CONTENTS

CHAPTER 1 GENERAL

CHAPTER 2 HANDGUNS

CHAPTER 3 SHOTGUNS

CHAPTER 4 RIFLES

CHAPTER 5 BLACKPOWDER

CHAPTER 6 CATALOG

COVER

Shown on the cover of the 1981 Annual are a Colt New Frontier in .44 Special with Art Jewel ivory grips (top), a Ruger Security Six stainless in .357 Magnum (middle), and a .45 stainless Safari Arms Enforcer (bottom). Photo by Pat Brollier.

1981 Guns & Ammo Annual 0-8227-3017-0

GUNS & AMMO SPECIALTY BOOKS STAFF

Editorial Director
Craig Boddington
Managing Editor
E.G. Bell, Jr.
Art Director
Martha Nelson
Senior Staff Editor
Payton Miller
Associate Editor
Michael Durr
Art Assistant
Bruce M. Morrison
Copy Editor
Patty Kenney
Associate Art Director
Jervis Hill

Vice President, Executive Publisher
Thomas J. Siatos
Publisher
Richard S. Fay
Executive Editor
Howard E. French

PETERSEN PUBLISHING COMPANY

Chairman of the Board
R.E. Petersen

President
F.R. Waingrow

Sr. Vice President
Robert E. Brown

Sr. Vice President
Dick Day

V.P., National Advertising Director
James Walsh

Publisher V.P.
Robert Macleod

Group V.P.
Thomas J. Siatos

V.P. Finance
Phillip E. Trimbach

V.P. Circulation Director
William Porter

V.P. Manufacturing
James J. Krenek

Assistant Director, Circulation
Jack Thompson

Director, Circulation Marketing
Nigel P. Heaton

Director, Production
Louis Abbott

Director, Research
Don McGlathery

Director, Graphics
Al Isaacs

Director, Photography
Bob D'Olivo

Director, Advertising Adminstration
Carol Johnson

Acknowledgements: Brass Rail, Los Angeles, CA; Martin B. Retting, Culver City, CA; London Guns, Santa Monica, CA; The Gun Shop, Lancaster, CA; Pony Express Sport Shop, Encino, CA; Pachmayr Gun Works, Los Angeles, CA; Reloading Bench, Panorama City, CA.

Chapter 1

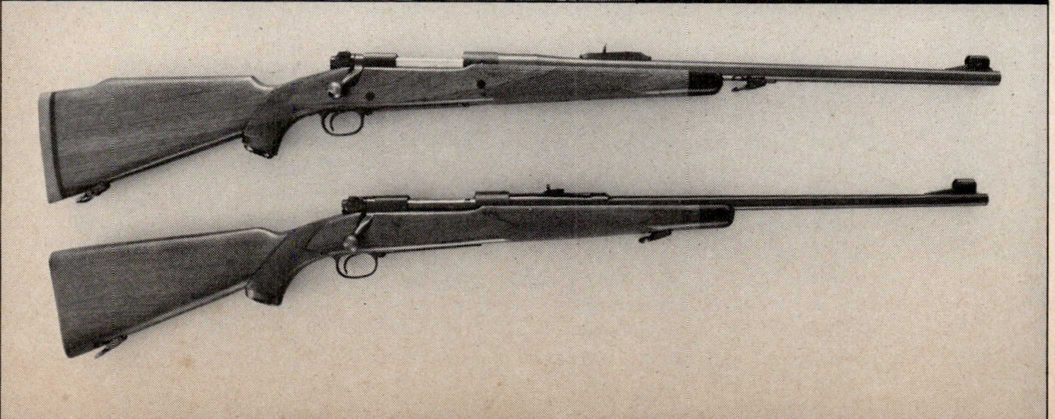

General

An Interview With John Milius

One of Hollywood's top director/screenwriters—the man who created "Dirty Harry"—speaks out on guns, shooting, and the movie industry.

G&A: John, you're extremely well known in the film industry and our readers are going to be familiar with your work, but they may want to know a little about your end of the movie business and how you might describe your particular job.

Milius: Well, I'm a writer primarily and a director out of self defense. I became a director because I had things that I had written done in ways I didn't like so I became a writer-director. I like being a writer more than anything.

G&A: How did you get started in the screenwriting end of the business? Were you a writer in the more conventional terms of writing originally or has your writing primarily been screenplays?

Milius: It's been totally screenplays. I went to cinema school when I was a lifeguard. When I got out of cinema school I was still a lifeguard and I wrote a script that summer. I said, "I'm going to give this a try," and the first script sold and I was on my way.

G&A: You're pretty well known as a shotgunner. Of course, your films reflect an interest in handgunning and military firearms. What kind of shooting sports are you interested in most?

Milius: Well, my wife and I enjoy trap and skeet shooting, We go every week. We can't live without firing four boxes of shells a week. My wife has become a fine shot—a fine trap shooter. I particularly like to shoot continental trap, which is a little different. I enjoy shooting skeet more than she does. I enjoy all the target games.

G&A: Down at the rifle range a little bit ago you shot a group with your Mannlicher that I thought was fantastic. Have you done a lot of rifle shooting?

Milius: I've done a lot of rifle shooting but I'm not a particularly good rifle shot. I need fantastically accurate rifles so that I can shoot as good as most people, but I think that was a lucky group. I probably flinched it.

G&A: What kind of handgun shooting have you done? You mentioned that you've been to the Jeff Cooper school.

Milius: Yeah, we did that because all my life I've shot handguns and lots of it, like most people who are interested in firearms. I'd decided I'd never really learned how to do it properly. I've never had any real instruction in it. I'd heard so much about this school and how good you became with it and I was really amazed at how much he can get out of people. My wife and I went and we really came back amazingly skilled. You know, we just couldn't believe how far we had progressed. The main thing about it that was terrific was that I'd never had a sense of safety with a handgun like I had after going to that school. You really feel totally confident—you know very, very well. how to handle a handgun properly. I've shot all my life and I can really recommend that people should undergo some sort of handgun training. I'm eager to go back and take the rifle course.

G&A: Have you stayed with the PPC combat shooting at all?

Milius: We go out and practice occa-

If you go for this kind of action...

Hunter, pilot, conservationist and nature photographer—Tom Siatos, Vice President, Executive Publisher of *Guns & Ammo* and *Hunting Magazine*, is a special kind of man.

He needs a special kind of watch! Carved from a solid block of stainless steel, engineered to withstand the hazards of freezing Arctic temperatures, built to endure and able to absorb constant shock, his watch is ready to go anywhere, anytime, to meet any challenge.

The watch Tom wears is the Rolex GMT-Master, a self-winding 30-jewel officially certified chronometer, which is individually tested and guaranteed to a depth of 165 ft.

Rolex . . . recognized the world over for excellence in craftsmanship.

The Rolex GMT-Master with matching bracelet is available in steel, steel and 14kt. gold and 18kt. gold.

you'll wear a Rolex

ROLEX

ROLEX WATCH U.S.A., INC.

Rolex Building, 665 Fifth Ave., New York, NY 10022

sionally. We really are shotgun shooters. I just do enough pistol shooting to hone up on it and to sort of remember how to do it. But what Cooper calls practical shooting applies so well to all kinds of shooting, so I'd really be interested to see what his theories are on the rifle.

G&A: You mentioned that you'd been shooting all your life; how exactly did you get started shooting? Was your father a shooter as well?

Milius: No, he wasn't, but a neighbor was, and I was really quite fascinated. I grew up looking at westerns on TV and of course I wanted to own a Colt .45 and a Winchester and ride around on a horse and be a cowboy or whatever, when I was young. I was always fascinated with guns, particularly with fine shotguns. I remember they always held some sort of romance of a good life or something, that you went out with your fine hunting dog and your fine shotgun and harvested the game on a beautiful fall day. I always sort of looked at that as a nice thing to do, one of the great bonuses, I suppose.

G&A: The comment you just made regarding a good hunting dog is the first thing you said that has anything to do with hunting. How do you view the hunting end of the firearms sport?

Milius: Well, I love to hunt. I don't hunt big game very much anymore. I used to fairly regularly go out and hunt big game, whether I shot anything or not because I just liked to get out there and ride around on a horse for a week and see some beautiful country. I don't do it so much now but I love to go bird hunting. I look forward every year to the first of September. I eat everything. I won't say I shoot it for food because I can certainly go and buy the food but I do enjoy the hunting. I enjoy the whole process. It's like growing something in your garden, I guess. You go out and you train the dog and shoot a limit of quail and bring them back and cook them for breakfast. I really enjoy hunting in Hawaii. Yeah, I still go over there every year and shoot a pig and have a luau. It's also fantastic bird hunting. You start at 7,000 feet and walk down through all these different layers, different flora and fauna.

G&A: In recent years the film industry, or at least a lot of people in the industry have seemed to have become extremely anti-gun and anti-hunting. Are you an exception, or are there a lot of folks in the film business who enjoy firearms and the outdoors like you do?

Milius: In the time that I've been in the business, in my contacts, there's almost no one, except Bob Stack. He and I have become good friends and we go duck hunting. But I've noticed that lately there's a lot more people interested in firearms for home protection. And out of this a few of them get interested in it as a sporting activity. There seem to be a few more people out shooting skeet.

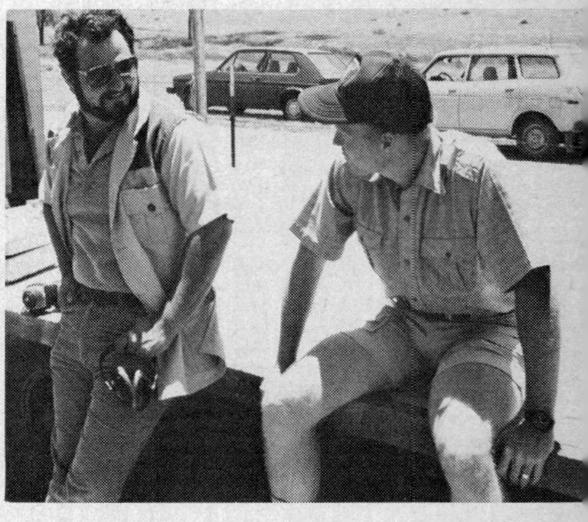

Milius owns two Steyr-Mannlicher SSGs in .308 Winchester. Although he doesn't consider himself an exceptional rifle shot, he was able to produce ½-inch groups at 100 yards.

G&A: Would you care to name some of these?

Milius: Oh, Arnold Schwarzenegger, who is going to be in my new film, is an avid skeet shooter and an avid hunter. He enjoys hunting and he certainly enjoys trap and skeet. He's a good shot, too.

G&A: You mentioned that you went to the Cooper school, the Gunsight Ranch school, with Jan-Michael Vincent.

Milius: Yeah, he enjoys it. He doesn't shoot shotguns that much, but he's a good pistol shot. He enjoys shooting the IPSC, the combat pistol thing. He's a pretty darn good shot.

G&A: So you think we've got a few friends out there and maybe we're starting to get a few more?

Milius: Yeah, I think that, at last the pendulum is definitely swinging back. Where it used to be, you know. I used to be looked at as kind of a colorful primitive, and I was tolerated that way, but I was still looked at as somewhat of a barbarian. I remember one time I was to have a lunch with a very famous critic in New York who has always disliked my work, and I invited her to lunch and she wouldn't come because she thought I was going to be armed.

G&A: Has your interest in firearms caused you any problems, not with the critics, but actually within the film industry itself? Some of your films have been quite actively pro-firearms.

Milius: No, I've never had any criticism within the film industry when I started out as a writer I was considered crazy. As a matter of fact, that's what writers are supposed to be. But they were never going to let me direct, because they said, "Milius will want to get out there and shoot somebody, you

Although primarily a shotgunner, Milius has completed Jeff Cooper's course at Gunsite.

Even though Milius is primarily a shotgunner and pistol shooter, he appreciates fine rifles—like this Anschutz .22 caliber sporter.

know, and so, he'll have a loaded gun and if an actor doesn't show up in time, he'll shoot him or something." Really, actually, they believe this in a way. People who don't know anything about firearms are so afraid of them, and so afraid of people that like firearms, that they really get kind of carried away.

I have never had it affect my work particularly, because Hollywood has long existed on violence. I would say that guns have paid the bills for many years.

G&A: It sounds like you certainly have never made any secret about your interest in firearms.

Milius: No, I don't believe in making any secrets about one's likes and dislikes.

G&A: It would be interesting to me to know, and probably most of our readers, when you plan a film, such as the *Dirty Harry* films or *The Wind And The Lion*, or *Jeremiah Johnson*, what kind of research is done to insure that the firearms used are actually what would have been used in the period, or is that the primary consideration when you make a film that involves firearms?

Milius: Well, for me it's always really a great consideration and it's really a lot of fun, because I love the study of firearms and it's really quite enjoyable to research what people did use and what would have been used tactically, in a military situation or something like that. Like in *The Wind And The Lion,* since it was 1904, we surmised that, remember that the Marine Corps had 6 mm Lee straight pull rifles for a while, and then they changed over to Krag in about '98, I think. We figured that a contingent of Marines would have the Krags in 1904 when that took place. And the machine guns that they would have would be those old Colt potato digger things. We had to make mock-ups and it was really quite easy to do it right. It was simply a matter of going out and asking them to do it right. It's often just as easy to do it right as to do it wrong. Often, when I see a western and it's an 1858 or 1868 movie and they're using '92 Winchesters. If it's 1868 or something, you say that you want something that was around in 1868. It's just as easy to do it right, but people avoid this all the time for various reasons. One is that it's easier to operate cartridge arms. If you have a western or something, you need a cowboy pistol, which to them, is a Colt single action, and it's also easier to work with a Colt single action than to have to fill the thing up with powder each time you want to shoot it. So there are various reasons why people avoid doing it right. The little extra attention you put into doing it right seems to always pay off. Every detail I try to get because it really adds up. It gives the whole thing a flavor after a while. It really seems to add to the depth of the whole thing. It's like a painting, you try to layer it as deep as you can. Detailing of the props really makes a big difference.

G&A: Your *Dirty Harry* films have probably done more for the handgun industry than any film that I can think of in a long time. Would you like to talk a little bit about the background for both the character Dirty Harry and also how you arrived at the Model 29 Smith & Wesson that he carries, which is definitely not a normal police gun.

Milius: Well, a Model 29 was really kind of an illogical choice from a practical point of view, because Jeff Cooper has brought me to task on this, that no one who is really well versed in the use of a handgun would use the Model 29 unless they wanted to shoot a car with it but the Model 29 is so dramatic that the idea of this guy carrying the biggest handgun he possibly could, knowing how to use it well, and he certainly did use it well, gave a certain aspect to that character. I'm really quite surprised that the whole *Dirty Harry* Model 29 identification came off as strong as it did. The gun became definitely a character in the movie. That very rarely happens. That's just something that happened in the case of those films, you know, the idea of Clint Eastwood, that

Here, Milius and two G&A staff members discuss one of the shotguns used during the interview—which included much shooting.

gun, the character named Dirty Harry. That character in particular is larger than the sum of its parts.

G&A: Do you have any film projects going now where firearms have a significant part, or what kind of things do you have in the works? Would you like to comment on that?

Milius: Well, I'm definitely going to do a few more things from American history. I want to do something on Jedediah Smith, and I'll eventually do that. I also want to do a movie about the early exploration of Kentucky. The firearms in those periods are very, very interesting, really much more interesting than the firearms of later periods. How big a part a personal firearm held in that existence! A Kentucky long rifle that a frontiersman had was a real part of him. He knew how the lock worked, the special way to load that particular rifle, powder capacity, the power of the rifle, the range of it, probably much more than people knew about their Winchesters. That one shot made a big difference, too. I sort of want to get this relationship with the characters. One of the things I want to do with the actors is have them spend considerable time learning how to shoot a muzzle-loader, flintlock, or caplock, loading it, firing it fast, using certain methods that they used until it really becomes second nature. In *Dillinger* it was very interesting. I trained the people. Rick Dreyfuss, who played Baby Face Nelson, didn't know how to shoot anything particularly well. And he was eager to learn but he had a terrible time with it. He'd come running out of the bank with a Thompson and he'd fire a burst and the drum would fall off the gun. These old guns, they had been dropped and used on many movies. Thompsons, by nature, just don't work in movies. The machine guns rarely work well in movies—you'll never see a machine gun in a movie fire very long. Most of them don't. You always see them fire and then you cut to what it's hitting, and then you cut back and it is still firing. The reason you cut is because the gun stops. The Thompsons are particularly terrible that way. And he came out of this bank firing the gun. It just wouldn't work and so I said, "Rick, what would happen if you really were Baby Face Nelson and you came out of the bank and your gun jammed. You'd have an extra magazine or you'd have an extra gun stuck in your belt or something. You know, you'd get to the car one way or another." He said okay and he went back in and we rolled again, and the next time he came out he was virtually clanking. He had a Thompson in either hand, a BAR over his shoulder, about six .45s in his belt, and magazines and drums all over him and he was determined to fire every one during the take. Everyone was laughing so hard by the end of it. But then, the point was made. It did work in that movie, people did change magazines, do stuff like that and it really looked good.

G&A: Going back to *The Wind And The Lion*, there were several scenes shot in the White House with Teddy Roosevelt and he had a range in the backyard of the White House and whatnot; how accurate a portrayal of Teddy Roosevelt was that?

Milius: I read a lot. Particularly, the letters in one of the Winchester books that Roosevelt wrote, and I quoted him directly about his ordering a Model 95 in caliber .405 and all this sort of thing. He was very concerned with his guns. He was very concerned that Winchester furnish him appropriate guns, that if he wanted to go to Africa he would be able to do it with an American gun. He did test the

Milius employs a two-handed grip to wring out a .357 Magnum Colt python—one of several handguns fired during the interview.

guns in the White House, too; it was quite alarming to the passersby. During the KretaKaris kidnapping, his first reaction was to select horses and rifles and go over and settle the thing himself with a battalion of Marines. That's true, that's on record. He said, "I'm Commander in Chief and I'm going to exercise that right."

G&A: Until now, I suppose myself and a lot of the readers felt that the directors and filmmakers just take any old gun and slap it into a fellow's hand and put him on the screen. I think that has happened in some films. Earlier today you were talking about your current film project, the *Conan* film, and I was really interested in the amount of trouble that you went to to find the proper sword for Conan to carry. Would you like to tell us a little bit about that?

Milius: Well, we had the swords made and they are real swords. You never see that in movies. They are always aluminum swords or sheet steel swords. Of course Conan, being a fantasy, we can make up any kind of sword we want, but we are making up a sword that would have been carried in this imaginary period, which is like a classic broadsword, but with a few refinements. We sort of made up the martial art to go with it, the way of using the sword, which is part Kendo and part real broadsword fighting. I've had people come in who were members of the Society of Creative Anachronism and these guys get together and get in armor and beat the hell out of each other with great big mock broadswords. They've really studied how these things were used; no one else had studied this. We have written accounts that are very sketchy of how men engaged each other with broadswords and no one teaches this anymore, there's nowhere you can go to watch it so we sort of had to reinvent it to try and figure out what they actually did do. So all the moves in the movie will be practical moves. All the weapons will be very practical weapons. They will have a certain dramatic flair in their decoration and everything, but the moves will adhere to a discipline the way they would in a Japanese samurai film. The thing they have is an ongoing tradition of Kendo that tells you how you really fight with a sword. We don't have anything except modern fencing, which is a completely different style, so we've had to decide what a battleaxe would actually weigh and if it weighed that much, how would it really be used. What materials would be used? How much armor? Where the armor would actually be worn. The style of fighting, this kind of thing. It's very, very interesting and complex. I'm really looking forward to staging some of these battles with all the information we have.

G&A: When do you start shooting on that film?

Milius: January 12th.

G&A: And that should be released?

JOHN MILIUS

Milius: About a year from then, or a little less than a year from then. Around Christmas.

G&A: When you write the script for a film, obviously that's not the end of the job for you. After you've written a screenplay and you know it's going to be a movie, what happens then and where do you go from there?

Milius: No matter how good a deal you have and everything else, the assurances from the studio, that the movie is going to be made when you write it, you have to cajole them and do what ever kind of pressuring you have to do, packaging, or getting people together, and there's a whole kind of process that goes on between the finishing of a script and the actual starting of a movie. Some movies come together inevitably; there's a movie that I know of now that they are going to make regardless of the fact that they don't have a script or a good script or even a particularly good idea. It sort of started to snowball and it will go ahead that way. Other movies seem to have a lot of trouble getting together and it seems to be just a process that goes on. I've gone through it enough times now that I know it's going to happen. The movie will get made and you just have to go through this kind of birthing process and then there is an enormous amount of preparation. You can never do too much preparation. Most of the disastrous stories you hear about people going way over budget and getting into trouble making movies are because of lack of preparation. If you prepare well and know what you want, there are very few contingencies to catch you off guard.

G&A: So your job doesn't necessarily stop when the script is completed.

Milius: Oh, no. If I'm going to direct it, then the real work begins. The most pleasurable work is writing because when I write it I see it for the first time and it's never quite the same.

G&A: If you're not the director yourself, you're still involved with finishing the project?

Milius: Yes. If I were just the writer, I would be involved in all the changes. In writing *Apocalypse Now,* I wrote for ten years. The first script, which is 90 percent of the movie that you see, was written in 1969.

G&A: You were involved with the project since 1969 until it was released?

Milius: Until it was released. I was still working on it until a month before it came out.

G&A: There's no question but that your films have done a lot to help us in our fight against the folks who would like to take our guns away from us, but how do you think we stand in our fight to retain our right to keep and bear arms?

Milius: Oddly enough, I think the wheel is really turning. I don't know whether any of us has done anything about it or not, or whether people's awareness that they need to have firearms has finally dawned upon them. The people that would want to take them away have probably made us more aware of how much we need them. It just seems to have gone that way. In the sixties, things looked pretty grim. Things look a lot better now. There's a lot more interest all over in the shooting sports. And that's just the tip of the iceberg. There is a great deal of interest in firearms. There are a lot of people now who are very interested in protecting their property with a gun. They are interested in protecting their property with a gun and the next stage is that that gun becomes a part of their life. This may not be the horrible thing that

One of Milius's favorite shotguns is a 12 gauge Spanish side-by-side. Although it is primarily a field or hunting gun, he likes to shoot trap with it as well. Both Milius and his wife are addicted to the clay target game—and both are very good wingshots.

they thought it was. Maybe they can think of taking that gun out and shooting a tin can. And if they can think of shooting a tin can, maybe a clay target. And then they have a completely different attitude. I think that you're going to see a lot more pressure on hunting simply because of the fact that there's less land to hunt on. The groups that are against hunting are using that. It's much more difficult to go hunting than it was in the '30s, let's say, when you could go out and shoot all the pheasants you wanted to. It's hard to find a place to go out and shoot pheasant. I don't know what's going to happen. I suppose it will come down to some sort of system like they have in

For Milius, fine shotguns have always had a special appeal—connoting the pleasures of fall birdhunting with a good retriever.

Germany where certain lands will be very carefully managed for hunting and that's probably a good thing. Some of the best hunting that I do is in Hawaii. That area where I go hunting is very carefully managed. The hunting is very sporting and very good. It is as rough as you like it. It's real hunting.

G&A: So you see a pretty bright future for shooting sports in America, then?

Milius: Yeah, I really do. I really see an enormously more enthusiastic reception by people to shooting sports. The way that some of these sports that you might have called oddities a couple of years ago, like combat shooting metallic silhouette, really attract a lot of people.

G&A: People who don't hunt are equally attracted by metallic silhouette shooting, I think.

Milius: Yeah, it's really fun. And the same with combat shooting. It's really fun to try and beat the clock and get your time down and get all your shots in. I see a tremendous future for shooting sports, much brighter than in the past 20 years, and I'm looking forward to doing a lot more shooting myself.

G&A: So are we!

"HELL, I WAS THERE!"

Excerpts From Elmer Keith's Autobiography

EARLY GUNS . . .

While living at 1012 Billings Avenue, we had several very happy years. My little brother Francis was born there. On Sundays we would drive to some of the streams like Prickly Pear or Beaver Creek for fishing. In the summer and in the fall we put in every Sunday hunting, either down to Helena Lake for ducks, or else go deer hunting up by Priest Pass or over on Nevada Creek.

I traded for an old Remington 1854 .58 caliber brass mounted smooth-bore musket, which was my first shotgun. It was only effective at very close range due to the extreme spread of the pattern. I did wreck a covey of green wing teal that came up the river past us one time, and was death on muskrats swimming down the river if I aimed about a foot in front of them. The concussion would kill them and they'd come up kicking and I'd have another muskrat skin to handle.

I began to realize that if I was going to get my share of the ducks that were so plentiful, I had to have a real shotgun. So I wrote the Ithaca Gun Company when I'd saved enough money for a No. 2 Ithaca double hammerless shotgun. I wanted Damascus barrels, but Lou Smith, the president then, wrote that during the war which had started then in nineteen and fourteen in Europe, he could no longer get Damascus barrels, and for that reason was sending me the higher grade No. 3 at the same price. It was my first fine shotgun and I did very good work with it. I shot that gun for seventeen years.

Father gave me a new model 1894 .25-35 Winchester long-barrel rifle. I killed my first deer with it on Nevada Creek. The first shot was three hundred yards up a steep mountain. The deer had seen me and I couldn't get any closer, so I raised the rear sight two notches, got a rest, and hit her square in the shoulder. It broke the shoulder, but the little 117-grain bullet never even went on into the chest. The deer pitched off down the hill in a run and I trailed her around the mountain for a half a mile, getting three more running shots before I got enough of those tiny bullets in the deer to put her down to stay. That was enough .25-35 for me for big game. Then I went to a .45-70 trap-

door Springfield carbine that had been given to me by an old stage driver. That and a case of 500-grain infantry loads of 70 grains of black powder and 500 of lead was a powerful rifle for a tiny kid then, but it did the job.

I asked to join the Helena Rifle Club, which was composed almost totally of much older men. The club members didn't want a kid in the outfit, but my friend, Bill Strong, sided with me and told them that I would qualify, and asked if they would let me in if I qualified expert over at the Fort Harrison military range.

They agreed, thinking it impossible that a skinny kid could handle a .30-06 and qualify expert. I bought a dollar-and-a-half Krag rifle through the N.R.A., practiced with it, and went over the course from two hundred to a thousand yards three times. Then I borrowed Bill Strong's Springfield and shot the full course with that. When the day came for the annual match, I managed to qualify expert with three points over, so they allowed me to join the Club.

That winter, the club shot in the National Indoor matches. They were mail order matches. Each club shot its own and mailed in the results. It was quite a lengthy match and we shot for weeks. At any rate, I wound up as high man for the Helena Rifle Club using an old .22 Winchester musket I'd bought from Holder Hardware Company for $10, and Mother loaned me another $10 to

WEIGH THE ADVANTAGES, AND YOU'LL WEAR BROWNING.

ADVANTAGE *Weight.*

You can take a pound off your feet by lacing on a pair of Browning Waterproofs instead of the competitions' waterproof boots. Multiply the steps you take in a hard day's hunting by a pound and the answer will be, ouch! You don't need an extra pound to carry around all day. But at the same time Browning gives you less weight, Browning gives you more quality.

ADVANTAGE *Open Cell Foam Insulation Versus Closed Cell.*

Browning Waterproofs are fully insulated from top to toe with 3/8 inch open cell foam that breathes to wick perspiration away from your foot. Most of our competitors use only 1/8 inch closed cell foam, which is like putting your feet in a plastic sock. Browning knows a dry foot is a warmer foot.

ADVANTAGE
Comfort Right Out of the Box.

Notice how soft and supple Browning's leather is, while other boots are stiff and bulky. That's one reason they break in easier and are more comfortable to wear. Another is the cushion your foot gets from the generous foam insulation and the smooth nylon tricot lining. Nylon tricot is lighter, as durable and more comfortable than leather lining. Moisture won't crack, rot or harden tricot. Browning does not use a weighty, poured midsole. Browning's trim midsole allows your foot to flex comfortably.

ADVANTAGE
Browning's Extra Durability.

Browning Waterproofs give you Vibram's best sole for firm support, bonded to the best top grain, silicone impregnated leather uppers. All seams and stitching are generously sealed with liquid latex and will not dry out or become brittle. Spring steel shanks and top quality insoles give you solid, yet comfortable arch support.

Quality is assured if your footwear bears the Browning name, and you pay no more for Browning's better boot. Weigh all the advantages. You'll pick a pair of Browning boots.

FREE CATALOG

Showing Browning's entire line of outdoor gear including 15 different hunting, chukka, western and pack boots for hunting, fishing, around camp use or casual wear. Send a postcard to Browning, Dept.20N , Morgan, Utah 84050. Distributed in Canada by Browning Canada Sports Ltd/Ltée.

In The Boot Race Heavyweights Finish Last.

BROWNING

buy a Stevens scope, which Johnny Linder put on for me free of charge.

ON PLACING THAT FIRST SHOT . . .

I believe it is time more people in this country hunted with single shot rifles, took their time and placed that one shot. If they don't get a good shot, they should not shoot. I am dead against the use of automatic rifles and spraying the brush in hopes of making a hit while shooting at too long a range.

Recently articles have appeared in *Guns & Ammo* Annual, and also in other gun magazines, of 1000-yard whitetail deer shooting across canyons in Pennsylvania. I am dead set against any such shooting.

To think of shooting at a big game animal at such range is simply beyond my comprehension. Furthermore if they are set up at a benchrest with a big spotting scope and Army range finder at a thousand yards, who is going to get across the canyon to where the deer was hit and find him in brush over there (where he probably can't see 50 yards) while across the canyon they show up plain?

It is inevitable that they must lose several deer for every one they kill.

I would prefer to shoot big game at ranges close enough so I could see their eyes. Whenever possible I stalk close enough to do just exactly that.

I have had to do long range shooting a few times in my life or forego the game that I wanted badly. I can remember killing one mule deer at 600 yards before about 30 people at Gibbonsville, on a bet. I didn't want to shoot at that distance, but they kidded me until I showed them it could be done. I also remember killing two mountain sheep at 600 yards, one at possibly 650. I also remember killing two coyotes, one at a measured 600, and the other one my father stepped at between six and seven hundred.

I do not believe that the average person has any business shooting over 300 yards at big game unless he is a trained expert rifleman with a background such as I have had of a lifetime of shooting, and then only from a rest or a prone position.

Otherwise more cripples are represented than clean kills.

I checked with different guides around this section of the country for several years. The consensus of opinion on elk is that the hunters coming in here and using rifles of .30 caliber, even the big magnums, lost at least two and possibly three elk for every one they bagged. Such shooting at long range, especially antelope, mule deer, sheep and goats, and the use of too small a rifle, is conducive to further elimination of all our big game.

The hunter training programs initiated by the National Rifle Association all over this country have done a great deal to help tyro hunters and others to realize the possibility of the arms and their own capabilities, and they should then limit the ranges at which they shoot to those capabilities.

When it comes to big game rifles I like to do all my hunting before I shoot. Many years ago I decided that for elk, moose, and our big bear, nothing smaller than .33 caliber or a minimum bullet weight of 250 grains at around 2200 feet or better, was adequate for these species.

This allows us a little .358 Winchester with 250-grain bullet, a Remington .350 Magnum with 250-grain bullet, a .338 Winchester Magnum, the .340 Weatherby, the .378 Weatherby, our line of OKH cartridges, the .333 OKH, the .334 OKH, and our great .338-.378 that Thomson and I designed. Also my .338-74 Keith from the 9.3 x 74 German cartridge, as well as other .35 caliber and .375 Magnum cartridges.

These are all adequate, and I consider them a bare necessity in an all-around rifle. Usually small deer can be shot with these big rifles with a heavy bullet at moderate velocity with far less meat destruction and cleaner kills than are accomplished by high velocity .30 caliber and smaller bullets that simply blow up on impact or inside the animal. Furthermore, you get a blood trail with heavy bullets of large caliber, while small high-velocity bullets go inside and blow up. The hide shifts on the outside when the animal moves, and you have no blood trail whatever.

ON THE TRAIL WITH ZANE GREY . . .

In the fall of 19 and 31 I had booked a party led by novelist Zane Grey for a two months trip over in the Middle Fork country. We had also booked a boat trip.

Cap ran the boat trip, and took another man to run the rear sweep, but they sunk the boat at Salmon Falls, lost everything they had, and had a hard time getting out on foot. We never lost a boat when Cap and I were running together, and that was the last boat that Cap lost. Cap had a habit before we'd come to a big rapid, of tying up, getting out on foot and looking it over. He'd see if any logs had lodged between the boulders where we had to go, or if the heavy high water current had shifted some of the boulders and closed the channel. Then we'd come back, Cap would kneel down, and we'd offer up a little prayer.

He'd always say, "Elmer, isn't that right (or left) side of the boat a little heavy?" It was always where we had a gallon jug of moonshine tied with quarter-inch rope. I'd get the jug, we'd have a little nip, cut the ropes loose, and we'd run the rapid. This was regular procedure on all the major rapids, and we made them all. This was during prohibition, and we'd buy a gallon jug of moonshine from Old Man Moore at North Fork that he and his boy Johnny had made. Then when we got down below Campbell's ferry an old Tennesseean had a ranch where he grew rye and made it all into whiskey to supply the people riding up and down the river. He made very good whiskey, much better than the Moores. We'd always stop there and buy another gallon and that would see the party through to the conclusion of the trip at Riggins. All our early trips started at Salmon, and went from there on down. The road ended at Shoup. From there on was nothing but a pack trail around through the hills. Then as civilization began at Riggins Hot Springs, a few miles up above Riggins, the rest of the canyon was devoid mostly of trails and civilization, just people squatting along the river living on game and fish and taking out a little gold each year. We'd always stop at their camps and they'd give us their little retort buttons. Usually they'd retort all the gold they'd got in a potato, burn the quicksilver off, and we'd take the gold out to Pelton's Grocery. He'd weigh it up and fill their grocery orders. The next trip we'd take their groceries down to them.

I remember that prohibition officers finally caught Old Man Moore in his still. They took him to Boise and put him in the calaboose for six months. When he got out I talked to his son Johnny. Johnny said, "Elmer, the old man was so dry I had to pour whiskey in him for three days before his hoops would tighten up so he could hold it."

During the summer of '31 we had some bad fires on the North Fork. Al Wheeler hired me to take charge of a crew of some twenty-odd men, and we fought one fire on the Nez Perce for three weeks, another one on the North Fork, and another small one back of the North Fork store up the steep mountain. John Kinney, the supervisor, had promised me a fire permit anytime I wanted.

I'd booked Zane Grey and his three-ring circus for a trip. Grey told me he wanted to go into wild country, and that they were old mountain people and would come light. However, he brought in three cars and a big truck loaded with gear. I had my pack string and my partner's, and we hired (all told with our horses) 57 horses and seven men. Then we had to make three trips to relay that three-ring circus. I heard Kinney talk to the forest supervisor over the phone. He informed Kinney not to let Elmer Keith and Zane Grey go in the forest at all. It was terribly dry, and he was going to close the forest and call out the National Guard to stop all traffic on all roads entering the forest. John Kinney told him that I was a very reliable man, that I'd handled a fire crew and trenched more fire for him than any crew he had that summer. But it was to no avail. In the meantime I learned that Williams Lake, part of it was in private ownership and well within the forest, had excellent rainbow trout fishing, so we moved the entire Zane Grey crew and all their belongings up to Williams Lake, and ensconced them in a nice little park at the north end of the lake, the day before the supervisor clamped on the lid. I was within the forest then, and he couldn't stop me from going out and bringing in grub and mail, and he couldn't get me off because I was on private ground. We had to sit there two weeks until they finally lifted the ban on the forest roads. Then I took the party over to Meyers Cove, went the eighteen-mile pack down Camas Creek to the Middle Fork, forded the river, and set up a nice camp there. We fished there for awhile, then went on down below the old Mormon Ranch. He wanted me to take them to the mouth of Waterfall, and cross there and go up Big Creek, then across over into Cottonwood Meadows, back up to Cole Meadows, down Crooked Creek and up Monumental Creek to Thunder Mountain, an old mining camp, and write a novel on it, which he did later.

While waiting for the forest to be opened up, I managed to get a permit signed by President Herbert Hoover allowing Elmer Keith to take himself and party within any National Forest in the United States. That stopped the supervisor cold, and we were able to go on our proposed two month's trip. From the Mormon Ranch down the Middle Fork there was no trail, but I'd heard several old timers say they had hunted sheep down in there and they'd found horse droppings and tracks, and knew the Indians had worked through some way or other. I told them anywhere an Indian could go with a horse, I could, so I set out with the party down the south side of the river. We managed to make it, though we had to climb up around the cliffs and back and forth. It was an all-day trip from the Mormon Ranch down to the Waterfall. There was a nice campground there. The trail had come in over the other side as far down as this campground, punched through by the Forest Service at some time or other, or the Indians.

The next day we forded the Middle Fork, or rather swam it. One mule drifted down to the last sandbar. We almost lost her into the box canyon below, but she finally got out. We took the north side of Big Creek. The cliffs finally crowded us down, when one of the pack horses we had named Toots, slipped off a ledge. I saw her belly flash in the sun twice as she turned over down the mountain and she stuck on a ledge down there, pack and all. It took us two hours to build a trail down to her, pack her packs up, get her back on the trail and repack her, but she apparently wasn't hurt.

The cliffs finally crowded me down into the river, but I had a fine saddle horse named Brownie; 17½ hands high. He'd swim back and forth across the deep pools. We came out below old Cougar Dave Lewis's camp on Big Creek a short distance from where there was a blacksmith shop and a forge in under a huge cliff. From there we went up to Dave Lewis's, camped there, and the old man came over and had supper with us and came again for breakfast. I remember picking up a spearhead and twelve arrowheads there on the flat that evening that I gave to the Grey party.

Next we went up to Coyote Springs, hunted there, and killed several mule deer. From there we went down on the lower Cottonwood and hunted goats. They got their goats and some more deer. We packed up through Cole Meadows, down Crooked Creek, and up Monumental Creek to Thunder Mountain. This had been quite a gold camp. The old cemetery was in ruins. Trees had fallen across the headstones. A big mud slide had come down Mule Shoe Creek and dammed off Monumental Creek, and this in turn had formed a lake. Some of the buildings and the old saloon was floating around out in the lake. The beaver had built a nice nest in the attic of the old saloon, as it was floating around there.

This was a hard party to handle.

Romer Grey, Zane Grey's son, and his son-in-law Bob Carney, were the first to run the horses. I told them no more running of any of that stock because they had an awful long trip to make over the roughest country on God's green earth, and a walk was fast enough.

They had a Japanese cook with them who was very agreeable and a fine person. I got along with him fine. They caught all the fish they could eat and then some, both up at Williams Lake in the two weeks there, and also on the Middle Fork.

From Thunder Mountain we went up Mule Shoe Creek, dropped over onto Marble Creek, went down it to where a trail cut to White's Gulch, and from there over into White's Gulch where we camped one night. I remember the next day I started counting mule deer when I gathered the horses.

From this camp on White's Gulch above the old White's Creek Ranger Station down to the mouth of the Camas, I counted 1000 mule deer that day, many of them within 30 or 40 yards.

We made it from there down to our old camp across from the mouth of the Camas, then back up to the Ramshorn Ranch where the party had left their cars, ending this two month's trip. They took off for California.

I had cached six packloads of grub in the cellar at the Ramshorn Ranch, and when we got back somebody had stolen all of it. We were out of grub, so my partner Gerry and I had to come on home on my six-gun alone, eating pine squirrels, ruffed grouse, one snowshoe rabbit and one little porcupine.

I believe I showed Zane Grey the first long range pistol shooting he'd ever witnessed. He and Bob Carney, his son-in-law, and Romer Grey, were shooting at a rock at 400 yards with a .30-06 Model 1895 Winchester

I laid down with my back to a log, took my old 7½-inch .44 Special single-action Colt and hit the rock repeatedly. Grey asked me why I packed a rifle. I told him I wanted the rifle for more power, the six-gun was just a handy weapon to use all the time.

I came home about ten pounds underweight that fall, and my big saddle horse had lost 200 pounds. In fact, everything we had was pretty well wore out in the way of pack stock. Zane Grey wrote the novel *Thunder Mountain* from this trip. He sent me a copy of it, autographed, and it was loaned to somebody.

One day when fording the Middle Fork this blond secretary was right behind me on the little buckskin mare. There was a lot of big flat rocks in the river bottom. The little mare got astraddle of them and went down. I whirled my old saddle horse around, reached down and I got her by the shoulder of the buckskin jacket. I lifted her up in the air until her old horse got its feet under it again, and dropped her back in the saddle. In his book, Zane Grey had a big rescue scene on the Middle Fork and told how the river had washed the horses away, as he grabbed the lady and waded out with her. That was utterly impossible, because a horse can hold the bottom long after a man hasn't a chance. Anyway, he wrote *Thunder Mountain* from that trip.

Editor's note: Elmer Keith's autobiography, "Hell, I Was There!" is available from Petersen Books, 6725 Sunset Blvd., Los Angeles, CA 90028, for $19.50 per copy plus $2.00 each shipping and handling. (California residents add 6% sales tax of $1.17.)

The Empty Chamber

From the earliest frontier days of the old west, the mechanical characteristics and limitations of the single-action revolver have been well understood, as the writings of the day amply demonstrate. The safety precaution of loading the "six-shooter" with only five cartridges and resting the hammer on the empty chamber was universally practiced. According to folklore, some cowboys habitually carried "burying money" or their last bank note rolled up in the empty chamber. These old-timers understood that the notches in the hammer provided only limited protection, and that an accidental discharge could result if a fully loaded revolver were to be dropped, or if the hammer were to receive a sharp blow.

Despite the lessons of history, there are still people who get themselves in trouble by ignoring the following common sense rules of gun handling:

1. The shooter should thoroughly understand the mechanical characteristics and rules for handling the particular type of firearm he is using.

2. The safest way to carry any old style* single-action revolver is with five chambers loaded and the hammer resting on the empty chamber.

**The Ruger "New Model" single-action revolvers are not subject to this limitation and can be carried safely with all six chambers loaded. Some other manufacturers have added various manual safety devices to old style single-action revolvers, but no manual safety can ensure against accidents if the shooter fails to use it properly. Remember: There is no such thing as a foolproof gun!*

FREE CATALOGS AND INSTRUCTION MANUALS WHICH CONTAIN APPROPRIATE WARNINGS AND ADVICE FOR THE USE OF ALL MODELS OF RUGER FIREARMS ARE AVAILABLE FROM THE FACTORY ON REQUEST

STURM, RUGER & Company, Inc,
82 Lacey Place
Southport, Connecticut 06490 U.S.A.

ALL RUGER FIREARMS ARE DESIGNED AND MANUFACTURED IN RUGER FACTORIES IN THE UNITED STATES OF AMERICA

HERCULES POWDER
SPORTING
Xtra-Range
No. 3½ COPPER PRIMERS
ARROW
U. S. PRIMERS
NO. 4
"CLIMAX" and "AJAX"
IMPORTANT
AMERICAN
GAME
BIRDS
The Parker Gun
The Parker Gun.
PARKER BROS., Meriden, Conn.
MASTER GUN MAKERS,
U. S. A.
WETPROOF
20 GA. Remington
KLEANBORE
SHUR SHOT SHELLS

PARKER— The American Classic

From 1868 to 1941, Parker set the standard for American gunmakers!

By Art Blatt

The name "Parker" on a shotgun is synonymous with American-made quality. Like Zeiss is to camera collectors, whenever the name Parker is mentioned within a group of gun enthusiasts, all ears perk up. All died-in-the-wool collectors are constantly searching for that rare Parker shotgun from some unsuspecting owner of one of these truly great guns.

Of course, unknowledgeable gun owners are becoming rare too. I once witnessed a humorous situation at a Dallas gun show. An old-timer walked up to a gun dealer displaying his wares and politely asked the dealer if he bought guns. Clutched in his weather-beaten hands was a brown paper sack. The dealer's eyes immediately lit up suspecting that the old man had a rare find in the bag. Casually, the dealer replied, "Sure do, what ya' got to sell?"

Well, the grey-haired veteran reached inside the sack and pulled out a dark walnut gun box about 12 by 15 inches in size. He gently laid the box on the table and let the dealer pop open the 150-year old hinges and lever the top back. There, resting between the two opponents sat a near pristine Colt Baby Paterson with a full complement of accessories basking in its faded blue velvet lining. Stifling all emotion and camouflaging his real interest, the dealer casually asked, "How much do you want for this old relic?"

Without any hesitation, the old man said, "$17,000 and not a penny less!" At the time, this was the full-bore retail price of this ensemble. Obviously, the old man had done his homework and wasn't about to be taken in by a city slicker.

The same is true regarding Parker shotguns. There's been reams of information carefully sorted out and presented to the gun collector that has created quite a few neo-experts.

There are presently three books available about Parker shotguns *Parker—America's Finest Shotgun*, by Peter H. Johnson and a brace of books by Larry Baer, Volumns I and II. Each of these books is informative, and provides the would-be collector with much-needed information.

The Parker shotgun was born and achieved greatness in the tumultuous span from the Civil War through the end of the blackpowder era, finally ceasing production at the end of the depression years. Whether plain or fancy, Parker's trademark was always quality.

Unfortunately, many published source materials contradict one another regarding production models, years produced and all grades offered by the Parker factory. There's a great deal of conjecture and speculation because out of nowhere an "unknown" Parker will suddenly appear at a major gun show, which throws all hard-fast rules out the window.

Why all the mysteries surrounding the Parker shotguns? Well, in the mid-to-late 19th century, bookkeeping and record controls weren't quite up to today's sophisticated methods. Computers in those days were ink and quill. Microfilm files were usually empty machinery crates stuffed with old job orders and shipping vouchers. Fires, which were frequent, destroyed much of this printed information.

Studying and comparing Parker catalog reprints along with a few original copies, there's evidence that the company made mistakes in their own actual count of models and versions. These old price lists varied greatly from year to year and no firm conclusions can be drawn from these "company records."

Today, there are quite a few "known" Parker gun collectors, and for every soliciting collector, there's an equal number of "quiet" collectors who adamantly demand and protect their anonymity. Their prized guns are rarely viewed by "outsiders" and it is a pity that these selfish collectors will not share the wealth of information they own as they constantly live in fear of being "ripped-off."

One of the most respected Parker collectors and historians is Frank Pachmayr. This gentleman owns over 200 Parker shotguns. These guns range in condition from poor to pristine, in all

APPROXIMATE NUMBER OF PARKER SHOTGUNS BY GRADE

Lifter Models	20,000
Hammer Models	10,000
Trojan Grade	48,000
V.H. Grade	60,000
P.H. Grade	8,500
G.H. Grade	28,500
D.H. Grade	40,000
C.H. Grade	5,000
B.H. Grade	13,000
A.H. Grade	5,600
A.A.H. Grade	350
A-1 Special	325
Invincible Grade	2

Like handcarved cedar decoys, the Parker Shotgun is a part of our hunting past. (Photo by PPC photographer Pat Brollier.)

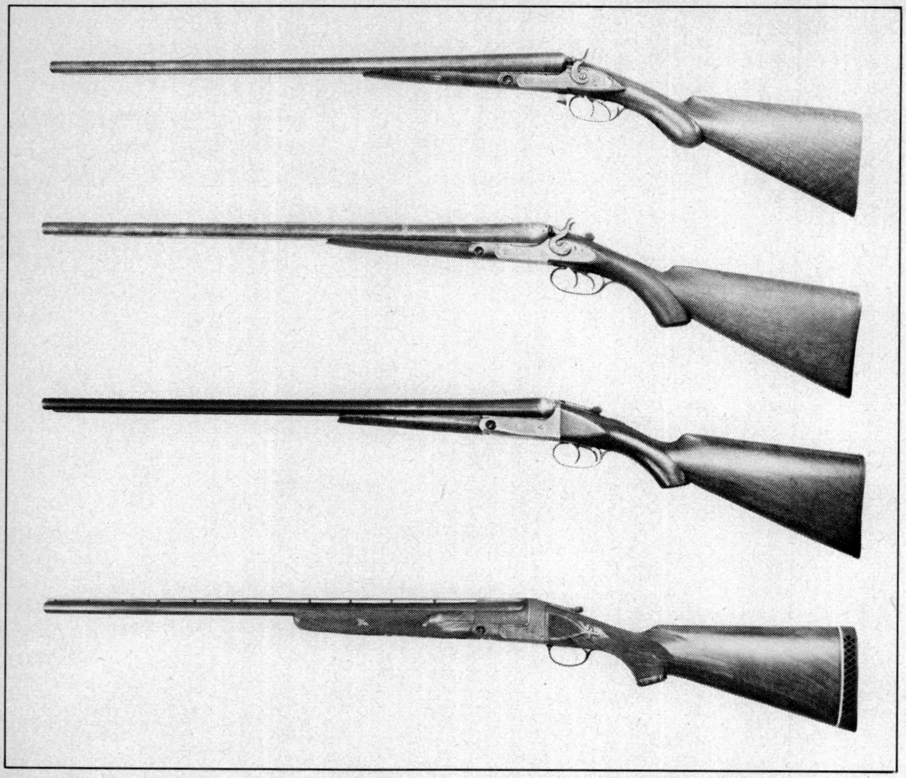

The first Parker shotgun was the lifter model (top). This was followed by the top lever action with exposed hammers (second from top) and hammerless models of early 20th Century. The single barrel trap gun (bottom) was one of the latest in the Parker line.

grades from "G"s and "P"s to A-1 Specials and in gauges from .410 to 8 bores.

We, at *Guns & Ammo* magazine, were privileged to study and photograph these magnificent guns for this story. Many of these guns have never been seen before and for this reason, we doff our hat to Frank Pachmayr for sharing his collection and expertise with us.

LIFTER MODELS

The very first Parker-produced product wasn't a gun, but rather a coffee mill manufactured around 1832. The first Parker built gun wasn't a shotgun but instead a repeating carbine for Civil War use in the mid-1860s. After the end of the strife between the states, Charles Parker decided to build sporting guns and combining his hindsight and excellent production methods, chose to build an improved version of an already accepted product—the double.

Charles Parker wasn't a gambling man, and he therefore preferred to take a successful product, improve on it and offer it for sale. Mr. Parker has been called many things, but innovative was not one of them. He was a shrewd Yankee trader and all of his products were sound, basic money makers.

The very first Parker shotgun—around 1868—was a 14 gauge double barrel with 29-inch barrels and utilized a super simple lifter bolting mechanism. This style and method of easy opening, locking and closing of the shotgun's action remained in production almost into the 20th Century, with the very last lifter action manufactured in 1899. In the late 1880's Parker began offering a top lever design and even a true "hammerless" gun was offered as early as 1889, yet Parker customers had a demanding preference for these strongly built and easy to operate lifter guns. And, as long as sales were still there, Mr. Parker would accommodate his customers.

The first model lifter guns were produced with a forend that locked to the barrel through the use of a tightly fitted wedge that had to be pried out from the right side of the forend. This was the same system as used on most muzzleloading rifles at that time. This forend lock remained in vogue for nearly 13 years and in 1880, Parker adapted the use of the Deeley & Edge forend latch and retained this method of securing the forend to the barrels until the last Parker shotgun was built in 1941—save of course for the Trojan spring-loaded forend.

In 1872, Parker realized that there was a definite market for higher-quality guns and offered different "grades." The upgrading of plain Jane models was a contributing factor to Parker's ultimate success as the affluent were able not only to own and shoot a Parker, but one with distinctive engraving and boasting finer figured wood. Parker, throughout their production, never lost sight of the "little man" and even in later years boasted that their lowest grade shotguns were always the finest available anywhere in the world in their price range.

In 1874, Charles A. King joined forces with the Parker company and contributed his great gun design expertise. King, at that time, was one of the foremost gun designers having worked many years for Smith & Wesson. During Mr. King's long tenure with Parker, he helped to develop other lifter-type actions, hammer models, hammerless models with external cocking levers, and finally the toplever hammerless model with automatic selective ejectors.

Damascus barrels in various grades were offered as follows: PH, the lowest grade with "plain twist" marked on the barrel. GH used two-iron stubs twisted and welded together to form the Damascus pattern. DH designated a three-iron barrel and finally, the esteemed CH, whereby the barrels were marked "Bernard" which was a very fancy ornamental design and, at that time, Parker's finest offering.

Interestingly, Parker never manufactured any of their own Damascus bar-

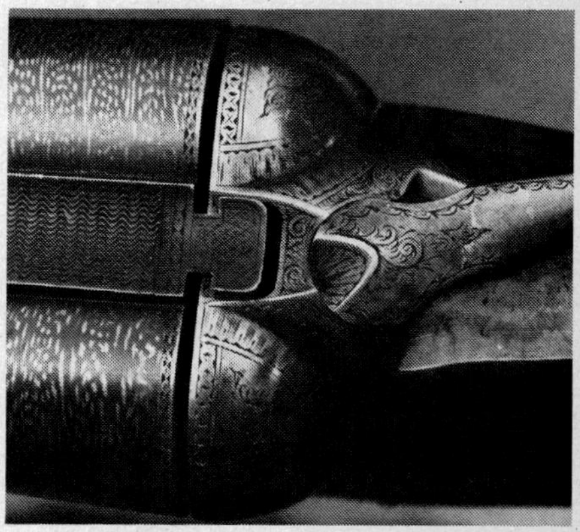

The Parker doll's head locking device on hammerless guns was extremely strong. It was tapered to compensate for wear.

Except for the opening mechanism, early Parkers were quite similar. The lifter model was often favored for its strength.

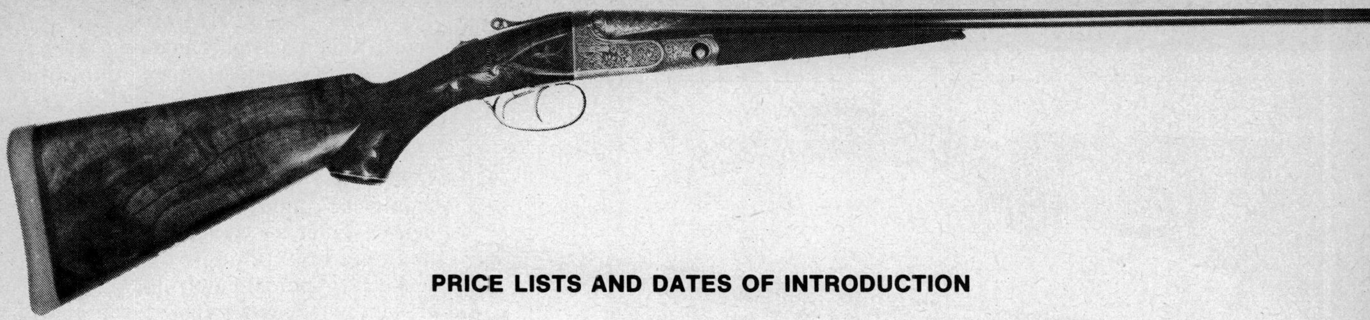

PRICE LISTS AND DATES OF INTRODUCTION

Grade	1874	1899	1912[1]	1912[2]	1915[2]	1917[1]	1919[1]	1930[3,4]	1934[3,4]	1937[4]	1940[5]
LIFTER MODEL											
various	$45 to $250	—	—	—	—	—	—	—	—	—	—
HAMMER MODELS											
AA	—	$400	—	$300	$300	—	—	—	—	—	—
A	—	$300	—	$225	$225	—	—	—	—	—	—
B	—	$200	—	$150	$150	—	—	—	—	—	—
C	—	$150	—	$112	$112	—	—	—	—	—	—
D	—	$100	—	$75	$75	—	—	—	—	—	—
E	—	$85	—	—	—						
G	—	$80	—	$60	$60	—	—	—	—	—	—
H	—	$70	—	$64	$64	—	—	—	—	—	—
R	—	$60	—	$45	$45	—	—	—	—	—	—
T	—	$55	—	$44	$44	—	—	—	—	—	—
HAMMERLESS MODELS											
A-1 Special	—	—	$394	$375	$400	$600	$725	$796	$875	$942	$942
AAH	—	$400	$319	$300	$325	$500	$610	$662	$728	$776	$791
AH	—	$300	$244	$225	$250	$340	$410	$457	$498	$547	$562
BH	—	$200	$169	$150	$169	$250	$302	$332	$361	$407	$422
CH	—	$150	$131	$112	$131	$182	$220	$245	$267	$307	$322
DH	—	$100	$94	$75	$94	$137	$166	$190	$207	$230	$227
EH	—	$85	—	—	—	—	—	—	—	—	—
GH	—	$80	$79	$60	$79	$107	$129	$143	$155	$163	$188
NH	—	$70	—	—	—	—	—	—	—	—	—
PH	—	$65	$67	$49	$67	$91	$109	—	—	—	—
VH	—	—	$56	$37	$56	$75	$90	$115	$124	$128	$168
Trojan	—	—	—	—	$27 +	$43 +	$55 +	$55 +	$89 +	$101 +	—
HAMMERLESS 8 GAUGE											
DH	—	—	$101	—	$101	—	—	—	—	—	—
GH	—	—	$90	—	—	—	—	—	—	—	—
PH	—	—	$79	—	$90	—	—	—	—	—	—
SINGLE BARREL TRAP											
A-1 Special	—	—	—	—	—	$550	$635	$625	$687	$725	—
SAA	—	—	—	—	—	$450	$520	$525	$577	$612	—
SA	—	—	—	—	—	$310	$358	$350	$385	$415	—
SB	—	—	—	—	—	$225	$260	$210	$231	$256	—
SC	—	—	—	—	—	$150	$173	$160	$176	$191	—
SPECIAL SKEET											
VHE	—	—	—	—	—	—	—	—	—	$168	$182
DOUBLE BARREL TRAP											
VHE	—	—	—	—	—	—	—	—	—	$200	$215

Legend
1. Hammerless guns with auto ejectors
2. Hammerless guns without auto ejectors
3. Hammerless guns with auto ejectors and single selective trigger
4. Prices from Remington-Parker catalog
5. Last published prices from Remington Parker catalog

\+ Without auto ejectors

Note: All published prices have been rounded off.

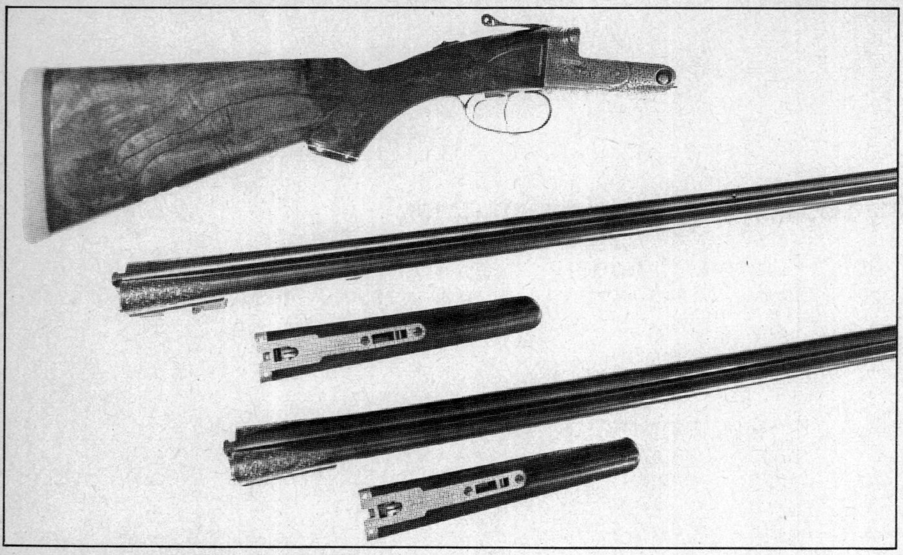

One of the rarest Parkers is this 28 gauge A-1 Special with two sets of barrels. Today, this A-1's value exceeds $50K.

rels and imported all shotgun tubes from Liege, Belgium. Total production of lifter model Parker shotguns numbered around 20,000 units in various gauges and configurations.

HAMMER MODELS

In 1882, the first Parker top lever action gun was introduced. This style made use of an auxiliary round bolt and a bolt lever that pivoted beneath the top tang. Now the Parker customer had a choice of two distinctive and different designs from which to choose. Ultimately, the top lever design won out and the lifter faded in 1899.

Throughout this era of producing the finest quality shotguns, Parker refused to use malleable iron castings to manufacture his guns. All parts were made of steel forgings; the frame, top lever, forend iron and hammers. Even then, this manufacturing process was slow and expensive, yet to keep Parker's reputation they felt it was absolutely necessary to build guns in this manner to make them absolutely safe to shoot and to last the average owner a lifetime—in fact, most Parker shotguns have outlived their original owners.

The Parker company was one of the first gun making firms to initiate "product liability insurance." Lloyds of London was contracted to protect the Parker company from suits against them in the event of burst barrels or injuries caused by a Parker shotgun. Lloyds was the only insurance company willing to sell Parker this type of coverage and the London based firm was never called upon to pay a single claim.

In 1899, there were 14 models of hammer guns available, which was five more than hammerless versions. Many extra features could be ordered like special barrel lengths, extra sets of fitted barrels, pistol grip or straight grip stocks and practically anything the buyer could imagine. Hammer guns were available in five gauges—8, 10, 12, 16 and 20 bore and this particular model remained in the Parker catalogs until 1915. After that, sales for the old "mule-ear" models dwindled and their demise launched Parker into their true "hammerless" models and sales skyrocketed. Approximately 10,000 top lever hammer guns were produced and today, these specimens are prized by collectors.

HAMMERLESS MODELS

From the end of World War I through the late 1930s were the golden years for Parker. The development of smokeless powder and fluid steel barrels helped to launch Parker's success story into the stratosphere. In 1915, the Trojan grade was first offered and nearly 50,000 units were produced. All Parker shotguns, with the exception of the Trojan grade, were available with a customer's choice of full-pistol grip, semi-pistol grip or straight stock, but all Trojan grades were only shipped with full-pistol grips. The buttplate on the Trojan model was always made of hard rubber with a milled surface. Recoil pads were

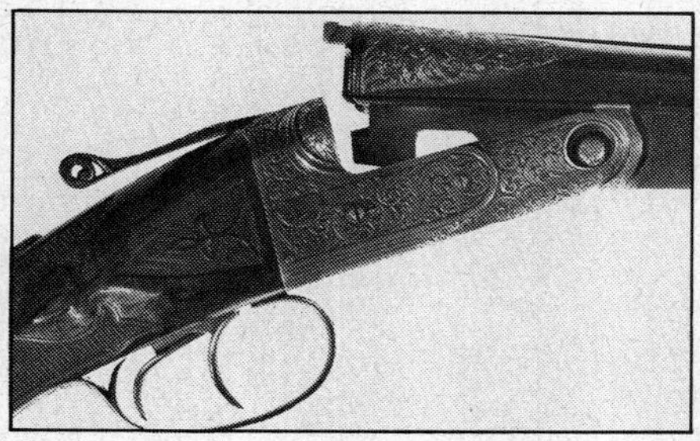

The finer grades of Parker were finished with an extremely handsome and well executed engraving as on this hammerless.

Parker was one of the earliest gunmakers to use the double locking screw arrangement to keep the gun securely together even after many years and thousands of rounds. This C-grade is adorned with fine scroll engraving and a panel depicting a bird dog on point.

PARKER FRAME SIZES AND NUMBER

Frame No.	Gauge	Action Width	Barrel Width
#1½	Light 12 & Heavy 16	1¾″	2¼″
#1	Light 16	1 23/32″	2 3/16″
#0	Light 20	1 21/32″	2 1/31″
#00	Light 28	1⅝″	1 23/24″
#000	Light .410	1½″	1¾″

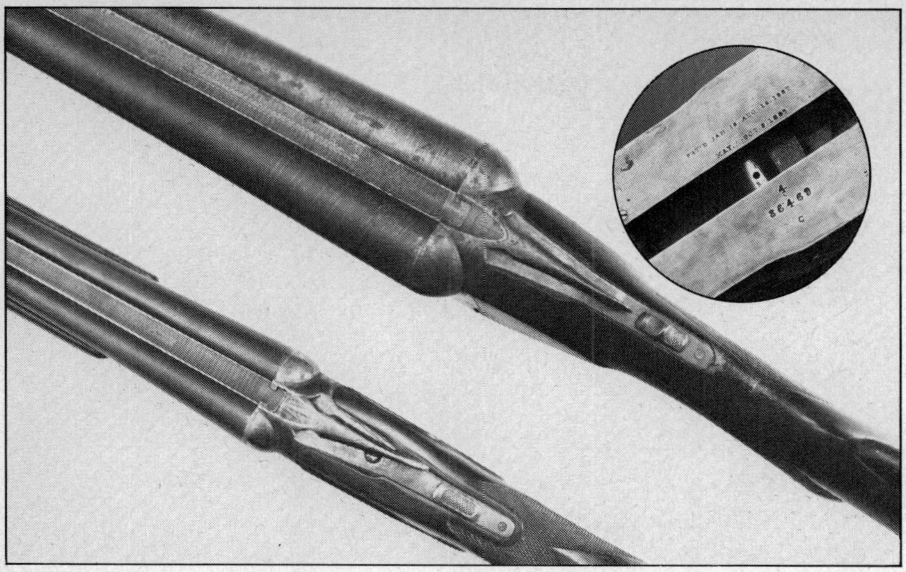

The difference in size between an 8 gauge and a 28 gauge is extreme, yet the Parker features are still present, notably the doll's head lock. All Parkers could be identified to grade, frame size, and often gauge by markings on the water table (inset).

not installed on Parkers until the late 1920s or early 1930s. Most of them were Hawkins pads.

The most popular Parker model—in numbers produced—was the VH grade with slightly more than 60,000 made from 1912 through 1941. The "V" grade Parker was the basic action for all Parker guns—even the higher grades. The same steel, quality workmanship and production methods used to produce the lowly V grade was the same used to produce the fabled A-1 Special—which incidently is incorrect, as the factory referred to this model as the "A No. 1 Special."

The single barrel Parker trap gun was introduced in 1917 and it was such an immediate hit among trapshooters that much of Parker's advertising claims revolved around this particular model. The single barrel was available in five grades—SC, SB, SA, SAA and A No. 1 Special. Barrel lengths were 30, 32 or 34 inches long and automatic ejector and raised ventilated rib was standard.

The early 1920s were a transition period for Parker shotguns as either Damascus or fluid steel barrels could be ordered. The late 1920s finally saw the disappearance of Damascus barreled offerings from Parker sales catalogs. In 1922, Parker offered their single selective trigger and one year later, the first beavertail-type forend appeared.

It wasn't until 1926, that the ventilated rib was first seen on a Parker double (the raised ventilated rib was standard on the single barrel trap gun) and in 1927 the first .410 Parker came off a craftsman's bench.

The mysterious Parker shotgun serial number 200,000 was produced in 1930 and this was one of two known "Invincible" grades. Truly the most magnificent shotgun ever factory produced, it carried an almost unheard of price tag of $1,500. In comparison, the best grade Purdey sold in the United States for about $750. The Great Depression was taking its toll and through the mid 1930s, Parker Brothers suffered along with the rest of the country. During these trying times, a variety of inexpensive double barrel shotguns started to erode Parker's sales. Offerings from L.C. Smith, Fox, Lefever and Baker helped lead to the demise of the great Parker name along with a host of repeaters from Ithaca, Savage, Winchester and Remington.

Finally, in 1937, Remington Arms Company bought out the great American tradition, made up whatever guns could be sold from existing parts and finally in 1941, the Parker bell tolled for the last time.

SIGNIFICANT DATES

1868 First Parker shotgun commercially produced. 14 gauge lifter model.
1872 First year "grades" were offered.
1880 Fore-end lock (Deeley & Edge) introduced.
1889 Adoption of hammerless lock.
1902 First year for automatic ejector.
1905 First 28 gauge offering.
1915 Introduction of Trojan grade.
1917 First single barrel trapgun produced.
1922 Single trigger available on double-barrel guns.
1923 Introduction of beavertail fore-end.
1926 First offering for ventilated rib.
1927 First year of production of .410 guns.
1930 Number 200,000 rolls off the line.
1934 Remington purchases Parker factory.

VARIOUS GRADES

Note: The use of the letter "E" after the grade markings indicated that the Parker shotgun was equipped with automatic selective ejectors.

V.H.—Lowest quality, simple border-line engraving about the receiver and at the base of the barrels. Vulcan steel barrels and in gauges—12, 16, 20 and 28.

P.H.—Heavier line-type engraving. Barrels of English twist steel and available in five gauges—10, 12, 14, 16 and 20.

G.H.—More extensive engraving with birds resting inside an oval on both sides of the receiver. Barrels of either Damascus steel in a colorful pattern or Parker Special Steel and in six gauges—10, 12, 14, 16, 20 and 28.

D.H.—Scroll-type engraving motif with the figure of a bird dog inside an oval. Barrels of figured Damascus steel or Titanic steel. No monogram shield and five gauges—10, 12, 16, 20 and 28.

C.H.—More elaborate engraving in scroll pattern with the figure of a dog inside oval on both sides of the receiver. A running deer was shown on the front of the receiver and the triggerguard was engraved. Damascus barrels were ornamental and marked Bernard steel. Fluid steel barrels marked Acme steel and in gauges—10, 12, 16, 20 and 28.

B.H.—More elaborate scroll engraving with figures of dogs inside oval on side of receiver and a scene depicting running dog and deer on front of receiver. Engraved barrel extension and trigger guard. Fore-end and stock panels finely checkered. Damascus barrel of extra-fine steel or Acme steel barrels in five gauges—10, 12, 16, 20 and 28.

A.H.—Heavier engraving encompassing more receiver area with a dog in woodland scene on the sides of the receiver—not in an oval—and two scenes on the front of the receiver, with the larger scene illustrating a dog standing over a fallen deer. Gold monogram shield imbedded into stock. Finely figured Damascus barrels or steel in gauges—10, 12, 16, 20 and 28.

A.A.H.—Opulent engraving completely covering receiver, receiver extension and top lever. Barrels engraved from breech extending approximately two inches. No scenes of any kind—completely scroll engraved. Lavish checkering on stock and fore-end and gold monogram plate imbedded into stock. A few Damascus barrels but mostly Whitworth fluid pressed steel along with a certificate of authenticity from the barrel makers—Sir. W.G. Armstrong Whitworth and Co. in four gauges—12, 16, 20 and 28.

A No. 1 Special—Engraving pattern and checkering design unequalled in a production gun at its time. Receiver and trigger guard completely covered with the trigger guard pierced with a diamond-shaped hole and two teardrop holes above and below. Grip cap checkered and engraving on barrel extends partway up from the breech. Trigger(s) heavily gold plated along with the monogram plate. Whitworth steel barrels and in four gauges—12, 16, 20 and 28.

THE "RIFLEMAN'S RIFLE"

In the days before and after WWII, the pre-'64 Winchester Model 70 reigned supreme as the ultimate sporting rifle.

By Bob Tatsch

When I was a youngster during the early 1930's, I used to visit our local hardware just to look at the guns. Not far from the front entrance was a large showcase of revolvers and automatics. The wall behind the showcase contained an alcove filled with new rifles and shotguns. On both sides of the alcove were ammo shelves.

On one special Saturday, noticing my rapt involvement, the proprietor asked if I would like to hold one of the rifles. He reached over the counter handing me a lever action Winchester. I remember being surprised by its weight and compactness. Born at that moment was the steadfast resolve to some day own as many of those rifles as I could afford. This resolve was not satisfied until many years later when I became both a shooter and a collector.

I encountered my first Model 70 Winchester in 1938. I was 18 at that time and worked every Saturday at the neighborhood gas station. The owner, Red Halburton, was a Texan who seldom if ever missed a deer season.

One Saturday he returned from lunch with a long slender cardboard carton which he unwrapped on the station workbench. Inside the box was a .30-06 Winchester Model 70 Standard Grade rifle with low comb stock and open sights. It was equipped with a leather sling. I remember all these details because I eventually acquired that rifle. All that remains of the original rifle is the receiver. I have had it re-barreled and re-stocked twice over the years. It is now chambered for the 7 mm Remington Magnum. I sincerely wish that I had left the rifle in its original state, perhaps put it away when the original barrel began to wear.

During Red Halburton's period of ownership the Model 70 was kept in the station tool locker. This provision gave me the opportunity to examine it whenever I got the urge—which was pretty often.

As the years passed I entered college, worked in the pre-war aircraft industry and in June of 1942 enlisted in the Navy. Following discharge in 1945 I spent a few weeks looking up old friends, one of which was Red.

At the conclusion of our visit he removed his Winchester Model 70 from the tool locker and gave it to me. I was very touched by the gift and thanked him with considerable emotion. I always wanted that particular rifle and I am sure he knew it. Not long after my visit with him he returned to Texas and thus ended our association.

Not far from where I lived in Los Angeles in those days is the Santa Monica mountain range. During the years following World War II, these mountains provided excellent hunting. They were well inhabited by all of the small game indigenous to Southern California and supported an impressive population of deer.

Opening morning found me at the mouth of Latigo Canyon, about a one hour drive from my home in Los Angeles. As dawn was breaking I left my 1938 Ford parked along the beach and started working my way up the mountain. It seemed strange hunting deer where I could hear the surf pounding on the beach behind me.

I soon topped out on an old fire access road well rutted and partially overgrown with native grasses and brush. Deer sign was everywhere. Brush rabbits bounded off ahead and small coveys of quail exploded at every turn of the old road. I unslung my Model 70 and continued to work my way toward more open terrain which offered better visibility.

Without any warning a large buck stepped out of the brush onto the road ahead. We stared at each other in total surprise, both absolutely motionless. He recovered before I did and cleared the upper embankment in one leap. As he bounded up the hillside I caught occasional glimpses of the tips of his antlers. It wasn't until he reached the top of the ridge that he came into clear view. I led him slightly and touched off the only shot I was going to get. All was quiet for a moment, then at a point beyond where I had last seen him, the brush began to thrash. Working my way up the hillside I found the point of impact, then a little farther on my buck. Three points on a side and quite large for a coastal blacktail. The shot hit him high in the shoulder breaking his backbone. He was the first of many bucks I was to take in those mountains above Malibu Beach. Red Halburton would have been pleased to know that his Model 70 was being well used.

During the period 1947 through 1955 I added more Winchester Model 70's to my collection of hunting rifles beginning with a pre-war .257 Roberts followed by a .250-3000 Savage. Both were Standard Grade with low comb stocks. Each was factory original with iron sights. My only concession toward improved sighting came when I installed a Lyman 48 receiver sight on the .30-06 received from Red Halburton.

In the late 1950's and early 1960's I began to add some of the larger calibers to my collection. The belted magnums opened up new possibilities and as the trophy class bucks were showing themselves at greater distances.

It wasn't until 1961 that I became seriously interested in obtaining pre-'64 Model 70's for purely collecting purposes. I began watching for Winchesters in the classified ads, gun shops, pawn brokers, gunsmiths and anywhere else Model 70's could be found. I am quite sure that for a period of time, collecting these fine rifles became an obsession with me.

Honest guns were not just floating around anymore. It therefore became very important for the collector to be on guard. As pre-'64 Model 70's became more difficult to find, "parts guns" or "made up" rifles were beginning to appear in increasing numbers.

When I started collecting there was very little written on the subject of pre-'64 Model 70's. Collectors shared information among themselves and learned the hard way, by making mistakes. We didn't have Dean Whitaker's fine book and guide *"The Model 70 Winchester 1937-1964."* This excellent volume is an absolute must for any collector, but especially the novice. The book is a wealth of accurate and reliable information expertly presented.

To a great extent the first Winchester Model 70 to be produced was a Model 54 with improvements. For this reason some knowledge of the Model 54 is

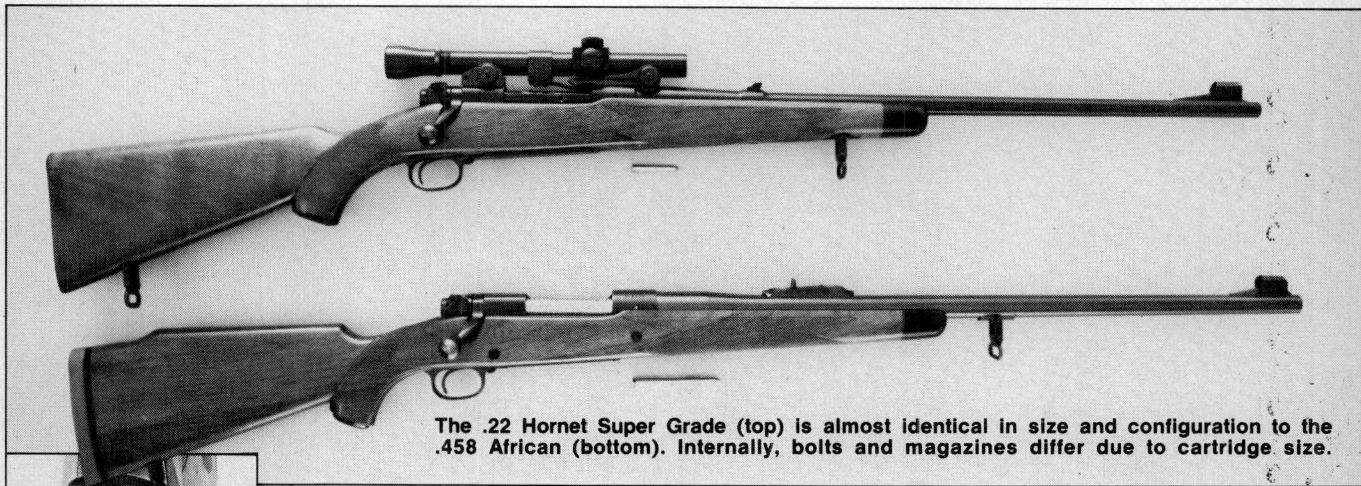

The .22 Hornet Super Grade (top) is almost identical in size and configuration to the .458 African (bottom). Internally, bolts and magazines differ due to cartridge size.

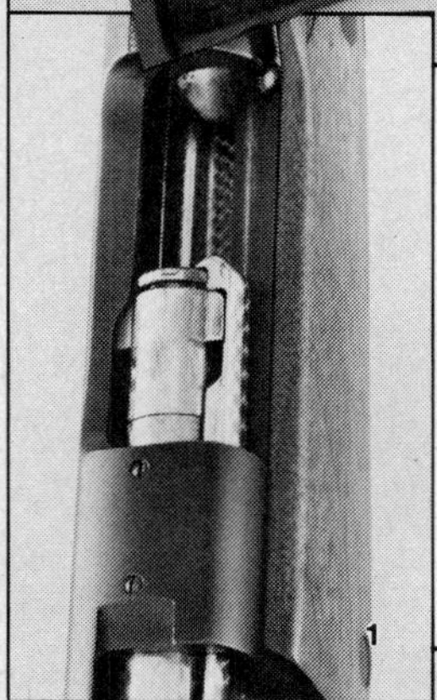

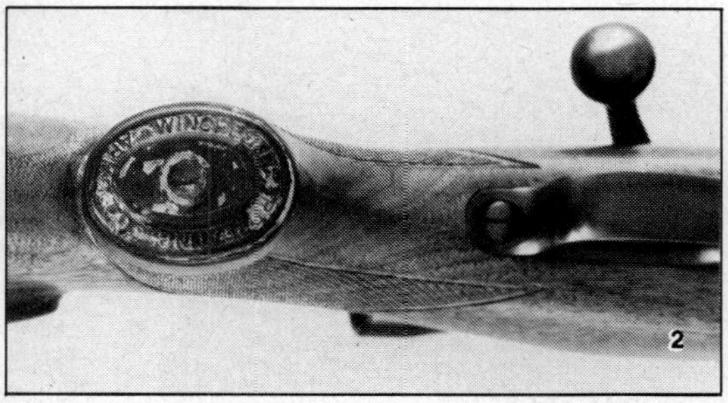

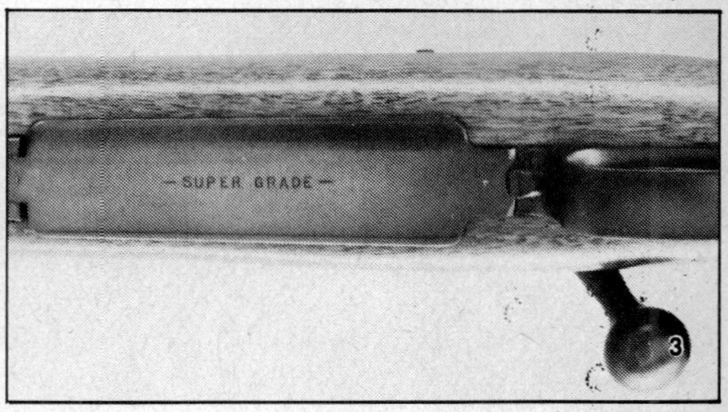

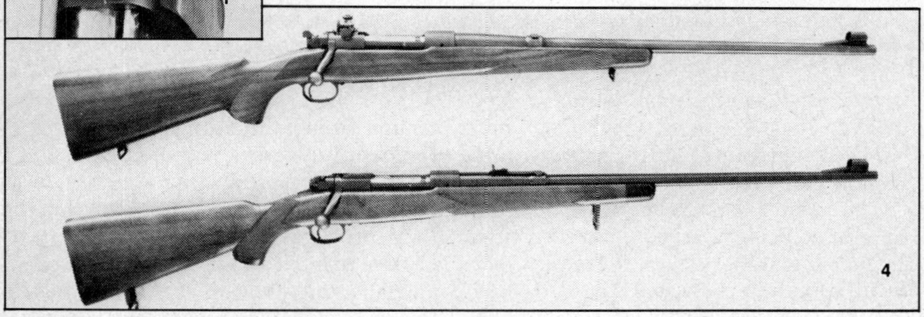

1) Super Grade actions of the later period sometimes have jeweling. 2 & 3) Photos show the distinctive grip cap and floor plate of Super Grades. 4) A major difference between the pre-war .30-06 (top) and the Super Grade transitional .270 is the safety lever. 5) The .458 African (bottom) needed recoil lugs to strengthen the stock, a feature not found on the .375 H&H Alaskan (top). 6) The Alaskan's sling swivels were mounted on the stock, similarly to other pre-'64 Model 70s. On the African, the swivel was affixed to the barrel.

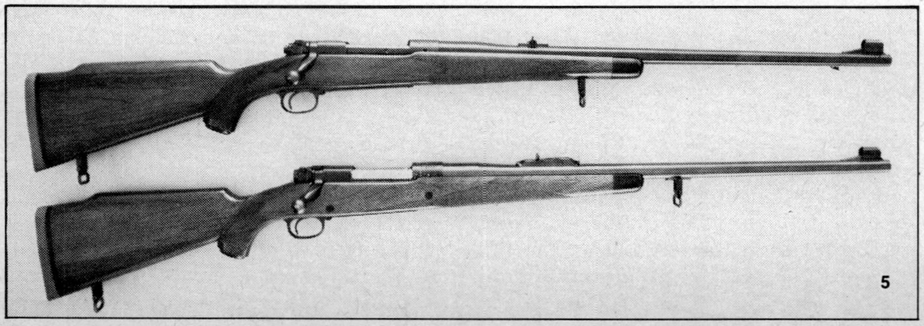

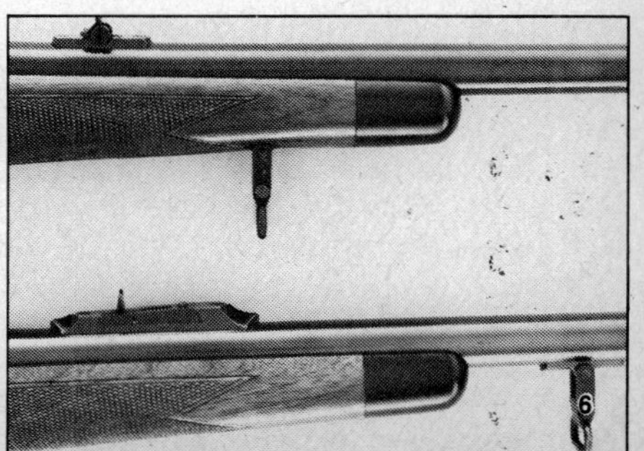

beneficial for the new Model 70 collector. If I were collecting today, I would certainly include some of these fine rifles with my early Model 70's.

This collectible occupied a substantial period of time in Winchester bolt action history and as such warrants some attention from serious collectors.

Publications are available describing in detail the history, variations, changes and markings of the Model 54 Winchester. The following will serve only to introduce the Model 54 as the forerunner of the Model 70 and provide some of the descriptive details.

The Model 54 was first introduced in 1925 and produced with periodic improvements and minor changes until 1937. Approximately 52,000 were produced in ten different calibers. The early Model 54 was the first high powered, bolt action, box magazine, solid frame sporter manufactured by Winchester.

The first introduced had a round tapered 24-inch barrel of nickel steel with integral front sight ramp and rear sight ring. The integral ramp and ring eliminated the necessity of cutting the sight slots into the normal contour of the barrel.

The stock was a one piece sporter with N.R.A. configuration and made from American walnut. Pistol grip and forearm sides were adequately checkered. Butt stocks were fitted with a shotgun type steel butt plate horizontally ribbed. There was no cap on the pistol grip.

The early Model 54 was chambered for the .270 W.C.F. and .30 Gov't '06 cartridge only. To withstand the pressures of these two cartridges bolts were long, well engineered, with two sturdy integral locking lugs at the forend. The bolt cocked with an easy cam action on uplift. The safety was the Mauser type located at the top rear of the bolt. The trigger was a two stage military type with corrugated finger piece. The bolt handle was round, solid, smooth and slanted rearward.

The sights for the early Model 54 were specifically designed for the rifle. The front sight was a Lyman Gold Bead and the rear was a Lyman 66 W folding leaf. The first leaf was for 200 yards, the second for 500 yards. Also furnished was the Winchester 22 G rear sight or the Lyman 48 W Receiver Sight. The receiver was factory drilled and tapped with two holes to accommodate the Lyman 48 W, which was an optional extra.

Early production carried serial numbers 1 through 8000. Caliber markings were placed on the left side of the barrel near the receiver. Winchester proof marks were located on the left side of the barrel and receiver, close together and level with each other.

Winchester factory records indicated that production of the pre-'64 Model 70 actually began in late 1936, however, the first rifles were introduced in 1937. During the following 27 years, the Model 70 pretty well dominated the domestic market as far as high powered center fire, bolt action rifles were concerned. The Model 70 provided the shooter with a reasonably priced quality firearm, well designed and accurate.

Production of the pre-'64 Model 70 was influenced by three distinct time periods of American history. The pre-war or first period of production extended from 1936 until the beginning of the second world war early in 1942, when regular production was halted in favor of war production. During the first production period approximately 65,000 rifles were produced. Collectors and shooters alike referred to rifles of this period as "pre-war" and they are held in higher esteem than those produced later. These particular rifles are examples of the best quality of workmanship, handfitting and finish of all Model 70's produced. The wood of these early guns was dense, high grade, American walnut sometimes resembling French walnut. Stocks were close fitting and generally heavier than those produced later. They were stained followed by a satin-sheen varnish both tough and durable. Checkering was the fish-tail

Robert E. Petersen (left), with Elmer Keith, took this lion using a customized pre-'64 Model 70 .375 H&H Magnum. The pre-'64 action is noted for its dependability. In particular, the Mauser-type extractor is very reliable, even handling heavyweights like the .375 H&H.

pattern, generous and well executed. To my knowledge all pre-war stocks were low comb attesting to the fact that this was the era of the iron sight.

The second period extended from late 1945 to 1951 and is referred to as the "Transition" period. Approximately 160,000 rifles were produced during this seven year period. Rifles classified as transitional are characterized by a low profile fluted safety lever as compared to the high profile rudder shaped lever of the pre-war period. I have seen transitional type safeties on rifles with the early clover leaf receiver tang as well as the late straight tang receiver. There was a time during the second period of production when leftover pre-war parts were being used by feeding them into second period assembly parts inventories. The result was a rifle of high quality but from a collector's standpoint, quite confusing. The overlapping of early and second period parts eventually ran its course, however, there are enough exceptions to the rule to cause the advanced collector to refrain from making any positive statements.

Receiver numbers below 80,000 were generally of the early clover leaf tang. Transition rifles usually carried serial numbers under 100,000.

During the third period of production, consumer interest in a smaller and lighter rifle brought forth the Model 70 "Featherweight." Cartridges for this new rifle were based on the 7.62 mm case and included the .308 Winchester, the .358 Winchester and the .243 Winchester. Later the "Featherweight" was produced in calibers .30-06 Springfield, .270 Winchester and .264 Winchester Magnum.

During the late 1950's and early 1960's consumer interest developed for the magnum cartridges. The Weatherby line of cartridges and rifles either developed this interest or answered it. I am

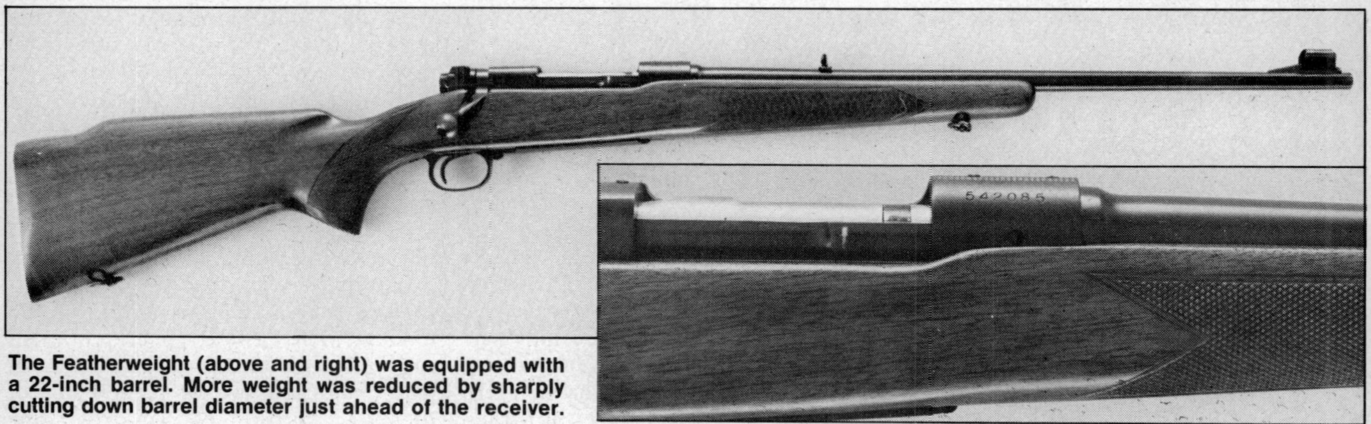

The Featherweight (above and right) was equipped with a 22-inch barrel. More weight was reduced by sharply cutting down barrel diameter just ahead of the receiver.

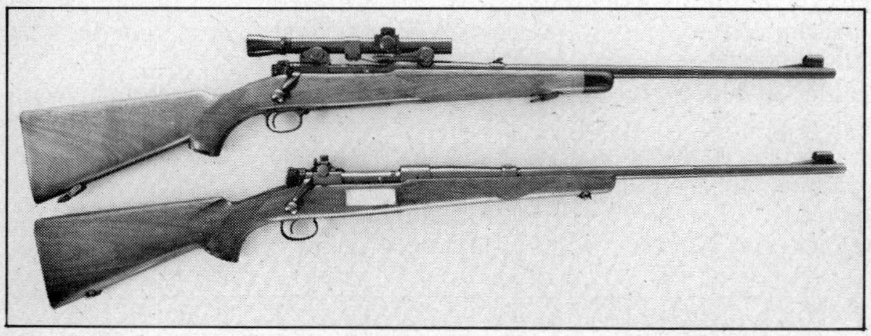

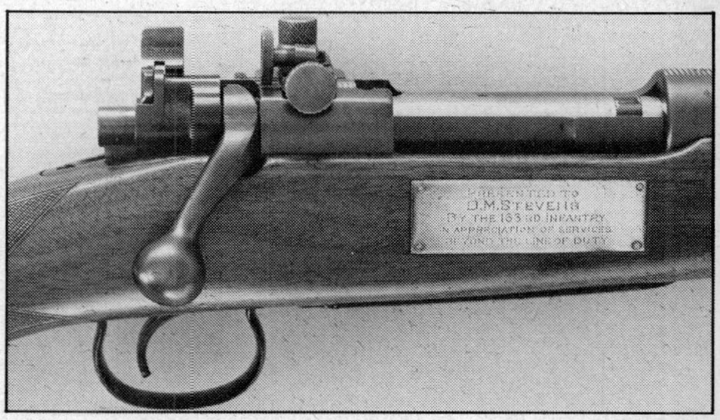

The Model 54 (below) was the forerunner of the pre-'64 Model 70 (above). One of the distinctive features of the M54 is the Mauser type safety (right).

The author, unlike many collectors, enjoys using his collection; few remain unfired.

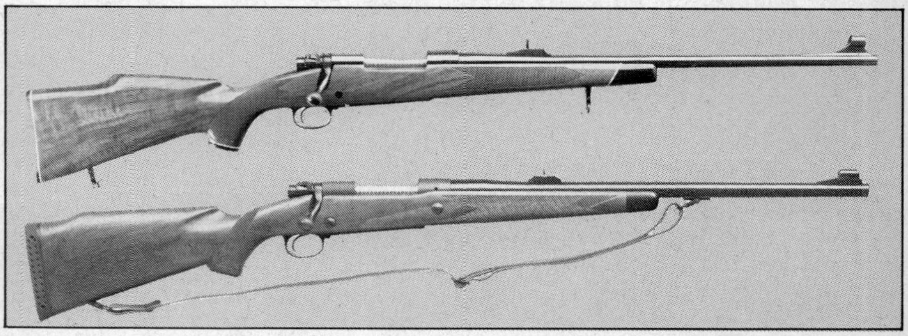

Today's Model 70 has undergone several modern changes and to collectors is a different rifle from the pre-'64. Above, top is a current M70 XTR; bottom, a current M70 African.

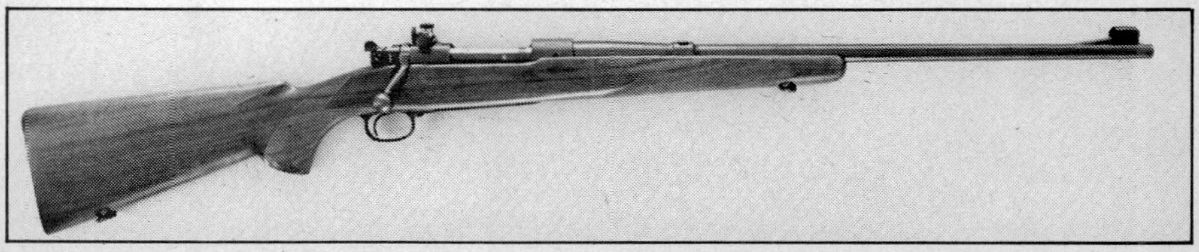

The pre-war Model 70 is considered by collectors to be the finest of the pre-'64 Model 70s, in terms of workmanship. Note the low comb, characteristic of iron sight era.

The legendary pre-'64 Model 70 was probably not as accurate as today's production rifles, due to improved manufacturing methods. In its day, though, it was tops.

not sure which it was. I used to drop by Weatherby's first shop during the early years following WW II and wonder about his beautiful rifles and powerful ammunition.

In answer to the increasing interest in Magnum loads, Winchester introduced a group of belted Magnum cartridges with rifles to handle them. These rifles were marketed as the .264 Winchester Magnum and .300 Winchester Magnum "Westerner," the .338 Winchester Magnum and .375 H&H Magnum "Alaskan," and the .458 Winchester Magnum "African."

Another rifle was introduced about this time and was named the "Varmint Rifle." It was produced in calibers .243 Winchester and .220 Swift.

In 1960 the "Supergrade" was discontinued. Serial numbers had reached the neighborhood of 500,000 and although the quality was still excellent, stocks were not as closely fitted, the checkering and wood density was not of earlier quality. To the eye of the experienced collector, other earlier refinements were also missing. No doubt the inroads of increased labor and material costs were responsible. It was a great disappointment to many when the decision to discontinue the production of these fine rifles was announced.

Production ended in late 1963 with approximately 583,000 rifles having been produced. This ended an era in which fine rifles were produced and marketed to match anyone's needs.

During the years of its production, the pre '64 Model 70 was manufactured in four basic types. During certain periods of its production it could be ordered in 20 different calibers and 14 separate configurations on a readily available basis. On special order the Custom Shop could, and did, turn out many additional interesting variations. The four basic types were the Standard, Competition, "Super Grade" and Special Order. Within each of the four divisions or types there were different styles and configurations. I found that the type of stock a rifle was fitted with made it easier for me to determine its classification. Rifles classified as "Standard" were the Standard Rifle, Standard Carbine, Standard Featherweight Rifle, Varmint Rifle, "Alaskan" and "Westerner."

All six of these rifles were fitted with stocks of the same general configuration. That is plain, except for checkering on each side of the pistol grip and forearm. There were minor variations such as steel, aluminum, and plastic butt plates and two or more kinds of rubber recoil pads. All had fixed sling swivels. Also, there were differences in barrel lengths and shapes due to type and caliber.

Rifles classed as "Competition" were all fitted with the heavier "Marksman" stock with the thick forearm and no checkering. These rifles were the Sniper's Match Rifle, National Match Rifle, Target Rifle and Bull Gun. All four of

Standard pre-'64 Model 70s were made with a sight ring boss to which the rear sight was attached. Sight is early design.

The pre-'64 Featherweight has no sight ring boss. To reduce weight, the barrel is shorter and also of smaller diameter.

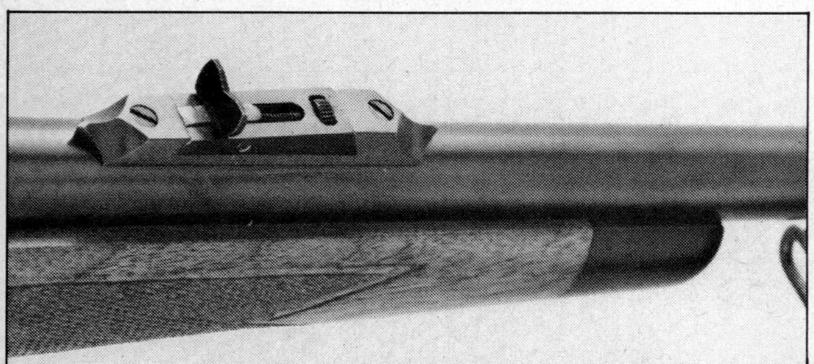

The .458 Win. Mag African was fitted with the heaviest of the pre-'64 rear sights, a modified express sight.

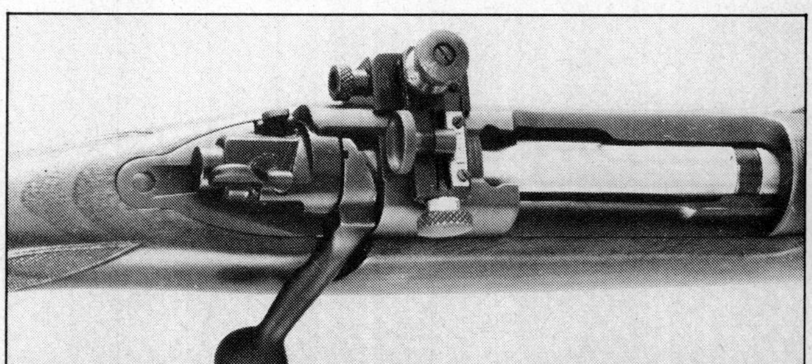

This pre-war .30 Gov't '06 Standard was equipped by the Winchester factory with a Lyman WJS 48 receiver sight.

Well before the pre-'64 Model 70 went out of production, the author was an admirer. Here he poses with M70 and two blacktails in post-war open spaces near Los Angeles.

these rifles differ from each other in other respects but basically share the same type stock.

Rifles classed as "Supergrade" again share a similar type stock. These deluxe rifles shared other more subtle similarities, however are easily recognized by the full wrap checkering on the forearm, plastic or steel pistol grip cap, black forend and distinct cheek piece, to mention a few.

Rifles classed as Special Order of Custom were the products of the customer's special requirements. These came in a variety of configurations not fitting the description of those routinely offered. Special Order rifles are quite rare and could be delivered with special stock dimensions, unique checkering patterns, engraving, custom sighting and other specialties ordered by the customer. Care should be taken before judging such rifles as not original or parts guns. Just because they do not fit catalog descriptions does not mean that they do not qualify as excellent collectibles. Quite the contrary, they can be more valuable as Winchester factory Custom Shop products than the more common production rifles.

For the benefit of the beginning collector the following paragraphs will generally describe each of the rifles produced under the four separate class-

MODEL 70 RARITY

During the twenty-seven years the Model 70 was produced chambering and configuration was strongly influenced by many factors. The end results, from a collector's standpoint, is that some rifles are rarer than others. This rarity makes itself felt in two distinct areas, configuration and caliber. For example, two rifles of the same production period, type and configuration will not be the same value if one is a .30-06 and the other a .35 Remington. The latter is many times more valuable because a lot less were produced. Therefore the following two charts will serve the beginning collector as a guide.

Style and Configuration Rarity
(Beginning with the most common)
Standard Rifle
Standard Featherweight Rifle
Supergrade Rifle
Target Rifle
Varmint Rifle
Bull Gun
Alaskan
Westerner
National Match
African
Standard Carbine
Supergrade Featherweight Rifle
Supergrade Carbine
Snipers Match Rifle

Caliber Rarity
(Beginning with the most common)
.30-06 Springfield
.270 Winchester
.308 Winchester
.243 Winchester
.300 H&H Magnum
.375 H&H Magnum
.220 Swift
.257 Roberts
.22 Hornet
.264 Win. Magnum (Westerner)
.458 Win. Magnum (African)
.338 Win. Magnum (Alaskan)
.250-300 Savage
7 mm Mauser
.300 Win. Magnum (Westerner)
.358 Winchester
.35 Remington
.300 Savage
9 mm Mauser
7.65 mm Argentine

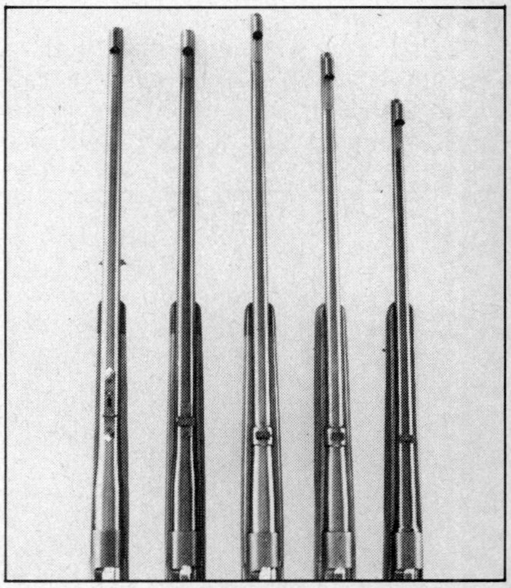

Photo shows variations in barrels and sights. From left: African, Super Grade .375 H&H, Super Grade .300 H&H, Standard .30 Gov't '06, Featherweight.

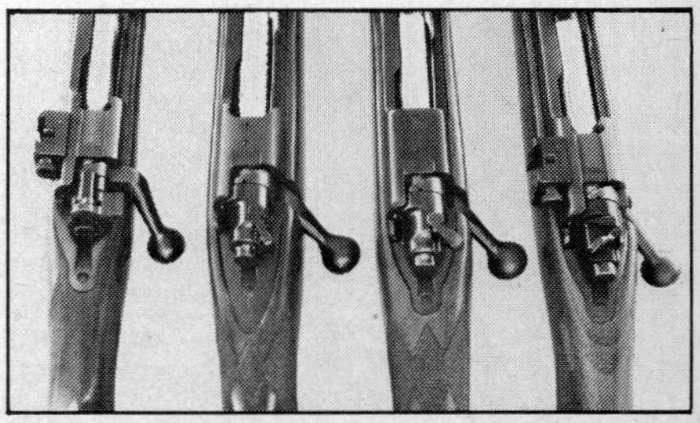

At left is Model 54 next to three Model 70s. M54 had a Mauser type bolt that would make modern low scope mounting impossible. Note early cloverleaf tangs.

ifications: Standard, Competition, Supergrade, and Special Order.

Standard Rifle

Introduced with a pistol grip N.R.A. type American Walnut stock with full well shaped forend. Grip and forend checkered. Box magazine, hinged steel floor plate, steel magazine follower, steel triggerguard, fixed sling swivels, front sight on ramp base with front sight cover. Steel or plastic butt plate except Magnums which in some cases were fitted with solid rubber recoil pads. Usually the .300 H&H Magnum came with steel butt plate. Barrel length 24 inches except for the .220 Swift and .300 H&H Magnum which were standard at 26 inches. The .375 H&H was standard at 25 inches.

Standard Carbine

The same as the Standard Rifle except that it was fitted with a 20-inch barrel.

Standard Featherweight Rifle

The same as the Standard Rifle except for a 22-inch barrel minus the rear sight base and stamped Featherweight, aluminum triggerguard and hinged aluminum floor plate. Earlier models furnished with oxidized aluminum butt plate, later models plastic.

Varmint Rifle

This rifle was provided with a straight tapered 26-inch barrel in calibers .220 Swift and .243 Winchester only. Walnut stock of same configuration as Standard Rifle, N.R.A. pistol grip and hand checkered. No sights were provided but blocks for telescopic sights were installed and the barrel was drilled and tapped for a front sight. I have seen late models with aluminum triggerguards and floor plates, however most models were all steel. Standard type fixed swivels.

Alaskan

This rifle was produced in calibers .338 Winchester Magnum and .375 H&H Magnum only. The latter was furnished was a 25-inch barrel and a special rear sight of two types, both folding leaf. Between 1960 and 1962 the .338 Winchester Magnum was produced with a 24-inch barrel. Both calibers were fitted with Standard Rifle walnut stocks.

Westerner

Produced only in calibers .264 Winchester Magnum and .300 Winchester Magnum. Barrel length for the .264 Winchester Magnum was generally 26 inches, however a 24-inch barrel was available from 1961 to 1963. Barrel length for the .300 Winchester Magnum was 24 inches. Stock and swivels the same as the Standard Rifle.

Sniper Match Rifle

The Model 70 Sniper's Match Rifles were generally fitted with the extra heavy 26-inch round barrel produced for the Model 54 in 1936 and had forged front sight ramps milled to accept a sight hood. Also, a few of these rifles were assembled from early production Model 70 barrels in both 24 and 26-inch heavy barrels.

When the stock of Model 54 barrels was depleted Winchester fitted a number of Sniper Match Rifles with heavyweight 25 and 28-inch barrels made from the first production run of Model 70 barrel blanks. The Sniper Match, unlike any other Model 70, was routinely provided with any of three separate stock configurations.

National Match Rifle

This rifle had the appearance of the Target Rifle in most respects except that it was fitted with a 24-inch Standard barrel. It was furnished with the plain Marksman stock, adjustable sling swivel bases and the receiver bridge was slotted for a clip. Front sight ramp was lower than the Standard barrel ramp and not slotted for a hood. Quite often the rear sight base was not slotted.

Target Rifle

This rifle was produced in a variety of calibers with medium weight straight shanked barrels generally 24 inches in length. Non-checkered Marksman type stock with adjustable sling swivel bases. Receiver bridge generally clip-slotted but occasionally not. Adaptable for a large variety of target sights but generally found with Lyman or Redfield "Olympic Model" receiver and muzzle target sights. Barrels are generally drilled and tapped for telescope blocks.

Bull Gun

This rifle came in two cataloged calibers only, .30-06 Springfield and .300 H&H Magnum. Other factory original

1

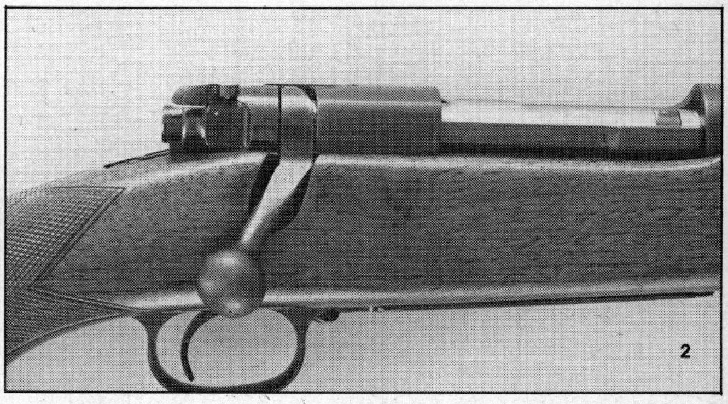

2

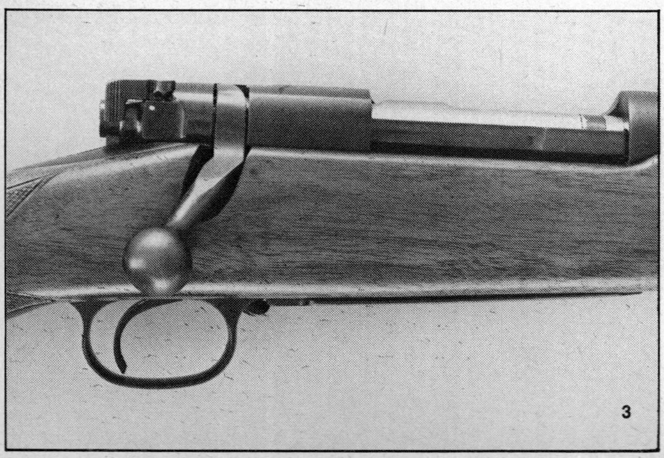

3

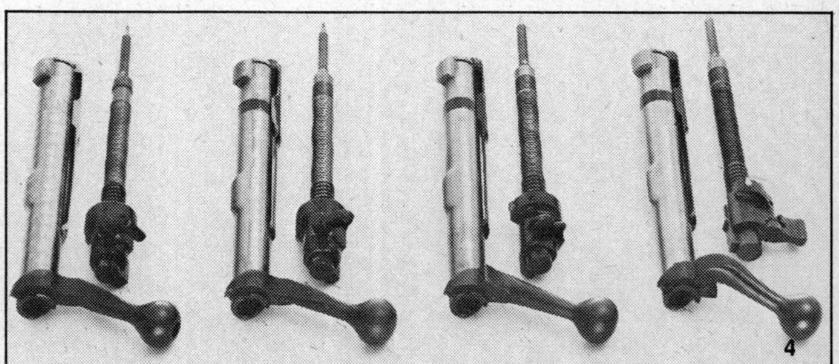

4

Variations in bolt safeties are visible in profile views: 1) a pre-war rudder type safety; 2) fluted transitional type safety; 3) latter period low profile safety. Photo (4) illustrates the similarities and subtle differences in bolt construction. From left: Super Grade .300 H&H, Transitional .270, Pre-war .30 Gov't '06, Model 54.

From its 1930's birth, the pre-'64 Model 70 was the premier American hunting rifle. As the iron sight era waned and optics became popular, the M70 was adapted for scopes.

calibers were produced on special order. Furnished with extra heavy 28-inch barrel, Target style stock with adjustable sling swivel bases. Clip feeding slots in rear bridge of most .30-06 Springfield bull guns, however omitted in late production guns. Bull guns weighed about 13 pounds.

Super Grade Rifle

These rifles were the deluxe production of the Model 70's. They differ from the Standard Rifle in that the stocks were specially selected wood, full wrap hand checkered forepiece with black forend tip, pistol grip cap and cheek piece. Detachable deluxe sling swivels with floor plate stamped "Super Grade." Prior to the late 50's many of the barrels were stamped "Super" on the bottom denoting that these units were specially selected. I have never seen a Super Grade with a plastic butt plate. These deluxe versions were similar to the Standard Rifle in every other respect except late production units were sometimes furnished with jeweled bolt shaft, extractor and follower.

Super Grade Carbine

It is my understanding that only a very few of these units were produced. They were the same as the Super Grade Rifle except for the 20-inch barrel length. Depending on caliber and wood density these weigh about 7¾ pounds.

Supergrade Featherweight Rifle

This rifle is the same as the Standard Featherweight rifle except for the stock which is the same as the Super Grade Rifle.

African

Produced in caliber .458 Winchester Magnum only. Barrel length 25 inches. Bolt Shaft, extractor and follower usually jeweled, however some are not jeweled or only one or two of the three parts are. Equipped with a special rear sight and special front sight bead. General specifications the same as the Super Grade Rifle except for front sling swivel which is mounted on a flat lug brazed to the barrel directly in front of the black forearm piece. From 1956 to 1960 floor plates were stamped Super Grade. From 1960 to 1963 both plain floor plates and aluminum floor plates stamped Super Grade were sometimes used. "Africans" were furnished with the Noshoc solid red rubber recoil pad with the molded Winchester trade mark.

Production of these fine rifles came to an end in late 1963 after a total of 583,200 rifles had been produced. Many loyal followers feel that this ended the age during which our best firearms were manufactured.

During the years the pre '64 Model 70 was being produced, North America provided great hunting opportunities. Most of the rifles purchased during that period were used for that purpose. When examining the marks and blemishes on one of these fine collectibles, I can't help but wonder about the story it could tell. Nothing to enter into the journals of history perhaps but certainly a story of hunting during an era that is gone for all time.

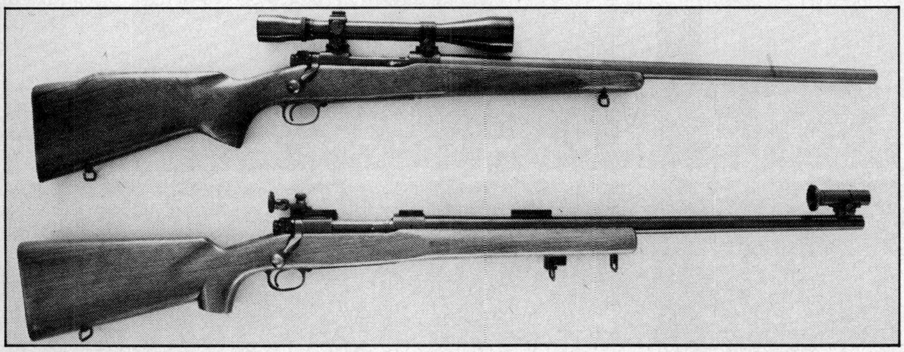

Top is a .243 Varmint and below it a .30-06 Target. Although they both have heavy barrels, only the Target models have the heavy stock configuration and scope blocks.

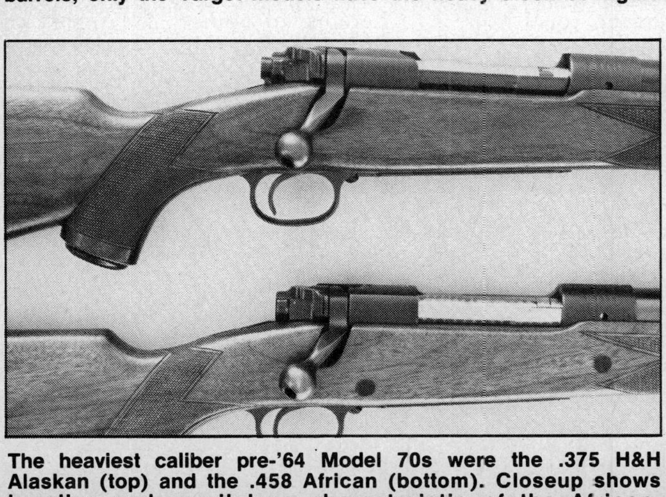

The heaviest caliber pre-'64 Model 70s were the .375 H&H Alaskan (top) and the .458 African (bottom). Closeup shows jeweling and recoil lugs characteristic of the African.

Nearly two decades after production ceased, the pre '64 Model 70 is still sought after by hunters for its rugged dependability. But today, hunters compete with collectors.

IS THE SPORTSMAN'S KNIFE A TOOL OR ART?

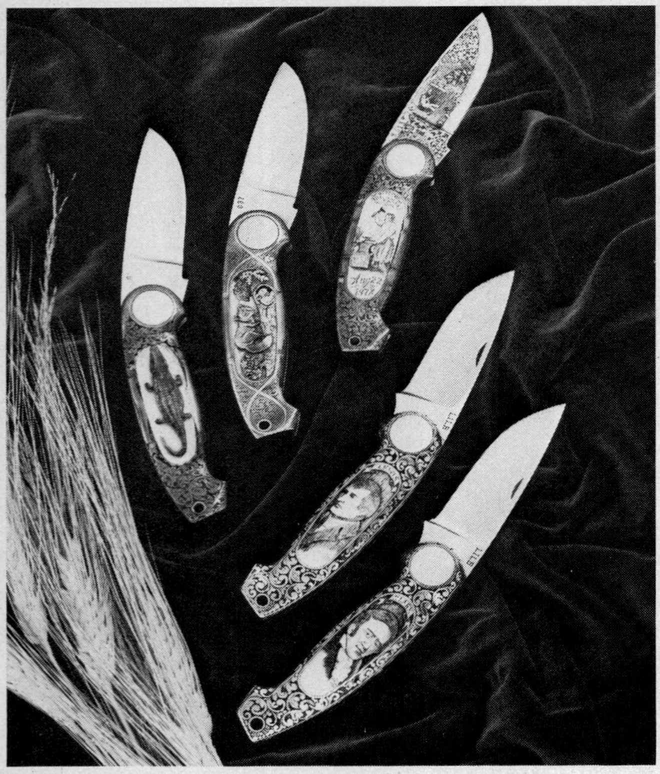

Here's an example of the knife as a work of art. These highly engraved Lile folders would lose a great deal of their investment value if they were closed or sharpened—much less actually used!

This is the opposite side of the coin; the knife as a mere cutting tool. For this purpose, the ability of the knife to take and hold an edge is more important than its actual appearance.

A custom knife commands a high price these days—do those extra dollars buy a better knife?

By Jon Sundra

My professional hunter worked quickly and with a deftness that comes only from repetition. The knife was a blur of movement yet not a swipe was misplaced as the cape of my Waterbuck came off as easily as a banana peel . . . and almost as quickly. I was helping as best I could but for the most part I was simply keeping tension on the skin.

Watching the whole proceeding was another hunter from camp, an Alaskan as I recall. A nice enough chap but one of those "equipment snobs." You know the kind—if Gucci had made jock straps and safari boots, this guy would have been wearing both.

Anyway, on completing the caping job my PH set his knife down, only to have it picked up by our spectator who then eyed it critically. I don't recall exactly what he said but it was some vaguely derisive comment about how well even crude instruments could work in the hands of someone who knew how to use them.

With that, my hunter politely reclaimed his knife.

"It cuts, don't it mate? What else is a knife supposed to do?"

Indeed, what else *is* a knife supposed to do . . . if anything? After all, when sharp, a ten-buck knife will cut as well as a bench-made one. The differences are in the length of time each will stay sharp, the quality of materials used, and how well those materials are combined in the end product. Which brings us to the most interesting question of all: "At what point do those added dollars cease buying a "better" knife, just a fancier one? To put it another way, when does a knife cease being a tool and become an *object d'art?*

There is, of course, a category of benchmade knives which are so lavishly embellished that only an idiot would take one afield or use it for anything other than letter-opening. Take the "Indian Chief" folder series of Jimmy Lile for example; sharpening or even closing one of those knives would detract from its collectors' value. Perish the thought of actually *using* one. But that's all right, too, because as man's oldest and most personal tool, a knife has always been a natural medium of artistic expression. The same can be said of firearms and armour.

Between that ten-buck number, then, and those $6,500 apiece Lile folders is this vast gulf of production, semi-production, benchmade and custom knives

A critical step in knifemaking is the actual heat treating of the blade. A slipup here can render a hunting knife useless!

Even mass produced knives, like these Case fixed-blade hunters, have several steps of hand work involved in their manufacture.

These highly engraved boot knives, by R.W. Wilson of Wierton, West Virginia, aren't for field use, but are fine collectibles.

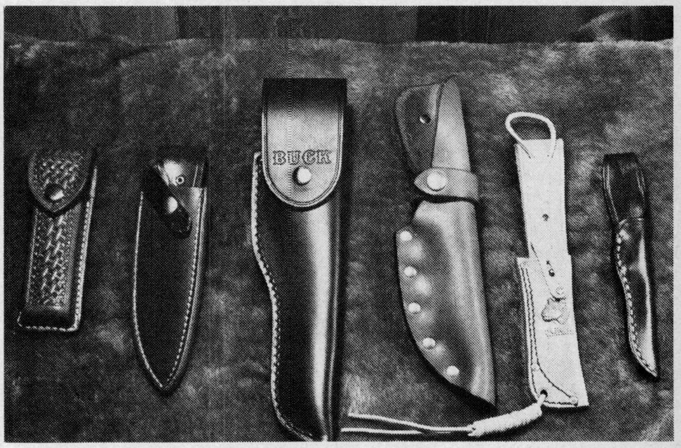

Most knifemakers include leather scabbards with their knives. (L to R); Browning scabbard, S&W, Buck, Rigid, Puma, Gerber.

that represent the major categories. Surely no knife belonging in those first two production categories could qualify as anything other than a working tool, a knife designed and meant to be used. Oh, one could have a production knife lavishly embellished by a Lynton McKenzie and thereby, by virtue of the artistic effort alone, transcend the medium itself, but sensible folks just don't lavish a grand or two of top-quality engraving on a ten buck knife . . . or a fifty-buck one for that matter.

So when we talk about a knife and at what nebulous point it ceases being a tool and becomes an art form, we're talking about at least a benchmade product, i.e., made by hand one-at-a-time. However, thanks to what could best be described as a renaissance in knifemaking over the past 20 years, features and styling once found only on a "custom" knife are now commonplace among production ones. So perhaps it is fitting here that we first look briefly at the current state of the mass-produced, production knife.

Don't let the term "production" mislead you; there's a surprising amount of handwork in even modestly priced knives, especially folders. Risky though it is to equate quality with a rigid price structure because there will always be exceptions, today's mass-produced blades generally retail for under $25.

A generation ago a typical "sheath knife" was characterized by a hidden tang covered with leather washers with an aluminum pommel. The hilt consisted of a stamped piece of nickel silver or brass sheet stock poorly fitted and held in place only by virtue of its being held captive by the washers behind it. Blade material was an AISI-1095 or a similar, easy-on-punch-dies steel and the configuration was invariably a Bowie or clip-type design with varying degrees of upsweep at the point.

Even less innovation was seen in folders. Lock backs were unheard of in the mass-produced product. A quality knife had solid nickel-silver bolsters, brass liners and bone scales. Period.

Today it's a much different game. The tremendous interest in benchmade knives and the new methods and materials pioneered by custom cutlers of the past couple of decades have resulted in dramatic changes in the industry. Today's consumer is far more knowledgeable, far more demanding than he used to be.

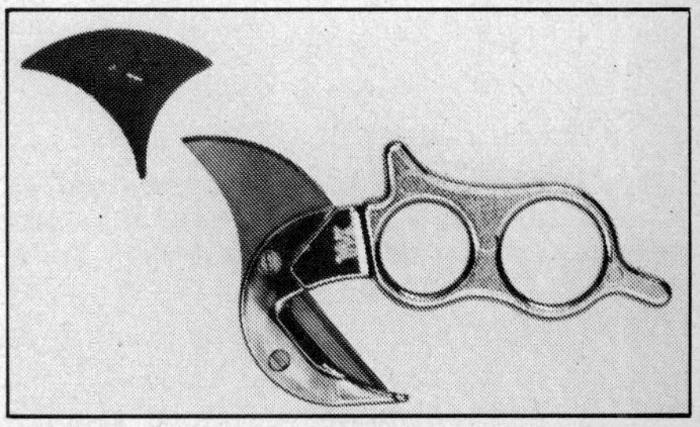

Despite its bizarre and unconventional appearance, the Wyoming knife is one of the most efficient skinning knives.

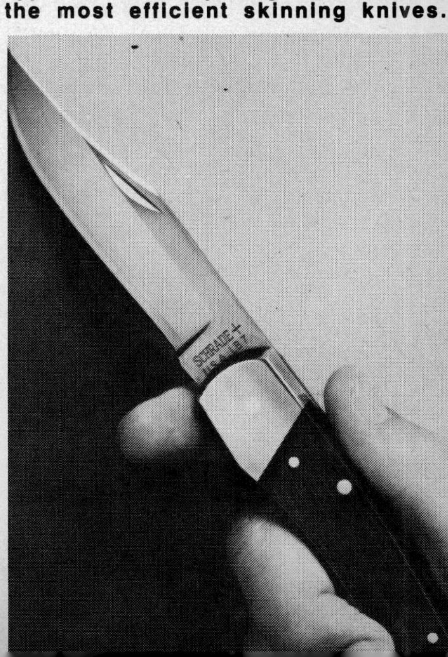

Heavy-duty, single-blade lockback knives are very popular with hunters. This big Schrade would be an excellent choice.

Professional outfitters put more mileage on a knife in a single season than most hunters do in a lifetime. Because of this, they're more interested in performance than appearance.

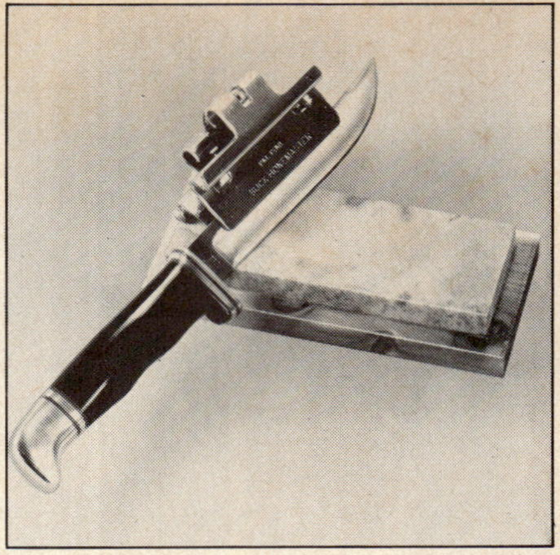

A dull knife is pretty worthless. The Buck Honemaster will keep the proper angle of blade to sharpening stone for a good edge.

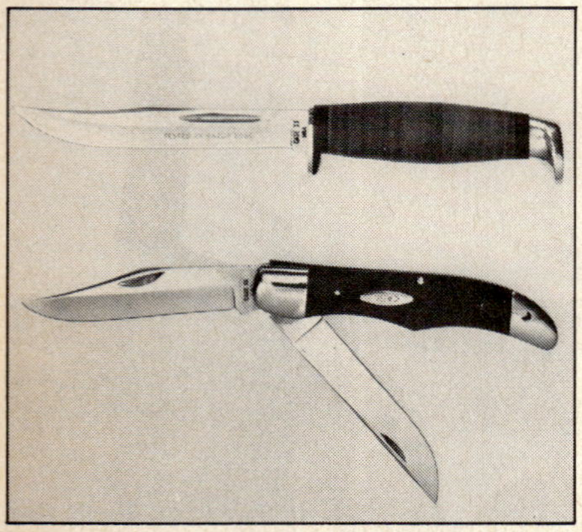

Case makes fixed blade knives (top) as well as folders. Although they're mass-produced, Case builds an excellent knife.

For that under $25 figure mentioned earlier, it is now possible to find fixed blade knives in 440C stainless, tapered tangs, superior scale materials like resin-impregnated wood and linen micartas; silver-soldered hilts, and blades with more functional shapes for both.

You're not likely to find all of the above features in a single knife selling for under $25, of course, but it's not uncommon to find one or two of them.

In folders it's the same story: the use of 440 and similar stainless steels is now commonplace, as is the use of micartas and other synthetics for scales. The drop point design,´ better for general use, is on its way to replacing the clip point.

So what *don't* you get that you could if you were to spend more money? Primarily, it's a matter of quality more than edge-holding. In the next price category, that of semi-production knives costing, say, $25-$70, your money is definitely buying you a better product but it is better more in terms of the quality, workmanship and durability. Most knives in this category use the same steels as found in less expensive knives. That is not meant to be disparaging of the semi-production category, but rather a compliment to the fine quality of steel used in the mass-produced product.

The semi-production knife is one of a set model and style but one on which there is more hand work done by fewer, more skilled craftsmen. Good examples of this category are the knives of Kershaw, Gerber, Buck, Al Mar, Cam III, Puma and the top-of-the-line models of Case, Camilus, Western and Shrade, to name but a few.

The two major things you still do not get in a semi-production knife is that unique, one-of-a-kindness that you get in a benchmade knife, and the more exotic steels which have better edge-holding characteristics—though not as dramatic as one might think.

In this category of knife it is common to find incorporated all of the features which once were found only in the benchmade product. The most prevalent steel is stainless, especially in folders, or at least a high chromium stain-resistant alloy. It is not so much a matter of stainless being best but it is the most practical. Don't get me wrong: today's stainless alloys are very good indeed. A 440C in the 57-59 RC range, for example, will take and hold an edge very well. But steels with less than a 15 percent chromium content tend to stain and discolor, to say nothing of rusting if really neglected.

The quality, materials and workmanship found in the upper end of the semi-production category can buy you a very well made knife indeed. In fact, some of them can hold their own with a lot of benchmade blades in terms of fit and finish, symmetry of bevels, sharp, well-defined facets and edges—all the criteria used to judge quality. Scabbards also exhibit comparable class and evidence more thought-out designs which not only look attractive, but allow easy access to the knife while still protecting it and the owner. If most or all of these characteristics can be found in the semi-production knife, why a benchmade one? Perhaps the best answer would be to simply draw an analogy to the gent who commissions a gunsmith to build him a rifle. The rifle may not be expected to actually *shoot* any better than a production Ruger, Remington or Winchester; it's simply a matter of wanting something different that isn't available over the counter. Surely a custom rifle is no more necessary than a benchmade knife but as one becomes more knowledgeable and discerning, whether it be in knives, guns, art or cuisine, it follows that one also becomes more demanding. The beauty of a benchmade knife is that you can indulge yourself; you can page through catalogs and cruise the aisles at shows and often discover that precisely what you want already exists. Or perhaps it's just a slight change in line here, a contour there, and a different handle material that's needed to come up with what you deem perfection. No problem. Just tell the maker what you want. Or maybe you want to start from scratch with a design all your own. Again no problem. As I see it that's the only difference between a benchmade and a "custom" knife: one is a more or less standard model of a particular maker; the other is made to the customer's specifications.

What I regard as the greatest attraction of all about the benchmade knife is the prospect of finding an unheralded maker whose work is close or equal to that of established "name" makers . . . at a fraction of the price.

For what is often just $10 or $20

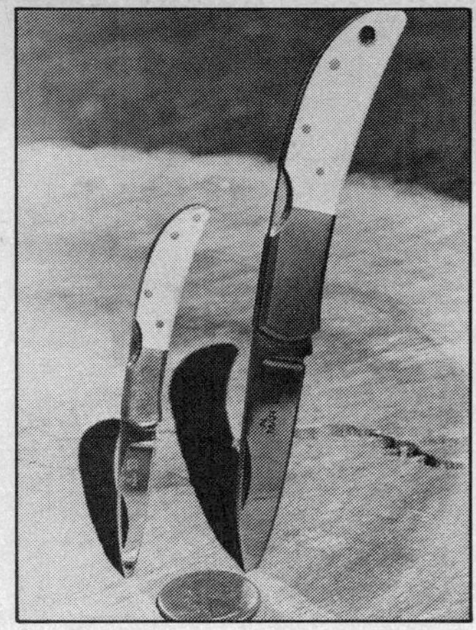

1

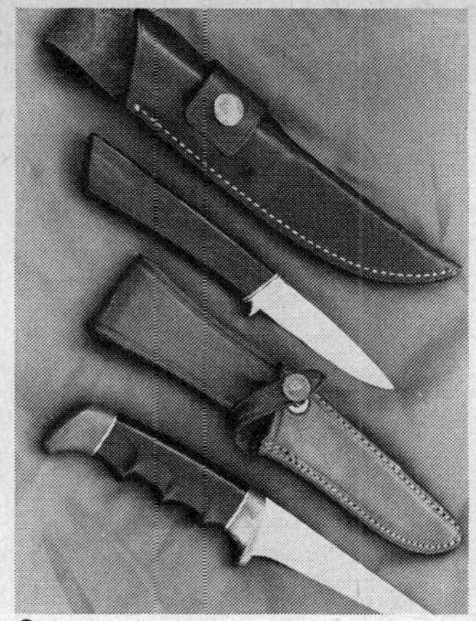

2

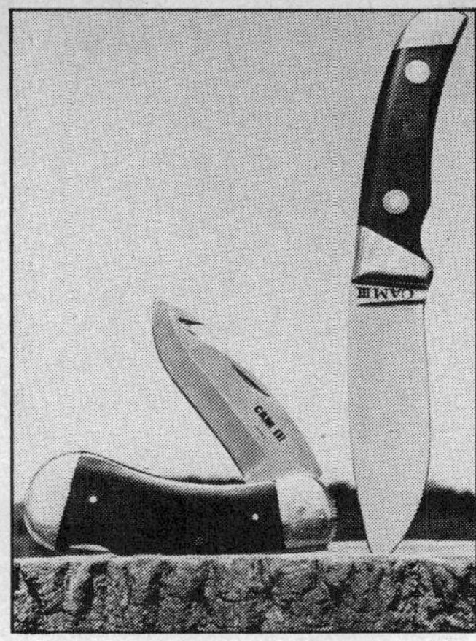

3

4

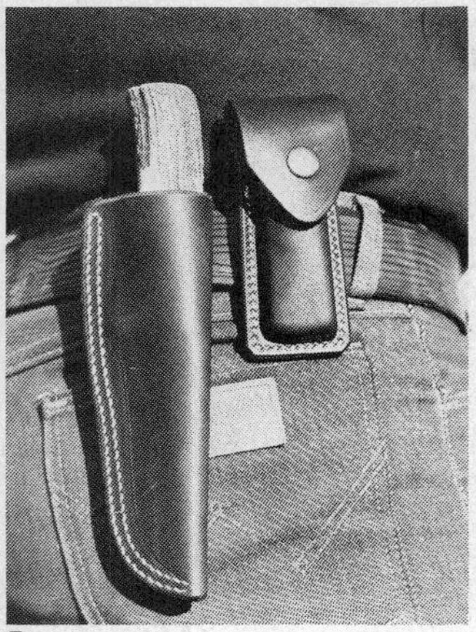

5

(1) Al-Mar lockbacks exhibit many features once found only on handmade knives. (2) Gerber (T) and Kershaw (B) both come with high quality scabbards. (3) Cam III lockbacks feature micarta handles and guthooks. They also offer sheath knives. (4) Both Gerber (T) and Kershaw (B) now offer folding lockbacks with rotary locks—a feature once found only in custom lockbacks. (5) Many hunters carry sheath knives and lockbacks—particularly if they anticipate heavy skinning and gutting (6) This bench-made skinner and caper were made by Vince Feragotti of Industry, PA and cost no more than most semi-production knives.

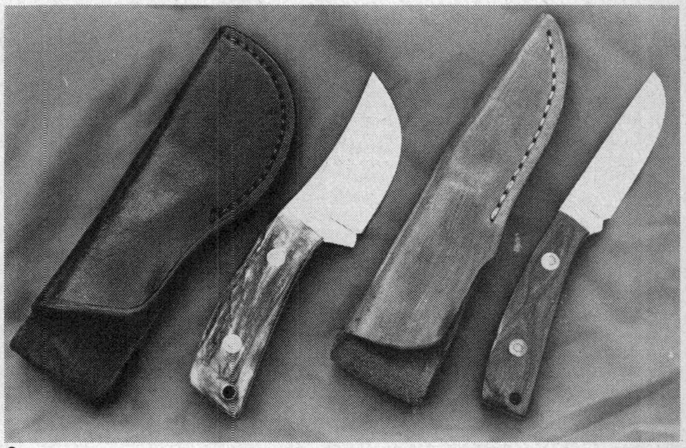

6

more than some of the better semi-production knives which you must take "as is," you can have the real thing. You can specify the steel; the blade design, size and thickness; the handle material and shape . . . anything you want. And the scabbard (or pouch if it's a folder), is custom made and wet molded to fit *your* knife, not just one that's similar.

Finding aspiring, talented makers willing to put reasonable prices on their work is not as difficult as one might think. You do have to look, however. But then that's the fun of it. Anyone can walk into a wine store, plunk down 50 bucks, and come out with a bottle of Lafite or Mouton that will be extraordinary. There's far more challenge and satisfaction, however, in discovering an unheralded chateau selling for five bucks which may be quite excellent.

And a true custom-made knife need cost no more than the maker charges for his standard models—if he even has "standard" models. That's assuming, of course, you're not specifying exotic materials and artistic adornment.

It should be mentioned here that about the only aspect of quality that can't be evaluated visually is that of the steel used and the proper heat-treatment of same—the latter being more important than the former. Here is where the reputation of the maker is your only assurance . . . and that's worth money. The alloy tool steels being used today, steels like 154CM, D2, A2, 440C require complex and highly accurate heat-treating procedures that go beyond the skills required to fashion the knife itself. Most cutlers do not have that knowledge or the sophisticated furnaces and related equipment required to do proper heat-treatment and must rely on outside facilities. It pays to inquire into this aspect of the knife making picture before committing to a purchase or a commission.

Returning now to the question posed at the outset: where do you draw the line? When does your money stop buying a "better" product and start paying for the reputation of the maker.

At the risk of stepping of some very soft ground, I'd say with the kind of inflated dollars we're spending in 1980, a hundred bucks is as good a figure as any. There are sure to be many exceptions, of course, but for the most part going beyond the century mark for a working fixed-blade knife—a bit more for folders—is going to buy you more reputation than product. That's fine with me; after all, the ability to command $250 or more for just a basic knife because a guy's a name maker is what the free enterprise system is all about. That's what any artist strives for—recognition of being numbered among the best.

There's no way a name cutler's work is three times better than a lesser-known craftsman, twice as good, or even 20 percent. Indeed, what distinguishes good from great can be so subtle that it may exist only in the eye of

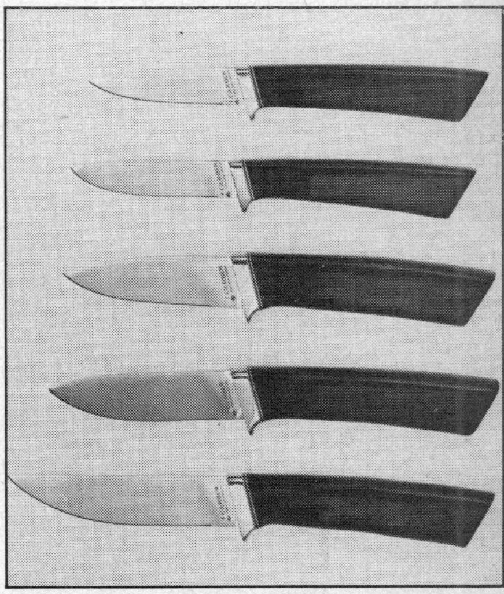

Gerber blades have a reputation for edge-holding characteristics. They offer blade sizes and shapes for every requirement.

the beholder because he wants to see it and the maker's rep *says* it's there. Then there's the single magazine article that may catapult an unknown cutler into the kind of national prominence that enables him to remove $75 price tags on his knives and replace them with $150 ones.

In the final analysis the relationship between quality, price and status is no different with knives than with anything else. You can whip out your Loveless in front of the guys at hunting camp and let it speak for you. Or you can let the product speak for itself.

During the siege of Kimberly in the turn-of-the-century Boer War, the British were saved by Yankee ingenuity in the form of a handmade, long range field piece called. . .

Long Cecil

By Jack Lott

War erupted between the two Boer republics of South Africa, the Transvaal and the Orange Free State, and Great Britain, on October 12th, 1899. Free State commandos invaded Britain's Cape Colony and Transvaal commandos occupied Laing's Nek, Natal. Thusly had Transvaal President Paul Kruger and Orange Free State President Steyn answered the ultimatum of England's Sir Alfred Milner, demanding "reforms," including the vote, to British subjects among the "Uitlanders" (foreigners) living in Johannesburg, the Transvaal's "City of Gold." Kruger had armed his republic to the teeth with the help of Germany's Kaiser Wilhelm, the overflow of weaponry spilling into Steyn's Free State, whose burgher (citizen) commandos joined with the Transvaalers in a common cause. Kruger had presented his own ultimatum to the British, confident in his sharp-shooting burghers with their new 7 mm Mausers, and the small but ultra-modern German-staffed Staats Artillery. Kruger demanded that the British pull their troops back from the borders of the republics but the British War Minister, Lord Wolseley, had made his plans for war and there would be no backing down from a struggle to achieve Britain's goal—total domination of the gold and diamond rich land.

Cecil John Rhodes, the great mining magnate and founder of Rhodesia, was in a social whirl at his estate "Groote Schuur" (great barn) on the outskirts of Capetown, and he kept his anxieties from his guests as he watched the ominous buildup of forces. Kimberley, the hub of Rhodes' diamond mining industry lay vulnerable to the north, only four miles from the Free State frontier. When the Free State publicly declared solidarity with the Transvaal, the fear of siege went through Kimberley like a veld fire, and Rhodes decided to leave the security of the Cape and share the risks of siege with his associates and employees of his De Beers Company. On the night of October 9th, Rhodes slipped away from the guests at Groote Schuur and took the last train to Kimberley. The train arrived a few hours after war was declared. Rhodes' mission: To ensure that all steps needed to defend Kimberley were taken.

In command of Imperial troops at Kimberley was Lt. Col. Robert G. Kekewich, commanding the 1st Battalion of the Loyal North Lancashire Regiment, arriving four weeks before war was declared to assess the military readiness of the garrison. The arrival in

Photos Courtesy of McGregor Museum, Kimberly, S.A.

During the siege of Kimberly, the imposing figure of Cecil John Rhodes was a common sight at the beleagured city's gun emplacements. This Boer-wartime artist's sketch shows the founder of Rhodesia inspecting his namesake, "Long Cecil."

Kimberley of the aristocratic Rhodes, with his dislike of professional military men and an ingrained inability to accept second place, spelled trouble. Rhodes had conquered Lobengula, king of the Matabele in 1893 in what is now Rhodesia, using his Chartered Company's forces, and continued to exercise overall leadership through the savage Matabele and Mashona rebellions of 1896 and 1897.

Colonel Kekewich was alarmed as he contemplated the inadequate Kimberley defenses manned by volunteers of the Diamond Fields Artillery, Diamond Fields Horse, and the Kimberley Regiment. Rhodes had raised the Kimberley Light Horse, subsidizing the over 300 mounted infantrymen. What Col. Kekewich found out was that the total volunteer forces available in Kimberley were but 540 men. The garrison's artillery were obsolete 2.5-inch rifled muzzle-loading mountain "screw guns."

By late September a company of the Royal Garrison Artillery, four companies and a mounted unit of the Loyals, plus Service and Medical Corpsmen, some 600 men in all, arrived in Kimberley. After the declaration of war, Kekewich's command included 1220 regulars and volunteers with 12 2.5-inch rifled muzzle-loaders.

The strength of Boer forces surrounding Kimberley included eight 75 mm Krupp field guns with a 5000 meter range and an ancient 9-pounder rifled muzzle-loader with a 3200 meter range. The nearly 5000 man Boer force was deployed effectively about Kimberley's perimeter with their guns emplaced solidly in rock and earth shell-proof emplacements, just out of range of Kekewich's muzzle-loading "screw guns."

The normally phlegmatic Kimberleyites awaited the approach of Imperial forces under Lord Methuen apprehensively, as he moved north to Kimberley's relief. In the path of the British advance lay the sharp-shooting commandos of "The Lion of the Western Transvaal," Boer General Koos Delarey, 3500 strong and all armed with the Model '95, 7x57 mm Mauser with its 173-grain steel-jacketed, flat-shooting bullet. When Methuen's men reached the Modder River, just south of Kimberley, they were met with a withering fire from Boer trenches dug into the south bank. Methuen was wounded and hundreds of his men killed and wounded in the 108 degree heat of the open veld. Delarey's eldest son was killed, but less than 100 Boer casualties resulted for the heavy British cost of crossing the river by a flanking movement after failure of a frontal assault. Delarey had proved his point; the day was done when infantry could prevail across an open plain against large numbers of well-emplaced accurate riflemen. The Boers withdrew, permitting the British to penetrate deeper into the perilous hinterland, alive with keen-eyed Boer riflemen. There remained one more barrier to the relief of Kimberley, the Magersfontein kopjes, a low range of rocky hills which General Piet Cronje chose as the site of a Boer "Thermopylae," where "they shall not pass."

On Dec. 11th after an appalling bombardment of the Kopjes, Scottish General Andy Wauchope's kilted Highlanders of the Black Watch, Seaforths, Argyll and Sutherlands and Highland Light Infantry, advanced silently on the Magersfontein at 4:00 a.m. What they didn't expect was that Gen. Koos Delarey had studied the terrain, and instead of relying on the rocky hills for positions, installed a long trench at the base of the kopjes level with the veld. Awaiting the advance of the resolute Highlanders were equally resolute Boers hidden behind a screen of thornbush and barbed wire. On the rocky kopjes he placed at varying intervals some blackpowder guns and riflemen to make visible smoke puffs to draw British fire. On came the Highlanders, fearlessly and relentlessly. When they were within point-blank range of the trenches, a withering hell-fire erupted from the muzzles of hundreds of Mausers, staining the land with the blood of laddies who would never again see their native Highlands.

Meanwhile the confident Boers surrounding the town played a waiting game. The people of Kimberley weren't going anywhere.

It was in this almost hopeless position that those well-attested Anglo-Saxon qualities of tenacity and ingenuity materialized almost miraculously. In Kimberley were some Americans, including Henry Labram, Chief Engineer for De Beers. Labram had resigned this position in August 1899 to take up another position in Johannesburg, but for some reason was still in Kimberley when the Boer noose closed tight. Labram had left the Silver King Mine in Arizona in 1893 to build a crusher in Kimberley, and had stayed on. During the early part of the siege he created a refrigeration plant for food and a much needed water supply system. He also designed much of the town's defenses, including minefields.

Relations between Cecil Rhodes, the "Colossus," and Lt. Col. Kekewich, the Imperial officer commanding Kimberley's garrison, became pathologically tense. Kekewich did his duty as he saw it, but Rhodes commanded far more resources with his mine labor, engineers, equipment, facilities, and a private army and arsenal.

By November's end the garrison had used up some 30% of its ammunition, though a large supply of blackpowder

Photos Courtesy of McGregor Museum, Kimberly, S.A.

On December 26, 1899, work was begun to form an artillery piece from this billet of mild steel stock. The gun was designed to reach the most distant Boer emplacements.

Photos Courtesy of McGregor Museum, Kimberly, S.A.

After machining of the barrel assembly, iron hoops were shrunk onto the breech for reinforcement. The craftsmen worked under constant threat from the Boer guns.

was available. With Rhodes' authorization, Henry Labram converted part of the De Beers workshop for the production of shells, fuses and charges for the 2.5-inch garrison guns. Once under way, the products of this enterprise were daily shot into the clear Kimberley air towards, but not capable of reaching the gun emplacements of the surrounding Boers.

Henry Labram had seen a large billet of mild steel in the De Beers' machine shop yard and proposed to Rhodes that a gun be designed and built capable of silencing the out-of-range Boer artillery. Rhodes gave his consent to the project on Christmas Night and work began the following day and continued on a 24-hour basis until completion. Labram had set about attempting to learn all he could about artillery and shells from the limited information sources available in Kimberley. He sought the aid of De Beers' Chief draftsman, Edward Goffe, A.M.I. Mech. E., and the pair scoured Kimberley for reference material, neither having any experience in artillery design. Some useful information surfaced, including the Encyclopaedia Britannica, articles in old engineering journals, a War Office treatise on ammunition, and a gunnery textbook belonging to an officer of the Diamond Fields Artillery. Labram discussed the idea with Lt. Col. Chamier, commander of the garrison's artillery, who reacted with stuffy skepticism.

Labram calculated that a 4.1-inch shell weighing 29 lbs. with a blackpowder charge of five pounds would achieve a range of over 7,600 meters, or enough to adequately bring to bear a bombardment of the furthest Boer artillery emplacement. A chamber with a capacity of some 2,785 cubic centimeters for the propellant was needed and would be 4.25-inches in diameter and some 12-inches in length. The breech pressure was calculated at about 22 long tons per square inch, or 50,000 psi. Was the mild steel billet up to such a strain? Labram decided that since ordnance steel had a higher carbon content that the barrel required reinforcement in the form of rings to be shrunken on the breech end while red hot.

The work progressed rapidly even as the gun was being designed. As drawings for one component were finished, they were sent to the workshop where materials were layed out and cut. The morning after Rhodes gave the go-ahead the big billet for the barrel was taken into the machine shop and set up in a huge lathe with a one foot swing and a 14 foot bed.

The billet was turned to a cleanup of the original surface to achieve concentricity, and a shoulder of ⅛-inch machined to position the trunnion ring. A long taper towards the muzzle was turned, and the breech end diameter was 10½ inches. By holding the breech end in a dog chuck, and muzzle end in a hardwood bearing, a 1½-inch twist drill was run right through, followed by a three-inch drill. The drilled hole was then bored to 3-15/16-inches by a boring bar and bit kept cool by a constant stream of soapy water through a hole in the bar connected to a pump. This turning and boring occupied a week, during which the reinforcing rings were forged, nine of them for the first row, and four for the outside row. These were made from 6x2½-inch Low Moor iron, forged into a circle and the ends welded together. The rough hoops were then turned to fit the breech end of barrel by shrinking them to fit very tightly.

It was decided by Labram that a gain twist would be easier on the gun, so this was included in the design and set-up of the rifling tooling. The rifling was designed for a 1 in 100 calibers right-hand twist, progressively increasing to a 1 in 32 calibers twist for a distance of 18 calibers, which then remained a constant 1 in 32 calibers for the final two feet of barrel. It was left to the Chief Draughtsman, Edward Goffe to design the rifling tooling, which he did by recalling a lecture he had attended by Sir William Anderson. So that any mistakes in grooving could not spoil the only available barrel stock, the rifling was begun from the breech end in the area to be enlarged for the powder chamber. After some preliminary problems with packing under the cutter the 32-groove rifling was completed in 20 hours.

A flat plane parallel to bore axis was machined for placing the clinometer and a cup-shaped gun metal casting bolted to the underside of breech to accept the ball end of elevating screw. It was later necessary to replace this with a hinged joint, since recoil tended to displace the ball and socket, making the screw fall down. A back sight was copied from that of the 2.5-inch screw guns and incorporated windage adjustment and set on a two degree angle to the right to allow for "drift" of the shell due to rifling angle. The front sight to be mounted on the trunnion was at first a small bead inside a tube, but was later modified to a "knife edge" minus the tube. At this stage the gun was completed and ready for mounting in its carriage.

The carriage frame was constructed from ¼-inch steel plates riveted together then cut to shape. The elevating screw was a 2½-inch diameter steel bar with a

Photos Courtesy of McGregor Museum, Kimberly, S.A.

"Long Cecil" and the men who made her! Neither the designers nor the workmen had any previous experience with ordnance, but they fashioned a gun that would fire accurately beyond 8,000 meters—easily the equal of the era's most modern artillery pieces.

square-threaded ¼-inch pitch operating in a nut pivoted between the side plates and worked by a hand wheel. The wheels were from a portable engine and bored out and fitted with gun metal bushes to fit the axles, and brass dust caps screwed on. Just 24 days had produced the gun, from Rhodes' "Go ahead," on Christmas Night to January 19th when it was rolled out of the workshop for testing.

The ammunition was made according to available blackpowder supplies, consisting of compressed cylinders, 1⅜-inches diameter by two inches, but was in limited supply, so it was decided to compress the finer "mealed" powder into blocks for slower burning. This process was not perfected and it was found that the available cylinders were adequate. A good grade of wool serge was used for the powder bags as made up by a local draper and bound with silk ribbons. Both "ring" and common shells weighed 29 pounds when filled with one pound of the bursting charge of blackpowder. Muzzle velocity was 1,710 fps with a charge of five pounds of "kidney" blackpowder. The shells were made of cast steel and each casing tested for flaws by steam pressure. The only ammo components not made at Kimberley were the fuse detonators, the blackpowder and the Mark III friction tubes. Each shell had "De Beers" and a diamond cast into the base, but many had "WITH COMP'TS CJR" (Compliments of C. J. Rhodes) stamped into the casing. This was the ultimate convincer that Rhodes never did anything impersonally!

Friday, January 19th, 1900, "Long Cecil" was removed to an emplacement for calibration and testing. Cecil Rhodes went along enthusiastically, accompanied by local functionaries and senior officers, including Lt. Col. Chamier, the garrison's artillery commander, who Rhodes invited to fire the first round. According to witnesses, Chamier stuffily declined on grounds that the gun was "non-regulation." At this point Rhodes asked Chamier to retire to a safe distance, then sent his horse and buggy to fetch Mrs. Willie Pickering, wife of the Secretary of the De Beers Company. Rhodes handed the lanyard to her and told her to pull it, which she did rather timorously. The result was almost as amazing to those present as it was to the Boer burghers in laager at the Intermediate Pumping Station 7200 meters away. Intercepted Boer letters told of the shock and alarm resulting when the first shell exploded in their laager. Before the party departed the gun fired 15 more rounds, including several by Rhodes.

On January 21st, "Long Cecil" was delivered to the Diamond Fields Artillery and a ten man detachment, including some ex-"blue jackets", was formed

Photos Courtesy of McGregor Museum, Kimberly, S.A.

An American, George Labram, left, and his associate, Edward Goffe, designed "Long Cecil." Labram was killed by Boer artillery.

Photos Courtesy of McGregor Museum, Kimberly, S.A.

Boers with a captured 2.5-inch British rifled muzzleloader, typical of the artillery used by the British Regulars at Kimberly.

Range Table: 4.1-inch Gun 'Long Cecil'
Projectile: 2 crh Common or Ring Shell: 29 lb.
Charge: 5 lb. 'Kidney' Powder
Muzzle Velocity: 1710 fs

Quadrant elevation	Time of flight	Remaining velocity	Range
degrees	seconds	fs	yards
1 00	2.0	1334	1000
2 45	4.5	1070	2000
4 49	7.5	937	3000
7 27	10.9	838	4000
8 57	12.8	796	4500
11 03	14.6	760	5000
12 38	16.6	723	5500
14 56	18.7	689	6000
17 09	20.9	659	6500
19 42	23.2	630	7000
22 21	25.6	602	7500
26 16	28.1	577	8000

Table of comparative data for 'Long Cecil' and equivalent British guns in service in 1899

Gun		BL 30-pr Mk I	BL 4-in Mk IV	Long Cecil
Weight of piece (nominal)	kg	1016	1320	1625
projectile (filled)	kg	13.6	11.34	13.15
charge (powder)	kg	1.13*	5.44**	2.26
Calibre	mm	101.6	101.6	104.14
Length overall	m	2.92	3.04	3.04
of rifling	cm	22.42	22.54	24.95
Chamber capacity	cc	4096	6833	2785
Muzzle velocity	m/s	503	579	521
Remaining velocity at 900 m	m/s	389	440	406
Elevation (maximum)		+16°	+25°	+26°
Range	m	5750	7050	7300
Rifling		24 PPS(mod); 1 in 120 increasing to 1 in 30	24 PPS(mod); 1 in 120 increasing to 1 in 30	32 PPS; 1 in 100 increasing to 1 in 32

Note 1.*—cordite charge only; 2.**—3½ lb (1.58 kg) cordite charge also used.

under Sgt. James Wheaton, an ex-Royal Navy man. Next day the gun and crew were placed under the command of Lt. H. M. Close, R.A., who was in charge of a section of Royal Garrison Artillery at No. 1 Searchlight Redoubt. The next day "Long Cecil" fired 18 rounds, and 57 the following day. The gun crew began the firing of the 29th, January with a cold gun and a charge of six pounds instead of the normal five pounds. According to Edward Goffe, "On Saturday night firing ceased as usual, Sunday being observed as a day of rest—from gun firing—by the Boers. But at daybreak on Monday morning, the first shot fired by "Long Cecil" was productive of an extra-loud and peculiar report, and the idea that something had gone wrong was general. A telephonic message came from the redoubt immediately afterwards and an examination showed that the second ring in the outer row had burst through the line of the vent hole." The gun was returned to the shop and the radial vent plugged and the damaged rings replaced. The copper tube lining the vent had allowed gas to escape around it in view of it not being tight enough, and the gas travelled under the rings, bursting them. An axial vent hole was installed through the obturator bolt for the friction tube and tested for safety with a feature to preclude the friction tube blowing out. It was also found that the powder chamber had bulged internally, and was 0.79 mm larger in diameter at center.

Meanwhile, Boer General S. P. du Toit, commanding Transvaal forces at Kimberley, requested and received a 155 mm Schneider-Creusot "Long Tom" for his emplacement at Kamfersdam some 6 kilometers north-west of Kimberley. It began shelling Kimberley the 7th of February, 1900, creating panic but little damage with its 94 four-pound shells. During these days Labram experienced a series of narrow escapes, and around 6 p.m. of February 9th was in his second floor room at the Grand Hotel shaving before dressing for a dinner with Rhodes. The last shell fired that day by "Long Tom" burst directly into Labram's room with an eerie accuracy that superstitious townfolk interpreted as an evil omen of Kimberley's doom. Rhodes commented grimly on the circumstances of Labram's death with "Well, what could a man do when God's been chasing him for three days?"

After Labram's death Col. Chamier reduced the firing charge to four pounds which cause little loss of range, and "Long Cecil" continued trouble-free until February 15th when Kimberley was relieved by Lord Roberts. Oddly, "Long Cecil" often shelled the "Long Tom" at Kamfers Dam but it was never shelled in turn. Due to "Long Tom's" effective fortification on the crest of a steep mine dump it apparently was never hit. Its last day in action, "Long Cecil" fired six rounds at "Long Tom" before the latter vanished to an unknown destination. In view of the disappearance of so large a gun it is

The untrained Boer citizen-soldiers proved the deadly efficiency of the then-new 1895 Mauser rifle in 7x57, and dealt a worse blow to British morale than even the darkest days of World War II. The *burghers,* left, proudly display the new Mausers that carried the field in the early days of the war. Above, the Boers had excellent leadership from such as General Delarey, "the lion of western Transvaal." The Boers knew the African bushveldt and used it to excellent advantage against the staid British.

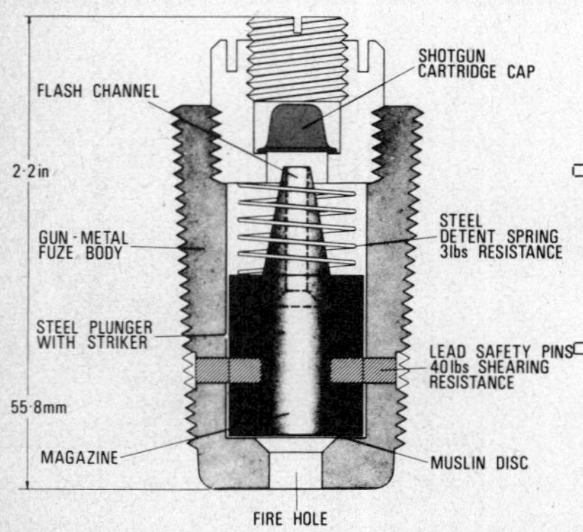

LABRAM FUZE

RE-DRAWN FROM ORIGINAL BLUE PRINT

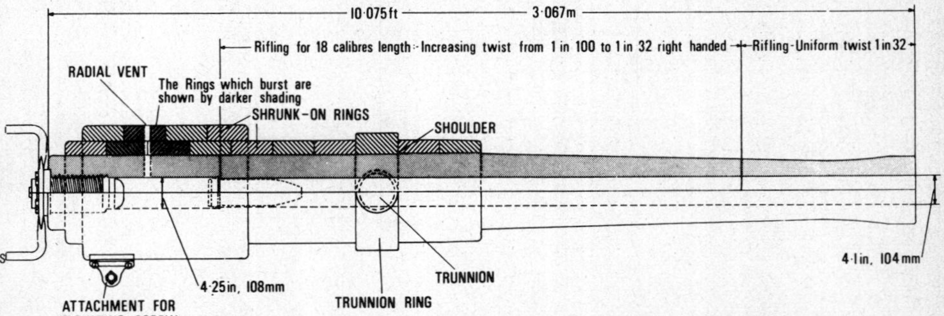

SECTION B.L. 4·1in GUN "LONG CECIL"

Re-drawn From Original Blue Print

The original blueprint for "Long Cecil," above shows the radial vent, screw breech, and gain twist rifling. Labram's fuse used a shotgun primer to detonate the 29-pound shell's explosive charge. The fuse never once malfunctioned.

Photos Courtesy of McGregor Museum, Kimberly, S.A.

Only two photos still exist of "Long Cecil" actually being fired. Here, gunners of the Diamond Fields Artillery engage distant Boer positions on Carter's Ridge.

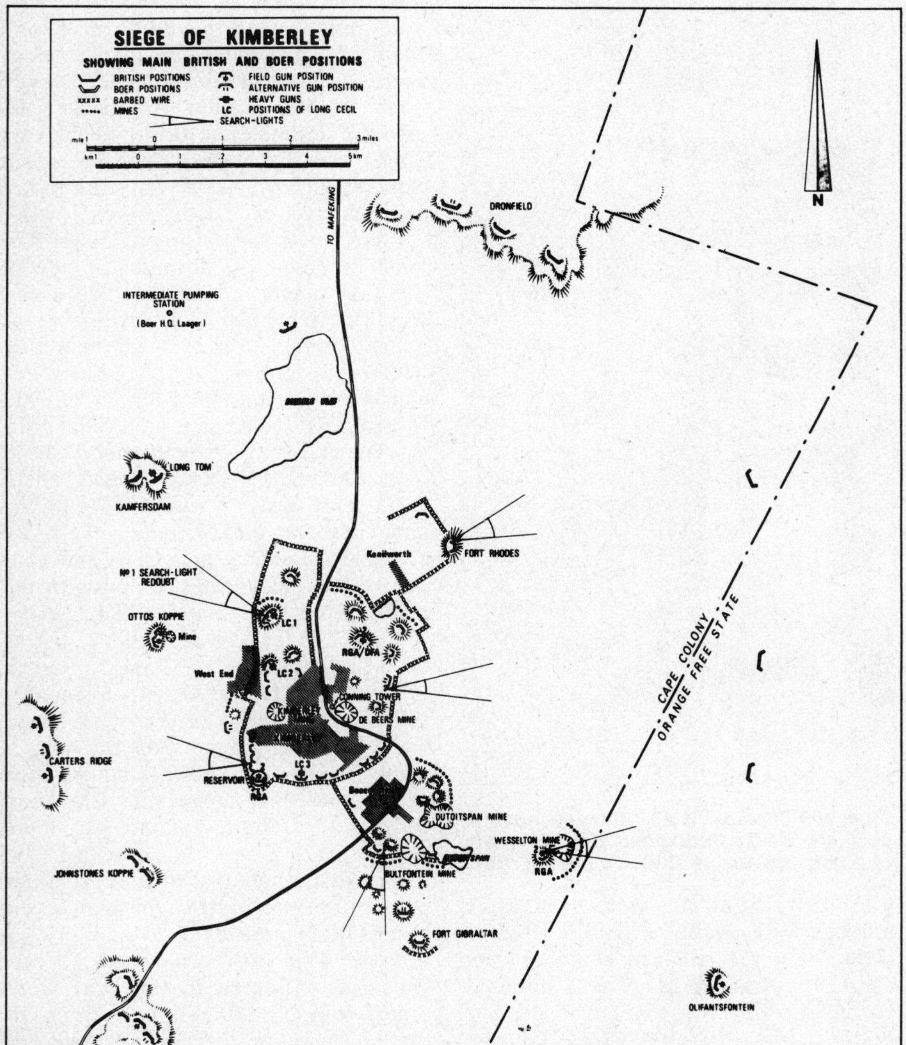

Diagrams Courtesy of South African National Museum of Military History

Boer warfare generally relied upon "hit and run" guerrilla tactics. Due to stubborn defense aided by "Long Cecil," the Boers were unable to end the Siege of Kimberly prior to the arrival of the relief column. After Kimberly, there were few set battles.

possible that "Long Tom" was in fact "killed" by "Long Cecil" and the Boers, not wanting to give Rhodes such satisfaction decided to destroy the gun.

"Long Cecil" fired 260 rounds; more than any other gun at Kimberley during the siege, a unique record for Labram's non-regulation gun. "Long Cecil" was to play another important role before permanent retirement to the Siege Memorial at Kimberley, including a trip to Capetown with its crew for the visit of his Royal Highness the Duke of Cornwall and York and his Duchess, who were later known as King George V and Queen Mary. On March 26th, 1902 Cecil John Rhodes, the great Imperialist and founder of Rhodesia died at his cottage in Muizenburg at the Cape. Though but 49, Rhodes had long been a sick man, his weak constitution and monumental intellect and character coexisting curiously. His heart had finally given out as the Boer War which he had helped start drew to a close under Kitchener's relentless campaign.

Rhodes was buried in his beloved Matoppo Hills of Matabeleland, Rhodesia atop the hill called "Malindidzimu" (Hill of the Spirits) by the Matabele. As his casket was lowered into a hole cut from the solid granite a horde of Matabele warriors raised their shields and spears, while shouting "Bayete," the royal Matabele salute, never before or since given to a white man. Many were veterans of the fighting of 1893 and 1896 against Rhodes' Chartered Company forces.

Kimberley honored the American engineer who contributed so much to its defense and comfort during the siege, by naming a suburb for him. De Beers paid an annuity to his widow and an educational grant for his son.

Both Rhodes and Labram as Victorians, were products of an age of individual initiative and excellence. It was an age that saw the conquest of the American and African frontiers and when neither doubts nor fears or red tape deterred the most daring intrepidity. No top-heavy corporate bureaucracies such as those which ponderously control our government agencies such as in aerospace spawned such achievments as "Long Cecil." Its story is that of the triumph of Man's brain-power over adversity, given freedom of action, minus the stultifying constraints of bureaucracy. Even with today's machinery and education it is doubtful that so few men with such limited information and experience could duplicate the creation of "Long Cecil" in so short a time.

The author wishes to express his appreciation for permission to use the excellent R. H. Wisehart drawings and ballistic tables from the South African National Museum of Military History's publication by the late Colonel D. E. Peddle, "Long Cecil."

THE AMERICAN SNIPER

Even in today's push-button age, our Scout/Snipers are better trained and better equipped than ever before, and draw on a legacy of rifle marksmanship dating back to the Revolutionary War!

By James A. Dunnigan

American history books have long perpetuated the myth that the 13 Colonies won their independence from Great Britain primarily by the accurate fire from Kentucky long rifles. The idea of buckskin clad frontiersmen hiding behind trees and mowing down the massed, scarlet-coated British sounds good, at least to Americans, but there is little basis in fact.

When the Revolutionary War started, the Americans had very few bayonet-equipped muskets available to them, and almost no well-trained militia. Instead, they had to rely upon civilian hunting arms, mostly rifles, in the early stages of the war. European military doctrine called for massed musket fire followed by a bayonet charge. Though the long rifles could inflict casualties at ranges far greater then any musket, they were slower to load. American officers quickly learned that their troops could not win a set battle with the British.

Rather, they adopted the guerrilla tactics of the Indians and fought a long-drawn war of attrition, meanwhile seeking foreign assistance in the form of muskets, bayonets, and professional soldiers to train the troops.

The tide of the Revolutionary War did eventually turn. The colonists received arms and military aid from France, and the troops were drilled and schooled by such European professionals as von Steuben.

Even though the rifle did not win the war, our Revolution was the first major conflict in which the rifled arm played a significant part. In fact, our modern concept of the sniper was born on the battlefields of Saratoga, King's Mountain, and dozens of other revolutionary battles.

While it became immediately obvious that the slow-loading rifles, *sans* bayonet, were not a front line weapon, the American army maintained rifle companies throughout the war, even after muskets were readily available. Used as skirmishers, the riflemen could fire accurately at ranges of several hundred yards, while the musket was all but done at less than 100 yards. George Washington himself espoused the idea of rifle companies, and even recommended buckskins as their uniform, saying they were practical, quiet in the brush, and camouflaged the rifleman. Washington also maintained that riflemen would not be used as frontline troops, but as snipers and skirmishers.

From WW I through Korea, the dependable, accurate .30-06 Springfield with a variety of sights served as the U.S. sniper rifle.

The most classic triumph of rifles over muskets was the Battle of King's Mountain, and it is this battle which probably gives rise to the general myth. There, well-entrenched American riflemen did decimate the British at extreme range, and the hapless Brits never got close enough for volleyed musket fire and the bayonet charge.

In one of history's great ironies, Colonel Ferguson, the inventor of one of the first practical breechloading rifles, was killed by American rifle fire at King's Mountain, and his excellent rifle became forgotten. At the Battle of Freeman's Farm, Tim Murphy of Dan Morgan's Riflemen shot and killed British General Frazier with one shot at over 200 yards. George Rogers Clark led a small army of riflemen from Virginia into the old Northwest Territory of Illinois, Indiana, and Ohio and conquered the small British outposts in the name of the 13 Colonies. When the Paris Peace Treaty was signed, it was through the efforts of Clark's riflemen that the new United States of America included the Northwest Territory. Thus, while the rifle did not win the war, it made its presence known, and the idea of accurate, long-range rifle fire from concealed positions—sniping—was here to stay.

The rifle again played a very minor part in the War of 1812, but rifle companies were employed effectively by both British and American forces. American riflemen in that war were armed with the 1803 Harper's Ferry rifle, the same arm that accompanied Lewis and Clark.

In 1836, at the Alamo, the Texans were armed almost entirely with rifles, and the terrible toll they exacted from the Mexican army is incredible to this day. American forces continued to employ rifle companies throughout the Mexican War, though the musket was

Photo Courtesy Vermont Historical Society

"Watching for Rebs!" Union scout Truman Head, known as California Joe, was a member of the elite Berdan's Sharpshooters. Photo was taken at Gettysburg in 1863.

The current U.S. sniper rifle as issued by the Marine Corps is the Remington M700 in 7.62 mm (.308). Bull barrel and Parkerized finish separate it from civilian model.

The ability to use cover and concealment is every bit as important to the modern day sniper as his marksmanship ability. Current U.S. military training stresses use of camouflage and the ability to cover long distances without being spotted by enemies.

still the standard arm. At Chapultepec the Regiment of Mounted Rifles and the First Mississippi Volunteers, all armed with the U.S. Model 1841 rifle, distinguished themselves and carried the battlefield. The 1841 rifle is still known to collectors as "the Mississippi Rifle."

By the start of the Civil War, the rifle had become the standard military arm, and the common foot soldier's effective range had increased even past the Revolutionary rifleman's 200 yards. At the start of the war there was no sniper unit or concept in employ, but certain free-thinking military men and statesmen saw a definite need. Snipers could pick off artillery crews and ranking officers at long-range and disrupt and demoralize an enemy force.

The motivating force behind the sniper concept in the Union Army of 1861 was Colonel Hiram Berdan. At the outset of the War Between the States, he petitioned the Secretary of War to form a regiment of "Sharpshooters," and equip it with the best rifle of the time: The .54 Model 1859 Sharps breechloading rifle using linen cartridges and firing a conical ball. The proposition was accepted by the Secretary of War on June 15, 1861. However, Berdan's dream of a unit of unmatched marksmen very nearly died here, the victim of petty bureaucracy.

The Assistant Secretary of War, Thomas Scott, joined the prevailing military opinions in scoffing at the idea of highly accurate, long-range rifle fire. The American military's "thinkers" still clung to the muzzleloaded arm as the status quo. Berdan's requests for breech-loaders bordered on lunacy. No rifles came. But the gritty, foresighted colonel was only temporarily obstructed.

When Mr. Scott and President Lincoln visited the regiment shortly thereafter, Col. Berdan personally proved the validity of his theory, using a James telescope rifle. Responding to a taunt voiced by Scott questioning his ability, Berdan agreed to fire at the silhouette of a Zouave 600 yards distant. As the officer prepared to assume the prone position, Scott spoke up again to ask whether an officer of Berdan's rank would deign to lie in the dirt before the Commander-in-Chief. Berdan chose to fire offhand. Once more Scott interrupted, to ask where the Colonel intended to aim. Without waiting for Berdan's reply, Scott blurted, "fire at the right eye." Skillfully cornered by the glib bureaucrat, Berdan raised the heavy rifle and fired. When the target was retrieved, Berdan had knocked out the right eye. Lincoln was delighted. Upon leaving, he instructed Berdan to" . . . come down tomorrow, and I will give you the order for the breechloaders."

Although stories recounting the escapades of both Union and Confederate snipers are legion, an enlisted man, Truman Head, serving under Col. Berdan became the celebrated sniper of the war, and a true American hero. Head, a professional hunter from California, never failed on a mission. None of his mates or officers called him anything but "California Joe."

Joe's prowess with a rifle during the Peninsula Campaign linked his name with Berdan's "almost inseparably together" in the press. Berdan's Sharpshooters became a sensation; Joe became a legend. A comrade said of him:

"Joe was one of those splendid creatures that made him a hero in spite of himself. Entirely free from brag and bluster, an unassuming man, past the

Photo Courtesy Vermont Historical Society

Col. Hiram Berdan, left, and Truman Head pose in this rare Civil War photo. Berdan organized the Berdan Sharpshooters, and is largely responsible for the modern scout/sniper concept. Truman Head and his Sharps rifle were popular heroes on the Union side.

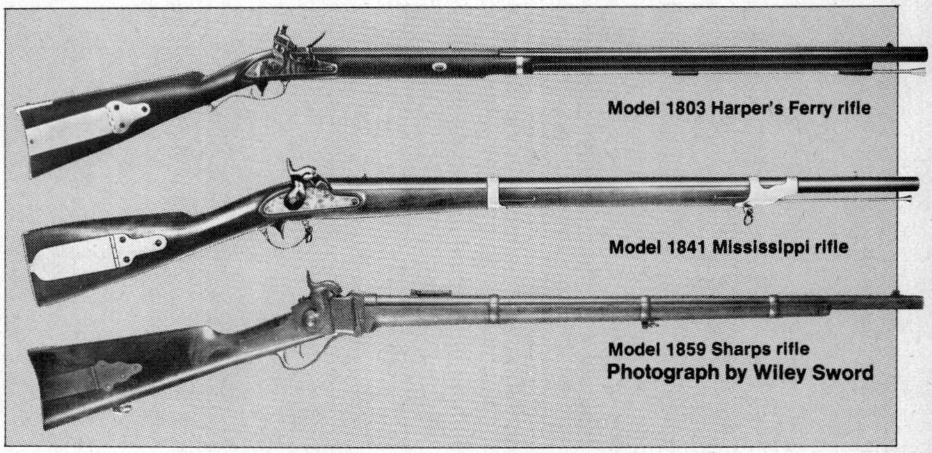

The Harper's Ferry flintlock rifle, top, was the first standard issue U.S. rifle. It was used on the Lewis & Clark expedition. The Model 1841 Mississippi rifle was used by regiments in the Mexican War. Breechloading Sharps' were used by Berdan's troops.

middle age, short in stature, light in weight, and a true gentleman in every sense of the word, he was always a special favorite with the entire command."

There were eventually two regiments of Sharpshooters, the 1st U.S. Sharpshooters commanded personally by Col. Berdan, and the 2nd U.S. Sharpshooters commanded by Col. Henry Post. It is said that these two regiments caused more casualties amongst the Confederates than any other unit in the army, and they accounted for quite a number of ranking rebel officers. Individual Sharpshooters also silenced gun emplacements, sometimes for days at a time, with witheringly accurate fire that decimated gun crews. In effect, a single sniper would capture a gun from hundreds of yards away.

During the Confederate retreat, two Sharpshooters were ordered to fire on a single rebel soldier simultaneously. Allowing a "couple of feet" for windage, the two men fired their .54 Sharps with the sights set for 700 yards. The fleeing soldier fell, wounded twice; in the right thigh and left hip.

A unit from Vermont surpassed that incredible feat by bringing fire to bear on a Confederate signal corps observation post 1,500 yards from their position! Since their Sharps sights would only extend to 1,000 yards, the snipers rigged twigs to extend the range on their rifles. After a few experimental cuttings, with an officer spotting and adjusting for them, they succeeded in driving the Southerners from their postage point.

But the Yanks too had their troubles with Confederate snipers. Berdan's counterparts in the Confederacy were known as "Lee's Sharpshooters." They used a different weapon, having access to British arms through the South's trade agreements. The Sharpshooters in Lee's command tested Minie rifles, British Enfields and Whitworths, various European models, Springfields and Mississippi rifles. Many finally preferred the .577 Enfield Rifled Musket, for its dependability and accuracy to ranges of 900 yards.

The ultimate proof of the Confederate snipers' skills came during the Battle of Spotsylvania Courthouse, May 8, 1864. The firing of Sharpshooters on both sides proved withering. But the Southerners persevered, striking Union staff officer Col. Fredrick T. Locke and wounding General William H. Morris. The final blow fell as a Southern sniper killed Major General John Sedwick, U.S.A. The general's death completely demoralized his troops and cast his ranking officers into total confusion.

Various Confederates claimed the shot that killed Sedwick, but an independent sniper, Sergeant Benjamin Powell, may have shot either General Morris or possibly Sedwick. Powell was armed with a .45 British Whitworth rifle equipped with a telescope.

Berdan's breechloading philosophy eventually sent shock waves throughout the post-war world: The breechloading system for firearms was so well received, muzzleloading soon became obsolete. Accuracy continued to improve.

When the Marines hit the soil of France during the First World War, they were invited to demonstrate their much publicized ability with firearms. The marines gladly obliged, generating incredulous outcries from the French marksmanship instructors as they routinely struck targets at ranges of 900 yards with their 1903 Springfields. The Germans were even more surprised.

The Second World War refined precision optical instruments and hence, accuracy. An old marine hand present on Saipan related that the Corps' accuracy experts hadn't lost their touch since France, 25 years before:

"We were pinned down on the beach at Saipan by a machine gun bunker. The pill box commanded a sweeping view of the area, and there was just no way we could get to it. Plenty of our boys had died trying.

"Finally, one of our 90 day wonders got on the horn and requested a sniper, of all things. A few minutes later, I saw these two old gunnery sergeants sashaying towards us, wearing shooting jackets and campaign hats! As soon as I saw those Smokey Bears bobbing over to us, I figured this could be some show. And it was.

"Those two old sergeants skinnied up to the lieutenant and just asked him to point out the bunker. Then they unfolded two shooting mats, took off their Smokey Bears and settled down to business. One manned a spotter's scope while the other fired a 1903 Springfield with a telescopic sight rig.

"That bunker must have been 1100 or 1200 yards away, but in just a few

Telescopic sights were employed by snipers on both sides in the Civil War, and casualties were inflicted at astonishing ranges.

These U.S. Marines at Da Nang, Vietnam, are equipped with pre-'64 Winchester M70 .30-06 sniper rifles with target scopes.

Marine snipers are currently trained as two man teams, with one spotting while his partner shoots. Range sessions are normally conducted in full camouflage gear.

Currently, a Redfield 3X-9X variable scope with Redfield mounts is the standard issue.

Can you find the Marine in the photo? During the six-week Scout/Sniper school, countless hours are spent learning the finer points of camouflage and using natural cover.

minutes, with three or four spotting rounds, this old gunny on the Springfield slipped a round right into the bunker's firing slit. One dead machine gunner. But their commander just stuck another man on that gun. Our sniper shot him, too. After the fourth man bit a slug, I think they got the idea. We moved up on their flank and destroyed the bunker while our snipers kept the machine gun silent. Then the two gunnys dusted themselves off, rolled up their mats and settled their Smokey Bears back on their heads. And just moseyed away."

Today, the American tradition of outstanding marksmanship on the battlefield, our legacy of accuracy, continues in the military's best rifleman: the sniper. Carefully selected and painstakingly trained as never before, today's sniper is a highly regarded specialist.

Recently, I was priviliged to visit nearby Camp Pendleton Marine Corps Base in southern California and observe the First Marine Division's Scout-Sniper

School. The present Marine doctrine calls for a two-man Scout Sniper Team to be assigned to each infantry company, giving the Marine Corps the widest possible use of the sniper's long-range capability. The highest traditions of American marksmanship are very much alive in today's Marine Corps, and no doubt in our other services.

While the Army prefers a match grade M-14 for sniping, the Marine Corps has carried the sniper concept a step further by equipping their personnel with what some marines feel is the best sniper rifle available. The Remington 700 in .308 that you or I might consider as a fine sporting arm undergoes a series of special modifications before becoming the Marines' M40A1 Sniper Rifle.

A heavy, stainless steel barrel is installed to enhance the already fine accuracy of the rifle and to prolong barrel life. The standard wooden stock is replaced by a Parkerized (camouflaged fiberglass) stock specially designed to resist weather variations affecting accuracy. The receiver is glass-bedded at the factory and attached to the stock by Allen bolts, which allows for a completely free-floating barrel. The triggerguard is also stainless steel.

The ammo used is 7.62 mm Nato Match. Although the Marine Corps and Army shooting teams have long experience with belted magnums for 1,000 yard competition shooting, sniper rifles have in this century always been chambered for the standard service cartridge, either the .30-06 up through Korea, or the 7.62 mm (.308 Winchester) today. A flat shooting belted magnum might add to the sniper's long-range capability, but ammunition resupply could become a problem. Worse, the sniper definitely does *not* want a rifle that sounds different from the standard service rifle or machine gun.

The sighting system used currently is a Redfield 3-9X variable scope. The Marines have another sighting system by Unertl which is currently being considered. It features a steel casing (highly applauded by the snipers I interviewed) and a trajectory compensation system which will significantly simplify sighting at all ranges.

Currently, the sniper zeros the Redfield scope at 500 yards and changes the point of aim as range changes. To strike the center of mass of a silhouette representing an enemy soldier six feet tall at 200 yards, the sniper sights on the crotch. For a 600-yard shot, he aims just above the helmet. Thus, as range changes, so must the point of aim, to compensate for the 500-yard zeroing.

The Unertl scopes, equipped with trajectory compensation, will be zeroed at 300 yards. Then, as range changes in combat, the sniper will simply read the range on an internal range scale, and adjust a range knob on the scope. Once range is set, regardless of distance, the sniper will always fire center of mass.

The newest wrinkle in U.S. sniper rifles is the addition of a camouflaged fiberglass stock. The new stock aids concealability, won't warp, and withstands combat conditions.

To graduate from the school, Marine snipers must be able to make one-round hits on stationary silhouette targets at 1,000 yards and on moving targets at 800 yards!

The goal of the sniper during training, his credo, is a simple yardstick of excellence: One round, one hit, one enemy soldier out of action. To be successful, the sniping team must maintain complete invisibility from the moment of insertion to final extraction. Armed with one bolt action M40A1, one service rifle and two .45 sidearms, the team is constantly vulnerable. Without the firepower or numbers of an infantry unit, the team must rely on keenly honed fieldcraft skills if it is to evade detection and attack after a successful strike.

The list of lessons each Marine sniper must successfully complete during training offers a testimonial to their skill:

1) A one-shot kill on a stationary target out to 1,000 yards
2) A one-shot kill on a moving target out to 800 yards (target measuring only 12 inches in width).
3) Stalking undetected for 1,000 yards (while under direct observation).
4) Judging distance to 1,000 yards with only ± 10% error.
5) Demonstrated skill at reading maps, aerial photos, and land navigation.
6) Observing, collecting, and reporting intelligence data.
7) Selecting, preparing, and manning sniper positions.

Even if you pride yourself on being a crack shot and stalking like an Indian, imagine attempting the Marine sniper's final exam: Inserted in an unknown area, each sniper must locate and stalk within firing range of an unknown observation post. During the stalk, the instructors on the observation post constantly sweep the area with spotting

Part of the Marine sniper's final exam is stalking undetected for 1,000 yards while under direct observation from a tower.

Marines selected for the elite Scout/Sniper school must already have qualified "Expert" with the rifle for two years.

Today's snipers have not only drawn upon their 200 year legacy of rifle marksmanship, but they've learned some things about camouflage from eastern turkey hunters! The "ghillie" suit includes head-to-toe green and brown burlap and netting for face.

The accurate Remington M700 is mated with precision-made 7.62 mm Match ammo. Recoil is mild and accuracy impressive.

scopes. Any sniper they detect fails immediately. Once in position, the sniper fires a blank round at the o.p. The instructors then move a radioman to stand within a ten-meter radius of the hidden sniper. If he still remains undetected, the observation post orders him to fire a second blank round. If they cannot detect movement or muzzle blast while looking directly toward the sniper's position, that sniper must then identify his target via observation of hand signals. If he fires at the right target, twice, while under direct observation and still remains completely invisible, he passes and becomes a Marine sniper.

Once trained and outfitted, the sniper is an invisible, long-range, deadly tool employed for maximum disruption and demoralization of enemy forces, in both offensive and defensive situations. Snipers support attacks by striking at key enemy personnel immediately prior to an offensive, which causes confusion and denies the enemy units time to bring up replacements before battle. Snipers also protect our armored vehicles and tanks by seeking out enemy anti-tank units.

Defensively, the sniper can engage the enemy beyond the range of the standard service rifle. The carefully selected fire of sniper teams has a marked effect on the security and morale of enemy troops, especially when focused prior to an enemy attack. In addition to the M40A1, our snipers have mounted .50 machine guns in static defense positions with 12X to 20X scopes. Firing single rounds from this configuration in Vietnam, kills have been made at ranges approaching 3,000 yards.

However, most sniper teams are deployed in pairs a spotter and a sniper—to pursue specific targets in a given combat zone. Intelligence reports play a central role in first identifying key enemy personnel. The snipers are highly trained in detailed observation so that they themselves can oftentimes provide the intelligence necessary to identify their targets. Once assigned, the sniper team is inserted in their area, well away from their suspected quarry. Wearing specially camouflaged, veiled uniforms called Ghillie suits, the snipers often must low crawl for thousands of meters, hours or even days to come within striking range of a target without suffering detection. A team may also construct a hide or blind from which to observe the enemy and await an opportune target.

Today, the Marine Corps sniper program is flourishing. They've come a long way since the days of California Joe. Refinements in rifles, sighting systems and camouflage have abounded. Tactics too have improved, as has schooling, but the core elements of the American rifleman remain.

For more than fifty years, Frank Pachmayr has been at the forefront of custom gunsmithing and firearms innovation.

By Art Blatt

THE PACHMAYR STORY

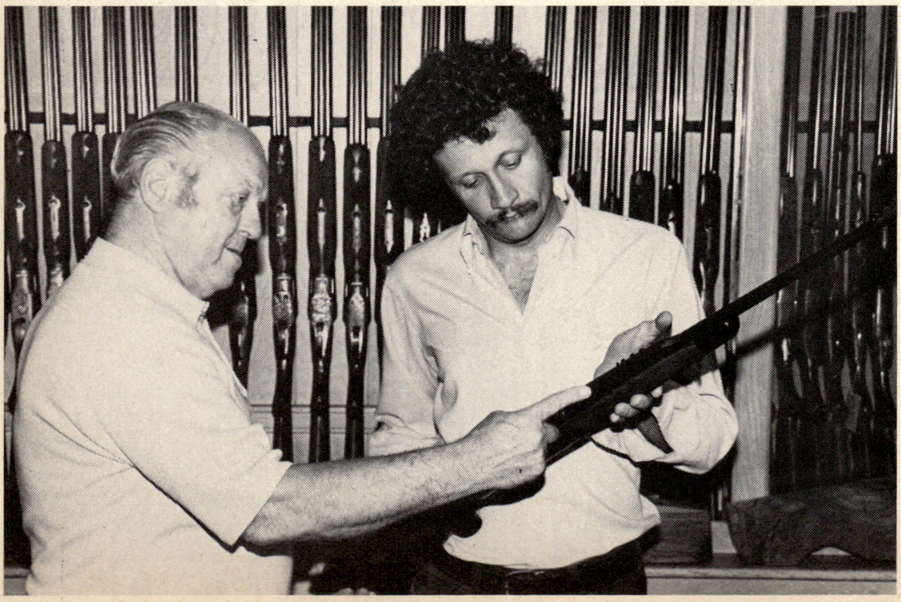

The old saw about "more complex—sooner dead" has been shot down by Pachmayr Gun Works of Los Angeles, California. For over 50 years, its founder, Frank Pachmayr, has thrived on versatility. Not only does this company produce a myriad of shooting-related products, but it has established itself as a successful and respected service-oriented company.

Since Frank Pachmayr opened the doors at 1220 South Grand in downtown L.A., this 74-year old gentleman has doggedly served the shooting frater-

The Pachmayr operation consists of a lot more than the retail sales of firearms and accessories. It runs the gamut from custom rifles and handguns to purchasing, curing and shaping custom stocks for shotguns and rifles. The store has been a mecca for shooters since it opened—serving generations of handgunners, and riflemen. It's not just a store—it's an institution!

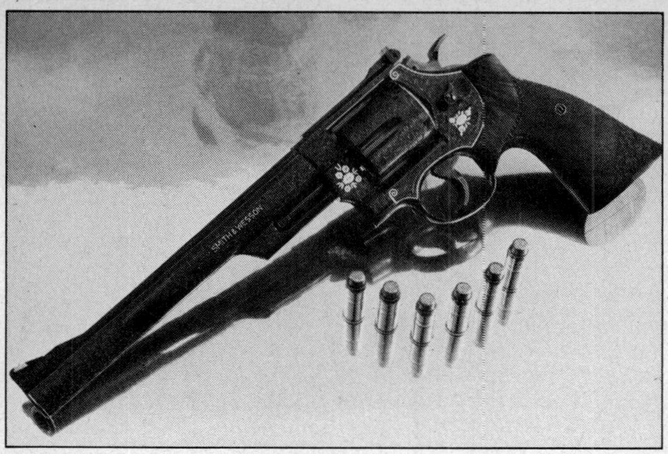

This beautifully engraved S&W M29 features a Pachmayr special—custom California Claro grips checkered by Richard Boucher.

This highly customized Pachmayr Signature .45 is truly the tool of the expert who can take full advantage of its capabilities.

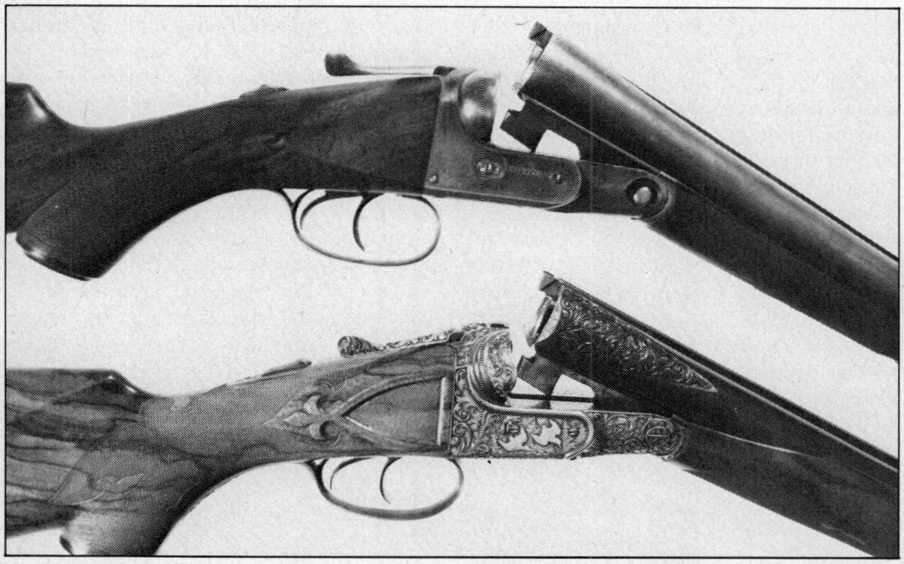

One of Frank Pachmayr's main interests is in collecting and upgrading classic Parker shotguns. The top Parker is an original; The bottom one has had the full treatment!

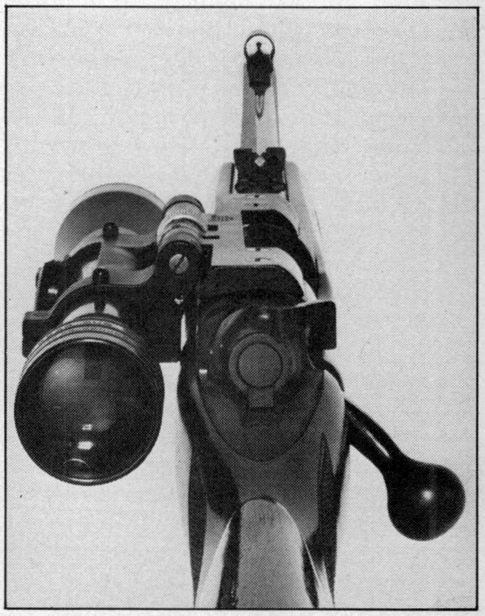

One of Pachmayr's most successful items has been the fine Lo-Swing scope mount.

nity with a wide variety of products and services. Frank is highly inventive, with over 100 patents to his credit, and Pachmayr-designed products have helped shooters rack up some impressive shooting scores and game trophies around the world.

If one were to ask Frank Pachmayr what was his single most important product, it would be an unfair question. All of the company's products are given equal billing, and for those of you who aren't familiar with these products and services, we'll list a few of them for you. First, the homegrown products are: Rubber cartridge packs, revolver grip adaptors, neoprene revolver and pistol grips, low swing scope mounts, pistol cases, sling swivels, and, of course, recoil pads. In the service department, there's the famous .45 accuracy jobs, custom built rifles from .22 Rimfire through .458 Magnum, plus shotgun restorations. This latter is headed by Pachmayr's famous Parker restoration and upgrading service.

Most recently, another entity has developed through necessity—a thriving gunstock wood business. Pachmayr owns a pair of lumber mills in northern California and is constantly searching for walnut groves that are past their prime. These stagnant trees are converted into some of the finest rifle and shotgun stocks ever seen.

Frank opened the gunsmithing store right in the middle of the Great Depression, and in spite of serious economic conditions, in a few years Pachmayr Gun Works was the leading gunsmith shop on the West Coast. A few years later, Frank was joined by his younger brother, John, and the Brothers Pachmayr were soon swamped with work. The boys' father, August Pachmayr, maintained his own separate shop and, in essence, was a limited competitor of his two sons.

The 1930s were good years for Frank and John. Between them they earned an excellent reputation and developed the nucleus for some of today's best-selling products. The first Pachmayr recoil pads were sold in 1931. Accuracy jobs for the Colt 1911 Auto were offered in the late 1930s. The grip adaptor was also founded in this decade, and, little by little, the general shooting public was beginning to see more and more Pachmayr products and services in the field.

The 1940s were hustle-bustle years for Frank Pachmayr. He was active in the war movement and the Pachmayr Corporation soon grew to almost 500 employees turning out specialized products for the war effort. Many of today's Pachmayr products are "off-shoots" from World War II products, like the flush-mounted sling swivels. These swivels, initially, were devised as a quick-detachable device to simply hold two aircraft parts together. After some limited modification, this device became the most successful sling swivel ever produced. It is standard equipment on Browning bolt action rifles and will soon be found on many H&K products.

Frank Pachmayr has always created new products by seeking a better way. All of the products are carefully tested by Frank under actual field conditions and, if they don't pass his personal inspection, they aren't good enough.

The Lo-Swing scope mount was developed in 1946 and is an inexpensive "insurance policy" guarding against a potentially disastrous hunting situation. There have been numerous situations in which a scope has failed the hunter at an inopportune time. Scopes can be susceptible to fogging, snow, rain, or may simply break at the wrong time. If the big game hunting rifle is equipped with a Pachmayr Lo-Swing scope mount, all the hunter needs to do is swing the scope and mount abruptly to the left and utilize the rifle's open sights.

When one thinks about recoil pads—for both rifle and shotgun—the obvious name is Pachmayr. Today's catalog lists close to 750 variations from which both shooter and firearms manufacturer can pick and choose. Remington, H&K, Browning, and many other gun-making companies use Pachmayr recoil pads under their own company name. Recently a shotgun was sent to Pachmayr Gun Works for some attention and the attached pad was of 1936 vintage—still pliable and totally useable.

Pachmayr's neoprene handgun grips are standard-issue on many police revolvers. One can hardly attend a combat pistol match without noticing that 80 to 90 percent of all combat autos are Pachmayr gripped. Pachmayr neoprene grips are being purchased today by many handgun manufacturers and are standard equipment.

Always a friend of the competitive pistol shooter, Pachmayr's famed Pistol Box has been around since 1946. It was first manufactured—mostly by hand—in wood and leatherette covering. Today, it is still made the same tried-and-true method and is still the pistol case that other companies emulate.

Since the early 1930s, many of the top-notch pistoleros have used and shot Pachmayr-prepared pistols. Col. Charles Askins was—and still is—a handgunner who ran up some impressive scores using a Pachmayr-tuned Government Model .45. Jeff Cooper, Ray Chapman, and a host of others are proud owners of these highly refined "old warhorses." With the upswing towards PPC and IPSC-type shooting events, Pachmayr .45s are eagerly sought and currently it takes close to 40 months to get a pistol returned after a "full house" job has been done to it.

An area of controversy has been caused by Pachmayr's upgrading and refurbishing of Parker shotguns. Many Parker purists claim that Frank Pachmayr shouldn't restore or upgrade these fine old shotguns. When Pachmayr Gun Works has completed one of their fabulous restorations, or upgrades, these individual guns are always marked in a manner to indicate that they were remanufactured by Pachmayr's. There's never a doubt that these are upgrades and not "fakes" to be sold to some unsuspecting buyer.

The dedication to "old world" craftsmanship was drummed into both sons by August Pachmayr and this spirit of producing quality at a fair price has remained the byword for Pachmayr Gun Works. Unless Frank is off on a hunting trip, every gun is inspected by him before it is returned to the customer. Complete customer satisfaction is always guaranteed regardless of who's "right or wrong."

The retail store is the gathering place for many notable hunters and shooters. Hundreds of guns are on display and for sale, ranging from everyday .22 repeaters through .600 Nitro Express double rifles. Even the store's inventory is versatile and spans the needs from plinkers through big game hunters and serious target shooters. All the store's personnel are hunters and shooters and accurately serve the customer's needs.

How does one encapsulate over 50 years of gun knowledge into a limited article such as this? It's impossible! But with Frank Pachmayr's leadership and total dedication to providing service and quality to his customers, the best way to describe the Pachmayr story is to invite all you readers into his operation. Pachmayr's is still at their traditional location at 1220 South Grand, near the heart of downtown L.A. Next time you're in Los Angeles, stop by Pachmayr Gun Works. It's firearms history in action, and it's a visit you'll never forget!

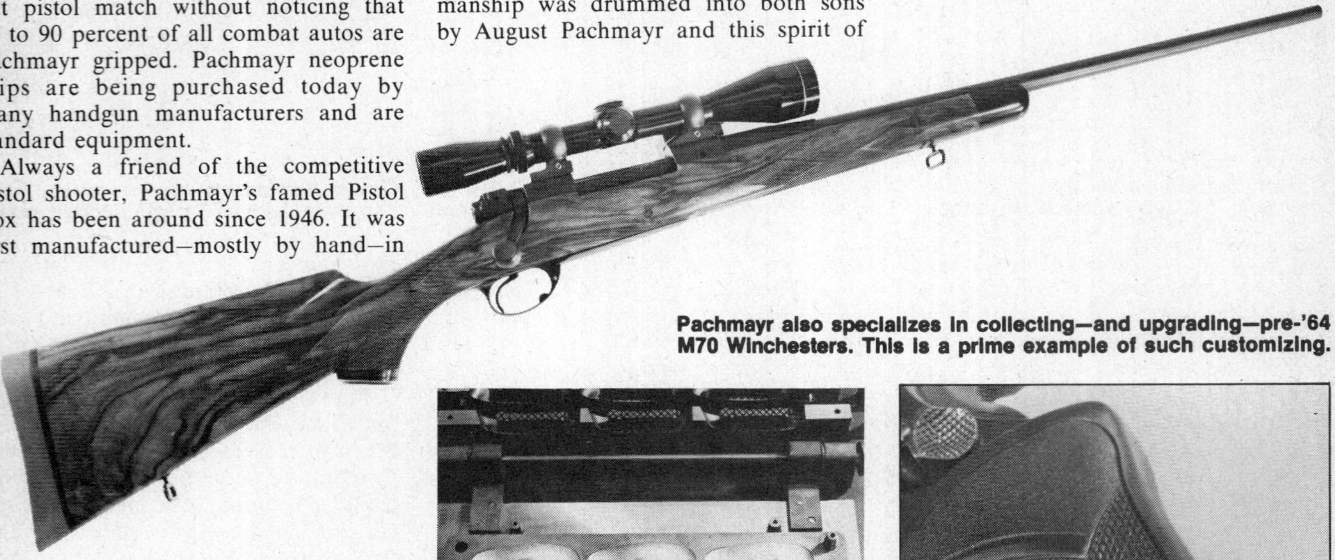

Pachmayr also specializes in collecting—and upgrading—pre-'64 M70 Winchesters. This is a prime example of such customizing.

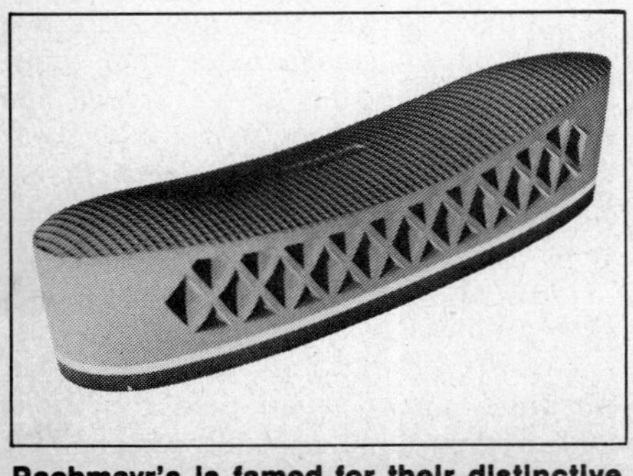

Pachmayr's is famed for their distinctive "white line spacer" rubber rifle recoil pad.

These molds for recoil pads attest to the high number made—and the high demand.

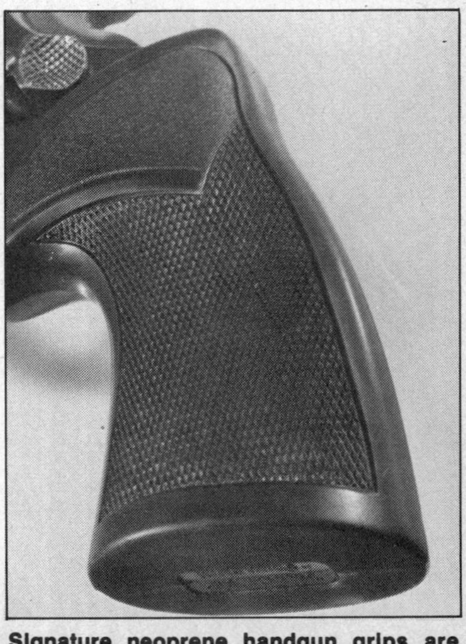

Signature neoprene handgun grips are preferred by combat shooters and police.

"Weaver's got more ways to help you shoot better than a ol' hound dog's got fleas."
"Now that I'm a big star, I can afford to indulge my passion for the sportin' life. I got a special pair of designer huntin' boots just for Sundays; a duck blind with a wet bar; and a hand-rubbed cabinet full of custom-made rifles from around the world. But when you get right down to the old powder and ball, it's a Weaver Steel-Lite II™ that calls the shots. Weaver makes a scope and mount for every kind of shootin' and every kind of shooter. Ol' Slim wouldn't trust anything less."
Slim Pickens
Weaver Scope
One tough scope... packed with quality.

Extremely popular in Europe, rarely seen in the U.S., this versatile three-barrelled gun may be best for your hunting needs!

By James E. Fender

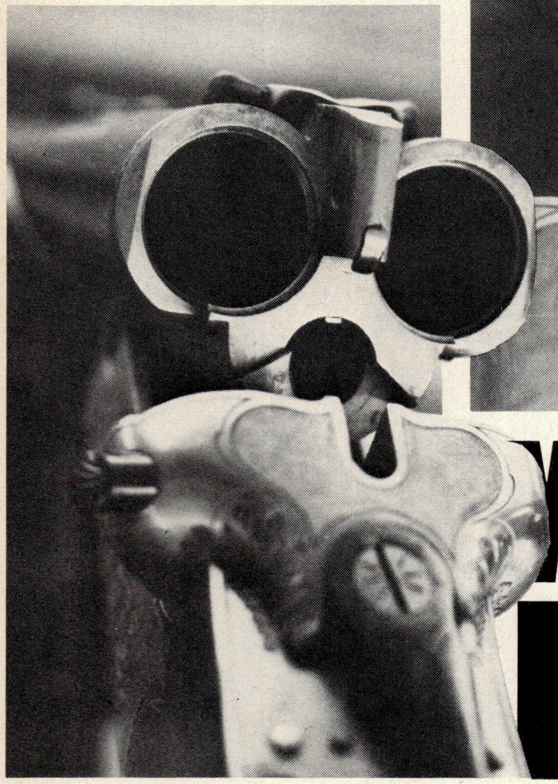

WHY NOT A DRILLING?

The drilling, or "drillinge" as normally spelled in German, evolved in central Europe during the latter part of the 19th Century. It was originally designed as a "poor man's firearm" because a jaeger could have the advantage of both rifle and shotgun without the expense of purchasing separate arms. More affluent sportsmen quickly saw the merits of the combination gun and the well-made three barreled firearm gained rapid acceptance among Continental hunters, particularly in Germany and Austria.

Hunting conditions and traditions differ vastly between Central Europe and North America. The drilling has never enjoyed a wide following in the United States or Canada, and most of the drillings brought into this country, either as trophies of World War II or more recent commercial imports, grace firearms collections more than they are taken afield by American hunters.

Is the drilling suitable for American hunting conditions or American shooters? Are the merits of having both rifle and shotgun incorporated in the same basic firearm outweighed by the expense of a current production drilling?

I have owned and hunted with a drilling extensively for the past 10 years. My drilling has accompanied me on European, African, American, and Canadian hunting trips, and it served me well . It has taken a wide variety of game. Consequently, my experiences may be of value to any American hunter who has ever been intrigued by the three barreled gun.

My drilling is a J. P. Sauer Model 3000 in 12x12x7.63 mm, or .30-06 as we North Americans know the cartridge. My drilling was one of the first Sauer drillings chambered for a rimless cartridge. The sales representative at the Sauer factory attempted to dissuade me from my choice of calibers and assured me that I would be much happier with a rimmed cartridge. Even though the 7x57 mm rimmed cartridge which was the factory representative's recommendation is a very effective cartridge, my long familiarity with the .30-06 led me to prefer that caliber. The main arguments given by the dealer and factory representative centered around potential extraction problems for a rimless case. Since extraction is less positive with a rimless case, care should be taken that only cartridges which do not exceed factory pressures are fired in a drilling chambered for a rimless cartridge. I have never experienced any extraction problems in my drilling.

The reticules of German telescopic sights have never appealed to me, though I recognize the extremely high quality of their optics. Consequently, I wanted to have an American made scope mounted on the drilling. This presented a problem since European made scopes are 26 mm in diameter while American manufactured scopes are normally 1" or 25.4 mm in diameter. The Sauer factory finally acquiesced to my specifications and fabricated special diameter rings for the Weaver V4.5 scope I sent to the factory. Since I am left-handed, the Greener-style shotgun safety was inletted into the right side of the small of the stock rather than the left. The factory representative told me when he delivered my drilling that this gun was one of the first Sauer Model 3000s to be chambered for the .30-06 cartridge, was certainly the first to have an American scope mounted on it, and to his knowledge was the first truly left-handed drilling Sauer had ever made.

The scope was carried in claw mounts which mated with lugs inletted into and alongside the barrel rib. The scope is

Hunting parties in France (above) wait for signal to begin hunting. Birds are the primary game on such a hunt, but roe deer may be flushed, a situation that has made the versatile drilling popular. At left, the author sights in his drilling.

very quickly detachable, yet the underside of the rings had been milled to provide a tunnel through which the standard iron rifle sights could be used while the scope is mounted. A test target with a three-shot group fired at 100 meters with the three shots less than 2″ on centers was included with the drilling when I received it. I have since fired numerous three-shot groups at 100 yards which halved this spread, though groups do begin to open considerably when firing is prolonged.

I had planned to take this drilling to Africa on a safari to Mozambique in July 1972, but as I was test firing Brennecke slugs to determine appropriate aiming points for both barrels the sear of the left shotgun barrel broke. I was living in Portugal at the time and there were no gunsmiths available locally who could repair the gun. Given mailing and customs delays, repairs could not be made by the factory in time for the drilling to accompany me. Needless to say, the fact that firing mechanisms for three separate barrels are incorporated in one rather compact frame makes the drilling somewhat more complex and prone to breakage than other firearms.

I was unable to use the drilling in Africa in 1972, but did use the drilling in Africa subsequently and it performed flawlessly. Since that one failure on initial test firing, the drilling has fired several thousand rounds through shotgun and rifle barrels at targets and game with never a moment's trouble. However, anyone who purchases a drilling (or any new firearm for that matter) with intentions of taking it on an extended hunting trip to a remote area should thoroughly test fire the arm beforehand. It has been my experience that if something is going to malfunction in a firearm the malfunction will occur, and can be remedied, within the first one hundred rounds.

My drilling performs a number of hunting tasks effectively. Whether it performs as well as a separate rifle or shotgun is frequently a subjective decision. What one man considers an aesthetically pleasing blend of rifle and shotgun in one compact and efficient package may be viewed by another as something neither "fish nor fowl," or a compromise that fails. However, no one should condemn the drilling without hunting with one—then forming his own conclusions.

Some would consider the single rifle shot feature of the drilling to be a disadvantage. However, every true hunter strives for the humane one-shot kill of any game animal.

The drilling balances well for carrying afield. However, the extra weight of the rifle barrel makes the drilling more difficult to track quickly in shotgun mode when the target is a gamebird flying at right angles to the hunter. The drilling mounts to the shoulder as quickly and as easily as any fine shotgun, but swinging rapidly enough to maintain the correct amount of lead can be a frustrating experience.

If a drilling has been fitted with detachable scope mounts, the slight protrusion of the bases along the sighting plane can be distracting to some hunters. Most American hunters who have any experience with double barreled shotguns are accustomed to finding the safety mounted on the tang just to the rear of the opening lever. On most commercially available drillings what appears to be a tang safety is actually the rifle barrel selector. The shotgun safety, being of the Greener-type, is mounted on the small of the stock. Many shooters, mistaking the rifle barrel selector for the shotgun safety, have been unable to get off a shot at a flushed gamebird, and have vented their frustration on the firearm, not their lack of familiarity with it.

This brings up a point concerning safety with the drilling. On all drillings, regardless of age or manufacturer, sliding the rifle barrel selector forward raises the rear rifle sight from its recess in the barrel rib. In and of itself this movement does not activate the rifle firing mechanism. The Greener safety on the small of the stock must likewise be pushed off "safe" position before the rifle barrel can be fired. All drillings' rifle barrels are fired by pressing the front trigger, the same trigger which fires the right shotgun barrel when the drilling is used in the shotgun mode. With the Greener safety disengaged and the rifle barrel selector pushed forward, the rifle barrel may be fired—but so also may the left shotgun barrel be fired! Just move the finger to the rear trigger.

Therefore, there is the danger of an unintentional discharge of a slug or shotshell in the left barrel when the shooter is attempting to use the rifle barrel. Such a feature could have unfortunate consequences if the shooter is unaware of this feature of the drilling. Many friends to whom I have explained this feature consider it to be undesirable. An American hunter would naturally consider the rifle barrel selector to control the rifle safety and the Greener side safety to control the shotgun. However, the drilling was designed for European hunting, and European users of the drilling overwhelmingly prefer to hunt with a Brenneke rifled shotslug in the left barrel while the rifle barrel is loaded with a standard cartridge when hunting game such as boar or European red deer. If game appears unexpectedly at close range the hunter would have the option of a quick choice between the cartridge or the slug. This feature was deliberately designed into the drilling, but the potential for an inadvertent firing of the left barrel should be kept firmly in mind by every hunter afield with a drilling.

The majority of drillings available today are built on the "Blitz" action with Greener crossbolt or some modification of the time-tested Anson shotgun action. The German firms of Suhler and J. P. Sauer use the "Blitz" action exclusively, while the firms of Heym and Krieghoff manufacture sidelock drillings as well. The Heym boxlock is a modified Anson action, while the Model 37 sidelock is fitted with unbreakable coil springs, and is a true sidelock action, though the sideplates are not normally hand-detachable.

For American shooters who are unfamiliar with a "Blitz" action this is an action in which the firing mechanism is placed on the trigger plate. The entire action is strengthened by a Greener crossbolt which locks through a hole in the rib extension. By placing the lock mechanism on the trigger plate the action body is almost a solid mass of steel, but this type of action does require that the stock grip be cleared out inside to the extent that there is less wood in the grip area, and prolonged shooting of heavy loads may cause the wood in the grip area to spread apart.

Most drillings have firing pin indicators which stand above the tang when

Ring-necked pheasant

Photograph by Leonard Lee Rue III

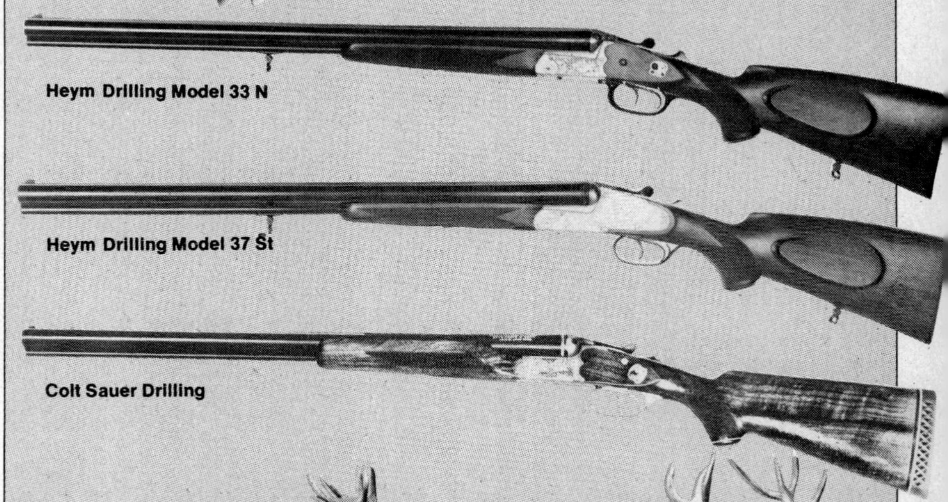

Heym Drilling Model 33 N

Heym Drilling Model 37 St

Colt Sauer Drilling

European Red Deer Stags

Photograph by Leonard Lee Rue III

the action is cocked and fall flush with the tang when the particular barrel is fired. No currently manufactured drilling has automatic ejectors—to incorporate this feature into the three barreled gun would make an already complex action even more complicated.

Some hunters have expressed doubts about the rigidity of the detachable claw mounts which are the only type of mounting systems useable on the drilling if the versatility of the firearm is to be preserved. A set of properly made and fitted claw mounts provides as firm a telescopic sight mounting system as any other system in use today. I have experimented with my Sauer drilling by removing the scope and its mounts from the firearm between rifle shots and then replacing the scope to determine if any change in zero occurred. Regardless of how frequently the scope was removed and replaced the zero held constant.

The drilling disassembles as does any conventional double barreled firearm into buttstock, barrels, and forend assemblies. I frequently fly a single-engine aircraft over rather foreboding territory in the northern United States and find the drilling an ideal survival arm. The drilling itself fits neatly into a 32-inch long rigid case, with room enough left over for 20 rounds of rifle ammunition and 25 rounds of shotgun ammunition. The case and its contents weigh only 15½ pounds and fit unobtrusively in the baggage compartment of

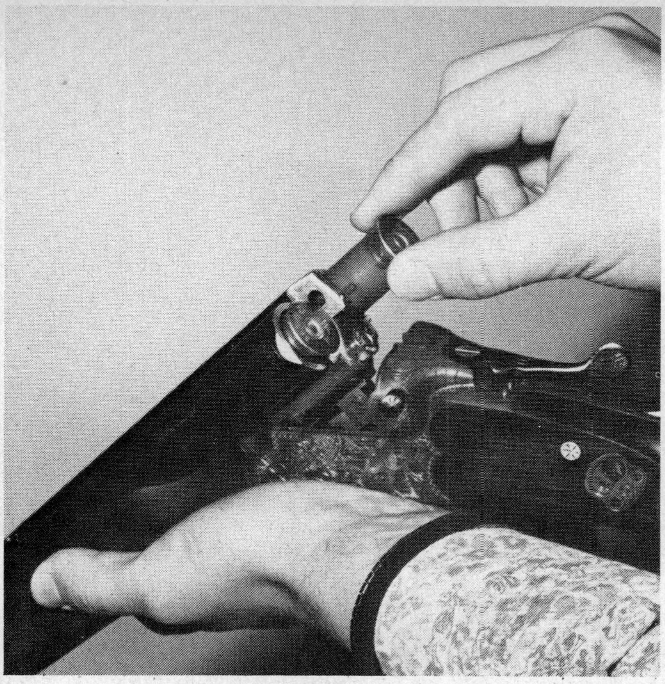

The Colt Sauer Drilling is available in .30-06/12 gauge or .243/12 gauge. For bird hunting, gun performs similarly to a side-by-side double barrel.

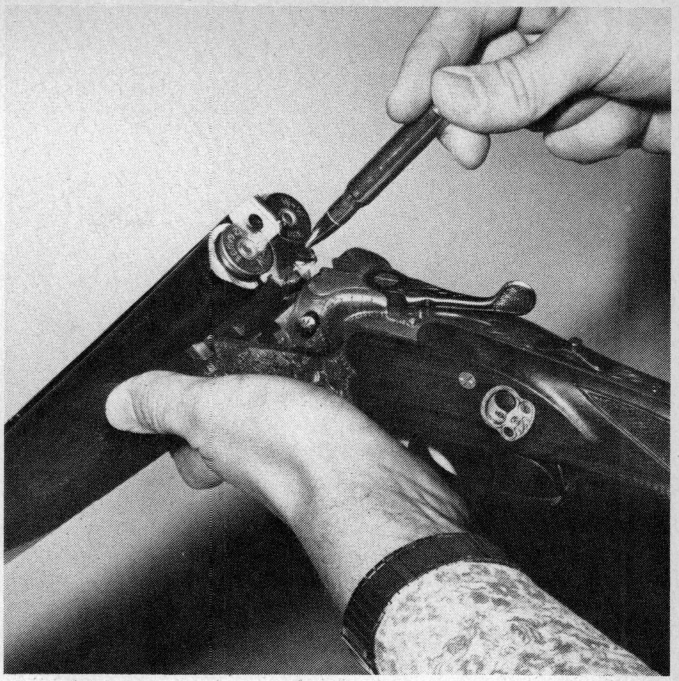

For big game, the drilling shows still more versatility. Lower rifled barrel carries the standard cartridge while the upper left holds a rifled slug.

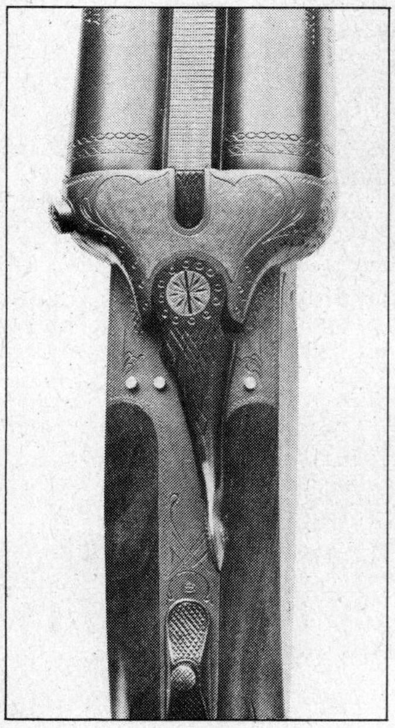

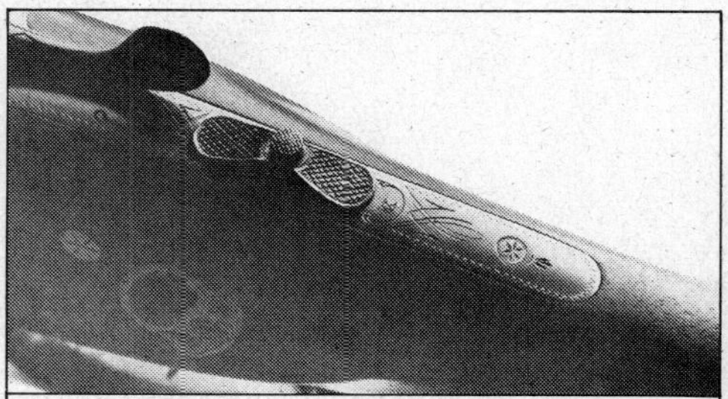

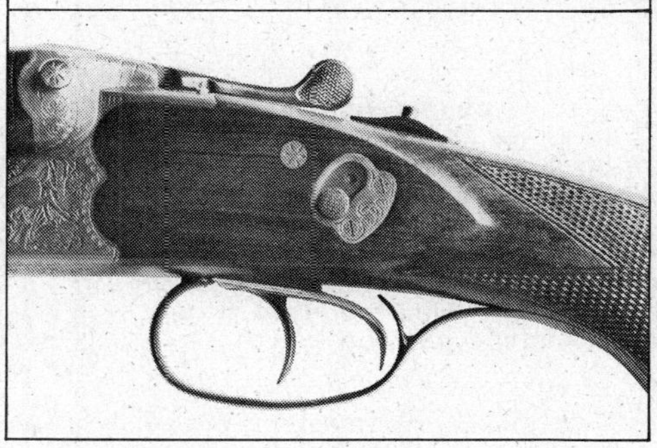

At left and above are closeups of the Colt Sauer Drilling action. The three raised studs in the photo at far left are cocking indicators for the three barrels. What appears to be a tang safety is actually the barrel selector. Sliding this forward raises the rear rifle sight from its recess. To fire the rifle barrel, the Greener side safety must also be pushed into firing position. With the selector forward, the front trigger fires the rifle barrel while the rear trigger will still fire the left shotgun barrel. To avoid confusion, it is understandably important to become familiar with the gun before taking it into the field.

my plane. A number of reputed survival authorities recommend the carrying of a pistol in such situations, but a pistol cannot match the versatility of the rifle-shotgun combination the drilling offers. My survival kit also contains a flare launcher, since there are small flares available which can be fired from a shotgun when a search aircraft is heard by a downed flier. I know of no flares which can be projected from a pistol. Additionally, a substantial portion of my flying is done parallel to the Canadian border, and occasional bad weather has forced me to divert to a Canadian airport to wait out a storm. Canadian officials take a dim view of pistols in the hands of the general public, including visiting Americans, but my drilling has never caused any difficulty as long as declared immediately to customs officials upon landing.

An earlier version of the Sauer Model 3000, the Model 30, was actually issued to certain pilots of the German Luftwaffe as a survival firearm during World War II. The short barrels of the Model 30 were 12 gauge and the rifle barrel was chambered for the 9.3x74 mm Rimmed.

Very little is known about the Model 30, either its actual use by Luftwaffe fliers or the total production, due to the paucity of records which survived World War II. Less than 100 of these unique firearms have found their way into American hands. They would certainly qualify as one of the rarest martial arms ever produced, and anyone who possesses a genuine Luftwaffe issue Model 30 drilling has an arm of historical and practical importance.

The drilling can be made even more versatile by the addition of an insert barrel for either .22 rimfire or .22 Winchester Magnum Rimfire. These insert barrels are made by Krieghoff and are available from Waffen-Franconia in Wurzburg, Germany. At one time the insert barrel was available in a .22 Hornet chambering, but with the advent of the .22 Winchester Magnum the Hornet chambering was dropped in favor of the .22 Winchester round. The insert barrel is fitted into the right shotgun barrel, and does not hinder the operation of either the left shotgun barrel or the rifle barrel. Thus, it would be possible to have a drilling set up for one shotshell and two different rifle cartridges. I have seen such inserts but have had no practical hunting experience with them. Some drilling users have reported little difficulty in getting the insert barrel to shoot to point of aim at ranges up to 50 yards by using the standard folding leaf rear sight. Other insert barrel users have never achieved proper correlation of rifle sights and point of bullet impact.

I have found my drilling's rifle barrel to be superbly accurate for the first three shots with any of a variety of handloads. While the rifle barrel is very thin, it is given great rigidity by being affixed parallel to the shotgun barrels. However, as the barrel begins to warm the shots begin to string. In my drilling the bullets begin to change point of impact low and right, though a 10-shot group often measures a respectable two inches. This is certainly acceptable hunting accuracy which extends to all practical hunting ranges past 100 yards.

In summary, the drilling occupies a very necessary niche in the firearms world. It is an admirable firearm to have where various game such as pheasant and deer may be hunted simultaneously.

The effects of American inflation are sadly reflected in the price one must now pay for a contemporary Continental drilling of good make. Ten years ago a Sauer drilling carried a retail price of less than $800, customs duty paid, delivered in this country. Current retail price for the Colt/Sauer drilling as it is now called by the importer, Colt Firearms, is $2,300 exclusive of scope or mounts.

Yet, even as our major American firearm manufacturers search for simplified methods of manufacturing to hold down production costs, more and more American hunters are turning to custom gunsmiths for firearms which are easily superior in appearance and function to commercial offerings.

There are large numbers of drillings available in America today, arms imported either many years ago or brought by returning service personnel. Many of these drillings are chambered for relatively obscure German metric cartridges. Many are in need of repair, or even though fitted with claw mount bases are missing telescope sights and claw mounts. These drillings are frequently transferred through inheritance or located in gun shops where they may be priced far below the cost of a current production drilling.

If a drilling requires repair, one of the most knowledgeable gunsmiths for this type of work is Victor Strawbridge, 6 Pineview Drive, Dover, NH 03820.

Rifle accuracy of drillings is good up to three shots. Beyond that the shots tend to string out. Rifle was dead on at 25 yards (above); slug from left shotgun barrel shot very close to rifle's point of aim.

The five shot group (above) was fired from a range of 50 yards with Brenneke slugs from the left shotgun barrel. When the rifle barrel selector is pushed forward, the rear sight (right) flips up. The left barrel and sight are regulated for slugs at close range, giving the shooter a rifle cartridge with the front trigger and a backup shotgun slug with the rear.

In a 10-shot group, several rounds elongated the same hole. As barrel heated up, shots began to string.

Mr. Strawbridge is not only a master in the repair of these fine arms, but can also fabricate claw mounts and fit a telescope to a drilling which may have lost its scope and mounts.

As mentioned earlier, Colt Firearms, Hartford, CT 06102, is now the importer of the Sauer drilling, which is available presently only in 12 gauge and either .30-06 or .243 caliber. However, to my knowledge, Colt is not able to furnish claw mounts and integral bases for the drillings it imports. The widest selection of drillings available in the North American market is provided by Mr. Nick Winter of Waidmannsheil Limited, Box 861, Bolton, Ontario, Canada L0P 1A0. Mr. Winter imports the excellent Heym drillings in 20, 16, and 12 gauges and a variety of rifle calibers. Necessary scope mounts and other accessories for the drilling are also available through Waidmannsheil Limited, and anyone dealing with Mr. Winter can be assured of prompt service.

Admittedly, the drilling has many negative features, such as high cost and slightly increased fragility, as it has positive aspects of ease of carrying and the convenience of having readily available both rifle and shotgun. But skeptics who have given the drilling an honest evaluation have become converts to the three barreled gun. This type of firearm does have a definite place in the American hunting scene. Each year will see more and more American hunters utilizing—and cherishing—this most useful of firearms. In the ten years since I first purchased my drilling it has become the arm of first preference on an increasing number of hunts.

Chapter 2

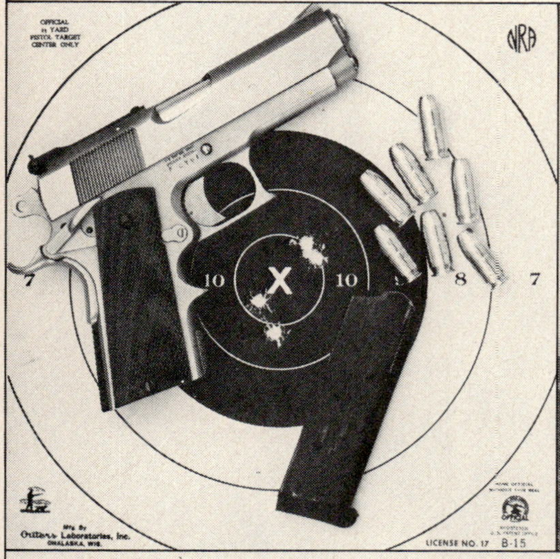

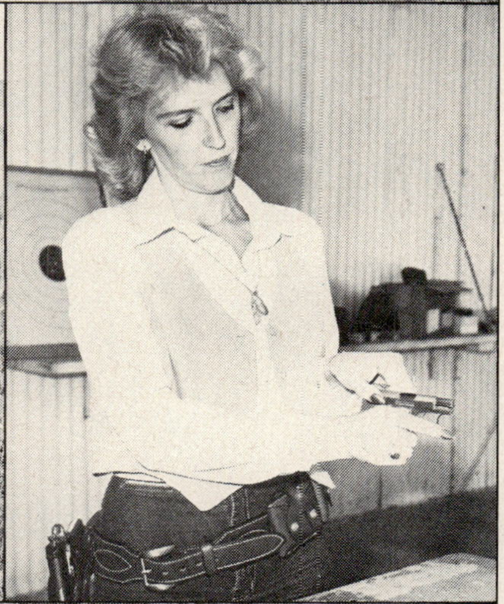

Handguns

9 mm Military Handguns

For nearly 80 years, the mighty 9 mm Parabellum and the pistols chambered for it have been the service sidearm for most of the world!

By Dr. Ralph C. Glaze

When it comes to mystique, the Luger is without a doubt the most famous of all 9 mm pistols. It was adopted by Switzerland (left), but was famous as a sidearm of the German Army (top). It saw extensive combat in World Wars I and II.

The most widely used automatic pistol cartridge in history is the 9 mm Luger, also commonly known as the 9 mm Parabellum. The latest and currently correct designation of this venerable caliber is 9 x 19 mm, a term that describes both bullet diameter and case length. The list of handguns made for the 9 x 19 is a long one, and includes many of the world's most illustrious designs. The 9 x 19 has become practically the universal military handgun caliber in spite of the fact that other, larger rounds are demonstrably more effective in combat situations. Its shortcomings notwithstanding, the 9 mm Luger is a good compromise in terms of acceptable effectiveness, light weight, accuracy, and felt recoil.

The 9 x 19 mm is one of those serendipitous results that sometimes occur when the simplest solution to a problem is sought. The 9 x 19 case is essentially the 7.65 mm bottleneck case without the bottleneck. George Luger merely omitted the last step in manufacturing the bottleneck case, and thus produced an ideal carrier for the 9 mm bullet. Since the body and rim diameter is identical to the 7.65 mm case, the same machinery could be used for making the 9 mm cartridge, and the Luger pistol needed no alteration to accept the new round except for fitting a barrel of the proper dimensions.

Since the Luger pistol was the first to chamber the 9 x 19, it seems only fitting that a discussion of 9 mm handguns should begin with the most famous of the 9's. The Luger design began in the mind of Hugo Borchardt in the late 1880s, culminating in the patenting of the Borchardt semiautomatic pistol in 1893. Hugo Borchardt, an American citizen, worked for both Winchester and Sharps. Although he is best known for his famous Sharps-Borchardt single-shot rifle, he also created a series of revolvers for Winchester that were never produced—not that they were bad designs, they were withheld as part of an agreement with Colt not to enter the handgun field, which is another story best told at another time. Inasmuch as Borchardt could not interest an American company in producing his revolutionary new pistol, he eventually arranged for its manufacture in Germany by the Ludwig Loewe Company of Berlin. About 3,000 of Borchardt's rather large and somewhat awkward pistols were produced between 1893 and 1899. The Borchardt was never adopted as a military sidearm, although it was carried by some officers as their personal choice.

George Luger was a salesman for the Loewe organization and, later, for DWM (Deutche Waffen und Munitionsfabriken) a new company formed by a merger of Loewe and the German Metallic Cartridge Company in 1896. While demonstrating the Borchardt, Luger became acutely aware of its shortcomings and set about a program of modifications to the original design that led to the development of the Model 1900 Luger pistol. The Model 1900, in 7.65 mm caliber became the first semiautomatic pistol to be adopted

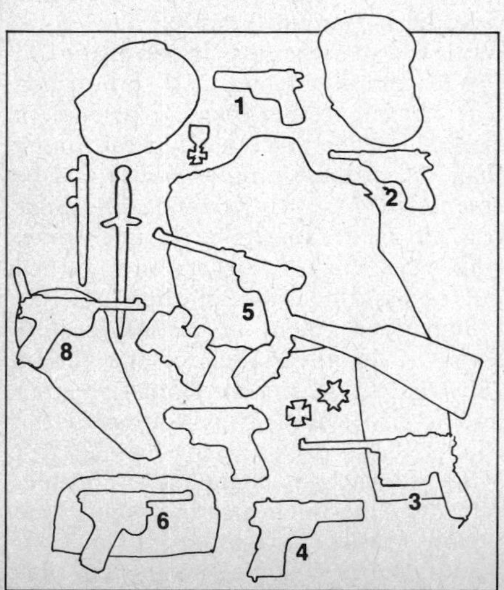

Shown on the opposite color page are several of the most famous military pistols in the widespread 9 mm Parabellum. (1) The Browning P-35, the long time champ of cartridge capacity, holds 14 rounds. (2) The P-38—standby of the German Army in World War II, and the first successful DA military sidearm. (3) The sinister looking 1896 Mauser in 9 mm—had a red "9" on grip. (4) The rugged Polish Radom featured several Browning features. (5) The artillery model Luger shown here, features a snail drum magazine and shoulder stock and was capable of high firepower beyond normal pistol ranges. (6) Italian Glisenti was chambered for a less power ful 9 mm cartridge than the Parabellum. (7) The H&K P9S is the most recent and modern of 9 mm military sidearms. (8) The Finnish Lahti boasted the strongest action of the 9 mm autos. Pistols shown are courtesy of Pony Express and Cole's. Military accoutrements courtesy of Bob's Military Antiques, Santa Monica, CA.

as an official military sidearm by any standing army of consequence when Switzerland ordered 3,000 of them for the officers of their army.

Both the Borchardt and Luger pistols use a locking system that is unique in auto pistol design. Based on a principle first used by Sir Hiram Maxim in his much earlier machine guns, the Luger depends on a recoiling barrel and barrel extension to raise an articulated toggle joint above its extended central position so that the joint assembly can flex, much like a knee joint, allowing the breech block attached to the forward half of the toggle to continue to the rear upon firing, ejecting the fired case and cocking the firing mechanism. The breech block is returned to battery, and a new round chambered by a recoil spring located in the rear of the grip. This system allows ample time for barrel pressure to drop to near zero by the time the breech block starts its rearward travel. It is a solid locking system, but is quite sensitive to variations in ammunition and, in its exposed positon, is severely prone to fouling by dirt, dust, and grime under combat conditions. A weak round will not impart sufficient velocity to the breech block for proper extraction and ejection of the fired case. In spite of these major deficiencies, the Luger remained in service in Germany through World War II, and was used in large numbers by the armed forces of many nations. In fact, it is still highly valued by many of the smaller countries of the world even at the present time. Among the nations that have used the Luger as a standard sidearm are Russia, Bulgaria, Portugal, Switzerland, Persia, Finland, The Netherlands, Latvia, and Austria. The United States tested the Luger, but rejected it in favor of the Colt .45 ACP, Model 1911.

The Luger was first adopted by the German navy in 1904, when approximately 1,000 of them were delivered in 9 mm caliber. In 1906, the German navy ordered 24,000 of an improved model. The Model 1906 has a coiled recoil spring instead of the flat spring used in the earlier versions. This change necessitated altering the dimensions of the frame and receiver, so that parts are not interchangeable with the early pistols. These modifications were the only ones ever made to the basic Luger design in its long production history, attesting to the "rightness" of the original concept.

In 1908, the German army chose the 9 mm Luger as its standard sidearm, and gave it the designation "Pistol Model 1908," or P-08 for short. The P-08 lacks the grip safety found in the first Lugers, and the safety works in the opposite direction. The standard P-08 has a 4-inch barrel, but at least 40,000 were made for the German navy with 6-inch barrels. To handle production of the vast number of P-08's ordered by the army, the government arsenal at Erfurt was licensed to manufacture them as well as DWM.

In 1914, a variation with an 8-inch barrel was introduced that was fitted with a removable shoulder stock, a 32-round rotary magazine, and adjustable tangent rear sight. The Model 1914 is basically a P-08 with a longer barrel, since all other parts are fully interchangeable. The Model 1914 was created to increase both firepower and accuracy in the hands of artillerymen who couldn't carry a rifle conveniently, but needed more accuracy than the P-08 provided for long-range shooting. The rear sight is adjustable to 600 meters, which may seem to be an optimistic range for a handgun. However, the author was able to score 3 hits out of 5 shots at a 4-foot diameter gong at 600 yards distance with his 1915-dated DWM Model 1914. This was done standing with the shoulder stock in place. As a boy, I watched a friend of my father's drop innumerable running jackrabbits with a 1902 12-inch barreled caliber 7.65 mm Luger carbine. All of which only proves that the long-barreled Lugers are capable of near rifle accuracy in spite of their thoroughly rotten trigger pulls. The Luger trigger has been described as being like pulling a brick through wet concrete—which is not a bad analogy.

Over the years, during two monstrous wars and an uncertain peace, Lugers have been manufactured by many companies and arsenals in addition to DWM and Erfurt. In 1918 production was started at the Spandau arsenal in Berlin, but the war ended before more than 100 of these rare versions could be assembled. Mauser, Krieghof, and Simson all made Lugers at a later time, while the English Vickers firm turned out several thousand of them for the Dutch government in the 1920s. The Swiss made a variation of the Model 1906 that has a straight front grip strap and a grip safety. It is known as the Model 1929.

The Luger had a long and useful life, but its rather delicate and vulnerable action (being also quite difficult to manufacture) gave way to the far sim-

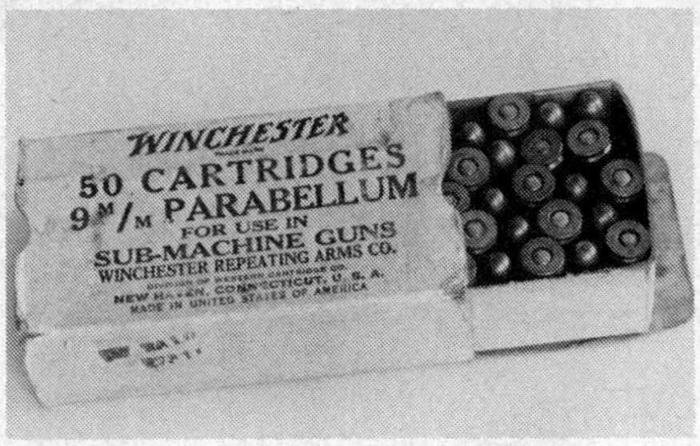

Ammo production for 9 mm arms has been a universal undertaking. Winchester made it (above), and of course, Germany—both with steel (L) and brass (R) cases.

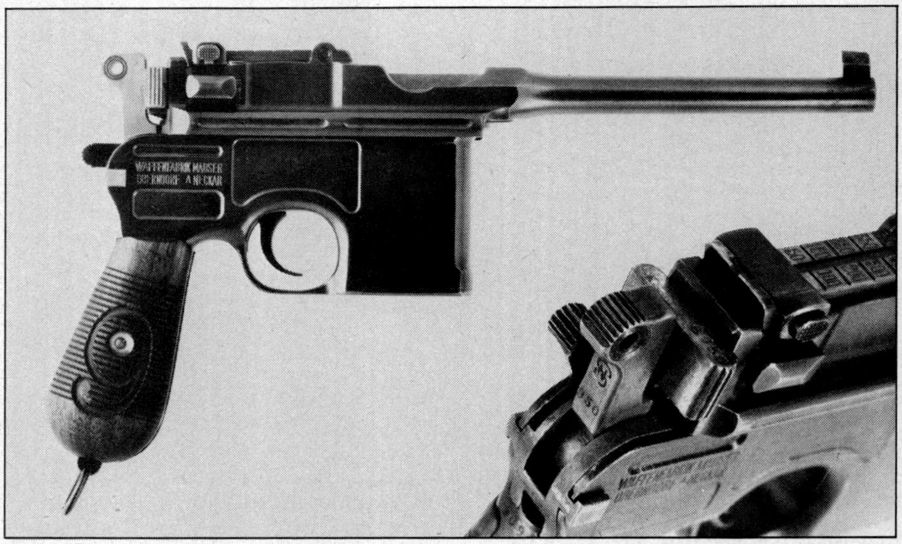

The Mauser was originally produced in 7.63 mm, but later models were available in 9 mm—as signified by the red "9" on the grip. The "cone" hammer (inset) replaced the spur hammer very early. Although never an "issue" arm, it was widely used.

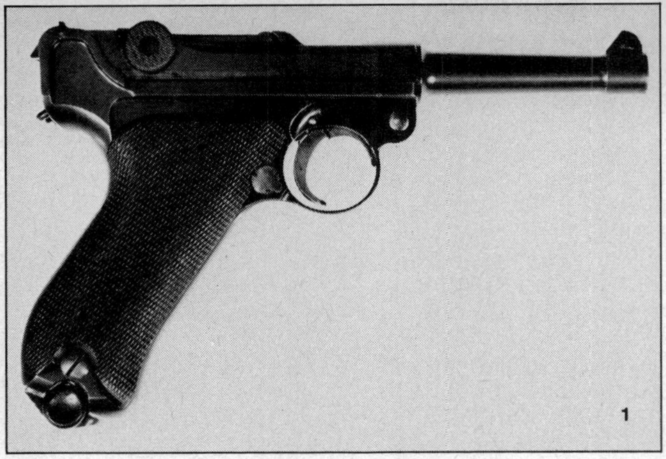

1

2

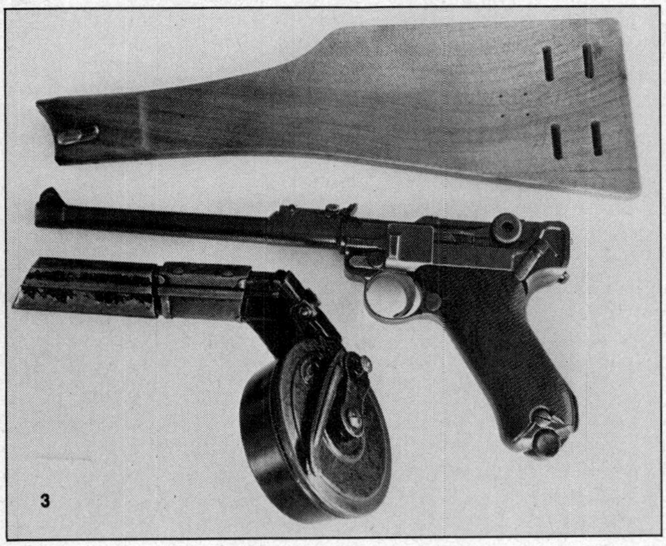

3

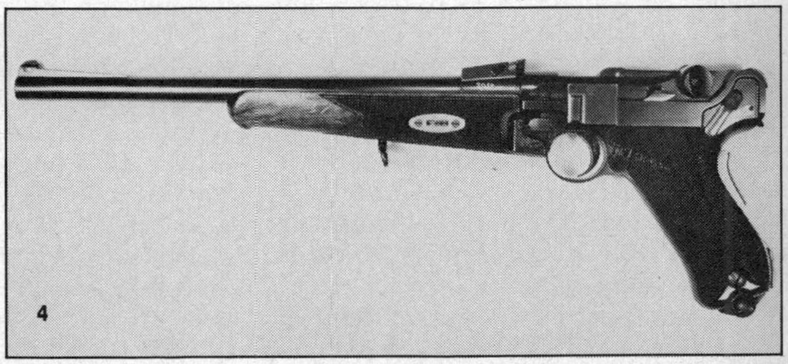

4

Lugers came in a wide array of variations. The standard military version (1) had a 4-inch barrel. The artillery model (2 & 3), had a 32-round drum, shoulder stock, and long range sights graduated to 600 meters—which is an optimistic setting! Several special Luger "carbines" were for hunting (4) and were used on small game and roe deer at short to medium range.

pler and more reliable P-38 as the first choice of the German army. Production of the Luger for military use consequently ceased in 1943.

A contemporary of the Borchardt was the Mauser Model 1896. The Mauser, called the "Broomhandle" because of its rounded grip, is a clumsy-looking handgun, with its funny grip shape and magazine ahead of the trigger. Fortunately, it shoots better than it looks. The '96 Mauser is rugged, well built, and superbly finished.

Although most Model 96's were in 7.63 mm caliber, the pistol is mentioned here because a considerable number were made in 9 mm Luger caliber for the German government during World War I as a secondary issue sidearm. Since many officers were carrying personally acquired 96's in 7.63 caliber, a large numeral "9" was burned into the wooden grip plates of the 9 mm versions to remind all parties to use the correct ammunition.

The Mauser Model 1896 was never adopted by any major power in 9 mm caliber as a first line sidearm, so it deserves only a brief description in this article even though the pistol has great historical value and evokes a vast amount of interest among collectors.

Developed in the early 1900s, the Glisenti was adopted by the Italian government in 1910 as their standard semi-automatic pistol. The Glisenti M10 was originally chambered for the 9 mm Luger cartridge, but testing soon revealed that the action was not strong enough to withstand the pressure generated by this relatively powerful round. The cartridge ultimately adopted for the Glisenti is dimensionally identical to the 9 mm Luger, but is loaded down to an acceptable pressure level. Even though it does not use the fully loaded Luger cartridge, it is one of the 9 mm family, nonetheless. The Glisenti uses a rotating locking block, one side of which pivots in a removable sideplate—not a very strong arrangement. The Glisenti is a rather complicated design, made of soft steel. In spite of its weaknesses, it remained an official sidearm of the Italians until the advent of the Beretta 9 mm after World War II. Many of the Glisentis in existence today have been fired using 9 mm Luger ammo and show considerable damage from the effort. Continued use of the too-powerful Luger ammunition will destroy the pistol, with perhaps disastrous results to the shooter.

Brief mention should be made of the Dreyse, a German substitute standard pistol used in World War I. The Dreyse was a simple blowback design. Its only claim to fame was a recoil spring that was so stiff that it was almost impossible to activate the slide. A unique device was employed that disconnected the recoil spring to allow a person of normal strength to cock the piece. Very few of these pistols were made, and the design made no lasting contribution to the handgun world.

In 1912, Austria-Hungary made the Steyr pistol their official arm. This excellent handgun was originally chambered for the 9 mm Steyr cartridge, which is similar to the 9 mm Bergmann-Bayard or 9 mm Largo. A large number of these pistols were later modified to accept the 9 mm Luger round and, in fact, were used by the Germans as a secondary issue arm in World War II. The Steyr (often referred to as the Steyr-Hahn, or hammer model) uses a rotating barrel locking system, somewhat similar to the American Savage semi-auto. The Steyr does not use a separate magazine, but loads from the top by means of a stripper clip, through the open action, into a fixed magazine in the pistol butt. Except for this awkward loading method, the Steyr is a fine pistol. It is well made, reliable, and accurate. It should be considered as one of the all-time great designs.

In the early 1920s, John M. Browning saw the need for a new 9 mm semi-automatic. Rather than merely adapt his existing .45 caliber M1911 design to a smaller cartridge, he developed an essentially different model, using a cam-controlled locking action, instead of the swinging link that he used on the M1911.

Introduced in 1935 by Fabrique Nationale in Belgium, this pistol is known as the P-35 and also as the Browning High Power (although of course it is no more powerful than any other pistol chambered for the 9 mm Luger round.) The P-35, with its 13-round staggered magazine, is a superb handgun, and has been accepted by more armed forces around the world than any other 9 mm pistol. It is, or has been, standard issue in Canada, The United Kingdom, China, Denmark, Holland, Argentina, Belgium, and a host of so-called Third World countries. Belgian-made P-35s were also used by the Germans in WWII as secondary issue, although the quality of these war-time products is generally quite poor.

The Browning P-35 is highly regarded by collectors and shooters alike, and is still in production at the present time. It has been made in a number of variations, the most notable of which is a model utilizing a shoulder stock that doubles as a holster. This model has a tangent rear sight that is graduated to 500 meters.

Although the Germans were painfully aware of the shortcomings of the Luger, economic and political conditions prevented the development of a replacement sidearm intil the 1930s, when a number of handguns were put through an exhaustive competitive program. The winner of the competition was a modified version of Walther's "Heeres Pistole," or army pistol, that was accepted for production in 1938. Known as the P-38, more recently re-designated the P-1, the Walther pistol has remained as the principal army sidearm up to the present time. Although the Luger was kept in production until 1943, the P-38 was the preferred army pistol of the second World War.

The P-38 is a rugged, reliable arm that can easily be made by mass-production techniques. The P-38 has a wedge-type locking block beneath the breech that is quite positive and virtually foolproof. The P-38 has a double-action trigger that allows the pistol to be carried with the hammer down. The first pull of the trigger cocks the hammer and fires the first round. While this type of trigger has some disadvantages, it does allow an under-trained, nervous, or tired soldier to fire the pistol quickly without fumbling with the mechanics of loading and cocking.

The P-38 was manufactured by several makers other than Walther. Among these are Mauser, Spreewerke, FN in Belgium, and Brno arsenal in Czechoslovakia. Quality of workmanship varies from factory to factory, and according to date of manufacture. Those made near the war's end were quite crudely finished by comparison with the earlier examples.

While not as simple in design as the Colt Model 1911 or Browning P-35, the P-38 was, and is, an excellent military autopistol. It has few weaknesses and no important design faults. The P-38 served its purpose very well, and remains as one of the all-time great military handguns.

In addition to the P-38, two other outstanding 9 mm semi-autos came on the scene in the 1930s—from rather unlikely sources. These are the Radom from Poland, and the Lahti pistol from Finland.

Fully automatic Mauser Broomhandles or "Schnellfeuers" were in 7.63 mm—and used by Germans in both World Wars.

The Radom, or VIS, combines features taken from the Browning Model 1911 and P-35 designs, plus some unique additions such as a simplified take-down lever on the left rear of the frame and a hammer-dropping lever on the slide that allows the pistol to be uncocked safely with one hand. The overall configuration resembles the Model 1911, while the locking system is by a barrel cam such as is found on the Browning P-35.

The Radom, named after the arsenal in which it was manufactured, was used extensively by the Germans during World War II under the designation "P-35(p)." The name VIS comes from the initials of the designers, Vilniewczyc and Skrzypinski.

The high magazine capacity of the P-35—which holds 14 cartridges—plus its reputation for dependability, made it one of the best 9's. Many military versions had long-range sights.

The Astra 600 (1) fires both 9 mm Largo and 9 mm Parabellum. Polish Radom (2) was used by the Germans in W.W. II. The Glisenti (3) was the service arm of Italy in W.W. II. French M1950 (4) is essentially a Browning 1911 with crossbolt safety. The Steyr-Hahn (5) used a rotating barrel locking system. The P-38 (6) was a rugged mass-produced double action auto.

The Radom in its better examples is an excellent pistol, rugged and reliable. Some of those made late in the war are lacking hammer releases and slide locks, and they are quite poorly made. Poland has produced many fine machines of various kinds, and a good Radom qualifies as one of the very best.

The Lahti pistol, named for its designer, Aimo Lahti, is undoubtedly the heaviest and strongest of the nine millimeters. It uses a large sliding locking block and incorporates an accelerator of the Browning machine gun type to insure positive operation in sub-zero climates. Introduced to the Finnish army in 1935, and known as the L35, the Lahti was also made in Sweden by Husqvarna beginning in 1940. In addition to the armies of Finland and Sweden, Lahtis were also used in Denmark. Lahtis are pleasant to shoot, highly accurate, and extremely reliable. The fact that they are bulky, plus being difficult to manufacture undoubtedly kept them from more widespread use. They are mechanically superior in every detail.

A little known German substitute standard pistol of World War II is the Astra 600, made in Spain in considerable numbers for German occupation forces. The Astra 600 is a shorter barrelled version of the Astra 400, or Model 1921, long a standard arm of the Spanish army. Both Astras are derived from the Campo-Giro pistol, developed in Spain in the early 1900s. Incidentally, although the Astra 400 is chambered for the 9 mm Largo cartridge, most of them have a tight chamber that allows them also to fire the 9 mm Luger round. The Astra 600 is a simple blowback design that has a very stiff recoil spring. Even though cocking the slide is difficult, the pistol is dependable and comfortable to shoot. Astras are generally well made, with good finishes, and are reasonably accurate.

Soon after World War II, several pistols of more or less conventional configuration were introduced in 9 mm Luger caliber. Among these are the French Modele 1950, the Tokagypt, the Beretta M1951, and the SIG 49.

The French Modele 1950 is essentially a copy of the Browning M1911 with a crossbolt safety and a magazine safety added, and the grip safety eliminated. The M1950 has no substantial merit as a separate design concept except that it was made in France for French consumption.

After years of making do with low-powered, substandard pistols, Italy was in sore need of a modern, adequately powerful semiauto. In 1951 Italy chose the Beretta as its military standard handgun. The Beretta M1951 uses a locking wedge system similar to the P-38. The M1951 was also used by Israel and Egypt. It is a good quality, compact and lightweight pistol that represents a considerable improvement over many older designs.

The Tokagypt is a 9 mm version of the Russian TT33 (Tokarev) made in Budapest by FES for the Egyptian forces in the late 1950s. The pistol is another virtual copy of the Colt-Browning 1911 with some notable changes. The best feature of the Tokarev concept is a sear-hammer group that is removable as a unit from the frame. This facilitates stripping the pistol as well as making sear adjustments much easier. Tokagypts are of better than average quality as regards both workmanship and finish. When found in good condition, they are desirable both as shooters and collectors pieces.

Switzerland always has been noted for excellence in things mechanical, and SIG pistols are true to this tradition in every way. The SIG 49 is similar to the Browning P-35 in its locking system, but instead of having the slide ride over the outside of the frame rails, it moves inside and between the rails. While this arrangement makes for a tighter slide and hence greater accuracy, it is somewhat more expensive to produce. Largely because of its high price, the SIG 49 is used only by the Swiss army and by Denmark. The SIG, due to its close manufacturing tolerances, is one of the most accurate of all the 9 mm semi-autos, and for this reason is used frequently by competitive target shooters as their first choice for centerfire matches.

The most recent developments in the 9 mm field include the Walther P-5, and the P-9 from Heckler and Koch. The Walther P-5 is a modernized version of the P-38, since it has essentially the same double-action mechanism. The P-5 differs from the P-38, however, in that it has a slide that fully covers the barrel—more in the style of the Browning pistols. The P-5 is more compact than the P-38, although weighing about the same. The Walther P-5 has been adopted by certain units of the German army and national police as their standard sidearm.

The semiautomatic pistols from Heckler and Koch represent the ultimate in design engineering and materials application. The P9S from H&K has a double-action trigger that cocks the internal hammer on the first pull if desired, although the pistol may be cocked by an external cocking lever. Polygonal rifling is used, producing slightly higher velocity than traditional cut rifling. The P9S is locked by a rolling block system that provides positive and consistent action. The P9 is a fine police and military pistol that has all the features one could desire in such a pistol.

The 9 mm Luger cartridge lacks the stopping power of the .45ACP, but has nevertheless become the almost universal military handgun caliber. Pistols for the 9 mm can be made a bit smaller and lighter, and a given amount of ammunition can be transported in a smaller space with less weight. Cost per round is also lower than for the .45 ACP—a fact to be considered when dealing with tight military budgets.

The 9 mm Luger has been called the "perfect" automatic pistol caliber—and with good reason. It holds enough propellant to give optimum pressure and velocity, and it has the correct ratio of length to diameter to assure efficient feeding in a large variety of autopistols. If the latter statement needs defending, consider the shorter 9 mm Kurz (.380 ACP) that frequently "stovepipes" in feeding through many pistols, and the longer 9 mm Bergmann-Bayard (9 mm Largo) that sometimes has a bit of trouble getting around the corner when going from magazine to chamber.

A 9 mm pistol has less recoil than a .45, and thus can be fired more comfortably and accurately by the average soldier-marksman. For these reasons, at least, the 9 mm is an attractive choice for many armed forces.

Military handguns in 9 mm caliber continue to proliferate and, in spite of much valid criticism, probably will continue to be standard fare in most armies of the world for a long time to come.

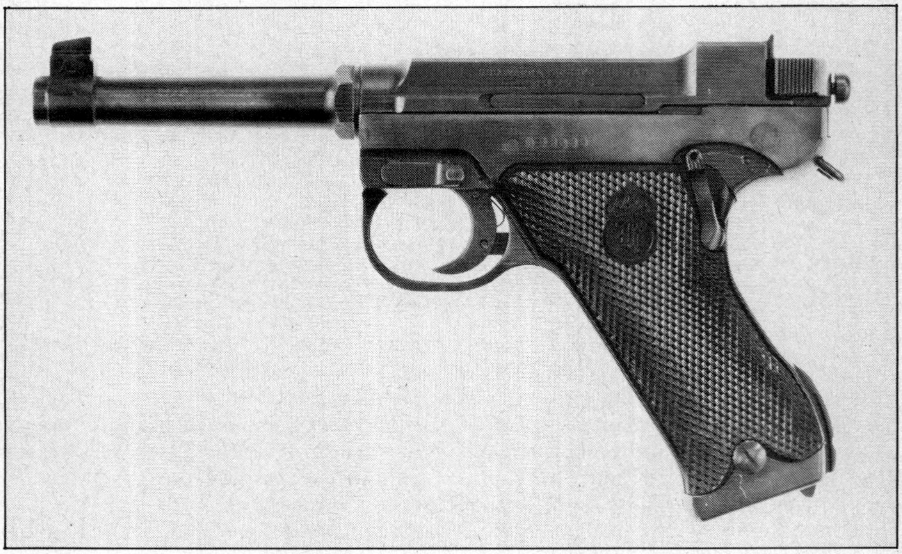

The Lahti was adopted by the Finnish Army in 1935. This exceptionally well-made pistol exhibits a high degree of accuracy.

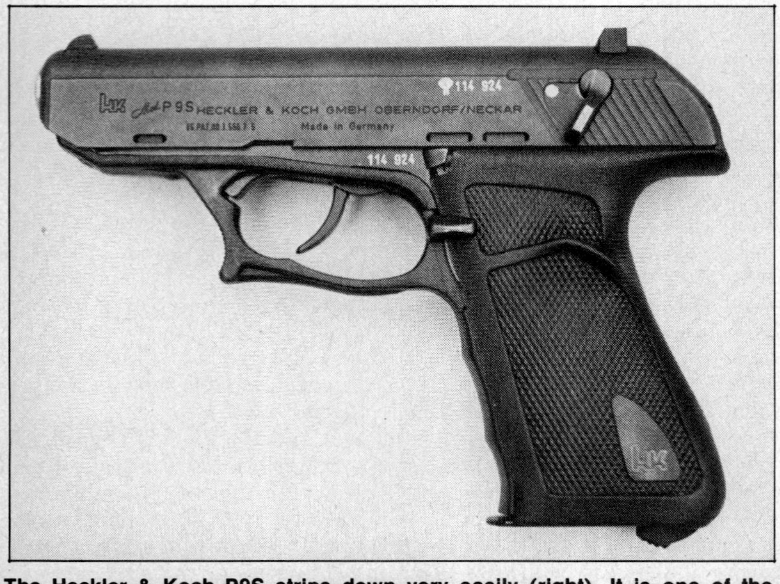

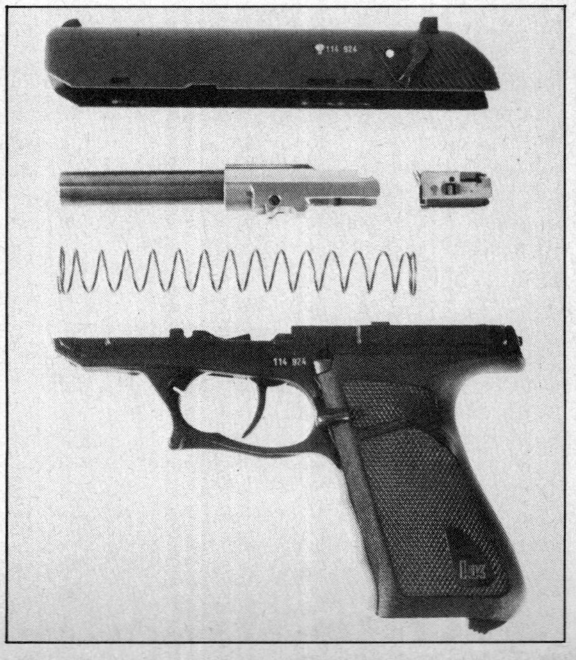

The Heckler & Koch P9S strips down very easily (right). It is one of the most accurate and easily controlled 9 mm pistols on the market. It has been adopted by many police agencies as well as anti-terrorist units.

THE XP-100 GOES BIG BORE!

With the brand new 7mm Bench Rest chambering, Remington's bolt action pistol is made to order for handgun hunters and silhouette shooters!

By Bob Milek

Remington's introduction of the XP-100 bolt action pistol in .221 Fire Ball chambering in 1962 represented the first major deviation from the norm in handgun design in this century. At the time the XP-100 was a very radical concept and shooters were slow to accept it. Not until long-range varmint hunters discovered the attributes of this pistol/cartridge combination did sales of the XP-100 pick up. Then benchrest pistol shooting and metallic silhouette competition came along and suddenly XP-100's were in demand—sometimes in their original form, but often so that the action could be rebarreled to a more powerful cartridge for big game.

In early 1980, Remington announced a new XP-100, this one with a 15-inch barrel instead of the 10½-inch used with the .221 Fire Ball, and the pistol was to be chambered for a new cartridge, the 7 mm Bench Rest. Apparently the demands of silhouette shooting were the force behind Remington's long overdue introduction of the XP-100 chambered for a new, powerful cartridge. The .221 isn't adequate for serious silhouette shooting and Remington was losing sales to other single-shot pistol manufacturers who offer a variety of powerful cartridges well suited to the silhouette game.

However, when I first read of the XP-100/7 mm B.R. combo, I saw it as a prospective hunting handgun. Here, I thought, is a handgun that might just be a super performer in the field—a rig with which to hunt both varmints and big game at ranges exceeding 100 yards.

Although I received my XP-100 in 7 mm B.R. chambering in early June, testing was held up for lack of Remington benchrest brass and dies with which to make 7 mm B.R. cases. But while I waited for these items, I went about familiarizing myself with the new pistol.

Its appearance differs from the older XP-100 only in that the barrel is 4½ inches longer, a little heavier and isn't cluttered up with the plastic rib and open sights common to the .221. There are no sights on the 7 mm version of the XP-100, but it is furnished with a dovetail base, secured to the barrel back 4½ inches from the muzzle by a screw, into which a variety of front sight fixtures can be set. The receiver is drilled and tapped on top for scope mounts and on the side of the bridge for installation of a micrometer receiver sight.

The stock is the old cumbersome nylon design that Remington has put on the XP-100 from the very beginning. I find this stock almost impossible to shoot comfortably, but this may be just a personal quirk. I know some shooters who find it quite satisfactory.

The bolt action is the 600 series around which Remington once built a number of excellent little carbines. It's nothing more than a scaled down version of Remington's famous 700 action—strong, smooth and reliable. The bolt face is counterbored, surrounding the head of the cartridge with steel, and the extractor and plunger-style ejector are the same as on the 700 action. The XP-100 sports a slender, graceful bolt handle whose knob is rounded on the top half, but ground flat on the underside and serrated to afford somewhat of a non-slip surface.

As on any bolt action pistol with a center-located grip, the trigger is positioned well forward of the sear housing—5¼ inches forward in the case of the XP-100. Remington connects the trigger to the sear housing by using a 5 ¾-inch long trigger link fabricated of heavy sheet metal. The trigger link assembly is attached at the rear to the sear housing and in the front to the recoil lug. The trigger can be adjusted for weight of pull and overtravel, but I didn't have to touch mine. It came factory adjusted for a 1¼ pound pull.

While I waited for cases and dies, I set about mounting a pistol scope on my XP-100. The scope I chose to test was the new Redfield 4X set in a Redfield bridge mount. My initial impression of this scope was that it was superb, but only time would tell. Too often in the past I've become fond of a pistol scope only to have it fall apart before my tests were completed.

Dies for the 7 mm B.R. were supplied to me by Redding. Dick Beebe, the company's president, made mine up from cartridge and chamber drawings furnished him by Remington. He had no fired cases to work with. Still, he got them right out to me—they arrived before I had any cases—and they were perfectly dimensioned. The die set consists of a number 1 form die, a trim die that doubles as the number 2 form die, a special ream die and reamer, a full-length sizing die and a seating die.

Cases for the 7 mm Bench Rest cartridge are made from Remington benchrest brass. Although this looks like .308 brass, it has thinner walls and uses a small rifle primer. Remington cautions against making 7 mm B.R. cases from .308 brass—as do I at this time. I'm sure that .308 brass can be used, but entirely different loads will be required owing to the large rifle primer and the thickness of .308 brass as compared to that of the special benchrest cases. The data shown in the enclosed load table is intended for use with cases made from Remington benchrest brass only! Time was short for completion of this piece, so I was able to develop only a few loads. However, those shown for bullets weighing from 115 to 145 grains are all extremely accurate. I was unable to develop loads for the 150 and 154-grain bullets that were satisfactory for accuracy.

The 7 mm Bench Rest has a maximum case length of 1.520 inches, just slightly longer than the similar .308x1½. The Bench Rest case as it comes from Remington is 2.05 inches long. Form die number 1 shoves the shoulder down to approximately the right location, but retains the same shoulder angle as is on the original case. This leaves a neck about .735-inch long. When this case is run through the number 2 form/trim die, the shoulder angle is changed from the original 20 degrees to 30 degrees. While the case is in die number 2, it can be trimmed with a hack saw and file. However, I prefer to trim the case in an RCBS lathe-type trimmer to which I attach an electric drill. This makes fast work of the trimming job and I trim to 1.510 inches. A word of caution, though. As the case is formed, the brass inside the neck develops a taper, or wedge shape. At the mouth the inside diameter of the neck is .317-inch, but just ahead of the shoulder/neck junction it's only .270-inch. If you decide to trim with a lathe-type trimmer, you must use a pilot no larger than .270-inch in diameter. A larger pilot will jam inside the neck as you near the proper case length.

After trimming, the inside of the case neck must be reamed to remove the wedge of brass, creating a straight neck wall, and to reduce the thickness of the brass in the neck to about .011 inch. Outside neck turning is useless in this instance because it will not eliminate the wedge on the inside.

Redding's ream die is easy to use. The case is just shoved all the way up into the die and the reamer is inserted into the die from above via the close-fitting pilot hole. To turn the reamer, use a T-handle or a carpenter's brace. After reaming, the case is run though the full length resizing die in preparation for loading.

This sounds like a lot of work just to make a case. Believe me, it is. However, it's absolutely essential that each and every step be handled perfectly, the reason being that the 7 mm B.R. chamber in the XP-100 is tight—built to benchrest dimensions rather than to those normally used for sporting cartridges. If the case isn't exactly dimensioned, it will not chamber. I belabor this point for one important reason. The minimum dimension in the neck portion of the chamber is .309 inch at the mouth, .310-inch at the shoulder/neck junction. Should the brass in the neck be too thick, greater than .011-inch, the neck will be too tight in the chamber to allow sufficient expansion to release the bullet. Should this occur, chamber pressures will jump drastically. When a .284-inch bullet is properly seated in a 7 mm B.R. case, the outside neck diameter should be no more than .307-inch. After reaming and bullet seating, the necks on my cartridges measure .3065-inch in diameter. Check this dimension after you have the cases loaded and check it again if a round doesn't chamber easily. Also, the brass tends to flow forward and thicken in the neck after a few firings, so repeated reaming will be necessary.

As I said earlier, my interest in the

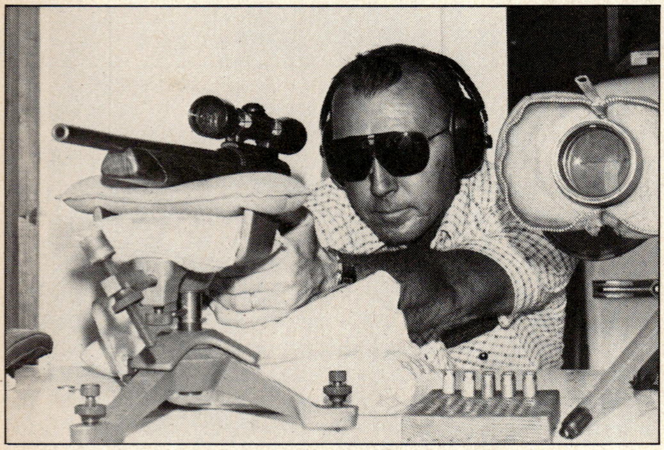

The author's pet load for the 7 mm BR will consistently group 5 shots into 2 inches at 100 yards. That's super pistol performance!

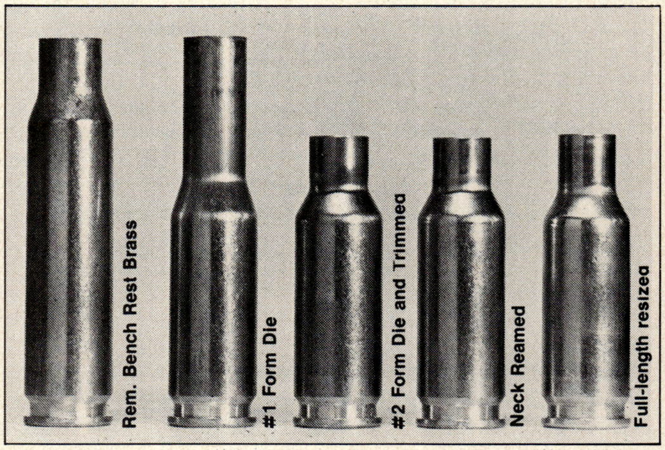

This sequence shows each step needed to form a 7 mm Bench Rest case. Remington warns against using .308 brass at any time.

This is the author's answer to control problems with the factory Remington XP-100 stock. He put on a custom thumbhole stock.

The new XP-100 in. 7 mm BR sports a 15-in. bbl, no open sights, and a nylon stock. Unlike the original .221 XP-100, it has no rib.

XP-100 7 mm B.R. is for hunting, so it was toward this end that my load development work was directed. As of this writing, all of my hunting has involved small varmints—prairie dogs and rockchucks—with the former providing most of the action.

Prairie dogs are small animals that present a tough target yet provide little resistance to help initiate bullet expansion. On these small animals, I found that only 115 and 120-grain bullets expanded rapidly enough and even these would occasionally fail to open up, particularly at long range. Bullets weighing 125 grains are OK for chucks, but anything heavier goes right through prairie dogs and chucks with little expansion.

As a result of my tests, I'd suggest that bullets weighing from 115 to 120 grains be used for varmint hunting with the 7 mm B.R. I was very pleased with the performance of these lightweight numbers in the field. For big game, bullets weighing from 125 to 145 grains will probably give the best results. If you anticipate having to break some heavy bone, I think I'd use bullets in the 139 to 145-grain weight class. They should hold together to smash bone and penetrate deeply. Many metallic silhouette shooters will undoubtedly prefer heavy bullets—150 to 160 grains—but I had trouble getting accuracy from my XP-100 with these heavyweights and I have a hunch that they wouldn't expand well on game. Of course, all of my recommendations for big game bullets in the 7 mm B.R. are based entirely on supposition. Not until after I get to use the cartridge on big game this fall will I be able to check out my theories.

Having taken well over 200 prairie dogs with the XP-100/7 mm B.R., I feel that my initial hunch about this pistol/cartridge combination is correct. It's destined to become one of our most popular hunting handguns. However, there are a couple of things that I feel must be addressed before it will receive the acceptance it deserves.

The first of these has to do with the stock. I can't shoot the issue version comfortably and from any position except a benchrest it's an abomination to control. My answer to this is a custom stock. There wasn't time for me to order one and have it inletted and finished before this article was due, so I used a different approach. A couple of years ago I had Dick Bryan stock my XP-100 .221 with Fajen's excellent thumbhole design. I removed the .221 barreled action and took the stock and my 7 mm B.R. to Thermopolis, Wyoming, gunsmith Ralph Hastings. Ralph opened up the barrel channel to accommodate the heavier barrel on the 7 mm, made a few minor changes in the action area, then glass bedded the barrel and action for me. This transformed my XP-100 from an uncontrollable ugly duckling into a handsome, nice-handling pistol that's a pure pleasure to shoot. Since the nylon stock is a common complaint handgunners have about the XP-100, I think Remington would do themselves and many shooters a favor if they would give us the option of buying either the complete XP-100 as now offered, or buying the barreled action only.

The other problem a hunter faces with the XP-100 is how to carry it in the field. It's much too large for any holster, even one of shoulder carry design, and too heavy and bulky to be carried around in your hand all day long. An experimental sling arrangement being tested by Michaels of Oregon may solve the problem of how to pack the XP-100 in the field. This uses an adjustable nylon web sling and "Uncle Mike's" detachable swivels. The rig is still in the prototype stage, but I look for it to be marketed before too long.

With the 7 mm Bench Rest XP-100, I feel that Remington has finally jumped into the thick of the handgun hunting scene. My favorite loads will consistently group five shots in 1¼ inches at 100 yards and the cartridge has the velocity to be effective on varmints and big game at well beyond 100 yards. I dropped a good many prairie dogs at ranges from 100 to 150 yards and even managed to down a few at 200 yards. Despite the accuracy of the pistol/cartridge, though, none of my long-range success would have been possible were it not for the new Redfield 4X pistol scope. For varmint hunting, 4X magnification is great, but because of the small field of view this scope offers, I believe I'll swap the 4X for a good 2X or 2½X before I take the 7 mm B.R. out after antelope and deer.

Now there is another cartridge for the XP-100 besides the Fireball—the 7 mm Rem. BR.

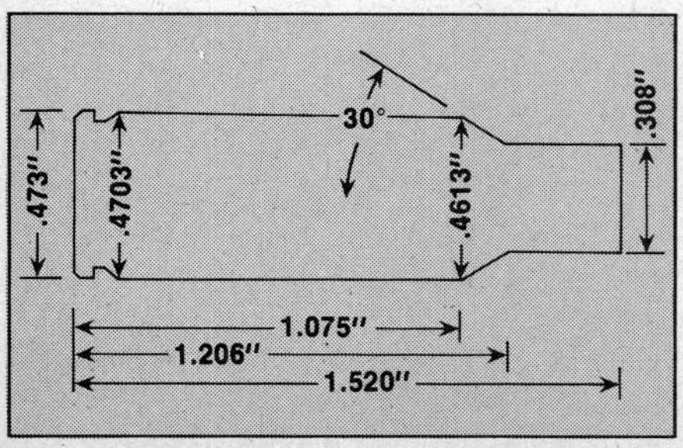

Shown above are the case dimensions for Remington's new 7 mm silhouette hotshot.

REMINGTON XP-100
LOAD CHART

Cartridge: 7 mm Remington Bench Rest
Test Gun: Remington XP-100
Barrel Length: 15"

BULLET	POWDER Type	Wt. Grs.	PRIMER	CASE MAKE	INSTRUMENTAL Velocity fps.
Speer 115-gr. HP	H322	30.0	Rem. 7½	Rem.	2505
Speer 115-gr. HP	BL-C (2)	34.0	Rem. 7½	Rem.	2359
Sierra 120-gr. Spitzer	H322	29.5	Rem. 7½	Rem.	2487
Sierra 120-gr. Spitzer	BL-C (2)	34.0	Rem. 7½	Rem.	2367
Rem. 125-gr. Pt. S.P.	H322	29.0	Rem. 7½	Rem.	2478
Sierra 140-gr. Spitzer	H322	27.5	Rem. 7½	Rem.	2318
Sierra 140-gr. Spitzer	748	31.0	Rem. 7½	Rem.	2253
Speer 145-gr. Spitzer BT	H322	27.0	Rem. 7½	Rem.	2310
Rem. 150-gr. Pt. S.P.	H322	26.0	Rem. 7½	Rem.	2109
Hornady 154-gr. Spire Pt.	H322	25.5	Rem. 7½	Rem.	2049

Note:
These loads were developed using Remington bench rest brass. In no instance should cases made from .308 brass be substituted!
All loads are safe only for my particular XP-100. For all other pistols the powder charges should be reduced at least 10 percent for starting loads.
Hodgdon's H322 powder appeared to give the best combination of accuracy and velocity with all bullets listed.

DAN WESSON MODEL 9-2

The Dan Wesson system is truly that—a system. It offers the wheelgunner a myriad of barrel length choices for a myriad of purposes.

By Jan Libourel

About ten years ago, the late Dan Wesson, scion of one of the most illustrious names in American arms-making, began production of the revolvers bearing his name. The most distinctive feature of these sixguns, as most readers know, is the fact that their barrels are readily interchangeable. Thus, a police officer who owns a Dan Wesson can use a 2½-inch barrel for undercover or plainclothes duty, a four-inch for holster wear in uniform, a six-inch for PPC shooting, and, perhaps, an eight or ten-incher for hunting or silhouette matches. By using the Dan Wesson interchangeable barrel system, he could, in effect, have four or more different revolvers—at a small fraction of the price he would pay for four separate guns.

I'll have to confess that I used to view the new system, for all its ingenuity and versatility, with considerable skepticism—anything that revolutionary and different just *had* to give trouble. Well, the Dan Wesson people proved me wrong. Their revolvers have stood the test of time. The barrel-locking system has been slightly improved over the years, and the company has offered their revolvers in an ever-increasing array of options, including their recently-introduced revolvers in .22 rimfire and .22 Magnum.

Our test gun was the Model 9-2—the .38 Special version of the basic Dan Wesson revolver; it came with a 2½-inch barrel. Since Dan Wesson .38 Specials are identical to their .357 Magnums except for the depth of the chambers, and most shooters prefer the .357 option, I had always assumed that the .38 Specials were intended for law enforcement customers who were prohibited from carrying .357s because of departmental policy. However, I was informed that the majority of Dan Wesson's .38 Specials are produced for export. It seems that in a number of foreign countries, among them Mexico, civilians can legally own .38 Specials, but not .357s. Obviously no country has a monopoly on irrational gun legislation!

Dan Wesson .38 Specials have the same dimensions as their .357 Magnums, and the steels and metallurgy used are exactly the same. The Dan Wesson .38s should thus have more than ample strength for any .38 Special +P loads or any sensible handloads.

Everything about this revolver reflects both high-quality workmanship and intelligent, progressive design. The fitting of all parts is excellent, and the bluing and polishing compare favorably with the finish on any revolver on today's market.

One of the features I have always most admired about Dan Wessons is that the frame has no grip straps; instead, the stock attaches to the frame by means of a through-bolt. Consequently, the grip on a Dan Wesson can be shaped to virtually any contour, and a grip suitable for the largest or the smallest hand can be fitted with equal ease. Dan Wesson offers several excellent grip designs as options, or you can get an inletted walnut blank from them and rasp and sand it until you have exactly the stock you want. Our gun came with their new "Speed-Loader" design—a large grip of handsomely-figured walnut with palm swells and finger grooves that closely resembles the designs favored by many of the top PPC competitors. I have always tended to look askance at grips with finger grooves, but this one fit my medium-sized hand to perfection. It did, however, seem disproportionately large and bulky for a 2½-inch barreled revolver, which is normally bought as a concealment gun.

Another very good feature is the cylinder latch located in front of the cylinder on the crane. The latch projects into a recess in the frame underneath the barrel and serves to give the cylinder and crane assembly a desirable fore-and-aft lockup. Many is the time my thumb has been lacerated by the cylinder latches on competitive revolvers under the recoil of heavy magnum loads, but this can't happen with the Dan Wesson. Its latch is safely located where it won't draw blood from your thumb. Some nit-pickers have objected to this system on the grounds that it makes it more difficult to reload one-handed in a hurry. Well, I don't claim to be any expert in gunfighting, but it seems to me that if you are facing hostile fire with one hand disabled and an empty revolver, the situation has probably deteriorated to the point where the location of the cylinder latch is of minor concern.

The rear sight is fully adjustable for windage and elevation. The sights are wide and easy to pick up. The front sight has a red inlay, and the rear a white outline. Front sight blades with yellow or white inlays are readily interchangeable with the original sight.

The sight adjustment and sideplate screws are of the Allen-head variety. The wisdom of this arrangement should be apparent to anyone who has ever burred a screw slot. The barrel wrench furnished with the revolver does double duty as an Allen wrench.

The trigger is of an ideal width for single or double-action work—⅜-inch. In accordance with current doctrine, it is smooth-faced. A small set screw in the upper rear of the trigger serves as an adjustable overtravel stop. Single-action pull was excellent—crisp, clean and just a shade over three pounds.

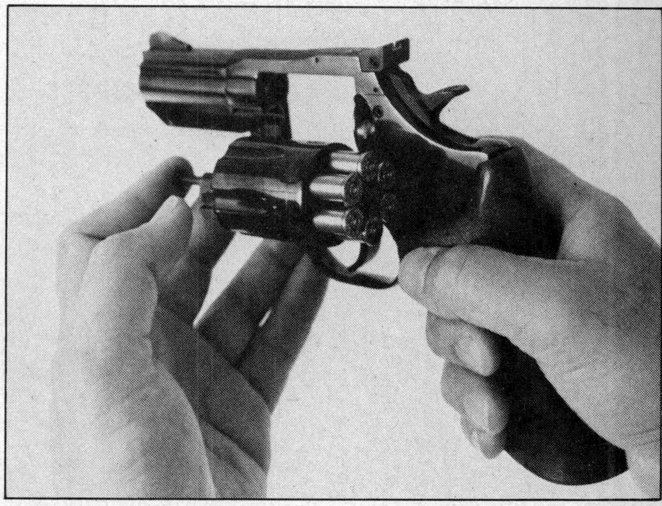

Contrary to what some shooters think, short-barreled wheelguns can be accurate—particularly if they have adjustable sights like the Dan Wesson. Author Libourel was able to get a 3-inch group at 25 yards firing double action (left). The revolver functioned perfectly. Ejection of empties was quick, smooth, and positive (top).

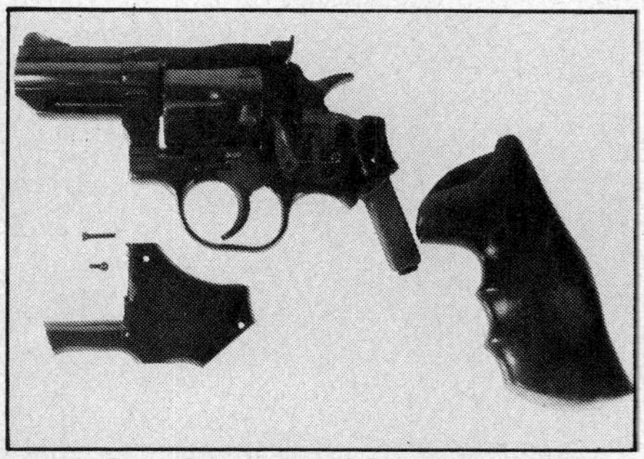

One nice feature of the 9-2 is that there are no grip straps. This means that the stocks can be shaped to fit almost any size hand.

I had always found the double-action pull on Dan Wessons somewhat disconcerting. Our test gun's broke at about 12 pounds, which is about par for any DA revolver of recent manufacture, but the Dan Wesson has a short, abrupt trigger-cocking action that just never felt "right" to me. I had seen some fine double-action work done with Dan Wessons in police combat matches years ago, so I knew these guns could be shot well in this mode, but I suspected that the Dan Wesson's "peculiar" system would be a handicap to me. I have since revised this opinion.

The lockwork on the Dan Wesson is simple, sturdy and well-designed. The cylinder bolt and hand are both hefty and should be long-wearing. The hand is of the two-stage design similar to Colt's; that is, it rises as the trigger is pulled to lock the cylinder into firmer alignment. Cylinder rotation is to the right, pushing the cylinder and crane toward the frame at the moment of firing—a system that many believe contributes to firmer lockup and greater durability. The safety system is the positive, reliable transfer bar, which is employed on virtually all revolvers of recent design. The lockwork is relatively easy to disassemble; instructions are found in the owner's manual.

Unfortunately, I was unable to experiment with the removable barrel prior to testing the gun because someone had filched the barrel wrench before the gun had reached my hands.

On the day of the shooting tests, I fired several hundred rounds of various types of .38 Special ammunition in the Dan Wesson. As expected, the gun performed flawlessly, with no misfires or any other malfunctions. The gun's weight and well-designed grips made it a pleasure to shoot with the hottest +P loads—something that cannot always be said of the smaller-frame .38 snubbies. Accuracy testing was rendered difficult because the wind had turned into a howling gale, and the first groups I shot ran about an unimpressive four inches. I doubted if the revolver itself was at fault, for Dan Wessons have an excellent reputation for accuracy, and this particular revolver had shot well in earlier tests.

When I turned to some standing, aimed double-action shooting, I got two surprises. That double-action that had seemed stiff and abrupt when the gun was dry snapped proved excellent for controlled DA shooting. Moreover, for some reason, I got better accuracy shooting in this manner than I had shooting single-action off the bench. With Federal 148-grain wadcutters, I put ten quick double-action shots into a group measuring just a shade over three inches. That convinced me of both the accuracy of our Dan Wesson and the excellence of its double-action system. Any time I can shoot a ten-shot group, double-action, in a high wind with a snub-nosed revolver that will fit inside the ten-ring at 25 yards, that gun is all right with me! When I tired some fast-as-you-can, aimed double-action bursts on a combat target at ten yards, I found that our test gun gave amazingly small groups.

I left this firing session with newly-heightened respect for Dan Wesson guns. These revolvers are competitive with any sixguns available. They are finely made, and even aside from their interchangeable barrels, they display many progressive design features that set them apart. Anyone in the market for a new revolver should take a long look at the Dan Wesson line. I know that I, for one, am eagerly awaiting the appearance of their .44 Magnum revolver in the not-too-distant future.

SPECIFICATIONS

DAN WESSON MODEL 9-2

Manufacturer:	Dan Wesson Arms 293 Main Street Monson, MA 01057
Caliber:	.38 Special
Barrel:	Interchangeable system; available lengths 2½, 4, 6, 8, 10, 12, and 15 inches
Action:	Double action revolver
Capacity:	Six shots
Weight:	32 to 69¾ ounces
Grips:	Seven styles available, all interchangeable
Sights:	Click graduated adjustable rear; serrated front with inserts
Price:	$232.90

Long before the advent of scoped handguns and metallic silhouettes, Elmer Keith was exploring the mysteries of . . .

Long-Range SIXGUNS

By Elmer Keith

Elmer's early interest in long-range sixgun shooting stemmed from his experiences with single-action Colts. He found that a good .44 Special load, out of a custom SA Army, like this one, was effective at long-range.

Today, most shooters think a six-gun is useless beyond about 50 to 100 yards. Nothing could be further from the truth. They do not question the fact a good artilleryman can hit a building three miles away and over a hilltop regularly with a 6-inch howitzer once the big gun is bedded down properly. For the ranges involved, its barrel is even shorter in proportion than that of a six-gun.

Just after the turn of the century, a revolver club in Kentucky regularly held 300-yard turkey shoots with revolvers. They used Smith & Wesson .38-44 and .44 caliber guns with 6½ to 7½-inch barrels. They made up and installed high rear sights on these target guns that were correct for their load and the 300-yard range.

During all those years, I wore a six-gun the same way as I wore my pants. Three years straight I killed 42, 41 and 43 grouse with a .36 Navy Colt. My six-gun accounted for all manner of small game and several deer and elk before I left Montana.

As a boy, I had an excellent six-gun coach in Samuel H. Fletcher, who had served through the Civil War in the 2nd Illinois Cavalry and was one of the few in that outfit who survived the war.

While herding sheep and also running wild horses, I often used the six-gun to turn them by raising dust in front of the sheep and to either side of a band of running wild horses we were trying to corral.

Dad would buy me Laflin & Rand black powder by the 25 pound keg, primers by the thousand and lead in 200 pound lots–that cost just 5 cents a pound in those days. I would spend part of the night reloading .45 Colt ammo to throw in front of those old black-face ewes the next day, to keep them from getting on our neighbors' property.

There is nothing mysterious about long-range six-gun shooting and any good shot who can keep and hold a sight picture can soon learn the art. Each summer I give Erv Malnarich's Packer and Guide School class a day's lesson in both short and long-range six-gun work, as well as rifle shooting. Many of them pick it up in one day's practice while others do not have the right temperament and will never learn.

Back in 1927, Harold Croft came out to our little Oregon cow ranch and spent a month with me, experimenting with six-guns. He brought along a suitcase of guns, all .44 and .45 caliber from the .44-40 and .44 Special to the .45 Colt. We nailed a bunch of boards together forming a 4x4 foot target and perched said target on a dry dusty hillside, 700 yards away.

Elmer's greatest feats in long-range sixgun shooting were accomplished—fittingly enough—with the cartridge he was instrumental in developing, the .44 Magnum. Here, he lines up on a distant target with his S&W M29.

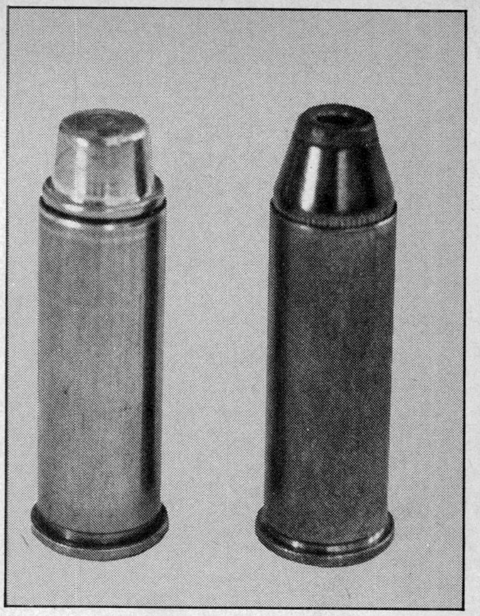

The .44 Magnum gives best results with either a Keith-type SWC bullet (L), or a jacketed hollow point (R).

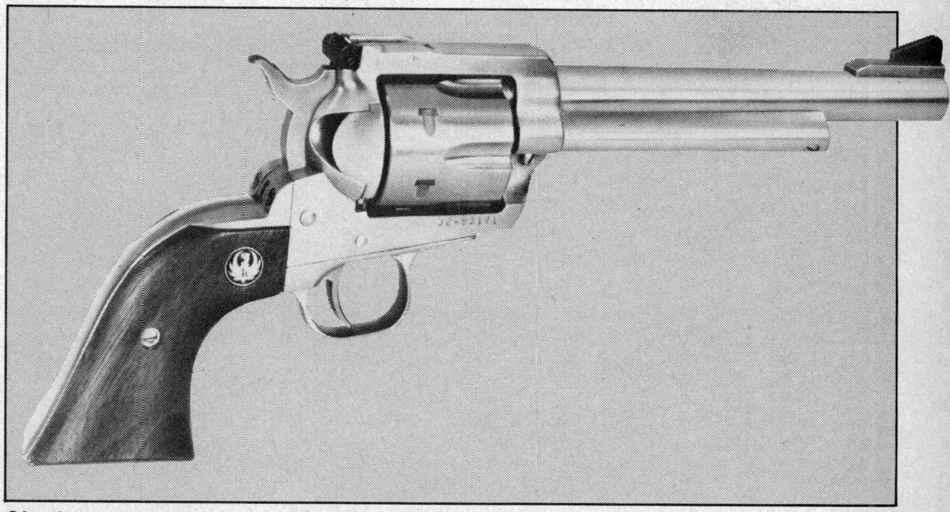

Single-action sixguns with fairly long barrels and good adjustable sights are still favored by many shooters for long-range. Ruger makes some of the best. Their Super Blackhawk in .44 Magnum (left) is a top choice for hunting and long distance shooting. Their .357 Blackhawk (above) can rightfully be considered a classic. Its simplicity, ruggedness and accuracy are nearly legendary. The Blackhawk is now available in stainless steel.

I hit that target with every gun he brought out, before it was empty, except a 2-inch barreled .45 Colt slip gun. I had to use 11 shots by aiming at a sage bush on top of the hill in back and off to one side—before I walked that short slip gun onto the target.

With the 3½ to 7½-inch barreled guns I had no trouble walking my slugs onto the target before the gun was empty. With the longer-barrel guns I got three hits out of five shots once I had learned that particular gun. This shooting was with both hands and the gun held between my drawn-up knees. My saddle was used for a back and head rest, while I shot from the shade of the trees out across the dry, dusty hay meadow.

The first requisite for long-range shooting is an accurate six-gun, preferably with 4-inch or longer barrel and an accurate, powerful long-range load. Wadcutters are out as they are not accurate beyond 50 yards in most six-guns. Many guns and loads will shoot well at 25 to 50 yards but are hopeless for long-range.

Big caliber, heavy bullets throw up more dust—it is easier to locate their strike and thus walk your bullet onto a distant target. The .22 Magnum rimfire in Ruger or Smith & Wesson guns is excellent for long-range practice over water, where safety will permit, but the tiny bullets seldom throw up enough dust to locate their strike.

The Lugers are accurate and are also good over water or real dusty ground for long-range shooting but the best is a good Colt or Ruger single-action or Smith & Wesson double-action in calibers .41 Magnum and .44 Magnum. With my 220-grain Hensley & Gibbs bullet, backed by 19 to 20 grains of 2400 in the .41 Magnum and my 250-grain bullet backed by 22 grains of 2400 in the .44 Magnum, from good Smith & Wesson or Ruger six-guns, you are in business. Also, some of the heavy factory full loads are excellent in these two calibers.

When two hands are used, the left, or nonshooting, hand should be brought up merely as a steadying rest for the shooting hand, but without any grip or pressure whatever on the gun from the non-shooting hand.

The only time I use one hand for long-range is in my old reclining position that so many new gun writers have now adopted. This is a very good position. The right knee is pulled up with the right hand and gun resting alongside and in the hollow of the bent knee, never over the top. Be damn careful to keep the butt of your gun away from your knee bone or you will soon learn the error of your ways. The back of the head can be supported by the left hand

The Colt Python, with its silky-smooth action and excellent sights is a superb long-range handgun. Although the .357 Magnum isn't as good a long-range load as a .41 Magnum, it can still be effective past 100 yards.

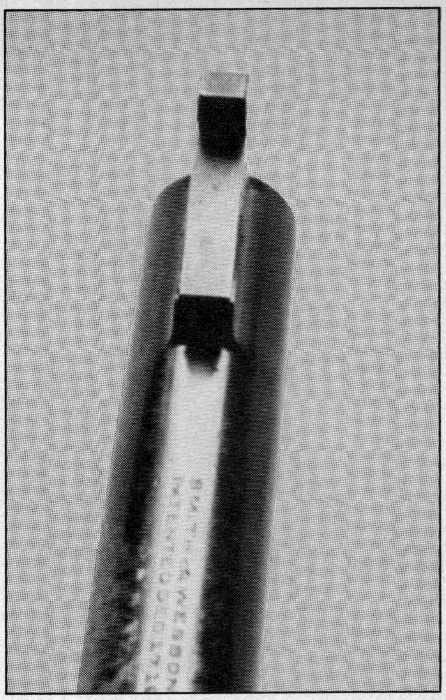

This undercut front blade sight is an excellent choice for long-range work, teamed with a Patridge rear.

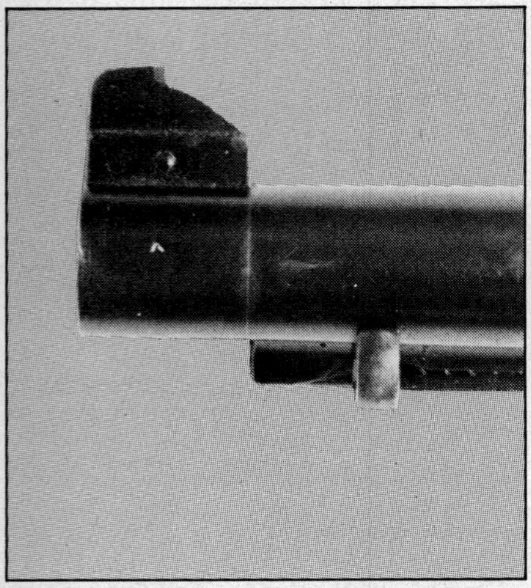

Keith's method of holdover requires a different type of front blade than normally comes with a stock sixgun.

This front sight, on one of Keith's old S&W Triple Locks features an even more radical undercut design.

The gold bead front sight on this particular sixgun of Elmer's permits a very fine hold at long-ranges.

or the left arm thrown back as a further steadying rest for the body. Once learned this is a wonderful long-range position.

About the best position of all, however, is with the back and head against a wall, post, tree stump, rock, or log and the knees drawn well up in line with the eyes and gun held between the knees, not over them, with two hands. The shooting hand gripping the gun, the other merely supporting the shooting hand. Both of these positions place the gun and sights well away from the eyes and give a sharp sight picture, which is necessary to any accurate long-range shooting.

When you are forced, through lack of time or any rest position, to shoot offhand, then face the target squarely and use both hands and arms fully extended. Shooting with gun alongside the right knee or between the knees is, by the way, not conducive to the longevity of your overalls, as the escaping gas at barrel and cylinder junction will burn and blacken them.

In game shooting if you do not have time to assume the reclining position or backrest positions and have to shoot offhand, then you can often rest the shoulder of the nonshooting hand against a tree, rock face, or any object to help steady your body. Always take advantage of any rest possible but never, never let the gun touch anything, neither the butt of the gun nor the side.

A big six-gun is as sensitive to a change of grip as are the strings of a violin, so always grip it in the same way and as high as possible.

Last but not least, assuming you can squeeze a trigger properly, comes the sight picture. Many years ago, in the early Thirties, I designed a long-range front-sight blade with three narrow gold bars. One at the top, one in the middle, and one near the bottom of the sight blade. It was designed for the D. W. King Gunsight Company of San Francisco. Smith & Wesson also furnished revolvers with these special Keith-King sights for a time, under Doug Wesson.

In the late Twenties I had also designed a three-leaf rear sight for the sixgun and Croft had one made up for a .44 Special flat-top single-action for me. Croft and I designed and had a good many flat-top single-actions made up with the forerunner of the ramp front sight as well as my three-gold-bar long-range front sight.

With the three bars, I had three definite aiming points on the front sight when these bars were held level with the top of the square-notched rear sight.

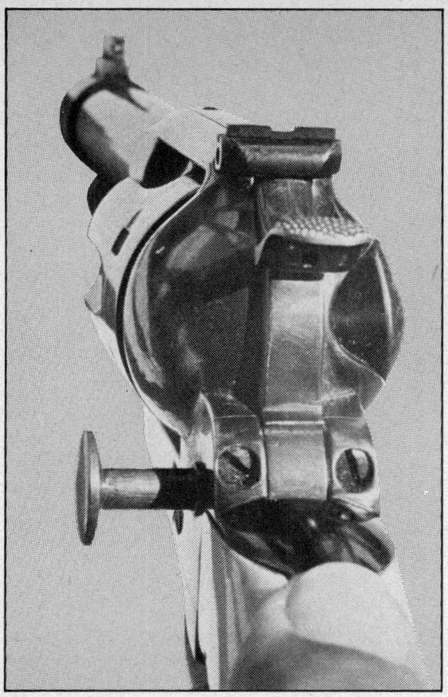

This particular long-range sixgun of Keith's features a large headed pin on the side of the frame to fit into a metal belt slot.

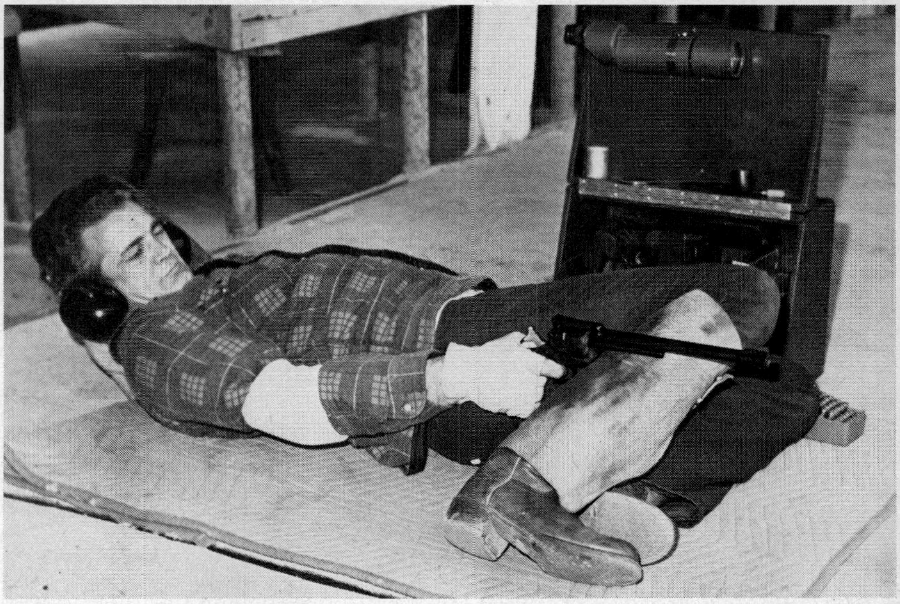

Shooting in a supine position, with the sixgun rested against the side of the leg is a technique used by many of today's metallic silhouette shooters. Elmer had been employing it long before the silhouette game.

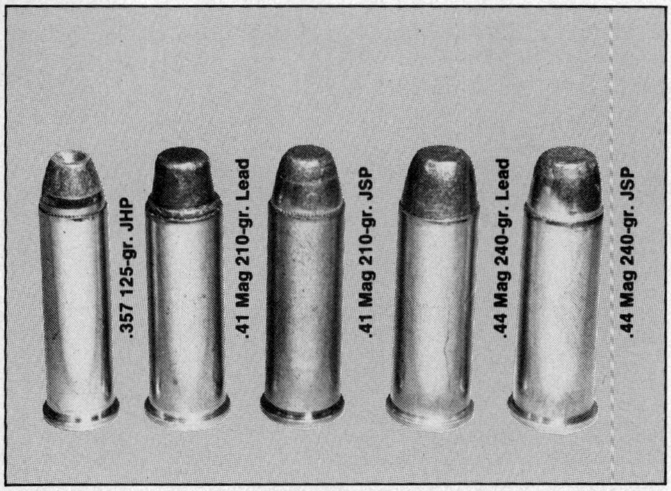

Magnum revolver cartridges suitable for long-range shooting have different bullet weights and styles from 110-gr. JHP on up to 240-grain jacketed SP's.

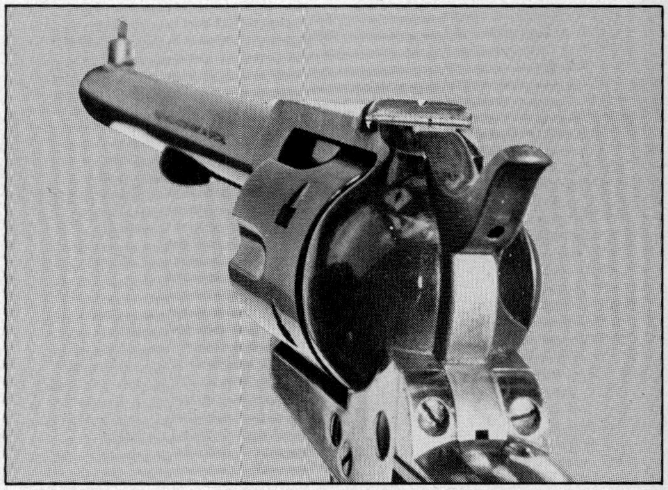

One of Elmer's pet sixguns features an unusual sighting arrangement — a gold bead front sight on a high ramp front post, and fixed, low profile rear with a U-notch.

For longer range, simply hold up more front sight. Then watch for the strike of the slug. If low, hold up more front sight and if high cut down the amount of front sight held over the rear sight bar. High front sights are a distinct advantage and the worst gun of all to do long-range shooting with is the .45 auto or similar low-sighted autopistols.

Just the same, at Ogden Arsenal I once put on a demonstration at 250 yards for a lot of army brass who thought a .45 Colt Auto was accurate only at short range. A snow drift some 18-inches high by 2 feet wide was located on the slope 50 yards beyond the 200-yard target butts. Holding the front sight of my .45 auto up so high I could see a bit of the slide between the front sight and the rear sight notch, I went over the snow drift the first shot, corrected my holdover of front sight and went low on the second shot, and then put the last five of the clip all into the snow drift. I then slipped in another clip and put all seven into what was left of that snow drift out at 250 yards. Then I asked any doubting Thomases to go and lie down near the snow drift and let me empty a clip at them, but got no takers. I used my old reclining position with right knee drawn up, supporting my body weight on my left arm and had my right hand holding the gun in the hollow outside my drawn-up right knee.

Many old cavalry officers of the late 1800s were fine long-range shots and even with the poorly sighted 7½-inch single-action Colts were able to make it a sporting proposition for an opponent.

Those who scoff at long-range shooting are simply ignorant of their subject and cannot be told, but you can take away their money on a bet with a little practice. I consider long-range shooting the finest test of any six-gun and ammunition. I have taken around 40 head of big game with a six-gun, one deer out to what would be a crazy long-range for a rifle, several at 200 to 250 yards, and some coyotes to 400 yards.

Long-barrel guns, up to 7½-inch maximum, are best for game shooting on account of their higher velocity and better downing qualities. The short guns, however, are handy to pack daily and almost as accurate. With the 4-inch barrel guns a very little front sight held up over rear sight bar will lob the slug to considerable ranges. However, with long 7½-inch barreled guns you have to hold up a great deal more front sight because of the longer sight radius.

Some years back, Hugh Duncan asked me to go down to the river with

Custom stocks, like the Herrett walnut troopers on these two N-Frame Smith & Wessons (left) help to distribute the recoil evenly, and can improve the long-range performance of a handgun. A consistent grip (above photo) is vital to good results at long-range.

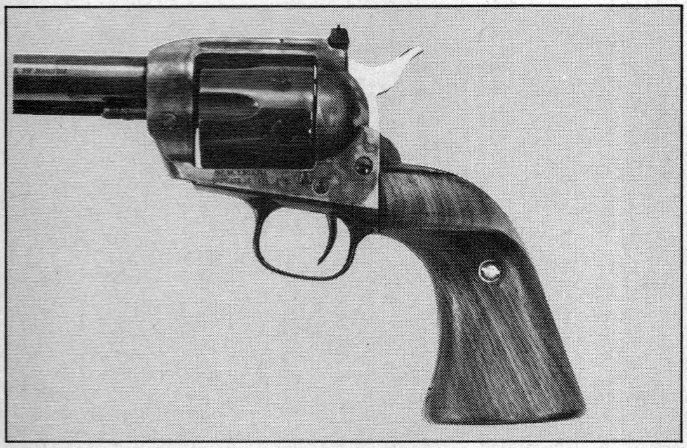

If handgun grips are too small or too large for your particular hand size, your control will be diminished. Thus grip size is extremely important. These fine Magnum Grips fit large hands.

No sixgun, regardless of its potential, will shoot any better than you can hold it. A two-hand grip or natural rest is one of the great secrets of shooting at long-range.

his boy Norman and his boy's wife for steelhead fishing. As we drove along. Hugh was telling Norman what a good pistol shot I was. I told Hugh to shut up or I'd probably have to prove it. Just as we rounded a corner, there was a nice mule buck standing 250 yards away, across a field, on a sand bar at the edge of the river.

Hugh said, "Elmer, there's your buck!" His boy stopped the car. I got out and crossed the road, trying to sneak to a fence post. The buck saw me and turned halfway around. I held the front sight higher than normal, and set the deer on top of it. The shot went just over the top of the animal's shoulders and splashed in the river. This, with a 4-inch .44 Magnum Smith & Wesson. I reduced the amount of front sight I held up for the next shot. The deer humped up and lunged into the brush. I knew I had hit him.

Then I saw his head twice in the jumble of logs and brush. I held on it as carefully as I could and shot twice more. I heard the bullets chunk into the logs, but I didn't hit the deer. Then the deer came out through the heavy brush and logs into the rye grass, I could see all the body except the legs.

My next shot hit squarely in the right shoulder and out behind the left shoulder. I was shooting my 235-grain hollow points, cast by G. E. Murphy, and copper plated. At that range, with a 4-inch gun, they didn't open at all. The deer jumped, kicked at his belly, ran in a circle, and piled up. I marked the deer's location from an old log with the bark peeled off. We went down across the field, and crossed the slough. When I found the log and got on it, I soon found the buck. We field dressed him and washed him out. Then we dragged him to the car, and went fishing in the stream nearby.

I have never been able to do as well with the 10 and 12-inch so-called Buntlines, as with guns of 7½-inch, or less, barrel length. The extra-long guns tend to whip out to the left and shoot to the left for me and I prefer barrels under 7½-inch length. For older eyes the short-barrel guns, with sights closer together, give the clearest sight picture as they are more in the same focal plane.

In all long-range shooting, remember you must hold and maintain a perfect sight picture with the front sight blade exactly in line with the center of the rear sight notch and the top of the front sight held up enough for the range involved. The target should always be perched on the top center of the front sight blade. I prefer square or Patridge-

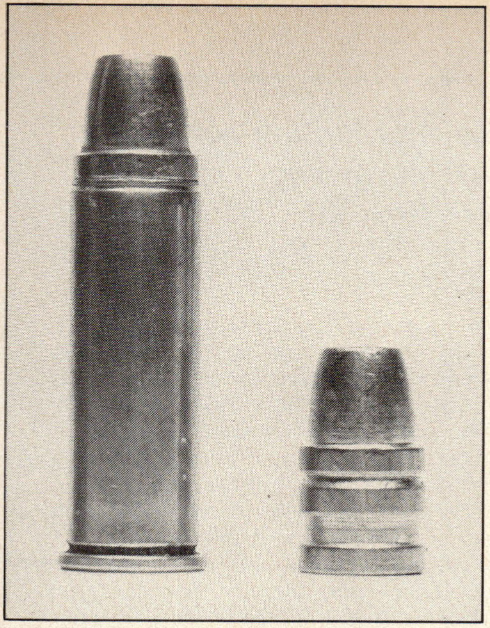

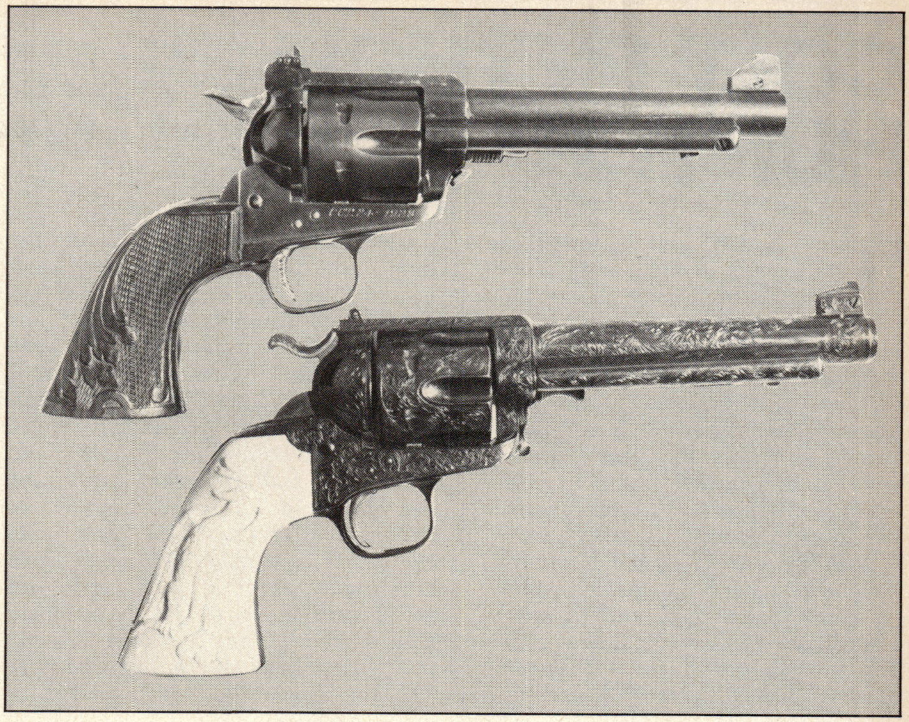

Keith has always preferred hard cast bullets for big game—feeling that they give superior penetration. (above). Two of the author's favorite long-range sixguns (right), are a .44 Special Colt SA Army with custom sights (top), and a highly-engraved .45 Colt Bisley which also features adjustable sights. Both serve as very effective gamegetters.

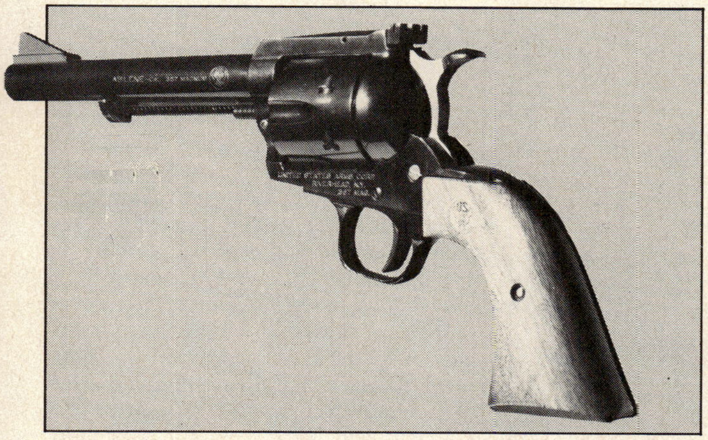

The author has seen many of today's companies improve their sighting equipment over the old SA "hog wallow" sights. The U.S. Arms .357 has adjustable Patridge sights.

One of the author's specially modified sixguns has a unique front sight adjustable for elevation by loosening the base screw. Rear sight must be drifted for windage.

type sights, never round beads and U notches. The rear sight notch must be wide enough to permit a good strip of light on each side of the front sight with the gun fully extended.

Since using the .41 Magnums in Alaska, I have now had a chance to shoot them at long ranges here in Idaho under average conditions of temperature and wind.

My first long-range shooting was at ranges from 400 to 450 yards and I found I was holding too high and shooting over everything. This .41 Magnum load is flatter than my handload for the .44 Magnum of 22 grains of 2400 and my 250-grain cast bullets that I have shot so long and am used to. When I used the same sight picture on 4-inch .41 Magnums I invariably shot high. When shooting Norma, Remington, and Winchester jacketed, soft point .44 Magnums the sight picture seems about the same for both the .44 Magnum and the .41 Magnum with these factory soft point 210-grain loads. My hand load for the .44 Magnum goes only around 1400 feet velocity but is so accurate I prefer it for all my shooting to the factory 1470 to 1500-foot load and especially as my alloy, lead bullets do not wear the barrel a fraction as much as a metal jacket.

At long range I noticed that the 210-grain, .41 bullet strike is harder to locate than that of the heavy 250-grain .44 Magnum handload, or the 240-grain factory load. The larger caliber, heavier bullet certainly throws up more dust, mud, or snow.

The pistol hunter must be a still-hunter of the highest order. He should strive to get as close to his game as possible. In many sections where whitetail and Pacific Coast blacktail as well as the Texas whitetails are hunted, you're hunting thick brush and your shots are usually anywhere from 10 yards to 40 yards. A 100-yard shot is a long one. This applies equally well to the little deer on the Alaskan southeastern coast. A man is just as well off with a six-shooter. If he's a good shot, he's better off than packing a rifle. Then when he makes his kill, he can dress his little deer out, put it on a pack board, and have both hands free to help climb hills, packing out his game. Likewise, for hunting wild boars and hogs in the brush, I'd just as soon have a heavy .44 Magnum, 6½-inch barrel as any rifle; it's handier.

Get a good six-gun and ammo and practice what I believe to be the most interesting of all six-gun shooting, long range.

COLT PYTHON HUNTER

Colt's new silhouette/hunting variation of their Python extracts maximum "downrange" accuracy and power from the .357 Magnum!

By Dave Hetzler

The year 1955 saw the introduction of the Colt Python, and since that time its popularity has grown to where now it is one of the most well respected .357 Magnums ever sold. For most of its life the Python has been offered with three different barrel lengths: 2½ inches, four inches and six inches. However, this year something new has been added—two more inches, making this new version a full eight-inch barrel handgun.

With the advent of the metallic silhouette game, and the growing popularity of the handgun as a hunting firearm, Colt found themselves a little short (no pun intended) with their longest barreled double-action revolver measuring six inches. The new version of the Python is the cure for that, and it is offered in two somewhat different forms. First is the bare bones model, if any Python could be called bare bones, and that one is nothing more than the basic arm. The second, and the most interesting, is the "Colt Python Hunter" model, wherein the eight-incher is decked out with rubber Colt/Pachmayr grips and a Colt/Leupold 2X scope mounted in custom Redfield rings. This whole package comes in a Zero Halliburton fitted aluminum case that also carries an MTM cartridge box with the rampant Colt on the lid, and a special cleaning rod with a beautiful wooden handle. This combination isn't cheap, listing for a suggested retail price of $995, but what the shooter receives is a .357 Magnum that will keep six shots in a three-inch circle at 100 yards with factory ammunition, from a sandbagged rest of course. That kind of accuracy from a .357 Magnum is outstanding, and we felt that it could be improved on with carefully worked up handloads.

Introduced in 1955, the Colt Python has become an American classic. The new 8-inch barrelled Python Hunter with Colt-Leupold scope should prove a boon to handgun hunters.

Upon receiving the revolver from Colt we spent the first few days just looking at it. Finished in the now traditional (for the Python) Royal Colt blue, the gun looked almost too pretty to shoot. But when the chance came for this writer to test the Python in the varmint fields of Wyoming with Bob Milek, our resident handgun expert, it was quickly decided that no matter how pretty it was, a few powder smudges weren't any big thing.

Upon arriving in Wyoming we took the Python to Milek's range and sighted it in. Once on the paper the fine scope adjustments were made, and it was then that we found the gun capable of firing those three-inch 100-yard groups, at least with Milek pulling the trigger. My own groups opened everything up about another half-inch, and I was more than pleased with that because I've never thought of myself as a long-range handgunner. Our particular Python was happiest firing 125-grain Remington JHP ammo, however it also shot most 158 grainers well also. The gun definitely didn't like the lighter 110-grain bullets, and consequently shot none of them well; five and six-inch groups were the norm with the lighter weight bullets.

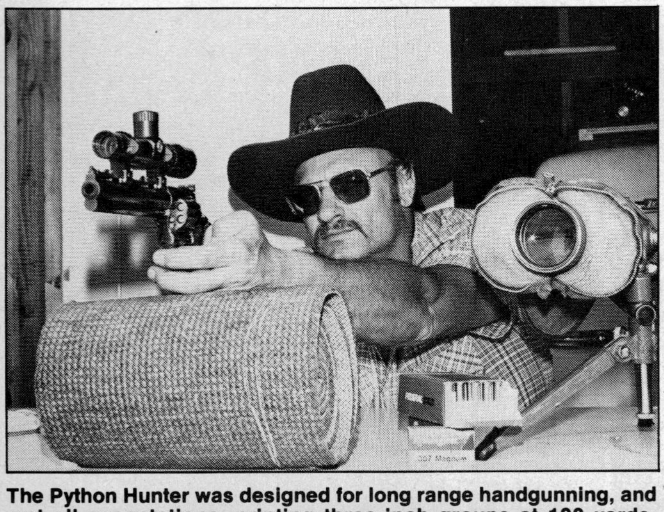

The Python Hunter was designed for long range handgunning, and met all expectations, printing three-inch groups at 100 yards.

The Python Hunter's action is super-smooth, with a crisp 3¼-pound trigger pull. Sideplate is easily removed for cleaning.

With the new system, author reckoned any prairie dog within 100 yards was in trouble!

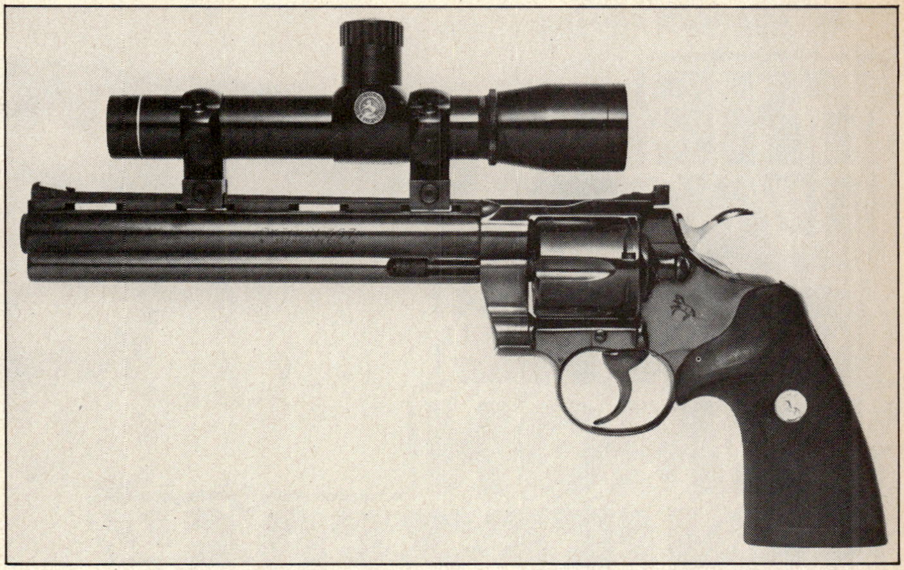

Aside from the new eight-inch barrel length, unique features include 2X Colt-Leupold scope on custom Redfield rings and neoprene Pachmayr/Colt grips with Colt logo.

Performance of the scoped Python, utilizing the 125-grain Remington ammo, was spectacular to say the least. Any prairie dog sighted at less than 75 yards was in deep trouble so long as the shooter did his part, and the ratio of prairie dogs hit at about 100 yards was much greater than one would expect. The longest shot scored was at a generously paced 110 yards, and honesty forces me to say that Bob was the man behind the gun. Needless to say, all shots at those kinds of distances were from a very solid, resting position.

Aiding the shooter is the kind of trigger pull that has become the norm with the Python. Single-action, the trigger breaks at 3¼ pounds, and double-action, the hammer drops when the scales reach 9½ pounds. Obviously, all hunting and target shooting was done in the single-action mode.

Although the 2X Colt/Leupold scope is standard equipment, the old standby iron sights are still in place, with the rear sight being fully adjustable for both windage and elevation. It's our opinion that a hunting handgun should have iron sights already fitted, just like any hunting rifle should have. If the scope should break, or be broken, there's no reason for the hunt to end if the gun is equipped with iron sights all ready to go.

The rubber Pachmayr/Colt grips, resplendent with the Colt logo, aids the shooter in putting his shots where he wants them. About the only complaint most shooters have with any Python is the grip shape—it's uncomfortable all the time, and almost unusable in rapid fire, admittedly something that most probably won't come up in a hunting situation. The rubber grips are shaped in such a manner that the average hand should have no trouble getting a good grip on a not-so-hard kicking .357 Magnum. Because of the overall weight of the gun/scope combo the felt recoil is tamed a great deal, and no one should have any trouble firing this revolver, especially in the hunting fields where the recoil is not normally noticed.

For Colt lovers who are interested in the iron chicken game, or are avid hunters, this new version of the Python should go a long way to guarantee that the rampant horse stays in their hands. With handgun hunting becoming more and more popular Colt was hurting for a double-action revolver to get the job done. The Colt Python Hunter package was meant to fill the bill, and for the proper sized game we think it will do just that. The .357 Magnum cartridge isn't a long-range, big game killer, but within the capabilities of the caliber itself, the new Python can guarantee accurate delivery.

SPECIFICATIONS
COLT PYTHON HUNTER

Manufacturer:	Colt Firearms 150 Huyshope Ave. Hartford, CT 06102
Action:	DA revolver
Caliber:	.357 Magnum
Capacity:	6
Barrel length:	8 inches
Weight:	Approx. 48 oz. (without scope)
Sights:	Colt/Leupold 2X scope with Redfield rings
Finish:	Blue
Grips:	Pachmayr rubber signature
Price:	$995 (with Zero Haliburton aluminum case)

Python Hunter comes in Zero Halliburton aluminum case with matched accessories.

The outfit isn't cheap, but it offers a complete package for hunters and silhouettes.

In production for 87 years, the unsung, oft-maligned Smith & Wesson Safety Hammerless is an American classic!

THE LAST OF THE "LEMON SQUEEZERS"

By Patrick F. Rogers

There is a handful of handguns we all regard as classics; the highwater mark of American handgun design. The Colt Single-Action, the Smith & Wesson Military & Police, and the .45 Colt Automatic would make most lists. They are classics because they are outstanding designs which solved the tactical problems of their times far better than their competitors. I would nominate one gun for the American Classics list that is seldom mentioned. Daniel B. Wesson's last design, the S&W Safety Hammerless.

This unusual and radical design was an outstanding success in its day. Its production life ranged from 1887 to 1974. In its final version as the Smith & Wesson Centennial model, it was probably the most practical, effective, and efficient .38 Special pocket revolver ever made.

Its effectiveness is such that if you need a .38 Special snubnose *for self defense,* and no other purpose, I would advise you to pass up all current models and buy any good Centennial you can find. Why? Because of the special characteristics of the original design, and the problems it was designed to solve—and did! When D. B. Wesson designed the .38 Safety Hammerless, he had a clear objective in mind. He intended to develop a revolver effective for close-range self defense and as near to being absolutely safe as a handgun can be.

Handgun safety was a matter of real concern in the 1880's. Accidental discharge was possible with many of the guns on the market. Most designs lacked the positive systems incorporated into today's designs. They would frequently fire if dropped when loaded. Wesson was particularly concerned about accidental discharges and resolved to prevent them.

The basic characteristics of the new gun were easy to determine. The most popular police and defensive cartridge of the day was the .38 Smith & Wesson. Wesson had designed the cartridge himself in 1875, and Smith & Wesson had produced a long and successful series of single-action and double-action revolvers in this caliber. All the previous designs had five-shot cylinders and incorporated the Smith & Wesson top-break, simultaneous ejection system. Public and police acceptance of these guns and the .38 S&W cartridge was widespread. It was logical to continue their basic features in the Safety model.

The safety features presented a more serious design problem. An ideal pocket revolver must have a number of special characteristics, some of which conflict with each other. It must be capable of being kept fully loaded, ready for instant use, for long periods of time. It should be safe from accidental discharge if dropped. It should instantly return to a safe condition after firing one or more shots, preventing additional, unintended shots from being fired in the panic mode. There should be no feature requiring conscious action by the user. All safety features should be working as the shooter grasps and points the pistol and go off automatically as the trigger is pulled.

These features are easy to list but very difficult to incorporate in one design. Nevertheless, Wesson did it. He completely enclosed the hammer inside the upper framer where it could not be touched by the shooter. Then he introduced the grip safety, a spring loaded lever automatically pushed in by the shooter's hand as he gripped the pistol. Attached to the grip safety was a hammer block acting against the rear of the hammer, locking it into place. Pushing on the grip safety raised the hammer block upward out of the hammer's path. This allowed the pistol to be fired with a normal double-action trigger pull. If the pistol should be dropped, the grip safety spring instantly pushed the grip safety back out, locking the hammer at the instant the pistol left the shooter's hand. It was the then-novel feature of the grip safety that gave the Safety Hammerless its characteristic nickname of the "Lemon Squeezer."

Wesson had achieved his objective. It is impossible to fire a Safety Hammerless, as they christened the new model, without simultaneously pushing in on the grip safety and pulling the trigger through the long double-action arc. Smith & Wesson were not exaggerating when they said, "even fully loaded it is as inert as a block of wood . . . while ready for instant action."

Smith & Wesson introduced their new revolver in their 1887 catalog as the "Hammerless Safety Revolver—Absolute Protection Against Accidents" and went on to say: "The great number of accidents recorded during past years, in the use and handling of the revolver, has made it evident that some decided change in this arm, by which these accidents shall be avoided is not only desirable but necessary. To this end much time and thought has been expended but not until quite recently has the objective sought after been fully obtained in the perfection of our HAMMERLESS SAFETY REVOLVER which we now offer to the public.

"It is well known to all familiar with the subject that a very large proportion of the accidents with revolvers arise from some unintentional manipulation of the hammer. Either it receives a blow, is allowed to slip off the thumb in cocking, is accidently caught on some foreign object and raised or is unintentionally left at full cock. The only other source of accident is the unintentional manipulation of the trigger.

"In the new HAMMERLESS SAFETY REVOLVER all these liabilities to accident are made impossible. First by placing the hammer of the arm entirely within the back frame so that no external force whatever can be applied to it, and second, by so arranging the trigger that it cannot be pulled except at the instant of deliberate firing, and only by this means. One very important feature of this arrangement is the safety of the arm in the hands of children, as no ordinary child under eight years of age can possibly discharge it. The new arm is also provided with a rebounding lock as an additional source of safety and protection.

Smith & Wesson introduced the scaled-down .32 S&W version of the Safety Hammerless in 1888. The new .32's were beautifully made, with the outstanding workmanship and finish that characterized S&W in the last century. They incorporated all the excellent design features of the .38's.

Smith & Wesson took an optimistic view of the .32. A pre-W.W.I catalog describes it as follows: "Although small in size the .32 caliber 'Safety Hammerless' is in no sense a toy. It shoots a cartridge which develops 30% more energy than the .25 caliber Automatic cartridge, and because of its larger diameter is very much more effective."

Still, there have always been people who like small calibers in small guns, and the 1880's were no exception. The .32 Safety Hammerless was offered in a wide variety of barrel lengths, ranging from 2 to 6 inches. A total of 242,981 had been produced between 1888 and 1937, when the .32 was discontinued.

Remarkably, the .38 S&W Safety Hammerless was seriously considered as a military service pistol. In 1891, the U.S. Army decided to replace the .45 Colt single-action with a double-action revolver of reduced caliber. Rapid fire and rapid ejection were considered

The double action-only firing mode of the Centennial, plus its grip safety and 19-ounce weight, takes a bit of getting used to. After a few practice rounds however, some excellent results were obtained (right hand photo). Using Federal 148-grain wadcutter ammunition, a five shot group at 25 yards came in at well under two inches—if you discount the single flyer! This type of performance is truly "above and beyond" the call of duty for a 2-inch belly gun. For defensive purposes, however, something hotter —and admittedly less controllable—than wadcutters would probably be used.

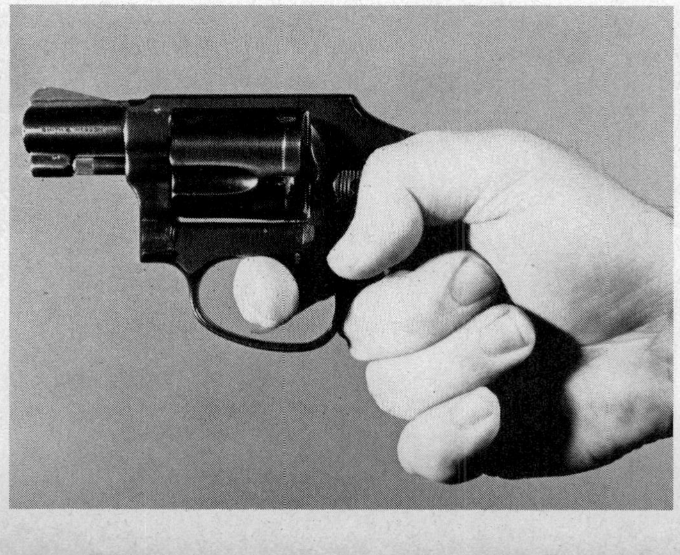

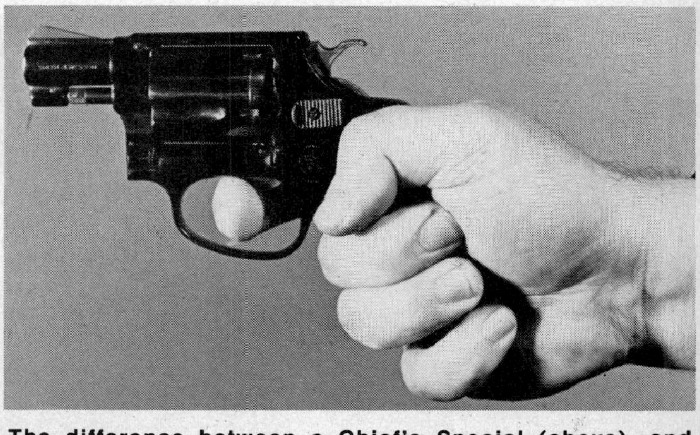

The difference between a Chief's Special (above), and the hammerless Centennial (left) in terms of control, is that the Chief's Special sits higher in the hand, with the line of recoil above the support point. The Centennial sits lower, and is easier to control in rapid fire.

THE LAST OF THE "LEMON SQUEEZERS"

essential. Two revolvers were selected for testing, a .38-caliber, solid-frame, swing-out cylinder, double-action Colt and the .38 S&W Safety Hammerless.

The decision to test the Safety Hammerless was due to the strong recommendations of two experienced Cavalry officers, Colonel Elmer Otis of the Eighth Cavalry regiment and Captain William P. Hall of the Fifth. Captain Hall was particularly enthusiastic. He regarded the Safety Hammerless as the best possible pistol for military service! In a letter to the test board he stated, "A revolver cocked with the forefinger has a great advantage over one cocked with the thumb. In cocking a revolver with the thumb, its position in the hand is rendered very insecure at the instant the thumb is changed from the hammer to the stock, and this insecurity is so increased in cold weather as to render a premature discharge probable, and greatly diminish the rapidity and accuracy of fire." Captain Hall went on to say that he had carried two .38 Safety Hammerless revolvers for more than a year, had fired 3000 rounds, and dry fired them over 5000 times without a single problem.

The test board was impressed and stated, "These points seem to be well taken, and certainly Captain Hall's reputation as an expert pistol shot should give his opinion great weight."

And so the Safety Hammerless was tested. It did well in the accuracy, penetration, rapid fire, and rapid reloading tests but came to grief in the rust test.

In this test, the revolvers were cleaned, all oil was removed, and the guns were severely rusted by being soaked in ammonium chloride and left to rust for 48 hours. After being subject to this test twice, the S&W suffered a broken main spring and a frozen barrel catch and was judged a failure. It was probably just as well. As a pocket revolver, the .38 Safety Hammerless was outstanding. As a cavalry weapon it left something to be desired.

Smith & Wesson remained enthusiastic about the Safety Hammerless. Their 1930's catalogs proclaimed it as, "THE QUICKEST, SAFEST, MOST DEPENDABLE, POCKET ARM EVER MADE," and went on to say, "The S&W Safety Hammerless has for nearly fifty years given the service for which it was designed and has earned the reputation of being the only perfect pocket firearm.

"The 'Safety' is not a target weapon although in the most practical kind of target shooting—rapid fire—it is hard to beat....

"The 'Safety' was designed to be carried in the pocket, from which it can be drawn without the slightest fear of catching; it was designed for quick work at short range, and it was made so that accidental discharge could not possibly occur, and so that it could not be fired by children.

"It can only be fired double-action; there is no exposed hammer and the safety lever on the back of the stock locks the hammer in the uncocked position until the arm is grasped correctly for shooting. No blow or fall can cause accidental discharge as the hammer is locked in position and cannot move."

The .38 Safety Hammerless was a popular gun in the Smith & Wesson line for over 50 years. It passed through five models. Each successive design variation incorporated minor improvements. The .38 Safety Hammerless was discontinued in March 1940 to allow Smith & Wesson to concentrate production on large quantities of .38 Military & Police revolvers for the British Government. Total production of the .38 Safety Hammerless had reached 261,493 pistols at this time.

That should have been the end of the Safety Hammerless design, but it was not. Knowledgeable revolver shooters appreciated the special features that only the .38 S&W Safety Hammerless could provide. There was a steady demand for the design in the early post-war years.

Noticing this, Smith & Wesson decided to introduce a new and improved version of the .38 Safety Hammerless. In the fall of 1950, they had introduced the .38 Special 'Chief's Special', a small, five-shot, .38 Special revolver built on an improved .32 frame. The Chief's Special was an instant success. A safety hammerless version of the new gun with solid frame and swing-out cylinder was designed and introduced in late 1952. Smith & Wesson had been founded in 1852. Although the new pistol was not a commemorative model in the current sense, the factory decided to name it the "Centennial" to honor the company's 100th year in business.

The new Centennial was produced in an all-steel version and an "airweight" model with an aluminum frame. Both versions were made with either blue or nickel finish with standard 2-inch barrels. The all-steel version weighed 19 ounces, the "airweight" model 13 ounces. When Smith & Wesson intro-

With a steady, two-handed shooting position, the Centennial is capable of good offhand field accuracy.

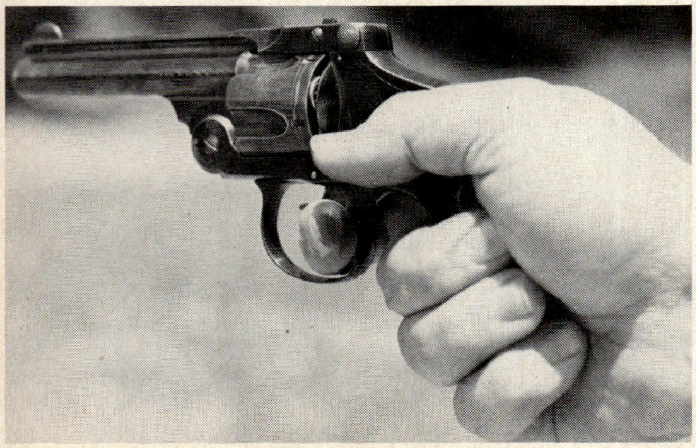

This is the hesitation trigger pull in action. The cylinder of this old Safety Hammerless has revolved and locked. A bit more pressure is all that's needed to fire.

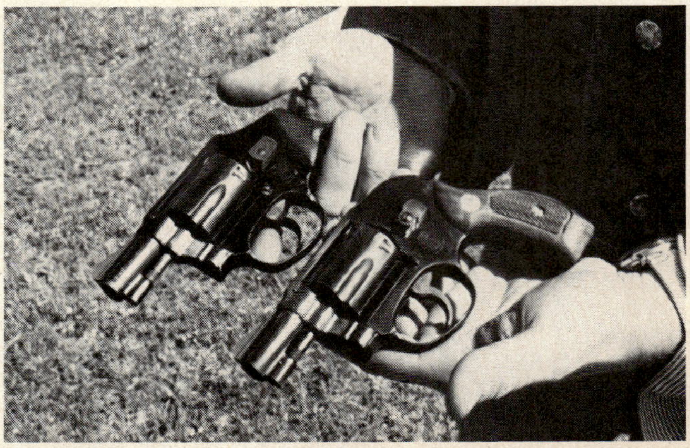

Smith & Wesson's current Model 49, right, slightly resembles the defunct Centennial, left. The M49 is not a true hammerless design though, having a small cocking piece.

duced their model number system, the all-steel Centennial became the Model 40 and the "airweight," the Model 42.

Pocket revolvers have to be small. Long barrels, heavy frames, and large, hand-filling grips defeat their basic purpose. A pocket revolver tends to be carried much and shot little. It is often carried as a second gun. The customer tends to demand light weight in addition to compactness. A survey of today's popular, high-quality pocket revolvers confirms this trend. The popular, fast-selling models have 2-inch barrels, small frames and grips, weigh 14 to 21 ounces, and are chambered for the .38 Special cartridge. These characteristics make rapid, accurate, double-action firing difficult.

Modern ammunition makes the problem worse. Shooters today know more about stopping power and bullet performance. When it comes to self protection, they demand high-performance ammunition. Since the 1960's, a wide variety of high-velocity, hollow point loads has been offered in .38 Special. These cartridges reduce bullet weight and load to maximum allowable pressure standards to obtain enough muzzle velocity to make their bullets expand. The modern .38 Special "+P" high-performance cartridges are more effective.

The choice is logical, but the results are unfortunate. When modern .38 Special high-performance ammunition is fired from a pocket revolver, all hell breaks loose!

The Centennial design goes a long way towards solving this problem. The saw handle design of the grips and the extension of the backstrap up into the area of the hammer arc on a standard revolver allow for an entirely different grip. Your hand is positioned behind rather than below the pistol. As a result, the Centennial kicks straight back into your hand, and it remains controllable even with the hottest .38 Special +P loads.

Notice that I said "controllable," not pleasant. The recoil is not reduced; it is redirected. After firing ten or fifteen fast shots double-action, you will feel it. But the Centennial is controllable, far more so than any other small .38 Special revolver. And as Elmer Keith once remarked on the same subject, "Recoil, of course, is very severe from such heavy loads in such light weight guns, but the guns will take it and any man would rather have a sore hand than a bullet through his middle." I would define acceptable accuracy for a pocket pistol as staying on a man-size target. Shooting carefully with a two-handed hold, a practiced shot can usually do this. In fact, when using +P high-performance ammunition, the Centennial may be superior. Its excellent recoil control and longer trigger pull help prevent flinching.

This leaves one final question. If the Centennial is really this good, why did Smith & Wesson discontinue it? The answer is simple—it did not sell well. Its best features cannot be appreciated by brief handling in a gun store or a few trial shots on a range. You must get one and practice steadily for a while to really see what it can do. Still, if you want a .38 Special snubnose for self defense, you can't find a better gun.

The Centennial is slowly becoming a collector's item. Mint, unfired specimens are already commanding premium prices. Most people who own Centennials intend to keep them. Still, every now and then a good, used one will turn up at a gun store or show. If you see one, grab it! But watch out! That large, greedy fellow trying to beat you to it is probably me.

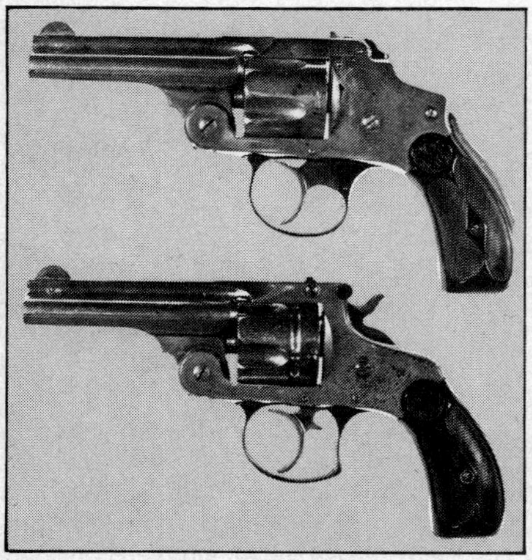

The early Safety Hammerless (top) and a standard .38 Double Action (below) were made at around the same time.

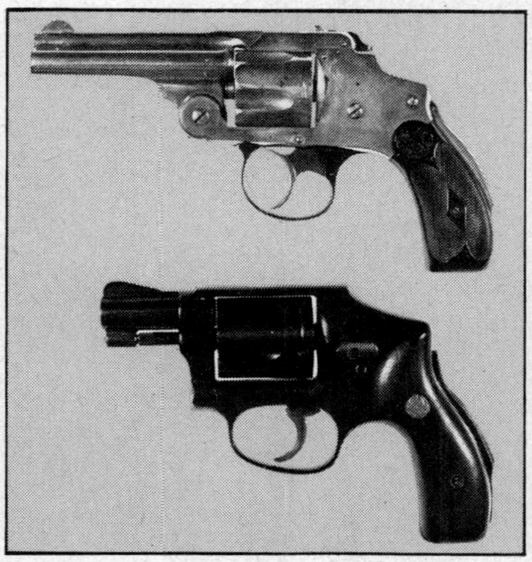

Evolution of the Safety Hammerless design is shown in these models. The Centennial was the final model.

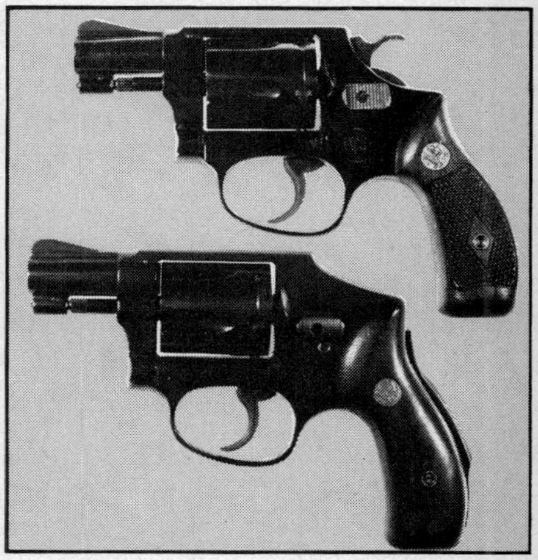

The totally different configuration of the Centennial (below) compared to the Chief's Special, is obvious.

The straight-line recoil of the Centennial is nowhere more obvious than in rapid fire. Five rounds of rapid fire full-power loads stayed well within the "K-Zone."

The top-break, automatic cartridge ejection of the original Smith & Wesson Safety Hammerless was a great tactical advantage, permitting exceptionally fast reloading.

SMITH & WESSON MODEL 29

In the category of magnum handguns, the undisputed "king of the hill" classic is still Smith & Wesson's original .44 Magnum.

By Craig Boddington

The first Smith & Wesson Model 29 .44 Magnum revolver was produced in early 1954. At the time it was the most powerful production handgun in the world, producing velocities around 1400 feet per second with a 240-grain bullet. The new gun and cartridge literally took the handgun world by storm, and opened up whole new vistas for the handgun hunter.

For many years—perhaps too many years—Smith & Wesson had the big bore hunting revolver market to themselves, unless the buyer preferred a single-action such as the fine Ruger Super Blackhawk. That has changed, and will change even more as the competition hits the market in numbers. Even so, a Model 29 in any barrel length or configuration rarely stays very long in a dealer's display, and at the guns shows it's not unusual to see a Model 29 sell for several dollars *above* retail, whether new or used.

It seems that most folks in the gun writing trade who have Model 29s have had them for some time, and tend to hang onto them. That plus their scarcity on the dealers' shelves added up to the fact that it had been a long time since we at *Guns & Ammo* had examined a *new* Model 29. With the current upsurge in handgun hunting, handgun silhouette shooting, and the large number of new extra-powerful handguns, we thought it might be interesting to take a new look at the current version of the gun that started it all—the Smith & Wesson Model 29.

The long-barrelled Model 29 lived up to its reputation as a premier hunting handgun. With the long sight radius and hand-filling grips, it's excellent for long range shooting.

Our test gun arrived in its wood presentation case lined with thick blue velvet, and sported the long 8⅜-inch barrel. Appearance-wise, it was easy to see how the Model 29 got its reputation. The finish was an excellent deep blue with the exception of the color case-hardened hammer and trigger and the serrated and sandblasted top strap and rib. The grips were nicely figured, deeply checkered.

The Model 29 is an imposing handgun, weighing with the long barrel some four pounds 3½ ounces! Overall length is 13¾ inches—not exactly a hideout gun! Finishing was excellent overall, with no visible blemishes or tool marks. The sights were, as expected, excellent. The rear sight is fully adjustable with a white-outline square notch. The front sight is a raised ramp with red insert, and on initial handling it was very easy to align the front sight into the notch.

The action is very smooth, and the single-action trigger pull is crisp and clean, with a 3¾-pound let-off. The double-action trigger pull was a heavy but very smooth 11½ pounds.

On the surface, the gun was every-

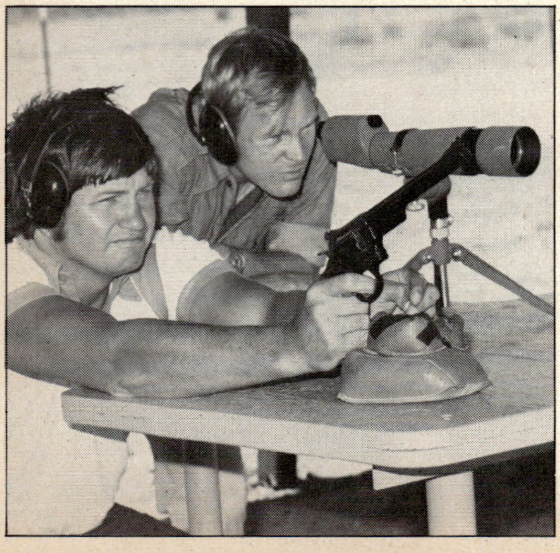

Rhodesian professional hunter Dave Joubert, left, wrings out the Model 29 from a bench rest. Recoil is definitely noticeable, but gun weight and good grip design make it manageable. Above left, the front sight is a raised ramp with red insert. Rear sight, above right, is S&W's adjustable sight with white-outline rear notch. Sights were highly visible and adjustments precise.

Using Federal 240-gr. factory loads, 25-yd. groups were in the one-inch class. 50-yard groups opened to three inches.

The "works" of the M29 were well-machined with a noticeable lack of toolmarks. Single action trigger pull was a crisp 3¾ lbs.

thing we expected and more, but to my mind it's a hunting handgun rather than a wall-hanger. I was fortunate to have the gun along on several shooting sessions, and to date have put nearly 500 rounds of assorted factory and full-power handloads through it. Good though my first impression was, this is a gun that grows on you! The timing has remained in tune, and we have had no problems. After initial sighting in, our test gun produced 25-yard groups in the one-inch class with Remington, Winchester, and Federal 240-grain ammo. Federal's 180-grain jacketed hollow point ammo fell a bit behind this in our particular pistol, but accuracy was still exceptional. I don't admit to any talent with a handgun, so I was pleased to keep 50-yard groups in the two to three-inch class.

Elmer Keith has long touted the .44 Magnum as an ideal long-range sixgun, and in his hands it will certainly do things that I can't do. However, at our range we have several 18-inch steel gongs at various ranges. With the Model 29's long sighting radius and excellent trigger pull, we found it easy to score hit after hit on the 125-yard gong, using a two-hand hold either offhand or from a rest. For me, that is long-range sixgunning! While we found it quite possible to hit the 250-yard gong, that was beyond my own capabilities for a sure one-shot hit.

The long-barreled Model 29—or any .44 Magnum,—for that matter—is not designed for target shooting or plinking. A pistol like this comes into its own for hunting or metallic silhouette shooting, and the Model 29 has an enviable reputation in both areas. We were not able to take the pistol big game hunting, but I did manage several ground squirrel shooting sessions with it. Now, nobody needs a .44 Magnum to dispatch ground squirrels, but it's my opinion that any handgun accurate enough and that handles well enough to make hits possible on a two-inch by six-inch ground squirrel ought to be a good deer gun if the cartridge is powerful enough! The Model 29 performed well enough that we reckoned any squirrel within 75 yards was in serious trouble!

The Model 29 currently comes in three barrel lengths—four inches, 6½ inches, and the long 8⅜-inch tube. In my experience the short barrel makes too much noise, although it is handier to carry. The 6½ incher is easier to carry than the 8⅜-inch model, and factory ballisitcs of 1470 feet per second with the 240-grain bullet and around 1700 feet per second with Federal's 180-grain load are quoted from the 6½-inch barrel. The extra long barrel on our test gun does have the disadvantage of making the pistol awkward to carry, at least for normal-sized folks like this writer. However, the long barrel offers a couple of very real advantages. Obviously, the extra 1⅞ inches of barrel milks a little extra velocity and resultant muzzle energy from the mighty .44. More importantly, from a practical standpoint, it adds recoil-dampening gun weight. That extra weight is added at the muzzle, which increases the shooter's ability to hold the pistol steady, and the sighting radius is also lengthened significantly. This last is very important for metallic silhouette shooters and handgun hunters, especially at long-range.

All in all, the Model 29 is today everything it ever was, and that adds up to a first class, fine looking handgun that shoots as good as it looks. Smith & Wesson may not have the double-action .44 Magnum market to themselves anymore, but its a sure thing that the Model 29 cannot and should not be overlooked by the prospective .44 Magnum buyer. Externally it's the same gun that Elmer Keith tested in 1954, but internally the action has been beefed up and improved over the years, and the current Model 29 lives up to its reputation as a latter day American classic.

SPECIFICATIONS

SMITH & WESSON MODEL 29

Manufacturer:	Smith & Wesson Springfield, MA 01101
Caliber:	.44 Remington Magnum
Action:	Double-action revolver
Capacity:	6
Weight:	51½ ounces
Barrel length:	8⅜ inches
Overall length:	13¾ inches
Finish:	Bright blue
Grips:	Concalo Alves checkered target style
Sights:	White outline adjustable rear, red ramp front
Price:	$388.00

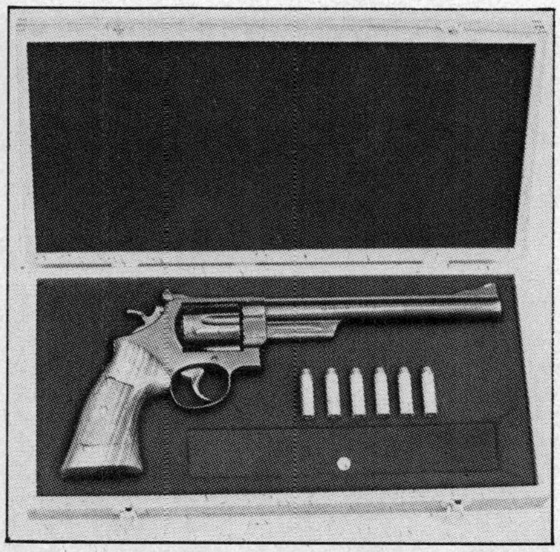
The S&W M29 comes with a wood case. Introduced in 1954, it was the first .44 Magnum and is still a first rate gun.

Fewer and fewer shooting sports are for men only. What do women think about breaking into what has always been considered a male pastime? Here are a lady's views on. . .

By Pamela Parriott

WOMEN AND COMBAT SHOOTING

Combat shooting! The very words strike horror and revulsion (or at least scornful disdain) into the hearts of most women. So what kind of woman is it who gets herself involved in such a thing?

Certainly, in this predominantly male sport there are very few women. Shooting in general is shunned by the majority of women for various reasons, perhaps the number one reason being that guns are viewed as tools of aggression, whether as a constructive outlet or a destructive one. And contrary to what the women's libbers may have you believe, little boys really are by nature more aggressive than little girls, and thus begins their childhood fascination with the sport. This reason, along with some other cultural ingraining, has led many women to think that there must be something unfeminine about it.

Another common occurrence which has given shooting a bad name in the minds of women are the many gun nuts who spend more time and money on their hobby than they do with their family, and thus through their irresponsibility create a situation where the wife (at least understandably, if not rightly) blames the guns for the problem, much as the anti-handgunners blame guns for crime instead of the criminals who misuse them.

Looking at the ladies who do get into shooting, most get started with it much later in life than the men do, and most are drawn into it by their husbands or boyfriends who are crazy about it. Funny though how once they try it, this shooting game gets into their blood and they become just as enthusiastic about it as their mates—or almost, anyway.

As for myself, I got started a bit differently than most women. Until about two years ago, when my husband and I were separated, I had no interest in guns whatsoever. While I was married, I had depended on my husband to take care of most facets of my life, but once I was on my own, I realized I had to start taking care of myself in a lot of different ways that I never really had given much thought to before. Looking around in the world, or even on the front page of the newspaper, and becoming aware of what was really going on, I came to the conclusion that the only sane thing for a woman to do to protect herself is to have a gun, and to know how and be willing to use it.

One day I happened to hear Mel Tappan on the radio discussing his book, *Survival Guns.* I bought his book and then went to my local small town

The four-inch .357 and snubby .38 are traditional male choices for combat. The author says women, though, may find the .357 revolver unwieldy and the recoil of the snub nose, excessive.

The author's choice for a combat arm is the .45 automatic. Its weight is rearward, aiding control, and the size of the firearm reduces recoil. Grip size on the auto fits most women.

library to see what they had on the subject. There was only one book on the shelf about handguns and, fortunately, it was by Jeff Cooper.

Reading through the two books was a real effort for someone who knew absolutely nothing about guns except that they went bang when you pulled the trigger. You gentlemen out there may not have noticed, but practically all of your books and magazines about guns take the approach that the reader grew up with guns and has been shooting since he was 12 years old. Of course, he already knows what it means when the author says to "put down the hammer" (are we nailing something?), or "drop the magazine" (but I haven't finished reading it yet), or "work the slide" (I did that at the playground when I was four years old).

Finally, after much head-scratching and perseverance, I managed to glean enough information from the books to make my decision: I wanted a .45 automatic because of its stopping power and ease of rapid reloading. I was a little worried about whether my hands would be large enough to hold the grips well, and if my nerves were good enough to endure the recoil, but I was determined.

I still remember the incredulous look on the sales clerk's face at the sporting goods store when I walked in, holding my little two-year-old girl in my arms, and told him that I wanted to rent a .45. (After all, their ad in the yellow pages said they rented guns, and I thought that would be a perfect way to try the gun out and see if I really could handle it before actually buying one.) After he informed me that they only rented rifles for their indoor range, I replied, "Well, in that case I guess I'll buy one." I picked out the Colt Combat Commander that I wanted, and while the clerk was filling out the paperwork he looked up at me dubiously and asked, "You do know there's a two-week waiting period on this don't you?" "Oh," was my classic reply.

I had my first shooting lesson a few weeks later from a rangemaster at a nearby range, and it was very encouraging. I even hit the target at 25 yards on my first shot offhand, and the recoil wasn't nearly as bad as I thought it would be. I had visions of the gun practically pulling my arm off and having to practice for weeks before I could even come close to hitting the target, but to my surprise, such was not the case. I found the recoil to be quite tolerable, and the gun as a whole to be very manageable. After that lesson I was on my own, shooting every other weekend. At first my interest in shooting was strictly functional–I wanted to practice to become accurate enough to defend myself well on the first shot, but somewhere along the line, I got to thinking that this was fun and good mental discipline as well.

I wanted to get into something more practical than just target shooting though, but I did not know of any place locally where such things were taught. As fate would have it however, one day at the range I happened to meet a person who had a lot of experience in just such a thing. From him I learned the basics of drawing from a holster and picking up the sights quickly, shooting two or three times on multiple silhouette targets, doing a quick magazine change on the run, clearing different kinds of jams quickly, and of course the old favorite, the beer can roll. If I thought target shooting was fun, this kind of shooting was a blast!

A few months later I found out about a "practical pistol" club that was forming in the area at Coto de Caza in Orange County, California and have been shooting at their matches for over a year now. I always enjoy the friendly, low-pressure competition and the challenging and clever set-ups of the courses. Although I find target shooting is still a valuable mode of honing a shooter's accuracy, it can also get to be a bore very quickly. This is where combat shooting can provide another facet to expand the shooter's interest in the sport, not to mention the practical aspects of it which test your judgment and teach you to shoot (or not shoot) under pressure. Not only must you aim the gun at the right spot, but you must also determine quickly which targets to shoot while sparing the "hostage" targets, count your shots so you don't fire the gun dry, recognize the kind of jam when it occurs and clear it quickly without stopping to call "time out," know when to reload and do it without dropping everything. On top of all that, you still have to be careful you don't trip over your own feet, point your muzzle into the crowd of spectators, fire off a round accidentally, or any such other nonsense which would immediately qualify you in the "Klutz-of-the-month" catagory. These types of things you just don't learn with straight target shooting. It's a completely different game and involves more of the total person, physically and emotionally.

Many ladies in this sport, or contemplating trying it, may be worried about doing something wrong and making a fool out of themselves in front of all these experienced guys who have been shooting for years. However, I've talked with many men who felt exactly the same way when they first started. Those guys out there may look cool, but let me assure you ladies, their hands shake just as much as yours do, so you needn't feel so silly about it.

In a sport such as this, where most of the competitors are men, one might think that there would be a certain amount of male chauvinism directed at a woman who gets into it, but I've found this to be quite untrue. Oh sure, there's a lot of good-natured kidding about "What's a little girl like you doing with a big gun like that," and "You'd better behave yourself around her. She packs a .45 under that jacket." However, practically all the men are very enthusiastic about having some women in the competition, and they really admire any woman who tries her hand at it. One of the nice things about shooting "with the boys" is that you will very quickly pick up from them some very clever subterfuges to get around your mistakes. It seems they have a whole list of intelligent sounding excuses as to why their gun or ammunition failed to function properly, when really, the problem could be more aptly attributed to operator error or brain-fade. In addition to these pearls of wisdom, they are also very willing to offer sincere advice or assistance to help out, and it is given in a spirit of mutual respect and equality. Very seldom in this sport have I met one of the "Here-little-girl-let-me-show-you-how-it's-done" types, who are not really as interested in trying to help me as they are in trying to prove how much more they know than I do.

There was one incident, however, that I have to laugh at whenever I think about it. I was at the range teaching a girlfriend to shoot a .45. During the cease fire period two guys came on the range. One of them put his holstered gun down on the bench, then drew it out and started waving it around down-range while everyone else was out there checking their targets. As I started ducking for cover, I heard the rangemaster yell out, "Put that gun down!" A little later, as I was showing my friend some pointers, this same chauvinist charlatan came and parked himself behind us to "check it (and us) out" and give us some over-the-shoulder advice. He tried to get me to shoot, saying, "I want to see how good you are," whereupon I informed him that I

In the correct "ready" position, feet are set comfortably. Pistol is held in two-hand grip. Safety is on, eyes are on target.

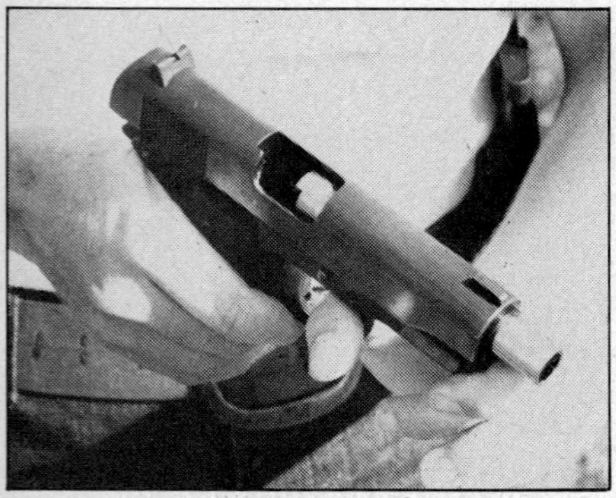

An easy way for small hands to pull back a slide is to hook thumb in triggerguard, index finger under barrel; then squeeze.

With small hands, it may be necessary to move right hand slightly to right for trigger contact. Left hand overlap is firm.

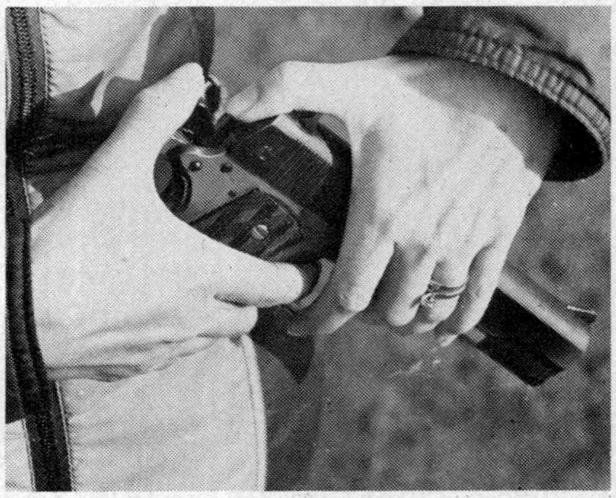

To lower hammer over live round, use left thumb as hammer block while right thumb lowers hammer. Practice with empty gun.

To unload the pistol, keep finger off the trigger and remove the magazine. Tuck magazine under little finger or in belt.

To complete unloading, chambered round must be removed. Push slide smartly to rear. This will eject round into hand.

Arm is "safe" when right thumb engages the slide lock. Locking the slide back allows the chamber to be visually checked.

Practice should be at ranges normally encountered in defense situations, up to 10 yards. Work on accuracy, then speed.

A combination of metal and cardboard silhouette targets are used in the practical handgun course. A cardboard target marked with a cross (at left) is the "hostage" target. The shooter must hit the metal target twice without nicking the hostage.

would shoot when I was ready. He then brought over his .357 Magnum and put it in my hand and said, "Here, try this." So I did, and I put a nice shot right in the ten-ring (thank goodness). Without a word I handed the gun back to him and he muttered, "Hm. That's pretty good," then sauntered off, never to be heard from again.

Contrary to popular opinion (subconsciously, at least), shooting is not a macho activity at which men are naturally gifted. Men are generally more adept than women at running and jumping and other such activities which employ what is known as "large muscle coordination." However, aiming a gun at a target and pulling the trigger at the right time requires a completely different type of "small muscle coordination" and good eye-hand coordination, something at which women tend to be particularly good. This may account for the many stories concerning some gun nut who has gone out every weekend for the last ten years to practice, and suddenly finds himself embarrassed to shoot next to the little lady who just picked up shooting a few months ago and only practices occasionally.

One area in which we women are at a disadvantage is that, due to the innate difference in our frame, we have much weaker shoulders and arms than a man. This can make it rather difficult for some of us to hold a fully-loaded gun out in front of us without it wobbling all over the place. It can also cause a problem in pulling the slide back fully on an automatic pistol. Those of us who have gotten used to carrying around a child all the time have developed some amount of strength in our arms, but are still no match for a man's natural strength in that area. As regards pulling the slide back, the usual method of doing this is to use the thumb and fingers on the grooved area, having the thumb pointing away from you on the right side of the slide. This method is very difficult for some women because you cannot get a good grip to pull on it. The fingers do most of the work this way, and if you don't have large strong fingers and a grip like a vise, this can be frustrating. An easier method would be to reverse the position of the hand on the slide with the thumb pointing towards you on the left side of the slide, the outside edge of the palm pushing in on the same side of the slide and the first joint of the forefinger resting against the rear sight so as to have something to push against and prevent the hand from slipping off when pulling the slide back. The last two fingers are kept up and out of the way in order to see into the chamber and to allow the cartridge to eject.

By the way, resist the tendency to point the gun into your stomach when using this method. While the left hand is pushing back the slide in a fast, hard movement (it's easier that way than trying to do it slowly) you can also use the palm of the right hand pushing against the back of the frame in a contrary motion to the slide. Pulling the hammer down first before you pull the slide back makes the whole operation much easier. A light recoil spring in the gun also helps. Another way to check the chamber is to hook the thumb inside the triggerguard (don't you dare touch

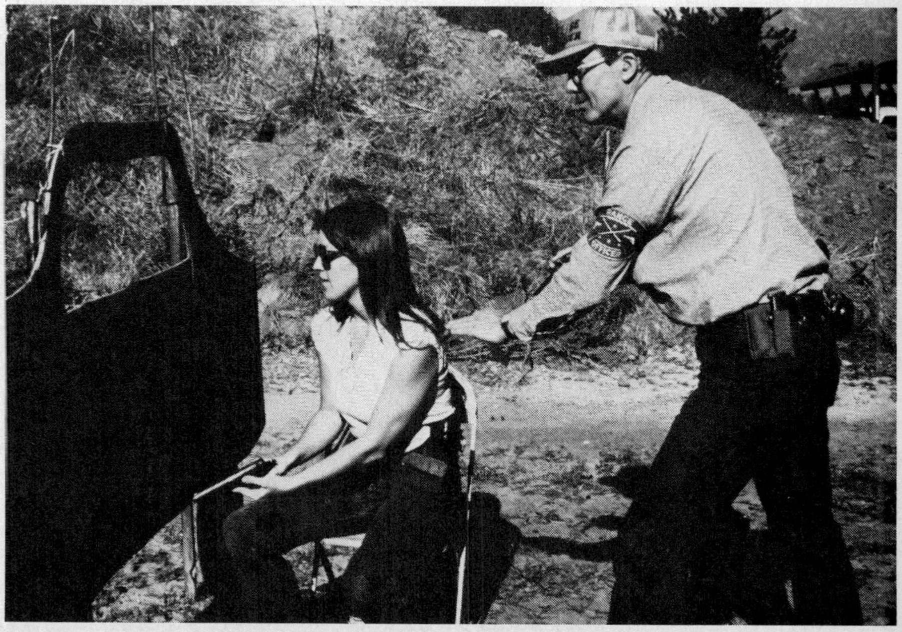

"Shooter ready?" asks the rangemaster, as the author gets set to run through the mock-up car stage of the course. The shooter must fire through the window of the "car" at several targets placed at various distances. First shots are from a seated position.

the trigger though!) and the forefinger around the front of the recoil spring plug, just under the muzzle, then pull the slide back with that same finger. This is a very easy, effortless method but is not to be recommended for beginners because of the safety factor.

Now as to the problem of holding up that heavy, fully loaded gun, there are several points to consider. First of all it may be wise, although not absolutely necessary, to do some arm and shoulder strengthening exercises like push-ups, or hold some three to five-pound weights in each hand out at arm's length until you can't stand it anymore. After that, holding a loaded .45 should be a piece of cake. Just as important as arm strength however is wrist strength, especially when you do any one-hand shooting (that goes double for the cursed "weak-hand only" stages they always throw in at the matches). A good exercise for that is to take a separate sheet of newspaper in each hand and crumple them into a ball as fast as you can. Then uncrumple the sheet and do it over again.

Another point along this line is that some women may find an automatic easier to hold than a revolver because most of the weight of the gun is being held securely inside the palm of the hand, while the bulk of the revolver lies outside and in front of the hand where it may seem a bit unwieldly and contribute to some wobbling. Getting a revolver with a short barrel doesn't help matters any either, because this makes the recoil much more unbearable and the accuracy considerably worse. I personally find a 2-inch .38 Special to have a much sharper recoil and lousy accuracy compared to 4-inch .357 magnum or a .45 auto. So think again, those of you who believe the little woman needs a little gun. In this case, bigger is better.

Another consideration which worries many women shooters concerning automatics, and especially the .45, is the relatively large size of the grips. This worry usually fades into obscurity after she actually tries it. I must say, I have yet to meet a woman who cannot handle the grip size reasonably well. I'm not saying that such women do not exist, but it just isn't as big a problem as one might think. There are a few things that can be done to make the grips more comfortable for small hands, such as replacing the arched mainspring housing

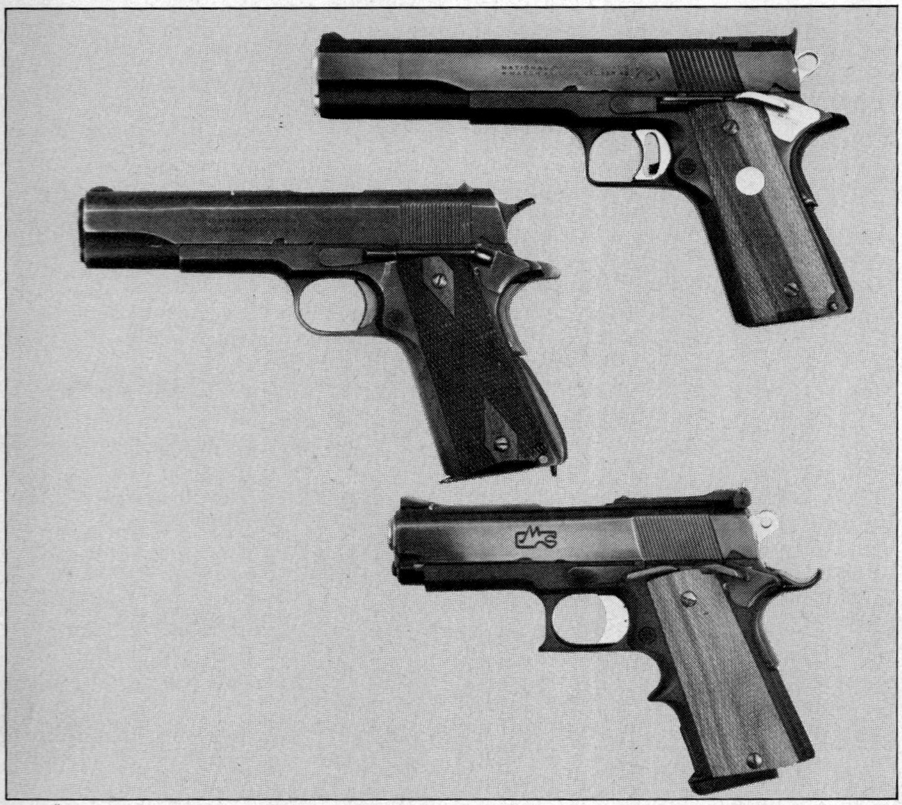

Combat .45s come in great variety, from customized Colt (top) to GI issue (center). Some production guns like Safari Arms Enforcer (bottom) have many custom features.

After firing from inside the mock-up car, shooter drops to a prone position (above) and repeats the firing sequence. Photo at right illustrates that running the practical handgun course actually means *running!* Shooter fires six shots from behind a barricade, then she must run up to close range and fire two more shots, reloading on the way.

on the back of the frame, which usually comes on it from the factory, with a flat housing to make the grip size smaller. The standard checkered wooden grips also tend to be a bit rough on delicate hands and can be replaced with smooth wood or plastic grips. For a more secure feel, the Pachmayr rubber grips are a good choice, but the front strap on these may need to be cut off, otherwise they will make the grips even larger than normal.

In addition to the aforementioned issues, it seems that many women have a nebulous fear of a .45 for some strange reason, no doubt in part because they have heard or been told outright that it's a "man's gun" or that it's hard to handle or that they won't like the recoil because it's so powerful or that automatics are tricky to operate (all of which can be considered pretty much groundless fears). To try and teach someone to shoot a .45 by first telling them all the supposedly negative factors involved is like teaching them to ride a motorcycle by saying, "You'll probably fall off at first." Well, they probably will fall off if someone insists on planting the suggestion in their minds.

Shooter is starting the "Jungle Walk." In this event, shooting targets and hostage targets have been placed on a trail. Most of the competitors shoot from two-hand crouch.

Jungle Walk targets may be on the ground, behind bushes, or in trees. There is no 10-ring; scoring is based on proximity to center and on time taken to complete course.

In this respect my complete ignorance of guns in the beginning was actually to my advantage. I just didn't realize that a woman isn't supposed to be able to shoot a .45 without a lot of preliminary practice using a .22 first, then gradually working up to "the big one." In my naivety I thought that I could just jump right in and do it. In fact I am still convinced that most women who want to can do the same. Starting with a .22 may be a very good idea for some, but in order to start right out with the big bores, a lady doesn't have to be a gutsy type with nerves of steel and the strength of an Amazon. All she really needs is a certain kind of attitude which encompasses a simple desire to protect herself and her family and a quiet determination that she will do it, no matter what it takes. Anyone going into it with a more frivolous attitude (just to please her husband or just to try it out) may not fare too well.

As far as holsters for women, just about anything seems to work and it's largely a matter of just trying out different types and seeing which one works and feels best, although those who are particularly well-endowed may want to stay away from shoulder holsters and crossdraws.

I know there are a lot of men who would like to get their wives or girlfriends to try combat shooting. I also know from talking to some women that there are quite a few of them who would like to try it but are hesitant to begin, one reason being that they don't know how to do it and want to learn privately before they get out in front of a whole crowd of male observers and make a spectacle of themselves. They may have a difficult time learning from their husbands (for reasons which are too involved to explain here). Also some women feel intimidated by "performing" in front of a lot of men even if they are pretty good and would feel more comfortable in an all-women's group. Such a group would not necessarily have to be run exclusively by and for women, however. Certainly any support, advice, and encouragement from the men would be necessary and welcome. This would open up some new possibilities to bring more women into a sport that I'm sure they will find fun, educational, and perhaps some day even crucial for their survival.

Editor's note: Ms. Parriott is starting a newsletter for women interested in practical pistol shooting, with the possibility of forming such clubs for women. Those who are interested may send a self-addressed stamped envelope to "Newsletter" c/o Pam Parriott, P.O. Box 1234, Brea, CA. 92621.

Test Report

RUGER SECURITY-SIX

If you're looking for the best in a combat magnum, take a look at the Ruger Security-Six .357 Magnum. Its rugged construction makes it one of the most dependable of today's revolvers.

By Michael Durr

What's the best combat magnum handgun?

It's an obvious, often asked question. Many people getting a gun for self-defense don't want one that's okay, adequate, good-for-the-money, or whatever. They want the *best!*

The answer to this question is that several guns are best in terms of combat performance. Any gun in this group will do the job and whittling down the choices becomes a matter of personal preference.

Certainly one of the best is the Ruger Security-Six. This double-action .357 Magnum has proven dependability, a design that incorporates many of the most desirable features in modern handgunning, and a relatively modest price tag.

Last year Ruger started putting oversized grips on the Security-Six to improve the controllability of the big magnum. Then a few months ago they changed the trigger surface from grooved to smooth. These alterations plus the general interest in this type of handgun seemed like good enough reasons for getting a test gun and taking a look at it.

The Security-Six was introduced in 1972. Since then it has found approval not only with the public but also with many law enforcement agencies, a market that had been virtually sewn up by "the big two."

Law enforcement has good reason to be cautious in accepting a new gun. Duty sidearms, without a doubt, take the most abuse of any handgun: they're dropped, thrown, filled with dirt in scuffles—you name it. Since they aren't always the property of the people who carry them, duty sidearms may not get much *TLC* other than an occasional barrel-swabbing.

Petersen Ranch foreman, Wayne Prindle, carries a Security-Six and used it to stop this coyote, a running shot at 110 yards!

Obviously, many guns get better treatment than I've described, but the point is that a duty sidearm should be able to take a lickin' and keep tickin'.

Ruger had this in mind when they designed the Security-Six. The frame is solid with no side plate. A top strap only slightly smaller than a railroad tie maintains unit integrity in what is normally the weakest place on a revolver frame. Even if it is dropped, the solid frame combined with the heavy top strap will probably prevent the frame from twisting or torquing.

Coil springs are used throughout the revolver, including the hammer spring. Our test gun had a four-pound trigger pull, single-action. Double-action, the pull was a little over 12 pounds, heavy but not excessive.

Ruger offers 2¾, 4, and 6-inch barrels on the Security-Six. The sample gun had a 4-inch barrel, a good compromise for quick handling, balance, and accuracy. The barrel, grooved sighting rib, and ejector rod housing are one piece.

The Security-Six has two safeties. The floating firing pin is housed in the frame. With the trigger released, the hammer rests on the frame, not the pin. The hammer never strikes the firing pin directly. A transfer bar, the second safety, moves into position between the firing pin and the hammer as the trigger is pulled. The hammer hits the transfer bar which in turn hits the firing pin. If the hammer falls without the trigger's being pulled, the transfer bar will remain in a lowered position and the force of the hammer won't reach the firing pin or cartridge. While not unique to the Security-Six, these two safety features are probably the best available for preventing accidental discharge.

Both blued and stainless steel Security-Sixes are produced. Our test gun was stainless, one of the most interesting firearms innovations of the past decade.

Investment castings, heat treated to a high quality of hardness, are a hallmark

Double action, the Security-Six has a smooth if somewhat heavy trigger pull. Shoulder holster is by Lawman Leather.

of many Ruger guns including the Security-Six. Ruger has pioneered in this area and it is a prime reason Ruger is able to keep the price of their guns down. Heat treatable alloys for stainless were not developed until the early 1970s and the stainless Security-Six came out a short time later in 1975.

With the exception of the front and rear sights, the entire Security-Six is stainless, inside and out. But, as Len Pardee at Ruger's Newport factory told me, this doesn't mean that you can leave a stainless Security-Six in the bottom of a fishing tackle box for six months and expect it to work like new. Stainless steel guns *will* rust and they need cleaning just like blued steel guns.

Cleaning—even a thorough open up, strip down cleaning—is a fairly simple matter on the Security-Six. Remove the screw in the grips and from there you can pull out the trigger assembly, the hammer, and the cylinder and crane.

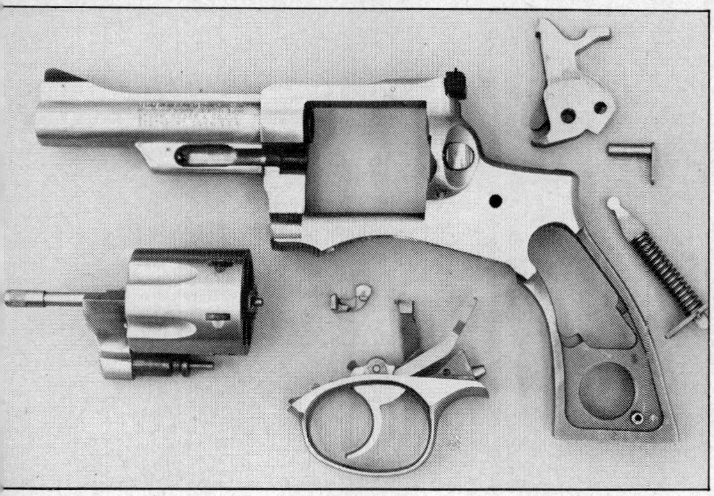

The massive solid frame and heavy construction of the Security-Six make it about as indestructable as a handgun can get. Takedown into major components requires only a small screwdriver (a coin will do) and a couple of minutes. And the rugged gun shoots well—2½ inch groups at 25 yards.

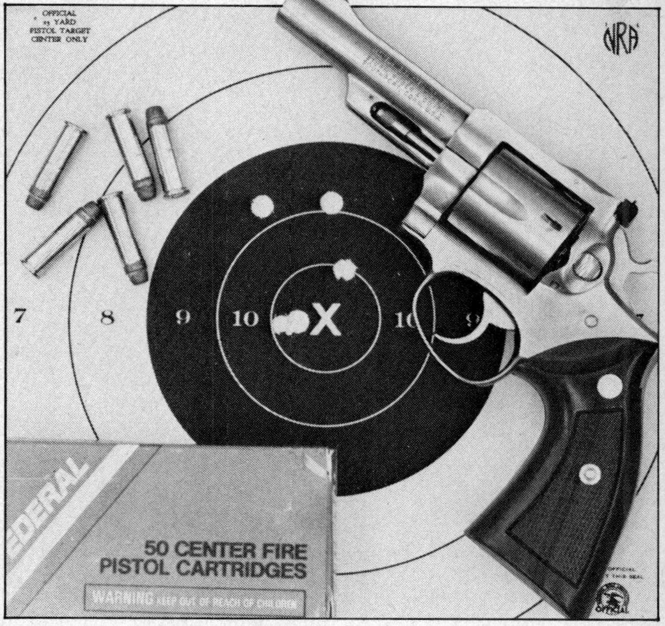

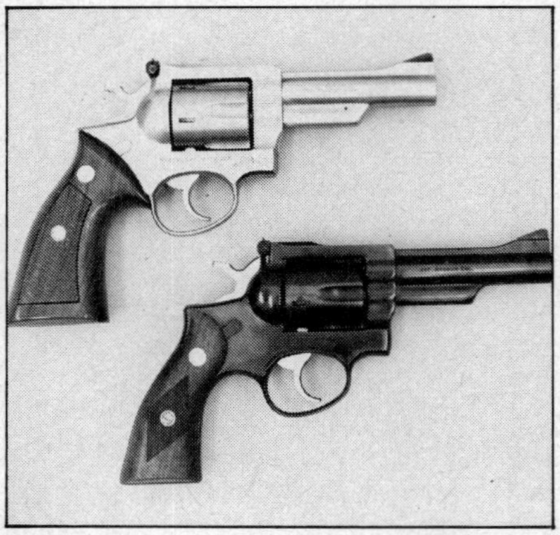

The quality finish on the blued Ruger is equalled in the stainless version. Big grips are a feature on all the new guns.

Following the instructions in the owner's manual, it took barely a minute to disassemble the sample gun.

For a stainless steel gun in particular, the metal finishing was good. Exterior surfaces showed a minimum of burnishing, something that mars the appearance of many stainless steel firearms. Metal edges and internal parts were quite free of burrs and rough edges.

Out at the range, the stainless Security-Six performed identically to a blued Security-Six which I'd picked up for comparison. There was no noticeable heating, a problem found in some earlier stainless alloys.

The five-groove, 18¾-inch twist rifling is supposed to handle long heavy bullets best, but after testing 110, 125, and 158-grain magnum slugs, there didn't seem to be any significant difference in accuracy. With magnums, the sample gun shot consistent four-inch groups at 50 yards. The Security-Six's best groups were with .38 Special semiwadcutters—two inches at 25 yards. These weren't outstanding groups, but certainly good enough for a gun designed for rugged dependability rather than match accuracy.

Accuracy could be improved by installing a finer notched blade in the rear micrometer sight. From a bench position, the standard notch allowed considerable light on either side of the front blade, fuzzing up the sight picture. The standard sights, however, are sturdy and changing them would be a waste unless extreme accuracy were required.

The recent changes in the Security-Six—the trigger and the grips—had been recommended to Ruger in the feedback the company actively solicits from handgunners around the country.

A smooth trigger surface is much more comfortable in double-action firing. More importantly, double-action pull tends to be smoother with the trigger finger riding, not stuck to, the trigger surface.

The new grips are designed for a medium to large sized man's hand and the circumference is ideal for maximum control. You can really appreciate these bigger grips if you alternately shoot the old and new style as I did.

The Security-Six is designed and made to be one of the best of the combat magnums, an elite group of handguns. As I said earlier, choosing from among the best is a personal matter, largely of look and feel. With its recent changes, the Security-Six is certain to increase its already large number of admirers.

SPECIFICATIONS

RUGER SECURITY-SIX

Manufacturer:	Sturm, Ruger, Inc. Southport, CT 06490
Caliber:	.357 Magnum
Weight:	36 ounces
Barrel Length:	4 inches
Sights:	Front: Serrated Rear: Adjustable
Grips:	Checkered walnut, semi-target style
Finish:	Brushed stainless steel
Price:	$227.50

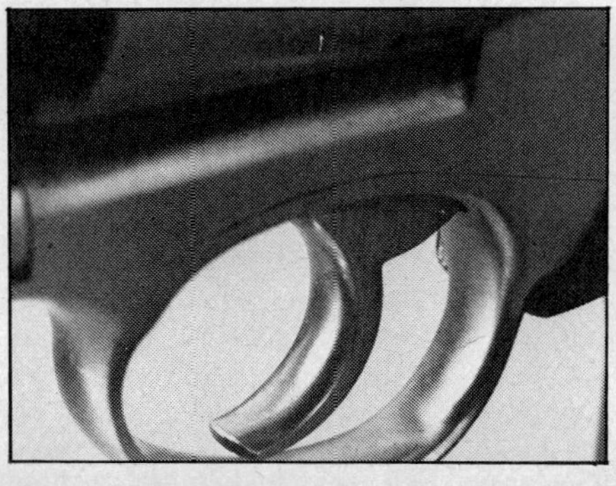

Security-Six now has a smooth faced trigger, the type most preferred for combat.

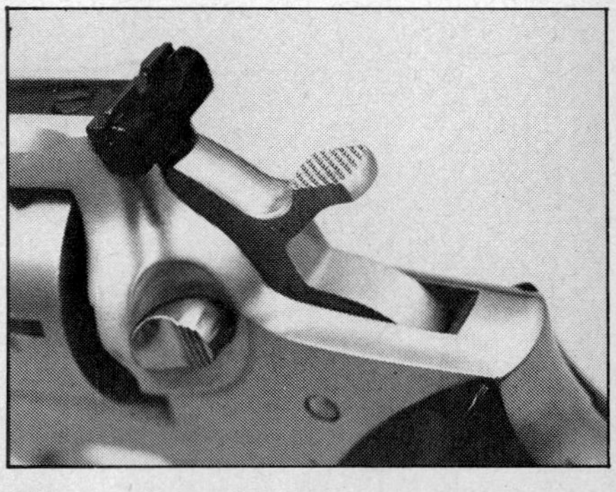

Micro adjustable sights are tool steel. Hammer is serrated and flat, perfect for SA.

The .45 Is Still The Best!

70 years young and still going strong, the Colt 1911 .45 Auto is tops for combat and defense!

By Greg Moats

"Why do you carry a forty-five?" I asked a police officer friend of mine.

"Because they don't make a fifty," he replied.

This simple statement epitomizes the attitude that those of us have who are devoted to the .45 auto. In fact, many consider the desire to shoot any handgun other than a 1911A1 to be at best a lapse of fidelity, at worst a vice.

At the 1980 NRA Convention, I stood listening to two handgun authorities, Ken Hackathorn and John Robbins, "discuss" the merits of the .45 auto vs the 9 mm Parabellum for our military sidearm. As I had heard (and participated in) identical "discussions" that went on and on and on . . . . ad nauseum, my mind drifted off to one of Murphy's laws: "when something works, don't try to fix it." The .45 auto has "worked" for quite awhile now; it is ironic that in it's 70th year the government is seeking to "fix it" by adopting a different handgun design.

Experience with the .45 usually forces the shooter into a dichotomy concerning it's value; those who like the .45 *love* it, while those who dislike the .45 *hate* it; there seems to be no middle ground. Those who hate the .45 generally acquired their feeling while they were in the military. Due to a negative experience with it, they classify the .45 (along with entrenching tools and barracks mops) as objects of misery. Since no one can be talked out of something that they weren't talked into, no amount of facts, figures or logic will ever sway their feeling. They believe that the .45 is inaccurate, kicks like a mule, jams frequently, is unsafe if carried in Condition One (cocked and locked), is too slow into action if carried in Condition Three (hammer down on an empty chamber, magazine in), and ugly. Sound familiar? Let's untwine some of this malarky that's been passed as Gospel about the .45.

"The .45 is inaccurate." Such rumors usually originate in the back of a barracks where a few disillusioned servicemen clean their pistols after having failed to qualify with them. It's undoubtedly true that some of the .45 autos currently in our military inventory (usually in training commands) can be "field stripped" by being grasped firmly and shaken violently from side to side. These are the exception rather than the rule. Since the mass production of the collet bushing, the .45 auto has provided more accuracy than most shooters can muster. Accuracy is, after all, a varying accomplishment; a handgun that will hold a two-inch group at 50 yards will produce "inaccurate" results even at close range in the hands of the inept or untrained. The Model 1911 design possesses a uniformity of action that, when subjected to the skill of a knowledgable pistolsmith, results in incredible accuracy. The .45 ACP cartridge is considered "inherently accurate," a concept that most people (myself included) don't thoroughly understand. An inherently accurate cartridge, combined with a uniformly functioning mechanism has caused the .45 auto to dominate NRA Bullseye competition, which is *the* game of supreme accuracy. Therefore, when a person brandishes a

Nowhere is the versatility of the .45 more evident than in the various modifications on 1911 design: gold plated .45 (upper right), heavy ribbed "bullseye competition" (upper left), modified combat .45 (lower left), "home defense" modified Commander (lower right).

For the maximum in combat situation accuracy, the "rollover prone" position with .45 is excellent if there is enough time.

Practical pistol shooting with the .45 is just that—practical. All manner of positions and barricades are used in competition.

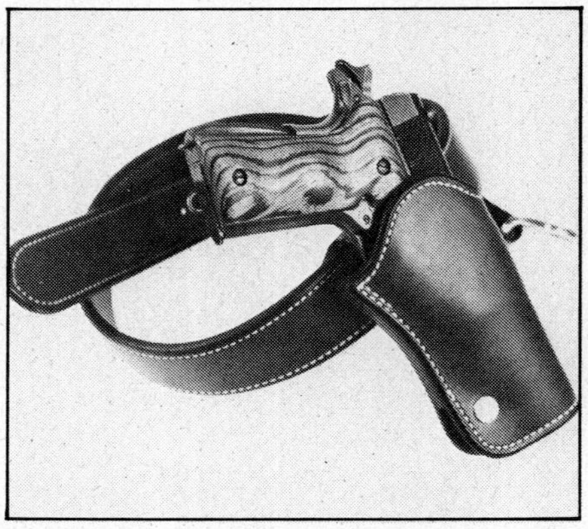

Custom grips like these contoured Magnum Grips, plus a holster and belt like this Davis, are used by top combat competitors.

Although the kneeling position is seldom employed in competition, things are flexible enough to allow its use if you want.

statement about the inaccuracy of the .45, that person is either extremely naive about handguns, or is comparing the .45 to a "minute-of-angle" benchrest rifle.

"The .45 kicks like a mule." Having never been kicked by a mule, it is hard to empirically compare it to the recoil of a .45. Recoil is an inexact phenomena. That is, the movement that actually takes place, and what the shooter *feels* take place are often two different things. For example, a person that has been firing a double-action revolver for years is used to a long trigger pull which initiates action (hammer retracts and cylinder turns) *while* he squeezes the trigger. After the hammer falls the only perceptible action is the movement caused by the expanding gasses of the cartridge. When that shooter fires a .45 auto, just the opposite phenomena is experienced! As he pulls back on the trigger, there is no perceptible action or movement. After the hammer falls, not only is there the movement caused by the expanding gasses but also the slide comes slamming back at him, an empty piece of brass is hurtled out into space, the hammer is recocked and the slide slams forward again. Since all of the action takes place *after* the hammer falls, it tends to give the unfamiliar shooter the feeling of uncontrollability. On top of that, if the shooter's grip is wrong and he gets the web of his hand "pinched" by the hammer, he immediately thinks, "my gosh, the .45 kicks like a mule." The actual recoil or movement of the .45 is noticeably less than any other major caliber handgun and occasionally less than some minor caliber handguns. Recoil controllability is, afterall, more a function of training than mechanical design, that is, you'll tend to get good at what you practice.

"The .45 jams frequently." To decry the superiority of the .45 on the basis that it jams frequently is akin to criticizing automobiles because they have a penchant for colliding. Historically, the .45 auto has proven itself to be the *most* reliable, battleworthy sidearm extant. The .45 is functional in its pristine state but becomes unreliable when: (1) poor quality ammunition is used, (2) magazine lips or followers are deformed, (3) the internal components of the handgun are altered (i.e. sears filed, slides tightened, springs cut, etc.). During my brief experience as a range officer in the service, 80% of the "alibis" given for malfunctions went to individuals who were shooting .38 Special revolvers. In the sport of practical pistol shooting, where no alibis are given, the .45 auto is used almost exclusively, attesting to its reliability.

"The .45 is unsafe if carried in Condition One" (cocked and locked). Had John Browning designed the Model 1911 with an internal hammer, a lot of paranoia would have been avoided. The .45 auto *is safe* while being carried in Condition One—it was meant to be carried that way! A loaded pump shotgun with the safety on is cocked and locked; however, it doesn't give the feeling of immediate threat that a Condition One .45 auto gives because it doesn't have an exposed hammer. Once an individual shooter gains a thorough understanding of how the .45 operates, this ceases to become an issue. However, since law enforcement agencies have Public Relations to contend with, many departments have adopted double action autos or double action conversions for the .45. Perhaps an easier solution and certainly a cheaper one, would be to put a leather hammer guard on the holster to conceal the appearance of a "cocked" handgun.

"The .45 is too slow if carried in Condition Three" (hammer down on an empty chamber, loaded magazine in). That's right, it is too slow. A .45 carried in Condition Three is just slightly more time consuming to get into action than a single-action revolver. Condition Three is only prudent for the shooter that has small children around the house. Ideally, most small children would have neither the strength nor the dexterity to pull the slide back and chamber a cartridge. Once those children are big enough to accomplish that, hopefully, they will have had the training and discipline to not stick the keys in the light socket, poke the dog in the eye, or touch the gun on the dresser.

"The .45 is ugly." I doubt that anyone takes this criticism seriously. As the saying goes, "pretty is, as pretty does."

What has made the .45 the unparalleled choice of handguns for defense, for combat competition and the choice of our military for the past 70 years? Let's examine the requirements of each of these classifications.

Many competitors prefer custom grips with finger grooves for their .45—like these fine walnut stocks made by Bullshooter's.

The .45's reputation for brutal recoil is undeserved, particularly when the shooter employs a proper two-handed technique.

A military sidearm must be simple to operate, disassemble and clean. It must provide parts interchangability but still be rugged enough to survive prolonged abuse. It must be relatively compact but need not be concealable. It need not possess "minute-of-angle" accuracy but must be easy enough to master to allow a less than refined marksman to hit a man sized target at 25 yards. It must possess the power to knock down a full grown adult assailant, regardless of that assailant's adrenalin flow or influence from drugs; however, it need not have a flat trajectory. It must be designed so that pieces of dust, sand, etc. will not render it inoperable; in short, it must shoot every time the trigger is pulled regardless of the circumstances. Since the handgun is a reactive tool, it must be easy to get into action quickly and provide enough firepower to give its bearer confidence. (Note: In a defensive situation, firepower *may* be necessary, but stopping power *will definitely* be necessary. To assert that firepower compensates for a lack of stopping power is like thanking your parents for getting married so that you wouldn't be illegitimate—it's a nice thought but makes too many assumptions.)

The .45 auto fulfills all of the above criteria better than any other handgun is current production. Revolvers are not easy to disassemble, do not lend themselves to parts interchangability, and do not stand up as well to abuse or neglect. The 9 mm Parabellum has not provided the stopping power that the .45 has historically demonstrated. It does provide better penetration, but that's a dubious virtue in a combat handgun. Although double-action automatics are in vogue, they are not new; they've proven to be notoriously hard to train with, subject to unintentional discharges, and most types are of comparatively delicate mechanical composition. If we therefore exclude these alternatives, we are left with the 1911A1 .45 as the quintessential military sidearm.

The requirements of a handgun for home defense differ slightly from its military counterpart. While it must also provide stopping power, it must do so without producing hyper-velocities that would endanger family members in adjoining rooms. It must be easy to get into action quickly and easy to handle in the dark. It must be controllable and easy enough to operate that almost all family members can achieve a moderate level of competence with it. Although not necessary, it would be nice if the action were slightly complex so that extremely small children couldn't inadvertently fire it. Finally, it must be relatively inexpensive and ammunition must be easy to come by.

The .45 auto fulfills all of the above criteria better than any other handgun in current production.

The requirements of a handgun for combat competition (or practical pistol shooting) differ very slightly from those of the military sidearm or the home defense handgun. The "practical handgun" must function flawlessly, it must be as accurate as possible within the parameters of perfect reliability, it must be extremely fast into action and easy to reload quickly. The practical handgun must be of major caliber (it must move a ballistic pendulum as far as a .45 ACP hardball round). For competition purposes, it should be easy to use with both hands and be controllable enough to not hinder multiple shot strings. At times the practical pistol may need to be concealed, so it should therefore possess minimum bulk and be fairly compact. The practical pistol must be rugged enough to endure the large amount of ammunition that will be poured through it in a short period.

The .45 auto fulfills all of the above criteria better than any other handgun in current production. The proof to this strong statement is that serious PPC competitors almost without exception use the Colt 1911.

The .45 has tabulated an incredible record in competition and combat by weight of years and sheer numbers in use. Although the .45 is one of the most abundant handguns ever produced, it is probably the most altered sidearm also. There are two general motives for most shooters having custom work engineered on their .45. They are: (1) to help the shooter accomplish a specific task more efficiently, and (2) to provide the shooter with the ego-reward that owning a "custom gun" (and having everyone notice it) gives. Just as some "dancers" go to a disco to strut their high fashion clothing, so some "shooters" enter practical pistol matches to show off their outlandish equipment.

For the person wanting to get started in the sport of practical pistol shooting and desiring to be well defended, it is suggested that a Colt Government Model Mark IV, Series 70, .45 ACP be employed. In practical shooting, a bad reputation is the kiss of death, where equipment is concerned. Due to the rolled steel pin that was used to hold the sights on, the Gold Cup gained a reputation for having the sights shoot off at extremely inopportune times. Although the reputation lingers, more and more Gold Cups are finding their way back into practical shooting events, the maladies having been corrected. The Lightweight and Combat Commanders are frequently passed over in favor of the heavier, more recoil resistent Mark IV. Although many like the feel of the Commander, the abbreviated slide and frame negate the asset of parts interchangebility with the abundant "full sized" .45 spare parts.

They say that imitation is the highest form of flattery. A brief look through the ads of the shooting periodicals will amptly show that there is no shortage of Colt "copies." There are few serious competitors that use anything other than original Colt, not that the quality control of the "others" is poor, it's a matter of reputation. Experience has shown that more problems exist per capita with non-Colt products, therefore most people judiciously play the averages and stick with Colt, which in the long run is probably the prudent choice.

What is necessary to make the Colt Mark IV "combat worthy?" Assuming that it functions, nothing! Almost everything that is done by way of modification is designed to make *You* more "combat worthy" with it! Some basic things that you might want to consider having done to a .45 are:

1. Polish the barrel throat and feed ramp to insure flawless feeding.

2. Install a set of high visibility, low profile sights (not necessarily adjustable). Proper sights are made by Bo-

Mar, Micro, MMC, Behlert, Kings, Smith & Wesson and others.

3. Have the trigger mechanism refined, a function generally called a "trigger job." You may want a long trigger with an overtravel stop installed, but this is a matter of preferrence not procedure. It may be best to shoot the pistol until the "newness" is worn off before you decide that you need a trigger job. Many "out of the box" .45s after having been "broken in" have quite acceptable trigger pull.

4. You may want to install an extended safety (or "speed" safety as it is often called) or an ambidextrous safety, especially if you're left handed. These modifications are considered "basic" and, with the exception #1, are designed to make the shooter, not the handgun, function better. Other modifications designed to make the handgun function better, such as precision fitting of the frame to slide, oversized bushings, links, etc. produce results that are only of benefit to the advanced pistolero. I've heard that there are over 150 modifications cataloged for the .45; rather than list 146 other alterations, let's look at a few things that are *seldom* found on the top competitors' guns.

Combat modifications include: extended slide release, adjustable rear sight, bobbed hammer, ambidextrous safety, beavertail grip, magazine pad, wide front sight blade, elongated trigger, oversized barrel link, recoil buffer, plus heavy duty recoil spring.

Double-action conversions

The 1911A1 was designed to be carried with a round in the chamber and the hammer cocked and locked. Double-action conversions are designed to convert the handgun to be carried with a round in the chamber and the hammer down, the inducement being fear of an accidental discharge. To believe that the slide safety and grip safety *won't* work accidentally, but that the trigger assembly and inertia-type firing pin *will* work coincidentally at the same time, seems somehow illogical. Since after the first shot with a 1911-type action, the hammer is automatically cocked, the double-action apparatus is only in effect for the first round. This produces two distinctly different trigger pulls between the first and second shot, making accurate rapid fire shooting a matter of considerable complexity. Since it makes shooting more difficult and only solves a non-existent problem, double-action conversions are seldom seen on competition guns. However, these conversions have proven themselves quite reliable and many police officers swear by them.

Extended/Ambidextrous slide releases.

In practical pistol shooting, it is considered poor tactics to ever shoot the gun "dry" (empty it so that the slide is locked to the rear). Therefore, to install an extended slide release on a competition gun is like a tire salesman giving away free bumper jacks with every set of tires he sells, its existence connotes something bad. Besides the fact that extended releases are not significantly faster to operate than the standard lever, they often interfere with proper holster fit.

This magazine release is not extended, but "spread out," in order to lessen the possibility of inadvertent magazine release.

An overly high ambidextrous safety can be unintentionally disengaged when holster's thumbbreak safety strap is released.

Extended magazine release buttons.

Extended magazine release buttons are good in theory but poor in execution. Their function was supposed to make it easier for the shooter to find the release button and increase the mechanical advantage the shooter has in depressing it. Unfortunately, they also have a tendency to be accidentally depressed by the holster rubbing against the extended portion. To draw your .45 and have the magazine fall out can be disconcerting, and could quite possibly be fatal.

These are a few of the currently available alterations that are seldom used by the top competitors. Many new shooters sit down with a gunsmith and mull over that rather exhaustive list of 150 modifications, seeking advice as to what's necessary. A better approach is to have the "basic" alterations done, go out and shoot for a year and then go back and *tell* the gunsmith what you'd like done; you will undoubtedly appreciate the result more.

Whether you're in the military, a practical shooting competitor or an individual concerned with protecting his life and property, the .45 Auto is the handgun for you. It is the only mechanical piece of equipment that has bridged the historical gap between the Indian Wars and Star Wars; it is truly greater than the sum of its parts.

Stainless Steel Handguns

Some shooters like it, and some still prefer the traditional blued steel, but it's a fact stainless steel is here to stay, with more "rustless" handguns available every year!

By John Lachuk

Smith & Wesson created quite a stir in 1965 when they introduced the first production stainless steel handgun, their Model 60 "Chief's Special," a compact 5-shot double-action revolver chambered for .38 Special. One of the early guns off the production line has been my more or less constant companion ever since. An inside-the-waistband holster, subject to constant perspiration, is not the best permanent home for any revolver. I've had others rust literally in *one day* of such use! Yet the tiny Chief's Special remains perennially shiny. (Perhaps a little *too shiny?*)

It can be argued that stainless steel is not truly *rust proof.* But it certainly is more *rust resistant* than ordinary blued steels! Back in 1966, Edward M. Yard, one of the most objective scientific minds among gun scribes, subjected a stainless steel Chief's Special to incredible abuse, finally succeeding in rendering it inoperative. However, the Model 60 did survive 72 days in a 100-percent humidity chamber at 112 degrees F with only mild signs of rust in non-strategic areas. The gun remained functional! Only after table salt was added to Ed's witches brew did the plucky little revolver succumb to serious corrosion on the hammer and trigger.

Stainless steel has complemented new designs like the Wildey Magnum above and beautifully engraved S&Ws on facing page. Engraving is by master engraver Gino Cargnel of Simi Valley, California.

Both the hammer and trigger are machined of a higher carbon alloy (thus more susceptible to rust) for greater hardness and wear resistance on the sears. Smith & Wesson uses several grades of 400 Series Chromium Grade stainless steels, with high resistance to corrosion, plus relative ease of machining, and the ability to heat treat to acceptable levels of hardness while retaining high mechanical properties. The use of slightly dissimilar alloys in working parts also prevents galling problems that sometimes crop up when identical stainless metals are brought together under conditions of high abrasion.

A telling testimonial for the Model 60 appeared in the "Letters to the Editor" column of the October 1969 *American Rifleman,* from a serviceman who had carried one for months in the highland jungles of Vietnam during the rainy season. "Once," he related, "I carried my Model 60 for 12 hours in a drenching rain. The holster acted like a sponge. When I finally retreated from the wet weather, I tossed the revolver, still in its wet holster on an open shelf. It remained there until the holster itself had dried out. A close inspection revealed no signs of rust or corrosion on the revolver." He stated further, "My Model 60 was subjected to constant exposure . . . carried in contact with my wet, perspiring body for days without being wiped off . . . forever sweaty hands and returned to a sopping holster without maintenance," all of this without rusting!

The instant popularity of the stainless version of the Chief's Special prompted Smith and Wesson to introduce a number of other models in the shiny silver-colored metal. The competition, including a number of smaller independents joined the stainless parade, thanks largely to emerging state-of-the-art techniques of precision investment casting, which provides raw parts of intricate forms, such as revolver or autoloading pistol frames in nearly finished dimensions, requiring a bare minimum of finish machining. Difficulties of machining have really held back the use of stainless alloys until recently. Stainless has a tendency to become gummy during machining, rapidly dulling milling cutters and drills, causing chatter and rough cuts. Development of "free machining" alloys has also helped popularize the metal with gun makers.

One of the more sophisticated stainless alloys is 17-4 PH, the choice of Harry Sanford, of "AutoMag" and AMT "Hardballer" fame. Billed as a precipitation hardening chromium-nickel grade stainless, 17-4 PH boasts corrosion resistance, " . . . superior to that of hardenable straight chromium grades such as type 410," plus, "excellent mechanical properties at high strength

Color photo by PPC photographer Pat Brollier

level," and possesses, "...good resistance to galling, seizing and stress corrosion." Hardened to 44 Rockwell "C," 17-4 PH attains a tensile strength of 200,000 psi.

Tensile strength is the most often quoted statistic in relation to steels, and rightly so! All other basic properties of steel are related directly to the tensile strength. Tensile strength is determined quite empirically, by tearing a test sample asunder! A huge machine grabs two ends of a bar with a square inch cross section and simply pulls it apart. At first the bar stretches in the middle like taffy, revealing its "yield point." Then it finally fractures, and the machine records the number of pounds pressure exerted at the breaking point.

Because stainless steels can't seem to

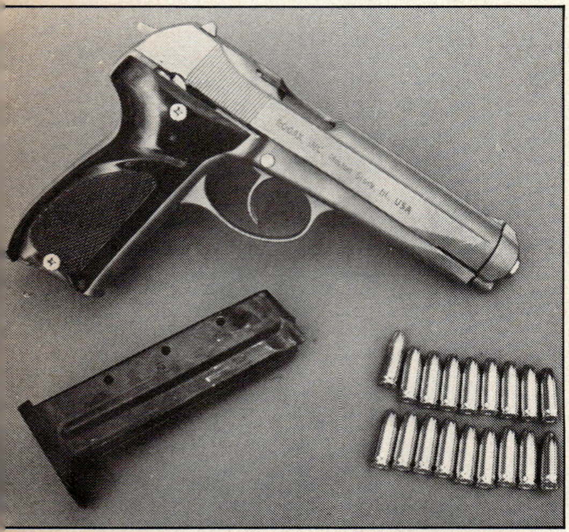

The L.E.S. P-18 has a gas-buffered blowback system and its magazine holds a prodigious 18 rounds of .380ACP or 9 mm!

match high carbon tool steels in terms of attaining high wear resistance and hardness together with maximum toughness, many makers such as AMT prefer to use tool steel for internal action parts. By using A-2 tool steel, Harry Sanford feels that he can get a more crisp trigger release on his Hardballer. Somehow, stainless seems to impart a slightly "mushy" feel to the trigger pull.

The fact that stainless can't be blued makes it a poor material for sights. Even sandblasted to break up reflections, there is enough glare to make sighting difficult against light backgrounds. Most makers recognize the problem by using conventional blued steel sights in the rear, but a number of them continue to use stainless up front because it allows simpler production procedures. One manufacturer, M-S Safari Arms of Phoenix, Arizona, solves the difficulty by offering either black or red plastic inserts in the silver sights up front on their remarkable "MatchMaster" and "Enforcer" .45 ACP autoloading pistols.

Buying an M-S Safari Arms' target-ready MatchMaster or combat-capable Enforcer is like getting a custom .45 ACP in a box, for about half the custom cost, and without the usual months or year-long wait. All of the features commonly associated with expensive custom pistols are original equipment, such as extended ambidextrous safety, extended slide release, broad beavertail grip safety, burr hammer, checkered neoprene mainspring housing and grips, finger grooves and checkered front strap, grooved long trigger with built-in backlash adjustment, fully adjustable micrometer rear sight, and rubber base pad on the magazine. Barrel bushings are threaded into the slide for a tighter and more stable fit.

The Enforcer is marked by its compact dimensions, with 3.8-inch barrel and the grip bobbed just below the bottom grip screw, reducing magazine capacity to six rounds. Despite the shorter grip, finger grooves and the spade grip safety offer improved recoil control.

M-S Safari Arms uses 17-4 PH stainless steel for the frames and Type 416 in the slides, thus placing two slightly dissimilar alloys in contact at the frame rails, where the greatest friction occurs, pretty well eliminating galling problems. Owner Don McNabb selected at random a gun off the line and put it through a degreasing bath. Then he fired 50 rounds through it absolutely dry of lubricant. *There was no apparent galling.* He followed that with 10,000 rounds after lubricating with "Break Free" he could only detect a slight polishing effect upon the frame rails. Every gun off the line is fired 50 times to confirm function and accuracy.

The .45 ACP "Hardballer" manufactured by AMT (Arcadia Machine & Tool) is a near-exact copy of the Colt National Match, save for one important attribute that has made it extremely popular—*it is made of stainless steel!* The slide has the full-length raised rib, grooved to reduce glare, with a tall, square post front sight backed by a Micro adjustable rear. Trigger, slide stop, safety, etc. are all in the Gold Cup image, with checkered walnut grips rounding out the picture. The slide-to-frame fit is approaches the firmness of a custom-fitted job, and accuracy is above average for an out-of-the-box pistol.

AMT offers some variations on the theme in the form of their shortened "Skipper" and stretched "Long Slide Hardballer," with two inches added to the barrel and slide. The Long Slide is rapidly becoming a favorite with combat competitors for long range matches where hits count as much as speed. The longer barrel length contributes to somewhat higher velocity, plus adding a bit of weight out at the muzzle where it dampens recoil and allows quicker recovery time. During our chronograph tests, the Long Slide recorded velocities more than 100 fps above those posted by the 5-inch barreled Hardballer. And that is no small matter with a .45 ACP! The AMT Long Slide has been around the *Guns & Ammo* working battery for some time now, and has garnered a reputation for never-miss function with full-bore loads.

Harry Sanford has this to say about rust on stainless handguns. "Heat-treatable stainless can develop corrosion under extreme conditions of dampness and exposure, mainly because the heat treatment brings carbon atoms to the surface. But you'll never get any deep penetration. I keep one of my Hardballers on my boat moored down at the ocean marina. I deliberately neglected it for a period of months, never cleaning or oiling it after use. Finally, it developed an overall exterior surface rusting. I took some fine steel wool and Hoppes and scrubbed it right back to its original finish! There was absolutely no pitting!"

Because demand is unflagging, AMT continues to turn out a limited quantity of AutoMag pistols. Increasingly popular is the AMT "Backup" autoloading pocket pistol, perhaps the smallest practical package to offer .380 power. The Backup doesn't rely upon a tight slide or bushing for accuracy, although toler-

Mossberg's stainless combat .45, produced by A.I.G., features a slide mounted safety and a compact four-inch barrel.

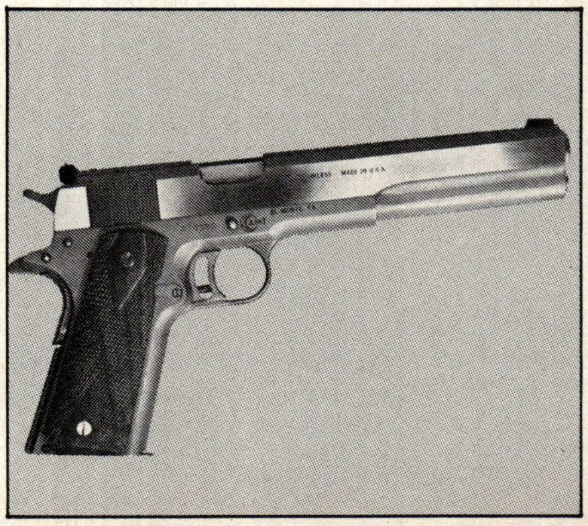

The AMT Long Slide Hardballer with its seven-inch barrel improves .45 velocity and dampens recoil for quicker recovery.

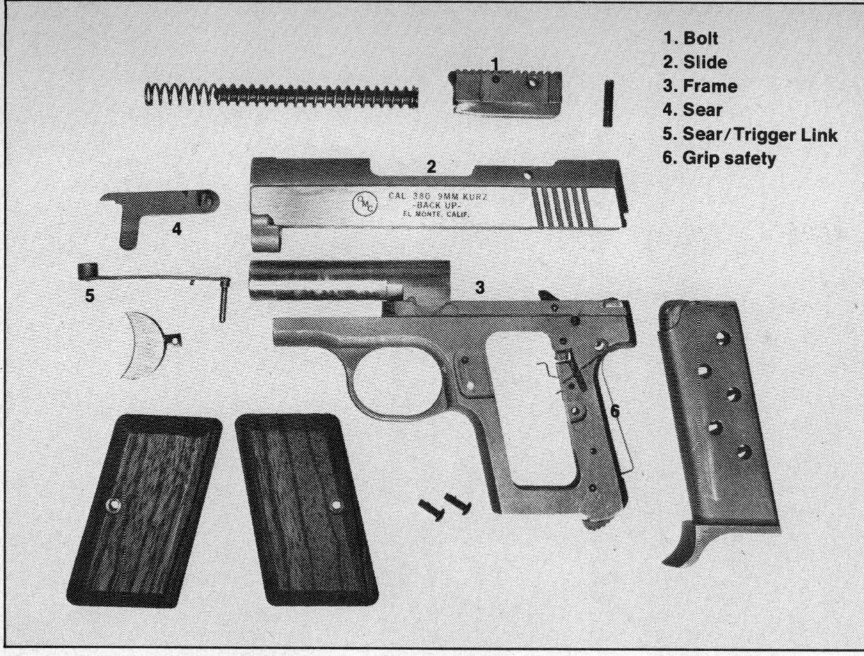

AMT is the manufacturer of the tiny .380 Back Up. This all stainless steel auto is a mere 4¾ inches overall. It features a rigid barrel on machined frame and is surprisingly accurate. The grip safety arrangement blocks the sear.

M-S Safari Arms uses one stainless alloy for the slide and another for the frame, thereby reducing friction and galling.

ances are laudably tight. The 2½-inch barrel is welded solidly to the frame, resulting in surprising accuracy for a gun so small. Sandbag five-shot groups ranged from two inches to 4½ inches at 25 yards. The solid barrel contributes to reliability as well, because the feed ramp is a continuous smooth groove, without the steps and sharp edges found in separate barrel designs.

Sometimes it's hard to remember that Bill Ruger was once a struggling independent, operating virtually in his garage. Today, Sturm, Ruger & Company, Inc. routinely challenges the goliath gun companies. Then we recall that Bill Ruger's forte was innovation—that it was he who first introduced precision casting of entire revolver frames—a process that Ruger utilizes in fabricating his several lines of stainless steel revolvers, beginning with his "New Model Super Single-Six" .22 rimfire.

Ruger revived the languishing single-action revolver in 1953. Ruger also revised the single-action's innards with modern coil springs and beefed-up parts, recently updating it even further by adding a transfer bar action that made the "five-shooter" of old into a true—and *safe*—"six-shooter! As currently offered in stainless steel, the New Model Super Single-Six is delivered with dual .22 Long Rifle/.22 Magnum cylinders, with optional 5½ or 6½-inch barrel, ramp-mounted Patridge front blade with micro-adjustable square-notched rear sight, and smooth walnut grip panels. Note that all major components on the Single-Six stainless *are made of stainless steel,* including the backstrap and triggerguard, and the ejector tube, all of which are alloy on the blued model. That alone makes the stainless rimfire revolver well worth the slight additional cost!

Duplicating all of the above features but scaled up to handle bigger bores are Ruger's New Model Blackhawk revolvers, massive frontier-styled handguns chambering the .357 Magnum. Barrel lengths are optional 4⅝ or 6½-inch. The Ruger "Old Army" black powder hogleg, essentially the same action and frame as the big bore cartridge single action, was stretched enough to provide space to seat picket bullets ahead of the cylinder, to be seated by the sturdy rammer latched under the long 7½-inch barrel. Offering it in stainless steel was an inspired move, as anyone who has shot black powder much could certainly attest. The cleanup after shooting is reduced to a few brisk wipes with an oily rag!

Ruger double-action revolvers are also available in series 400 stainless, in .38 Special and .357 Magnum. The "Security-Six" with 2¾, 4, or 6-inch barrel, has a squarish beefy frame roughly reminiscent of the Blackhawk, but with a compound-curve pistol grip in a single unit with the solid-sided

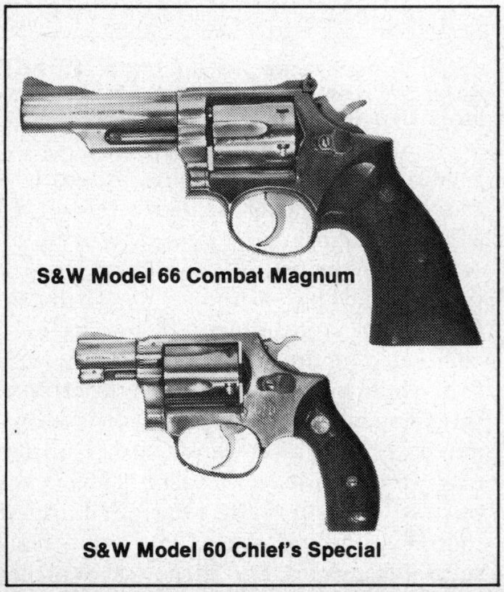

Smith & Wesson pioneered in the field of stainless handguns, introducing the Chief's Special (bottom) in 1965.

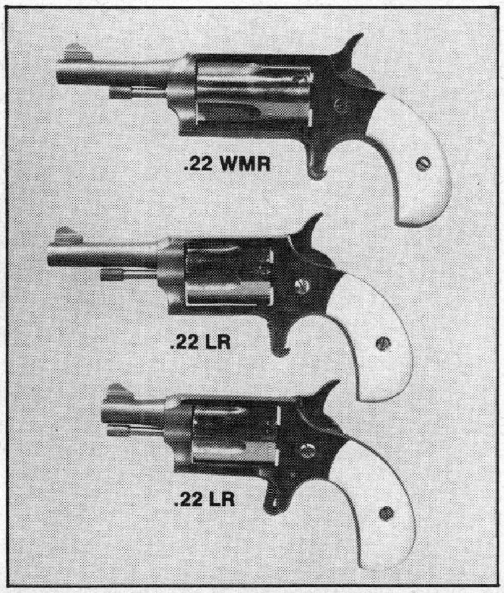

These diminutive stainless .22s are from Freedom Arms. Cardtable accurate, they are nonetheless attractive and functional.

frame. Disassembly involves removing the mainspring and hammer/pin, then popping the trigger guard out the bottom, along with the bolt, trigger, hand, and tranfer bar, thus eliminating any weakness that might result from using a removable sideplate, *a la* Smith & Wesson. The six-shot swing-out cylinder responds to a firm push inward on a button left/rear, unlatching front and rear. The cylinder rotates counter-clockwise. The front lock resides in a fully-enclosed shroud integral with the barrel. Topside is a grooved full-length ramp, rising at the front to accept a black blade sight.

At the rear is Ruger's standard offering on single or double action revolvers, an aluminium sight body with black steel blade square notched without outline, click adjustable for windage and elevation, pinned in a groove in the top strap, protected from side assault by thick ears. The sight picture is adequate but subject to some annoying highlights.

The single-action release on our four-inch barreled sample released crisply at three pounds. The smooth curved stainless steel trigger cycled comfortably through double-action without any hitches in the "gitalong!" The finish is "natural," with a satin silver look. Fitting is good, with a minimum of gaps between mated parts such as the yoke and frame. Ruger's "Police Service-Six" (square butt) and "Speed-Six" (round butt) lack only the target sights.

Big news among big bore enthusiasts is introduction of the Ruger "Redhawk," a brutish 3¼-pound revolver designed to dispense .44 Magnum slugs with utmost efficiency! The 7½-inch barrel makes the gun look somewhat huge for a handgun, but next to the popular AutoMag, it handles like a ping-pong paddle! Basic design resembles the Security-Six, but *bigger!* For hunting and silhouette shooting, this gun should be a premier performer.

Ruger doesn't have a monopoly on front-loading stainless steel revolvers—there's some formidable competition from Navy Arms in their .44 caliber "New Model Army" replica of the ever-popular Remington cap-n'-ball. This all-over silver six-shooter with the bright brass triggerguard is a faithful reflection of the original, save for the high chromium/nickel alloy.

In the modern military vein is Navy Arms' 9 mm "Mamba." Looking like a misbegotten mating of a P-38 and a Browning Hi Power, it is investment cast of solid stainless steel, and features such assets as double-action on the first shot, ambidexterous safety, and overwhelming 14-round magazine capacity! Takedown is similar to a Model 1911, as are the fixed sights. At two pounds six ounces with a barrel just shy of five

The Ruger Security-Six .357 Magnum with four-inch barrel chronographed in the same range as the Smith & Wesson Model 66, about 1400 fps with full house magnums. In a Ransome machine rest, the Security-Six drilled 25-yard five-shot groups under 1½ inches.

Stainless revolvers come in old and new designs. Here a traditional SA frontier design is compared to modern DA combat magnums.

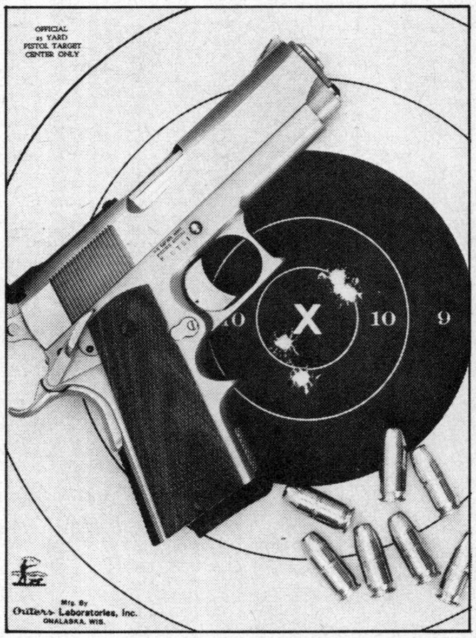

Safari Arms makes the accurate .45 ACP Enforcer with many custom combat features like an ambidextrous thumb safety.

inches, the Mamba should make a functional service type sidearm.

Also in military configuration is the L.E.S. P-18 9 mm, a unique stainless steel creation with 5½-inch polygonal barrel solidly welded to the stamped steel frame. The massive slide is gas-assisted to the rear, to eject and strip a new round from the 18-shot clip! The first shot is double-action. The safety is located at the rear of the slide. An optional single-action version is also offered. Despite the European look, the P-18 is American-made.

Rather more refined is the Mossberg ".45 Military-Combat" completely fabricated of stainless steel, save for the blued front and rear sights. Boasting a batch of custom features, the .45 Military-Combat is manufactured by A.I.G. Corporation of North Haven, Connecticut, based upon a design by Bo Clerke. In appearance, it resembles a channeled and chopped .45 ACP, with four-inch barrel and abbreviated six-shot magazine. The grip safety is eliminated and the safety lever shifted to the rear of the slide, with thumb levers for port-siders as well as right-handed shooters. Pulling the safety lever rearward disengages the trigger and blocks the firing pin as well. The serrated single-action trigger is adjustable for both weight of pull and overtravel. In

Ruger Redhawk .44 Magnum DA

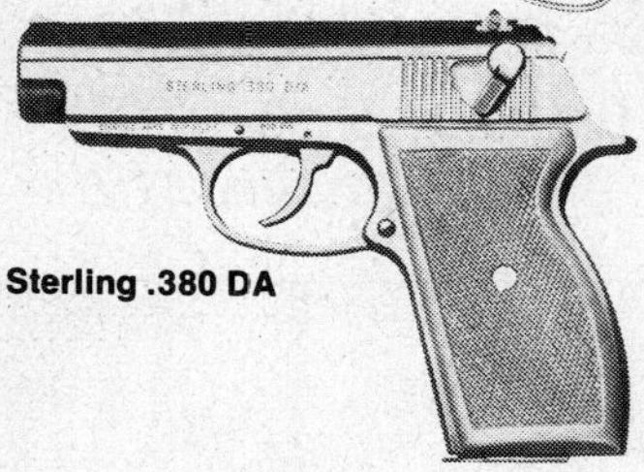

Sterling .380 DA

Smith & Wesson Model 63 .22 Kit Gun

Each of these three stainless steel handguns has enough size and weight to make for comfortable, controllable shooting.

Detonics Mark VI is one of the smallest .45 ACPs. No more than palm size, the seven-shot Mark VI is easily concealable.

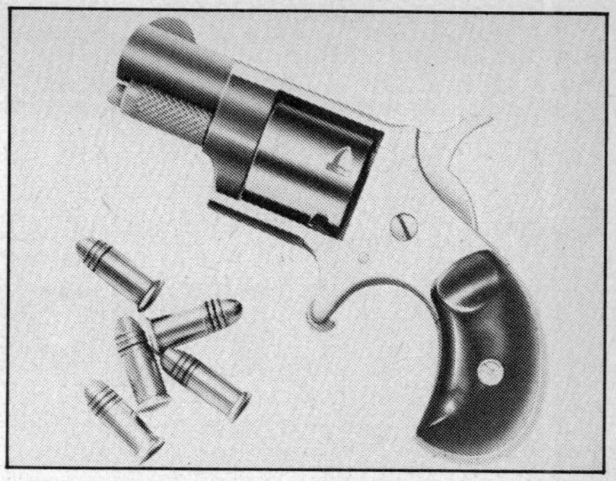

A push button latch on the base pin permits easy cylinder removal for reloading the North American Arms Mini-Derringer .22.

the combat image, the .45 Military-Combat features an extended slide release and a spur on the triggerguard for a firm two-hand hold. The barrel bushing is immovably secured by a threaded nut, for greater accuracy. All of the corners have been planed off for smooth release from the leather, giving this black-handled satin-silver pistol no-nonsense look of competence!

Another product of the Mossberg/A.I.G. partnership is the rugged-looking flat-top stainless steel single action "Abilene" revolver, available in .357 or .44 Magnum, in barrel lengths from 4⅝ to 8½ inches. A tall ramp-mounted black blade up front is backed by a broad-blade micrometer rear sight set deep in the thick top strap. The six-shot cylinder is fluteless. Smooth walnut grips further the frontier image.

Apparently the smallest of the chopped .45-firing palm-pistols is the stainless steel Detonics .45 Mark VI, hardly larger overall than the Smith & Wesson Chief's Special! That allows for about a two-and-a-half-finger grip on the smooth front strap. One would expect the recoil to be brutal with full-house .45's from so small a package. Surprisingly, it's bearable—although few would choose the Detonics as their favorite plinker. The cunning compactness makes the 29-ounce, 3¼-inch barreled Detonics a favored concealed handgun for law enforcement. plus popular with the masses as well. Because internal parts and frame openings are identical to the .45 ACP, a standard magazine can be substituted for the six-shot clip.

Another palm-punishing pocket pistol is the 21 oz. 2½-inch-barreled stainless steel .357 Magnum from Security Industries. Only the addition of a set of Pachmayr semi-soft neoprene grips makes the tiny five-shot revolver controllable. Also available in stainless is Security Industries .38 Special on the same frame, but with 1⅞-inch barrel.

The impressive number of small silver-colored pistols and revolvers made bespeaks their unprecendented popularity. A slightly heftier handgun is the 26 ounce, 3¾-inch barreled Sterling 400S MK II, double-action seven-shot stainless steel autoloader, with slab-sided checkered walnut grips, low-profile adjustable sights, and slide-mounted thumb safety. Somewhat larger than the smallest .45 ACP, the Sterling MK II handles the modest-powered .380 ACP. However, what it gives up in terms of power and compactness, it more than makes up in ease of handling—especially important to a woman—and controllability during sustained fire.

The generally accepted "woman's gun," the .25 ACP pocket pistol is typified by the Bauer Automatic Pistol, prepossessing in appearance with its satin silver look and mother of pearl grips—almost a jewel to wear—but difficult to aim and control.

Also under the heading of handy but marginally adequate defense handguns are the trio of tiny .22's from Freedom Arms. The stainless steel single action revolvers are offered with .22 Long Rifle five-shot fluted cylinder, in one-inch and 1¾-inch barrel lengths, or with 4-shot fluted cylinder and 1¾-inch barrel only for the .22 Magnum cartridge, weighing approximately 4½ ozs. The mini-derringers come semi-matte silver, with tiny "bird's head" grips ivory or ebony, sheathed serrated triggers, and tall-spur hammers for easy cocking. Blade and notch sights are virtually useless because the hammer obscures them when the revolvers are cocked. Loading requires removal of the base pin and cylinder—a tedious process!

Similar in design is the North American Arms Mini-Derringer .22 Long Rifle stainless steel revolver, available with five-shot cylinder and 1⅛-inch barrel only. Loading is simplified by a positive-locking push-button latch at the front of the base pin.

Expected to take the worlds of handgun hunting and silhouette by storm is the stainless steel gas-operated autoloading Wildey Magnum Pistol in either 9 mm Winchester Magnum or .45 Winchester Magnum. Weighing just over 50 ounces, it nonetheless is rather compact in overall dimensions—actually smaller than the Colt .357 Python! A five-lug rotary bolt lock-up contains the working pressures in excess of 50,000 psi! Winchester has the new rimless ammo in ample supply to meet shooter demand.

Lest it be forgotten that Smith & Wesson started the stainless steel stampede, note that they currently offer S.S. versions of their incomparable .22/32 Kit Gun with four-inch barrel; the Chief's Special, the fixed-sight Military & Police with two or four-inch tube, and the four-inch Combat Masterpiece, all in .38 Special; plus the Military & Police four-inch Heavy Barrel and the Combat Magnum with optional 2½, 4 or 6-inch barrel in .357 Magnum caliber. All of the S&W revolvers are marked by Smith's superb workmanship and attention to detail, upholding a quality level that all other makers strive to emulate!

Among the more impressive efforts to duplicate the Model 1911 A-1 in National Match configuration, is the stainless steel "Vega Combat Match .45" from Pacific International of Sacramento. The slide has a full-length raised ramp with smooth top. Overall finish is satin silver, with the sides of the slide and frame both high-gloss polished. Either fixed or adjustable sights are available. Overall workmanship of test sample appeared to be excellent, and accuracy better than average for an out-of-the-box .45 ACP.

All stainless steel pistols share that magnificent asset of being highly rust resistant through and through, without the aid of any protective coatings. Contrary to popular opinion, they *do not* wear better than high grade high carbon steel! However, few of us will live long enough to notice the difference. And the peace of mind engendered by that ageless satin silver luster, untarnished by time or neglect, is certainly well worth the slight additional cost!

Test Report

COLT NEW FRONTIER

The Colt New Frontier—thanks to its adjustable sights—is one of today's premier field guns even though it's a 107-year-old design

By Payton Miller

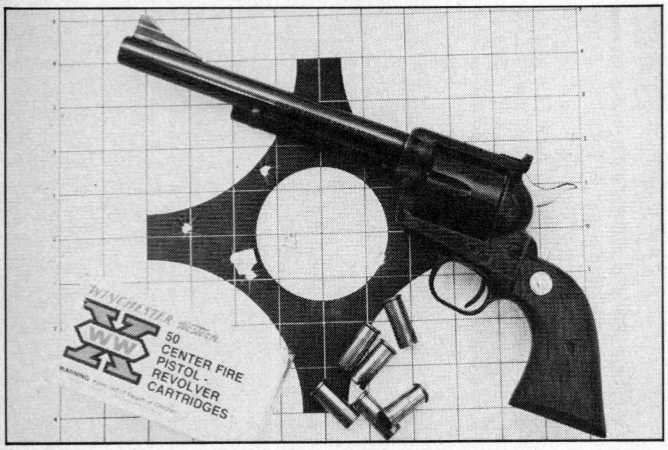

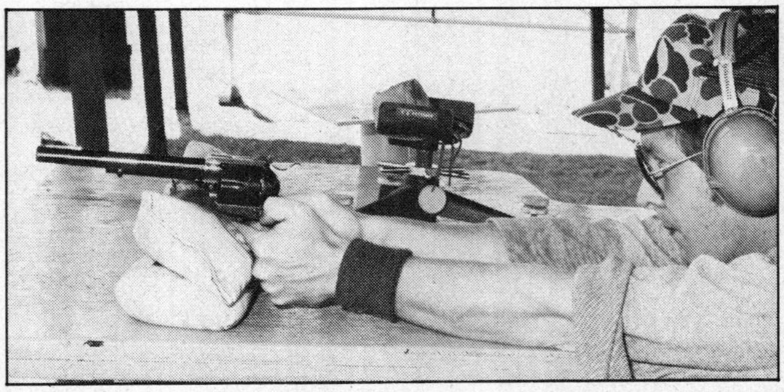

Although few single action revolvers are noted for real target-type accuracy, the Colt New Frontier, thanks to its adjustable sights is capable of adequate hunting and field accuracy. 25-yard groups of around 2½ inches were possible, using Winchester 246-grain ammo. Handloaders could improve on this figure!

Single-action revolvers—even in this age of PPC double-action wheelguns and customized .45 autos—are going as strong as ever. Colt, Ruger, United States Arms, Mitchell, Navy Arms and Interarms all offer versions of their own variations of big, plowshare handled, large caliber hoglegs.

In any assessment of this breed of handguns, the original Colt version has to be reckoned with. Since its birth in 1873, the Colt Single-Action Army has been the yardstick by which all other single actions have been judged. Plain and simple, it's a classic, and is to single-actions what the Mauser is to bolt actions, the Colt 1911 is to autos, and the Winchester Model 12 is to pump shotguns. While it is no longer a tactically viable combat gun—and never was seriously considered a target arm—the SA Army still fills a niche.

Although it probably reached its peak of modern popularity during the fast draw craze of the 1950s, most shooters today employ it at ranges quite a bit beyond pull-and-point distances. Because of this, the New Frontier adjustable sight variation makes more sense for serious shooters. Collectors will probably always have a soft spot for the old grooved top strap, hog-wallow sighting arrangement of the standard version, as will filmmakers and any surviving fast-draw fanatics. For test purposes, however, we wanted the New Frontier version, and we wanted it with a long—7½-inch—barrel. It's also available with a 5½-inch tube.

We chose to go with the .44 Special, a cartridge long known for exceptional accuracy and controllability. In fact, until the advent of the .44 Magnum, handloaders were getting near-magnum performance from this oldtimer, something that isn't nearly as safe to attempt with the .45 Colt.

When the New Frontier arrived, we were pretty impressed with it. It's not exactly a cheap piece of ordnance, running around $478 a copy! The Colt runs on the old single-action system, meaning *no* transfer bar. With the exception of better materials, it's exactly the same mechanically as it was in 1873. Whether or not this has any merit is totally a subjective judgment on the part of the shooter. From an objective safety standpoint, it does mean that carrying the gun with an empty chamber under the firing pin is recommended.

Appearance-wise, the New Frontier was extremely attractive. The blueing was very deep and lustrous, as befitting a premier-priced, top-of-the-line sixgun.

The trigger pull on our test sample broke at a shade over five pounds—which is just a bit on the heavy side for a target gun, but since the SA Army has never fit that definition, that's rather an irrelevant complaint.

Actual firing of the New Frontier was done at the Petersen Ranch. It was an exceptionally windy day, so we decided to stay at 25 yards to group the gun. We used Winchester 246-grain factory ammo. While the round nosed shape isn't the best choice for a field load (an SWC or JSP would be better), it was the only available ammo we had.

Shooting the factory .44 Special out of a single-action revolver with a 7½-inch barrel weighing over 40 ounces is extremely pleasant. The micro adjustable rear sight and ramp front are easy to get a good sight picture with.

Our best group—excluding one called flyer—was around 2½ inches. We felt that the gun was capable of better, but the conditions just weren't conducive to finding out just *how* much better.

To be perfectly honest, we're not quite sure what type of niche a 7½-inch barreled .44 Special single-action fits into. A collector would probably opt for the standard, fixed-sight version. A non-handloading hunter would probably pick a cartridge that develops a bit more steam in factory loadings. It would seem that the best use then, for this big sixgun, would be for the hunter/handloader. From the 7½-inch barrel, a .44 Special could be handloaded to pretty respectable energy levels, although it would be rather optimistic—and potentially hazardous—to try to approach low-end .44 Mag. power.

Still and all, this is an excellent, well-made, pleasant-to-shoot sixgun, and it's a pleasing blend of nostalgia and modern sighting improvements.

SPECIFICATIONS

COLT NEW FRONTIER SA ARMY

Maker:	Colt Firearms 150 Huyshope Rd. Hartford, CT 06102
Caliber:	44 Special
Barrel:	7½ inches
Weight:	Approx. 40 oz.
Finish:	Blued and casehardened
Stocks:	Smooth walnut
Sights:	Ramp front, target rear
Capacity:	6
Price:	$478

Chapter 3

Shotguns

"If you're going to start 'em, start 'em right!" Here are some tips on picking . . .

The Beginner's Shotgun

By Chuck Adams

A youngster's first shotgun is a mighty important purchase. Author stresses that care must be taken to make the correct choice.

At least half the good hunters I know started out with wretched shotguns—firearms that handicapped their progress as shooters and caused them grief in the field. I was a typically sorry example. I started hunting pheasants, quail, doves, and other feathered game at the tender age of seven years old—a perfectly legal age in my home state of California—and because I was growing like a weed, my dad borrowed my grandad's old .410 bolt action shotgun for me to use. This gun was relatively light in weight and had a conveniently shortened stock, but was otherwise one of the most poorly suited choices any unfortunate beginner could have been saddled with.

I suffered with that old .410 for two years before I scraped together the dough from odd jobs to buy my first decent shotgun—a well-used but solid Winchester Model 12 pump in 16 gauge. The old .410 did teach me to shoot precisely because if I didn't the little pencil-like shells with their 11/16 ounces of shot always missed clean as a whistle. I should say *nearly always* missed clean. That .410 did lose me some cripples, the direct result of not enough pellet hits when my shots were slightly off center.

My first shotgun had other severe drawbacks besides limited firepower. It was an awkward, cumbersome tool, as most bolt action shotguns are. I don't know who invented the bolt action scattergun, but he certainly made a mistake. My particular gun also had a sticky wig-wag safety above the tang—a safety that didn't work worth a hoot in chilly weather when fingers tend to be numb and mechanisms tend to stiffen.

The little 16 gauge pump was a remarkable improvement . . . the direct result of reading *Guns & Ammo* magazines and paying close attention to the tips of experts. The gun was light enough in weight for a youngster to handle—a touch over six pounds—and had a wonderful safety that worked crisply every single time. The 16 gauge shell with its one to 1¼ ounces of shot threw larger, neater patterns, especially since the choke on my particular gun was modified instead of the standard .410 full. And when I occasionally hit a bird that didn't drop, I had only one quick, smooth pump between me and a finishing shot. On top of all these good things, a gunsmith cut off the stock to the proper length for my size and added progressively thicker butt pads until my growing streak was over. I had some great times with that old 16, and never touched the .410 clunker again.

A beginner's shotgun should seldom be the same gun an experienced hunter would use. The exception to this rule is when the beginner is an adult man of normal physical size. In this case, any standard issue 12 gauge shotgun of reputable make will do the trick, provided of course that choke, barrel length, and action type are chosen to suit some general hunting needs. However, when most of us think of beginning shotgunners, we think of youngsters and women. Although the occasional adult man somehow breaks into shotgun hunting cold, the vast majority of beginners start as adolescents between the ages of 6 and 16. In addition, a fair share are women who somehow latch onto a hunting man and want to start shooting to please him and/or to have some new fun of their own. Because youngsters and adult women tend to be smaller in size than adult men, and because these beginners have not yet formed solid shooting habits, their first guns must be chosen to match special needs.

One of the biggest mistakes a person can make when choosing a beginner's

Proper stance and correct positioning of the gun to shoulder are essentials for both accurate shooting and to minimize recoil. Held incorrectly, even a small gauge shotgun can have considerable recoil, especially for a youngster on his first time out.

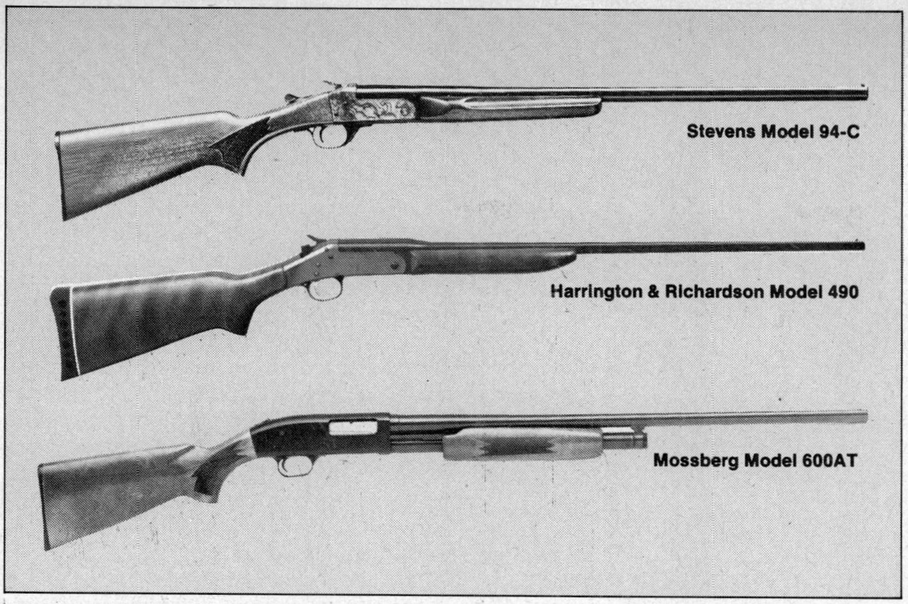

The break-open single shot shotgun is by far the most common gun selected as a beginner's shotgun, and there is a wide variety from several manufacturers to pick from. Author feels that a good, dependable repeater also makes a sensible "starter" gun.

Regardless of action, it is absolutely essential that any shotgun selected as a beginner's first shotgun have a safety that locks positively and is easy to reach for small hands. Training must stress use of the safety at all times, but should also impress upon the beginner that safety is not to be relied upon. It is possible for even the best safety to fail, so guns must be pointed in a safe direction at all times.

shotgun is deciding that the smaller the gauge the better. The end result of such thinking is the conclusion that a neat little .410 shotgun makes a bang-up beginner's gun. Nothing could be farther from the truth. A .410 is really just the opposite—an expert's tool which cripples or misses game unless the shooter can place tight patterns with pinpoint accuracy at relatively close range. The average .410 is light in physical weight, but invariably has a full choke because it must concentrate its light shot load for killing patterns. A .410 should not be given to any beginner because using such a gun will always be a demoralizing experience, leading to many missed shots and more than an average number of cripples. This is true no matter what birds are being hunted, but fast, high fliers like ducks and doves are *least* suited to the little .410.

The 20 gauge shotgun is ideal for most beginning hunters, especially since modern 20 gauge guns are now chambered to accept the potent 20 gauge 3-inch Magnum shell in addition to standard 2¾-inch fodder. A single 20 gauge gun will do it all with factory ammo, from taking close-range doves to dusting spooky Canada geese. Loadings readily available range from low brass field shells with ⅞ ounces of shot to potent 3-inch magnums packing a full 1¼ ounces of chilled lead. A 20 gauge shotgun with 2¾-inch shells produces recoil and muzzle blast mild enough not to spook the rankest beginner, and doesn't hurt his firepower options as he becomes a veteran. Many experts use the versatile 20 with excellent results.

Although a 16 gauge gun served me well in my youth, this gauge is rapidly losing ground today in favor of the 20 gauge with 3-inch chamber. As a result, it is not available in a wide range of factory loadings, and does not represent the good choice it happened to be 25 years ago. On the average, a 20 gauge shotgun is a bit lighter in physical weight than a 16, a plus for the slightly built youngster or lady who is prone to become tired of lugging a cannon about the fields.

A 12 gauge shotgun is the most popular choice of mature hunters, and a husky lad or full-sized man can easily handle the recoil, muzzle blast, and physical weight of the average 12 gauge gun. The 12 is also available in a maximum number of loadings to suit any purpose, making it a decent choice for the mature beginner. However, kids and women will generally do better with the lighter, milder-mannered 20.

No matter which gauge the beginner ends up with, his particular shotgun should be chosen with care to match his physical size, strength, and limited outdoor experience. The oft-accepted notion that a single-shot gun makes the

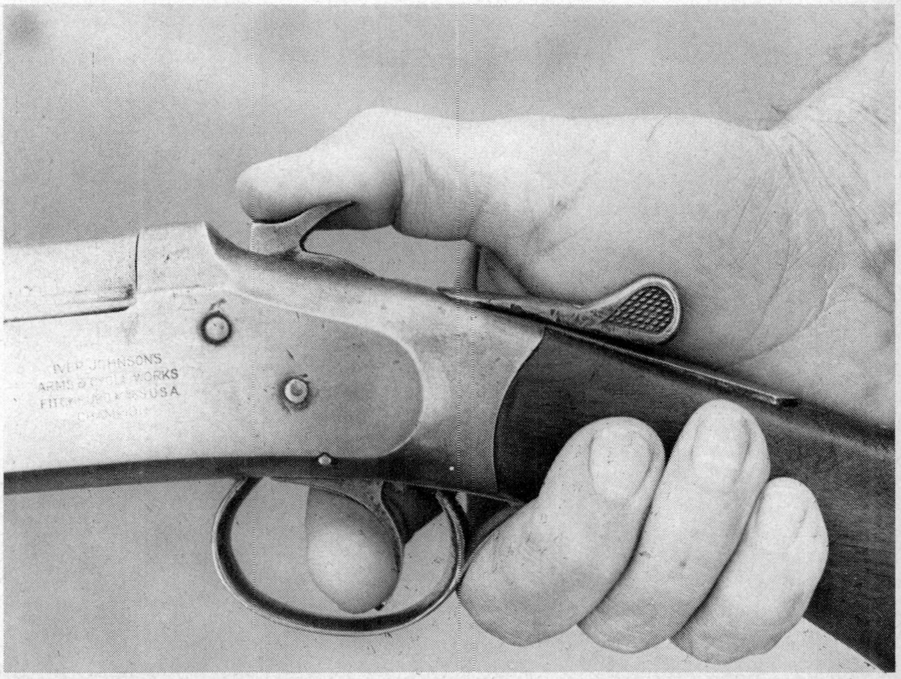

The single barrel, exposed hammer, single shot break open shotgun has long been considered one of the safest designs for beginners. It's simple, and very strong. However, extreme caution must always be used in lowering the hammer on a live round.

It takes time and a little practice for a youngster to learn to hit flying targets. The first step is to teach proper technique, and then allow him or her to shoot enough to get the hang of it. The biggest obstacle is usually hitting the first one!

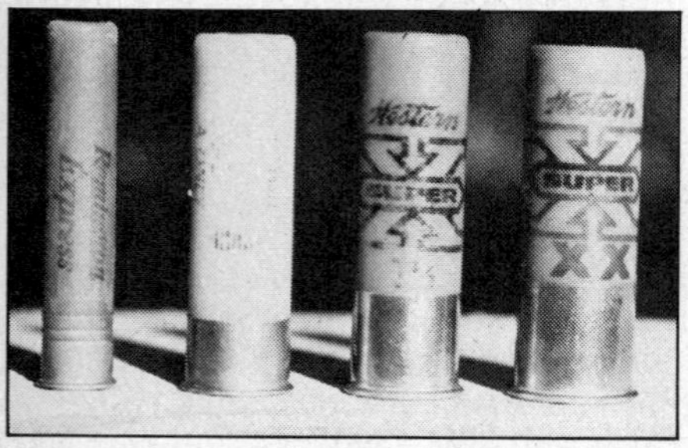

The .410, left, is usually picked for a youngster. Author feels either the 20, 16, or 12 gauge are better, as it's much easier to score confidence-building hits with them.

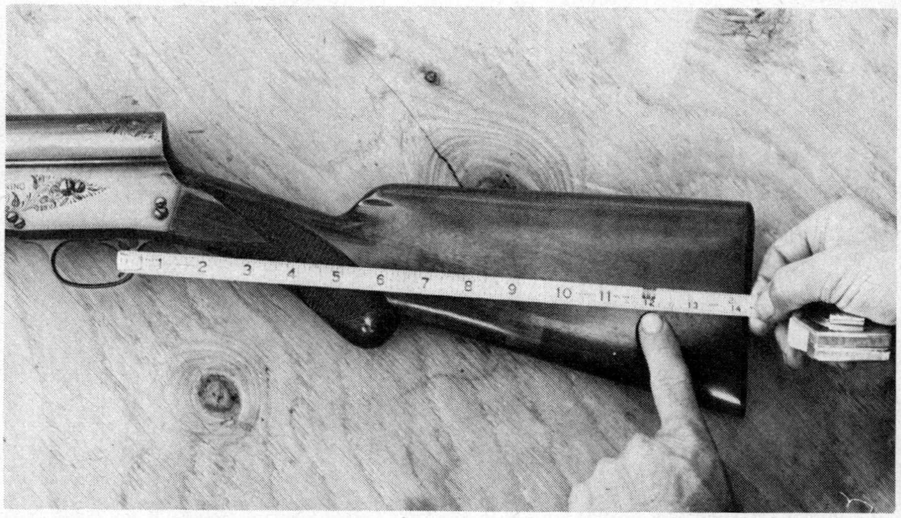

Most full-size shotguns have a length of pull of around 14¼ inches, far too long for most kids and many women. It's a simple job to shorten the stock, and as the youngster "grows into the gun," succeedingly thicker recoil pads can be installed for proper fit.

best beginner's gun is pure, unadulterated bunk. Sure, such a gun is light in weight, and sure, it can only be fired once for safety purposes. However, the beginner needs second and third shots every bit as often as the expert does, giving him the chance to double on covey rises, finish off cripples, and drop birds missed with first or second shots. In all but the smallest gauges, a single barrel arm is apt to knock the stuffing out of undersized beginners because of its excessively light weight . . . which does not result in confident shooting. In addition, a single barrel gun with a hammer can be the ultimate danger in neophyte hands because one slip of the thumb at half cock can blow a companion's leg off or at least scare the beejesus out of everyone in the general vicinity. A single barrel gun is quite cheap when compared to other designs—a plus for many beginners who don't have the money or the keen interest in shotgunning to fork out more dough. However, such a scattergun is not adequate for any beginner.

Although a lightweight side-by-side or stackbarrel shotgun in 20 gauge makes a nifty, well-balanced beginner's tool, the generally high cost of such a gun is likely to deter most neophytes. A well-made double weighing about six pounds and sporting barrels in the 25-inch category makes a delightfully compact beginner's gun, providing snappy performance, two shots, and the choice of two chokes like modified/full or improved-cylinder/modified. However, in our economically depressed times, a good three-shot gun makes a better beginner's choice because it features a much lower price tag without handicapping the tyro who wants to excel.

Both semiautomatic and pump action shotguns make decent beginner's guns, provided they weigh around six pounds and sport relatively short barrels between 25 and 28 inches long. Of the two action styles, I much prefer the pump for beginners—for two reasons. First, a pump action gun requires the shooter to physically eject a spent shell before firing again—a safer situation than being able to simply pull the trigger repeatedly. Second, an auto invites the beginner to rapidly empty his gun at anything that jumps up in hopes of knocking down game with sheer firepower instead of careful, pinpoint shots. The habit of emptying a shotgun tommy-gun style at birds seldom takes game, always wastes shells, and can be completely avoided by using a pump which makes every shot a separate, deliberate event. Those who favor autos over pumps are quick to point out that the average auto recoils less than the average slide action gun because part of the auto's potential recoil is used to eject the spent shell. This point is well taken, but I believe the merits of the pump tend to override the auto's slightly reduced recoil.

The best all-around choke for the beginner's shotgun is modified because it represents a happy compromise between the hard-to-hit-with full choke and the close-range improved cylinder. However, a beginner who intends to hunt a particular species under particular circumstances can easily fine-tune choke choice to meet these individual needs. For instance, the kid whose dad is an inveterate honker hunter will probably take the most birds with a long-reaching full choke, and the southern belle whose husband shoots bobwhite quail over good pointers will do best with an improved-cylinder choke or a double gun choked improved-cylinder/modified. The beginner intent on a variety of game birds like pheasants, doves, quail, and ducks should stick with the modified choke because it tends to forgive slight shooting errors.

Several modern shotguns are billed as beginner's guns, featuring cut-down stocks and lighter-than-normal physical weight. However, I believe a school kid or woman is best off buying a quality adult sized shotgun and having a gunsmith shorten the butt stock to match his or her shorter-than-standard length of pull. This allows the youngster to grow into his gun without having to buy another later on, and makes possible a wide choice of excellent hunting hardware. As a young beginner grows physically, his gun's buttstock can be lengthened with progressively thicker butt-pads, and eventually he can simply discard the original stock in favor of a new, full-length replacement.

A beginner's shotgun should not be a second-rate piece of equipment. If anything, the beginner needs easy-to-use, smoothly functioning gear even more than the expert because he does not possess the experience to make do successfully with a less than ideal gun. As

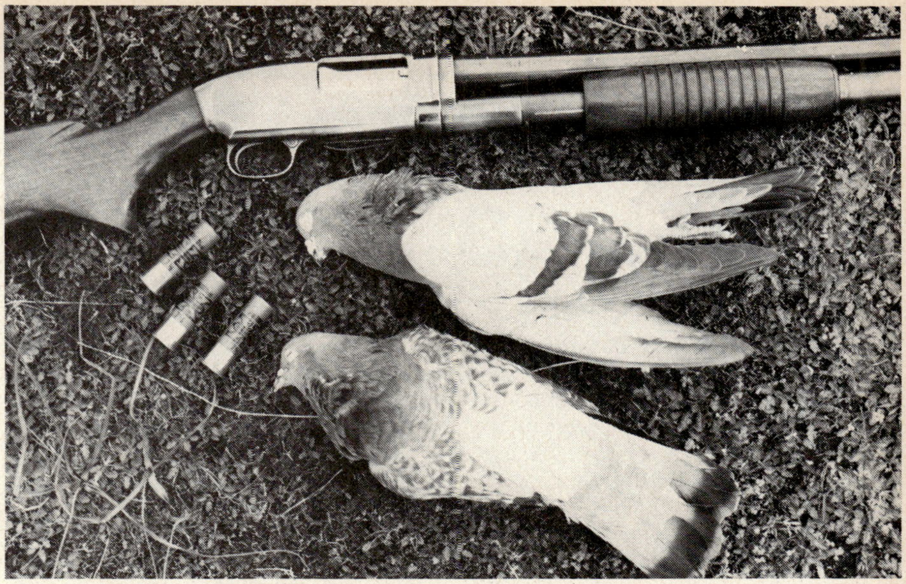

Like many young shooters, the author spent his first few hunting seasons with a .410—and bagged few birds. After "graduating" to a Winchester Model 12 in 16 gauge, above, his shooting average improved dramatically, as did his interest in shotgunning!

A beginning shotgunner, whether old or young, male or female, should be allowed as much practice shooting as possible before going afield. A few sessions with a hand-operated clay target thrower does wonders towards building confidence and skill.

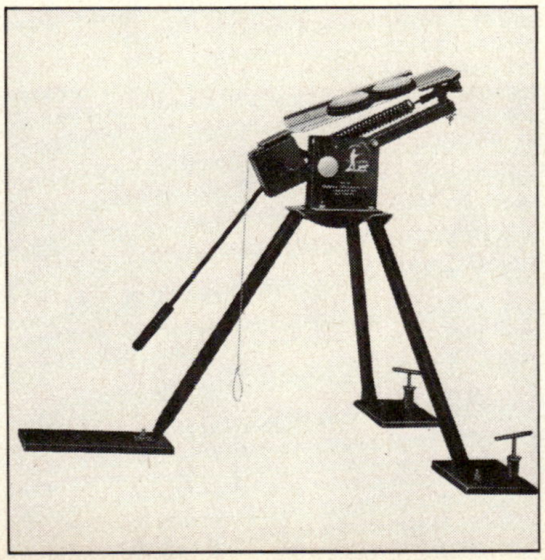

Formal trapshooting has become expensive, but the small investment in a hand operated thrower can be worth the outlay.

a result, inexpensive "off" brands and guns with weird action styles, safety catches, or gunstocks should be avoided at all costs. Standard, dependable guns like the Ithaca Ultrafeatherlight 20 gauge pump (with a 25-inch barrel and a 5-pound weight), the Remington 870 20 gauge pump (26-inch barrel, weight about 6 pounds), and the Winchester Model 1200 20 gauge pump (26-inch barrel, weight about six pounds) all make typically good beginner's guns when cut down to size because they work smoothly and. have positive, easy-to-use safety catches. Some pleasant time spent browsing through the catalog section of this *G&A Annual* will turn up several shotguns that nicely fill the beginner's bill.

The would-be shotgunner who is lucky enough to start with a practical gun is likely to enjoy his sport far more than the poor unfortunate who is forced to fight an inadequate gauge or gun design. Many of us have grown to love shotgunning *in spite of* the first guns we used, but it sure does help to start with practical equipment. The results are more pleasant shooting, better shooting habits, and considerably more game in the bag!

With a properly fitting shotgun and plenty of practice, most youngsters can shoot with experts—and maybe outshoot them!

Test Report

WINCHESTER MODEL 23

This "affordable" side-by-side double from Winchester is a first rate field gun—with style!
By Bill Jordan

It was during the 20's that Winchester's technical staff designed the Model 21 double. It was placed in production in 1929, with first warehouse deliveries in 1930 and announcement in Winchester's price lists of 1931. A less auspicious time could hardly have been conceived!

Although without question, one of the finest double guns ever made, the Model 21 was, for some strange reason, hard to shoot well. Furnished in Skeet and Trap grades, (pre-1959), it was never popular with the competitors in those sports. As an example, the only top rated skeet shot I have ever known who shot the 21 was Billy Perdue, the great Mobile, Alabama "Flyer" shooter. Billy used the 12 gauge 21 to win many important national and regional tournaments and to make the All American Skeet Team several times. I owned a Skeet Model 21 which I used on doves and quail. Although usually I wound up doing quite well with it by the end of a season, it always took the first two weeks of shooting to "learn" to point it.

Elevating the 21 to the Custom class put it out of the reach of most american double gun lovers and left Winchester with no twin tubed gun to take its place, unless one could count the "Steelbilt." While one of the most rugged and dependable guns ever built, the Steelbilt could hardly be mentioned in speaking of fine doubles. It kept shooting and made a fine club if needed to finish off wounded game. The need has finally been filled by the introduction, this year, of their Model 23 XTR Pigeon Grade double barrel shotgun in both 12 and 20 gauges.

The Model 23 was unveiled at last year's (1978) Winchester Seminar, held at Nilo Farms. Apparently some manufacturing difficulties kept it off the market for a full year, production finally getting under way this year. We shot it at the 1978 Seminar on quail, duck, pheasant and chukar. I was impressed by the fact that, unlike the trouble I had always experienced in adapting to the 21, I was very comfortable with this gun from the start. Checking back to refresh my memory to a writeup of the 1978 Seminar, I found that, speaking of the 12 gauge, I had written the following comments in the March 1978 "Snap Shots" column of Petersen's *Hunting* magazine:

"For hunting Nilo's upland game, the writers were furnished with Winchester's slick new side-by-side double, the Model 23. We found that it handled beautifully. My gun, with 28-inch improved cylinder-modified barrels, was a natural pointer which allowed me to get on flushed birds as quickly as any gun I have ever shot. For years I used a similar gun in Model 21 Skeet grade, for both doves and quail. Every year I started out shooting poorly with this gun and only began to score consistently with it after a period of getting accustomed again to the way it shot for me. There was no such problem with the 23."

Pleased to note, that in this instance at any rate, my memory proved also consistent, I unboxed the test gun forwarded to me with genuine anticipation. Arriving opportunely just before the start of the hunting season, it was a 20 gauge chambered for 3-inch shells and had 26-inch improved cylinder-modified barrels.

The overall picture was of a strikingly beautiful piece, with its dark, French walnut wood and highly polished, deep blue barrels set off nicely by the silver gray satin finish of the receiver and other metal parts.

Checking out the dull, but important matter of specifications, there was an overall length of 42¾ inches; 14 inch pull; 1½ inch drop at comb; 2½ inch drop at heel; and a weight of six pounds, nine ounces.

Stock lines are graceful and the gun comes up smoothly and quickly. Catching the eye is a deviation from usual Winchester pistol grip appearance. The 23 sports a round bottomed European style semi-pistol grip reminiscent of the old Browning autoloader. Checkering on grip and forearm is fine line and of plain, but pleasing, design. The butt plate, apparently of a black plastic, is set off with a thin white line spacer. Removing and replacing this plate with a nice rubber pad would eliminate a feature which is inconsistent with the quality of the gun in general.

The tapered ventilated rib is serrated and, in the model sent me for testing, has both a front and central white bead, although the guns description in Winchester's brochure mentions only one, the front sight. The words, "Pigeon Grade" appear on top of the rib in an oval shaped depression just forward of the chambers. Markings on the side of

the left barrel further designate the gun as a "Pigeon Grade" as does the figure of that bird outlined in engraving on the action's bottom.

On the right barrel, the words, "Made in Japan," would have struck a discordant note a few years ago. Today, it has come to be expected of any double gun, whether side-by-side or over/under, requiring painstaking hand fit-

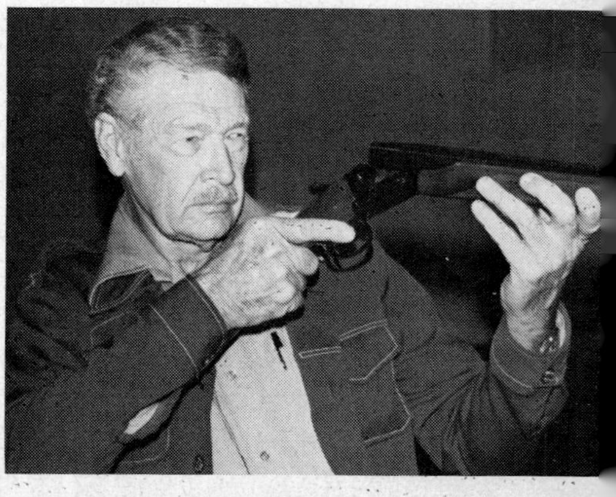

Chrome lined bores were but one of the features writer Jordan liked in the M23.

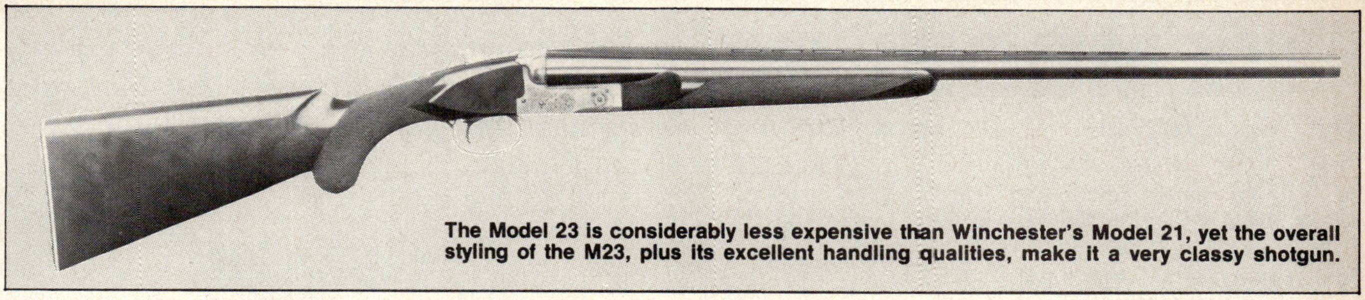

The Model 23 is considerably less expensive than Winchester's Model 21, yet the overall styling of the M23, plus its excellent handling qualities, make it a very classy shotgun.

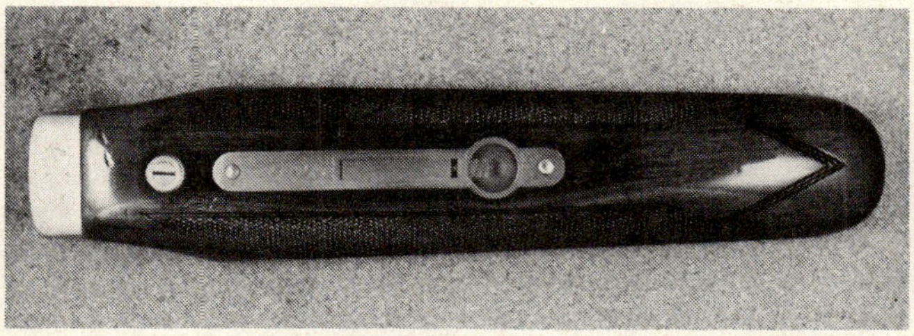

The broad beavertail forearm of the M23 shows fine hand fitting. Forearm release is an engraved flush-set lever.

The action of the M23 is a boxlock type with silver satin finish. Delicate scroll engraving surrounds a pigeon figure.

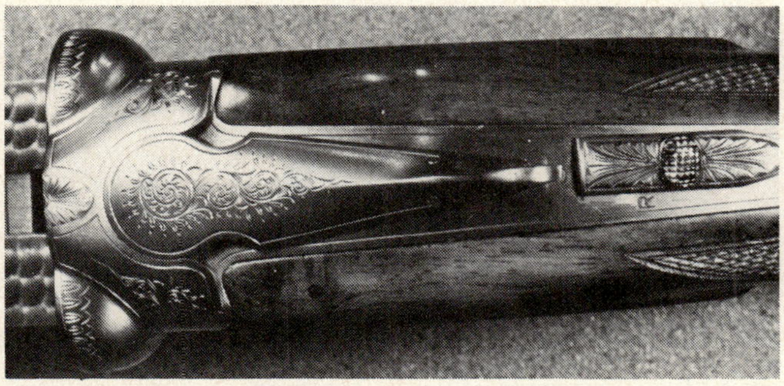

The safety and barrel selector are on a single slide located on the tang. Action cocking is mechanical, not recoil activated.

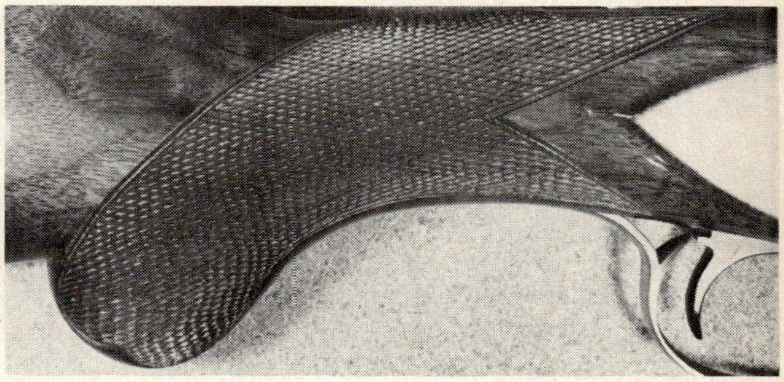

The pistol grip of the Model 23 has an unusual European style rounded bottom. As on forearm, the hand checkering is fine line.

ting at an affordable price.

There is a moderate amount of fine line engraving on the receiver, done tastefully, and notably without the oriental look typical of earlier Japanese made firearms. This treatment extends to the thumb latch and to the combined barrel selector and safety positioned on the upper tang. The only blued parts of this shotgun were the barrels, rib and trigger, all others having been given the silver-gray satin finish.

I have never cared for recoil operated actions. Being tall and fairly slender, I give easily to recoil and often there is not enough jolt to activate the action for the second barrel. I found this to be, insofar as I am concerned, a criticism applying particularly to the Browning Over/Under. No problem here with the 23. Not because of any feeling of excessive recoil, but because its single trigger is both selective and mechanical. i.e., it does not require recoil to cock the action.

Other niceties noted in this firearm are chrome lined barrel bores which resist corrosion and fouling, and selective, automatic ejectors. The barrels are made of high strength chromium molybdenum steel.

The first field test of this little 20 gauge came on my annual pilgrimage to Nebraska for opening of the pheasant season. Hunting with niece, "Bill" (her name is really Mildred Anne, but shortly after learning to talk she announced that it was "Bill," and being a determined tot, that took care of that); her husband, Dr. Vernon Glenn; long time host, Harley Truscott; and Winchester Sales Manager, Jerry Lambert; there were plenty of the big gaudy birds, plus a goodly number of quail coveys to give the gun a thorough workout. It took off just where its 12 gauge counterpart had stopped. That is, right on target. I scored with it as well as I had shot the 12. Three-inch shells took those tough pheasant cocks with ease and, when working on the frequent quail coveys, field loads were just as impressive on the little brown bombshells.

The 23 XTR not only fills the cost gap in Winchester's line caused by their promoting the 21 to millionaire status, but it seems to me a better field gun than its aristocratic cousin.

By no means cheap at a suggested retail of $980, the Model 23 Pigeon grade seems a bargain at today's inflated prices. To anyone wishing to add a bit of class to his upland gunning without quite moving up into the ranks of the thin lipped, belted back Model 21 set, this new Winchester could be just what you have been waiting for. Take a good look at it!

SPECIFICATIONS

WINCHESTER MODEL 23 XTR

Manufacturer:	Winchester-Western 275 Winchester Avenue New Haven, CT 06511
Gauge:	12 or 20
Action:	Boxlock, side-by-side
Triggers:	Single Selective (mechanical)
Ejectors:	Automatic selective
Barrel length:	26 or 28 inches
Choke:	M/F or IC/M
Weight:	6½ to 7 pounds
Stock & Forend:	Select walnut
Price:	$980.00

Test Report

FOX MODEL BE

The economical Fox side-by-side doubles seldom get much attention, but they rank among the most dependable and popular of today's shotguns.

By Dave Hetzler

Quality is an interesting word. When related to firearms, in the context of, "I've always wanted a high-quality gun," one immediately thinks of a Best Grade Holland & Holland, or a finely stocked and customized pre-64 Winchester Model 70. Or if your tastes are more toward shotguns, perhaps the mind thinks of a beautiful Perazzi trap gun, or a Famars 28 gauge field gun that's highly engraved.

The fascinating thing about this is that invariably the thought of quality firearms brings to mind a very high dollar gun. And rightfully so, because the rule of thumb is that the more an item costs the better the quality.

But there is another definition of quality that should be of more interest to those of us who are of average means. Take, for example, the Fox Model BE 20 gauge shotgun that's the subject of this test. In this instance quality means a solid shotgun that delivers good, well distributed patterns directly to where the gun is pointed, and will do it year after year, hunting season after hunting season. To our way of

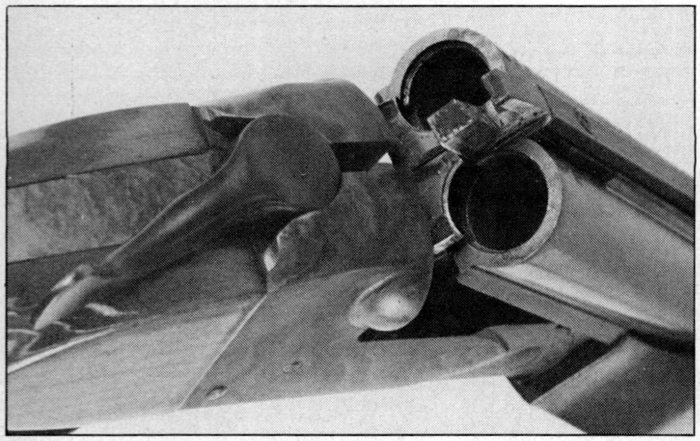

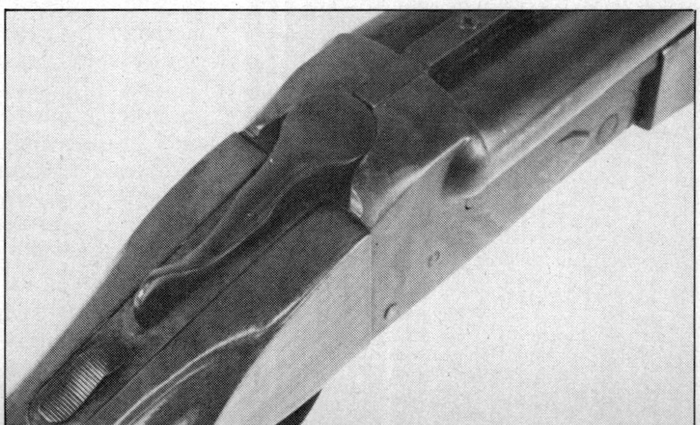

The Fox model BE features selective automatic ejectors, top left, as opposed to the Model B which has extractors only. Top right, upon opening the action the ejectors also lift unfired shells out of the chambers for easy removal. Above, the breech of the Fox double is beefy for plenty of strength and durability—the hallmark of Fox doubles. Safety, right, is sliding tang type.

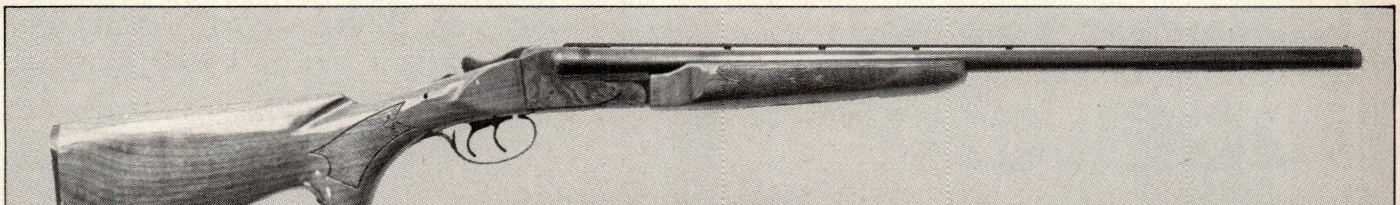

The Fox Model BE is a double trigger boxlock side-by-side with added features such as case-hardened receiver, raised ventilated rib, beavertail forend and ejectors.

thinking, that's also a quality gun.

The basic Fox Model B shotgun has been around since 1939 when the Savage Arms Company first introduced it. In 1930 they bought out the A.H. Fox Gun Company, which had been making fine quality firearms for many years. Today the only arm marketed under the Fox name is the basic Model B, the Model B-SE, and new this year, the BE. The primary difference among the three models is that the B is the "Plain Jane" version without automatic ejectors, and it's fitted with two triggers. The Model B-SE has the automatic ejectors, and a non-selective single trigger. The new model, the BE, has automatic ejectors and double triggers. All three versions share a ventilated rib and a color case-hardened receiver.

The stock and forend on the BE are what you would expect on a moderately priced side-by-side; good, honest American walnut that shows a decent grain pattern that ran, on the stock, straight through the pistol grip area. The piece of wood selected for the forend showed the same grain pattern as the stock. Checkering on both the stock and forend is impressed, and looks like the average impressed checkering job. The checkering on the pistol grip covers a small area, not enough to aid a shooter with cold hands in getting a good grip on the gun. However, that found on the forend is generous and the left hand can get a good hold. Dimensions of the stock are: length of pull, 14 inches; drop at the comb, 1½ inches, and drop at the heel, 2½ inches.

Fit of the wood to metal was fair in that there were few, if any, major gaps between the two. However, in some areas the stock protruded out past the metal, although certainly not enough to disturb anyone to any great extent.

The barrels themselves on the test gun measured 26 inches in length, meaning they were bored improved cylinder and modified. In 20 gauge there are also 28 and 30-inch tubes available, both lengths bored modified and full. All Model BEs are chambered for the three-inch magnum shells in all three of the gauges they're available in.

As stated earlier, the BE is fitted with a ventilated rib, something you don't normally find on a moderately-priced side-by-side shotgun. And the rib itself is fitted with two beads instead of one, definitely a plus feature to this shooter's mind. Dual beads, while perhaps a split second slower to line up on the target, will give better repeatability in getting the same sight picture time after time.

The rest of the gun is what you'd expect on a side-by-side shotgun. The easy portability, because of the quick takedown characteristics of the breed; the easily reached automatic tang safety that goes on every time the gun is broken open; and the extractors/ejectors that lifted the unfired shell about ¼-inch out of the chambers, throwing the fired hull about four feet behind the shooter. The ejectors on our test gun were interesting in that the one for the right barrel worked when it wanted to, while the left side operated perfectly every time. Dismantling them showed that some heavy packing grease had semi-hardened and was causing the one to malfunction from time to time. Cleaning it up cured the problem.

Other than that single item the BE performed perfectly. Hunting season wasn't upon us during the testing period, but the arm should still be in our possession when dove season breaks. With the Fox BE, and a little luck, the eating should be mighty fine!

In the hunting field, it's easy to see why the Fox is the world's most popular double.

SPECIFICATIONS

FOX MODEL BE

Manufacturer:	Savage Arms Emhart Ind., Inc., Westfield, MA 01085
Action:	Boxlock side-by-side
Gauge:	20 gauge (12 and .410 available)
Triggers:	Double triggers
Ejectors:	Selective automatic
Stock:	American walnut
Choke:	Imp/cylinder & modified
Barrel length:	26 inches
Weight:	6 pounds, 15 ounces
Price:	$278.85

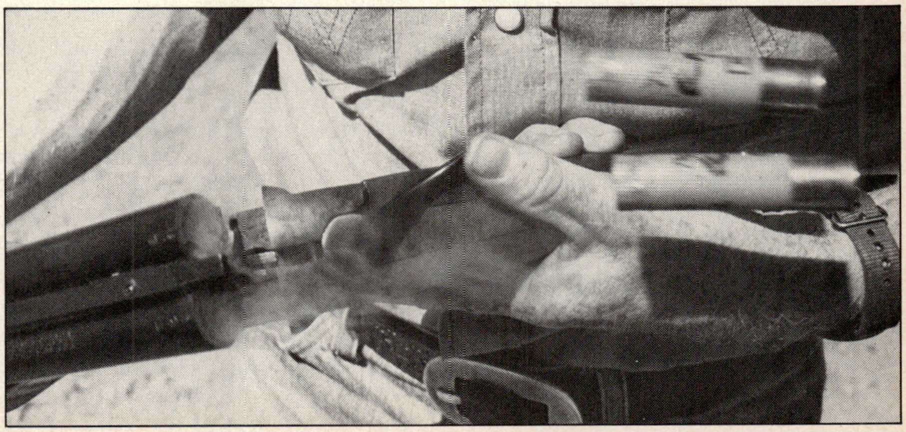

Upon initial testing, one of the ejectors "hung up." Hard packed grease was the cause, and after cleaning the ejectors functioned perfectly, kicking the hulls well clear.

Test Report

REMINGTON MODEL 870

Remington's "Wingmaster" has become today's classic pump shotgun!

By Craig Boddington

The slide action is one of America's greatest contributions to the evolution of modern firearms. While pump action rifles are not quite as popular as they once were, it's a fact that the pump shotgun is still the most popular action type with American bird hunters.

It's mighty fashionable in some circles to use a good double gun, whether over/under or side-by-side, and there's something about the feel of a double that few guns can match. However, when it comes time to get serious about bird shooting, that's the time I reach for my old pump gun! Like many shooters, I learned to shoot a shotgun with a slick old Winchester Model 12. Fashionable or not, it's a fact that I shoot a pump gun better than I shoot a double or autoloader.

There are some good reasons for this. The longer sighting plane of a pump or auto enables, at least for me, more deliberate swing and follow-through. For repeat shots, the act of "shucking" the action causes just enough hesitation to correct any sighting errors on the first shot. More importantly, as the hand moves forward to complete chambering a round, it acts as a finger pointing towards the bird or clay target. Of course, all this happens very quickly. Those who shoot a pump gun a great deal (and I'm one of them) will argue that a slick, well-broken-in slide action is every bit as fast as an autoloader.

Unfortunately, yesteryear's classic pump gun, the Model 12 Winchester, is gone. Today, though, there's another pump shotgun that has become the "old standby" for American shotgunners—the Remington Model 870.

Introduced in 1950, the Model 870 Wingmaster needs little introduction. In 1978 Remington turned out the *three-millionth* 870, and it continues to be a dependable, fine-shooting shotgun.

The Model 870 is available in various configurations and in all five gauges—12, 20, 16, 28, and .410. Our test gun was Skeet model in 28 gauge, as slick a little shotgun as you'd ever hope to see. The Lightweight 20 gauge and the 28 and .410 versions have a scaled down receiver, making the whole package lightweight, slim, and trim. Although seen very little away from the skeet field, the 28 gauge with its short shot string is a highly efficient bore size. In experienced hands for close in bird shooting, it makes a fine quail and dove gun, especially when mated with a fine-handling shotgun like the 870.

The test gun was a 28-gauge Skeet model with ventilated rib. It pointed and handled well, and the slide action was smooth and trouble-free, as expected of the famed 870.

Our test gun came up smoothly, and the stock fit was excellent. In appearance, the gun displays Remington's usual excellent finishing. Metal parts are blued, and the stock was of good grain with the glossy Du Pont RK-W finish.

Of course, the test of any shotgun is how it shoots. Not being much of a skeet shooter, I figured if I could break targets with the little 28 gauge, then it *had* to be a good-shooting gun!

Well, I didn't break 'em all, but I wouldn't have with a 12 gauge. After several rounds, scores averaged in the 21 or 22 out of 25 range, and the little gun handled well and swung smoothly. The action is of the twin bar design to prevent binding, and on doubles the second shot was there and ready without me being conscious of working the slide. That's the way a pump gun should work!

The Skeet models come equipped with a ventilated rib and a two-bead sighting arrangement; field models come either plain barrel or with rib at an added cost.

We were mighty impressed with today's version of the time-tested 870. It has the reputation of being a dependable "workhorse" of a shotgun, and it's not bad in the looks department, either! With five gauges to pick from, right or left hand versions, field models, trap and skeet models, engraved models, rifle-sighted slug barrels, and a wide variety of barrel lengths and chokes, there's an 870 to fit any shooter's needs!

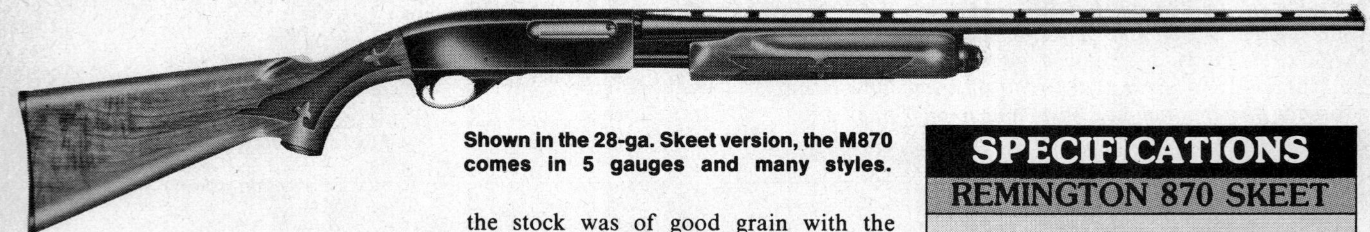

Shown in the 28-ga. Skeet version, the M870 comes in 5 gauges and many styles.

SPECIFICATIONS

REMINGTON 870 SKEET

Manufacturer:	Remington Arms Company, Inc. Bridgeport, CT 06602
Gauge:	28 (12, 16, 20, .410 also available)
Action:	Slide action
Capacity:	Five rounds
Sights:	Ventilated rib, twin beads
Weight:	Six pounds
Barrel length:	25 inches
Choke:	Skeet
Overall length:	45 inches
Stock:	Checkered American walnut
Price:	$316.95

Chapter 4

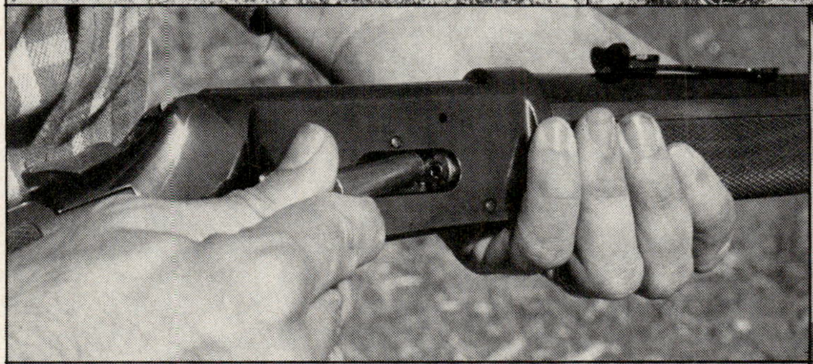

Rifles

Like groceries and gasoline, guns have gone up in the past few years! **Guns & Ammo's** *"rimfire expert" tells how to buy . . .*

.22 RIFLES ON A BUDGET

By John Lachuk

During these recession-ravaged days, many of us are obliged to search out the greatest value for our rapidly shrinking dollars. Time was, three ten-spots ransomed a fair to middlin' smallbore longarm. Today, three times three portraits of poor Hamilton barely do the job!

However, a rundown of recession rimfire rifles priced near or under the century marker brings to light some highly functional firearms. Admittedly, that old dictum, "You get what you pay for," applies with equal vigor in the field of firearms, but the sacrifices are mostly cosmetic. The percentage of plastic per rifle increases, usually in the areas of the triggerguard and sights. Metal surfaces display little polishing, and tool marks are abundantly evident! Perhaps exemplifying the best metal finishing to be found among budget priced rimfire rifles is the Marlin/Glenfield line, with their high-finish ground barrels, accented by a deep blue-black.

When advertising includes such phrases as, "genuine walnut *finish*," (italics by the author) rather than simply, "genuine walnut," likely the maker is attempting to pass off birch as walnut via tricky semantics. Often tell-tale patches of pale yellow give it away by peeking through the deep brown finish where careless buffing cut through on corners and edges. If in doubt, remove the receiver/barrel assembly from the stock and look into the inletted channel where no finish masks the truth.

The fact is, birch need make no apologies in terms of utility! Although it lacks the rich color and contrast of walnut, birch is close-grained and hell for strong! Generally, "grain" in the decorative sense is nonexistent, but two of the Mossberg rifles sent for testing actually had very attractive grain in their birch stocks, reminiscent of hard rock maple widely used by frontier gunsmiths on Kentucky rifles. Birch is perfect for machine turning because it cuts cleanly, without chipping or tearing as walnut so often does. The sanding and finish on most low-priced rimfire rifles is regrettably spotty, whether walnut or birch. Some of the stocks appear to be finished without any effort at sanding to remove tool marks. Happily, this is an area where any enthusiastic do-it-yourselfer can work wonders with sandpaper, stain and varnish.

Nearly all budget rimfire rifles can benefit greatly from some amatuer gunsmithing. They lack most of all the human touch! Many of them look as if they were extracted from the tape-controlled screw machines still dripping cutting oil and crammed into a box for shipment. Turning bolt actions often grate like a hand-cranked coffee grinder and autoloaders with tubular steel receivers are apt to function as if fighting a fistfull of sand. Both can be corrected by wrapping some medium-grit carborundum cloth around a wooden dowel large enough to almost fill the hole in the receiver, and working out the worst of the circular tool marks. Follow up with wet-or-dry paper of 400 then 600 grit. Secure the bolt or breechblock in a bench vice and polish all accessible working surfaces, inside and out.

Trigger pulls are apt to be disappointing on low cost rimfire rifles, exhibiting excessive takeup, creep, and backlash. Luckily, most trigger troubles yield readily to some talented tuning. You'll need a triangular fine-cut Swiss file, a triangular medium India slip stone, and a knife-edged hard Arkansas, to true and smooth the sears, lightening the weight of pull and removing any roughness. However, the long creepy trigger release will remain. Creep can

Semiautomatic rifles are preferred by countless plinkers, and two of the best are the Ruger 10/22 (right) and Remington's Model 66.

With a scope like Leupold's Compact 2½X, the tubular magazine semiauto Marlin Model 990 is an effective small game rifle.

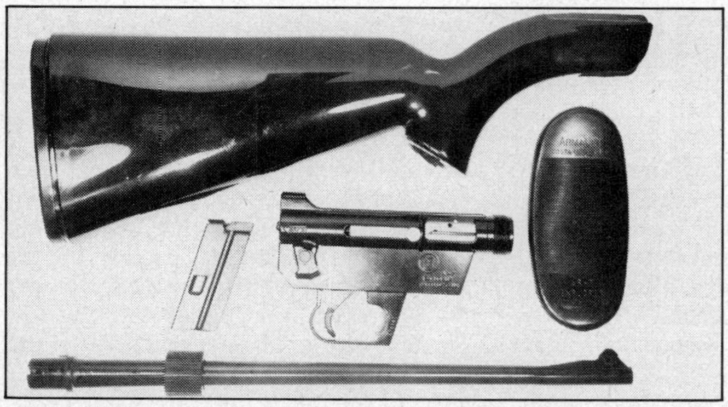

For backpackers, boaters, or pilots, the Charter Arms AR-7 makes a lot of sense. It breaks down into three units—receiver, barrel, and magazine—and the parts stow into a waterproof stock that floats. The total weight of this semiauto is a mere 2½ lbs.

be reduced by minimizing the depth of sear engagement—*consistent with the dicates of safety!* Don't remove so much metal that the safety is crippled, or the gun fires when jarred.

Exemplifying some of the best fit and finish to be found among Ford/Chevy class rimfires are the Marlin/Glenfield rifles. The 5½-pound Model 60 Glenfield with 18-shot under-barrel tubular magazine, and the 4½-pound Glenfield Model 70 with chrome plated seven-shot clip, share the same autoloading action, housed in a streamlined satin-black alloy receiver marked by its extra large ejection port and full-length male dovetail for tipoff mount scopes. The large brightly polished square breechblock carries an equally ample cocking handle with a large frontal area for easy-on-the-fingers operation. With the bolt retracted, a push inward on the cocking knob locks the bolt open—a vital safety feature on any auto action! No need to polish the inner workings on these guns. The finish is fine!

The bores of the Glenfield Model 60 and Model 70 must be cleaned from the muzzle. Disassembly for cleaning the action requires the removal of two screws through the stock, fore and aft of the receiver. The satin black alloy triggerguard with chrome plated trigger remains in the stock. Four more screws secure the action assembly to the receiver. Removal allows lifting out the breechblock for a thorough cleaning.

All Glenfield rifles share stocks of sturdy birch, tinted to walnut hue, and well finished, shaped with full pistol grips and Monte Carlo combs. The rifle length Model 60 with 22-inch barrel, is adorned with an impressed carving pattern on forend and pistol grip. The carbine length Model 70 with 18-inch barrel, has the carved squirrel and groove pattern on the grip only. The forend is accented with a barrel band and sling swivel.

All Glenfield guns retail well under the mythical $100 barrier, including the bolt action 5½-pound Model 20 with seven-shot clip and the single-shot Model 15, and I'll personally guarantee you more than your money's worth with any one of them!

If you aspire to loftier heights, the Marlin Model 990 and Model 995 closely resemble the Glenfield Model 60 and Model 70, with the added assets of Marlin's famed "Micro-Groove" barrels, gold rather than silver triggers, ramp mounted flat-faced gold bead front sights under oval "Wide-Scan" protective hoods, backed by high quality semi-buckhorn folding leaf sights with elevator adjustment, plus genuine American black walnut stocks graced by classic checkering patterns.

For low cost entry into the field of the .22 Magnum, it's hard to beat the six-pound Marlin Model 783, turning bolt action with cock-on-opening internal striker, housed in a grooved steel receiver. The full-length tubular magazine under the 22-inch medium weight Micro-Groove barrel unfailingly feeds 12 .22 Magnum rounds.

The action on our test unit was notably smooth and easy working. The attractively swept-back bolt handle offered easy uplift and retracted without any grating. Internal finish of the tubular steel receiver and the outside of two-piece bolt were of a gratifyingly high order! Lockup is effected by a massive rectangular lug at the base of the bolt handle.

The trigger on the test rifle broke cleanly at 4½ pounds, without takeup and with minimal backlash. The separate sear design results in crisp release. A handy thumb lever right/rear of the receiver firmly locks the trigger and sear.

The genuine American black walnut stock on the Marlin Model 783 offers eye-appeal and good fit, with man-sized proportions and crisp lines, from the rakishly-angled forend to the white-lined black butt plate. The year 1980 witnessed a sensible shift to classic checkering patterns front and rear, although our test unit still had the old-style carving pattern. The 783 also boasts sling swivels and a handy leather carrying strap! Marlin's "Mar-Shield" semi-gloss stock finish provides adequate protection and accents the attractive walnut grain.

In terms of design excellence and overall finish on both wood and metal, the Marlin Model 783 represents a real bargain on today's market! During testing, over a period of months with many rounds fired, it never failed to function perfectly. All components save for the alloy triggerguard and brass inner magazine tube were reliable steel! Accuracy was totally outstanding!

Marlin should be lauded for paying attention to the small, often overlooked needs of the shooter. For example, Marlin hardens the trigger sears to prevent wear that could ruin the pull, or worse, make it dangerous. They punch a hole in their seven-shot clip to show when it is fully loaded. And they place a spring-loaded collar at the tip of their tubular magazine to avoid rattles and keep it firmly locked.

Long famed for their broad line of modestly priced rimfire rifles, O. F. Mossberg & Sons, Inc., continue dominant in the field, offering their novel Model 353V with Tenite fold-down forend and unique Model 377 "Plinkster" with right-hand-only thumbhole stock molded of polystyrene structural foam, surfaced to resemble the color and texture of walnut. Both autoloading rifles share the same steel tubular, full-length grooved receiver with polished steel breechbolt and internal striker, feeding .22 Long Rifles (only) from a seven-shot clip and 15-shot butt stock tubular magazine, respectively.

The 5½-pound Model 353K has been in the Mossberg line so long as to be a tradition. The black tenite two-position stock extension offers an optional military style grip, especially handy when firing in prone position. A military type black web carrying strap attaches to fixed swivels on the left side of the birch stock, with Monte Carlo comb, impressed checkering, and white-lined black butt plate. In keeping with the carbine concept is the abbreviated 18-inch barrel, topped with ramped black bead front and step-elevator adjustable rear sights. A lug dovetailed into the barrel about ¼-inch ahead of the receiver holds the stock and action/barrel assembly together.

During testing, the Mossberg Model 353K functioned unfailingly, and delivered a high level of accuracy with its "AC-KRO-GRUV" rifling. The action was smooth-working, and locked open with ease by pushing in on the cocking knob. Happily for this southpaw shooter, the action threw back but little flak and unburned powder. Cases ejected forward at 45 degrees, thus sparing the shooter at the bench to my right from hot brass under the collar. The trigger release was an only slightly creepy 4½ pounds.

The Mossberg Plinkster shares the same basic action, but loads through a trough low on the right side of the butt stock into a tubular magazine that has given us some problems. The two units that we have tested over the passed several years have both attempted now and then to deliver cartridges to the chamber in pairs rather than singly. We traced the problem to a sluggish cartridge cutoff and were able to correct it with some tinkering.

Hefting 6½ pounds with a full 13½-inch length of pull, the Plinkster is certainly no "kid's gun!" It is an intriguing adult's toy, for informal target practice or small game hunting. The 20-inch barrel of the Plinkster lacks open sights because the rollover Monte Carlo comb is too high. Mounting a scope on the grooved receive with ordinary tipoff mounts results in a line of sight too low for the comb. Mossberg corrects the problem by shipping a stilt-like double dovetail mount to raise the scope. Alas, they also include a ¾-inch tube 4x15

Budget rimfires tested included (left to right): Charter Arms AR-7, Glenfield Model 20, Marlin Model 783, Marlin Model 990, Marlin/Glenfield Model 70, Mossberg Model 640K, Mossberg Model 340B, Mossberg Plinkster, Mossberg Model 353T, and Remington Model 581.

mm scope that is somewhat inadequate for the job! The forend of the Plinkster is molded with sharp diamond checkering and QD sling swivel studs are included. The rather creepy triggers on the two test rifles weighed 6 pounds and 6½ pounds.

Some of the most accurate .22 Magnum groups in my experience issued from the budget-priced muzzle of a Mossberg Model 640K, with turning bolt action that incorporates a rotating hammer in the manner of a "hammerless" autoloading pistol, cocking as the bolt is pulled back to extract the empty case. New rounds feed from a five-shot clip. The one-inch diameter grooved steel receiver is capped in plastic, to round off the rear and act as a bolt stop. To remove the bolt from the receiver, the single slotted-head screw that secures wood to metal just ahead of the receiver must loosened and the stock separated, exposing the trigger/hammer mechanism contained in a stamped steel housing bolted under the receiver. The trigger connects with the rotating hammer via a horseshoe-shaped strut that functions as the sear, offering a reasonably crisp release that weighed 6½ pounds.

The fully-proportioned birch stock

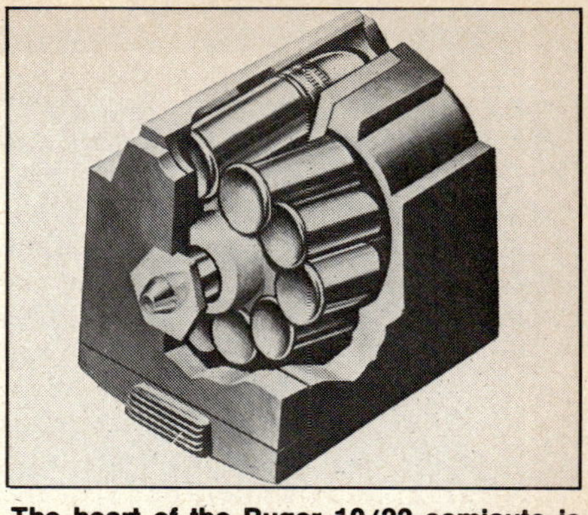

The heart of the Ruger 10/22 semiauto is the unique 10-shot nylon rotary magazine.

RECESSION RIMFIRE RIFLES ACCURACY CHART

GUN	AMMO "A"	AMMO "B"
Charter Arms AR-7	1⅝	⅞
Glenfield Model 20	1¾	1
Glenfield Model 60	1¼	1½
Glenfield Model 70	1 1/16	⅞
Marlin Model 990	⅞	1⅛*
Marlin Model 783 (.22 Mag.)	⅜	½
Mossberg Model 340B	11/16	⅞
Mossberg Model 353K	1½	1⅜
Mossberg Model 377	1	1
Mossberg Model 640K(.22 Mag.)	5/16	5/16
Remington Nylon 66	13/16	1
Remington Model 581	¾	⅝*
Ruger Model 10/22	⅝	11/16
Stevens Model 125	1⅜	
Stevens Model 246	1⅛	1
Stevens Model 887	1⅜	

Half a dozen various brands of ammunition were tried in each of the test guns. The groups listed were among the best delivered, regardless of ammo brand. Thus it became convenient to merely list them as "A" and "B".
** Five-shot groups. All others were ten shots, fired from a bench rest at 25 yards.*

For a little extra cash, the Ruger 10/22 is available in a sporter version with a shotgun styled buttplate, checkering and swivels.

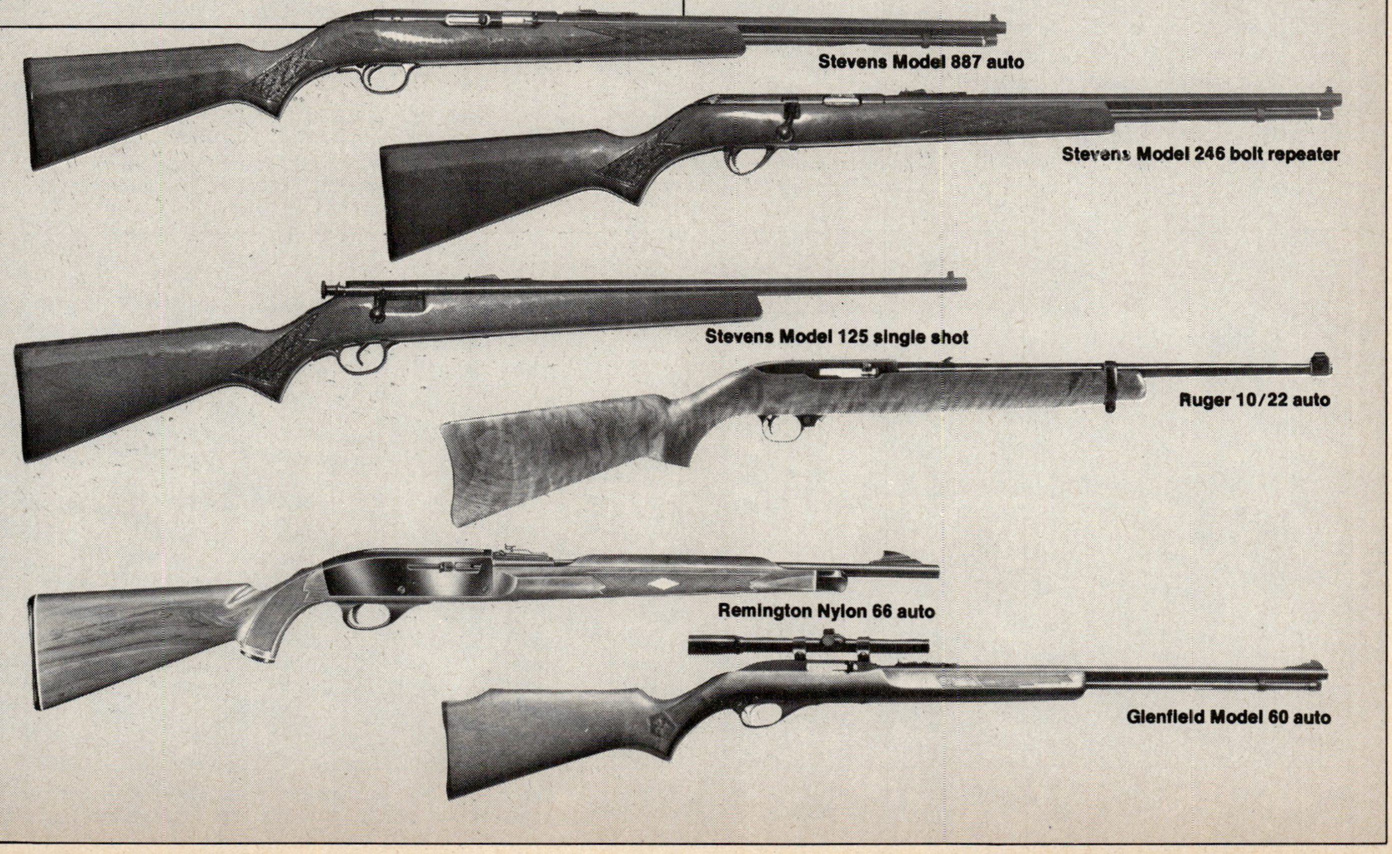

Stevens Model 887 auto

Stevens Model 246 bolt repeater

Stevens Model 125 single shot

Ruger 10/22 auto

Remington Nylon 66 auto

Glenfield Model 60 auto

boasts impressed checkering on forend and pistol grip, white-lined cap and butt plate, and fixed sling swivels, plus scope-height Monte Carlo comb. Conventional open sights top the medium heavy 24-inch barrel. Dubbed the "Chuckster" by Mossberg, the Model 640K is a man-sized six pounder, well fitted for field or trail.

The Mossberg Model 340BD echos most of the features of the 640K, sans checkering and white-line spacers, chambering .22 Long Rifles only from a seven-shot clip. Unique among field-styled rimfire rifles are the hooded ramp front sight offering a choice of aperture or post that rotate into position and lock with a small screw, and the Mossberg S330 micrometer rear sight with finger adjustable quarter-minute windage and elevation knobs.

As of this writing, the Remington autoloading "Nylon 66" MB sells for just a nickel under $100.00, and the bolt action Remington Model 581 clip-loader and Model 582 with under barrel tubular magazine cost more than one hundred depreciating dollars. However, these .22 Long Rifle chambered guns represent real bargains in terms of design excellence and endurance!

The full-length stock of the all-but-indestructable Nylon 66 is molded of DuPont "Zytel" structural nylon, grained to simulate walnut, with sharp diamond checkering on forend and pistol grip. The only metal parts on the four-pound rifle are the 19⅝-inch steel barrel, internal action parts, and the stamped steel action cover, which affords a groove for mounting rimfire scopes. The triggerguard, serrated shotgun-style safety and trigger are black plastic. An outer-space-styled front post is backed with an adjustable open rear blade. Disassembly of the Nylon 66 is not recommended because it's harder to get back together than grandpa's double spring alarm clock! The trigger on our test rifle offered a mushy 4½-pound pull. Despite that, accuracy was far above expectations. Light weight makes the Nylon 66 a good .22 trainer for smallfry. Its ruggedness makes it a good emergency arm for pickup or boat! And it's hard to beat for scaring hell out of local rabbits! The stock-enclosed tubular magazine holds 14 Long Rifle cartridges.

The Remington Model 581 is a straight forward no-nonsense bolt action, with possibly the best trigger to be found on any medium priced rimfire rifle! Our test sample was free of take-up or backlash, and released crisply at just 4½ pounds. The big bore-styled mechanism takes the weight of the firing pin spring off of the trigger with a separate sear, allowing a shallow engagement and crisp release with complete safety. The thumb safety doubles as the bolt release, freeing the trigger of that chore, allowing zero backlash! Also, the lever *locks the sear* rather than just the trigger for greater security.

Remington's unique interrupted thread artillery-styled locking system places three sets of double lugs about amidships on the bolt, plus matching shoulders in the forged steel receiver, resulting in one of the strongest .22 rimfire actions ever made! The smooth-faced bolt totes dual extractors. The counterbore to enclose the case head is in the barrel. A firing pin fall of less than ¼-inch provides lock time under two milliseconds, further contributing to accuracy.

The full-sized walnut stained hardwood stock of the Model 581 sports a Monte Carlo comb and full pistol grip. Totally lacking in adornment, it is beautiful in its simplicity! With 24-inch barrel, and weighing approximately five-pounds, the five-shot clip-loading Remington bolt action certainly merits serious consideration from adults as well as aspiring young nimrods.

First conceived as a "spittin' image" .22 trainer for the Ruger Model 44 Carbine, the Model 10/22 Carbine in .22 Long Rifle caliber soon outdistanced its big bore brother in popularity. Just over a yard long overall, and weighing but five pounds with its 18½-inch tapered sporter barrel, the Ruger 10/22 is handy and quick swinging! Designed with famed Ruger ruggedness, the 10/22 is a

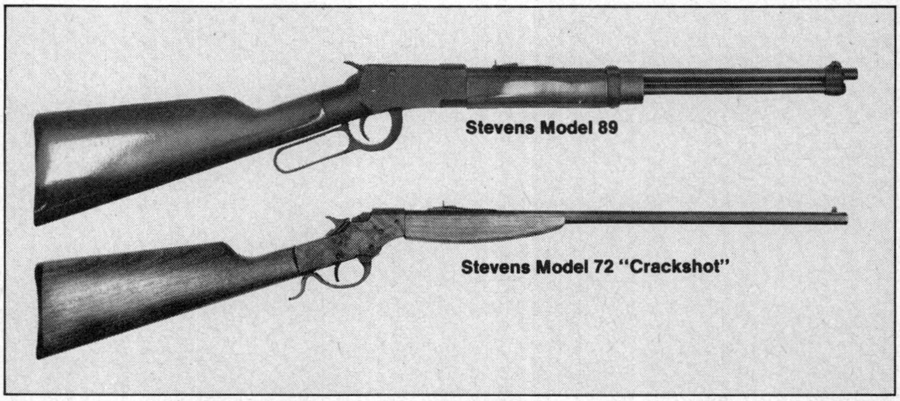

Savage/Stevens offers two economical single-shots, perfect for training youngsters. The Model 89 (top) looks like a lever-action saddlegun. The Model 72 is a falling block.

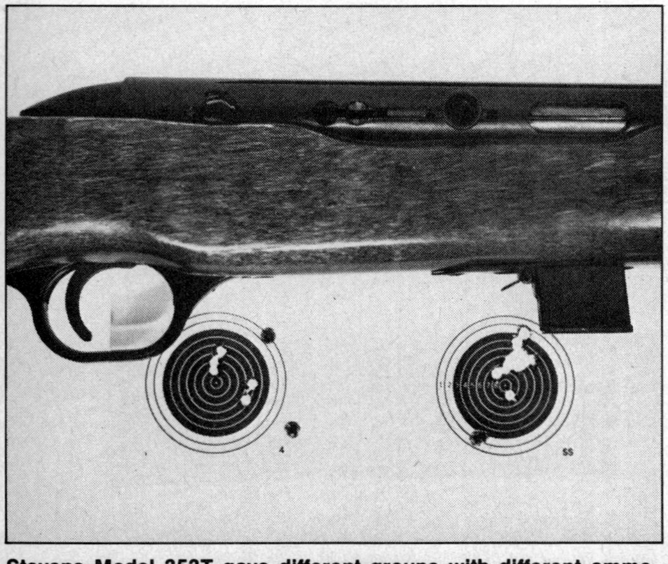

Stevens Model 353T gave different groups with different ammo. The best group—at right—was shot with Remington High Velocity.

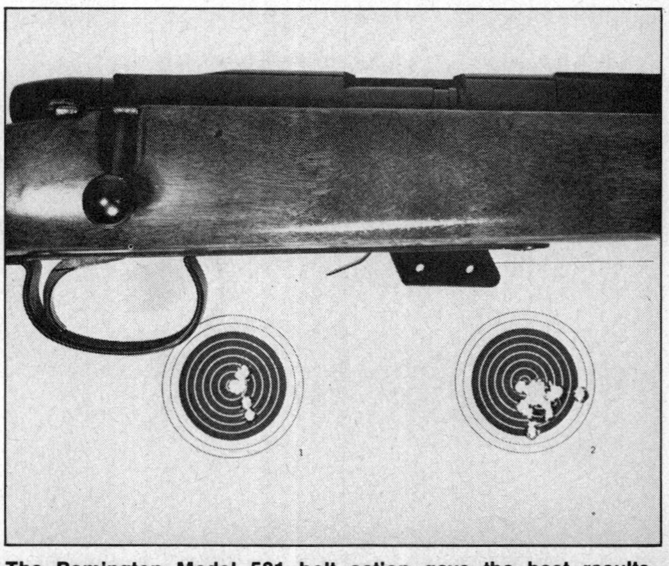

The Remington Model 581 bolt action gave the best results. 10-shot groups at 25 yards were more than adequate.

autoloader that seldom sees a repair shop! A unique rotary magazine reminiscent of the Mannlicher-Schoenauer allows cramming ten rounds into the plastic detachable magazine without making it jut out beneath the stock. The alloy receiver is not grooved, but it is drilled and tapped for a dovetailed scope mount included with the rifle. A post front sight and folding leaf rear top the tapered barrel. A crossbolt safety rides the alloy triggerguard ahead of the trigger, with an ingenious bolt-hold-open latch immediately ahead of that.

In its original version with barrel-band forend and crescent butt plate, the Ruger 10/22 still sells near the century marker. The newer 10/22 Deluxe Sporter with shotgun-styled butt plate, on a hand checkered American walnut stock with scope-height comb and sling swivels, sells about $25 higher, but is well worth the difference. Our test sample had a better-than-average trigger pull weighing 4½ pounds. Although not a high-gloss piece, the Ruger 10/22 is well finished inside and out!

Noted for bringing to America the superb line of Anschutz target and sporting .22 rimfire rifles, Savage covers the other end of the smallbore spectrum with their wide array of Stevens sporting rifles. Currently enjoying a resurgence in popularity is the recreation of the famed Stevens "Crackshot" falling block single-shot, with 22-inch octagonal barrel and turn-of-the-century classic styling, distinguished by its color case-hardened receiver plus soft-luster oil finished genuine walnut low-comb stock and abbreviated forend. The Stevens Crackshot is an open-sights-only rifle, as charming and appropriate in our modern melange as Tiffany lamps!

Another Stevens single-shot is the Model 89 lever action with Martini-styled breechblock and automatic ejection. Styled in the Western image with action lever and dummy tubular magazine under the barrel, the Model 89 has wide appeal with budding young "cowboys." The super-safe manual cocking hammer makes it a fine teacher rifle for beginners.

Both Savage single-shots handle the entire line of .22 rimfire ammo, from BB Caps to Long Rifles, and both are priced well within range of the limited budget. The 5½-pound Stevens Model 34 loads .22 Shorts, Longs or Long Rifles from a five-shot clip, by means of a turning bolt action with a rotating hammer which is cocked by retracting the bolt rather than on the upstroke. Conventional open sights top the 20-inch barrel, backed with a steel receiver grooved full-length for tipoff mounted scopes. The birch stock boasts Monte Carlo comb and impressed checkering.

The Stevens Model 246 is identical save for the tubular magazine which dispenses 22 shorts, 17 Longs, or 15 Long Rifles. Also, it has the classic comb rather than a Monte Carlo hump at the back. The same classic stock configuration and 20-inch barrel is shared by the Stevens autoloading Model 887, with under-barrel tubular magazine holding 15 Long Rifles only. Its somewhat longer scope-grooved tubular steel receiver houses a conventional blowback bolt with internal striker. Despite its utter simplicity, the action contrives to have a better-than-average trigger pull with negligible takeup and backlash, and a relatively crisp six-pound release. Slots on both sides abaft the chamber deflect much of the cheek-stinging shrapnel dispensed by many autoloading rifles. The action locks open by pressing in the large plastic cocking handle. The Stevens Model 125 bolt action single-shot weighing five pounds with 22-inch barrel gets back to basics, with a manual cocking knob and rebounding safety lock. Obviously designed as a small fry rifle, it is also available as the Model 125 (youth) with an inch trimmed from the butt.

Despite the frugal prices attached to the Savage/Stevens line, which is aptly billed, "super values" by Savage, the rifles exhibit deep bluing over a medium-gloss polish on steel surfaces. Alloys and plastics are held to a minimum. Internal working surfaces are smoother than one has a right to expect in this price range. The birchwood stocks feature simple impressed French checkering in good taste, and are well sanded to a smooth surface. The coating completely fills the pores, resembling a fine furniture finish.

Discerning smallbore buffs can step up from Stevens to Savage by opting for the Model 65-M .22 Magnum with five-shot clip, resembling the Stevens Model 34, but with higher gloss steel finish, improved open sights, and genuine American walnut stock. The Savage Model 80 .22 Long Rifle chambered autoloader resembles the Stevens Model 887, also upgraded as above.

Although designed primarily as a survival rifle, the Charter Arms AR-7 Explorer serves ably as an all-around plinking small game gun, and it still retails under $100! With alloy receiver, steel-sleeved alloy barrel, and eight-shot steel clip stowed in the high impact cycolac plastic stock capped by neoprene butt plate, the 2½-lb. gun makes a large but light lump in your packsack. Should the need arise, the AR-7 assembles in seconds!

In general, it pays to buy the best that you can afford. Having more money to work with allows firearms makers to enforce more stringent quality controls and use better materials. However, most of the budget-priced rifles tested by the author proved to be entirely functional—lacking mostly cosmetic touches. If you're burdened with a "beer budget," you won't suffer any great deprivation by shooting a sturdy, reliable "beer budget" rifle!

Bolt actions, like the Marlin M781, are excellent for youngsters. A good, low-cost rimfire will still serve them when they get older.

The fold-down forend of the Mossberg Model 353T is very handy. It's an excellent feature in prone and offhand shooting position.

Test Report

REMINGTON MODEL 788

Remington's budget tackdriver—the Model 788—has been given a new lease on life, with an altered design and a highly efficient new cartridge! By Craig Boddington

The Remington Model 788 is the kind of rifle that grows on you—the longer you fool with it, the more you like it. I tend to have a split personality when it comes to my taste in rifles. I dearly love a classic-style bolt gun with good wood and fine checkering, yet, being first and foremost a hunter, I have a built in admiration for a functional, no-frills working rifle.

Thus it was with mixed feelings that I unwrapped the newest version of Remington's popular Model 788 bolt action. I trust my friends at Remington will forgive me when I say the rifle wasn't pretty. It had a nice blued finish, and the stock had a good straight grain and clean lines, but it wasn't a rifle that immediately drew gasps of appreciation. It *did* look extremely efficient and functional. I lifted it from the box, inserted the bolt, and hefted the rifle to my shoulder. The wide forend and stout pistol grip felt good, and the short 18½-inch barrel swung easily. The rifle felt more than a little good. In fact, it felt real good, and I started to get a real itch to try it out.

Our test gun was chambered for the new 7 mm-08 Remington, basically a .308 Winchester case necked down to the 7 mm (.284) bore diameter. It's a fairly common saying these days that we have enough cartridges as is and don't need any more, and I'd be the first to admit that we've got a lot of rounds on the market that duplicate each other's performance. Still, Remington's new cartridge has a lot going for it. For one thing, the 7 mm is in itself an extremely efficient caliber. The bullets in that bore diameter offer excellent sectional density and ballistic coefficients. Better, actually, than the .30 caliber bullets we Americans are so fond of. The new cartridge, then, can offer the proven performance of the 7 mm coupled with the short .308 case capable of working through the shortest actions.

The only current factory loading is a 140-grain Pointed Soft Point Core-Lokt, with factory ballistics quoted at 2860 feet per second from a 24-inch barrel. Remaining velocity at 300 yards is a respectable 2189 f.p.s. Sighted to be "dead-on" at 200 yards, the bullet will print two inches high at 100 yards and only eight inches low at 300 yards, making the new cartridge an excellent medium range deer-getter.

Of course, our test rifle didn't come with a 24-inch barrel. Instead, it had an 18½-inch carbine length barrel, and in fact that's the only barrel length available in that caliber on the 788. Needless to say, one loses some amount of velocity when you remove 5½ inches of barrel. However, that short case is very efficient, and no doubt Remington is loading a fairly fast-burning powder. The velocity loss is not as great as one would expect; our chronographing revealed velocities averaging 2760 f.p.s. That's still stepping right along, and the difference in trajectory is negligible.

The iron sights on the test rifle were certainly adequate, and our initial groups fired with the iron sights at 50 yards were quite respectable. However, a flat shooting cartridge like that deserves a scope, and the rifle itself deserved the chance to deliver its best accuracy. We topped the rig with a Weaver K4 scope with duplex crosshair, mounted with two-piece Weaver bases and rings, and now the outfit looked like a complete hunting rifle.

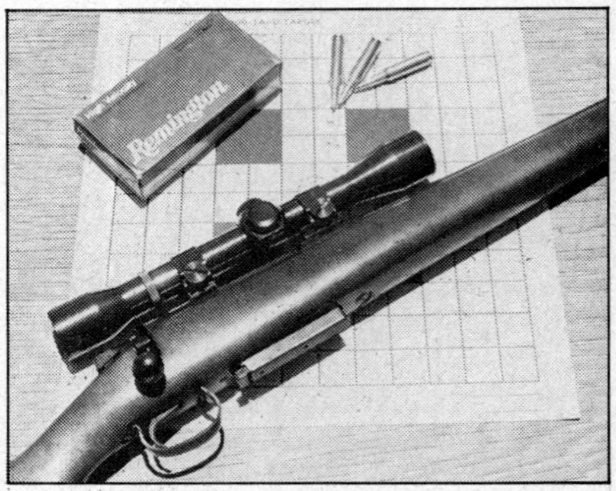

The shortened version of the Model 788 is now available in 7 mm-08, a necked-down .308. Accuracy was quite acceptable.

With the 140-grain factory loads, recoil off the bench was very mild. The rifle does not have a rubber recoil pad, but it doesn't need one—it's very comfortable to shoot as is. After a few shots to zero the rifle, we sat down to see how well it would group. And that proved to be very well indeed! Three shot groups consistently stayed between ¾ and one inch.

By now I was getting mighty impressed with this little "workhorse" of a rifle, and I decided to try a little offhand shooting. The gun handled so well that I was curious to see if I could shoot it off my hind legs. On our test range we have an 18-inch iron gong hung out at 250 yards, and it's a pretty good trick to hit it consistently offhand.

The Remington Model 788 features a detachable magazine and rear-locking bolt. The safety is positioned for easy access.

Topped with a Weaver K-4 scope and Weaver mounts, the 788 was accurate; the 7 mm-08 round was pleasant to shoot.

The carbine-length 788, right, was a joy to carry in the field and proved fast-handling for brush shooting. Above, Don Pine, left, and the author pose with a hefty California wild pig. The 788 in 7 mm-08 should prove a top combo for deer-sized game.

Still, the rifle feels naturally steady, and the four pound trigger pull is crisp and clean. I missed the first shot, and then rang the gong four times in a row.

After all of that, I was anxious to take the rifle where it really belonged—hunting! If I had to class the 7 mm-08 cartridge, I'd call it a deer cartridge. Being late June, there wasn't much chance to hunt deer with it. However, wild pigs are open year 'round in California, and offer challenging, exciting hunting in the coastal mountains.

Durwood Hollis, head of the Magnum Country sporting equipment, and I left our homes in Los Angeles about midnight and drove north to some private land in the Intercoastal Mountains, where we met California Fish and Wildlife Game Manager Don Pine at dawn. I loaded up the Remington's three-shot detachable magazine and we hiked into the hills.

The underbrush was thick, predominantly dense juniper, and I was immediately grateful for the shortness of the Remington.

Just when the sun was getting uncomfortably warm, we started into a steep draw and out jumped three deer. One, a little fawn, bounded across the draw and started up the opposite side only to jump smack on top of a bedded pig. The pig rose out of his bed like a whale surfacing, and the fawn was so startled it fell over backwards. Our pig faded from sight before we could shoot, but we knew he had stayed within a small circle of brush.

Eventually the pig smelled a rat and left his bed, taking three steps to my right. He stopped behind a screen of brush, and I was reluctant to shoot without an absolute certain shot. Don had him dead to rights, and he did the sensible thing. That is, he dropped him with a perfectly placed shoulder shot from his .30-06.

And that, unfortunately, was as close as I got to shooting anything with the 7 mm-08. At least so far.

So shoot something with the new rifle we did not. We did, however, take it hunting, and I remain convinced that the 788 is as fine a hunting rifle as can be found. It's not fancy, and certainly not ornate. But I'll change my initial statement. It is pretty in a simplistic, functional sort of way, and after hauling it around the hills a bit, it looks a whole lot better to me. As to the cartridge, it's not necessary to shoot a lot of game with it to know it will do the job. The caliber is proven, the bullet weight is adequate for deer-sized game, and the velocity is quite sufficient. And the Remington Core-Lokt design has an enviable reputation in the hunting world. The new 7 mm-08 will do the job, and mated with the shortened 788 and it's newly designed stock, it's our opinion that Remington has a winner.

SPECIFICATIONS

REMINGTON MODEL 788

Manufacturer:	Remington Arms Co., Inc. Bridgeport, CT 06602
Caliber:	7 mm-08 Remington
Barrel Length:	18½ inches
Overall length:	38⅛ inches
Weight:	7¼ pounds
Trigger Pull:	3¾ pounds
Action:	Bolt Action
Magazine Capacity:	3 plus 1 in chamber
Sights:	Blade front, adjustable rear
Finish:	Blue
Price:	$209.95

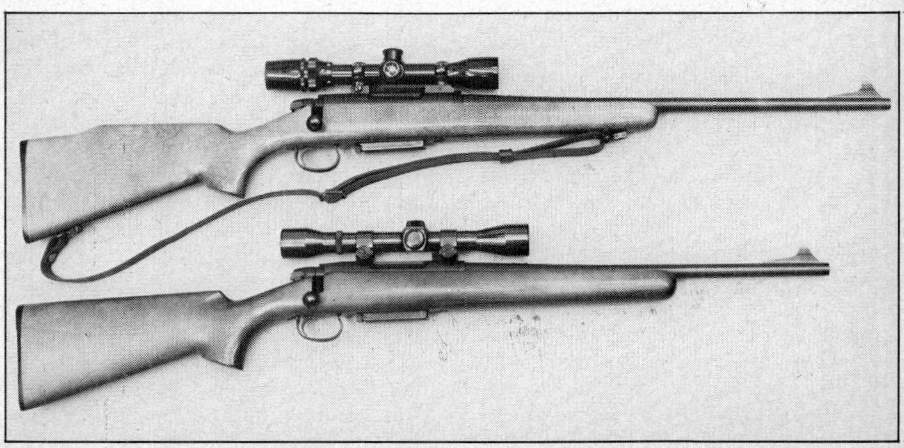

The main differences between older Model 788s, top, and the new version, below, are a redesigned stock with thicker forend and the 18½-inch barrel on big game calibers.

The North American Magnum

The term "magnum" means different things to different people. Most shooters, especially big game hunters, see all belted cases as magnums, but this isn't completely correct, either as a designation or performance. There are magnum shotgun shells with the term sometimes referring to the shell, and sometimes only to the load within the shell. There are magnum handgun cartridges that do not have belts, but are slightly larger than their brethren in the same calibers, and a lot more potent. There is the .22 Winchester Rimfire Magnum that is a true magnum in the .22 RF class, but there is no belt. And there are a number of belted cases that have "magnum" headstamps that do not fit the name. There are also cases that have magnum powder capacity and give magnum performance in rifles that do not have a belt and/or the magnum headstamp.

Both Webster and Thorndike define the word magnum as meaning a wine bottle with a capacity of roughly two quarts, with no mention whatever of the cartridge. Perhaps this is why we refer to magnum as "big jug" cases.

The cartridges discussed here are all of the belted magnum variety, but, as will be noted in some instances, they do not give magnum performance when compared to so-called standard capacity cases. One other point that should be cleared up right at the start, is that even if a cartridge is a true magnum in performance in its caliber, this does not necessarily give it superior performance to standard cartridges of other calibers. As an example, the .240 or .257 Weatherby Magnums will not perform as well in some respects as standard numbers such as the .270 Winchester.

As this rundown of magnum cartridges covers only those suited to hunting North American big game, the .240 Weatherby is the logical place to start. Since the advent of the .243 Winchester and the .244/6 mm Remington, there is little room for argument against them as good performers on deer-size game. When Roy Weatherby came along with the .240 Weatherby Magnum in the neat little belted case that added some 10-gr. of powder capacity and 200 fps to the 6 mm Remington cartridge, it was a true magnum development that boosted the efficiency of the .243 bore to a new performance height. This velocity boost did not increase cartridge performance greatly as far as game heavier than deer is concerned, but it did make it easier to hit the deer, antelope or whatever in a vital spot at longer ranges, and the same bullets

When it comes to stopping power and flat trajectory at long-range, the modern magnum rifle is the answer on this, or any other continent! By Bob Hagel

packed more punch for surer kills when they got there.

I had killed a good many head of deer-size game with the 6 mm Remington cartridge and various big game bullets before the .240 Weatherby came along, then shot some with the cartridge under similar conditions. As expected, anything the 6 mm Rem. could do the .240 Weatherby magnum did better at ranges out to 200 yards or so, and from there on the performance gap widened even more.

The .25 caliber cartridges in commercial form have a much wider spread than the .243s, everything from the old .25-20 to the .257 Weatherby Magnum. When we consider this wider range, including the old .25 standbys like the .250 Savage and .257 Roberts, the .25-06 comes close to being in the magnum class, but it is far surpassed by the .257 Weatherby, a true magnum in every sense. The .257 Weatherby is one of the better long-range cartridges for game like the pronghorn antelope that undoubtedly requires more long-range shooting than any other American game animal. It is also a superb cartridge for hunting mule deer, sheep and Rocky Mountain goat.

Another small notch up the bullet diameter line is the 6.5 mm or .264 caliber. Strangely, the only commercial American cartridges of that caliber have a magnum headstamp—one, the .264 Winchester Magnum is a true magnum, the other, the 6.5 Remington Magnum is on the same powder capacity level as the .30-06 based cartridges, even though it has a belt around its head. Except for the magnum designation, it can be completely discounted as a "magnum." The .264 Winchester is completely magnum from every standpoint. Case capacity and performance are similar to the .257 Weatherby with bullets up to 100-125 grains, but for shooting game larger than mule deer, the .264 has the advantage of bullets of much heavier weights being available. The .264 Winchester ammunition was originally loaded with a 140-gr. bullet, and nearly all bullet manufacturers make a 140-gr., with some producing extremely long slugs of up to 165 grains. These heavyweight bullets theoretically put the .264 Winchester in a class suitable for hunting game from the weight of elk up, because of their ability to dig deep into the vitals of heavy animals. There are a couple of bugs in the broth here that are not always considered, however. First, even though the heavy .264 bullets do give deep penetration (if properly constructed), they have a rather small frontal area when expanded correctly for deep penetration, which doesn't do as much damage to vital tissue and

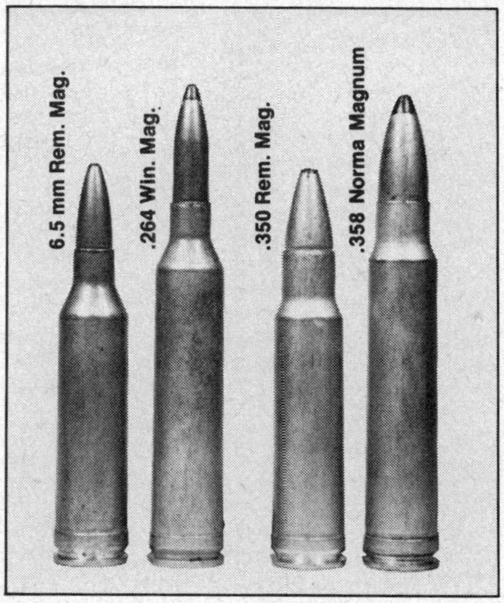

A belted cartridge case does not always mean extra power. The two short magnums above are magnums in name only.

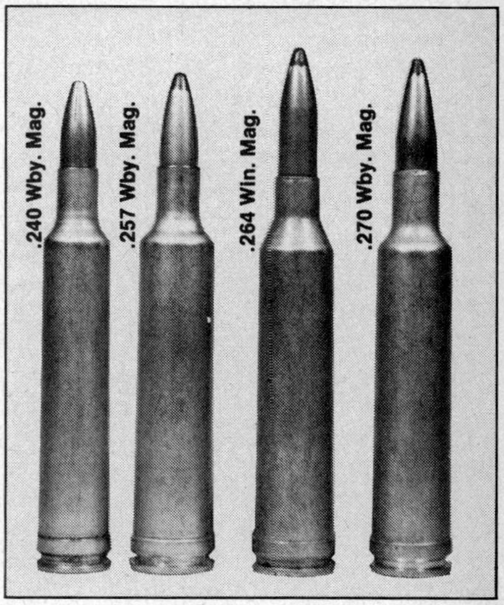

The cartridges above are best termed "small bore" magnums. They are suited for deer sized game at extremely long-range.

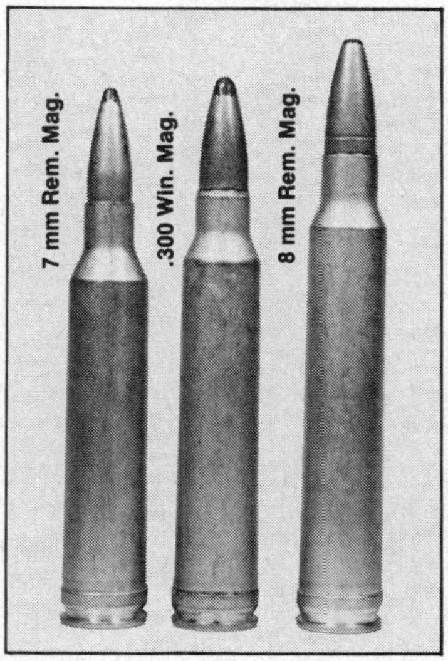

These cartridges are considered by many the top choices for all around North American hunting.

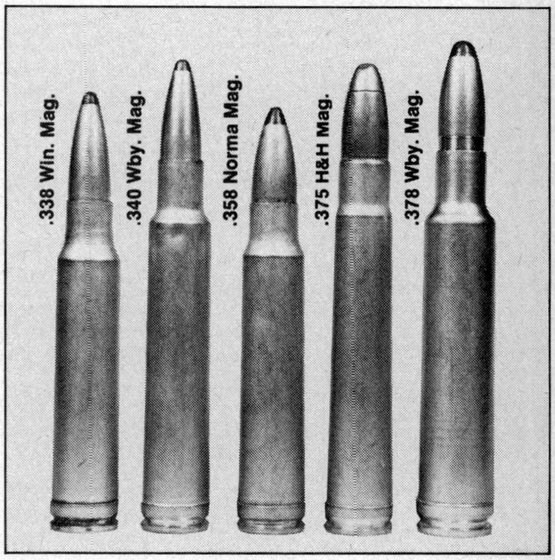

The big bore magnums all have considerable recoil, but they are effective on any game under the worst of conditions.

The Weatherby Mark V is available in a variety of belted, high-velocity cartridges, all designed for maximum power and flat trajectory at long-range. Depending on the size of game you're after, there's a Weatherby magnum suited for any type of hunting.

nerves as larger caliber bullets. Also, when you saddle the .264 bore with the 160-gr. bullet, velocity is not spectacular, even with the slowest powders like H870—on the order of 2900 fps.

Considering the popularity of the .270 Winchester, it seems rather odd that Weatherby is the only rifle maker that has come along with another cartridge in that caliber, either large or small. And while the .270 Weatherby Magnum is a true magnum any way you look at it, it never gained any great popularity. And this is not because of lack of performance of the cartridge. Anything the .270 Winchester will do, the .270 Weatherby will do better to the tune of some 200 fps. What this really amounts to, is that the .270 Weatherby will kick the 150-gr. bullet along at the same ve-

While caribou are not particularly large nor hard to kill, their wide-open habitat usually means long-range shooting—perfect for the modern flat shooting magnum. As well, moose and grizzly may be encountered, so magnum power is good insurance.

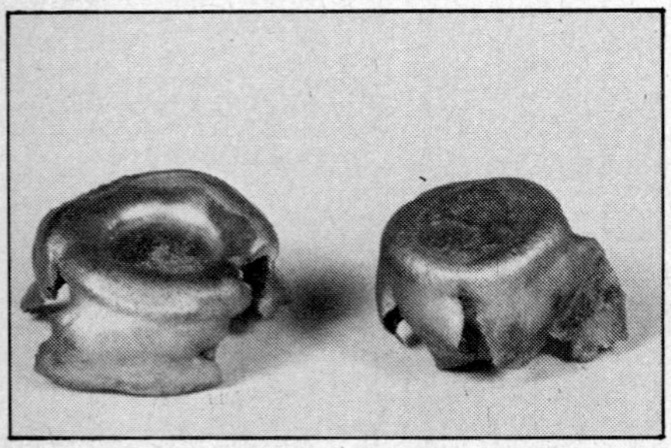

Choice of bullet is critical to magnum performance. The Nosler at right penetrated 24 inches into Duxseal, while the conventional bullet penetrated only 15 inches.

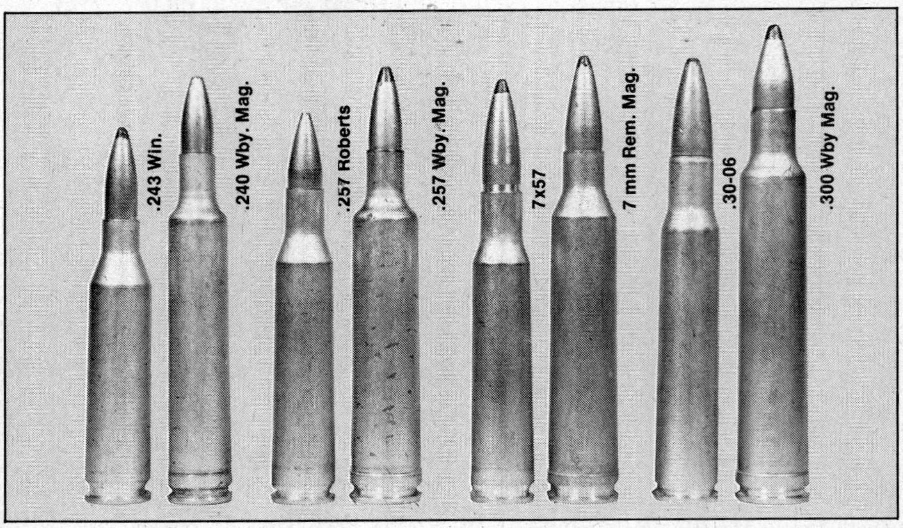

Above, various belted magnums are compared with their non-magnum counterparts. While it is true that magnum power is not always needed, the idea behind a magnum is that it will do everything a standard round does—and do it at much longer range.

locity the .270 Winchester will give the 130-gr. at similar pressures. In fact, if the best powder for the Weatherby case is used with the 150-gr. bullet, which is H870, the Weatherby will increase that by another 50 fps or so, or to 3200 fps. If I were to hunt the largest American game with the .270 Weatherby Magnum, my choice for a bullet would be the 160-gr. Nosler Partition, which can be boosted along at 3150 fps without undue pressure pains with the right powders from a 24-inch barrel.

Americans learned something about the efficiency of the 7 mm (.284) bore back during the Spanish-American War, but for some reason the caliber didn't catch on here very well until the advent of the 7 mm Remington Magnum in 1962. A few domestic made rifles were chambered for the 7x57 Mauser, but the cartridge never became very popular, and neither did the .280 Remington. Roy Weatherby came along with the 7 mm Weatherby Magnum 'way back in 1944, and while it did gain a good deal of popularity in the Weatherby line of cartridges, no other commercial firm chambered it. The 7x61 Sharp & Hart, imported in the Schultz Larsen rifle, did little better ten years later. It took the efforts of Warren Page, Shooting Editor of *Field & Stream* at the time, to promote the big seven with his 7 mm Super Mashburn. The fact that Warren Page and Remington's Mike Walker were close friends, probably had a lot of influence on the conception and birth of the 7 mm Remington Magnum—a cartridge that was to become one of the hottest ammunition sales items of all time.

For much hunting of American big game, the 7 mm magnum cartridges approach the ideal. A great array of bullets in every shape and structure, and in weights by nearly all manufacturers running from the 115 to 120-gr. varmint class to the 175-gr. heavy game bullets, and Barnes Bullets even making a 200 grainer. Nosler makes the highly efficient, time-tested partition jacket in 140, 150, 160 and 175-gr. weights, which certainly adds to the performance level of 7 mm magnum cartridges for all North American big game hunting. With one or the other of those bullets I have killed everything from pronghorn to Alaskan moose and grizzly with one or another of the big seven cartridges, and I can't recall I ever felt being either over-gunned or under-gunned. Few cartridges shoot flatter with the 140-160-gr. spitzers, and few give deeper penetration than the 7 mm magnums when used with the 175-gr. Nosler. There are other cartridges that pack more punch at all ranges, and certainly some I feel better with when following a brownie along an Alaskan salmon creek, but the 7 mm magnums are among the top all-around cartridges.

We have long been a nation of .30 caliber fans, and we certainly have not slighted the .300 magnums. It really all started back in about 1913 when Charles Newton developed the .30 Newton. No, the Newton did not boast a belt, but magnum it was from rim to bullet point. The next big .30 was not chambered in an American commercial rifle until Winchester put the .300 H&H in the Model 70. Since then, we have seen the .300 Weatherby, the .308 Norma, and the .300 Winchester magnums, all belted. There isn't too much difference in the performance of these .30 magnums, but the .300 Winchester and Weatherby cartridges do hold the edge. Like the 7 mm magnums, the big .30s come very close to being ideal for all North American big game hunting if the *right* bullet design and weight is used for the game and conditions. Again, I've killed everything from deer-size game to moose and grizzly with

The modern belted magnum cartridges are available in a wide variety of excellent production rifles. The typical magnum is a bolt action with a fairly long barrel but various single shots and the Browning BAR are also popular as magnum rifles.

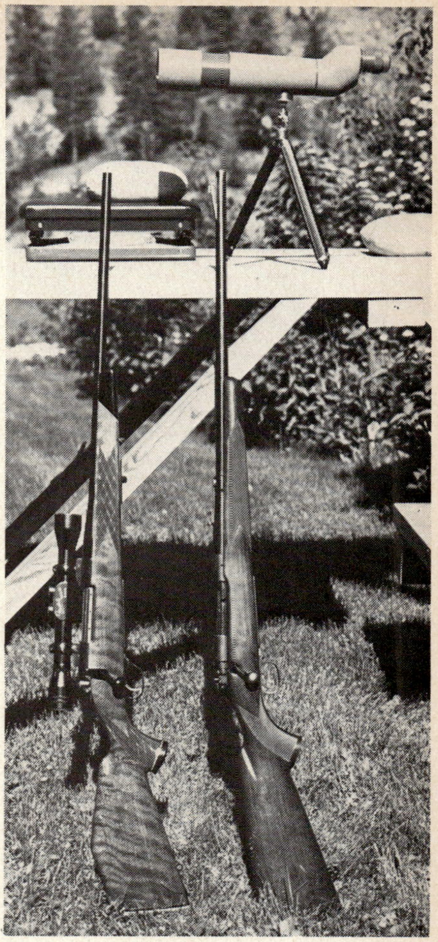

There is little difference in size and weight between a Winchester M70 in .270 Win. and a Weatherby Mk. V in .270 Wby. Mag., left, but the Weatherby cartridge far outclasses the standard .270 Win.

one of the .300 magnum cartridges, and I always feel comfortable with one no matter what the game. Also, like the 7 mm, .30 caliber bullets are made in every imaginable style, and in varmint weights of 110 to 130-gr. to 200 and 220-gr. heavyweights, with Barnes making a 250-gr. There is a .30 caliber bullet design and weight for shooting anything from jackrabbits to Kodiak bear, and if you use the right bullet the big .30s will handle them all.

Over on the other side of the Atlantic, the 8 mm (.323) bore has been very popular for big game cartridges since long before the turn of the century. On this side of the puddle we didn't get very excited about that bore diameter, and, strangely enough, the first American 8 mm cartridge to be chambered in a commercial American rifle, is a magnum. There seems to be little doubt that the reason behind Remington's thinking in bringing out the 8 mm Rem-

Many experienced big games hunters consider the .300 Weatherby Magnum the ultimate cartridge for mountain hunting. It offers a significant edge over the .30-06 in power and trajectory, and a slight edge over the .308 Norma and .300 Win. Magnums.

Photograph by Jon Cates

The performance of a flat-shooting magnum can mean the difference between success or failure on the trophy of a lifetime.

ington Magnum, is an answer to their success with the .244 Remington being rejuvenated by the 6 mm headstamp, the astonishing success of the 7 mm Remington Magnum, and to try to grab off a little of the medium bore magnum cake going to the .338 Winchester and .340 Weatherby magnums.

The 8 mm Remington Magnum cartridge is *impressive* to look at, having even slightly more boiler room than the .300/.340 Weatherby case. It also looks quite good in the ballistic tables, but somewhere someone slipped a little, because factory ammunition falls far short of quoted velocity when fired in factory rifles. Even worse, bullets loaded in factory ammunition are inclined to blow up at close ranges, and so far no bullet maker has come along with bullets that are any better. A situation that makes the cartridge less than dependable for heavy game at close quarters under adverse hunting conditions. There is nothing wrong with the cartridge if we just had good bullets to use in it.

The .338 Winchester Magnum and the .340 Weatherby do not suffer from the same problem. There are bullet weights ranging from 200 to 300 grains, and some of them are of the finest design for sure expansion at the longer ranges and deep, bone-shattering penetration close in. Nosler makes a spitzer 210-gr. that delivers astonishingly flat trajectory at all practical game ranges, and a 250-gr. for the big stuff at any range under any condition. After shooting a great variety of heavy game with various .33 caliber cartridges, I find that the performance of the best bullets in .338 caliber is perhaps the most versatile of any caliber for the heavier game from elk up at all hunting ranges under all conditions. It is no great problem to clobber an elk at 400 yards with the knowledge that the lighter spitzer bullets will get there with no more drop than the magnums with smaller bores, and expand adequately to do a lot more damage. And when you have to go into the brush after a bull moose or brownie, the heavier bullets like the 250-gr. Nosler, Bitterroot or Speer Grand Slam will drive deep into the vitals from any angle, while packing enough wallop to raise merry old hell with his innards along the way.

If we look back to the early 1900s, the .35 caliber cartridges were quite popular. As with the .30 magnums, we also had a .35 magnum in the form of the .35 Newton. There have been a lot of smaller .35 caliber cartridges chambered in American commercial arms, but no .35 magnum. True, the fat little .350 Remington Magnum does say "Magnum" on its belted head, but a true magnum capacity cartridge it is not. The old .35 Winchester did nearly as well back in 1903, and the wildcat .35 Whelen on the .30-06 case is as potent. The only true .35 magnum cartridge in commercial form, and it has never been chambered in American-made commercial rifles, is the .358 Norma Magnum. The .358 Norma is a good cartridge but, like the 8 mm Remington, there are no really good bullets made in .35 caliber that will stand up under the velocity the big case churns up. Nosler has never made .35 caliber bullets because of lack of sales. The Bitterroot Bonded core was made in .35, but is not available as of this writing. Either bullet brand would make one hell of a good cartridge out of any magnum .35, but without such bullets the .35 magnum cartridge is far less reliable on heavy game than either the .338s or the reliable old .375 H&H.

There is really little to say about the .375 H&H that hasn't been said dozens of times. There is little doubt as to the performance it gives on any class of American game, but even with that cartridge the right bullet design is the an-

The author feels the magnums shine in two areas—sheer power for dangerous game and flat trajectory for long shots. This mountain grizzly was taken with a single 200-grain Nosler from the .300 Win. Mag.—a classic example of the right gun for the job!

Photograph by Leonard Lee Rue Jr.

Game such as mule deer can definitely be taken easily with non-magnum cartridges, but big bucks come hard these days. When shots are long or a raking shot such as this must be taken, the extra power and penetration of a magnum can make the difference.

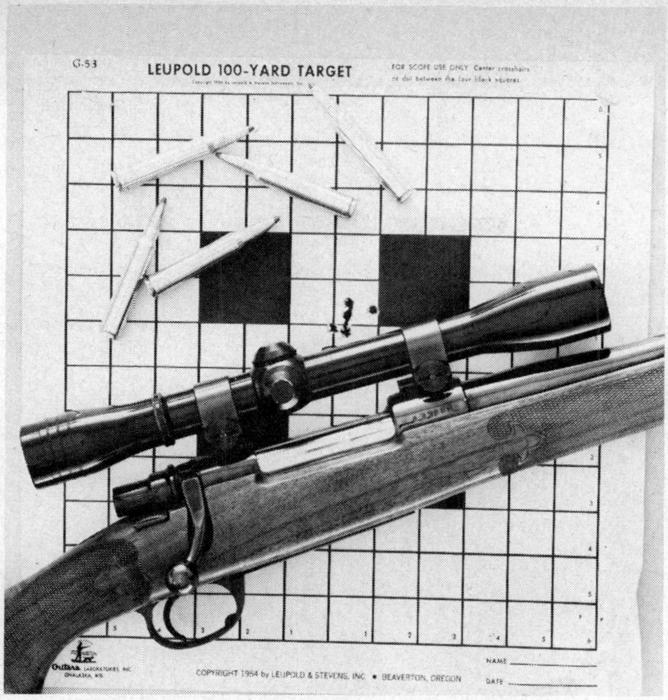

This five-shot 100 yard group was shot with a Mauser Mark X using Federal Premium 200-gr. .300 Win. Mag. ammo. Magnums have more recoil, but modern magnum cartridges can be very accurate.

The .257 Wby. Mag. is the hottest .25 caliber cartridge, delivering phenomenal velocity and punch. In spite of its power, it has relatively light recoil and is pleasant to shoot even from a bench.

swer to success. I've had bullets in the 270 to 300-gr. range blow up and fail to give adequate penetration on heavy animals with full-power .375 H&H loads. These bullets had jackets too thin to take the velocity on contact with big animals, but with 270 or 300-gr. bullets with controlled expansion design, the .375 is completely reliable. It shoots quite flat for a big bore, but does not compare to the .338 and .340 magnums for long-range heavy game shooting.

The .375 Weatherby is just a little more of the same thing, but few rifles chambered for it are still around, and Weatherby no longer furnishes ammunition for it. The mammoth .378 Weatherby Magnum really has little place in American big game hunting, but it does add about 300 fps to the top velocity of the .375 H&H. Needless to say, if you can shoot it well enough, it is very effective on the largest game at the longest practical hunting ranges. The big disadvantage of the .378 Weatherby, is that the rifle must be overly heavy for American hunting, and even then, few hunters can handle the recoil well enough for shooting at longer ranges.

Big billy goats can be tough, and are often taken at extreme ranges. This fine trophy was taken with a 7 mm Remington Magnum, an excellent choice for any mountain hunting. The 7 mm Rem. Mag. is the most popular magnum cartridge on today's market.

Of course recoil is one of the problems connected with shooting magnum cartridges in most calibers. There is little doubt that some of this is psychological—many hunters start mentally flinching when they look at the big cartridge—but if the recoil and/or muzzle blast of a magnum rifle bothers you enough to cause flinching, you *must* either learn to overcome it or use a cartridge you can handle. A .243 bullet in the lungs of a bull moose is more effective than a .375 that breaks a leg!

But there is the unquestionable fact that in the same calibers the magnum cartridges outperform their standard cousins. It has often been said that the animal hit in the lungs with a .30-06 doesn't know the difference between the jar it received and the jolt from a .300 Weatherby. That may well be correct, but it's when the bullet misses the lungs a little or just brushes them that the extra tissue destruction and wallop count. The hunter doing the shooting *will* know the difference if the animal is on the far side of a wide canyon; both in ease of hitting a vital spot, and by the greater punch the bullet packs when it gets there.

Someone else brings up the point that the magnum adds only 200-300 fps, "which really isn't so very much." If we all felt that way we'd still be shooting nothing better than the .30-30.

There is also the fact that the .30-30 is plenty for a deer in the brush at 100 yards and a .300 magnum is more than needed. True, but the .300 will do the job very well at 100 yards or out to 400, where the .30-30 is useless.

Test Report

MARLIN 375

By Rick Hacker

Marlin has a new variation in their classic 336 line of brushbusters. It's chambered for the powerful new .375 Winchester.

When John M. Marlin's firearms firm introduced their first lever action back in 1881, they were unknowingly setting the stage for a long succession of strong, popular, and fast-shooting repeaters that has culminated a century later with the new Marlin Model 375. As might be supposed from its numerical designation, this latest lever action is chambered solely for the .375 Winchester Big Bore cartridge introduced in 1978.

Basically, the Marlin 375 is a rechambered version of their popular and time-proven Model 336, which has been in the line since 1948 and which, in turn, was an updated and improved version of Marlin's Model 36, which traces its unaltered lineage back to the Model 1893, the first year the side ejecting lever gun was introduced. Needless to say, this mainstay of Marlin's has been around long enough to prove itself admirably over the years as a popular design for hunters. Unlike other lever action designs which had to be beefed up in order to handle the potency of the .375 cartridge, Marlin's solid steel receiver easily adapted to the Big Bore chambering without any external alterations.

Before one can fully appreciate the advantages of the Marlin 375, it is first imperative that we realize the potential and characteristics of the new round for which the rifle is chambered. Interestingly, the .375 is simply a "modernized" version of the old .38-55, a once-popular short range cartridge known for its accuracy and stopping power. The rimmed cases of the two cartridges are nearly identical, but there the similarities end; the .375 Big Bore in its 250-grain version contains enough powder to send it thundering out of the bore at 2200 fps with a muzzle energy of 2150, compared with the 255-grain .38-55's muzzle velocity of 1320 and a ME of 985. The conclusions are obvious: **(1)** the .375 Winchester is a vastly improved variation of the tried-and-true .38-55, giving hunters substantially more knockdown power for short-range shooting, and **(2)** it is totally unsafe and an admission of insanity for anyone to chamber the new .375 cartridge in a firearm designed for the .38-55; the cartridge and chambering will fit together, but you and your rifle won't after pulling the trigger! Therefore, if you enjoy using both hands when counting from one to ten, *only fire the .375 Winchester in rifles expressly made for that caliber.*

Presently, four rifles are chambered for the .375 Big Bore—Winchester's 94 Big Bore, Ruger's #3 Carbine, the Savage 99-375, and the Marlin.

And speaking of rifles expressly made for the .375 Big Bore, none combines the power, ruggedness and fast-handling qualities demanded by the round more efficiently than Marlin's Model 375. Because the .375 Big Bore is essentially a short range big game cartridge, Marlin's 20-inch barrel and butter-smooth lever action are ideal for fast forest shooting. When zeroed in for 100 yards, the 200 and 250-grain bullets both drop approximately five inches at 200 yards. At 300 yards the slugs drop very radically, as well as losing more than half of their killing energy. Thus, the .375 has an effective range of up to about 150 yards, and for this reason, Marlin's semi-buckhorn rear sight and beaded-post ramp front sight are ideal for the close-in taking of any North American game up to and including moose. However, for those hunters with focusing problems, or desiring to add a little more light gathering ability during the twilight hours, the receiver is drilled and tapped for a scope or receiver sight and the rear sight is hinged to fold down out of the way. Like all of Marlin's lever actions, the Model 375 features side ejection, permitting the scope to be mounted low and over the receiver. Personally however, I much prefer easy-to-see open sights for short range.

I have always had some apprehension about Marlin's side ejecting feature inhibiting southpaw shooters. However, being right-handed, I never had the opportunity to explore this concern until this field test, which was conducted on the big bore line of Angeles Shooting

The new Winchester .375 (right) is an updated version of the old .38-55 (left), but offers increased velocity and energy.

The author's benchrest tests of the new Marlin 375 demonstrated the inherent brush-range accuracy of the combo. Using Winchester 200-grain ammo at 50 yards with open sights, he shot a one-inch group, excepting a called flyer that opened it to 1½ inches.

Range in Tujunga, California. A fellow shooting on the bench next to me was using the wrong hand (as we right-handers like to say). Naturally, I could not pass up the opportunity to ask him to fire a few rounds with the 375 and tell me how he liked it. At the same time I could casually observe him to see if I detected any brass cases flying past his line of sight. Being a life member of the NRA and a man with a few hunting seasons under his belt, he was only too happy to make the sacrifice and touch off a few rounds for "the cause." The Marlin cleanly kicked the empties seven feet to the man's right, barely disturbing his aim. So much for my worry.

Firing at a 50-yard target right handed, the Marlin 375 performed extremely well for me. Surprisingly, when firing both the 200 and 250-grain cartridges from the bench, recoil was negligible. Using the 200-grain bullet, and trying for a grouping rather than a score, my first shot landed at four o'clock in the eight ring. The next shot touched it.

"Hot damn," I thought, "I'm hot today!" Buoyed with this self confidence, I promptly scratched the nine ring with the third shot at three o'clock, opening up what had been a ½-inch group to 1½ inches. A final shot put a hole back into the 8 ring, next to the first two. Firing at 100 yards scattered a five round tube full of 250-grain shells over a four-inch spread in the target, certainly adequate accuracy for a big game hunting rifle.

Adding to the Marlin 375's desirability as a hunting rifle is the fact that it comes with a leather sling and quick-detachable swivels. The walnut stock of the Marlin is devoid of checkering, which adds to its businesslike looks. Likewise, the top and bottom of the receiver is matted, which aides tremendously when sighting the rifle in bright sunlight. The hammer, as on all of Marlin's lever actions, is equipped with a thumb extension to make cocking easier, although this device really comes into its own when used in conjunction with a scope. With all this attention to details and hunting practicality, it amazes me that Marlin still insists on "dandifying" an otherwise traditional looking gun with their gold plated trigger. Yes, the 375 has it, just like most of Marlin's other offerings. I can see this feature on a commemorative, but not on a working gun, which the Model 375 most certainly is.

However, other standard features found on all Marlin firearms have made their way to the 375 and add to its desirability: 12-groove Micro-Groove rifling, Pachmayr rubber recoil pad, and an easy-carrying weight of 6¾ pounds make it an ideal brush buster for hunters wanting more knockdown power than the traditional 30-30, but without the kick of the .30-06. I do not in any way consider the short range limitations of the cartridge or the gun a detriment, as more than two-thirds of the big game animals I have harvested were taken at less than 100 yards. For fast-handling big game use in close quarters, as a super saddle carbine or even as a light, handy back-up gun for those far-away hunts when you never know what you will find, the Marlin Model 375 is an ideal addition to a lever action legacy that began 100 years ago.

SPECIFICATIONS

MARLIN MODEL 375

Manufacturer:	Marlin Firearms Co. 100 Kenna Drive North Haven, CT 06473
Action:	Lever action
Caliber:	.375 Winchester
Capacity:	6 (5 in magazine)
Stock:	American black walnut
Sights:	Adjustable rear and hooded bead front
Overall length:	38½ inches
Barrel length:	20"
Weight:	6¾ lbs.
Price:	$210

Marlin includes a hammer extension on their rifles to facilitate thumb-cocking if a scope or receiver sight is employed.

The factory-issue rear sight of the Marlin 375 is a step-adjustable, folding semi-buckhorn, which is adequate for close shots.

Test Report

BROWNING B-78

The Browning B-78 combines accuracy, simplicity and dependability with good looks. It's the heir apparent to the classic Winchester High Wall.

By Rex Thomas

We see it almost everywhere today, in almost every aspect of our daily lives—a return to basics. More and more we seem to realize that the simple things in life have more lasting value than what modern technology can mass produce. Firearms enthusiasts, whether they be nostalgia buffs, benchrest fanatics or just the kind of guys who really enjoy the "challenge" of a hunt, are the breed of shooters who have a genuine passion for the "oneshooters." They value and respect a simple, good shooting rifle.

In 1974, Browning introduced the B-78. The design of the modern B-78 is based on John Moses Browning's first successful rifle which he developed in the year 1878. One year later, he patented his single-shot rifle and five years later sold the manufacturing rights for it to Winchester. In 1885, Browning's original patent was reintroduced as the now famous Model 85 Winchester "High Wall." Many still regard the "85" as the finest single-shot rifle ever made in the United States.

Browning's modern B-78 is probably most accurately described as a modernized version of the original patent.

Overall, the B-78 measures 42 inches which means this single-shot is one to three inches shorter than the average bolt action rifle. Barrel length of the particular rifle I selected for testing was 26 inches which was an aid in getting all the velocity possible from the particular caliber I had selected—.22-250 Remington. I had a choice of either octagon or round barrel and chose the round barrel because it was about 12 ounces heavier. I wanted the extra stability and edge in accuracy for long-range, western style rockchuck hunting. "Chucks" in this part of the country are a very real threat to the farmers' crops so I try to help out whenever I can.

The B-78 is one of the few modern single-shots around that is easily adapted to optical gear. It is drilled and tapped to accept scope mounts and bases and rings were furnished to simplify adding good glass. I chose to mount a Weaver T-10 to my particular rifle which proved to be a deadly combo, but more on that later.

The B-78 is not a particularly light rifle for toting around on your shoulder all morning. Without scope, mine weighed 8 pounds 4 ounces but I didn't feel overburdened carrying it in rockchuck country. In fact, I'm quite certain barrel weight was a positive factor in making some of the long shots.

Cosmetically, the B-78 is the kind of rifle a hunter can really sink his teeth into. The buttstock and forearm are crafted from high grade select walnut with excellent figure. The Monte Carlo comb is carefully shaped and helped put my eye in perfect alignment with the scope. A black rubber recoil pad with white-line spacer is installed and the pistol grip is tastefully capped with a silver plate to accommodate the owner's initials. The forend is actually semi-schnabël in design and is attached to a special hanger which projects from the receiver rather than being fastened directly to the barrel. Two screws can be adjusted so the barrel is left free floating. Buttstock and forearm are crisply handcheckered 20 lines-to-the-inch and as the icing on the cake, a high-gloss finish is applied lending a real touch of class. Low profile sling swivel studs are furnished that will handle rifle slings up to 1¼ inches wide.

Wood to metal fit could be rated somewhere between good and excellent and all metal parts have a deep, rich blueing that is greatly enhanced due to the meticulous polishing of the metal prior to the blueing process. The receiver has been forged and milled from a single block of high strength steel in the famous High-Wall profile. The action is operated by means of an underlever which when lowered, pulls down a breech block within a mortise. Returning the lever raises the breech block, seals the chamber and cocks the exposed hammer. A half-cock safety is the recommended carrying position. Another unique feature is a "cartridge stop" that can be adjusted to deflect spent cases left or right or it can be set to retain cases for easy removal.

A trigger pull gauge indicated the factory pull weight had been set at just under four pounds. That seemed a bit stiff for me so with the Allen wrench provided, I reset the *externally adjustable* trigger to about three pounds. There was absolutely no creep in this particular rifle and if my sample was indicative of the rest of the line, then the B-78's have some of the finest factory triggers on the market.

Accuracy tests were impressive. From a shooting bench at a distance of 100 yards, five-shot groups could consistently be placed within a one-inch square. I tried two different loads to see if there was any variation. Federal factory ammo grouped well averaging about one inch groups—if the shooter did his part consistently. But some of my handloads fashioned after the exact recipe that was Harvey Donaldson's favorite proved to be a bit more accurate. The load consisted of 34 grains of 4064

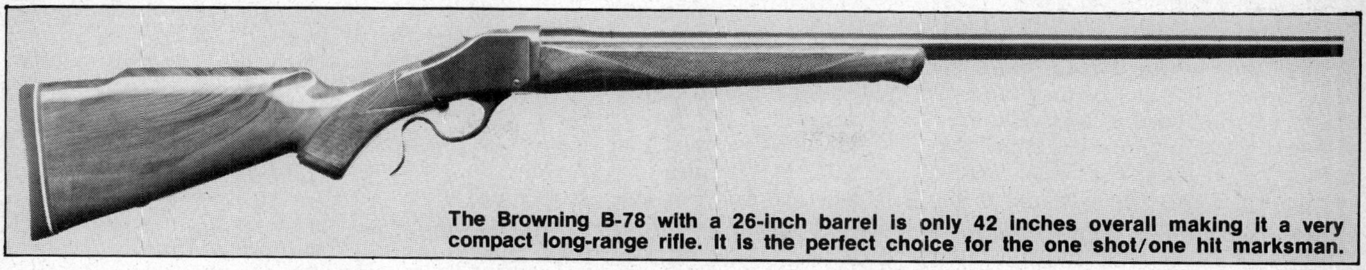

The Browning B-78 with a 26-inch barrel is only 42 inches overall making it a very compact long-range rifle. It is the perfect choice for the one shot/one hit marksman.

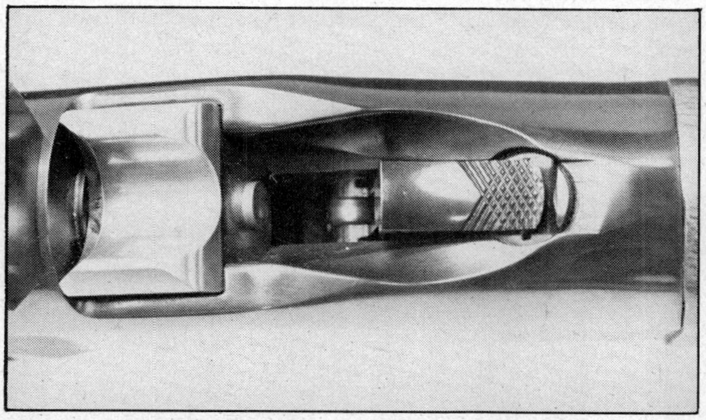

At the range, the .22-250 B-78 was an MOA shooter with factory ammo, and even better with handloads. The same gun (photo on facing page) was deadly on Idaho rockchucks, taking 25 varmints with few misses and no cripples. The author feels that this type of single-shot rifle is a good conservation tool as it demands careful shooting. Photo above shows closeup of action with breech block locked solidly within mortise.

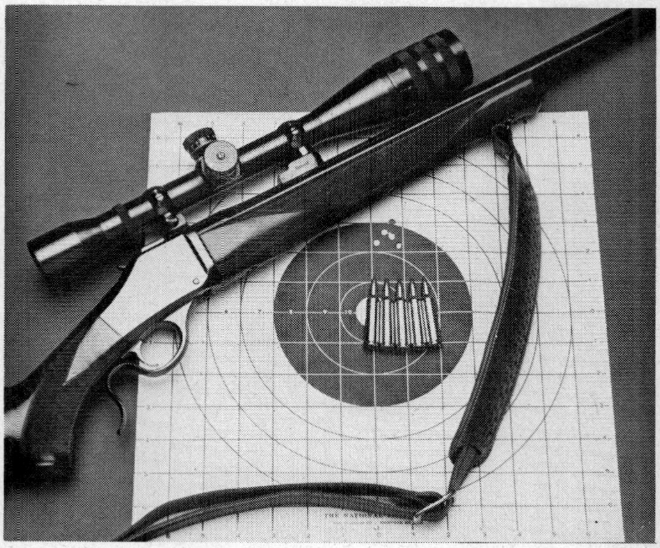

powder packed behind a 53-grain Sierra bullet. Groups averaged ⅞ inch. That is excellent accuracy and also proved to be a deadly load for the rockchucks mentioned earlier. On one morning's hunt near Soda Springs, Idaho my partners and I managed to take about 25 chucks. The efficiency of the load left no cripples and I can also say that there were very few misses. All of our targets were within 90 to 230 yards although I must admit that most of the confirmed hits were on the shorter end of the scale. I am convinced that the fine accuracy the B-78 is capable of can be partially credited to a psychological handicap the shooter consciously accepts with a single-shot rifle. When you have only one single round, you tend to really concentrate on making it count.

I am also convinced that the single-shot rifle could be utilized as a conservation tool both for reducing the number of wounded animals and giving the anti-hunters less to complain about—if all hunters had to use one. If we all only had one shot to consider, we would be more selective of when to shoot as well as what we shoot at.

Most of the single-shot enthusiasts of today accepted the challenge for one of two reasons. They either were not satisfied with the accuracy obtained from rifles with other actions, or they were dissatisfied with the challenge in their hunting experiences. You may consider joining the ranks for either of those reasons or maybe you are a candidate because of nostalgia. There is also real value in a B-78 for a firearms collector whether he be amateur or professional.

Regardless of the game you seek, or your reasons for considering a single-shot as part of your firearms battery, there is surely a B-78 for you. There are six calibers to select from including .22-250 Rem., 6 mm Rem., .25-06 Rem., .30-06 Sprg., 7 mm Rem. Mag. or .45-70 Govt. All are available in your choice of round or octagonal barrel except the .30-06 Sprg. and the .45-70 Govt. They both have octagonal barrels only. The .45-70 is the only offering with a straight grip stock, curved steel butt plate and iron sights.

I've discovered there is a bit of snob appeal wrapped up in a single-shot rifle. The hunter who carries one tells others, merely by carrying one, that he has enough confidence in his shooting capabilities to accept an additional hardship or challenge.

(Author's note: At the time of this writing, I have just been advised that the B-78 has been officially discontinued in the Browning line. There will be no more made as of this date but inventory does exist. Contact your Browning dealer or write Browning, Route 1, Morgan, Utah, 84050.)

SPECIFICATIONS
BROWNING B-78

Maker:	Browning Route 1 Morgan, Utah 84050
Action:	Falling block
Caliber:	.22-250 Remington
Capacity:	1 round
Barrel Length:	26 inches
Overall Length:	42 inches
Sights:	None
Stock:	High Grade Select walnut; Monte Carlo comb and Schnabel forearm.
Weight:	8 lbs. 4 ozs.
Price:	$399.95

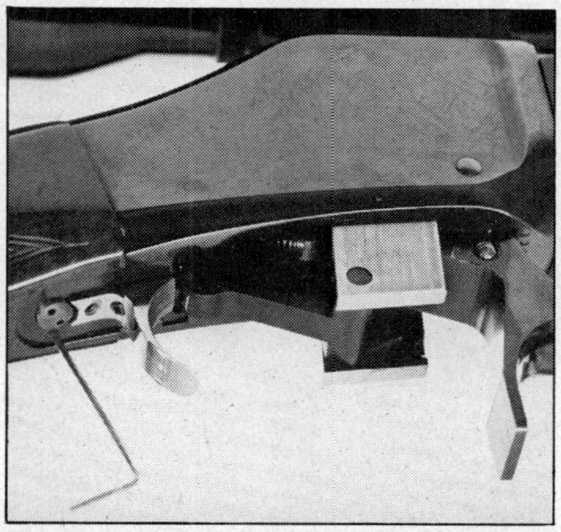

The standard B-78 factory trigger is fully adjustable *externally!* Author was able to adjust pull down to a smooth three pounds.

The term "varmint" means a lot of different things to different people although it is generally intended to encompass such critters as the coyote, fox, crow, woodchuck, rockchuck, prairie dog, jackrabbit, ground squirrel, etc. It would seem then that any rifle used for any of these animals would be considered a "varmint rifle." However, such is not the case; the "varmint rifle," as most of us have learned to consider it, is a heavy barreled gun with a high-magnification scope and chambered for a high velocity, flat-shooting cartridge. The gun is designed to be fired primarily from a rest at stationary targets.

However, since the advent of high pelt prices for predators such as coyote, fox, and bobcat, there has been the need for an entirely different type of rifle for these particular "varmints"—a varmint rifle for profit. I prefer to call it a "fur gun."

But before we get really serious about making money, let's have some fun. The varmints I go after, strictly for enjoyment—prairie dogs, rockchucks, woodchucks, ground squirrels, and crows—aren't worth a thing, moneywise. Actually, they have a negative value to farmers and ranchers which is so much better for the shooter. But in terms of sport and a challenge to the precision rifleman, they can be worth a lot. With these, part of the challenge comes from procuring a super-accurate rifle and working up a load to match the gun's ability. Then, the real excitement comes in trying to consistently hit the small targets at frequently extreme ranges.

Here, the traditional stereotype of the varmint rifle fits nicely. A bolt action is the usual choice and it's normally fitted with a heavyweight barrel. The pudgy barrel does several things which prove advantageous to the long-range varmint shooter. First, it isn't generally as persnickity about load variations—the heavy barreled rifle will often shoot a variety of components well. Secondly, it isn't as sensitive to heating up in sustained fire as a lightweight barrel. Thirdly, the heavy barrel makes a steady hold from a rest a bit easier due to the inertia of its weight. Besides all this, higher velocity cartridges with heavier bullets can be used with less apparent "kick" because of the added recoil-damping weight from the fat bar-

VARMINT RIFLES FOR FUN AND PROFIT

Year 'round hunting plus the high value of pelts makes "varminting" more popular than ever. Here are an expert's tips on picking the right rifle and cartridge!

By Rick Jamison

rel. Besides a large diameter barrel, most varmint shooters also prefer a long one, something on the order of a 24 or 26-incher in order to derive all the velocity from the cartridge they can. A good portion of the manufacturers supply heavy barreled, bolt action "varmint rifles" which are well-suited to the taking of varmints.

A high-magnification scope is the general rule for the fun gun—from 8X to 24X is ideal for the small targets at long-ranges. Often, in varmint shooting, the critter might not expose any more than the very top of his head, offering a target measured only in fractions of an inch. In order to define such a target well, which also tends to blend into the background, a high-magnification scope is necessary. However, in extreme mirage conditions, the high magnification scopes can be a real pain in the neck for they will reveal more apparent distortion from mirage—to the extent that it can be difficult to see the target. Therefore, using a lot of scope power is somewhat of a trade-off.

Almost any cartridge from the .17 to .25 calibers can be used on these "fun" varmints but the wise varmint hunter, particularly one in the wide-open spaces of the west, will select his with an eye towards high velocity. The faster a bullet, the less time there is for wind to alter its course over a give range. The faster bullets also fly flatter so that precise range-doping is not quite as critical as with the lower-velocity cartridges. For this reason, ideal all-around cartridges would include the likes of the .17 Remington, .22/250 Remington, .220 Swift, 6 mm Remington, and .224 and .240 Weatherby Magnums.

The .25's, such as the .25/06 and .257 Weatherby Magnum might also be included, but they're a bit much for most varmints. Cartridges larger than the .25's will, of course, work but most long-range varminting is done from the prone position and shooting larger, harder-kicking calibers from this position can become uncomfortable, particularly considering the large number of rounds often fired. I can remember prairie dog towns that contained so many uneducated pests that a rifleman could fire all the ammo he had, just about as fast as he wanted. In fact, frequently when shooting prairie dogs, I take two rifles and switch to the second when the first heats up. Other varmints such as crows, woodchucks, and rockchucks, are not normally found in these concentrations.

There are situations where a high velocity, flat-shooting cartridge is not ideal. For instance, woodchucks, compared to the smaller varmints, are quite large and are not generally shot at extreme ranges due to the hilly terrain and often heavy foliage where they're found in the east and midwest. For this

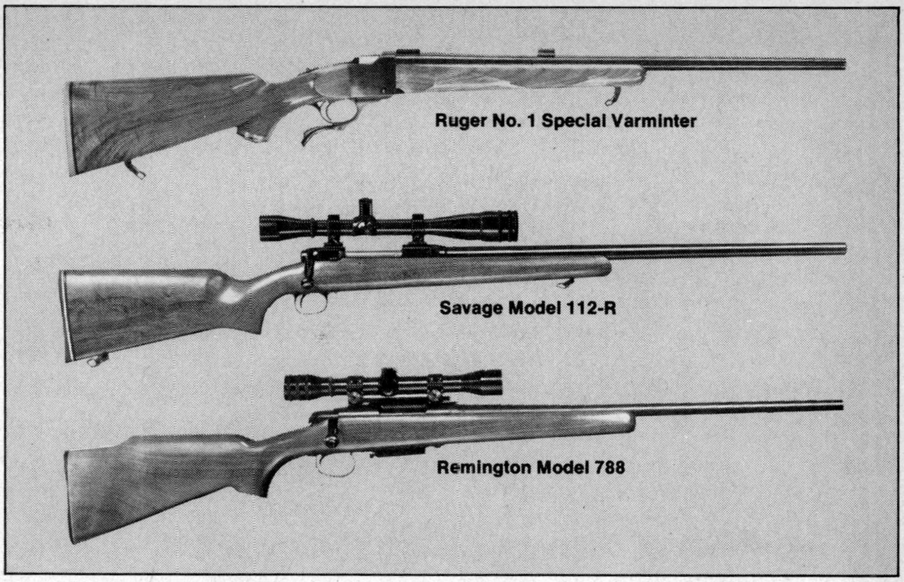

Heavy barrelled, super accurate rifles such as the Ruger Special Varminter and Savage 112-R are preferred for continuous long-range shooting at prairie dogs, woodchucks, and such, while lighter, fast-handling rifles such as the Remington 788 in appropriate calibers work best for fast-breaking situations like coyote calling.

The varmint caller places more importance on skill with call to lure varmints in close rather than the precision marksmanship of the long-range varmint gunner.

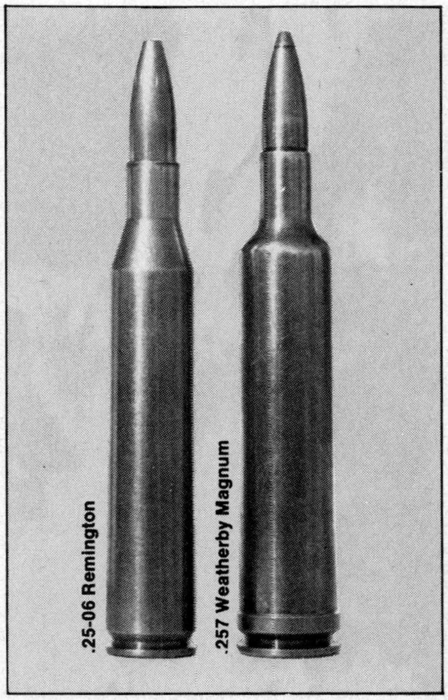

Author puts the .25 calibers at the upper end of the varmint rifle spectrum. Though effective, recoil is excessive for prolonged shooting, and pelt damage is extensive.

Varmint callers often get coyotes to come in very close, requiring a rifle that handles well and a cartridge/bullet combination that minimizes damage to the hide.

reason, a cartridge such as the .22 Hornet can be a good choice—particularly where noise is a problem, for the Hornet emits a relatively mild report. Another solution is to simply load down the larger cartridges.

Two of the most popular varmint cartridges, the .222 Remington and the .223 Remington, have so far gone unmentioned. I consider these "middle-of-the-roaders" for they do not offer extremely high velocity nor do they offer mild report like the .22 Hornet. Yet, all things considered, these two cartridges will probably take care of 90 percent of the varmint shooting done in North America. In some cases, such as in gopher shooting in parts of Montana and Wyoming, the targets are so small that any hit at nearly any range is a good one. Here, few shooters take shots beyond 300 yards and for this particular situation, the mild, accurate .222 or .223 would be good.

For any of these situations, it's generally best to use a fragile bullet (all but the heaviest, in varmint calibers) for a couple of reasons. First, the more fragile, highly explosive bullets are less prone to ricochets. Ricochets become more of a problem at low velocity—such as would be the case at the extreme ranges often encountered in varmint shooting. Ricochets have always been the subject of concern in the heavily-populated east although I think that every shooter owes it to the sport to be concerned about them no matter where he is. Houses are continually being built and where there was nothing over the hill last year, there may now be a half-dozen family dwellings. Secondly, the more fragile bullets are more apt to do a varmint in with a poorly-placed hit. Precise shot placement on these small animals at extreme ranges is next to impossible; therefore it's best to have a bullet that will do the job quickly, even with a poorly placed hit.

The varmint rifle to be used for profit via the accumulation of predator pelts is an entirely different matter. First of all, the most consistent way of taking these valuable furbearers such as the coyote, fox, and bobcat, is by using a call. In this case, particularly with the canines, the animal is often spotted at close range and moving fast. In such a situation, the heavy barreled varminter with high scope magnification would be a decided handicap. Can you imagine trying to pick up a running coyote with an 18X scope at 30 yards? There are times when ranges can be long, but even here the heavy barrel is not needed since the quarry is larger, usually moving, and seldom are enough shots fired to heat a featherweight barrel.

Even for the long shots often encountered between predator calling stands there are plenty of lightweight sporters that will shoot just as well as a heavy barrel gun for two or three shots. This is particularly true if the shooter is a handloader and has the opportunity to work up an accurate load, tailored to the vagaries of that particular barrel's vibrations. Like with the heavy barrel, the close-range calling situations demand a turnabout in scope selection—a wide field of view offered by the low magnifications is practically a must. For this dual-purpose, long or short-range situation, the variables can really shine; something on the order of a 3X-9X does it all. I generally turn the magnification down to three when calling and crank it up to nine when driving between stands.

However, I have one restriction on the use of a variable scope for the serious fur hunter. Generally, with the current pressure from fur hunters, either coyotes or foxes spotted between calling stands will be running; once in awhile you'll get a single standing shot, but only if you're lucky. I would guess that 90 percent of the coyotes I've taken,

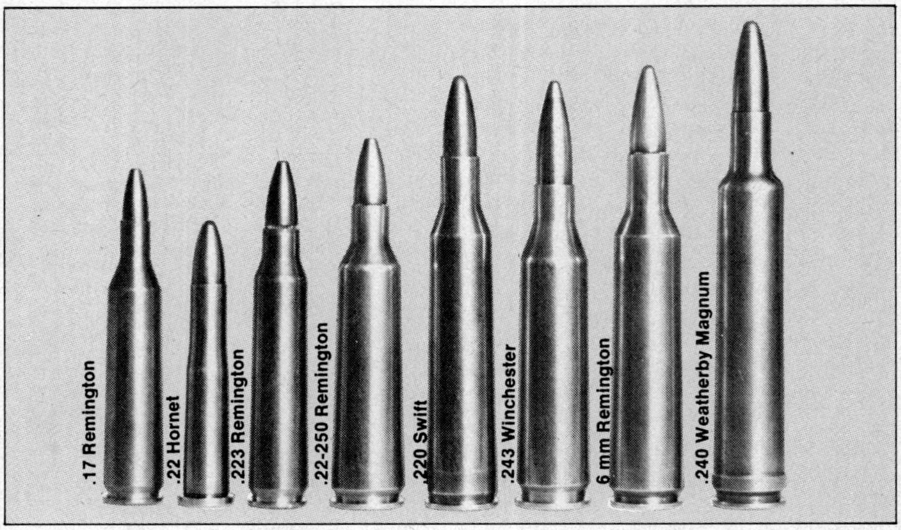

Cartridges shown above represent some of the more popular varmint cartridges in use today. The old Hornet is limited to short range work, while the .17 and .223 Remington cartridges are excellent medium range choices. The .220 Swift and .22-250 are good all-around choices, while the bigger 6 mm rounds shine at extreme long range.

Sitting jackrabbits are fine for long-range rigs, but running bunnies call for low-power scopes and fast-handling repeaters.

The Weatherby Mark V in .240 Weatherby Mag is an excellent choice for dropping coyotes at long-range in windy, open areas.

aside from calling, are shot loping or running—usually away from me. For this reason, it's critical that you be able to estimate range *through the scope*. Doing it otherwise just doesn't work out. By the time you judge range over the top of the scope on a running coyote and then find him in the field of view, he can easily be 50 or 100 yards farther away. With continued daily use of a single scope and magnification the hunter almost unconsciously determines range through the optics, then figures holdover and lead, as the coyote is moving. Switching magnification on a variable scope makes this almost impossible. A higher-magnification glass makes the animal appear closer, and consequently you'll be fooled into thinking that less holdover and lead are required. It will also foul up your judgment of his running speed. To solve the problem, I use 9X for all my shooting, particularly the long shots, except when calling; then I know the range will be short and there is no need of range or lead calculations; here the 3X is best.

Selection of a cartridge for the "fur rifle" is much more critical than with the varmint rifle built for fun. The man wanting to make a profit from the pelt would be wise to avoid anything that would tear a big hole in the hide. In my experience, I have found the 6 mm and .25's to be a bit damaging on pelts. I have used full metal jacket bullets in the 6 mm with mixed success. Either the animal isn't dispatched cleanly where he gets away in brushy country, or the high velocity projectile hits bone for excessive pelt damage from the secondary fragments.

I think that something on the order of a .240 Weatherby Magnum with either the 90-grain Speer or the 80-grain Hornady FMJ loaded to maximum velocity would be about ideal for coyotes and foxes at extreme ranges either in open country or on snow. Here, the high velocity 6 mm would have advantage in making hits easier while the soft-footed animal could generally be recovered by tracking on snow or by simply watching the quarry's direction in open country. Coyotes hit solidly with a FMJ will not generally go far, but even 100 yards can cause a loss in brushy country. A good friend of mine uses the 90-grain Speer FMJ in a .243 for coyotes all the time. However, he also has a good dog to trail up those that aren't killed cleanly.

I consider the .22 Hornet or .17 Remington to be about ideal for shooting strictly from varmint-calling stands. The fragile bullets from either cartridge will generally not exit at close range—an ideal situation as far as pelts are concerned. However, when the ranges become extreme, I consider these two a bit lacking for predators the size of coyotes. Likewise, I would not select the .222 Remington nor the .223 Remington for there are better cartridges at the long-ranges. The hunter really serious about getting fur wants everything in his favor. Why give an inch in terms of velocity and trajectory?

Either the .220 Swift or .22/250 Remington provides the best for all situations, on the average. Like with the .17, fragile bullets at high velocity in these calibers won't exit a coyote at close range. Yet, they have enough energy at the longer ranges to do a decisive job. Of course, as with any load, there are drawbacks—if a coyote or fox is hit somewhere around the periphery of his body outline, a large rip in the pelt will result. Sometimes, with the diminutive grey fox, even these .22's are too destructive. Still, I have found the hot .22 centerfires, with proper loads, to provide the best compromise between long-range hitting, killing power, and minimal pelt damage.

In order for the "profit rifle" to become most effective, it's necessary for the shooter to use it a lot. In fact, this is much more important than rifle or cartridge selection. In my opinion, nothing can replace actual "on the job" pelt hunting—either for becoming effective with a rifle, or just having fun.

Jackrabbits may be hunted year 'round in many areas, and offer real sporty shooting.

Shooting crows with a rifle calls for pinpoint accuracy and quick-expanding bullets.

Long-range prairie dog gunning is a specialized skill requiring accurate heavy barrels.

Serious long-range varminters want the most accurate loads possible, and opt for special precision loading dies like the RCBS Competition dies, far right, or the Bonanza Benchrest dies, right. Above, varmint shooters take the extra step to turn case necks for precise chambering.

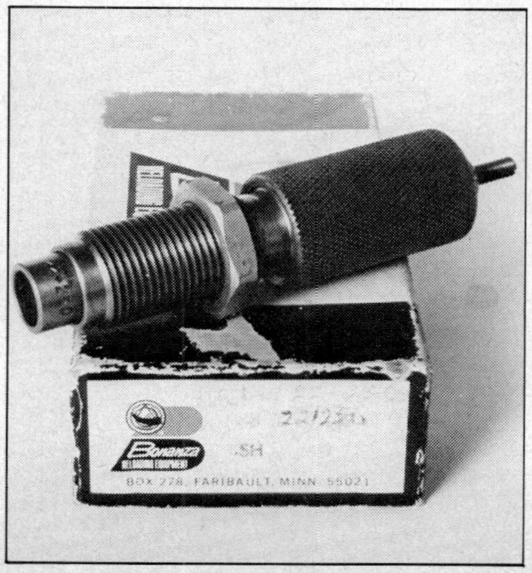

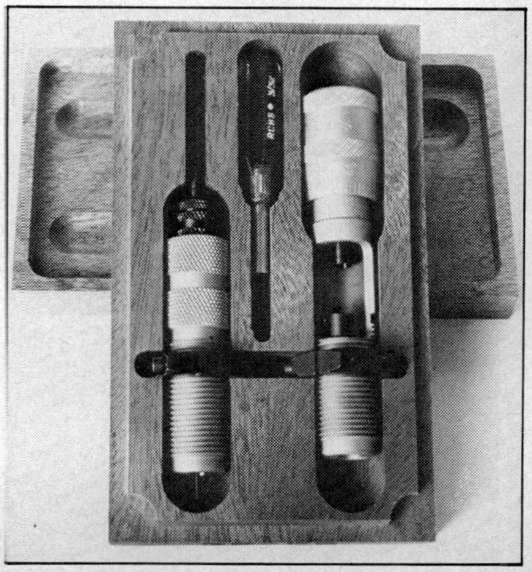

Test Report

KIMBER MODEL 82

At last, here's an American-made .22 sporter with class!

By Craig Boddington

Howard French, the editor of *Guns & Ammo,* once told me that the measure of a serious rifle nut was whether or not he owned a top quality .22 rimfire. I don't know if I'd agree with that strong of a statement, but it's a fact that the one most important, most often-used rifle in any shooter's battery is the .22. The ammo is cheap, recoil is non-existent, noise level is minimal, and a good .22 is amazingly accurate for small game, pests, and just plain plinking fun. Fun is the key to the .22's popularity.

There are dozens of accurate, well-made, functional .22 rifles available in all action types, and most of them are excellent for the uses a .22 has—plinking, target shooting, and hunting. Good though the run-of-the-mill .22 is, many of us want that most-used rifle to be something special. Not only functional, but attractive and unique. While the centerfire rifle fan has long had available any number of fancy-stocked, embellished, highly attractive production rifles to pick from, the rimfire enthusiast in search of something special has found slim pickings indeed. For years the realm of high-grade .22s has been limited to a few models—Remington's fine 541S, the superb German-made Savage-Anschutz line, Browning's high-grade .22 autos, Weatherby's Mark XXII, and Harrington & Richardson's Model 700 Deluxe. Well, one more fine .22 doesn't expand the field all that much, but the newest "top of the line" rimfire should answer a lot of shooters' demands for "something special."

I first saw the new Kimber Model 82 .22 Long Rifle at the Shooting, Hunting, and Outdoor Trade (S.H.O.T.) Show in San Francisco last January, and I'd have to say that it was one of the most exciting of all the new products unveiled there.

The best word to describe the Kimber is "classic." The hand-checkered black walnut stock on our test rifle had exceptional grain and color and just a trace of fancy figuring. The rifle is loaded with nice touches that show a lot of care and thought in its design. The pistol grip cap is polished blued steel, and the butt plate is checkered steel in a traditional crescent shape. The triggerguard and floorplate are one-piece blued steel, and would do justice to a fine centerfire rifle. Quick-detachable sling swivel studs are installed, and the five-shot magazine sits flush so as not to mar the clean lines of the rifle. Metal parts are blued and free of tool marks, and the stock has an attractive epoxy finish, resembling hand-rubbed oil but purportedly more durable.

Our test rifle was factory fitted with a Leupold M8 4X Compact scope with duplex crosshair. The two-piece mounts are designed for the Kimber and fit onto the dovetailed receiver. Our rifle's barrel was plain with no provision for open sights, but the rifle is also available with a Williams hooded ramp and bead and either a Williams or Lyman rear sight.

The rifle weighs a comfortable seven pounds, and it hefts like a fine sporting rifle, whether centerfire or rimfire. The stock feels comfortable against the cheek, and our rifle comes up with the crosshairs lined up and on target. It's grossly unfair to compare a brand new rifle with an old rifle, but if the new Kimber calls anything to mind, it would be the old Winchester Model 52 Sporter, long the king of American rimfires and regrettably discontinued in 1958.

However, the Kimber is not a Model 52, and although the gun is totally American made, the action dates back several decades and comes from Australia. Years ago, Jack Warne of Sporting Arms Limited developed and marketed the Sportco Model 62 bolt action .22 rifle.

The old Sportco action, redesigned, is the basis for the Kimber and it's a strong and efficient design. The receiver is beefy, and the cock-on-opening bolt is locked by two horizontally opposed locking lugs. The safety is conveniently located at the right rear of the bolt handle, and it locks both trigger and sear.

The trigger is wide, ⅜-inch, and is completely adjustable for weight of pull and overtravel. Our test gun had a crisp pull of three pounds, so we elected to leave it alone—it was just fine!

Jack's son Greg Warne wanted to produce the ultimate .22 sporter for those who demand quality and looks as well as utility. At a price tag of $569 for the scoped version, several times that of a well-made functional .22, the buyer has a right to expect a lot for his money. In

Mounted with a Leupold M8 4X Compact scope, the new Kimber Model 82 is designed for the rimfire hunter/shooter who demands the ultimate in looks, accuracy, and reliability.

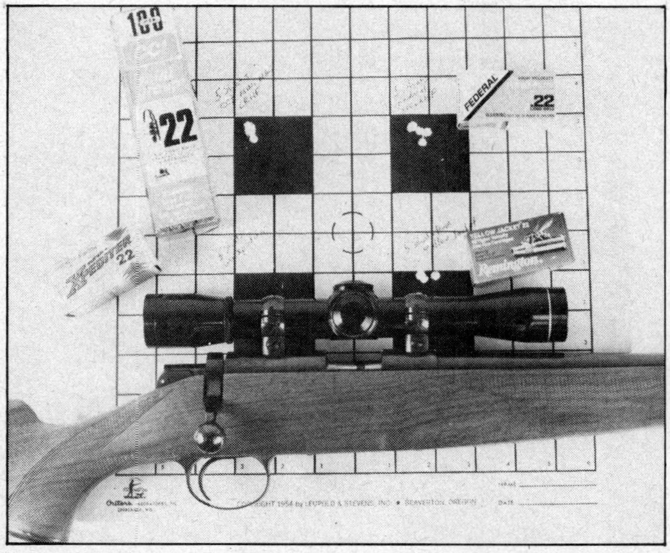

The Kimber produced excellent accuracy with all ammo tested, but grouped best with high velocity .22 LRs, as fits a hunting rifle.

The Kimber handled well under hunting conditions and proved easy to shoot. Author found it a dandy outfit for varmints and small game.

the looks department, we feel he gets his money's worth.

Needless to say, a .22 with a tariff like that should shoot as well as it looks. With that in mind, we repaired to the range to do some serious grouping with the new rifle. It seems that whenever we test a new gun, we're constantly apologizing for the high winds. Well, all I can say is that the wind blows a lot in California's high desert, and this day was no exception. With gusts hitting from 20 to 30 mph, we made no effort to group the rifle at 100 yards. Rather, we confined our testing to 25 and 50 yards, using eight different types of ammo, ranging from match ammo through high-velocity hollow points to the three ultra-velocity .22s—CCI Stingers, Winchester-Western XPediters, and Remington Yellow Jackets. Functioning was smooth and trouble-free with all types of ammo.

The results were a bit monotonous. At 25 yards, all shots touched in all five shot groups! At 50 yards, groups averaged from an exceptional ½-inch to an acceptable one inch. At that distance, the wind seemed to have a definite effect on the lower velocity ammo, while the best groups came from the three hot rimfires.

Of course, the Kimber is a sporting rifle, not a serious target rifle, and it's a rifle that should be taken to the squirrel woods, the rabbit warrens, and the prairie dog towns. The fact that it liked the hotter .22 loads was just fine with me, because our next test was to match it against the local ground squirrels. These little rodents are extremely common in our area, often to pest proportions, and in southern California they're known to be carriers of bubonic plague. Thus it's both desirable and necessary to keep their population in bounds.

The Leupold scope as supplied by Kimber is specially adjusted to be parallax free at 75 yards for optimum shooting at .22 caliber ranges, and we found both the rifle and scope a deadly varmint/small game combo. In fact, we found it difficult to miss a ground squirrel at any distance under 100 yards!

On first glance, I liked everything about the Kimber except the price. After shooting it, I'd say it's well worth the money. It's not a rifle that will be in everybody's budget, but many of us don't get too upset over paying that kind of cash for a high quality centerfire rifle, a rifle that most of us will shoot less than a hundred times a year. Why not, then, pay a little extra for a rifle that we can enjoy year 'round, whether for head-shooting cottontails and squirrels or just for busting tin cans on a lazy afternoon? If your taste runs to top quality and top performance in an appealing package, then the Kimber Model 82 is for you. And if what Howard French said is true about a serious shooter needing a fine .22, then the Kimber is a gun to get serious about!

SPECIFICATIONS

KIMBER MODEL 82

Manufacturer:	Kimber of Oregon, 9039 S.E. Jannsen Clackamas, OR 97015
Caliber:	.22 Long Rifle
Barrel Length:	22 inches
Overall Length:	40½ inches
Weight:	7 pounds
Trigger Pull:	3 pounds
Action:	Bolt action
Magazine Capacity:	5, 10
Sights:	Bead front, adjustable rear, or scoped
Finish:	Polished blue
Price:	$388, plain barrel $413, open sights; $569 with Leupold M8 4X scope.

Regardless of the intrinsic accuracy capabilities of any hunting rifle, shooting from a field rest must be done correctly. This is to say, that the support hand must be between the barrel and any hard surface. If this isn't done, the rifle will "shoot away" from the surface, and the result will be a wildly erratic shot.

Worse things can go wrong with a rifle than deterioration of accuracy; even inaccurate rifles are usually capable of bagging game or doing self-defense duty at close ranges. A broken firing pin or mainspring is more serious, but fortunately easier to remedy, providing one has brought along spares. In this case the cure amounts to no more than stripping the bolt and installing the part. There is no question about what is wrong or the remedy if the mainspring or firing pin breaks, since it is an abrupt transition from functional to nonfunctional status.

Deteriorating accuracy, on the other hand, is hard to detect and cure. Accuracy has been known to disappear as mysteriously as a frightened wraith or ghost with business elsewhere. Usually it happens on a new rifle as the still green wood dries out or otherwise warps, but it can happen on a trusty standby hunting rifle which has performed steadily and reliably for years. Many shooters with years of experience, myself included, have been mystified at times by the elusive cause of inaccuracy in certain rifles.

I well remember the most trying experience I ever had with an inaccurate rifle. It was a Super Grade Winchester pre-war Model 70 with a factory custom stock. The original barrel had been replaced with a Pfeiffer Premium Grade 7 mm Weatherby Magnum barrel of medium weight. The 7 mm Weatherby is a fine short magnum cartridge in the same category as the 7 mm Remington Magnum, and I envisioned excellent short to long-range performance.

Armed with a variety of handloads in 150 and 175-grain bullet weights, I visited the range, assuming that I would not get the tightest groups with these first loads, but only average ones until I found out which loads performed best.

When I settled down to serious benchresting, with the rifle sand-bagged at front and at butt, the best groups I could obtain averaged around five inches for five shots at 100 yards. I scrutinized every detail of the rifle, scope, and mount, to see if I could locate something loose, such as a mount base or a loose reticule. I checked the guard screws and the big middle screw under the hinged floor plate, but all was in order. I held the rifle up and pulled the forend away from the barrel as a friend slipped a business card in just behind the line where the forend tip joined the stock. None of this made any appreciable difference in the horrendous group size.

Now there is no doubt that I could

have taken game with the rifle within 200 yards, which is the average limit anyway, but I was not about to accept such inaccuracy from a rifle of such quality and a cartridge of such long-range capability. I tried 140-grain Hornady Spire Points to see if the rifle would shoot with lighter bullets. I checked the twist and found it 9½ inches per turn, satisfactory for most medium to heavy bullets.

Finally, after trying every load and bullet I could conceive of, I decided to see if rechambering the rifle to 7 mm Remington Magnum would improve its performance. I took the rifle to Arnold Capone of King's Gun Works, and he set the barrel back slightly and rechambered it. Then the process of load testing continued with the same negative results.

I then epoxy-bedded the receiver and the first 2½ inches of the barrel. The rest of the barrel I scraped free except a small "pad" at the bottom of the forend channel, just behind the juncture of the forend tip and the wood. This did nothing to improve the shooting. I tried free-floating the entire barrel channel. I changed scopes. And the inaccuracy persisted with remarkable consistency.

I thought that if my group sizes held so uniformly to around five inches, through all those caliber, scope, and bedding changes, that perhaps I had stumbled onto some new principle, but I could never learn what it signified.

What was the cause of the inaccuracy? I haven't the foggiest idea. Whatever it was existed in the barrel in the form of stress-related harmonics I would assume. The twist, groove diameter and chamberings were within proper tolerances. Some barrels simply won't shoot with any loads; they have erratic vibration patterns which no corrective bedding or load variations will compensate for. Admittedly this is rare, but I have encountered it in three rifles in my life.

Stock bedding problems are probably the most common cause of inaccuracy, aside from inherently inaccurate barrels. This assumes that the problem isn't the result of poor shooting. The only way to find out if it is your shooting ability or your rifle's, is to do all test firing with the rifle's forend solidly cradled in a sand bag or shot bag rifle rest and with the butt solidly bagged to form a precise sight alignment without hand support.

Remember that the purpose of such testing is to determine the accuracy of your rifle and not to display any target shooting ability. Your personal shooting ability must be left out of the equation as much as possible. If you are not confident of achieving this kind of controlled, immobilized benchresting then ask the range master to help. Usually there are a few experts at any range who enjoy doing this sort of thing.

When a new rifle refuses to group, or groups in a formerly accurate rifle begin to spread, perhaps the most common cause is loose screws. These may be either guard screws or scope mount base screws which have been installed without Loc-Tite or other cementing.

To properly remove or tighten a gun screw, the rifle or gun must be held solidly, which means in the padded jaws of a vise, unless in camp or where you have a friend hold the rifle as you work the screwdriver. Some find that they can hold the rifle in their lap, but if you slip while doing this you can damage more than your rifle!

The key to proper use of the screwdriver is to hold the tool absolutely as vertical as possible to the screw head. Use the stronger hand for gripping and turning only and the opposite hand for pressing down hard on the handle to prevent the blade from slipping out. If a screwdriver slips out of the slot, it is either because the blade is wedge-shaped, or it was not held down tightly enough.

In cases of very tight screws, a sharp rap on the screwdriver with a hammer or mallet will sometimes break the rust or oxidized oil cementation in the threads preventing removal. I prefer to wrap a piece of sandpaper around the handle, grit side in, paper side out. This will give us a non-slip grip. If the screw doesn't come out with such efforts, with rifle held solidly in a padded vise, take the rifle to a good gunsmith.

Sometimes action screws on a bolt action rifle can bottom out and fail to clamp the receiver to the wood. Usually this can be noted by placing the rifle in padded vise jaws around the barrel shank so as not to squeeze the thin part of stock around magazine, and shaking the barrel up and down. If not clamped solidly down to the inletting, the barrel and receiver will display a rocking independent of the stock. To cure this rocking effect, find out which screw offends and grind one or two threads off the end and rechamfer the screw tip.

When installing your barreled action in its stock, it is incorrect to merely install and tighten each screw in turn. If this is done without regard to the effect on positioning the receiver bottom surface evenly and flat relative to the inletting, a bending of the receiver tang or a strain can be created. The screws may be tight but the receiver will not be sitting down flat in the inletting.

To properly bed a barreled action, I prefer to press or tap the receiver down to a normal relationship with the inletting, and then to snugly draw up first the front guard screw, then the rear, while holding the receiver firmly in place. If there is a forend screw, this can be snugly installed before installing the action screws so as to hold the barreled action in relation to the stock. The better and closer the inletting of a stock, the easier it is to assemble the barrel and action in the correct and proper relationship.

Stocks with ample free-floating barrel channels require care in installing action screws since sometimes tightening the front guard screw first will pull the barrel down tight in its channel and lift up the tang so that it bends when its guard screw is tightened. In addition, many stocks have irregular bottom surfaces or rounded ones where the receiver bottom beds down, permitting the receiver to tilt one way or the other, depending on which guard screw, front or rear, has been tightened first.

Do you know what to do when your favorite rifle goes haywire? Here are an expert's . . .

RIFLE ACCURACY REMEDIES

By Jack Lott

When the receiver has been placed straight into the stock and the forward guard screw snugged up enough to hold the barreled action in alignment with the inletting, then snug up the rear guard screw and tighten the front screw firmly but not heavily. Then do the same to the rear guard screw. At this point it will be safe to tighten the front guard screw tightly and then the rear, but the main thing is never to tighten one guard screw before the other is firmly snugged up to prevent tipping or straining the receiver, causing distortion and sometimes bolt-binding when both screws are tight.

The condition of irregular non-fitting receiver mortising should be corrected as soon as possible, by a stockmaker or by epoxy bedding. It will continue to cause trouble if not corrected and the receiver bottom, flat or round, is the base area for proper location of the heart of our barreled action—the receiver and its bolt. Any stock can be precisely bedded with epoxy as made by several gunsmith supply houses, such as Bob Brownell's "Accra-Glass." Brownell's and other "glass bedding" kits have adequate instructions for accomplishing a first class job. The only area that requires a solid floor of "glass bedding" is the receiver inletting, but personally I find it advisable to also bed the first two to three inches of the barrel shank. If you do not want to do the work yourself, it can of course be taken to a good gunsmith.

Once the receiver is solidly bedded without strain, it is then time to check the barrel channel. It is not always necessary to free-float the barrel for best accuracy, but it is the cheapest way and, on a production rifle or a military conversion, probably the best. This is to a large extent because the more porous production commercial and military stocks do not lend themselves to solid wood bedding since they are more subject to warping than hand-inletted, dense and well seasoned stocks.

Personally, I do not like a wide gap around my sporter barrels, and I find that free-floating is best when there is but a slight clearance between the wood and barrel and the major clearance below the stock line. I test this by passing an ordinary piece of typing paper between the wood and the barrel, but such free-floating is insufficient for porous wood and is only intended for more expensive hand-inletted stocks of hard stable wood. This method also excludes the ever-present debris that can and will accumulate while hunting in the woods or brushy areas. Some rifles seem to shoot better with full contact in forend between wood and barrel when a solid pressure point at forend is constant. Others do better with only cosmetic contact at the barrel line of forend and cleared slightly below. Again, these last mentioned methods of barrel inletting apply only to top stocks of high quality, stable woods.

Some think that lightweight barrels are best when free-floated, but I have found that when very light they often need much more contact with the wood than do heavier barrels, so as to dampen their excessive vibration. This is why I have had such good accuracy in lightweight full-stocked sporters including several Mannlicher-Schoenauer originals in 6.5, 8 mm, 7 x 57, 9.5 and .270, all having unusually fine accuracy and all being full-stocked carbines with short light barrels. Much misinformation is bandied about concerning the alleged inaccuracy of full-stocked carbines and rifles, based largely on assumptions. My standby 24-inch barreled Holland & Holland .375 Magnum Mauser is stocked in this manner, using a second piece of walnut for the forearm which is glued on a long bias and doweled to the butt stock in the center of the forearm. There is a stud projecting from the middle of the barrel into which a sling swivel stud attaches via an engraved diamond-shaped plate. The muzzle is held in a constant relationship to the stock via a close-fitting, engraved steel muzzle cap.

When the stresses of still uncured wood react under atmospheric changes as with some production rifles, the result is often an erratic pressure on the barrel, but sometimes it warps away from the barrel. It is always better to have consistent rather than erratic or intermittent contact between wood and barrel, but with production quality porous walnut it may be better to simply free-float it or scrape the wood away under the topside plane of barrel channel and add a small "pad" of wood or business cards so as to apply pressure at the forend until adequate dampening effect has been achieved.

The business card or cards, depending on how many are required for the approximately ten pounds of upward pressure, can be cemented with epoxy or other good cement and then waxed for weatherproofing. It is amazing how stock problems can sometimes be cured simply by slipping business cards under the forend until enough pressure exists to dampen vibration. If this fails, then it may be time to free-float, but don't overdo it and clear out a space big enough to provide shelter for snakes, spiders and other squatters.

As I have stated, the correct bedding of the stock is the basic therapy for bringing the rifle to its best shooting potential, but if inaccuracy persists then you must look elsewhere. At least you will then start with a solidly bedded receiver and a barrel channel which is

If there's one golden rule to checking the accuracy of any rifle, it would be to remove as much of the human element as possible during firing. The shooter above, testing a Ruger No. 1, has rested the forend, as well as his elbow to achieve steadiness.

either closely contacting or free-floating.

It may sound as if I am a Dark Ages reactionary when I say that scope sights are a mixed blessing, but insofar as accuracy is concerned, scope sights and their mounts can be a source of inaccuracy that defies all but the most careful investigation. For one thing, iron sights, particularly receiver mounted peep sights, or those long-range micrometer tang sights such as the British use for service rifle shooting up to 2,000 yards, do not suffer from parallax. Parallax is a condition which exists when the image converges either before or behind the telescope sight reticule. Parallax does not occur when the eye is exactly on the optical center of the telescope, but since such precise positioning is not normally possible, any scope with parallax at the basic range of 100 yards will produce errors of aim, depending on the degree of parallax.

Unfortunately, there is no practical way that the shooter can know when his eye is exactly on optical center in view of the fairly wide latitude of eye positioning relative to the reticule and target images. Parallax can be easily detected by positioning the rifle in a normal shooting position on sandbags or a rifle rest so that the reticule of scope is centered on a bull without hand support. Once it is thusly positioned, move the head upwards and downwards as well as back and forth sideways, and if the reticule appears to "dance" while the bull stays still, then you have parallax. If excessive and if you do not position your eye at approximately the same position with each shot, then the result is the same as if you used a slightly different aiming point with each shot, thereby creating a much larger group than the rifle's accuracy would otherwise make. In hunting scopes of low to medium power with no parallax adjustment, the remedy is either a factory repair or a new scope.

Guns & Ammo **Executive Publisher, Tom Siatos tests a Shilen DGA benchrest rifle in 6 mm Remington. The butt of this heavy barreled tackdriver is firmly bedded on a sandbag, and the result will come as close to machine rest precision as humanly possible.**

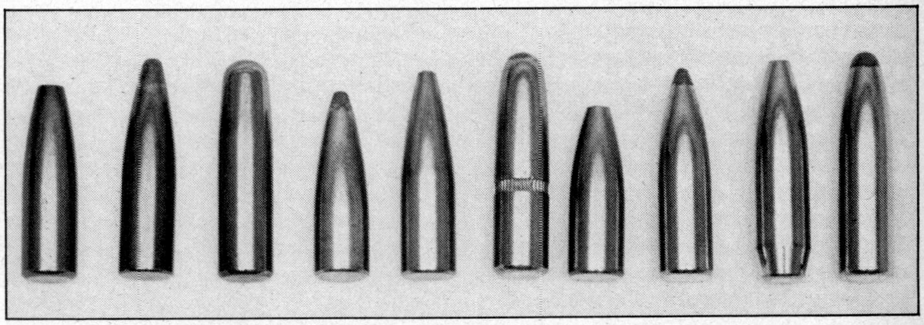

Of all the variables affecting the accuracy of a given rifle in a given caliber, the types of bullet styles—and their weights—are among the most important. All the above bullets are for the 6 mm,—from 105-gr. RN to 75-gr. HP. Some will give better results.

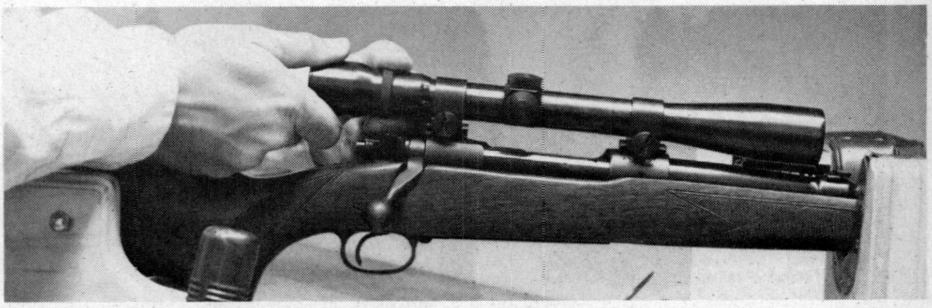

Final adjustment of the scope—with regard to such factors as eye relief and focusing are among the most basic—and simplest—steps to ensure that the rifle will perform well.

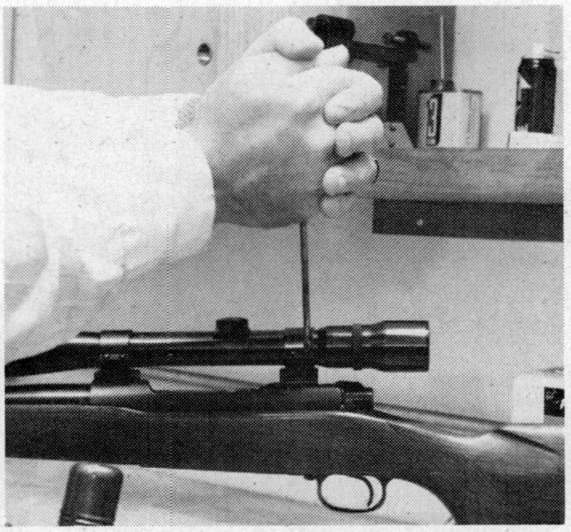

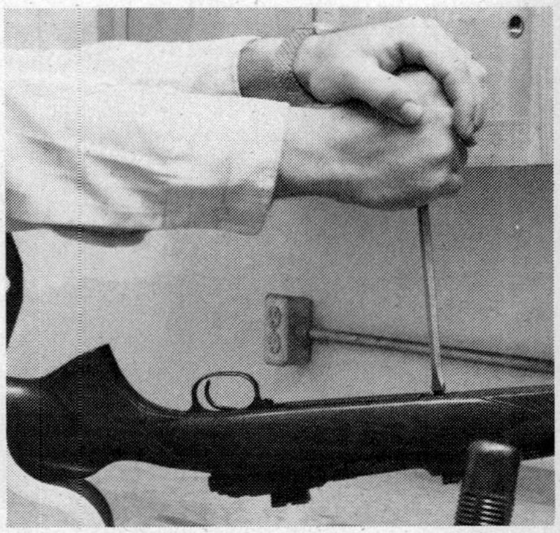

There are quite a few remedies available for improving the accuracy of a rifle. One of the most effective, is to glass-bed the receiver (left). This will provide uniform contact between the action and the stock, and can go a long way towards shrinking the size of those groups! Other steps that can be taken are even simpler, and because of their obvious nature, are often overlooked. Mount screws can loosen (above) and cause inaccuracy. This is one of the first things to check if your rifle is not performing. Another often-ignored culprit is the action screw (above right). This should be tightened snugly before you even take the rifle to the range for initial zeroing.

Another source of scope inaccuracy is the failure to tighten eye piece lock rings, which also causes an inaccurate optical center which changes with recoil causing the loose eye piece shift on its threads.

Loose scope mounts are often the cause of erratic performance and should be kept tightened up with screws cemented with semi-hard Loc-Tite, shellac, or cement, after de-greasing threads. The poor fit of scope tube to rings is also a possibility, though a rare one. I have seen scope rings which failed to clamp the tube properly due to a poor fit, or in cases where a 26 millimeter ring set is used on a one-inch tube, which is about .020-inch smaller in diameter. In such a case a well fitting paper or metal shim can be used to take up the difference.

A loose reticule can sometimes occur, and it is not always possible to see it move except in gross examples.

An incorrect twist for the bullet can sometimes be the cause of keyholing inaccuracy in custom barrels where it was caused by error or was a special twist for a heavier or lighter than normal bullet. If so, it can usually be made to shoot with the proper weight bullet, if this is satisfactory to the shooter. Such conditions occur most often when we purchase a second hand custom-barreled rifle .

Old primers in old or recently handloaded ammo can cause erratic ignition with resultant inaccuracy, and if any suspicion of unreliable priming exists, such as a frequency of hang fires, they should be discarded.

Shooting with the barrel resting on a hard or semi-hard object will cause the bullets to throw away from the point of barrel contact. The rule is never to shoot with anything but the forearm resting on the forward hand or something soft like a sandbag, hat, or rolled-up jacket.

Rifles with newly-installed quick detachable sling studs in forearm can cause erratic performance when the screw part projects even slightly into the barrel channel and contacts the barrel. To determine the existence of this condition, remove the barreled action, smoke the bottom of the barrel, and reassemble. And if upon the subsequent disassembly a bright contact spot shows, remove the screw and grind or file enough off to clear the gun's barrel and a bit more.

Throat erosion on hot magnums is a frequent cause of inaccuracy, and this can sometimes be cured by re-throating without setting the barrel back. If severe, barrel setting-back may be needed, along with re-chambering and re-bedding.

Poor or erratic ignition caused by a dragging striker, a weak striker blow or excess headspace causing the firing pin to drive the case forward, thereby reducing the blow on primer, is a source of inaccuracy. A trigger which hangs up, creeps excessively, or releases erratically permits the rifle to wobble a bit before ignition. Sometimes this is due to insufficient clearance between wood and trigger, which most gun owners can correct. An internal trigger problem should be fixed by a gunsmith or the trigger returned to factory or replaced.

Excessive pressure caused by overloading or case lengthening causing the case mouth to jam against the end of chamber thereby pinching the bullet is another cause of inaccuracy. By too much pressure, I don't mean for safety, but best accuracy is usually obtained with 85–90% of full power loads. We obtain the last 50 fps, up to full power, by sacrificing accuracy.

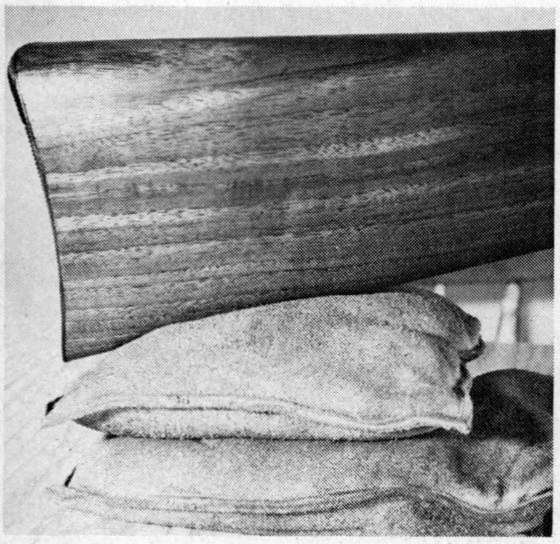

Closeup shows the "sandbag under the butt" method of bedding a rifle for maximum steadiness. Use it when possible.

Sometimes specialized equipment like an adjustable rest isn't available and you have to make do with what you have. Here, the author had to crisscross sandbags to test a rifle.

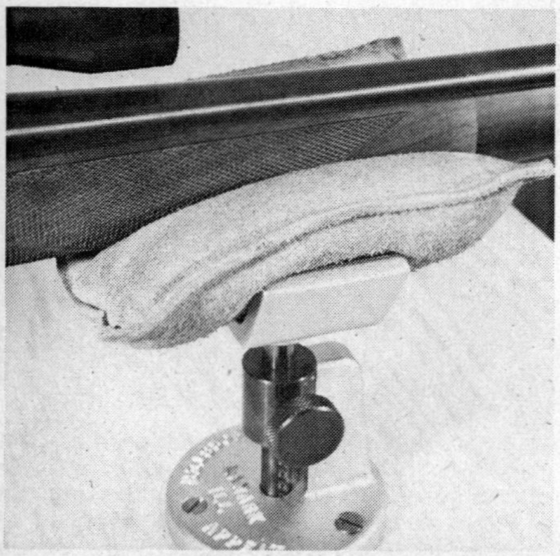

An adjustable forend rest that locks at the desired height and is padded, is an essential tool for the serious rifleman.

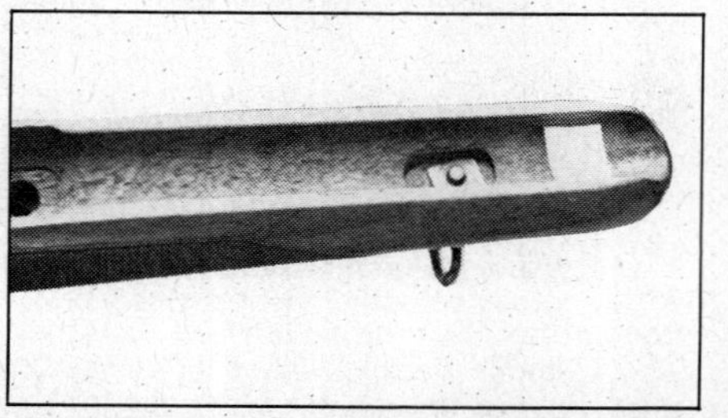

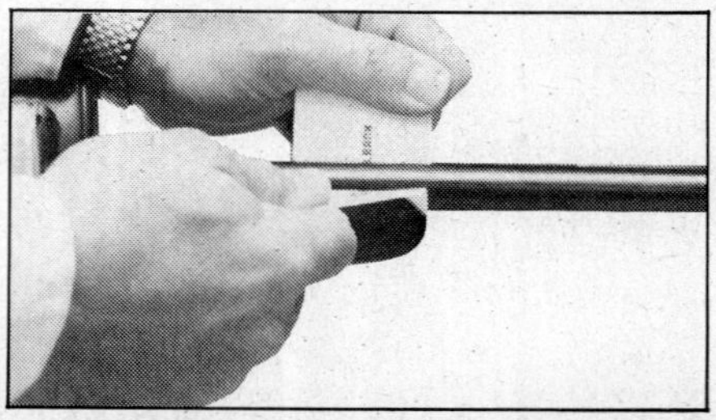

The relationship between the forend and the barrel is critical insofar as accuracy is concerned. If your rifle doesn't group well, try taking the barrel and the action out of the stock and slipping a business card in. Sometimes it will improve things.

Even ultra-sophisticated, specialized benchrest rifles won't shoot better than the sum of their components. Scopes have to be mounted correctly, screws must be tightened, and loads have to be tailored to the individual bore. The same principle applies to sporters.

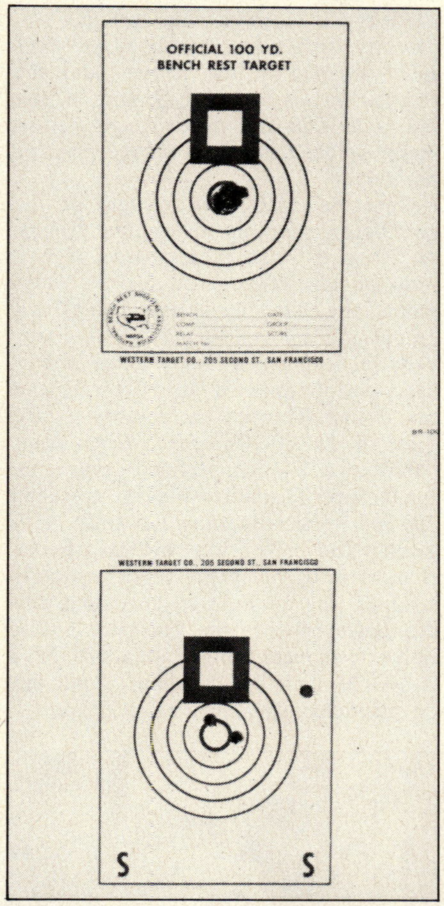

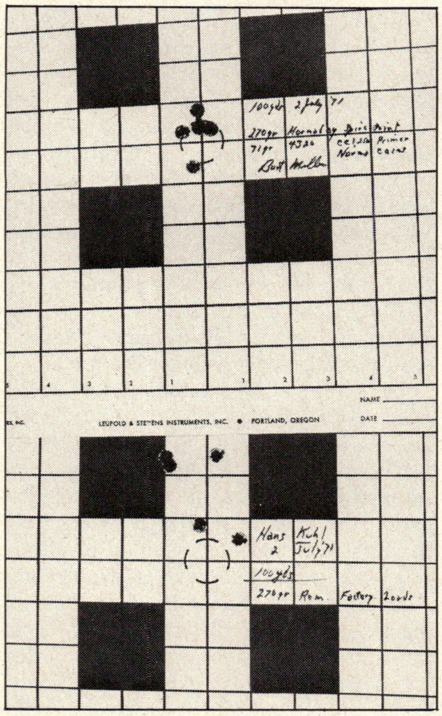

The above photo illustrates the difference in results that glass bedding can make. Top group was shot with 180-grain .30-06 ammo after bedding. Bottom group was shot prior to bedding. Probably the most effective way to improve the performance of any rifle is to experiment with handloads. Above right photo shows two groups fired at 100 yards with a .375 H&H. Top group was shot with 270-grain handloads while the bottom group was shot with 270-grain factory ammo. The difference is obvious! Photo at right shows what 180-grain bullets from a .30-06 will do after the rifle was sighted in with 150-grain bullets. The lesson is plain. Always hunt with the exactly the same weight bullets that you originally sighted in your rifle with.

When tightening up forend screw, or any other screw for that matter, a proper screwdriver won't chew up heads as shown.

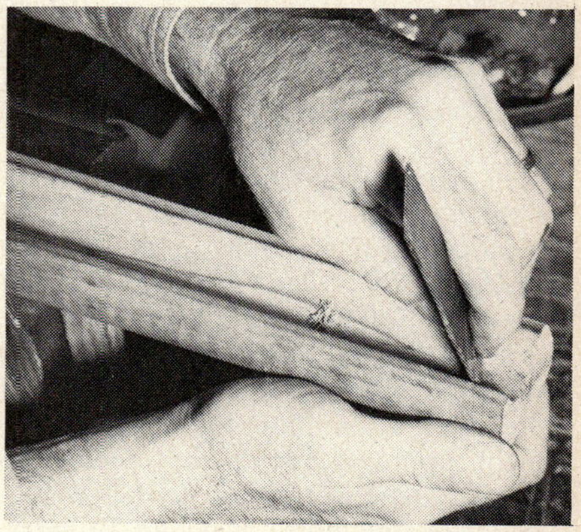

Proper harmonics from barrel vibration are essential for accuracy, and the forend/barrel relationship is a critical one.

Whereas I understand the appeal of short barrels in such cartridges as the 7x57 Mauser or the .308, the best recommendation I have is not to use barrels shorter than 24 inches on magnums. The volume of gas released in such magnums requires even longer than standard barrels to obtain magnum velocities, meaning 26 inches for less muzzle blast and full velocity.

Bullets are a major cause of inaccuracy, either from dimensional errors, nonsquare bases or an unbalanced, incompletely filled core. I also assume that eccentric jacket wall thickness sometimes causes it.

Rifle inaccuracy must be extreme before it can result in a practical inability to kill game at normal ranges, but few of us will tolerate a scope-sighted rifle which will only group inside three or four inches at 100 yards.

If the problems I have listed are the only type of accuracy problems you are having then you may well be on your way to obtaining tighter groups. But I am certain I have left out many more than I have listed, such being the mysterious nature of many accuracy problems not connected with individual marksmanship. The worst part comes when you have your pet rifle all tuned-up to produce one-inch five-shot groups at 100 yards, and then you return home empty handed after missing that big buck. I can offer no all-purpose alibi. Sometimes there are things better left unsaid!

WINCHESTER MODEL 94 TRAPPER

Winchester has trimmed the barrel of their classic M94 down to 16 inches—and the result is an exceptionally handy little brush gun—the Trapper!

By Michael Durr

Imagine a car or any other piece of machinery that would still be in production, still popular, 86 years after its introduction. It's unlikely, to say the least. Technology advances, tastes change, and "old" gives way to "new."

Yet, here I am writing a test report on the Winchester Model 94 Trapper, the most recent variation in the Model 94 series.

Brought out back in 1894, the Winchester M94 has been in continuous production all these years. As of 1980, more than 4.7 million M94s had been bought up by a public that seems to never lose its attraction to the old 94.

In 1894, the horse was the primary means of transportation and, for almost anyone who lived west of New York City, a saddle or scabbard gun was a handy item. The ideal saddle gun had a thin, flat receiver with no protrusions—bolt handles or cocking devices—to snag on the scabbard. It had to have a short barrel so the rider could remove it quickly and smoothly. The M94 fit these qualifications to a tee and became, deservedly, an immediate success.

Times have changed but the M94 stays with us. As horses have become less a part of our lives, the little lever action is just as likely to be seen in the gun rack of a pickup truck as in a saddle scabbard.

The M94 has appeared in many different configurations since its introduction. Winchester has produced M94s with straight grips and pistol grips, full magazine and half magazine, fancy finishes and plain ones, and various barrel lengths.

The new version, the Model 94 Trapper is very similar to the standard M94 Carbine currently produced. Like the standard, the Trapper is a top-ejecting lever action. The lever operates through a non-geared 90 degree sweep, long but not intolerable. A full length tubular magazine rides under the barrel.

The most significant difference between the Trapper and the standard 94 is the barrel length. The Trapper's barrel has been shortened down from the standard 20 inches to a minimal 16 inches. This shorter barrel reduces weight from 6½ lbs. down to 6⅛ lbs. With an overall length of 33¼ inches, the Trapper is the shortest and one of the lightest big game rifles around.

The finish on the Trapper is strictly utilitarian: plain, oil-finished walnut; semi-matte blued finish on the metal.

The Trapper is chambered for .30-30 Winchester, a cartridge that has been around just about as long as the M94 itself. While the .30-30 has long since stopped being a favorite of ballistics fanatics, the rimmed cartridge works well in lever guns and at ranges up to about 150 yards it will handle deer-sized game.

An important fact about the .30-30 is the availability of different loads. You can get .30-30 ammo in several bullet types, from the fast expanding variety to the more controlled expansion type bullets. Weights range from 125-grain up to 170-grain. And, Remington even offers a 55-grain, flat-shooting Accelerator version of the .30-30. All these loads give the .30-30 cartridge a whole lot more versatility than is usually attributed to it. A hunter can load his .30-30 Trapper for varmints, pelt hunting, small deer and javelina, or big mule deer and black bear!

Right out of the box, the Trapper needed a little work. Our test gun apparently hadn't been grouped or even rough aligned at the factory and the rear sight was way off. I had to fire into a hillside and make rough adjustments before the bullets started printing on paper.

Once the sights were brought in, the gun began shooting acceptable groups. At 50 yards with factory iron sights, semi-buckhorn rear and square blade front, the Trapper printed 2½-inch groups.

Barrel length is a major factor in ballistic performance of any firearm, and the 16-inch Trapper barrel does reduce some of the .30-30's speed. When compared to a 24-inch barreled lever gun, the Trapper's muzzle velocity averaged 200-300 fps slower using 125-grain, 150-grain, and 170-grain bullets. Since changes in barrel length result in nearly straight-line changes in ballistics, the difference between the Trapper's ballistic performance and the standard M94 is 100-200 fps. That's about 7 percent or, practically speaking, insignificant.

In the .30-30 Trapper, Winchester has come up with a tough little utility carbine that can handle a variety of chores. A 16-inch barrel keeps the overall length down to 33¼ inches.

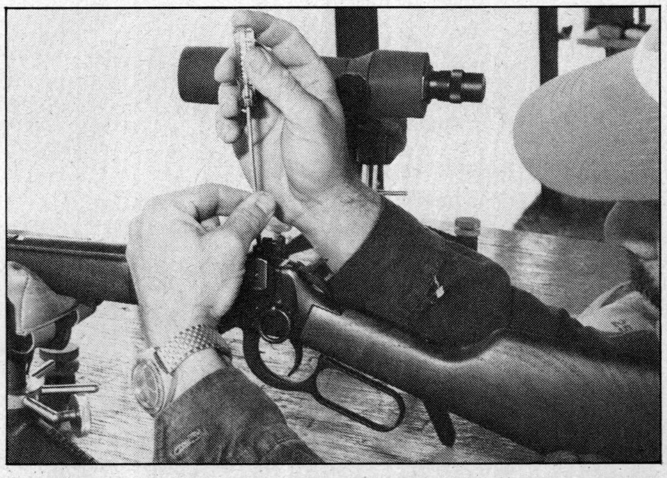

The Trapper is factory drilled and tapped for receiver sights and the installation of a Lyman Model 66 was simple and quick. Note saddle ring on receiver; it's the Trapper's only frill.

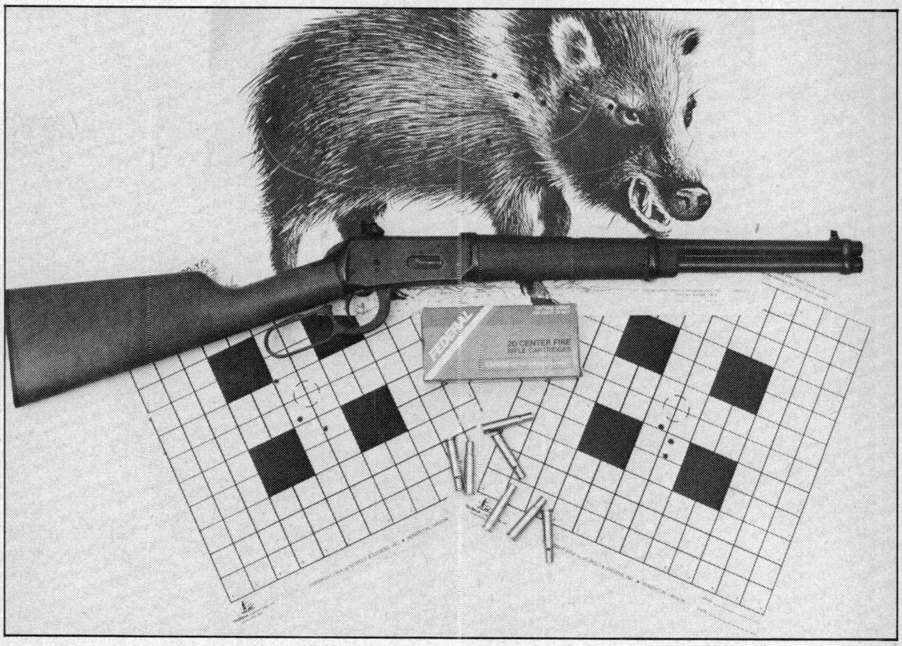

The Trapper functioned smoothly and proved quite accurate. With iron sights, it punched 2½-inch groups (above, left); a receiver sight cut the group size in half (above, right).

I have distinct reservations about most open rear sights. The Trapper is supplied with semi-buckhorns, no less accurate and somewhat stronger than those furnished on most other sporting rifles. Their sturdy build is an important consideration on a knockabout rifle, but they are so coarse that they restrict the shooter and don't really show what the rifle is capable of.

Having test fired the Trapper "as is," I next installed a Lyman Model 66 receiver sight. The gun is also tapped for scope mounts but putting a scope on the Trapper would seem contrary to the purpose and style of the gun.

The holes for the receiver sight are pre-drilled at the factory and the operation of removing two screws and fitting on the receiver sight takes all of three or four minutes. I set the receiver aperture as close as possible in line with the open sight, then removed the semi-buckhorns.

The receiver sight cut the group size by better than half. We began shooting groups just slightly over an inch at the 50-yard target. That's a respectable group size for any metallic-sighted sporting rifle!

As a repeater, the Trapper holds five rounds in the tubular magazine plus one in the chamber. The spring cover of the magazine was stiff and it took firm pressure, pointing toward the base of the receiver, to load the magazine. Invariably, a rimmed case hangs up at the loading port of a tubular magazine; the Trapper proved no exception. It took an extra shove to insert the cartridge completely.

This was off-season so we didn't have the opportunity to take the Trapper hunting. We did try the gun out shooting offhand at some running boar targets. It was fun shooting, the action worked flawlessly, and all of our shots would have resulted in pork barbeque if the target had been a real javelina or wild pig. The recoil of the Trapper is sizable, given the light weight and stock design, and by this time, having run several boxes of ammo through the test gun we were definitely feeling the shots. A recoil pad would be a sensible and welcomed addition.

With the Trapper, Winchester has come up with a useful variation to a proven winner, the Model 94. The Trapper loses a little velocity because of the short barrel, but not enough to worry about. Overall, the rifle is small, light weight, and surprisingly accurate. As a backpack gun on a sling, the 16-inch barrel would reach about neck level with little likelihood that it would snag or get in the way. As a saddle gun, utility rifle for a pickup, or woods and brush rifle, it should be just the ticket. And, at a price of $176, the Trapper is one of the most economical centerfires on today's market.

.30-30 BALLISTICS (Federal Ammo.)

Bullet Wt.	Muzzle Velocity (fps) 16 in. bb.	24 in. bbl.
125-gr. Hollow Pt.	2371	2693
150-gr. Hollow S.Pt.	2158	2415
170-gr. S.Pt.	2055	2266

SPECIFICATIONS

WINCHESTER MODEL 94 TRAPPER

Manufacturer:	Winchester-Western 275 Winchester Ave. New Haven, CT 06511
Caliber:	.30-30 Winchester
Barrel length:	16 inches
Overall length:	33¼ inches
Weight:	6⅛ pounds
Stock:	American walnut
Trigger:	4½ pounds
Sights:	Dovetailed blade front Semi-buckhorn rear
Magazine capacity:	5
Price:	$176

Saddle carbines today are more likely to be seen in the gunrack of a pickup truck. The short-barreled Trapper fits perfectly.

7x57
20 Büchsenpatronen Center Fire Rifle Cartridges
High Velocity
Remington
Weatherby
7 mm
WEATHERBY MAGNUM
FEDERAL
RIFLE CARTRIDGES
norma
SUPER 7x64
Western
SUPER X
284 WINCHESTER
Remington
7mm-08 Remington
Western
SUPER X
20 CENTER FIRE RIFLE CARTRIDGES
7mm REMINGTON MAG.
7 x 64
KYNOCH
7 m/m MAUSER
(7 x 57)
CARTRIDGES
JOHN RIGBY & Co.
Remington
KLEANBORE
280 REMINGTON
Hi-Speed
WARNING
KEEP OUT OF REACH OF CHILDREN
RIGBY ·275
CARTRIDGES
7x65R
20 Büchsenpatronen
20 Center Fire Rifle Cartridges
20 Cartouches à balle de grande chasse
Kegelspitz-Geschoß 8,0 g
Dynamit Nobel
7x65R

With short, standard, and belted cases, .284-caliber cartridges are gradually overtaking the .30 caliber in popularity. Accurate and efficient, there's nothing like . . .

The Magnificent 7 Millimeters

By Jerome Knap

The Golden Eagle, imported from Japan, is one of many fine rifles chambered for the 7 mm Rem. Mag. The "seven mag" has become *the* most popular belted magnum.

One of my favorite cartridges is a modest and, by today's standards, a rather ancient creation, the 7x57 mm. In 1892, when smokeless powder was new, the Spanish army adopted the then new 7x57 mm cartridge in equally new Mauser rifles as their service cartridge. Other European armies followed suit.

Americans first learned what the cartridge and the Mauser rifle could do in the Spanish-American war. The 7x57 mm and the Mauser rifle was clearly superior to the .30-40 Krag, and Krag-Jorgensen rifle, which was then the standard rifle of the U.S. infantry. Some U.S. troops still carried the .45-70 Springfield trapdoor, and it was pitiful compared to the Mauser.

The U.S. military brass knew almost immediately that their troops were outgunned—in both cartridge and rifle design—and set to design a better cartridge and rifle. The result was the .30-06 (actually .30-03 at first) and the Springfield rifle, both of which were copies of the 7x57 and the Mauser 98.

The British ran into the 7x57 in the Boer War of 1899-1902 and they too found the 7x57 superior to their .303 which is very similar to the .30-40 Krag.

Over the past 90 years the .284/7 mm bore diameter has spawned an incredible variety of cartridges and sporting rifles. The beautiful .275 Rigby (7x57), below left, is a typical World War I era sporting rifle. Ruger's M77 is typical of today's sporters, and has the distinction of being chambered for three 7 mm rounds—7x57, .280 Remington, and 7 mm Remington Magnum. Photo by PPC Photographer Pat Brollier.

They looked for a new cartridge and came close to choosing a cartridge of 7 mm caliber. But ultimately, they decided that they had too much money invested in the vast stocks of .303 Lee-Enfield rifles to make a switch. Instead, they opted for a new lighter, pointed boat-tailed bullets which improved the ballistics of the .303 British.

The fact that rifles are still made for the 7x57, almost 90 years later, clearly shows how good a cartridge it really is. But it wasn't long before some shooters and big game hunters realized that the 7 mm was an excellent caliber for the creation of the ideal long-range big game rifle.

Perhaps the first man to do so was Sir Charles Ross, a Scottish nobleman and inventor, who was an enthusiastic deer stalker in the highlands of Scotland. The red deer stags that Sir Charles stalked invariably called for long-range shooting. Sir Charles felt that the 7x57 mm was less than ideal for this type of hunting so he created the first 7 mm magnum—the .280 Ross.

The .280 Ross had the pre-World War I gun buffs jumping for joy. It was loaded with pointed 145-grain and round nosed 180-grain bullets. The 145-grain bullet delivered a muzzle velocity of 3050 feet per second with 26-inch barrels. This may seem pretty modest by standards of today but in 1910, when the cartridge was introduced, 3000 feet per second velocity was something phenomenal. Only the .250-3000 Savage came close and it did not come out until 1915 and then only with an 87-grain bullet. Even the 180-grain loading in the .280 Ross was nothing to sneeze at with 2,600 fps. The .30-06 delivered only a bit more and that was with a bullet with lower sectional density and hence less retained energy and velocity.

It is almost humorous to read the early accounts of the .280 Ross on game. But in a way, the accounts were similar to those about the 7 mm Weatherby in the early 1950's, and even about the 7 mm Remington Magnum in the early 1960's. Perhaps the world was simpler then.

I suspect that we would still be shooting the .280 Ross today had it been originally chambered in a Mauser-type rifle. The straight-pull Ross rifle was well made and superbly accurate but it was not suited for military use. The Canadian army abandoned it late in World War I. That spelled the demise of the Ross rifle and the .280 Ross cartridge although it was revived in the 1920's in Germany as the .280 Hagler. Some of the beautifully-made Hagler rifles using Obendorf Mauser actions were exported to the U.S. in the 1930's and sold for the princely sum of $1000.

The .275 Holland and Holland magnum came out on the heels of the .280 Ross in 1912-13, actually called the .275 Belted Rimless Magnum Nitro.

Why the .275 H&H Magnum did not catch on is a mystery to me. It surpassed .270 Winchester ballistics some 10 or 12 years before the appearance of the .270 Winchester. In fact, ballistically it equalled, if not surpassed, a number of 7 mm wildcat cartridges that had sundry gun writers writhing in ecstasy some 40 to 50 years later. The .275 H&H Magnum is still a good cartridge and a modern looking one at that. It is built on a shortened .300 H&H case—as is the 7mm Remington Magnum. The only logical explanation that I can think of why the .275 H&H Magnum did not catch on is that no modestly priced rifles were chambered for it. All of the rifles were expensive British jobs. The era of good and relatively inexpensive American-made bolt action rifles did not really start until the early 1920's.

The German ballisticians and cartridge designers were also active in the 7 mm field. Wilhelm Brenneke developed his 7x64 mm (and its rimmed sister cartridge 7x65R for break-open guns) in 1917. The 7x64 is ballistically the same as the much later American wildcat 7mm-06. The 7x64 clearly shows that the Germans were a way ahead of us in 7mm cartridge development during the World War I era.

The 7x64 instantly developed an outstanding reputation as a big game cartridge. Part of that reputation may have been because of the very dependable Brenneke Torpedo bullet but by any criteria the 7x64 mm is an excellent cartridge. Had any of the American firearms manufacturers picked up on it and called it .280 Brenneke, it might have been a runaway success.

Interest in the 7 mm wildcats in America grew. The first of the 7 mm wildcats was the 7 mm-06. The excellent but little known .276 Newton was an offshoot of this interest. The first commercial cartridge in that vein was the .270 Winchester. It appeared on the scene in 1925. Winchester seemed to have deliberately avoided the true 7 mm caliber. Why, I don't know. Perhaps Winchester wanted its own entity. Like the .280 Ross, the .270 Winchester is so close to the 7 mm bore diameter that there is virtually no practical difference between the .270 and the 7 mm. The .270 has a bullet diameter of .277. In practical terms, the .270 can be thought of as a bastard-size 7mm.

The basic difference between the 7 mm and the .270 is that there is a better selection of heavier bullets in a variety of styles and designs on the market for the 7 mm. So, to the handloader, the 7 mm is a more versatile bore size for big game hunting.

The mid 1920's produced another interesting American 7 mm cartridge—the .276 Pedersen. This cartridge was developed expressly for the U.S. Army which was looking for a cartridge that would be suitable for semiautomatic rifles and at the same time have less recoil without losing the stopping power of the .30-06. Ballistically, the .276 Pedersen was similar to the 7x57. It fired a 125-grain pointed boat-tail bullet at around 2700 feet per second at the muzzle.

So pleased was the U.S. Army Ordinance with the .276 Pedersen that they recommended its adoption in the Garand rifle. But Army Chief of Staff General Douglas MacArthur ordered the Garand in .30-06. Apparently, the army had many million rounds of .30-06 ammunition and many serviceable Springfield rifles in .30-06. General MacArthur did not want duplication.

The fact that the U.S. Army failed to adopt the .276 Pedersen did not dampen the interest in the 7 mm to wildcatters. The decades from the 1930's to 1950's saw many 7 mm magnum wildcats. It was in 1944 that Weatherby introduced the 7 mm Weatherby, first as a wildcat and later on as a proprietry cartridge.

The next 7 mm factory cartridge was half American and half European—the 7x61 mm Sharp and Hart. The cartridge was designed by Americans, Philip Sharp and Richard Hart but made and loaded in 1953 by Norma in Sweden. Even the rifle for which it was chambered was the Danish Schultz and Larsen.

The 7x61 mm Sharp and Hart is a very fine cartridge. Factory ammo delivered muzzle velocity of 3100 per second with 160-grain bullets out of 26-inch barrels. While most of the 7 mm magnum wildcats were based on shortened but blown out .300 H&H magnum cases, the 7x61 has slightly larger rim and base diameter than the .300 H&H. Unfortunately, the 7x61 mm never caught on. Perhaps, like with the .275 Holland and Holland magnum, the small selection of rifles chambered for the cartridge was the main reason.

By then it was a foregone conclusion that an all-American 7 mm cartridge would come out. And in 1957 Remington introduced the .280. This cartridge

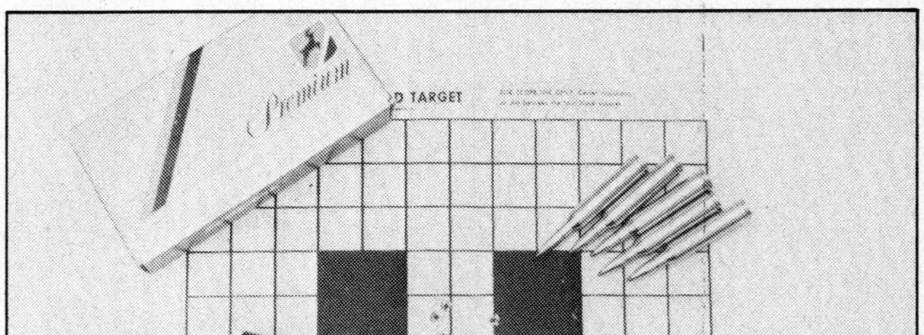

The Austrian-made Mannlicher Professional Hunter's Rifle, above, comes in several 7 mm rounds and features synthetic stock, rotary magazine, double set triggers and detachable scope. Right, the Savage 110-CL (left hand), using Federal Premium ammo, shows the fine accuracy the 7 mm Remington Mag is known for.

The high velocity, good sectional density, and long range capabilities of the modern high-performance 7 mms have made them a top choice for hunting in western mountains. This excellent New Mexico bull was taken with one shot from a 7 mm Remington Magnum.

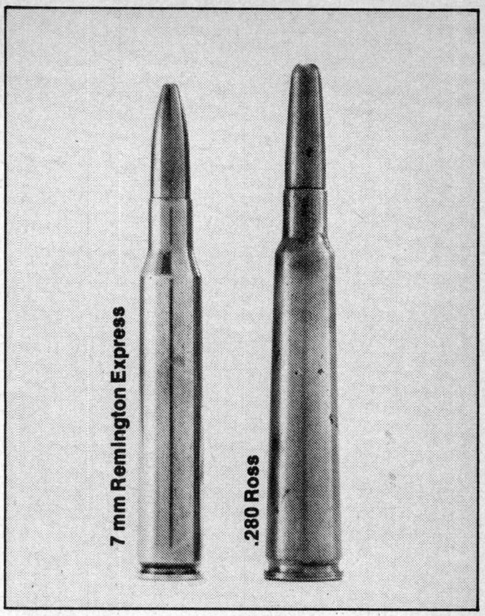

.280 Ross was first high-performance 7 mm—3170 fps with 140-gr. bullets.

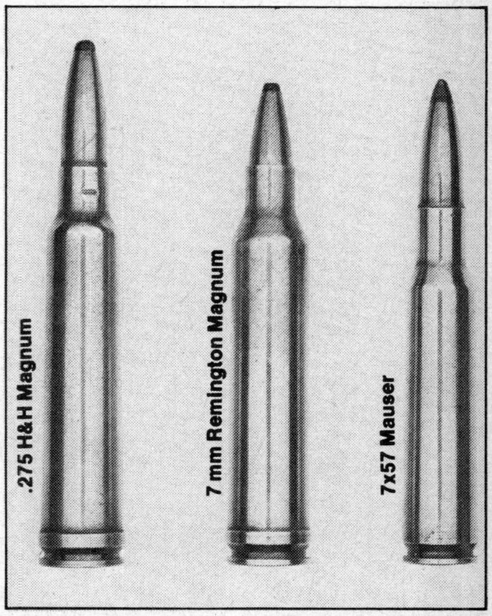

Introduced in 1912, the .275 H&H approximated 7 mm Rem. Mag. some 70 years ago!

is nothing but the 7 mm-.284 bullet in a .30-06 case necked down. And like all of 7mm-06 wildcats, it is an excellent all-around big game cartridge.

But unfortunately, Remington short-changed the .280. Initially, the .280 Remington was chambered for the Remington 740 autoloader (later 742) and then the Remington 760 slide action rifle. And finally the model 721 and 725 bolt action rifles. In order to make sure that the cartridge functioned reliably in the Model 740 and 760 rifles, Remington purposely loaded the cartridge down to muzzle velocity of 2890 feet per second with 150-grain bullets and 2820 feet with 165-grain bullets. Only by handloading for a bolt action rifle could the .280 Remington deliver its true potential. I firmly believe that it was this underloading that prevented the .280 Rem. from achieving more popularity.

The .280 Remington did not satisfy the craving on the part of many riflemen for a factory 7 mm magnum. In 1958, Winchester brought out the .338 Winchester and the .264 Winchester which of course is 6.5 mm. No doubt Winchester missed the boat by not bringing out a 7 mm instead of the 6.5.

Remington quickly capitalized on Winchester's mistake and in 1962 brought out its now famous 7 mm Remington Magnum and at the same time, introduced the fine Model 700 bolt action rifle. The new rifle was chambered for the new cartridge.

There is no question that Remington's step was long overdue. It took an American firearms manufacturer nearly 50 years to realize what the British knew in 1912—that a 7 mm short magnum came close to being an ideal cartridge for shooting of thin-skinned big game at long ranges—for the vast majority of shooters.

But even then, Remington did not come out with the ideal long range loading for the 7 mm Remington magnum. It introduced the cartridge and still loads it with a rather indifferent 150-grain bullet with a rather indifferent ballistic coefficient rather than the classic 7 mm bullet, a 160-grain spitzer boat-tail.

It is possible to safely handload the 7 mm Remington Magnum with 160-grain spitzers to 3200 feet per second at the muzzle. This velocity, along with the excellent ballistic coefficient of the 160-grain spitzer boat-tail, makes the 7 mm Remington a superb cartridge for anyone who wants the flat trajectory of the .300 Magnum with 180-gr. spitzer type bullets but with substantially less recoil.

As good a cartridge as the 7 mm Rem. Mag. is, it could hardly be considered new or revolutionary—even 18 years ago. It is largely a factory version of a number of short, belted 7 mm wildcats such as the Dubiel, Mashburn, Ackley, Luft, ICL and others. Its virtue lies in the fact that it is a commercial cartridge and available in moderately priced factory-made rifles.

Actually, the Europeans were again ahead of us in the development of a modern 7 mm magnum. Walter Gehmann introduced the 7 mm Von Hofe, sometimes called the 7x66 in 1962. Stoeger carried Swedish Husqvarna rifles chambered for this cartridge in the early 1960's. But even prior to World War II, the Germans had a couple of other 7 mm magnums—the 7x73 Von Hofe, a belted rimless cartridge with dimensions close to the .300 H&H and the 7x75R, a rimmed sister of the rimless 7x73.

The American development of 7 mm cartridges did not stop with the 7 mm Rem. Mag. In 1963, a year after the 7 mm Rem. Mag was introduced, Winchester brought out the .284.

The .284 Winchester is a short cartridge—2.170 inches long—shorter than the 7x57. It is the first American cartridge with a rebated rim that is actually smaller than the body of the case. The .284 was designed expressly for Winchester's Model 88 lever action and Model 100 semiautomatic rifles. Later, for a few years, Savage chambered the Model 99 for the cartridge as well. The usefulness of the .284 Win. rests entirely on the fact that it is short enough for lever action rifles. Ballistically, the cartridge is similar to the 7x57.

Unfortunately, Winchester no longer makes the Models 88 and 100 and Savage no longer chambers the Model 99 for the .284 Win. The .284 is doomed for obsolescence unless some unforeseen development comes to its rescue.

The next development in the 7 mm cartridge saga was the revitalization and renaming of the .280 Remington. Actually, Remington has done this before by beefing up the loads of the .244 Remington and changing its name to 6 mm Remington. The new loading and name gave the low luster .244 a new lease on life. Remington hopes that the revitalization of the .280, now called 7 mm Remington Express, will do the same.

The muzzle velocity of the 7 mm Remington Express has been beefed up over the .280. It is 2970 feet per second with a 150-grain pointed bullet out of a 24-inch barrel.

Actually, I am very pleased by this Remington development. I was afraid that the .280 Rem. was headed for the boneyard which would have been a pity because the .30-06 case has the ideal powder capacity for the 7 mm caliber. But I will have some more to say about this later.

The newest 7 mm cartridge is the 7 mm-08 Remington. As the name suggests, this cartridge is the .308 Winchester case necked down to 7 mm. This development is also not really new or revolutionary. There have been wildcats of this sort ever since the .308 Winchester was introduced in 1952. The 7mm-08 wildcats came into prominence with the introduction of the Winchester Model 88. And as I outlined earlier, the .284 Winchester was designed to fill the need for a short 7mm cartridge.

Thus far, the only hunting rifle that Remington is chambering for the 7 mm-08 is the 788. The 7 mm-08 is also chambered in the Model 700 "Varmint Special" for silhouette shooting. Shilen also chambers its DGA Silhouette rifle in this caliber.

With a muzzle velocity of 2860 feet per second with 140-grain pointed bullet, the 7 mm-08 Remington is a fine medium sized game cartridge but in the hunting field it can't do anything that the almost 90 year-old 7x57 can't do as well, in fact, better. That brings us to the thorny and debatable question of comparing the different 7 mms. Where do they all fit in? In fact, do they fit in? Do we really need all of them?

I'll begin answering these questions with a reference to the newest 7 mm cartridge—the 7 mm-08 Remington. If the 7 mm-08 has a useful role play in hunting then it is the same one that the .284 once had—it's a cartridge short enough for a lever action or a short bolt action. In the hunting field it duplicated the 7x57.

Basically, we have three classes of 7mms—the 7x57, the 7x64 and 7 Remington Express and the 7 mm magnums.

The fact that the 7x57 has not only survived 90 years but has grown in pop-

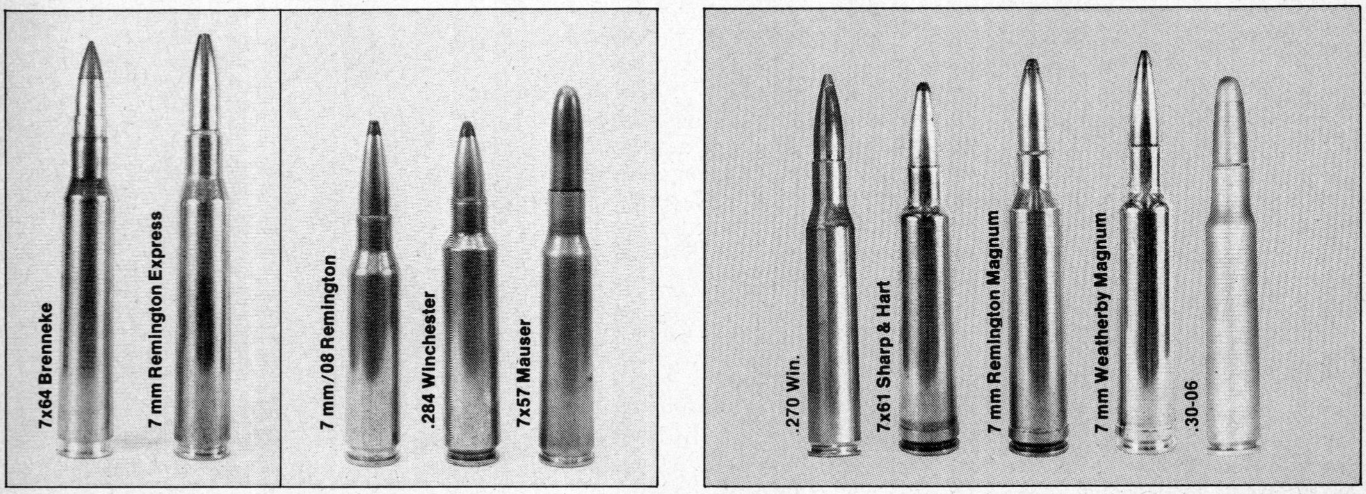

The author divides modern 7 mm cartridges into three classes: Left, general purpose cartridges suited for the majority of big game hunting; center, short range rounds for use in lightweight, short action rifles; right, magnums for exceptionally long range.

7 MM CARTRIDGE COMPARISON

Cartridge	Bullet Wt.	MV FPS	100 Yds.	200 Yds.	300 Yds.	ME Ft. lbs.	100 Yds.	200 Yds.	300 Yds.
7x57	139-gr.	2660	2400	2150	1910	2280	1850	1490	1190
7x57	175-gr.	2470	2170	1880	1630	2410	1830	1400	1100
7 mm Exp.	150-gr.	2970	2699	2444	2203	2937	2426	1989	1616
7 mm-08	140-gr.	2860	2625	2402	2189	2542	2142	1793	1490
.284 Win.	125-gr.	3140	2810	2512	2238	2736	2191	1751	1390
.284 Win.	150-gr.	2860	2595	2344	2108	2274	2243	1830	1480
.280 Ross	140-gr.	2900	2665	2445	*	2620	2210	1860	*
.280 Ross	160-gr.	2700	2395	2110	*	2600	2040	1580	*
.275 H&H	160-gr.	2700	2505	2320	*	2600	2230	1920	*
7x61	160-gr.	3100	2927	2757	2595	3415	3045	2701	2393
7x64	103-gr.	3572	3110	2685	2283	2930	2230	1670	1190
7x64	139-gr.	3000	2570	2260	1980	2780	2040	1570	1200
7x64	162-gr.	2960	2603	2375	2200	3150	2440	2030	1740
7x64	177-gr.	2880	2665	2490	2325	3270	2820	2440	2130
7 Rem. Mag.	125-gr.	3310	2976	2666	2376	3040	2458	1972	1567
7 Rem. Mag.	150-gr.	3110	2925	2748	2575	3221	2849	2515	2208
7 Rem. Mag.	165-gr.	2860	2710	2560	2420	3000	2690	2410	2150
7 Rem. Mag.	175-gr.	2860	2718	2580	2443	3178	2870	2586	2319
7 Wby. Mag.	139-gr.	3205	2909	2633	2374	2130	3170	2611	2139
7 Wby. Mag.	154-gr.	3075	2811	2561	2329	3233	2701	2242	1854

** Figures unavailable*

PETERSEN PRESENTS: Open Edition Prints

These prints are printed to the same exacting specifications that are used in the production of our limited edition prints. Each print is published on 80 lb. Vicksburg vellum paper, but is not limited to a specific number nor is it signed by the artist.

Each of these fine prints is offered either unframed or tastefully framed with genuine walnut-finished, wooden frames and double-matted with glass.

These prints are an excellent opportunity to fill those decorating needs where a signed and numbered print is not desired.

"BOBWHITE QUAIL" by Douglas Van Howd
No. 771 Size: 16″ x 20″
$7.50 unframed $34.00 framed

"BROKEN STILLNESS (RAINBOW TROUT)" by Russell Johnson No. 791 Size: 16″ x 20″
$7.50 unframed $34.00 framed

"SIX POUNDS OF TROUBLE (LARGE MOUTH BASS)" by Russell Johnson No. 792 Size: 16″ x 20″
$7.50 unframed $34.00 framed

"AUTUMN WOODS" by Lee LeBlanc
No. 793 Size: 16″ x 20″
$7.50 unframed $34.00 framed

"CANADA GEESE" by Douglas Van Howd
No. 772 Size: 16″ x 20″
$7.50 unframed $34.00 framed

"MALLARDS" by Douglas Van Howd
No. 773 Size: 16″ x 20″
$7.50 unframed $34.00 framed

"PHEASANTS" by Douglas Van Howd
No. 774 Size: 16″ x 20″
$7.50 unframed $34.00 framed

"BACKWATER MALLARDS" by Lee LeBlanc
No. 794 Size: 16″ x 20″
$7.50 unframed $34.00 framed

"THROUGH THE TREES" by Lee LeBlanc
No. 782 Size: 16″ x 20″
$7.50 unframed $34.00 framed

Please mail coupon to: PETERSEN PRINTS 6725 Sunset Blvd. Suite 316
Los Angeles CA 90028 (213) 657-5100 Ext. 431

Print Title	Stock No.	Framed	Unframed	Quantity	Price

(US delivery ONLY) Shipping and Handling add $3.00 ____

Add 6% sales tax for California delivery ____

TOTAL ____

Signature ____

SHIP TO: (PLEASE PRINT)

NAME ____

ADDRESS ____

CITY ____ STATE ____ ZIP ____

CHARGE TO:

MASTER CHARGE ☐ VISA ☐ AMERICAN EXPRESS ☐

Account Number ____ Expiration Date ____

For Master Charge only. List four digits above your name ____

PETERSEN SPORTING PRINTS

A DIVISION OF PETERSEN PUBLISHING COMPANY

ularity during the past decade would repudiate any argument that it is not a useful cartridge. The 7x57, along with the .30-06 and the .375 H&H, has become one of the truly international cartridges. I know several people who, because of low recoil tolerance, use the 7x57 as their all-around big game rifle. They not only have cleanly killed such dangerous beast as lions and grizzlies with it, but such huge herbivores as moose and eland as well.

The 7x64 and the 7mm Remington Express are virtually two peas in the pod, at least ballistically. They, too, fill a niche and that being their ballistic versatility. Both cartridge cases are of optimum powder capacity for the 7 mm caliber. Even with today's slower burning powders the 7 mm Remington and Weatherby magnums are still slightly over bore capacity.

But at present, the .270 Winchester is a better bet for North American hunting for the non-reloader than either 7x64 or the 7 mm Remington Express. Coming from a 7 mm fan, this may sound a bit like a blasphemy but there is more to selecting a hunting cartridge than its ballistics.

Neither the 7x64 or the 7 mm Remington Express is as widely distributed as the 270 Win. The 7 mm Remington Express is only available in one factory loading, with a 150-grain pointed bullet while the .270 Winchester is available with 100, 130 and 150-grain bullets in several different styles and designs. True, the 7x64 is also available with even a selection of bullets in several designs and weights but its distribution is poor. There just aren't enough rifles around in the 7x64. In fact, only three makes, the Steyr-Mannlicher Model M, Heym SR20, and Tikka M65 are imported into the U.S. in this caliber.

But a handloader would find either the 7 mm Rem. Exp. or the 7x64 excellent hunting cartridges anywhere. Either cartridge can be loaded to give muzzle velocity of a little over 2700 feet per second with 175-grain bullets. In fact, RWS factory load with 173-grain H model bullet gives a muzzle velocity of 2790 feet per second. This brings either cartridge into the .30-06 class. Actually, because the sectional density of the 175 grain bullet in 7 mm is greater than with the 180-grain .30 caliber bullet, the 7x64 retains its velocity better and is a little flatter shooting than the .30-06. The 160 and 140-grain bullets give the 7x64 and the 7mm Rem. Exp. excellent versatility for all mountain or plains game shooting.

There is, of course, a solid niche for a 7 mm Magnum cartridge. The commercial success of the 7 mm Remington Magnum is a solid proof of this. The 7 mm Remington Magnum is one of the most popular cartridges of the post World War II era. Only the .222 Rem. the .308 Win. and possibly the .243 Win. can surpass it in popularity. The fact that rifles for the 7 mm Remington Magnum are made by some 22 makers—Browning, Colt-Sauer, Dumolin, Golden Eagle, Harrington & Richardson, Interarms, Kleinguenther, Mauser 66, Steyr-Mannlicher, Parker-Hale, Remington, Ruger, Sako, Savage, Smith & Wesson, Tikka, Weatherby Vanguard, Winchester, Wickliffe 76 and such semi-custom rifles makers as Champlin and DuBiel, is further evidence of the 7 mm Rem. Mag.'s vast popularity. Only the shooter after a .30-06 or the .270 has a better selection of makes and models to choose from.

Game, however, is not killed by ballistics alone. It is killed largely by well placed bullets that perform well. In its first 90 years of existence, the 7x57 has killed much more game than the 7 mm Remington magnum ever will. Sure, the 7 mm Rem. Mag. shoots flatter, hits harder and theoretically should kill a bit quicker than the 7x57—if all things are relative. But let me illustrate what I am talking about with an example or two. A few years ago, a friend and I hunted moose with Louis Waswa, the Ojibway Indian outfitter out of Fort Hope in northern Ontario. I had my 7 mm Rem. Mag. while Jim Bowen, recovering from an eye operation, had a 7x57 because of its low recoil. We both shot smallish bulls that were called up by Louis to within 50 to 60 yards. Both moose were lung shot. Jim's moose travelled some 60 yards before collapsing while mine went twice as far.

In both cases the bullet placement was good. It was the bullet performance that played the key role in the kill, not the energy of the bullet. A novice hunter could easily come to a conclusion that the 7x57 is a better moose cartridge. Yet, we all know that this cannot be true.

I can recall an article by the late Jack O'Connor on the 7x57. O'Connor's wife, Eleanor, had 7x57 as her light rifle on an Angola safari while Jack had the then-new 7 mm Rem. Mag. They both shot about the same number of antelope and other plains game with their respective 7 mm rifles and both rifles killed equally well. I suspect that in this case, bullet performance again won over energy as it will every time and that is why the 7x57 performed as well as the 7 mm Rem. Mag.

The 7x57 enjoyed an illustrious reputation in the game fields of Africa long before any of the magnums. It was there at the turn of the century when African plains teemed with game in such numbers that today's Africa pales in comparison. It was during this era that W.D.M. "Karamojo" Bell killed over 1000 elephants with the 7x57, not-to mention several hundred with the 6.5x52mm. And, in effect, proving that bullet placement and performance is the key to quick and humane kills on all game.

A scoped 7 mm magnum like Remington M700 Classic is a most versatile big game rifle, equally at home in Africa or the U.S.

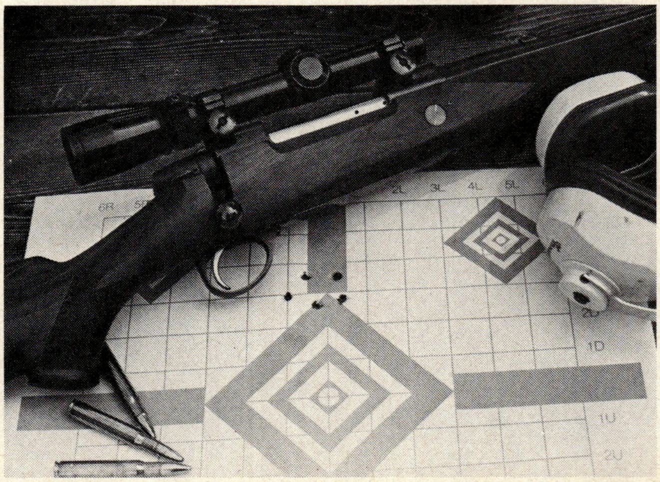

The new Sako Classic is yet another example of the wide variety of quality "7 mags." This Finnish import is very accurate.

The carbine length Remington M788 is currently the only rifle chambering the newest 7 mm—the 7 mm/08 Remington.

Famed ivory hunter, Karamojo Bell, shot over 1,000 elephant with Kynoch 7x57 solid bullets. Though inadequate for such game by today's standards, the 7x57 still ranks as an excellent hunting cartridge.

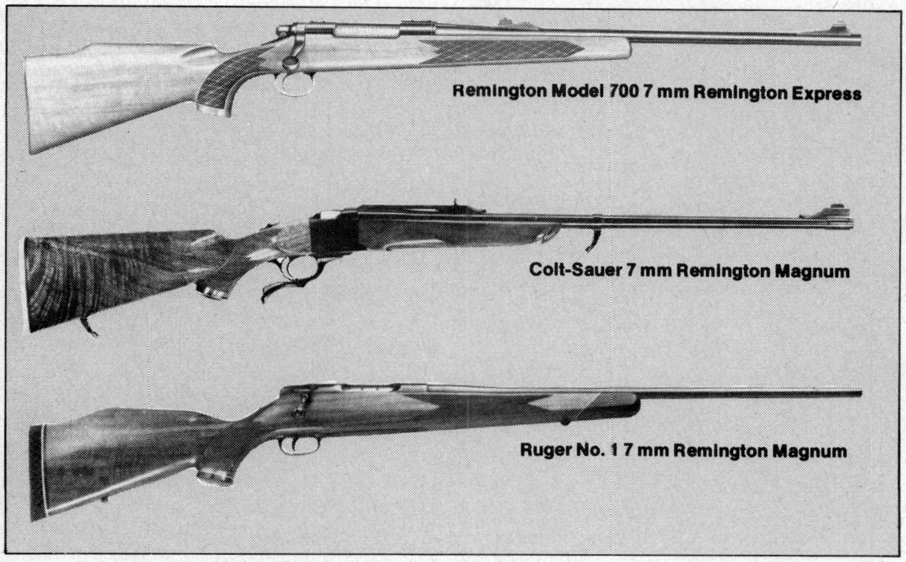

The 7 mm Remington Magnum rivals even the venerable .30-06 in the number of rifles chambered for it. The 7 mm Rem. Exp., author feels, will also gain very wide acceptance.

Even today, Brian Smith, a Zambia-based professional hunter with no small amount of experience and reputation, uses the 7x57 as his sole rifle—even for stopping the various big beasts that his clients have wounded. The fact that the Smith has yet to be trampled, gored, clawed or bitten despite many charges, further proves my claim on what really kills big game animals—of all sizes.

So where does this place the more potent 7 mms—the 7x64 and 7 mm Rem. Exp. and, of course, the 7 mm magnums? There can be no doubt that the 7 mms with greater cartridge capacity shoot flatter, hit harder and should theoretically kill faster than the 7x57—if all other factors are equal.

Frankly, I've had no experience with the 7 mm Rem. Exp. but I have shot a few animals with the similar 7x64 and have seen even a bigger number of animals killed.

The RWS 123-grain pointed hollow point is probably the most lethal bullet on such animals as pronghorn, impala, sprinbuck, Thompson's gazelle and the like that I have seen. I've seen standing impalas collapse in their tracks with almost any kind of a chest hit with this bullet. The exit holes are palm-sized. The lungs are a mush. But again, it's the bullet's performance, not the caliber, that is primarily responsible.

I, like many other shooters, was first fired up about 7 mm magnums through the writings of the late Warren Page. As shooting editor of *Field & Stream,* Page got interested in the 7 mm Mashburn Super Magnum wildcat and had much to do with the 7 mm magnum mania of the late 1950's and early 1960's.

But there is an irony to all this, which I grasped only after I acquired my 7 mm Rem. Mag. Page had expounded on the 175-grain bullet for all big game hunting with the 7 mm magnum but yet at that time there was not a single pointed 175-grain 7 mm bullet on the market. There was, in fact, only one semi-pointed 175-grain bullet on the market and that was the efficient Nosler partition.

This meant that Page was touting the 7 mm as an ideal long range cartridge when actually it was not a significantly better cartridge for truly long range than the old .270 Win. because there was no true long range bullet made for the 7 mm. The 7 mm Magnum only becomes a good long-range big game cartridge when loaded with proper bullets. Otherwise the hunter might as well stick with a rifle in the .270 Win.—7 mm Rem. Exp. class.

What are the best bullets for the 7 mm magnums. Well, any of the 160-gr. spitzer boat-tails are a good choice. They have enough sectional density for penetration and high ballistic coefficient for flat trajectory. For example, the 160-grain Sierra boat-tails can be loaded in the 7 mm Rem. mag. so that it has a muzzle velocity of 3200 fps and delivers over 2100 fps of energy at 400 yards. The new 175-grain spitzer boat-tail is even better. It has an amazing ballistic coefficient of 9654 and at 400 yards it retains over 2300 fp of energy. To boot, it has been my experience that the 175 grain 7 mm Nosler partition bullet gives better penetration than the 200-grain .30 caliber Nosler partition bullet fired in the .300 Win. Mag.

When the 7 mm Rem. Mag. first came out, a rash of overenthusiastic articles broke out in the various gun magazines. According to some of the writers, its bullets killed like lightning. Only an atom bomb was more lethal.

It took almost a decade for me to discover the superiority of the 7 mm Rem. Mag over my favorite .270 exists at 350—400 yard ranges, the maximum range that anyone should ever shoot at any big game and only when the cartridge is loaded with proper loads. Up to the 300 yard mark, the range at which nearly all big game is killed, the 7 mm magnums just don't have any significant killing superiority over the 7mm Rem. Exp., the .270 Win. or the old .30-06. Sure, the 7 mm magnum has superior ballistics on paper but the extra foot pounds of energy or the extra feet per second of velocity over the .270 or .30-06 did not seem to kill that much better.

On top of this, the rifle was heavier than my favorite .270 sporter. There was also more recoil and muzzle blast. Factory ammo costs more as well.

However, the 7 mm Rem. Mag. remains one of my favorite cartridges. But now I have no illusions about it. I use it only on hunts when long shots are a probability, not just a possibility. For all-around big game hunting, I consider the 7x64 or 7 mm Remington Express a better choice.

The irony of the 7 mm caliber saga is that we have almost come full circle. We started out nearly 90 years ago with the 7x57 and our new 7 mm cartridge is the 7 mm-08 which ballistically not quite equals the 7x57. In 1917, we had the 7x64 and last year we got its ballistic equal in the 7 mm Rem. Exp.

Are the ballisticians and cartridge designers trying to tell us something? Maybe. If there is a message here then it's that the 7 mm magnums are good but that for most of our hunting, the smaller 7 mms are all we really need.

ROTARY MAGAZINE RIFLES

Several of the most successful rifle designs available are built around the rotary magazine concept —and with good reason!

By Payton Miller
And Dr. Ralph C. Glaze

The history of the development of firearms is for the most part documented quite well, with surviving examples of significant design achievements being readily available in the collections of a great number of museums and private individuals. It is most unfortunate, however, that the history of the people involved with the evolution of firearms is not equally as well preserved. All too frequently, the stories of the people behind certain firearm creations have remained untold and, with the passing of time, probably can never be brought to light. Much uncertainty exists as to just who might have first thought of a particular idea, or who made it work for the first time. Pinning down exact dates in firearms history is equally difficult, with various sources often disagreeing on the date of a specific event by as much as several years. The story of the origins of the use of the rotary magazine in repeating arms is blurred into virtual obscurity. Some sources attribute the development of this familiar feed system to von Mannlicher. Others credit its beginnings to Schulhof or Schoenauer. Several "authorities," who obviously are unfamiliar with European firearms history give credit to Savage. It is quite probable that all are wrong.

The concept of a rotary magazine very likely came into being as an extension of the principle behind the revolving cylinder, such as is found in modern handguns, and that at one time achieved some popularity in rifles and shotguns. If a revolver's cylinder is reduced in diameter until each chamber is merely a trough containing a single cartridge, and a mechanism is then provided to push a loaded round into the barrel and to subsequently remove the fired case—we have all the elements of a rotary magazine system. This line of reasoning must surely have been followed by the inventor of the first successful rotary magazine.

In 1866, Sylvester H. Roper, an American, patented a rotary magazine system that was actually produced in the following year in a repeating shotgun. The Roper used reloadable steel cases fitted with a percussion cap. The modern brass primed cartridge was only coming into existence at this time, and it is entirely possible that Mr. Roper was unaware of its potential when he was developing his repeating scattergun. The Roper later was adapted to fire self-contained shotshells of the type still in use today. In the Roper system, the bolt extended thru the center of the carrier sprocket and moved fore-and-aft to load and unload the firearm. A spring rotated the carrier each time the bolt was pulled to the rear. The spring was re-wound as cartridges were individually inserted. This is the same operating principle that is found in all rotary magazines of later manufacture. Rather low-powered rifles also were made by the Roper Repeating Rifle Co. over a period of about 15 years prior to cessation of production in the 1880s. Roper was also noted for having joined forces with Christopher M. Spencer of Civil War rifle fame to produce a series of somewhat unsuccessful designs.

The next fully documented use of the rotary magazine was by the Waffenfabrik Steyr of Vienna in their Model 1879, a full 12 years after the appearance of the Roper. Although differing slightly in configuration, it is interesting to speculate just how much Steyr engineer Antonin Spitalsky was influenced by Roper's efforts when he created the M79 magazine. During that period in history, due to the pressures of the Civil War, most of the new ideas in firearms design originated in America, and all developments of American makers were closely inspected by European designers. Either the Spitalsky magazine was ahead of its time, or there were serious problems encountered in its production or operation. In any event, the Model 1879 Steyr did not become a fast-selling item. The Spitalsky magazine was not forgotten, however. It was merely shelved for awhile.

In 1880, Ferdinand Ritter von Mannlicher came to the Steyr Works. Over the next 23 years, until his untimely death, von Mannlicher supervised the development and introduction of over 150 firearms, many of which had an everlasting effect on firearms design. In 1883, Mannlicher resurrected the Spitalsky magazine and joined it to the same bolt system that had been used in Mannlicher's Model 1882 tubular magazine repeater. This was a two-piece bolt with rear locking lugs. After the usual development period, this rifle was patented in 1885 and later issued as the Steyr Model 1887 turnbolt. As this rifle was chambered for the Austrian Model 1877 11 mm cartridge, the magazine protruded beyond the stock on both sides and on the bottom in order to provide ample room for this fat, black-powder round.

The Model 1887 magazine drum was a light, hollow cylinder having 8 ribs. The drum is tapered and rotates on a shaft inclined at an angle to the bore. A spiral coil spring is located in the forward section of the drum, and is wound up by the action of loading individual rounds.

At the same time the Steyr Model 1887 was being perfected, Josef Schulhof of Vienna was producing a similar rotary magazine for this Model 1885 rifle. This also used the 11 mm (.43 caliber) cartridge, and had a magazine larger than the stock surrounding it.

Neither the Schulhof Model 1885 nor the Steyr Model 1887 was manufactured in any significant quanity.

While Steyr was bringing out the Model 1887, there was a parallel project in the works that combined the straight pull action of the Steyr-Mannlicher Model 1884 with the Spitalsky rotary magazine. For this project, Mannlicher made some small changes in his action design, and Otto Schoenauer, director of the Steyr Works, improved on Spitalsky's magazine mechanism. The result was the Model 1887-88 Straight Pull Rifle.

The Schoenauer-designed magazine did not extend beyond the lines of the stock, since it was intended for the new 8 mm Austrian cartridge that was considerably smaller than the 11 mm round it replaced. In addition, the Schoenauer magazine could be loaded with a charging clip as well as with individual rounds. The reduction in loading time afforded by clip-charging is a significant factor in a military rifle. The Schoenauer magazine may be unloaded just as quickly. Pressing a release button causes the magazine spool to rotate, expelling the loaded rounds upward through the open action.

The Schoenauer improvement of the Spitalsky design is essentially the same magazine as is found today in rifles made by Steyr-Daimler-Puch, successors to the old Viennese Steyr Works. In 1903, Greece adopted the Steyr Model 1900 in 6.5 x 54 mm caliber as their standard army rifle. The Greek "Mannlicher-Schoenauer" was produced in greater numbers than any other military version of the rotary magazine design. Many of these guns were imported into this country as surplus in the 1950s, and are still encountered occasionally, either in original form or as sporter conversions.

At least two other applications of the rotary magazine deserve mentioning. The first of these was a gravity-feed magazine attached above and to the left

For the rifleman with an interest in rotary magazine rifles, there's one to fill almost every hunting requirement. The Steyr-Mannlicher and Mossberg bolt actions will serve admirably for big game hunting, while the Ruger 10/22 is superb for small game.

Ruger's successful rotary magazine for their 10/22 fits in action, and detaches as quickly as a conventional clip magazine.

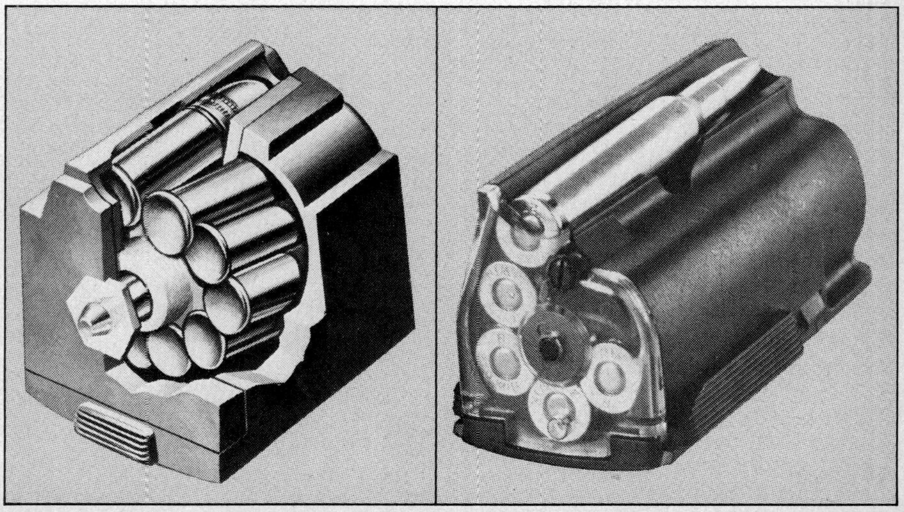

The detachable magazine of the Ruger 10/22 holds 10 rounds of .22 Long Rifle ammo (left photo). The Steyr-Mannlicher magazine is for centerfire rifles. Current versions are made of a synthetic—Makralon—and are every bit as reliable as the old steel ones.

of the receiver of the Martini-Henry British Army rifle. This attempt to increase the firepower of a singleshot rifle was made in 1886, but was abandoned. The British Farquhar-Hill automatic rifle, under development at the time of the first World War, utilized a detachable rotary magazine of rather large dimensions. In fact, it resembled a truncated kitchen sieve in terms of general size and shape. This rifle was never perfected due to the exigencies of war.

In 1895, Arthur Savage had perfected his lever-action, rotary-magazined rifle to the point where the United States Navy considered it quite seriously as their new military standard rifle. The Navy did not adopt the Savage, however, but it was later placed on the market as a sporting rifle and, in its 1899 configuration, it has been in continuous production to the present day with only minor changes. The Savage magazine differs somewhat in detail from the Mannlicher-Schoenauer, but is quite similar in operation.

The Savage Model 99 is one of our most popular big-game rifles. It's available in calibers from .243 to .358 Winchester, and is the only rotary magazine lever action made.

The latest military rifle to utilize a rotary magazine was the Johnson. This rifle was perfected in the 1930s by Melvin M. Johnson, then a Captain in the U.S. Marine Corps. It was offered to the U.S. Army also as a replacement for the Springfield Model 1903, but after a long and bitter struggle the Army elected to adopt the Garand instead. The Marines bought a few thousand, and some of these saw service in World War II. The Johnson magazine held ten .30-06 rounds and could be loaded with 5-round charging clips or with single cartridges. It could also be loaded or unloaded with the action closed and the piece on "safe."

At the present time, in addition to the Savage M 99 and the Mannlicher-Schoenauer, the Mossberg RM-7 and the Ruger 10-22 are the only other rifles available with rotary magazines.

That rotary magazines have been in use for such a long time would seem to indicate that they have certain advantages over more conventional types. This is quite true. Similarly, one might question why, if they are so good, the rotary magazine is not in more common use. There are good reasons for this, too.

The rotary magazine presents a cartridge in a straight line with the bore, so that feeding is smoother and more positive, especially with rimmed cases. Separate grooves in the spool keep cartridges from touching each other and this reduces the possibility of damage to cartridge cases.

It is largely a matter of economics that keeps the rotary magazine from more general use. They cost more to manufacture than a simple box magazine. A rotary magazine also must be a bit wider than its box counterpart, which might result in a weakening of the stock in the magazine area of a bolt-action rifle. Capacity is limited with the rotary design as well.

Rotary or drum magazines of the Thompson and Lewis types have been used in several machine gun designs. These are not considered in this article, however, since they are not of the same design pattern or used in the same way as the spool-type rotary magazines discussed here. The drum magazine indeed might be the primary subject of a separate article.

To temporarily escape from the task of researching the historical and mechanical aspects and applications of rotary magazine rifles—as well as to get some much-needed photographs for this story, we decided to take a couple of currently-produced representatives of the breed out for a field test at the Petersen Ranch.

Our first order of business was to decide on which member of the Mannlicher line to lay hands on. This isn't as easy as it sounds. There are nearly a dozen models and variations to choose from, and everyone on the staff had intransigent and aggressive opinions as to which one to get.

After interminable bickering, we (or rather, the editor) decided upon the Model M Professional Hunter's Model. His rationale was that he was intrigued by the idea of a Cycolac-stocked sporter with double-set triggers. A call to Robert Koumjian at Steyr-Daimler Puch of America, and the gun was on its way. Bob's only question—"OK, what caliber do you want it in?"—provided another editorial dilemma. Would we play it safe with a .30-06 (for which we had plenty of ammo), or go for some exotic European metric import? We opted to get the rifle in 7x64 Brenneke for two reasons. (1) According to published

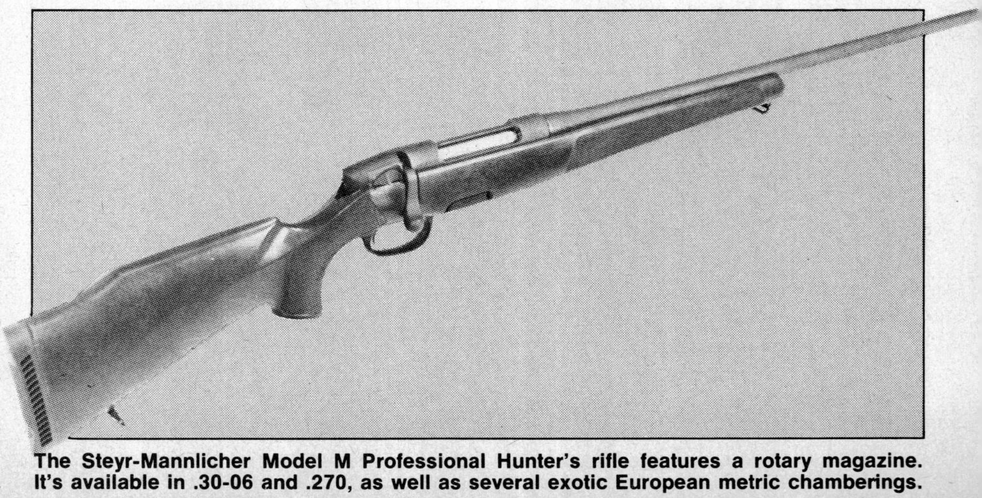

The Steyr-Mannlicher Model M Professional Hunter's rifle features a rotary magazine. It's available in .30-06 and .270, as well as several exotic European metric chamberings.

Norma and RWS ballistics tables, the 7x64 is a hot number—somewhat more energetic than the .270 or 7 mm Remington Express in terms of performance. (2) Nobody on the staff had ever seen or shot one—a situation which we felt to be intolerable.

When the rifle arrived, sporting a 2-7X Mannlicher scope (made by Burris, incidentally) on QD claw mounts, reactions to the rifle's visual appeal were mixed. The best term for this rifle, is *stark.* Some staffers found this appealing. Some of the more traditional-minded found the olive-drab synthetic stock, sandblasted metal finish, and no-frills configuration to be "as ugly as homemade sin." Since the rifle is a working hunting rifle in the purest sense of the word, the consensus was to see how it shot before jumping to any rash conclusions.

Firing our first 5-shot group over a sandbagged rest with 150-grain Norma ammo gave us our first lesson on the proper handling of the Professional Hunter. *Do not touch the front trigger—once it has been set—unless you want the gun to fire!!!!* It is light—around 8 ounces. Without being set, the trigger pull is a shockingly crisp 2½ lbs. For the vast majority of hunting situations, using it "unset" would be preferable. Having the set option is reassuring, should you ever have to take a long shot at a relatively stationary target from a dead steady rest.

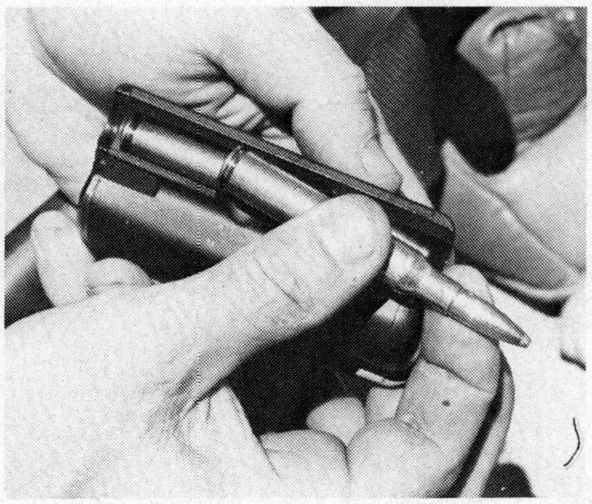

As far as practical hunting accuracy goes, forget all about the term "practical" when describing the Professional Hunter. At 100 yards, the best 3-shot group was a shade under ½-inch. Other 5-shot groups were a bit larger, to be sure. Some of them got as large as 1-inch. For a 7½ lb. sporter with factory ammo, this type of accuracy is almost indecent. Switching to RWS 150-grain ammo produced results in a similar pattern.

In our growing enthusiasm for the rifle, we very nearly overlooked the considerable merits of the cartridge itself. It's tough to find in this country, only available at present through Norma and Eastern Sports International (RWS). Factory bullet weights are 103-grain, 139-grain, 150-grain, 162-grain, and 177-grain. The RWS line also includes the excellent TIG, TUG, Cone-Point, and Torpedo point bullets. 7 mm handloaders can have a field day with commercial domestic bullets in this persuasion as well, once they've amassed a supply of brass.

Functioning of the rifle—and the Makrolon rotary magazine was flawless. The bolt action was a little tight at first, but smoothed up as the session went on. Felt recoil was noticeable, but firing 40 or 50 rounds of a bench isn't what the rifle was designed for. Incidentally, the cartridge capacity of the rifle is 6 rounds, which is two or three more than you get with most in-line magazines. Not really necessary in a hunting rifle, but what the hell! If you've got $467 to lay out for a rifle that won't reflect light, suffer stock warpage or splitting, and shoots like a house afire, you might look into the Model M Professional.

One of the latest domestic offerings in the bolt action category happens to have a rotary magazine as well. It's the Mossberg RM-7, currently available in .30-06 and 7 mm Remington Magnum. The .30-06 version (which we tested) has a 22-inch tube. The weight is identical to the Mannlicher—7½ lbs. The rifle has a 3-position safety, the middle position allowing you to run cartridges through the action while the gun is on safe. If you're really in a hurry to empty the RM-7, you can hit a release catch and regurgitate all the cartridges instantly.

The trigger pull on our test sample was a bit over 5 pounds and could benefit from some work. For testing, we

Despite its light weight—due to a synthetic stock—the Steyr-Mannlicher Model M is an exceptionally accurate hunting rifle.

Loading the rotary magazine of the Steyr-Mannlicher Model M is a relatively fast and simple process (top left). The test rifle was in 7x64 Brenneke, and the capacity was five rounds—more than most domestic rifles with "in-line" magazines. Throughout the tests, functioning of both rifle and magazine was perfect.

mounted a Redfield 3-9X Widefield Variable on the rifle. Using 150-grain Remington and 180-grain Federal ammo, average 5-shot groupings at 100 yards were consistently under 2¼ inches. That doesn't qualify the RM-7 as a tackdriver, but since we didn't have time to tailor handloads to it, there's no reason to doubt that it's capable of a tighter performance.

The RM-7 is a solid, functional looking rifle stocked in the resurgently-popular "classic" style. It's a bit beefy in the middle, thanks to the rotary magazine, but balances nicely. At $245 per copy, it's a good bargain.

Without a doubt, the most popular and widely used rotary magazine rifle in the United States isn't a bolt gun at all. It's the lever action Savage Model 99. The basic rifle has been produced in a myriad of versions since its inception, with some detachable box magazine models cropping up in recent years. The one we went with was the Model 99-358, a brush-busting powerhouse chambered for Winchester's impressive .358 round. With a 200-grain bullet at 2530 fps, and a 250-grain bullet at 2250, the .358 is a real stopper. Since the rifle is a woods gun, we elected to mount a Williams receiver sight on it instead of a scope—although a low powered scope would still be ideal for the rifle's intended purpose.

The rotary magazine of the 99-358 holds 6 rounds, and in view of the type of terrain the gun is tailored, for the speed of fire made possible by a lever action—this seems to be an advantage.

During firing tests, the 99-358 functioned perfectly. Magazine feeding went without a hitch. Accuracy, from the bench at 50 yards was quite good—around 1½ inches for 5-shot groups using Winchester 200-grain Silvertip ammunition. The trigger pull was a bit heavy at 5 pounds—but that should be easy to correct. The balance and shotgun-like "feel" of the 99, plus its straight-grip stock configuration makes quick, offhand shots easy. For elk, bear, or deer in heavy timber, the 99-358 should be a natural. List price is $257.

One of the handiest and popular rotary magazine rifles available today isn't a bolt or lever action. In fact, it isn't even a centerfire. It's the Ruger 10/22 auto, and is deservedly one of the most respected .22 rifles around. As far as we can tell, there are three reasons for this—(**1**) it's reliable, (**2**) it's economical, and (**3**) it's highly accurate. This little 10-shot repeater is offered in two versions, deluxe sporter or standard. Price for the standard version is $84, and the sporter goes for $103.

Not having one ordered specifically for this test, we "borrowed" G&A staffer Dave Hetzler's. Dave had installed a Weaver K-4 scope on it to use as a small game rifle. Using Federal Standard velocity Long Rifle ammo, the Ruger stayed under ½-inch at 50 yards all day. Even more importantly—considering the small game hunting qualities of the rifle—it grouped nearly as well with various brands of Long Rifle hollow point ammo. With over 200 rounds fired through the 10/22 there wasn't a single hitch. Not one jam or stutter. The little rotary magazine is tough, positive, and inserts and extracts as easily as a box magazine. The safety is of the crossbolt variety, and the top plane of the receiver is smooth. The 10/22 handles like a little shotgun. All in all, it's a first-rate choice for hunting and plinking.

Despite the production costs, limited capacities, and receiver width necessitated by the use of rotary magazines, their reliability is, for the most part, superior to in-line magazines. In all the rifles we tested—Mannlicher, Mossberg, Savage, and Ruger—feeding problems were nonexistent. That alone is sufficient justification for their existence, and hopefully, their existence will be assured.

Although the recoil of the Steyr-Mannlicher Model M was a bit more noticeable than in a wood-stocked rifle of a similar caliber, it was not unpleasant for a light sporter!

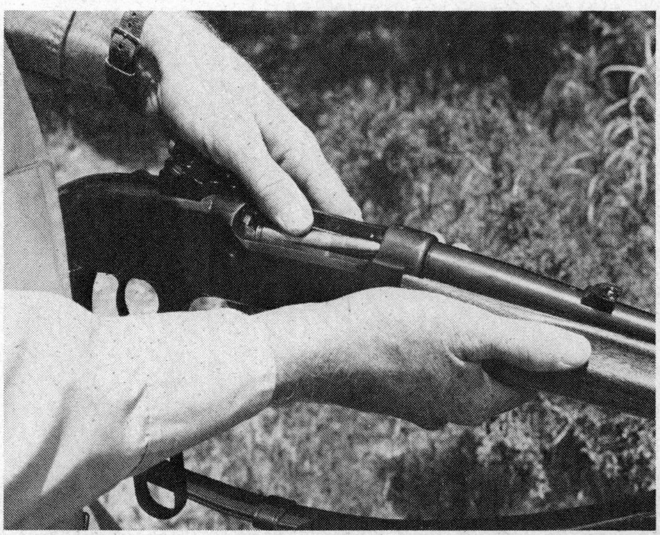

Unlike the Steyr-Mannlicher, our Savage Model 99 in .358 did not have a detachable magazine. The cartridges load through the breech. In .358, the rifle held six cartridges, which is a high capacity for a rifle of that caliber. The M99 features a cartridge counter showing how many rounds are left.

Chapter 5

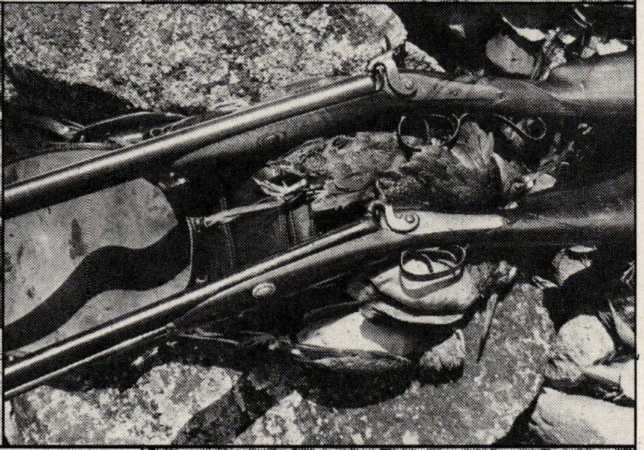

Black Powder

Tuning Blackpowder Rifles

Just like modern centerfires, muzzleloading hunting rifles can be easily adjusted and loaded for increased accuracy and performance!

By John Wootters

Serious hunting with a muzzleloading rifle is a different world. Please note that word "serious" it is not meant to apply to knocking off game with a frontstuffer just to say you've done it, nor to "blooding" such a rifle as some sort of ritual return to an earlier incarnation.

It is meant exactly as it sounds—the seeking of game with a black powder firearm with the same confidence, competence, and seriousness of purpose with which we hunt with modern rifles, not as a stunt but as a special way of homage to the game sought.

This means, among other things, the acceptance and management of the inherent limitations of the muzzleloading rifle. All hunting arms, from the Masai lion spear to the semiautomatic magnum rifle, have limitations. These differ largely in degree, once the hunter has become familiar with them. That Masai lion spear, for example, is an incredible weapon in the hands of a Masai warrior. African professional, John Hunter, tells of facing a wounded leopard with his double express rifle flanked on each side with a Masai, spear leveled. When Hunter's dogs badgered the cat into a short-range charge, he stated that *both* Masai spears passed *through* the leopard's body before he could get his rifle to his shoulder! In your hands or mine, one of those things would be not much better than a cattle prod; the difference is knowing what it will do and how to do it.

It is no different with a muzzleloading rifle. Knowing what it will do and how to do it simply requires more experimenting and practicing than a modern rifle. It will, however, almost invariably do much more than most modern hunters give it credit for. In fact, a well-handled muzzleloader of good quality will do so much so well that, in my opinion, the term "primitive firearm" is pretty shaky; with one, a hunter is at very little handicap, if any, in dealing with game normally shot within 100 yards or so.

The author feels that a muzzleloader can be an accurate and reliable hunting tool, provided, of course, that the owner spends enough time with it to learn its peculiarities.

The muzzleloader does enforce its own rhythms upon its master, however. You can't rush a blackpowder rifle. To get the best from it, you simply have to exercise patience and loving care, coaxing the gun along at its own pace. A few respond quickly, but the vast majority want to be wooed at a more leisurely pace. If you aren't in the mood to slow down, leave your muzzleloader at home. If your temperament is well-geared to the hectic pace of life in the Space Age, better not even buy the smokepole in the first place. It won't like you and you probably won't like it. These comments are not idle; they are the bedrock truth on which the remainder of this article is based.

Because of the versatility inherent in a rifle in which you can vary your powder charge from shot to shot, the distinctions between small game and big game muzzleloading rifles become a little blurred. You can use a .54 caliber rifle with a round ball and a half-charge to take a squirrel's head off as cleanly and with as little fuss as it can be done with a smaller caliber. However, the smaller bore may not be suitable for bigger game simply because of the lesser weight of lead it throws, regardless of the charge.

Let's say, then, that a small game muzzleloader will (except for a few custom guns) be of not smaller than .36 caliber and not larger than .45, and will be used on animals from squirrels up to and including bobcat, fox, coyote, wild turkey, and javelina, animals requiring only moderate penetrating ability in the ball. The big game gun must penetrate, above all, in order to get to the vitals of a deer, bear, or other such creature on a shot from a bad angle, or on an elk even from a good angle! This is usually considered to dictate a bore size of not less than .44 or .45 and on up to at least .58 or larger.

Experience has taught me to be a little dubious even of the .44 to .45 caliber class for game larger than small whitetails. Most modern American muzzleloaders seem to regard caliber .50 as about minimum for reliability on big game, and I agree.

The bore size dictates the overall weight of the gun to some extent, too, and few of today's buckskinners see

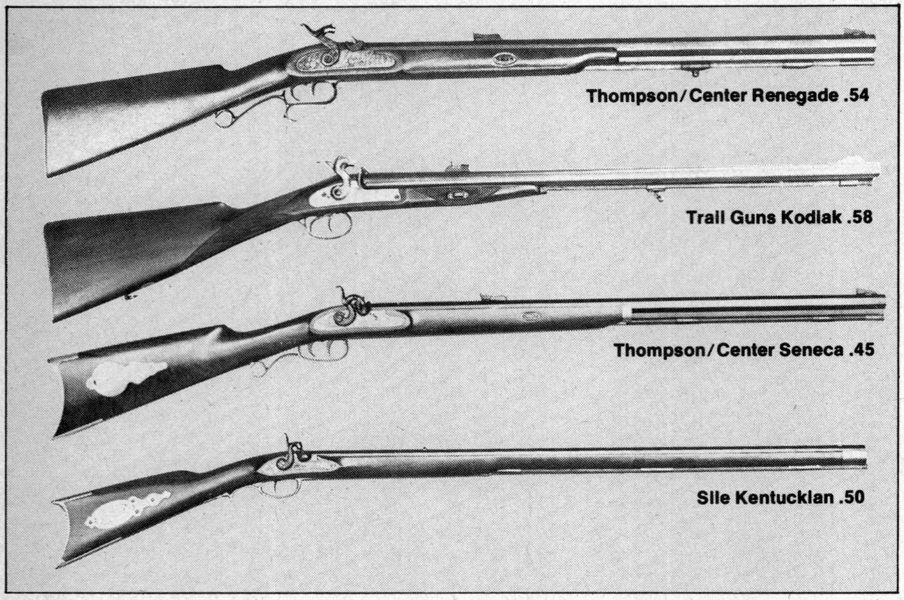

There are more than enough makes of muzzleloading rifles on the market to satisfy anyone. All the above are caplocks, and are in calibers from .36 to .58. A .36 or .45 caliber is OK for small and medium game, but a minimum of .50 caliber is best for deer.

Uniform loads are critical for hunting success, as is uniform seating force with the ramrod. Good loading technique is vital!

much point in toting an 11-pound, big bore mountain rifle around in the squirrel woods. Something in the .36 to .45 caliber class can be built light and dainty enough to be a joy in such service without sacrificing ballistic efficiency *or* loosening your molars with recoil.

The next question concerns rifling twist and pattern. In a muzzleloader, exactly as in a modern magnum, a longer, heavier bullet requires a quicker rifling pitch for stability in flight. The patched round ball, being the shortest, lightest practical projectile in any given bore size, may give excellent accuracy with a very slow twist, say, 1 turn in 56, 66, or even 72 inches, but that same barrel is very unlikely to shoot well with a heavy Maxi or Minie ball. Theoretically, a quicker twist should overstabilize the round ball, and it's true that barrels with a 1-48″ or quicker twist seldom offer top round ball grouping. But there's more to it than that; a barrel made for patched ball shooting usually has rifling grooves .010 to .012 inches deep. The quicker twist barrels of modern maunfacture, however, are almost invariably intended for use with conical bullets rather than balls, and are cut with grooves only about half those depths.

When such a barrel has its grooves deepened, even though the twist remains the same relative quick rate, it often groups much better with the patched ball. The conclusion would seem to be that what we call "overstabilization" wasn't the problem in the first place. In dealing with modern rifles, I sometimes wonder if there is indeed any such thing as over-stabilizing a bullet, and see no reason why a muzzleloader should behave differently in this regard.

Theoretically, too, one should be able to avoid such over-stabilizing of a round ball merely by reducing the powder charge, thus the muzzle velocity, and thus the RPMs of the flying bullet. But I haven't been very successful with this approach to shooting the patched ball in a factory muzzleloader barrel with a 1 to 48 inches or quicker twist. Again, deepening the grooves and lapping the bore, if needed, have quite often done wonders for patched ball accuracy, even at full velocity.

On the other hand, deepening the grooves may prevent modern elongated muzzleloader bullets, including Minies, from sealing the bore, leading to gascutting and wild shooting. For the small bore, small game muzzleloader, however, the round leaden ball is ideal, and accuracy is of paramount importance. A serious squirrel rifle must be capable of hitting a half-dollar every time at 25 to 35 yards, and really should be expected to do better than that to allow some margin for the hunter's wobble. Great power is not needed, but one-hole groups at squirrel-shootin' ranges are.

If your modern-made, brand-name smallbore delivers such groups as is, fine. If not, you may find it profitable to have the grooves deepened and the bore simultaneously smoothed by lapping. As one example, the sleek little Thompson Center "Seneca," with its good trigger and adjustable sights, can make a lovely little small game rifle in either .36 or .45 caliber with this treatment, despite its rapid twist.

For the big game rifle of caliber .50 and larger, the problem is not so easily solved. If it's a slow-twist (1 to 56 inches or slower) bore, as many of the more popular guns are, it's probably going to be a round ball rifle only, with one possible exception to be mentioned below. If the twist is faster than 1 to 50 inches, it will most likely perform well with conical bullets, which are the best choice for large game hunting in any barrel in which they'll group sufficiently better.

The reason for that sure-to-be-controversial statement is the better ballistic coefficients of the elongated projectiles. The BC of a .495 round ball of pure lead is .070, whereas that of a 370-grain maxi ball in the same caliber is .095, a 35 percent improvement. The maxi, therefore, loses its initial velocity very much more slowly than the round ball, and it penetrates more deeply in an animal's body. Although the lighter, faster round ball can be launched at significantly higher muzzle velocity, it slows down so rapidly from air resistance that, at 100 yards, the maxi is car-

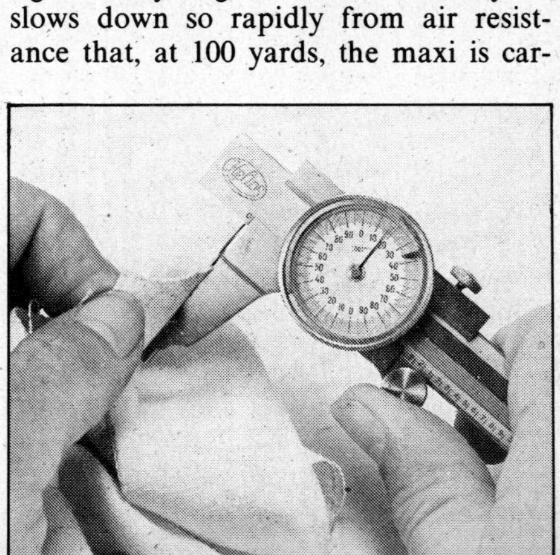

The size and thickness of the patch is a crucial factor with a round ball rifle.

rying as much or more kinetic energy.

This isn't to say that a .50 round ball will not kill a deer (I've killed too many with it to make that claim) or that it hasn't enough penetration for such game, at least on the broadside rib-cage shot. It does . . . and will often drive through and exit. This isn't the criterion, or at least isn't *my* criterion, however. No hunting rifle should be chosen for its adequacy on the *easiest* likely shot, but for the toughest one. In deer hunting, that's when the animal is at a bad angle, so that the ball must penetrate very deeply to reach the vitals, or must break heavy bone and dig through heavy muscles to get there. If you're loaded with a round ball, my advice is to refuse such shots, even if it means going home without venison. With a maxi, or something similar, you can make the shot with confidence.

A round ball often will expand very nicely in flesh, whereas the lower speed conicals seldom will. This becomes somewhat academic, however, when the *un*expanded projectile is carving a half-inch hole through whatever it hits. Likewise, the flatter trajectories possible with round balls have little meaning at ranges of 100 yards or less, where most muzzleloading game is collected. Accuracy counts, and so does penetration; expansion and flatness of trajectory are secondary in the realities of black powder hunting.

The buckskinner who regards the use of anything less traditional than a patched round ball as a sacrilege against history is not going to be impressed with these arguments, and I have no dispute with him. He gets his kicks his way, and I get mine another way, and so be it. If you *are* impressed with the discussion above but happen to own a rifle which simply will not group maxi balls well enough because of a slow rifling twist, there is still one trick to try, and that is the Lee R.E.A.L. bullet in its lighter weight. This is a solid-base conical with driving bands intended to be engraved by the lands during loading. In caliber .50, for example, its weight is 250 grains in pure lead, midway between the 180 grains of a round ball and the 370 of a maxi ball. It has a wide body and is, therefore, relatively short for the caliber. Since stability is a function of length, where caliber and materials are the same, this bullet will often stabilize at reasonable velocities in a slow-twist bore. I find it extremely accurate in two of my round ball rifles (1 to 56 inches and 1 to 66 inches) and an excellent compromise between the two extremes, combining some of the virtues of both round ball and maxi. It is also fast and easy to load in the field if carried in a loading block, already greased. Lee R.E.A.L. moulds are available in .45, .50, and .54 calibers.

Having made the all-important bullet

Zeroing from a benchrest position is just as necessary with a muzzleloader as with a modern rifle. In fact, because of the variations in loads, it's perhaps even more vital!

The author trophy hunts just as seriously with a muzzleloader as he does with a centerfire rifle. This Texas axis deer is No. 1 in the record book for muzzleloaders.

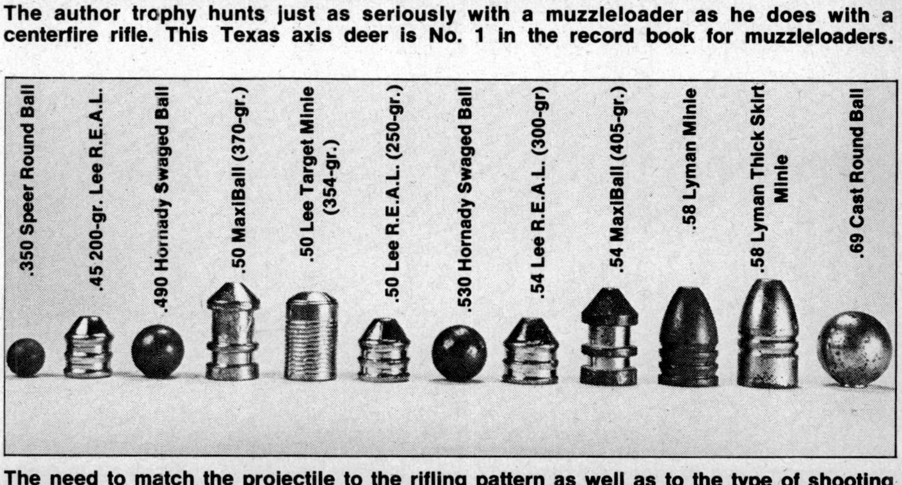

The need to match the projectile to the rifling pattern as well as to the type of shooting requirements, has resulted in a wide variety of projectiles that differ greatly in shape.

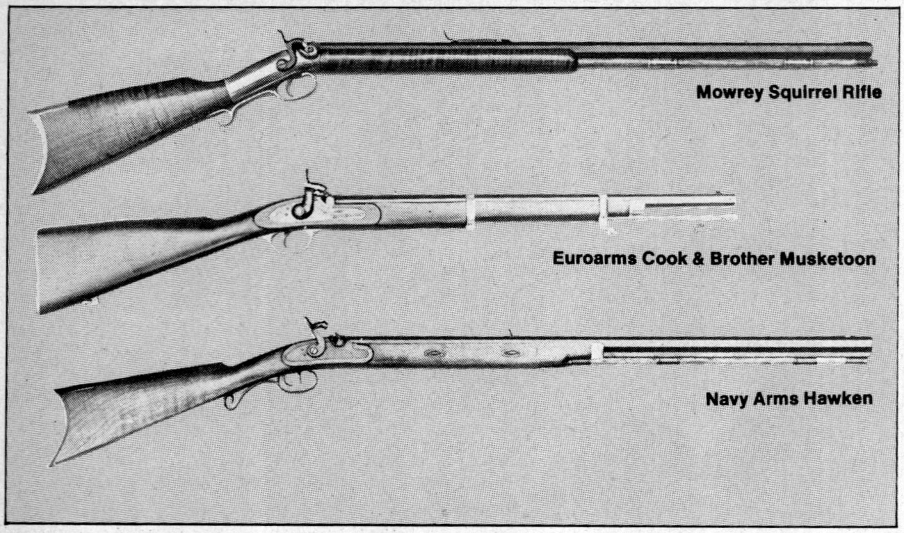

The Mowrey Squirrel Rifle (top) comes with a brass or steel frame, and is suitable for small game. The Euroarms Cook & Brother Musketoon doubles as a hunting/military replica piece (center). Hawkens (bottom) in .50 can up are the most popular for big game.

Learning how to add powder to the flash channel to "boost" out a misfire is a technique best learned during range practice.

selection, you have a great deal of shooting before you to determine the best combination of powder charge, lubricant, patching thickness, and cap, if yours is a percussion rifle. A general velocity level must be determined, bearing in mind that identical charges(by volume) of GOI, Curtis & Harvey, and Pyrodex will deliver markedly different velocity and energy levels, and, sometimes, different degrees of accuracy.

We have already discussed the accuracy standards for a small game loading to be used on squirrels. Most modern-day black powder hunters are satisfied with much poorer grouping for a big game rifle than can usually be attained. Many will settle for a load with which they can keep every bullet in a paper plate at 100 yards, hoping for a little luck, a rest, and a 35-yard shot when the big chance comes. With a little (or a lot) more experimenting, they might find that their rifles are capable of not worse than four inches for five shots at 100 paces, accuracy which many modern-day lever action carbines cannot regularly exceed. Some guns can be coaxed into two to three-inch groups—very good for a muzzleloader, but not really rare in the hands of an expert.

Two-inch groups from a benchrest at the 50-yard range at which most of us do our targeting should be the standard. If your hunting rifle will not at least approach that standard, regardless of what is tried, there's probably something wrong with it—an off-center bore, bent barrel, rod wear at the muzzle, or a rough or tight spot in the bore. Some of these conditions are not terribly uncommon in production muzzleloading rifles, sad to say, but most of them can be corrected at reasonable cost by a custom black powder gunsmithing shop. Muzzleloaders are not as sensitive to bedding problems as modern rifles, simply because their heavy barrels and low velocities do not produce much vibration, but they are by no means immune to them. A warped stock is a possibility which should be investigated, especially when a rifle's grouping goes suddenly to hell with the same load with which it formerly did well.

In round ball rifles, patching can make a big difference in accuracy, and the effects of changing patch thickness or material cannot always be predicted. Only patient testing will provide the real answers. Patch lubricant also has an effect. Where we once had the choice between tallow and spit, we now have an array of Space Age materials, including some promising dry lubes, with which to experiment. Teflon, in various forms, seems to have great promise in a couple of my own rifles in which I've been playing around with such products as Tri-Flon, Lubri-Lon grease, and Bullet Master lube. Teflon-

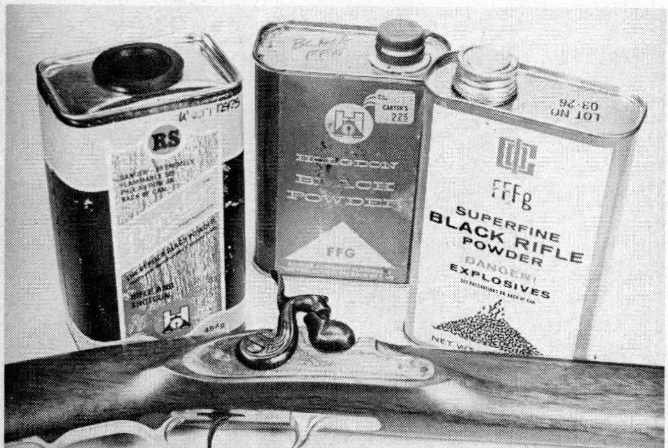

As with modern rifles, muzzleloaders may respond differently to various types and brands of powder. Tuning requires careful tests with as many of them as possible before going to the field.

Sprues are visible on the home cast balls (right) and can impede speedy loading and accuracy. Sprue-less swaged balls (left) are available from Speer and Hornady, and are superior for hunting.

based lubes seem to offer beneficial bore-conditioning effects which are well worth further experimenting. Bores conditioned with this stuff appear, in early trials, to deliver measureably better accuracy and fewer flyers, and have fewer fouling problems as well.

The quality of projectiles, as always, is a major accuracy factor. Availability of the Speer and Hornady swaged round balls has pretty well eliminated ball-related accuracy problems, but they still exist—and always will—with cast balls and bullets. Maxi balls are especially difficult to cast consistently and should be check-weighed and sorted for serious target and hunting purposes. The aluminum Lee moulds for the aforementioned R.E.A.L. bullets are, on the other hand, very easy to get good slugs from. They should still be inspected and weighed carefully, especially in the larger calibers, but I find very few rejects falling from them.

Many newcomers to the game appear to feel that muzzleloaders lack sufficient precision in general to make experimenting with small increments of powder worthwhile. It ain't necessarily so; certain barrels (especially the good ones) can reveal surprising discrimination between charges varying by only three or four grains (measured by volume, of course). With a frontstuffer that shows signs of wanting to group well, time spent in varying charges by small increments is likely to be very well spent, indeed.

Most novices in the black powder game seem to believe that you cannot overload a charcoal burner. Believe me, you surely can, and we're seeing more and more good quality barrels damaged as a result. There's no sense in it, either, since heavy loads are almost invariably unsatisfactory. For every barrel/ball combination, there is always a point beyond which more powder produces more recoil, more fouling, and poorer accuracy, and very little more velocity. This point is more easily identified with elongated bullets, it seems, than with round balls, but it can be noticed by the watchful shooter with anything. A chronograph, of course, reveals it plainly, as increasing charges of powder produce less and less additional velocity.

Admittedly, the very best accuracy is often found with subnormal charges, and the big game hunter must make some compromise between accuracy and power, but there is a delicate balance which can be found with enough trial shooting. Reference to the loading tables in Lyman's "Black Powder Handbook" can provide a useful shortcut to finding it.

In the four or five caliber .50 guns which I currently shoot, for example, I use an average of 80 grains (by volume) of Pyrodex or FFFg black behind a maxi ball, 90 grains with a 250-grain R.E.A.L., and 90 to 100 grains with a round ball. In every case, a 25 percent heavier powder charge delivers about a five percent increase in muzzle velocity—plus the bad recoil, fouling,

"Bullet Master Lube" was intended for cast bullets in cartridge guns, but does a great job as a patch lube and for conical slugs.

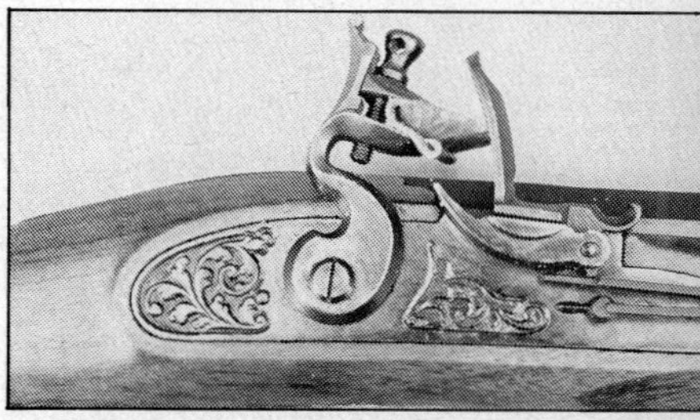

Flintlocks pose an additional handicap on blackpowder hunters because of slower ignition. Shown is a Lyman Plains Flintlock.

and accuracy effects mentioned. Those results should be telling an intelligent muzzleloader something.

Identifying this "breakover point," or whatever you wish to call it, as well as all the other performance characteristics of any given black-powder rifle contitutes an art, and not a science. It is not an art mastered overnight. The gun itself, however, never gives the wrong answers—*if* you ask it the right questions!

The proof of any worked up blackpowder load will ultimately come in the field. Author Wootters took this trophy boar javelina with a .50 caliber Maxi-Ball. The rifle is a Lyman Plains caplock.

Although Wootters took this whitetail doe with a .50 caliber round ball from a CVA Mountain Rifle, he feels that a conical projectile is superior for hunting, except for well-placed broadside shots.

Test Report

PARKER HALE VOLUNTEER RIFLE

This new English replica offers classic styling and first-rate target accuracy!
By Garry James

The British and French alliance during the Crimean War (1854-56) had been an uneasy one, so it was not surprising that a scant two years later relations between these two traditional adversaries would again start to deteriorate.

Emperor Napoleon III had been building his army into one of the most formidable fighting forces in the world. This, coupled with one or two awkward international incidents and a thinning of British regulars at home due to the Sepoy Mutiny in India, caused the English to view their neighbor across the channel with increasing distrust.

During the Napoleonic wars at the beginning of the 19th century, scores of volunteer units were formed to protect the English homeland from a feared French invasion. By mid 1858 this idea had been resurrected and a second volunteer movement began.

For the most part these units, like their counterparts in America, were little more than social clubs and shooting societies. While their uniforms were supposed to be predominately gray, depending upon the group's finances, many hues and flamboyant styles were adopted.

Their arms too, often matched their uniforms. While the rifles used by these units were generally modeled after the famous .577 caliber Pattern 1853 Enfield Minie Rifle, embellishments often took a decidedly elegant turn.

Another category of arms used by both civilian and military sharpshooters were the .451 smallbores. Perhaps the most famous arm in this category was the hexagonal-bored Whitworth.

British marksmen using these guns led the world in international competitions, and both their arms and shooting postures were widely emulated.

Original British smallbore match, sporting, and military-style target rifles are now sought-after collector's items. Many modern shooters, especially in Britain, still prize them for black powder competitions.

The supplies of these guns, particularly on this side of the Atlantic, have been sparse enough to drive the prices well into the thousands of dollars

Some years ago the respected British arms firm of Parker-Hale Ltd. introduced its splendid series of .577 Enfield style rifled-muskets. Many current shooters consider these to be the finest replica military style muzzle-loaders available.

Now they have capped off this achievement with their new .451 "Volunteer Rifle." Stylistically, the P-H Volunteer Rifle hearkens back to the military target rifle of the early 1860s. Its basic configuration is that of the popular Pattern 1856 two-banded Enfield Short Rifle—with a difference.

The hitherto plain stock of the P-56 is hand-checkered on the Parker Hale Volunteer rifle, and each barrel is glass bedded at the factory for improved accuracy. The 33-inch, .451 barrel features a patent breech (reduced-size chamber) to speed up powder burning, and is rifled with shallow lands and grooves, one turn in 20 inches, in the style of famed Irish gunsmith, John Rigby.

The adjustable rear sight also of the Enfield variety, comes with several interchangeable leaves to allow one to fire at varying ranges without having to raise the "ladder." The front sight is of the hooded globe type popular with shooters well into the 20th century.

Other special features on Parker-Hale's new "smallbore" include an adjustable detent lock, brass-tipped ramrod, small false muzzle to protect the rifling while ramming, and a copper beryllium nipple.

With the exception of the lock, which is case-hardened, all other steel parts on the gun are deeply blued. The Volunteer Rifle's buttplate, triggerguard, nosecap and lock escutcheons are of brass.

A precision arm such as this obviously needs special loading, firing and cleaning attention. Parker-Hale has supplied an accessory kit to make sure the

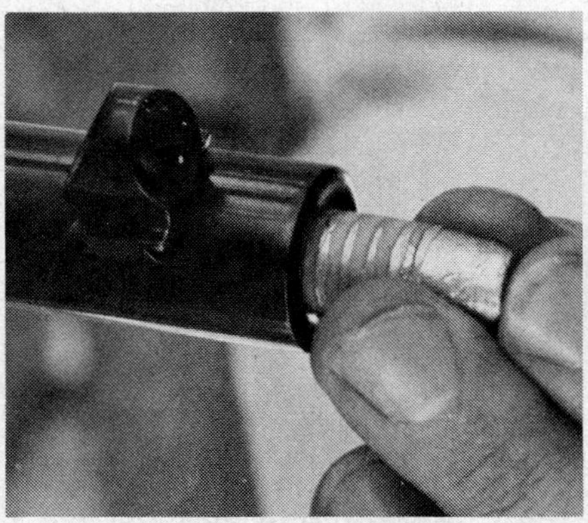

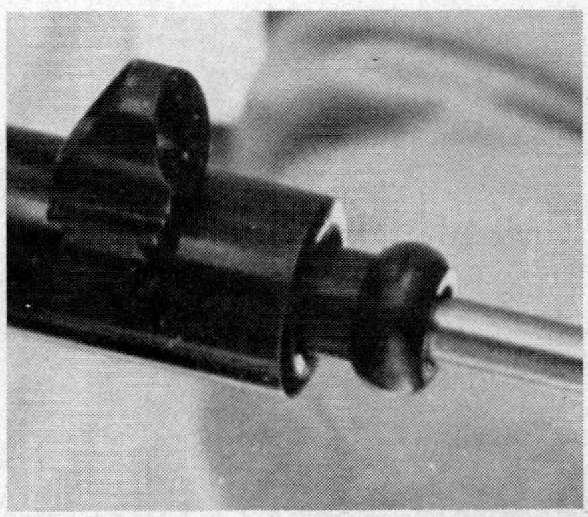

The Lyman mold supplied with the Parker-Hale Volunteer rifle casts a 545-grain conical bullet, above left, with eight annular grease rings. After initial seating of the bullet, a false muzzle is used to protect the rifling as the bullet is rammed home.

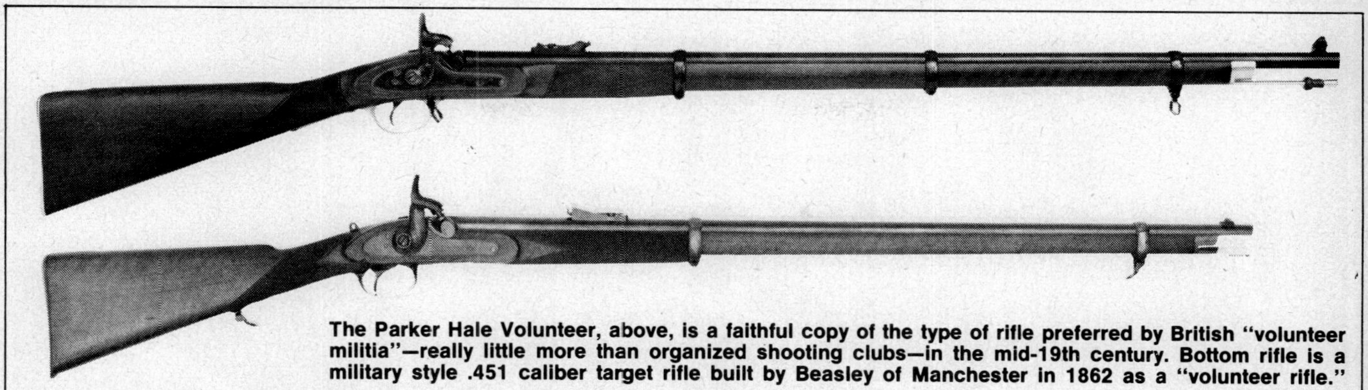

The Parker Hale Volunteer, above, is a faithful copy of the type of rifle preferred by British "volunteer militia"—really little more than organized shooting clubs—in the mid-19th century. Bottom rifle is a military style .451 caliber target rifle built by Beasley of Manchester in 1862 as a "volunteer rifle."

shooter gets off on the right foot. It includes a special Lyman mold for a .451, 545-grain bullet, a bullet sizing die and punch, a .451 wadcutter, an extremely well-designed nipple wrench, one spare nipple, three rear sight blades, two phosphor bronze cleaning brushes and three wool swabbing mops.

Perhaps the first step in firing the P-H Volunteer Rifle is casting the bullets. To begin with, always use pure lead, as this style of arm relys upon upsetting to expand the bullet into the rifling. After properly heating and "seasoning" the mold we cast several dozen of the Lyman #457121 bullets.

Our next step was lubricating the bullets. We melted a mixture of about 80 percent beeswax and mutton tallow, dipped each bullet in it to the depth of the top most of eight annular grease rings and set them aside to solidify.

Once this was completed it was necessary to size the bullets. To do this one must *carefully* press each bullet, base first, through the sizing die with the punch, taking care not to deform it in any way. This is an extremely important operation. Any bullet showing the slightest degree of eccentricity should be tossed back into the melting pot, for it simply won't provide reliable accuracy.

Prior to loading the Volunteer Rifle, we first snapped a couple of caps on the nipple to clear the channel of oil or other obstructions. Following that time-honored procedure, we then carefully poured 70 grains of FFg black powder down the bore (after experimenting with several loads, we found that 80 grains gave us the best accuracy), and topped it off with a lightly-tamped waxed wad.

Parker-Hale recommends that at this point the shooter attach a wool mop to the ramrod, dampen it slightly with water, lower it down on top of the waxed wad and mark the rod about a quarter of an inch above the point where it exits from the barrel. This ensures that when the gun is swabbed out, the mop will not be introduced into the powder chamber where the moisture could spoil the burning and ignition of the powder in the next load.

The bullet must be inserted carefully into the muzzle and *gently* pushed down on top of the wad. It is a good idea to mark the ramrod at this point as well, so that each bullet is seated to the proper depth.

We removed the ramrod and false muzzle, capped the nipple, aimed—and fired. Our first 100-yard shot struck about six o'clock in the eight ring. The next round punched a hole about two inches to the right of its mate, however subsequent 10-round groups (using the 80-grain powder charge) measured just over 1½ inches. I will readily admit that I am no competition shooter and am sure that any accuracy variances rested with the human, rather than mechanical, element of the battery.

The Parker-Hale Volunteer Rifle, available from the Navy Arms Co., 689 Bergen Blvd., Ridgefield, NJ 07657, sells for $575—not much more than many production Hawkens and Kentuckys now on the market.

If your taste runs to military arms (as mine does) or to match rifles, the Parker-Hale Volunteer Rifle comes as close to fitting both qualifications as one can expect. It's a good gun to have in your collection—even if you are not worried about a French invasion!

SPECIFICATIONS

PARKER HALE VOLUNTEER RIFLE

Manufacturer:	Parker Hale
Importer:	Navy Arms Co., 689 Bergen Blvd., Ridgefield, NJ 07657
Action:	Single-shot percussion muzzleloader
Caliber:	.451
Barrel:	33 inches, round
Sights:	Adjustable rear, globe front
Finish:	Case hardened receiver, all other parts blued
Stock:	Glass bedded, hand-checkered walnut
Weight:	9½ pounds
Price:	$575

Parker Hale's new Volunteer has hand-checkering at wrist and forend and comes complete with all the accessories needed to fire or service the arm, including an extremely practical nipple wrench, lower left, a punch for over-the-powder wads, flask, powder measure, and a specially designed Lyman bullet mold for accurate .451 cal. 545-gr. conical bullets, not shown here. The whole package is both attractive and functional, offering superb accuracy, fine workmanship, and good looks.

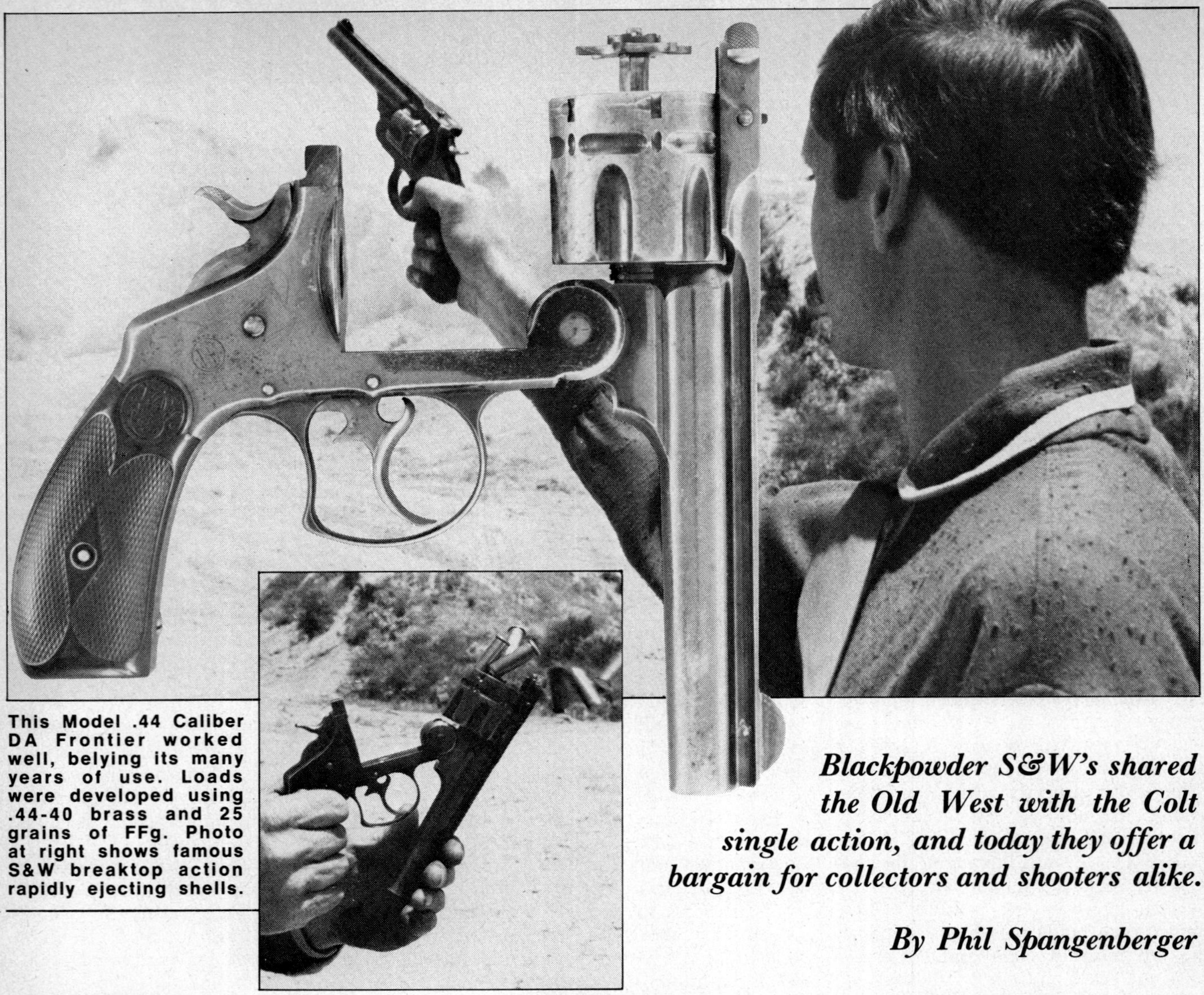

This Model .44 Caliber DA Frontier worked well, belying its many years of use. Loads were developed using .44-40 brass and 25 grains of FFg. Photo at right shows famous S&W breaktop action rapidly ejecting shells.

Blackpowder S&W's shared the Old West with the Colt single action, and today they offer a bargain for collectors and shooters alike.

By Phil Spangenberger

Shoot That Old Smith & Wesson!

With the rising interest in shooting blackpowder cartridge guns, many arms that until recently were relegated to wall hangers are now being turned into shooters. Firearms buffs are rapidly becoming aware of the vast assortment of usable firearms, both handguns and longarms, that are available in this field. Presently, because some of these arms haven't yet caught on with collectors, they are still priced at an affordable figure for most of us with a moderate pocketbook.

One of the leading manufacturers of the 19th Century, whose arms presently fall into the above mentioned category is Smith & Wesson. S&Ws were used widely around the world and have as colorful American frontier history as any Colt, Sharps or Winchester could ever boast of. Today, a budget minded blackpowder buff can afford a shootable S&W blackpowder cartridge revolver for just a little more than some blackpowder cartridge replica sixguns command.

Bill Evans, a fellow western enthusiast and blackpowder shooter from Redlands, California, recently picked up a couple of interesting S&W six-shooters. Through a little reloading research and a few pleasant hours at the reloading bench, Bill has turned these old Smiths into some mighty fine, and fun, sidearms. The guns are: Model .44 Caliber Double-Action Frontier and a Model .44 Caliber Double-Action, First Issue, also called the New Model Navy No. 3 revolver. Both guns are in fine

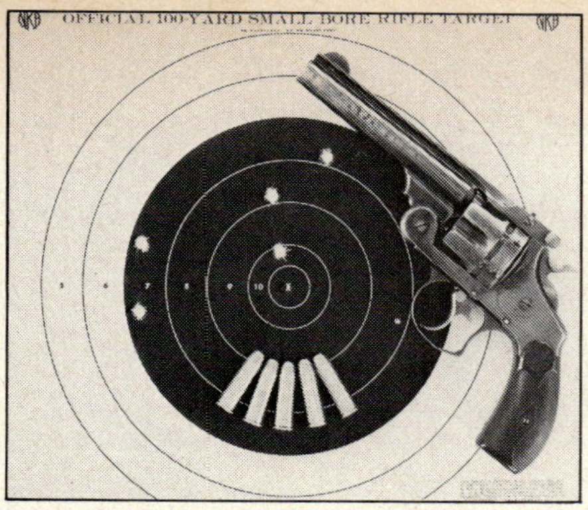

At 25 yards, the .44 DA consistently put five shots within a 4½-inch radius—fine accuracy for a fixed iron sight handgun.

A stylish and well made handgun, the DA Navy is of a quality unequalled on today's production guns.

Chambered for .44 SWR, the DA Navy shot 4½-inch groups. The .44 SWR was a popular target round of its day.

shape, showing most of their original finish.

Bill's D.A. Frontier model is a five-inch barreled, blued revolver. It sports the original factory hard rubber checkered grips. A total of 15,340 of this model was manufactured from 1886 through 1910.

Originally designed to handle the .44-40 (.44 WCF) which used a 200-grain lead bullet in front of 40 grains of blackpowder, Bill decided to load his ammo on the lighter side. Using a 205-grain Lyman cast bullet (mold #427980), along with 25 grains of FFg (2F), blackpowder and large pistol primers, Bill was able to get an average velocity of 720 feet per second. Accuracy-wise, Bill was able to group five shots within 4½ inches at a distance of 25 yards while using a two-hand, offhand hold. This is excellent accuracy for a five-inch barreled .44-40 with fixed iron sights.

Incidentally, this may be a good time to point out that many shooters complain that they find it next to impossible to load 40 grains of FFg blackpowder into a modern .44-40 case. And they are right, it is next to impossible, but not impossible. The old-timey cases were constructed with a balloon-head base and could easily take the prescribed charge of 40 grains. However, the newer cases are made with a thicker base and are not even made to accommodate a 40-grain blackpowder loading. Rather, they are designed to handle a comparable smokeless load. I have found that by loading the cartridge case about three-fourths full, then tapping it gently to settle the powder, the full 40 grains can be put in, although it is certainly tightly packed, sometimes causing the case to buckle during bullet seating. A better cure for this problem is to reduce the charge to about 35 grains of FFg blackpowder. In this way, you'll get a healthy charge and save yourself some aggravation.

The other revolver, the New Model Navy was chambered for the .44 S&W Russian cartridge, one of the most popular target rounds of the late 19th Century. However, because of the plain, non-adjustable iron sights on this handgun, it is obvious that this gun was not intended for target work.

For the initial shooting sessions with the D.A. Navy model S&W, we ventured to Angeles Shooting Range in Tujunga, CA, Bill utilized a loading of a 240-grain Hornady semi-wadcutter lead bullet backed by 18 grains of FFg blackpowder. This loading produced a muzzle velocity of 680 fps. Bill's load was actually a little on the light side. The old standard load for the .44 S&W Russian (.44 SWR) was a 246-grain lead round-nosed bullet and 23 grains of black powder. It can be safely assumed that FFg (2F) was the grade of blackpowder used, since this granulation was fairly common in big bore pistols during the late 19th Century.

At 25 yards, while shooting offhand with a two-handed hold, Bill was able to place five shots within 4½ inches. Not bad, however we must remember that the .44 SWR was a favorite target cartridge—even if this was not a target revolver.

Unlike the .44-40, Bill was not able to go to the local gun store and purchase a box of factory primed cases for his loads. So, Bill decided to cut down .44 Magnum cases to the appropriate length (.44 Special cases will also work well). Large pistol primers are used with .44 SWR loads.

Although Bill used the Hornady semi-wadcutter bullets, there are a couple of other bullets that will also do the job nicely and still maintain the original bullet shape of the .44 SWR. Lyman's mold #429184, which is a 235-grain round-nose bullet will serve, as will their mold #429383 which is a 245-grain bullet. Interestingly, in Frank C. Barnes' excellent reference work, *Cartridges Of The World Fourth Edition*, published by DBI Books, Inc., Northfield, IL, six out of seven loads listed utilize a 246-grain bullet. These loads produce muzzle velocities ranging from 695 fps to 900 fps.

Both of Bill's "forty-fours" are fine shooters and performed admirably, not only on paper, but in the field as well. Plinking with these guns was lots of fun. Both guns exhibited a high degree of workmanship with their smooth functioning performance. This pair of Smith & Wesson frontier-era revolvers are typical of the many varieties of firearms that are available to today's collector/shooters. For the money, you'll get a gun that is superior to any of today's production pieces, and with proper care, you'll have an investment for the future. There are many reasons one could justify the purchase of an old-time sixgun. However, the most important and simple reason is "just for fun." After all, that's why most of us are in this shooting game, and as far as this blackpowder buff is concerned, shooting a pair of 19th Century sixguns, after working up your own loads, or duplicating the old factory loads, is a great way to spend those precious leisure hours. So, the next time you see an old blackpowder cartridge revolver for sale at a gun show or store, check it out thoroughly. If it is in good shape, don't pass it by simply because the chambering is no longer available in factory fodder. You could be missing out on some firearms fun.

The Hall Breechloader

Years ahead of its time, the Hall Breechloader was the controversial forerunner of a new system. How good was it?

By Garry James

The idea of a breechloading firearm is not a new one. At least as early as the late 15th Century we find examples of chamberloading cannons and matchlocks. Even Henry VIII's arsenal contained ingenious breechloading handguns which protruded through small shields.

The advantages of a firearm which could be charged from the breech were obvious. A man could load such a gun faster with less exposure to the enemy or quarry, and a slightly oversized bullet could be made to fit a rifled or smoothbore tightly without using troublesome patches, increasing both accuracy and range.

Today's firearms enthusiast is treated to a large variety of antique breechloaders—some practical, some bizarre—and though the system was not particularly unusual at any time during the past 400 years, it still did not achieve widespread popularity until the early part of the 19th Century.

The arm responsible for this revolution was patented by a Captain John H. Hall in 1811. While this chamber-loading flintlock rifle was not unique (the system had been tried by Italian and British gunsmiths several decades prior to Hall's invention) its simplicity and durability caught the imagination of the American Ordnance Department.

Hall's version of the chamber loader called for a breechblock, containing the action and centrally mounted cock and frizzen, to be attached to the receiver by a rear mounted screw/hinge.

To load the arm, the soldier merely pressed back on a spring-loaded trigger-like catch beneath the stock and rotated the block upward. He then poured in a measured charge of powder and pressed the ball on top of it with his finger (later, paper cartridges were introduced to facilitate this operation). The block was snapped shut, the pan primed and the arm was ready to fire.

With practice, the Hall rifle could be loaded several times faster than the standard smoothbore military muskets and patch-and-ball muzzle-loading rifles of the period.

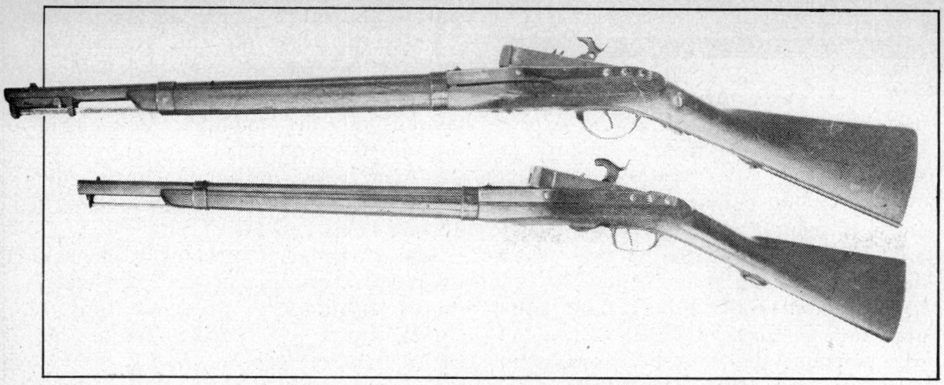

The Model 1836 Hall Carbine, top, was several inches longer than the 1843 Model. More subtle changes included replacement of sliding bayonet/rod for a standard ramrod and stronger hammer/nipple combo.

The first Hall rifle barrels were 22.75 inches long, .52-caliber, with 16 grooves .200-inch deep, twisting one turn in eight feet. The rifle's main drawback, according to contemporaries, was the escape of gas at the gap between the block and barrel. How severe this really was, we will see later on.

The initial government contract called for 1,000 Hall rifles. Hall himself was hired as assistant armorer at the Harpers Ferry Arsenal to supervise the manufacture of the arms, with a salary of $60 per month plus a $1 royalty on every gun made. The arms featured iron furniture similar to that used on the standard Model 1812 U.S. Musket. All metal parts (with the exception of the case-hardened block) were finished with a durable brown lacquer. Because of a number of revolutionary machines devised by Hall, his rifle was the first arm with truly interchangeable parts. Components such as the block, barrel bands, and action parts could be taken off one gun and fitted to another with little difficulty.

In 1833 the United States adopted a percussion carbine version of the Hall breechloader for its Regiment of Dragoons. This arm had the distinction of being the first percussion arm officially accepted by the United States Army.

The carbine action, with the exception of the replacement of the flintlock cock and frizzen with a percussion hammer and nipple, was virtually identical to its predecessor. The carbine itself employed much of the rifle furniture, though it did have some unique features. Its .52 caliber (1000 M-33 carbines were also made in .579), 26-inch barrel was smoothbore, and it incorporated a rather curious sliding triangular rod bayonet. The rear portion of the triggerguard was hinged to provide a compartment for a wiper and combination tool, and a carbine sling ring was also affixed at this point.

In 1840 two versions of a .52 caliber, 21-inch barreled carbine were produced—one with an odd elbow lever, replacing the M1819-style catch, and another with a "fishtail," replacing the elbow lever. The bayonet was replaced in favor of a standard ramrod, and the ring remounted at the rear of the triggerguard.

In 1841 a percussion version of the long rifle was introduced. While it retained its old silhouette, it too employed the fishtail lever.

A Model 1842 carbine with brass fittings and a sling bar on the left side of the receiver was the last of the Hall breechloaders to be manufactured at Harper's Ferry.

Finally, in 1843, the improved Hall carbine incorporating a spoon-shaped side, lever designed by Simeon North was adopted into the service. It was to be the last of its line, for in 1847 the government ceased its production in favor of a muzzle-loading, .69-caliber musketoon.

The barrels of the Hall carbines, though smoothbore, tapered slightly from the breech to the muzzle, causing the ball to compress and give a good tight fit. Accuracy, even sans rifling, was reported to be quite good. Because of the centrally mounted hammer, it was necessary to position the rear sight notch slightly to the left. The front sight was affixed at an angle on the muzzle so that it would align with the rear notch and be visible to the shooter.

The Hall carbine had a long, distinguished service, seeing action against the Seminoles in the late 1830's and during the Mexican War. It even saw some usage during the Civil War by hard-pressed Confederate troops and a few Federals. This diminutive breechloader was even the subject of a scandal when, in 1862, General John C. Fremont attempted to purchase some 6000 newly-rifled Hall carbines for his troops at what the Congress considered to be the exorbitant price of $22 each.

One attribute of the Hall that made it popular with soldiers was the ability of the block to be easily removed and used by itself as a handgun. Because the mechanism was completely contained in the breechlock, it made a handy, if somewhat unwieldy "pocket pistol" for close-in work. Sam Chamberlain, in his memoirs, *My Confession,* described a fracas during the Mexican War in a local cantina where his Hall "chamber" came in quite handy.

"I thought my time had come, but resolved not be rubed [sic] out without a struggle. With a bound I sprang behind a large table used for a bar, drew the chamber of my 'Hall's Carbine' (that I always carried in my pocket), said a short prayer and stood cool and collected, at bay before those human tigers, *guerrillars.* There was one grizzly old fellow who seemed more ferocious than the others, he had but one eye that glared on me with the fierceness of a wild beast. He rushed for the table as if he would spring over, when the sight of the little iron tube pointing straight for his solitary optic caused him to pause. A few tallow dips cast a feable [sic] light on the savage faces in my front, cries of *'muerte! muerte! los Ame-*

The artist's conception of Captain May's 2nd Dragoons charging the Mexican guns at Resaca, Mexico is 1846 shows the troopers armed with sabers. The 2nd Dragoons were armed with Hall carbines—both they and their Halls accounted well for themselves.

ricano ladron que muerte los Yanque burro' came from all parts of the room, but none offered to strike, 20 brigands were held at bay by the strange weapon I held, they seemed to know it was sure death to one, and none seemed willing to be that one."

Hall rifles and carbines (depending upon model and manufacture) were usually stamped on the top front of the breechblock with either a four line "U S, S. NORTH, MIDDLTN CONN., (date) or line J. H. HALL, U S (date) or H. FERRY, U S, (date) designation.

Modern critics have been quick to condemn the Hall for what appears to be the obvious weakness in its design—a fairly large gap between the barrel and block, which allowed a gas leak.

Of course, prior to the invention of the self-contained cartridge this was a problem with all breechloaders. Some percussion arms such as the Maynard, Burnside and Smith attempted to remedy the difficulty by employing brass or rubber cartridges that would effectively seal the breech.

My own personal experience in firing a good number of paper cartridge percussion breechloaders is that they all leak to a more-or-less disconcerting degree. For instance, the Sharps and Cosmopolitan burns one's wrist and hand, and the Terry and Leech blasts gasses out of the side across the shooters' face. Perhaps the two best caplock breechloaders were the Green and the Westley Richards, but the former was complicated and cumbersome, and the latter introduced so late that it was rapidly overtaken by improved metallic cartridge single-shot and repeaters.

Halls in any shape are getting harder and harder to come by, but for our test we were fortunate to obtain a Model 1836 (dated 1839) in good shooting condition. The bore and action were about perfect, however the exterior exhibited the nicks and dents one would expect the gun to receive from a century of storage.

At this point perhaps it would not be amiss to caution the reader. Hall rifles and carbines are valuable collector's items. Depending upon condition and model, many Hall carbines and rifles are worth in excess of $2000, so it is not recommended that they be subjected to extensive firing. Also, before one shoots any old gun, it is imperative that he takes it to a qualified gunsmith to make sure that it is in serviceable condition.

Keeping both of these precautions in mind, we determined that our specimen was mechanically perfect and decided that if proper care was taken, it would not be detrimental to fire a string of 50 test shots out of it.

Our first surprise came when we tried to introduce a .680 ball into the chamber of the breechlock—it rested at the chamber mouth, refusing by several hundredths of an inch to be seated upon our 70-grain, FFg powder charge. Different bullets were tried, with a .662 diameter found finally fitting perfectly. The bore at the muzzle was miked, and it proved to be closer to .64 than .69. From this, we can surmise that the story of the M-1836 being able to use issue musket balls is probably untrue.

Using the .662 ball and previously mentioned powder charge, we made up ten potassium nitrate soaked tissue paper cartridges. The remainder of the shots we decided to fire with loose powder and ball.

The first thing we noticed was that the nipple was quite small. The "top hat" musket cap we are familiar with today simply hadn't been invented in 1839. This Hall used the regular size 11 rifle and pistol caps of the day, although those manufactured after 1842 were equipped to handle the larger caps.

Pressing back on the catch, we snapped the block up in the loading position. A charge of powder was poured into the chamber, and a ball pressed down on top of it. The block was then closed, the nipple capped, the piece brought to full cock, aimed at a 50-yard gong, and then fired.

Almost immediately we heard the satisfying reverberation of the thick metal target. Bystanders could see only a small spurt of smoke from the chamber/barrel gap.

The process was repeated five more times, with five more hits. Paper cartridges, though a bit faster to load, elicited no change in the overall shooting effect or the accuracy.

Remembering trooper Chamberlain's experience with the *"guerrilars,"* I thought it might be amusing to fire the breech by itself. It was easily removed from the receiver and despite its angularity and absence of grips, was not uncomfortable to hold. The trigger is almost two inches long, so it was possible to wrap three fingers around it for a more positive pull.

We reduced the charge to 40 grains of FFg, loaded a ball, capped the nip-

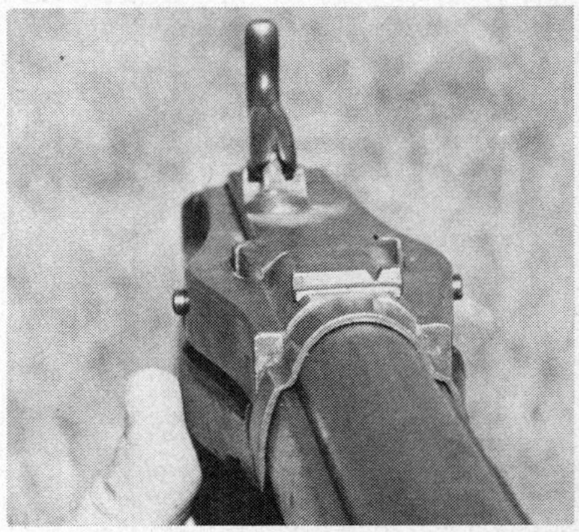

Because of the centrally mounted hammer on flintlock and percussion models, front and rear sights were offset to the left.

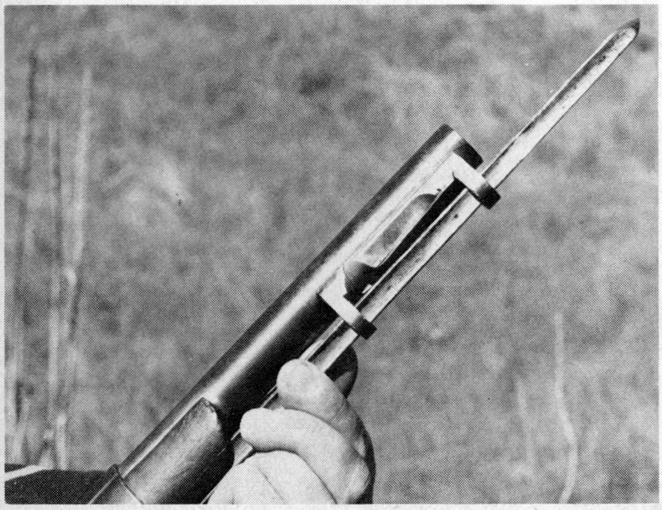

The Model 1833 and 1836 Hall carbines featured an unusual sliding ramrod/bayonet combination. Eliminated on later models of Halls, the same arrangement was used on 1884 Springfields.

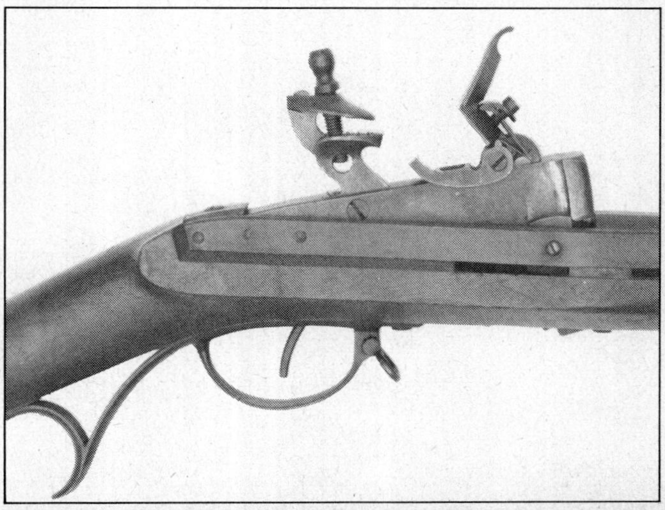

The breechloading operation of the early flintlock Hall rifles was identical to that of the later percussion models. The soldier so armed could load rapidly without standing to expose himself.

ple, and squeezed the trigger. The report was quite loud, but the recoil not prohibitive. We missed our 25-yard target by about a foot and a half, and subsequent shots peppered the area around the bull at distances varying from six inches to two feet. All in all, it was an interesting experience.

A number of shooters took turns with the carbine while others examined the piece for gas leakage during firing. There is no doubt that the Hall did leak, but certainly no worse than a Sharps. Not one shooter complained about discomfort from the gasses.

Accuracy was surprisingly good. Although we did not group the shots, the gong rang 37 times out of 45 rounds—not bad for any smoothbore.

Our recent test seems to indicate that the Hall has had "bad press" for a number of years without any real justification. Most of the early damning reports came from ordnance officers who viewed and fired guns well past their prime. This breechloader, in its various modes, remained in service for over a quarter of a century with few complaints from the soldiers to whom it was issued. One could hardly ask for a better testimonial than that.

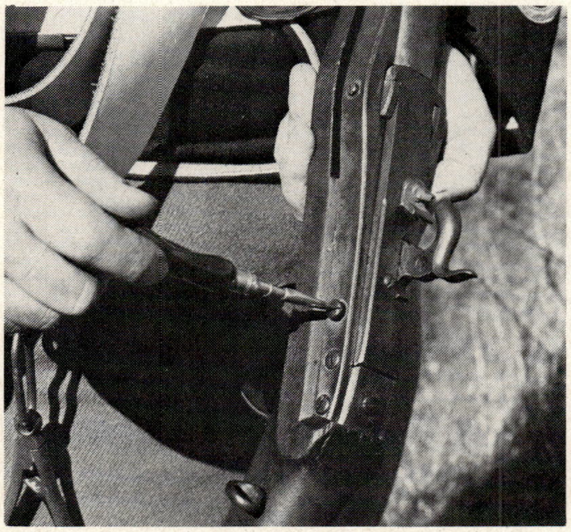

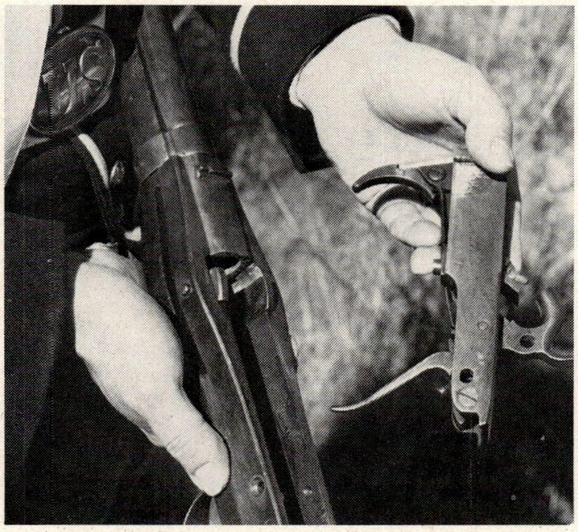

The breechblock of a Hall rifle or carbine, whether flintlock or percussion (shown), can easily be removed by simply unscrewing the rear pivot pin. The original concept for this unique feature was due to the interchangeability of parts with fairly minor fitting, and of course cleaning was made quite simple. Innovative troopers discovered that the breechblock could be carried as a small pocket pistol! carbine also exhibits the later saddle ring at the small of the stock.

Although understandably inaccurate, a Hall breechblock was deadly at close range. The practice was unauthorized, but the breechblocks were carried concealed by many an off-duty Mexican War trooper. Although very awkward, recoil is not excessive when fired in this mode, above. Left, a loaded and capped Hall percussion breechblock fits neatly into the palm of the hand.

Test Report

COLT WALKER

Reissue of the Colt Walker offers the chance to try out that massive firearm which was so much a part of our frontier history.

By Phil Spangenberger

The Colt Walker revolver was for many decades the biggest and most powerful mass-produced handgun ever built. Even today, it ranks as a magnum among cap-and-ball revolvers and can hold its own, at least on paper, with many of today's powerful combat handguns, including the venerable 1873 Colt Peacemaker and the "King" of the autopistols, the 1911.

What the Walker had going for it during its heyday was the .44 caliber soft lead ball, the massive cylinder whose chambers could take up to 60 grains of FFFg black powder, producing a whopping 1,100-plus feet per second, (fps), muzzle velocity. That's a good 300 or more fps faster than the 1860 Colt percussion revolver, any of Colt's massive Dragoon models, or the .45 Colt caliber 1873 Peacemaker. In fact, this approaches the muzzle velocity of the 1873 Springfield .45-70 Government "Trapdoor" rifle or carbine. With power like that in the mid-19th Century, it's no wonder that the Walker was thought of as the ultimate in a revolving pistol. Even today, the Colt Walker is considered somewhat of a blackpowder magnum.

Originally designed through the joint efforts of Col. Samuel Colt and Capt. Samuel Walker, a former Texas Ranger then assigned as a company commander of the newly created U.S. Regiment of Mounted Riflemen, the Walker Colt combined the power of the single-shot horse pistols then being issued to mounted troops, with the multi-shot capabilities of Colt's earlier invention, the Paterson five-shooter. Although the Colt Walker revolver entered the Mexican War during its latter phases, and played but a small part in the overall scene, it did gain rapid popularity with those who used it and those who saw it used. Issued in pairs to the Mounted Rifles, the Colt Walker was used effectively and through its favorable performance—in spite of drawbacks both in design and the fact that they were made of a poor grade of iron causing several to blow up in service—the Walker paved the way for generations of Colt produced revolvers, and automatics, in the U.S. military.

In recent years, Colt Firearms Division of Colt Industries, 150 Huyshope Ave., Hartford, CT 06102, have begun remaking several of their legendary black powder percussion revolvers. The Walker model is one of those that has been resurrected. It is Colt's intention to bring out an authentic looking and functioning "reissue" of their percussion revolvers, complete with each arm's idiosyncrasies. By offering these historic reissues of handguns that would otherwise be unobtainable, much less shootable, to most of us, Colt provides not only shooting recreation, they also offer the opportunity to study first hand the development of these six-shooters, enabling us to understand why each successive model was improved in its own way. With this point in mind, let's look at the 1980s version of the 1847 Colt Walker revolver.

While the original Walker weighs in at a hefty four pounds, nine ounces—unloaded, the reissue versions we've inspected at *Guns & Ammo* seem to run about two ounces lighter. Perhaps this is due to superior 20th century metallurgy. The overall dimensions are identical to those of the old-time Walkers, so the same 60-grain charges of black powder can be packed into their behemoth chambers. The nine-inch barrel which sports seven grooves, is blued, as is the cylinder, trigger, backstrap, wedge and all screws. The hammer, frame and loading lever assembly are color casehardened and the triggerguard is polished brass. The cylinder has the early-style circular bolt stops and the roll-engraved scene commemorating the famous "Hays' Big Fight" where 15 Texas Rangers, Sam Walker among them, all armed with Colt Patersons, defeated 80 Comanches in 1844. The stocks on the Walker are one-piece walnut and are nicely figured. Probably the biggest noticeable difference in this new-made Walker as compared with the original, is the fact that Colt is using their modern blue-black dull gloss finish on their firearms as opposed to the rich peacock blue of the 19th Century vintage arms.

During the test sessions, I had the opportunity to handle three specimens of this newest addition to the Colt black powder series. While they were all good looking six-shooters, I did encounter some minor gunsmithing problems with them all. They were the sort of problems that, while not very big, can be a real nuisance to the shooter/buyer. The cylinder bolt stop has been left unfinished, and in two of the three Walkers I tested, it left a light scoring line around the cylinder after a few rotations. Of course, the shooter can polish this himself quite easily, but I feel that this is Colt's responsibility. By smoothing off any minute rough spots on the stop before the arm leaves the factory it could save the marring of the gun's cylinder. In one of the guns,

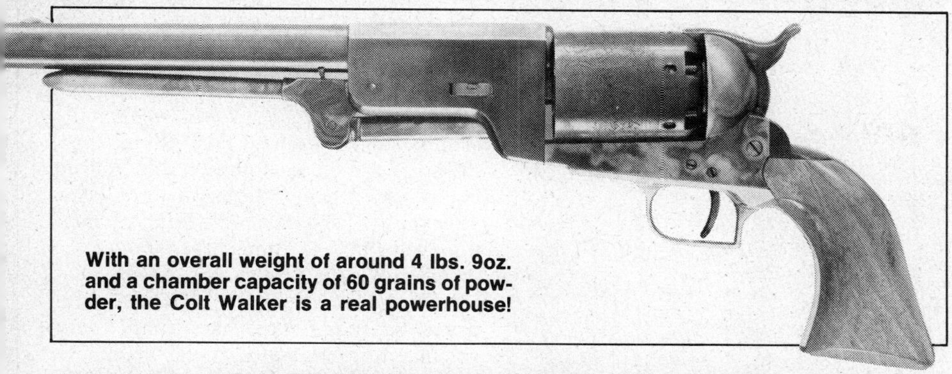

With an overall weight of around 4 lbs. 9oz. and a chamber capacity of 60 grains of powder, the Colt Walker is a real powerhouse!

The loading procedure for the big Colt Walker reissue calls for pouring the powder charge into the cylinder (left), seating the .44 caliber lead ball (middle) and then ramming it down with the loading lever (right). All that's left is to cap each nipple.

the wedge retaining spring (located in the slot of the wedge itself) had not been peened properly and it came loose when the wedge was removed from the barrel assembly. This same revolver had a loose mainspring stud in the backstrap assembly and a refitting would be necessary for smooth cocking. Dismantling the gun in the field can also be a problem as two of the three revolvers had frame/barrel assembly pins fitted so tightly that a bench vise was needed to pry them apart. A call to the Colt factory revealed that they are fitting them tightly on purpose, so they won't wear loose as fast. This writer disagrees

SPECIFICATIONS
COLT 1847 WALKER

Manufacturer:	Colt Firearms 150 Huyshope Avenue Hartford, CT 06102
Caliber:	.44
Action:	Percussion
Capacity:	Six shots
Barrel:	9 inches 7 groove, right hand twist
Weight:	4 pounds, 7 ounces
Frame:	Color case hardened steel
Sight:	German silver
Backstrap:	Blued steel
Triggerguard:	Brass
Grips:	One piece walnut
Price:	$447

with this practice. As Editor Howard French stated in his review of the Walker in the August, 1980 issue of *G&A*, "You can have too much of a good thing." Personally, I'd rather be able to dismantle the gun by hand—at the risk of having the pins wear faster. That black powder residue gums these cap-and-ball sixguns up enough without some design engineer's help!

While the Colt instruction manual states that No. 11 caps should be used, two of the three revolvers inspected required the No. 12 caps. These negative points can be attributed to quality control at the factory. Colt has had a good reputation for many years for producing top-quality handguns and with a little closer inspection of these percussion revolvers, they will maintain their well-deserved reputation of leadership in quality.

With these minor smithing problems ironed out, the Walker performed admirably, as would be expected of any Colt. Accuracy within the revolver's intended ranges of 25 to 50 yards was acceptable. Four to six-inch groups are easily obtainable with this gargantuan sixgun at these ranges and at the 100-yard range, Colt's Walker is still able to not only put the shots where they count, they still pack plenty of power. I might add that a two-hand hold is the preferred method of shooting this barrel-heavy revolving horse pistol, as the old-timers called it. The Walker was designed as a close-range combat arm to be fired from a plunging horse during the heat of battle, and this it did well. But when you heft that revolver while in a relaxed position at the firing range, *whoa* . . . that's a different story. Actually, snap shooting with the Walker, the way it was intended to be shot, can be lots of fun. Try it yourself. Set up some cans a few yards out, turn and face away at a 90-degree angle from your targets, with your revolver raised in the classic duelling position, and quickly turn toward the targets and let fly one or two rounds.

Our test loads consisted of both 50 and 60-grain charges of FFFg black powder, a .457 Speer swaged round ball, Ox Yoke Originals' "Wonder Wads" which take the place of a lubricant, and the appropriate sized caps. During the testing we used a variety of caps, including CCI-Speer, Navy Arms and Remington-made percussion caps. The massive weight of the Walker absorbed any recoil.

Ignition was fast and sure, and while my first impression of the mainspring was that it was a tad on the weak side, this is not the case. I suppose I expected a stronger pull. Actually, the cocking of this gun is quite comfortable and positive. All in all Colt's newest Walker .44 sixgun is a joy to shoot.

If you are looking for a true black powder magnum, Colt's reissue Walker is just for you. The Walker played an important part in the history of firearms and shooting this authentic version, by the same company that made the originals, gives an added flavor to this big, big six-shooter. Priced at around $446.95 it should make a nice addition to any shooter's collection.

Muzzleloading smoothbores have long been the most low-keyed of the blackpowder sports, but no more! Old-time guns and brand new equipment have combined to start . . .

By Rick Hacker

The Renaissance Of Blackpowder Shotgunning

Of all the black powder shooting sports, perhaps none combines the elements of adventure, nostalgia and convenience as well as shotgunning. Ever since its beginning in the 1750s, when the first flintlock double was perfected, on through the Victorian years of the market hunter, bird and small game hunting with muzzleloading scatterguns enjoyed an undisputed realm of popularity for both survival and recreation; its aficionados included the Western rancher, the farmer of the Midwest, and the coat and tie businessman of the East, as well as sportsmen throughout the world. Unfortunately, all this came to an end by the 1880s, when the fixed-shell breechloading shotgun was finally accepted by hunters.

It took 100 years for the muzzleloading shotgun to regain its rightful place in the shooting sports. Even though the "rebirth" of muzzleloading is generally acknowledged to have started in 1954, the scattergun has, until recently, lagged behind its rifled brethren in popularity. The reasons for this unfortunate situation were at one time, based in fact: a scarcity of safe, shootable and affordable shotguns, a lack of shooting paraphernalia required by the frontloaders, the inconvenience of carrying all of this equipment into the field, and the mistaken belief that the muzzleloading shotgun was not as effective as its modern, fixed-shell counterpart.

Today, all these negative notions have evaporated like a cloud of FFG smoke in an autumn wind, thanks to two important factors. The first is the growing number of shooters discovering the fun, effectiveness and economy of muzzleloading. Shot for shot, it costs less to shoot a muzzleloader than it does "store bought" ammunition.

The second impetus to the front-loading shotgunner's resurgence upon the shooting scene is the increased number of manufacturers who have entered the black powder hunter's camp. As a result, an amazing variety of products have come to light in recent years. Some are continuations of original equipment our great grandfathers might have used; others are new and innovative creations designed to make the black powder shotgunner's sport more enjoyable than at any other time in the history of the sport.

I began shooting muzzleloading shotguns back in the 1950s; my first scattergun was an 1863 U.S. Trenton musket whose .58 bore had been shot out to an undeterminable gauge. I used to pack that old Union Army tube with 75 grains of black powder and a handful of Arizona dry wash gravel (no one sold lead shot in my neighborhood and I could not have afforded it if they had), all held in place with Republic & Gazette newspaper wads. This home-brewed load was obviously less than perfect, but occasionally I got lucky and managed to bag a few rabbits or doves.

Even in those days, when replicas were non-existent and originals normally sold for give-away prices (by today's standards), good quality muzzleloading shotguns were hard to come by. Fortunately for the sport today, that situation has entirely changed, thanks to the dedicated efforts of a few manufacturers and importers. The black powder shotgunner can choose from a number of scatterguns, most patterned after two styles of original 1850-70s era arms;

both single and double barreled shotguns are available.

My personal preference for a muzzleloading shotgun leans heavily towards the traditional side-by-side double, even though very little of my furred or feathered game has ever been taken with that second barrel. Still, it is psychologically reassuring to have that extra tube loaded and instantaneously ready, should you ever need it, and on a few occasions, I actually have scored with the second barrel after missing with the first. But all this rationale not withstanding, I simply happen to *like* the way a good double looks and feels in my hands.

In the twin tube category, I feel best values for the money are the classically styled Dixie Double (Dixie Gun Works, Union City, TN 38261) which sells for $175, and the Navy Arms Delux Upland (Navy Arms Co. 689 Bergen Blvd., Ridgefield, NJ 07657), costing $225. Both guns are available in kit form for $160 and $210, respectively, if you want to save a few dollars and have some extra time on your hands. However, I should mention that the factory-assembled guns are well made and for the small difference in price, I would rather purchase a completed shotgun and spend my extra time shooting it. Both makes of scatterguns are 12 gauge, and sport 28-inch tubes, European walnut stocks with hand-checkering, and feature stamped engraving on lock, hammers and triggerguard.

There is also a very well made, high power 10 Gauge Duck & Goose Gun recently introduced by Trail Guns Armory (1634 E. Main, League City, TX 77573). Unlike most of the other muzzleloading doubles on today's market, this large bore scattergun features longer (30¾-inch) barrels, a cheek piece on the left side of the stock, twin bead front and middle sights, and sling swivels. The Trail Guns Armory percussion weighs in at a hefty nine pounds to help absorb recoil. Ornate engraving adorns all metal parts with the exception of the barrel. This handsome, heavy-duty shotgun sells for $379.79. For those hunters who already own a Trail Guns Armory double rifle, a set of 12 gauge shotgun barrels can be purchased for $267.50.

For shooters with Navy Arms' Zouave and Buffalo Hunter Rifles, that firm also offers pre-fitted shotgun barrels in 12 and 20 gauge for the Zouave ($75 per barrel) and in 20 gauge for the Buffalo Hunter ($65), thereby providing the opportunity to change your .58 caliber rifle into a smoothbore shotgun with a minimum of effort and expense.

Shotgunners preferring the lighter weight and ease of pointability of a single barreled scattergun have a limited number to choose from. (Although there are a number of well-made muzzleloading single barreled shotguns on the market, many feature brass receivers, which I consider unsuitable for hunting due to the reflective qualities of the metal.) One of the best single-shots I have hunted with is the Magnum Cape Gun, imported by Euroarms (14 W. Monmouth St., Winchester, VA 22601). This 12 gauge copy of an 1850

While vintage muzzleloading shotguns in shootable condition are becoming rare, the number of new muzzleloaders is increasing. Above is a Dixie Gun Works 12 ga. double.

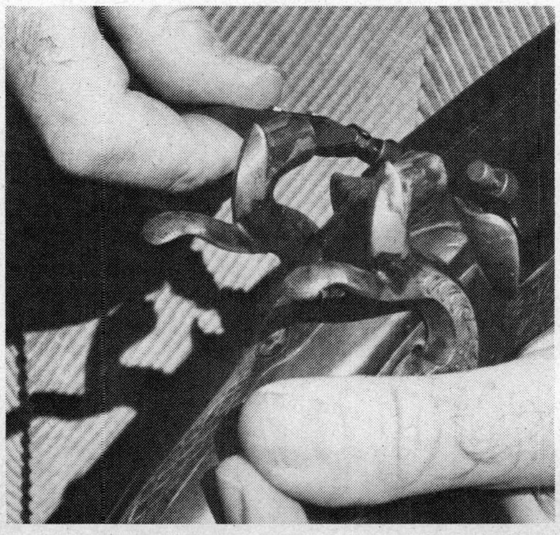

In the loading sequence at left, powder is poured from a pre-measured container, in this case a Butler Creek Speed Loader. After the powder, a thick cork overpowder wad is forced into the muzzle taking care not to split the cork. The load must then be rammed firmly. Consistent pressure ensures consistent patterns. Shot, carried in a Dixie Gun Works shot pouch is measured, then poured in. After a thin overshot wad is rammed, the shotgun is capped with a straight line capper. With practice, the whole loading process should take no more than 30 seconds per barrel!

style English sporting fowler has a 32 inch barrel with clean-out screw on the left side of the breech, a handy feature should you experience clogged or moist powder or—more embarrassingly—forget to pour your powder *first!* The Cape Gun also features a nicely-finished, hand checkered walnut stock. Light, tasteful engraving accents the hammer, triggerguard and lockplate, which is polished to a dull, silver matte sheen. All other metal parts are blued, making the gun extremely attractive for its $179.95 pricetag.

With the exception of the Trail Guns Armory double, all of the above-mentioned scatterguns weigh in at a scant 6 to 7½ pounds. In my humble, abject, and non-arguable opinion, this is much too light for any kind of enjoyable shooting. The fault lies not with the manufacturers, but with Nature herself. Unfortunately, thanks to the growing demands of our society, there is a scarcity of close-grained, high density (i.e., heavy) wood. As a result, more and more firearms firms are turning out lighter guns, which does little towards dampening recoil.

The black powder big game hunter rarely has the problem of recoil to worry about, even though he may be shooting bone-breaking loads; chances are he will not touch off more than one or two shots at his quarry all season. But the black powder shotgunner has the fast-flying opportunity to expend 20 or more FFG charges in a single day, if he has the good fortune to be stationed in a productive blind or in a heavily inhabited cornfield. As a result, many hunters have trouble shouldering their scatterguns after a few muscle-pounding shots. And with good reason.

One obvious solution to help diminish recoil is to wear heavier clothing. This is fine for late fall and winter forays, even though body movement is hampered to a degree. But in my region most bird hunting is done in the warm days of early fall, not to mention small game hunting in the spring. This mild weather hunting tends to turn the heavily-clad sportsman into a walking sauna, and bulky clothing can become more of a hindrance than a help. Adding a recoil pad to a muzzleloading shotgun is another solution to the problem, but to my purist mind, is akin to putting a scope on a caplock rifle; if you want to "go modern," then get one of those "new" self-contained cartridge shotguns. For the black powder fraternity, much of the satisfaction lies in shooting a scattergun that *looks* like it came out of the past. Thus, the problem remains: what to do about recoil?

Like many great ideas, this one was not thought of by me; it came from fellow writer and gun fix-it-person Toby Bridges, who hangs around Dixie Gun Works a lot. We had been discussing some of the problems inherent with the preceding paragraphs (today's shotguns being too light, too much recoil, etc.) when Toby suggested, "Why don't you install one of those Edwards Recoil Reducers in that front stuffer of yours? It's something I've always wanted to try but never got around to doing. Maybe it would work."

A quick look through my gun writer's reference book (i.e., the out-of-state yellow pages) quickly brought me a pleasant long-distance phone conversation with inventor Jesse Edwards. His recoil reducer has been around for quite a while, although the word is just now getting out concerning its effectiveness. Basically, the reducer consists of a patented spring-loaded metal tube that is inserted into a hole drilled in the wooden buttstock of a shoulder arm. The hole is then concealed by the gun's buttplate . . . ideal for muzzleloading shotguns! Jesse's anti-recoil principle works on an inertia effect: for every force forward, there is an equal force expended in the opposite direction. I decided to give his recoil reducer a try; to my knowledge, this was the first time the Edwards Recoil Reducer had ever been used on a muzzleloading shotgun. The service was unbelievably fast. I shipped the stock portion of my shotgun to Edwards via UPS on Monday. That friday, the gun was delivered to my house with not one, but two reducers installed. (Jesse promises same-day service, but in this case, he must have had his workshop in the back of a delivery truck enroute to my home!) The end result? The Jesse Edwards' product reduced my black powder shotgun recoil by 40% And according to my Blue Shoulder test, after spending a half day test-firing loads of 3½ drams of FFG backing a 1½ oz. charge of shot, there was absolutely no shoulder discoloration. The Dixie double I was using was not only a lot more enjoyable to shoot, but the added 12 ounces of weight in the stock caused by the two recoil reducers actually improved the gun's balance, shifting it over the locks and increasing its pointability. If you, like me,

A stalking gun case, like this from Uncle Mike's, gives weather protection plus easy accessibility when the gun is needed.

Stock design and metal buttplates can make blackpowder shotguns kick a lot. Author's answer was Edward's recoil reducer.

A bag comes in handy for carrying powder, shot, wads, and caps. Author carries this haversack when hunting upland game.

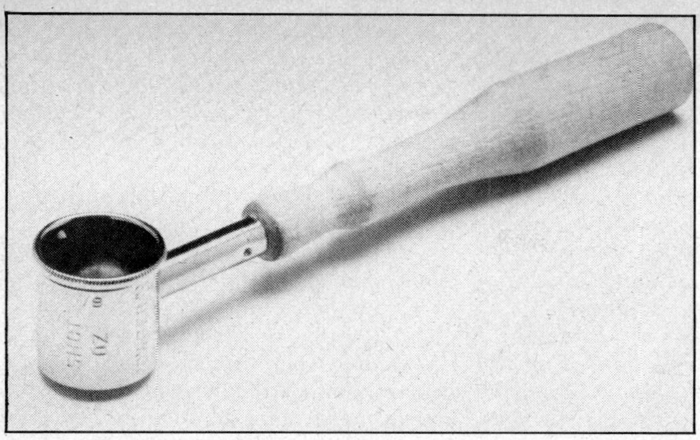

This 1880s-style dipper is made by Lyman and can be used to measure blackpowder or shot. The tool is adjustable, measuring ⅞ to 1⅝ oz. of shot or 2¼ to 3¾ drams of blackpowder.

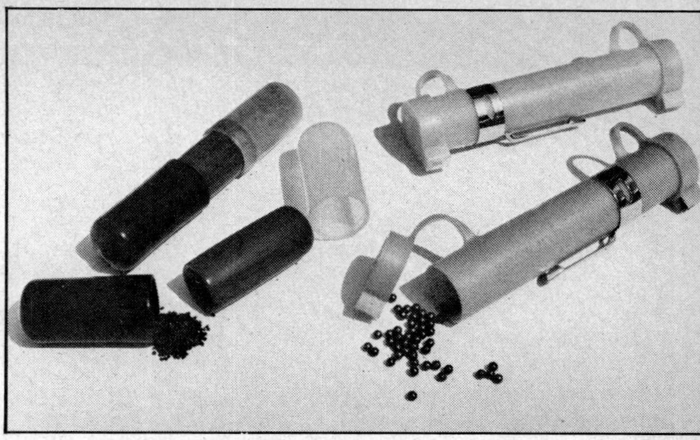

For fast reloading, pre-measured loads are the answer. Powder and shot charges can be conveniently carried in Winchester Sutler's Shotgun Tubes (left) or Butler Creek's Quick Loaders.

prefer shooting heavy, feather-penetrating hunting loads out of today's muzzleloading shotguns, I strongly recommend that you send the breech and stock portion of your gun to Jesse Edwards (Edwards Recoil Reducer, 269 Herbert Street, Alton, IL 62002) along with a check for $35 for installation of one reducer or $70 for two.

Another relatively new product which isn't always associated with black powder scatterguns is the much heralded Uncle Mike's stainless steel Hot Shot nipples. Prior to their invention, percussion shotgun shooters using a double gun would often get a face full of hot gases and semi-charred hands when firing the lock closest to their face. It is somewhat akin to a right-handed shooter firing a left-handed gun. But the new Uncle Mike's Hot Shot nipples have put an end to all that; they effectively dispel the gases, with absolutely no discomfort to the shooter. Another bonus is that the nipples, when installed on any black powder shotgun, will prevent those high inner-breech pressures from blowing your hammer back to half-cock after each shot, a sear-wrenching experience at best. I also found that the nipples increase velocity slightly, so after installation, you may find you'll be using less powder in your favorite load to achieve the same pattern. And of course, the stainless steel construction insures that the nipples will last as long as your gun.

In my conversation with other scattergunners (muzzleloading and otherwise) I am always surprised to learn that the practice of patterning a shotgun is rapidly becoming a lost art. A shotgunner's failure to pattern his gun is tantamount to a big game hunter not sighting in his rifle prior to a hunt. But evidently, in the mind of the shotgunner, there is a mistaken notion that because he is blasting forth with a cloud containing hundreds of pellets, no quarry is safe from its lethal effect. Absolutely false! I've seen shotgun patterns that were so open a flock of geese could fly through it sideways, without ruffling their down. Conversely, these same shotguns, worked up (or down) to the proper load, have been made to pattern so tightly even an anemic snake would be assured of getting a deadly dose of pellet poisoning. The secret is in shotgun patterning, and muzzleloaders have the distinct advantage of being able to tailor their loads in the field and after every shot to achieve maximum effectiveness. It's simply a matter of finding the best combination of black powder and shot that is right for your particular gun. And make no mistake about it, although mass-produced, each muzzleloading shotgun will shoot differently. It's up to you, as a responsible and effective hunter, to find out what works best. It's easy to do.

Patterning a muzzleloading shotgun is not only necessary, it is simple . . . and fun! All you need is a 4x4 foot sheet of butcher paper tacked onto a board. From a distance of 40 yards, aim for the center of the paper and fire your charge. Using a cardboard template cut in the shape of a 30-inch diameter circle, cover the largest concentration of your shot. Then trace the circumference of the circle onto your patterning paper. By evaluating the location of the circle in relationship to your aiming point, you can tell just where your shotgun is

Around the turn of the century, many fine breechloading shotguns were made for blackpowder. Pictured is a Remington (left) and an L.C. Smith. Conversion Arms makes stainless steel shot shell cases that can be reloaded and primed in the field.

Ignition, velocity, and gas dispersion were all improved greatly by installing Hot Shot nipples on the author's shotguns.

shooting. (Contrary to popular belief, most of today's muzzleloading shotguns do not shoot directly to point of aim.) Next comes the trial-and-error method of varying your powder and shot combination to produce the densest loads for your gun. Remember, a dram of powder is equal to 27½ grains. I've found my best hunting loads fall in the 2½–3 dram vicinity, behind a one to 1½ ounce charge of shot. When patterning your shotgun, it is wise to wear shooting glasses and ear protection. And always use FFG black powder or Pyrodex "RS" . . . *never* smokeless powder. The guns can't take it and consequently, neither can you.

Once you've got your shotgunning load established, you're ready to take to the field. However, therein lies another situation that has been a stumbling block for shooters, up until recently–in-the-field reloading. Not because it is difficult, but due to the fact that your shotgun requires a variety of paraphernalia. The scattergun hunter must carry percussion caps, powder, overpowder wads, shot and overshot wads, not to mention nipple pick and nipple wrench for emergencies, and of course, a game bag for both the sure-shots and the optimists! Happily, with the exception of the game bag, all of these elements can easily be transported by one of two methods.

The first method is the proverbial bush coat, a longtime favorite of mine. With its many, accordioned pockets, this one item of wearing apparel can carry every item a shotgunner might need. All that remains is for the hunter to decide which pockets will carry what, so that with a little practice he can reach instantly for the desired item without a moment's hesitation.

For those shooters not having, liking, or wanting to buy a bush coat, there is another option open: traditional Mountain Man "possibles bags" are normally too small to carry every shotgunning item needed. But there is a desirable alternative to black powder shooters. Dixie Gun Works sells a Civil War Haversack for only $5.50, which is an exact copy of the canvas, over-the-shoulder carrying pouch used by both sides in the War Between the States. It also makes an excellent, rugged and inexpensive shotgunner's bag. This one, sturdy item, with its button-down flap, will hold everything listed above. You might even consider obtaining an extra haversack to use as a game bag.

For safety reasons as well as weight, I dislike carrying large amounts of powder with me in the field. Therefore, I always opt for premeasured loads when hunting. One of the handiest, most versatile items in this category is the Supreme Quick Loader, which sells for $4.50 for a package of three and are available from Butler Creek Corporation (P. O. Box GG, Jackson Hole, WY 83001). Although originally designed for muzzleloading rifles, these two-compartment, waterproof plastic containers are ideal for the shotgunner. All you have to do is to reverse the standard procedure and put your loose shot in the "powder" side (it will hold more than 2 oz. of shot) and pour your premeasured powder in the "ball" compartment, which will easily carry 3½ drams of FFG . . . the heaviest load you're likely to use. These handy loaders also have compartments on each end for carrying spare caps.

Another new entry in the premeasured shotgun load category is the Shotgun Tube from Winchester Sutler (Bloomery Star Rt., Box 61, Winchester, VA 22601). These three-part black, red, and clear plastic tubes are watertight and can carry up to 1¼ oz. of shot and three drams of powder.

Black powder shotgunners desiring to "pour their own" can carry either powder or shot in the new leather and wood Shot/Powder flask, manufactured by Uncle Mike's and selling for $11.95. And it's waterproof!

Obviously, muzzleloading shotgunning has come a long way in just the last few years. In fact, it is somewhat ironic that today the sport is beginning to attract more shooters than at any other time in its 200 year history. And no wonder: there are more black powder shotgunners today than ever before, and in many cases the hunting has never been better; through the efforts of national organizations such as Ducks Unlimited and local sportsmen's associations, many game species have actually increased in number, making shotgunning all the more productive. In addition, during these times of tight money and questionable gasoline supplies, the muzzleloading scattergunner doesn't have to spend a week away from home or hire a guide in order to enjoy his sport. His charcoal-burning fun can be as near as the back 40, or an open field just a few hours drive away. Additionally, his gun and equipment, while evoking the memories of the past, have never been safer, or more up-to-date.

Perhaps then, *these* are the fabled golden years of black powder shotgunning; the celebrated era of 150 years ago merely set the stage for today, when the muzzleloading shotgunner can realize the full pleasure of this historic sport. Indeed, we are fortunate to be living in both a time and a country that permits us to enjoy muzzleloading shotgunning to its fullest.

These blackpowder cleaning kits, made by Hoppe's (top) and Dixie Gun Works, contain the essentials for shotgun care.

Chapter 6

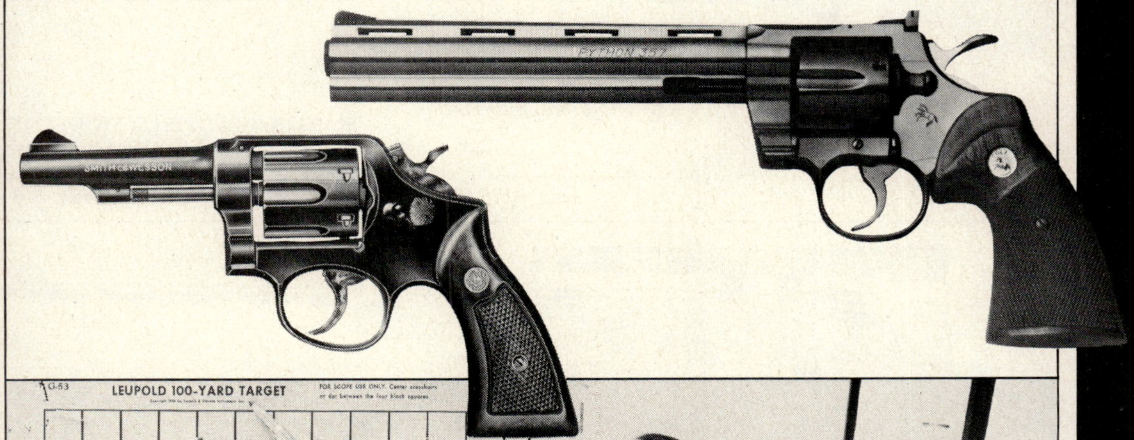

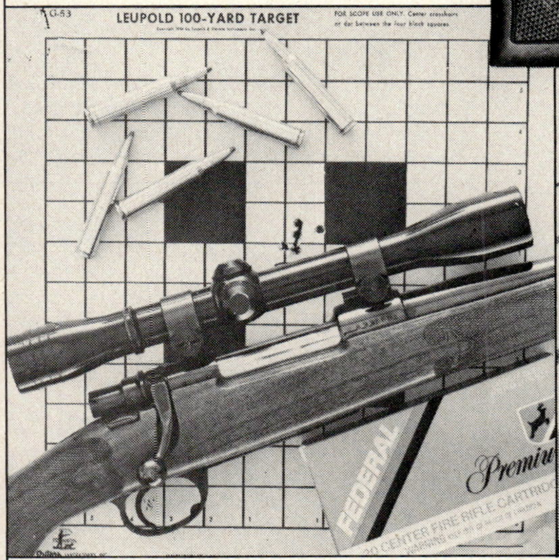

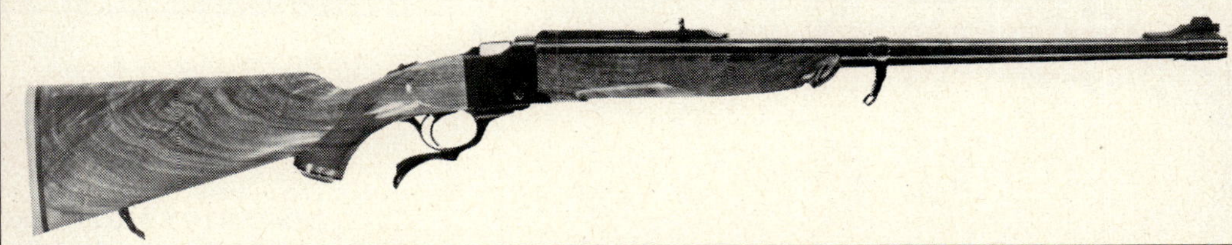

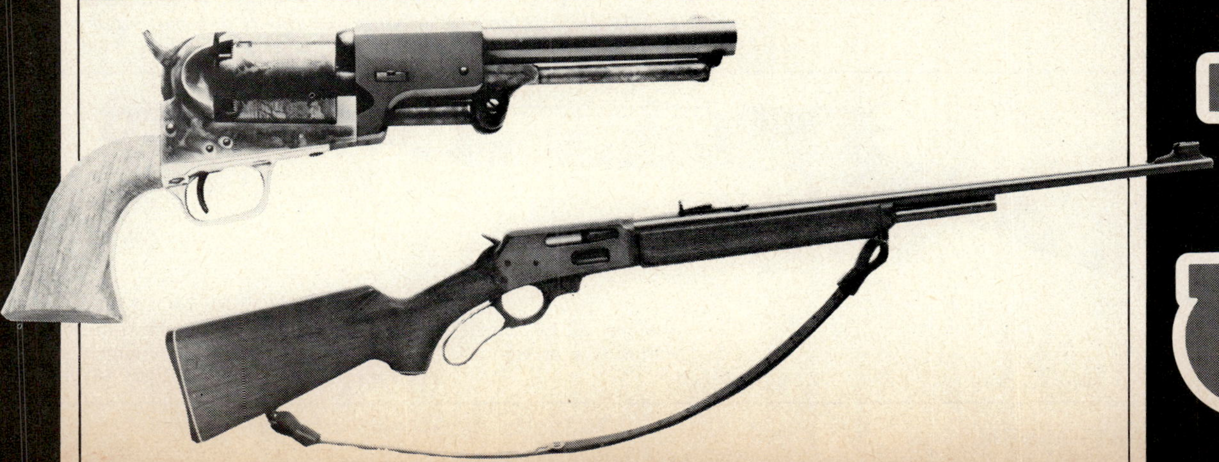

Catalog

RIFLE CATALOG

Single-Shot Rimfire

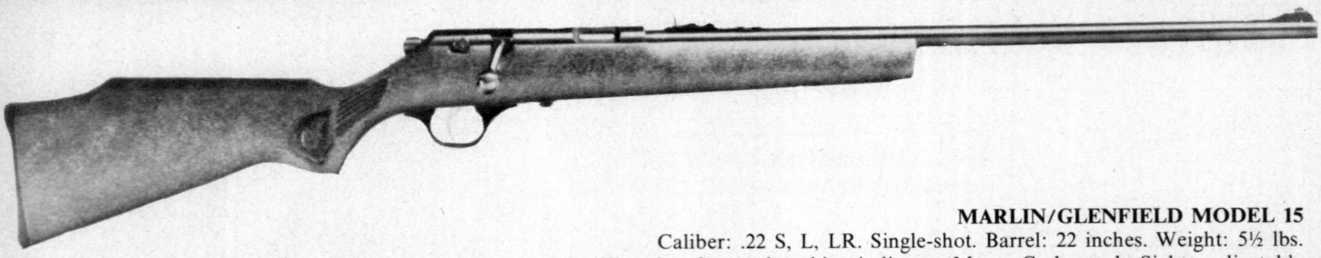

MARLIN/GLENFIELD MODEL 15
Caliber: .22 S, L, LR. Single-shot. Barrel: 22 inches. Weight: 5½ lbs. Thumb safety, red cocking indicator. Monte Carlo stock. Sights: adjustable rear, ramp front; grooved for tipoff scope mount. **Price: $68**

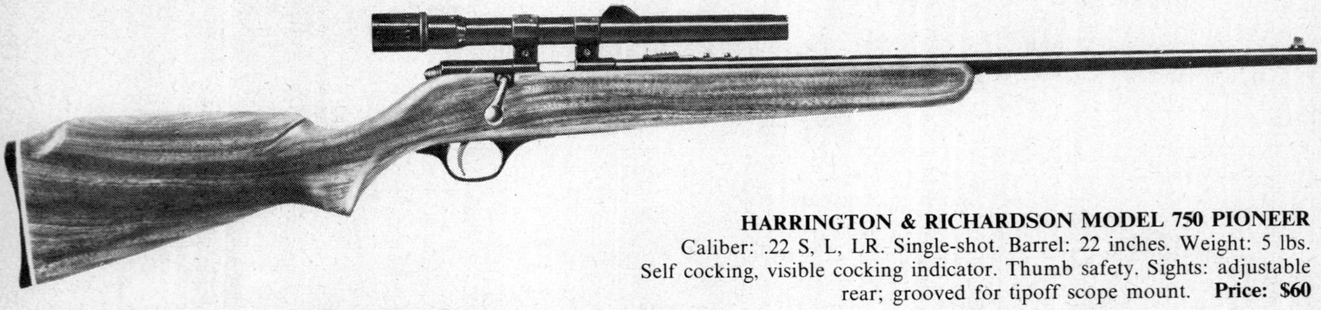

HARRINGTON & RICHARDSON MODEL 750 PIONEER
Caliber: .22 S, L, LR. Single-shot. Barrel: 22 inches. Weight: 5 lbs. Self cocking, visible cocking indicator. Thumb safety. Sights: adjustable rear; grooved for tipoff scope mount. **Price: $60**

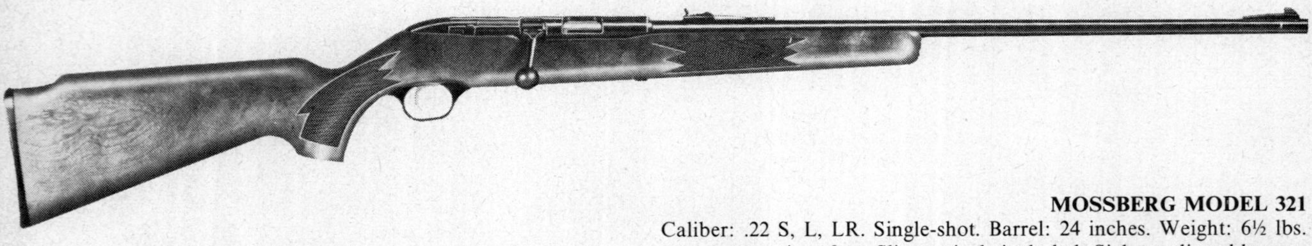

MOSSBERG MODEL 321
Caliber: .22 S, L, LR. Single-shot. Barrel: 24 inches. Weight: 6½ lbs. Automatic safety. Sling swivels included. Sights: adjustable rear, bead ramp front. **Price: $79**

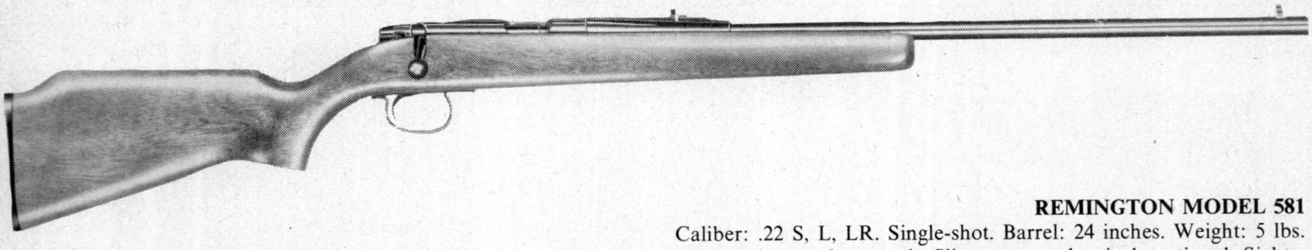

REMINGTON MODEL 581
Caliber: .22 S, L, LR. Single-shot. Barrel: 24 inches. Weight: 5 lbs. American walnut stock. Sling strap and swivels optional. Sights: adjustable open rear. **Price: $100**

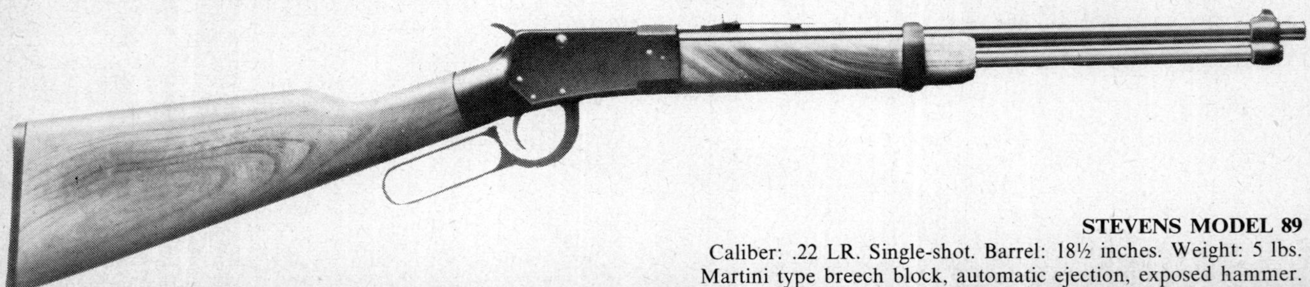

STEVENS MODEL 89
Caliber: .22 LR. Single-shot. Barrel: 18½ inches. Weight: 5 lbs. Martini type breech block, automatic ejection, exposed hammer. Sights: open rear, sporting front. **Price: $67**

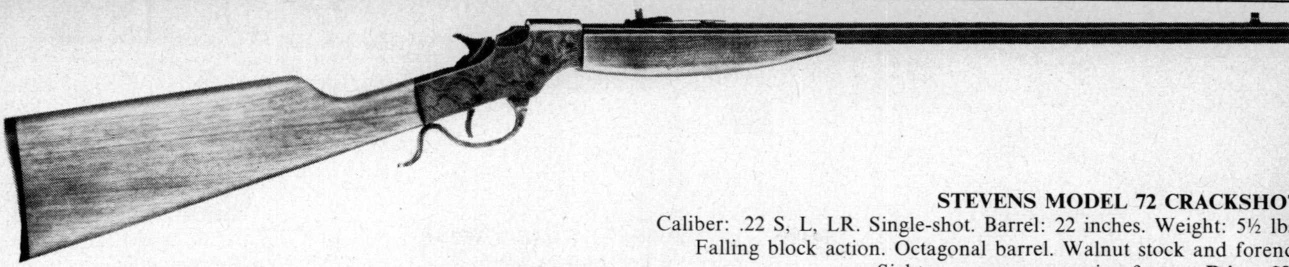

STEVENS MODEL 72 CRACKSHOT
Caliber: .22 S, L, LR. Single-shot. Barrel: 22 inches. Weight: 5½ lbs. Falling block action. Octagonal barrel. Walnut stock and forend. Sights: open rear, sporting front. **Price: $95**

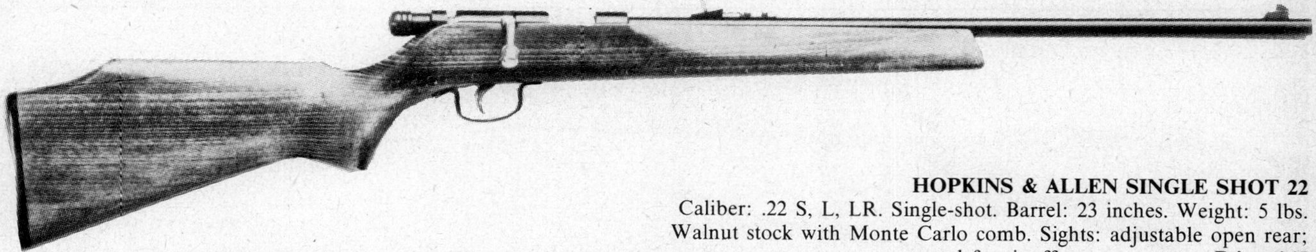

HOPKINS & ALLEN SINGLE SHOT 22
Caliber: .22 S, L, LR. Single-shot. Barrel: 23 inches. Weight: 5 lbs. Walnut stock with Monte Carlo comb. Sights: adjustable open rear; grooved for tipoff scope mount. **Price: $45**

Bolt Action Rimfire

HARRINGTON & RICHARDSON MODEL 865
Caliber: .22 LR. Detachable magazine, capacity: 5. Barrel: 22 inches. Weight: 5 lbs. Dual extractors. One piece hardwood stock. Sights: adjustable open rear; grooved for tipoff scope mount. **Price: $70**

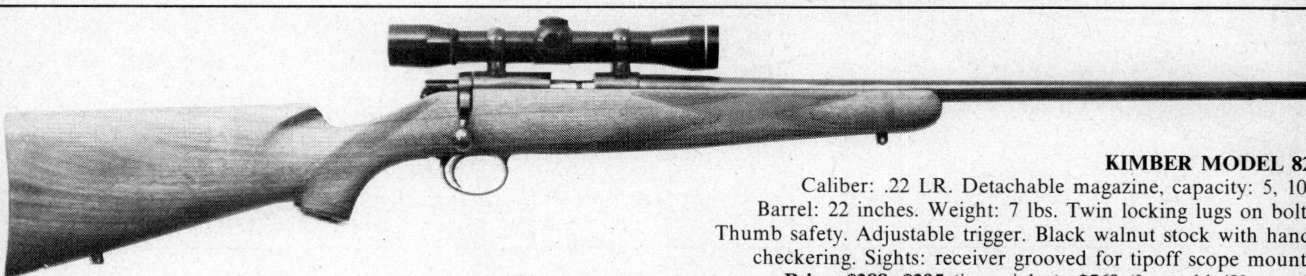

KIMBER MODEL 82
Caliber: .22 LR. Detachable magazine, capacity: 5, 10. Barrel: 22 inches. Weight: 7 lbs. Twin locking lugs on bolt. Thumb safety. Adjustable trigger. Black walnut stock with hand checkering. Sights: receiver grooved for tipoff scope mount. **Price: $388; $395** (iron sights); **$569** (Leupold 4X scope)

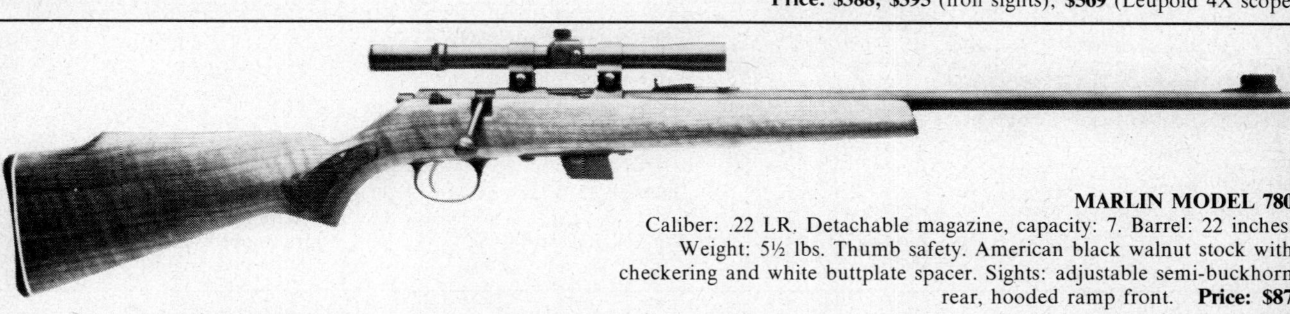

MARLIN MODEL 780
Caliber: .22 LR. Detachable magazine, capacity: 7. Barrel: 22 inches. Weight: 5½ lbs. Thumb safety. American black walnut stock with checkering and white buttplate spacer. Sights: adjustable semi-buckhorn rear, hooded ramp front. **Price: $87**

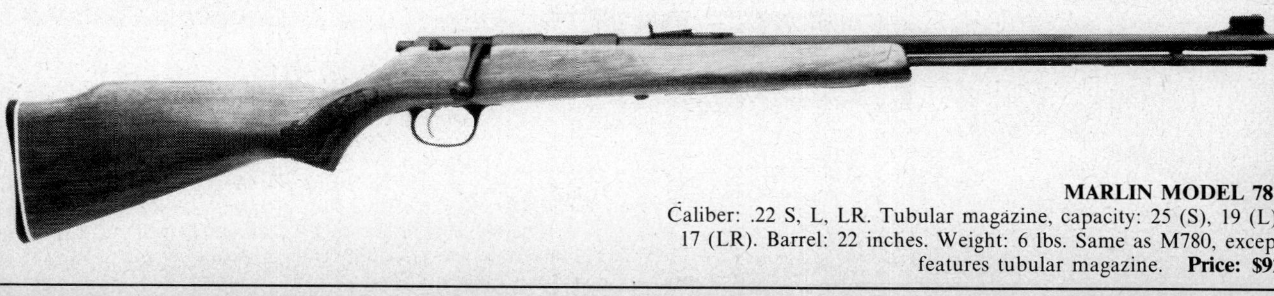

MARLIN MODEL 781
Caliber: .22 S, L, LR. Tubular magazine, capacity: 25 (S), 19 (L), 17 (LR). Barrel: 22 inches. Weight: 6 lbs. Same as M780, except features tubular magazine. **Price: $92**

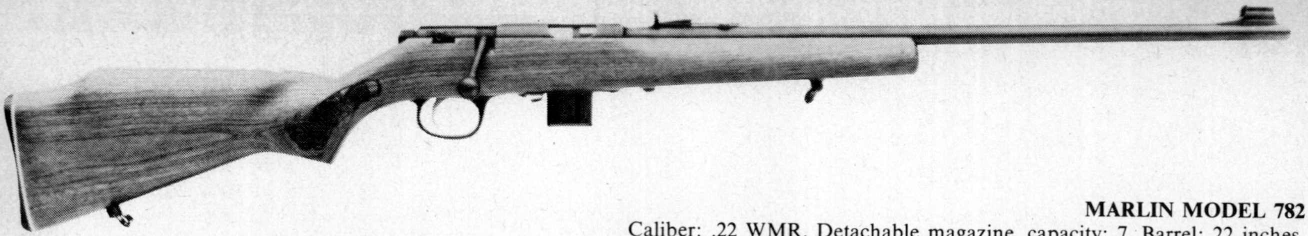

MARLIN MODEL 782
Caliber: .22 WMR. Detachable magazine, capacity: 7. Barrel: 22 inches. Weight: 6 lbs. Same as M780, except chambered for .22 magnum and includes sling and swivels. **Price: $97**

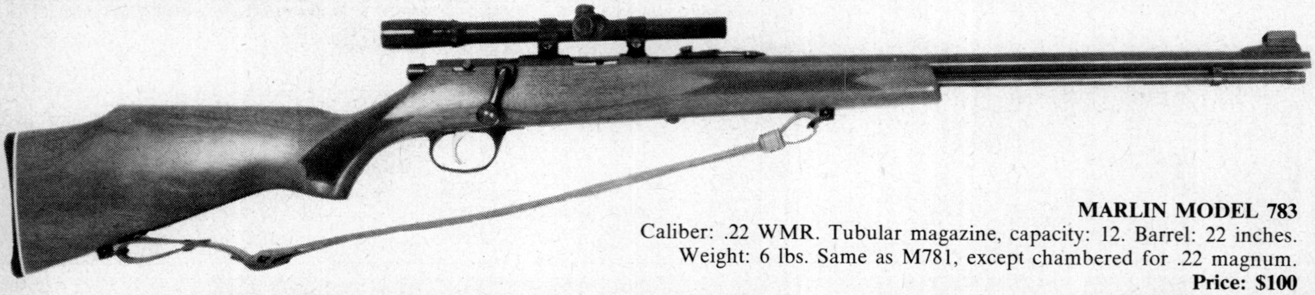

MARLIN MODEL 783
Caliber: .22 WMR. Tubular magazine, capacity: 12. Barrel: 22 inches. Weight: 6 lbs. Same as M781, except chambered for .22 magnum. **Price: $100**

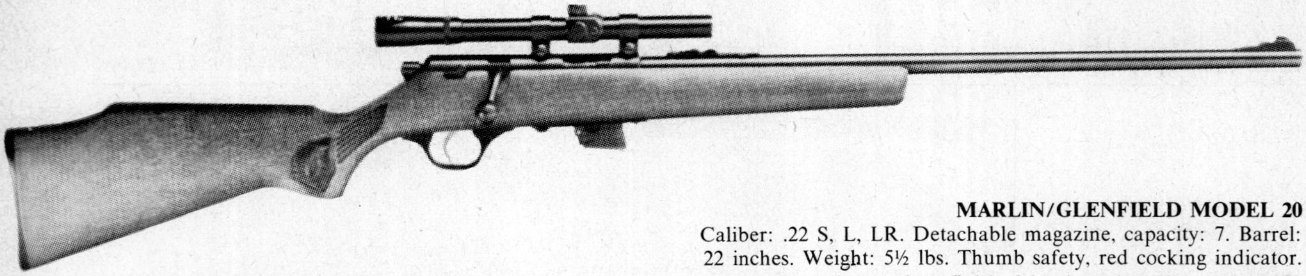

MARLIN/GLENFIELD MODEL 20
Caliber: .22 S, L, LR. Detachable magazine, capacity: 7. Barrel: 22 inches. Weight: 5½ lbs. Thumb safety, red cocking indicator. One piece walnut finished hardwood stock. **Price: $72**

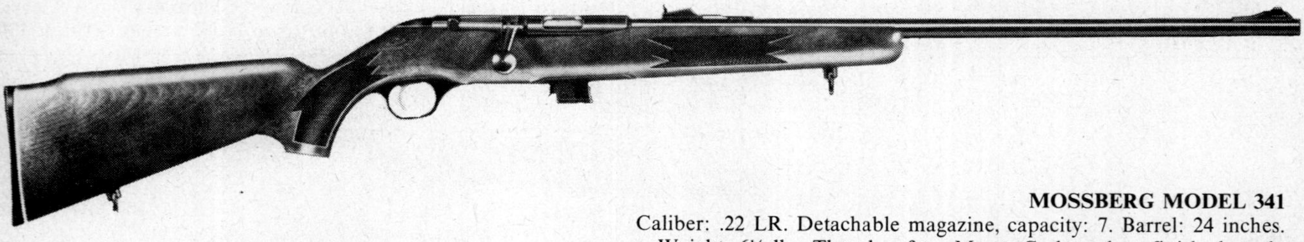

MOSSBERG MODEL 341
Caliber: .22 LR. Detachable magazine, capacity: 7. Barrel: 24 inches. Weight: 6¼ lbs. Thumb safety. Monte Carlo walnut finished stock. Sights: adjustable rear, bead front on ramp. **Price: $90**

MOSSBERG MODEL 640 CHUCKSTER
Caliber: .22 WMR. Detachable magazine, capacity: 5. Barrel: 24 inches. Weight: 6¼ lbs. Heavy duty receiver and bolt, double extractors. Walnut finished Monte Carlo stock. Sights: tapped for scope mounts or receiver sights. **Price: $102**

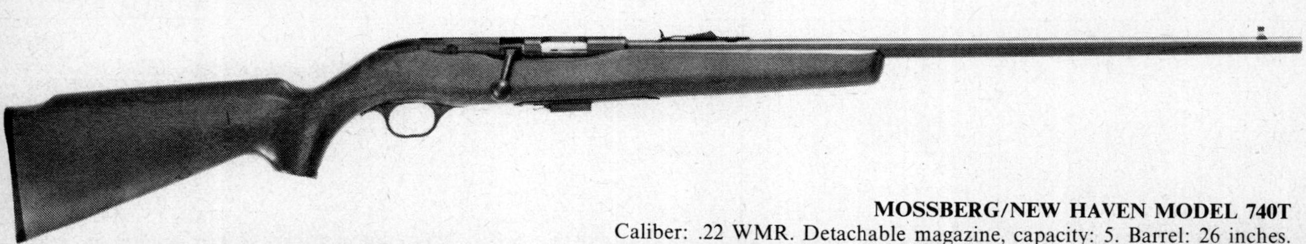

MOSSBERG/NEW HAVEN MODEL 740T
Caliber: .22 WMR. Detachable magazine, capacity: 5. Barrel: 26 inches. Weight: 7¼ lbs. Walnut finished Monte Carlo stock with sling swivels. Sights: adjustable rear; grooved for tipoff scope mount. **Price: NA**

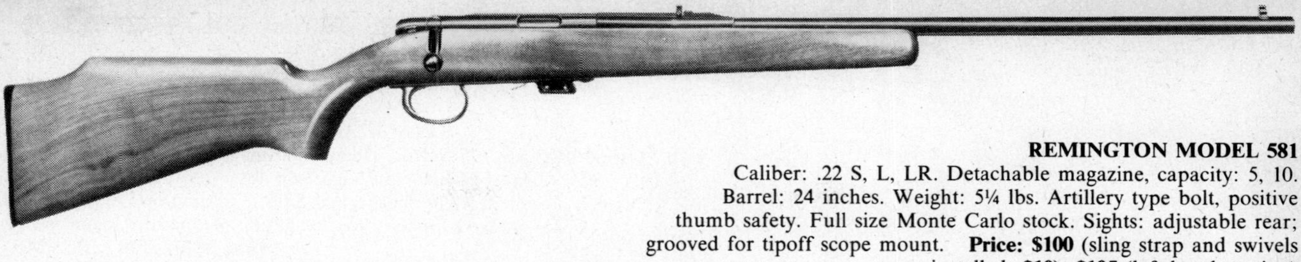

REMINGTON MODEL 581
Caliber: .22 S, L, LR. Detachable magazine, capacity: 5, 10. Barrel: 24 inches. Weight: 5¼ lbs. Artillery type bolt, positive thumb safety. Full size Monte Carlo stock. Sights: adjustable rear; grooved for tipoff scope mount. **Price: $100** (sling strap and swivels installed: **$10); $105** (left hand version)

REMINGTON MODEL 582
Same as M581, except features tubular magazine. **Price: $120**

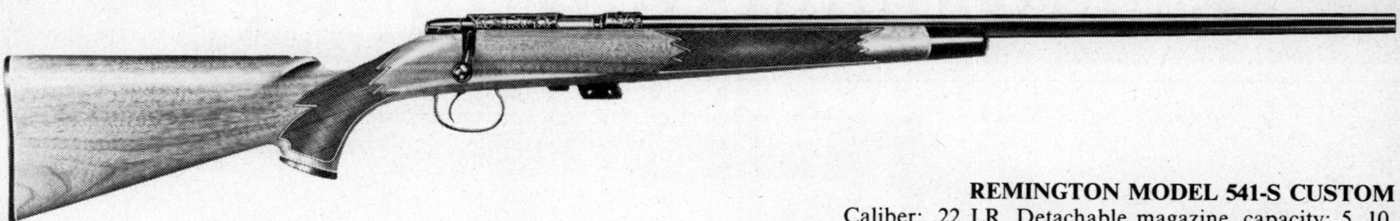

REMINGTON MODEL 541-S CUSTOM
Caliber: .22 LR. Detachable magazine, capacity: 5, 10. Barrel: 24 inches. Weight: 5½ lbs. Engraved receiver, adjustable trigger. Rosewood toned forend, pistol grip cap, buttplate. Sights: tapped for scope mount. **Price: $261**

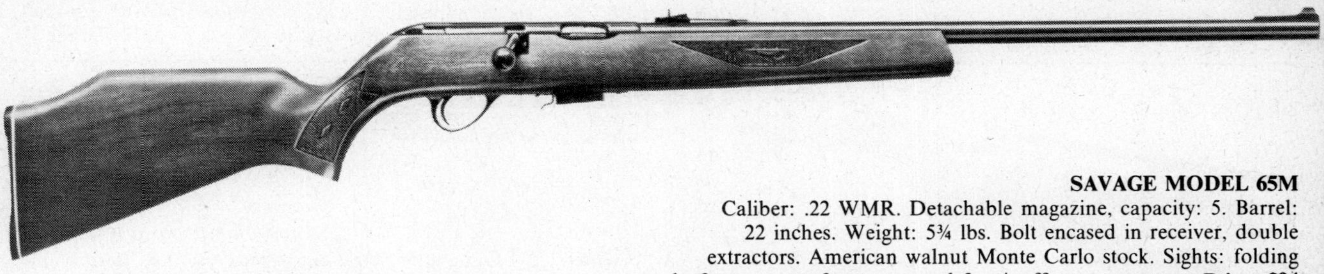

SAVAGE MODEL 65M
Caliber: .22 WMR. Detachable magazine, capacity: 5. Barrel: 22 inches. Weight: 5¾ lbs. Bolt encased in receiver, double extractors. American walnut Monte Carlo stock. Sights: folding leaf rear, ramp front; grooved for tipoff scope mount. **Price: $94**

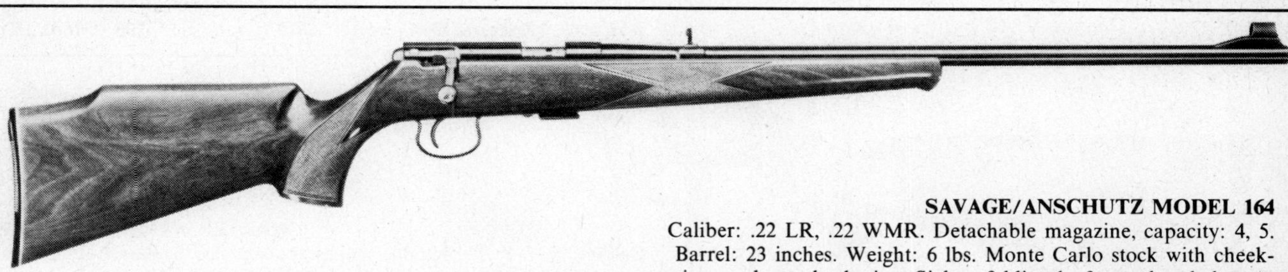

SAVAGE/ANSCHUTZ MODEL 164
Caliber: .22 LR, .22 WMR. Detachable magazine, capacity: 4, 5. Barrel: 23 inches. Weight: 6 lbs. Monte Carlo stock with cheekpiece and cut checkering. Sights: folding leaf rear, hooded ramp front; grooved for tipoff scope mount. **Price: $326; $334** (.22 WMR)

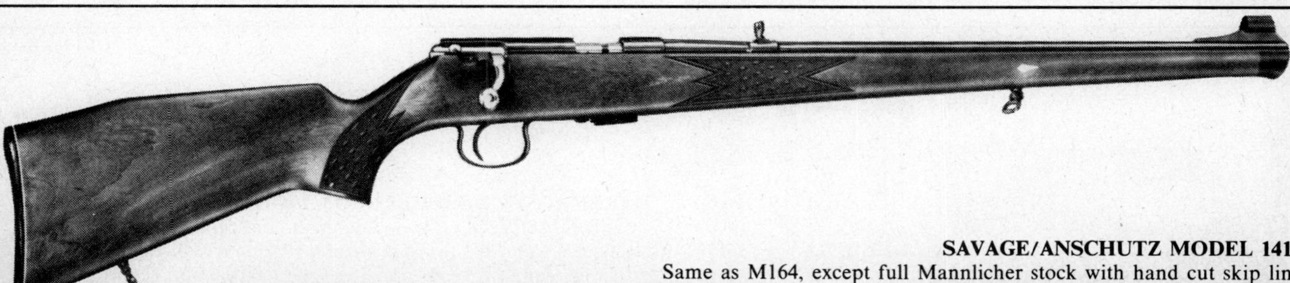

SAVAGE/ANSCHUTZ MODEL 1418
Same as M164, except full Mannlicher stock with hand cut skip line checkering. **Price: $463; $471** (.22 WMR)

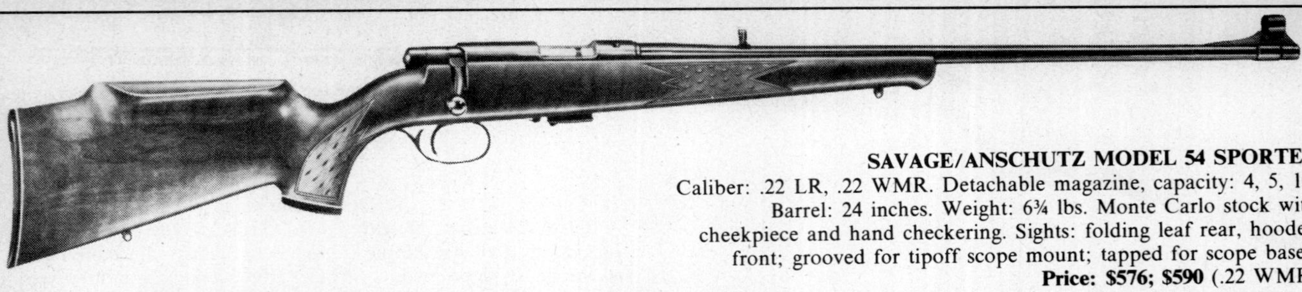

SAVAGE/ANSCHUTZ MODEL 54 SPORTER
Caliber: .22 LR, .22 WMR. Detachable magazine, capacity: 4, 5, 10. Barrel: 24 inches. Weight: 6¾ lbs. Monte Carlo stock with cheekpiece and hand checkering. Sights: folding leaf rear, hooded front; grooved for tipoff scope mount; tapped for scope bases. **Price: $576; $590** (.22 WMR)

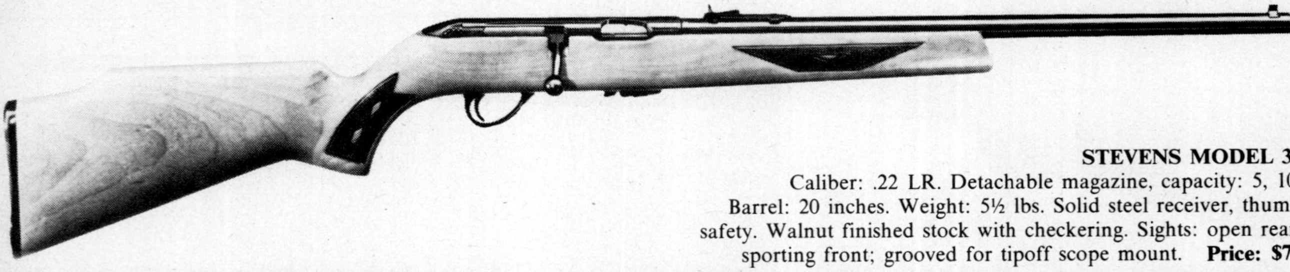

STEVENS MODEL 34
Caliber: .22 LR. Detachable magazine, capacity: 5, 10. Barrel: 20 inches. Weight: 5½ lbs. Solid steel receiver, thumb safety. Walnut finished stock with checkering. Sights: open rear, sporting front; grooved for tipoff scope mount. **Price: $74**

Semiautomatic Rimfire

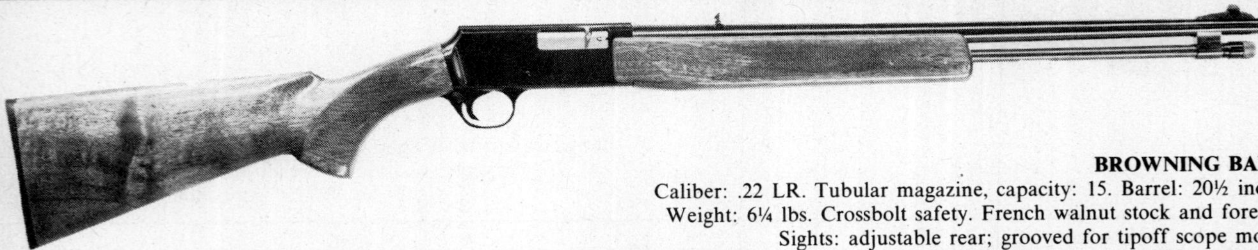

BROWNING BAR-22
Caliber: .22 LR. Tubular magazine, capacity: 15. Barrel: 20½ inches. Weight: 6¼ lbs. Crossbolt safety. French walnut stock and forearm. Sights: adjustable rear; grooved for tipoff scope mount. **Price: $210** (Grade I); **$310** (Grade II)

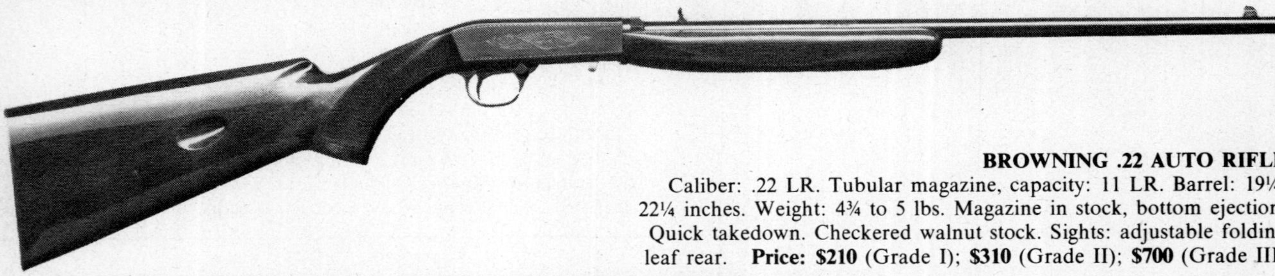

BROWNING .22 AUTO RIFLE
Caliber: .22 LR. Tubular magazine, capacity: 11 LR. Barrel: 19¼, 22¼ inches. Weight: 4¾ to 5 lbs. Magazine in stock, bottom ejection. Quick takedown. Checkered walnut stock. Sights: adjustable folding leaf rear. **Price: $210** (Grade I); **$310** (Grade II); **$700** (Grade III)

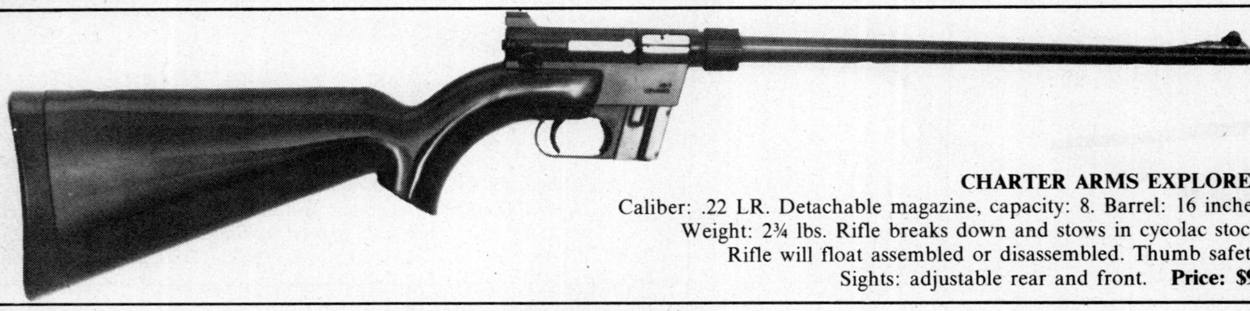

CHARTER ARMS EXPLORER
Caliber: .22 LR. Detachable magazine, capacity: 8. Barrel: 16 inches. Weight: 2¾ lbs. Rifle breaks down and stows in cycolac stock. Rifle will float assembled or disassembled. Thumb safety. Sights: adjustable rear and front. **Price: $98**

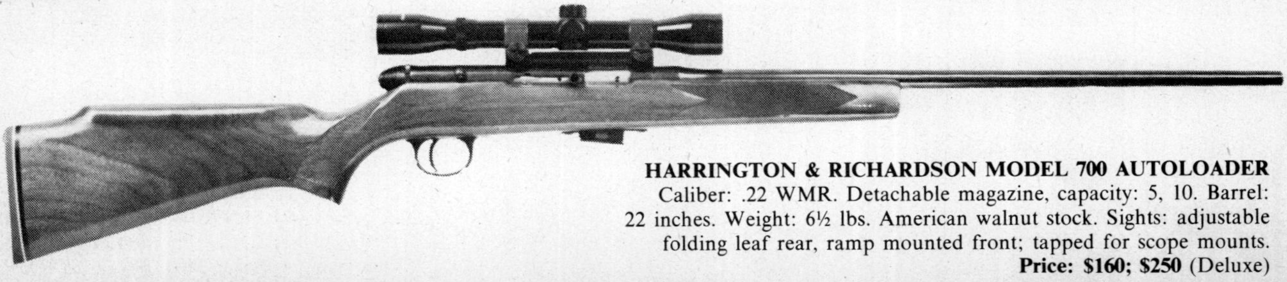

HARRINGTON & RICHARDSON MODEL 700 AUTOLOADER
Caliber: .22 WMR. Detachable magazine, capacity: 5, 10. Barrel: 22 inches. Weight: 6½ lbs. American walnut stock. Sights: adjustable folding leaf rear, ramp mounted front; tapped for scope mounts. **Price: $160; $250** (Deluxe)

HECKLER & KOCH .22 MAGNUM AUTO
Caliber: .22 WMR. Detachable magazine, capacity: 5, 10. Barrel: 19.7 inches. Weight: 5½ lbs. Single stage 3½-lb trigger. European walnut Monte Carlo stock. Checkered pistol grip and forearm. Sights: adjustable V-notch rear; tapped for scope mounts. **Price: $260**

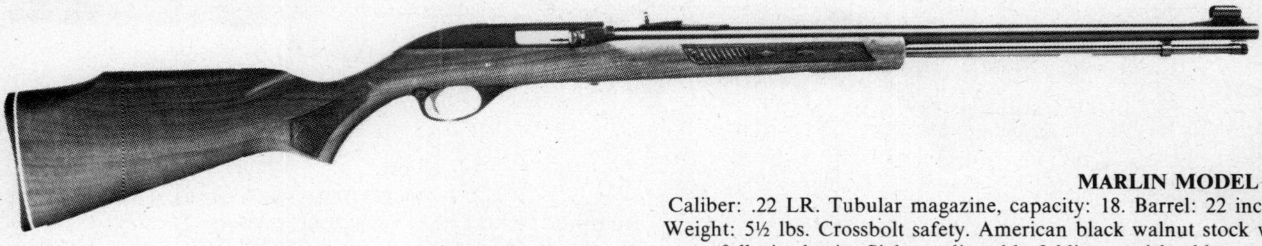

MARLIN MODEL 990
Caliber: .22 LR. Tubular magazine, capacity: 18. Barrel: 22 inches. Weight: 5½ lbs. Crossbolt safety. American black walnut stock with full pistol grip. Sights: adjustable folding semi-buckhorn rear, hooded ramp front; grooved for tipoff scope mount. **Price: $94**

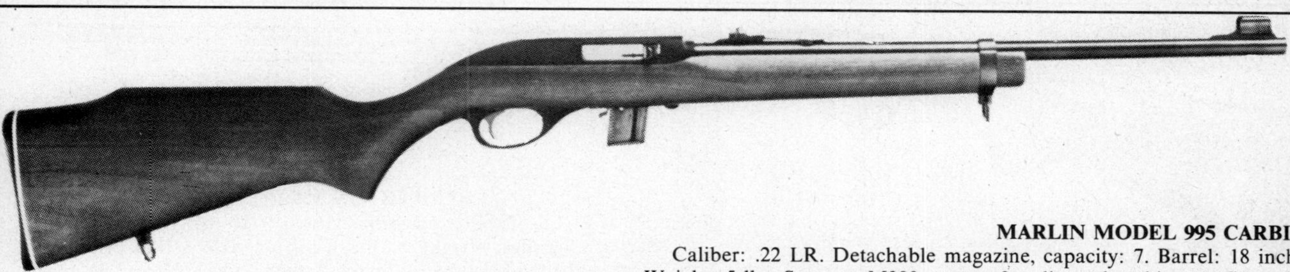

MARLIN MODEL 995 CARBINE
Caliber: .22 LR. Detachable magazine, capacity: 7. Barrel: 18 inches. Weight: 5 lbs. Same as M990, except for clip and carbine configuration. **Price: $88**

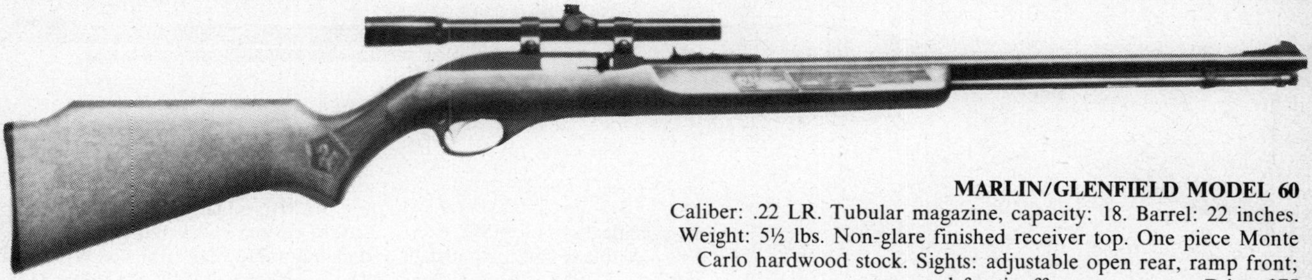

MARLIN/GLENFIELD MODEL 60
Caliber: .22 LR. Tubular magazine, capacity: 18. Barrel: 22 inches. Weight: 5½ lbs. Non-glare finished receiver top. One piece Monte Carlo hardwood stock. Sights: adjustable open rear, ramp front; grooved for tipoff scope mount. **Price: $75**

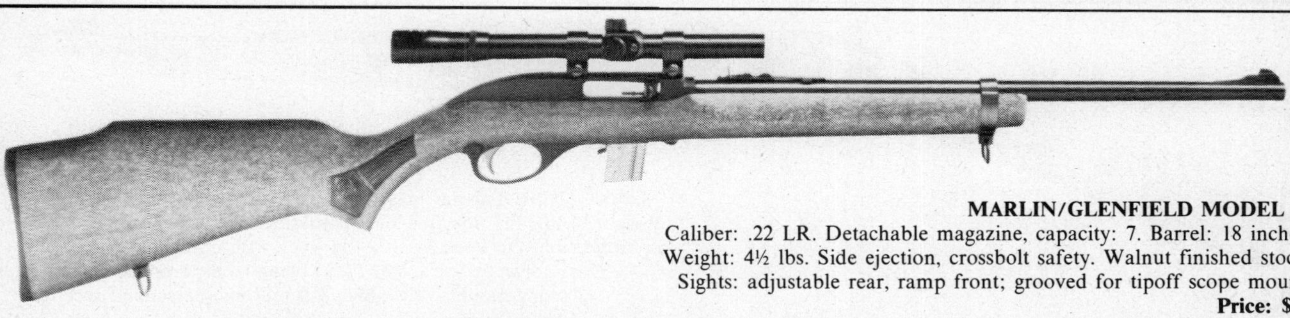

MARLIN/GLENFIELD MODEL 70
Caliber: .22 LR. Detachable magazine, capacity: 7. Barrel: 18 inches. Weight: 4½ lbs. Side ejection, crossbolt safety. Walnut finished stock. Sights: adjustable rear, ramp front; grooved for tipoff scope mount. **Price: $75**

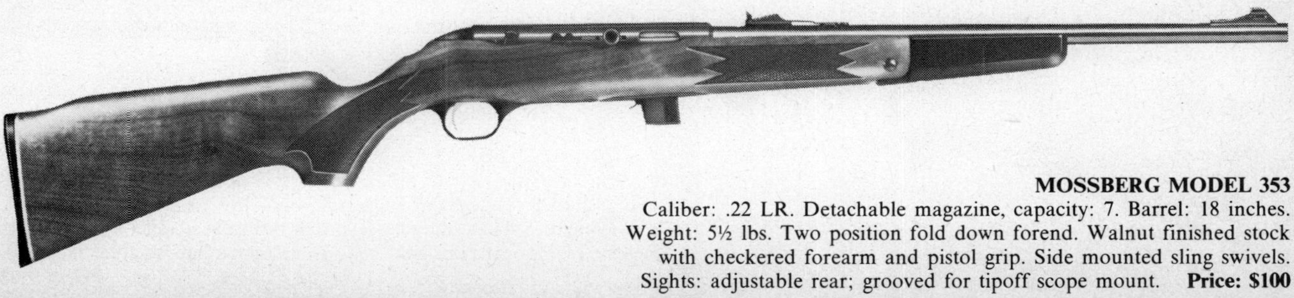

MOSSBERG MODEL 353
Caliber: .22 LR. Detachable magazine, capacity: 7. Barrel: 18 inches. Weight: 5½ lbs. Two position fold down forend. Walnut finished stock with checkered forearm and pistol grip. Side mounted sling swivels. Sights: adjustable rear; grooved for tipoff scope mount. **Price: $100**

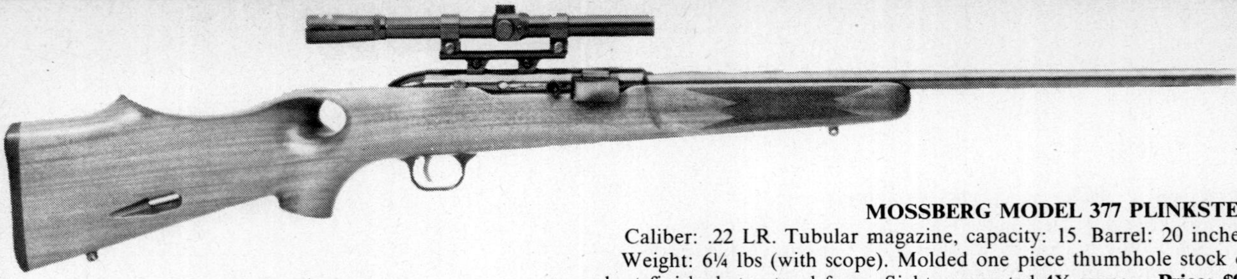

MOSSBERG MODEL 377 PLINKSTER

Caliber: .22 LR. Tubular magazine, capacity: 15. Barrel: 20 inches. Weight: 6¼ lbs (with scope). Molded one piece thumbhole stock of walnut finished structural foam. Sights: mounted 4X scope. **Price: $91**

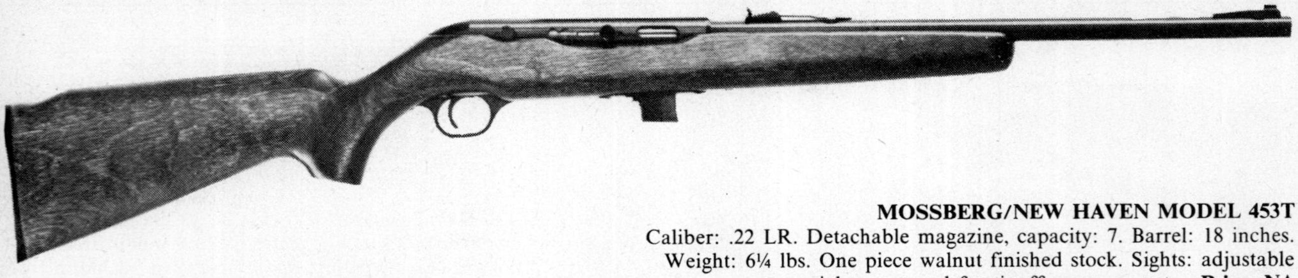

MOSSBERG/NEW HAVEN MODEL 453T

Caliber: .22 LR. Detachable magazine, capacity: 7. Barrel: 18 inches. Weight: 6¼ lbs. One piece walnut finished stock. Sights: adjustable open sights; grooved for tipoff scope mount. **Price: NA**

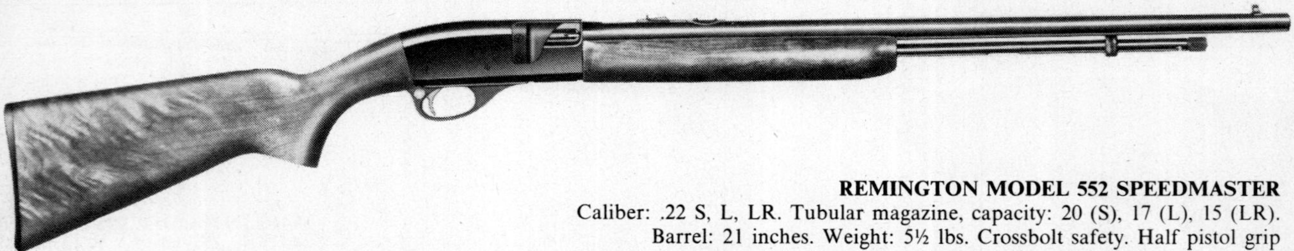

REMINGTON MODEL 552 SPEEDMASTER

Caliber: .22 S, L, LR. Tubular magazine, capacity: 20 (S), 17 (L), 15 (LR). Barrel: 21 inches. Weight: 5½ lbs. Crossbolt safety. Half pistol grip stock and semi-beavertail forearm. Sights: adjustable rear; grooved for tipoff scope mount. **Price: $115**

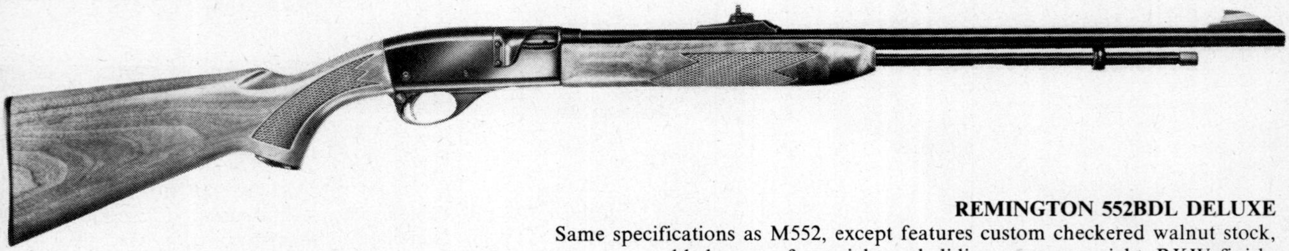

REMINGTON 552BDL DELUXE

Same specifications as M552, except features custom checkered walnut stock, blade ramp front sight and sliding ramp rear sight. RKW finish. **Price: $131**

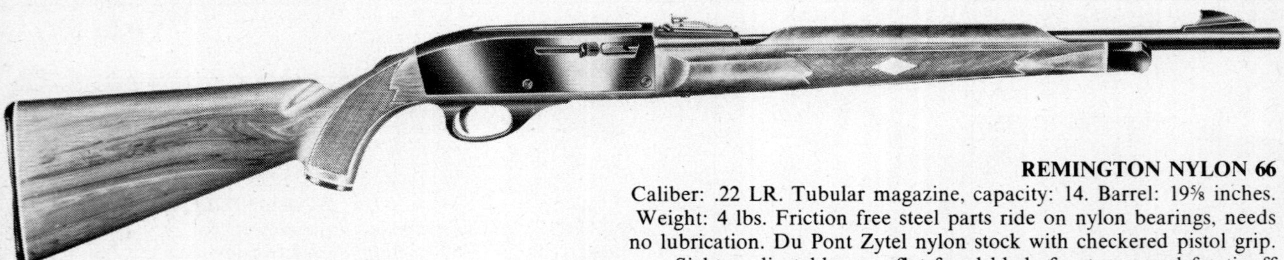

REMINGTON NYLON 66

Caliber: .22 LR. Tubular magazine, capacity: 14. Barrel: 19⅝ inches. Weight: 4 lbs. Friction free steel parts ride on nylon bearings, needs no lubrication. Du Pont Zytel nylon stock with checkered pistol grip. Sights: adjustable rear, flat faced blade front; grooved for tipoff scope mount. **Price: $94; $99** (chrome barrel and receiver)

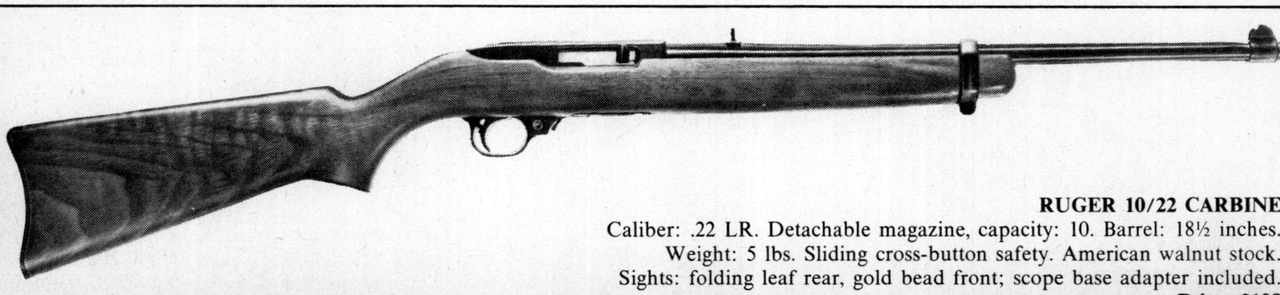

RUGER 10/22 CARBINE

Caliber: .22 LR. Detachable magazine, capacity: 10. Barrel: 18½ inches. Weight: 5 lbs. Sliding cross-button safety. American walnut stock. Sights: folding leaf rear, gold bead front; scope base adapter included. **Price: $108**

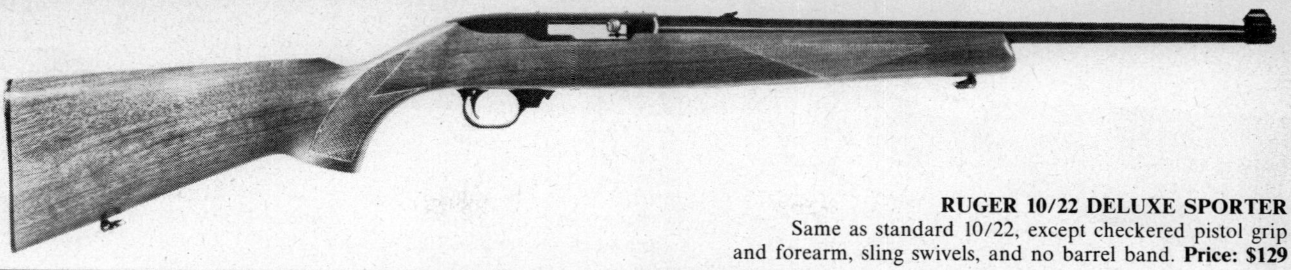

RUGER 10/22 DELUXE SPORTER
Same as standard 10/22, except checkered pistol grip and forearm, sling swivels, and no barrel band. **Price: $129**

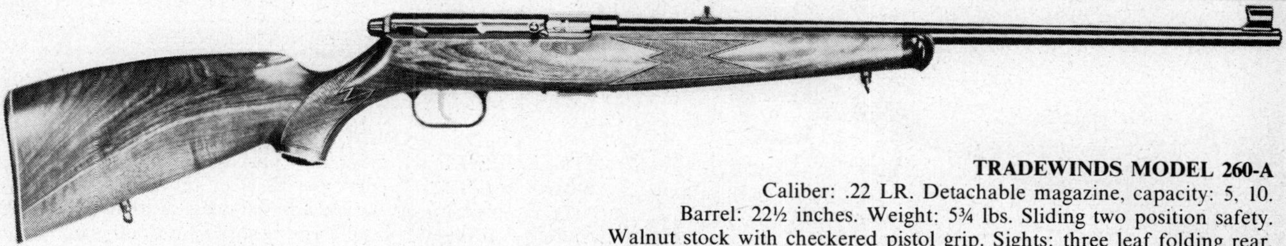

TRADEWINDS MODEL 260-A
Caliber: .22 LR. Detachable magazine, capacity: 5, 10. Barrel: 22½ inches. Weight: 5¾ lbs. Sliding two position safety. Walnut stock with checkered pistol grip. Sights: three leaf folding rear, ramp front; tapped for scope mounts. **Price: $185**

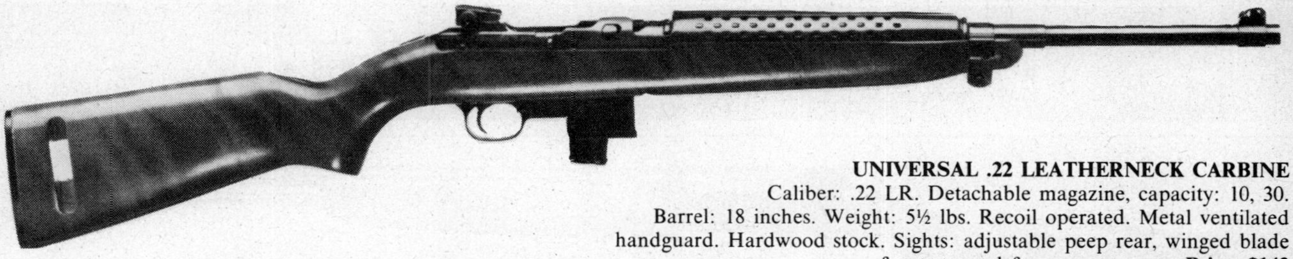

UNIVERSAL .22 LEATHERNECK CARBINE
Caliber: .22 LR. Detachable magazine, capacity: 10, 30. Barrel: 18 inches. Weight: 5½ lbs. Recoil operated. Metal ventilated handguard. Hardwood stock. Sights: adjustable peep rear, winged blade front; tapped for scope mounts. **Price: $143**

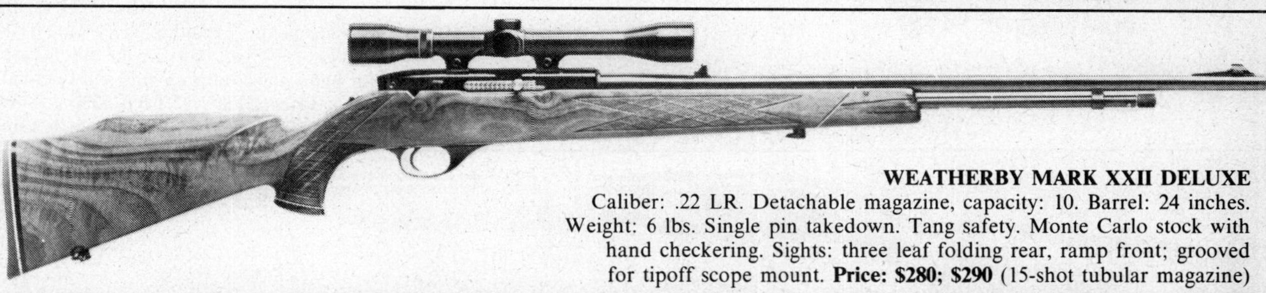

WEATHERBY MARK XXII DELUXE
Caliber: .22 LR. Detachable magazine, capacity: 10. Barrel: 24 inches. Weight: 6 lbs. Single pin takedown. Tang safety. Monte Carlo stock with hand checkering. Sights: three leaf folding rear, ramp front; grooved for tipoff scope mount. **Price: $280; $290** (15-shot tubular magazine)

Lever Action Rimfire

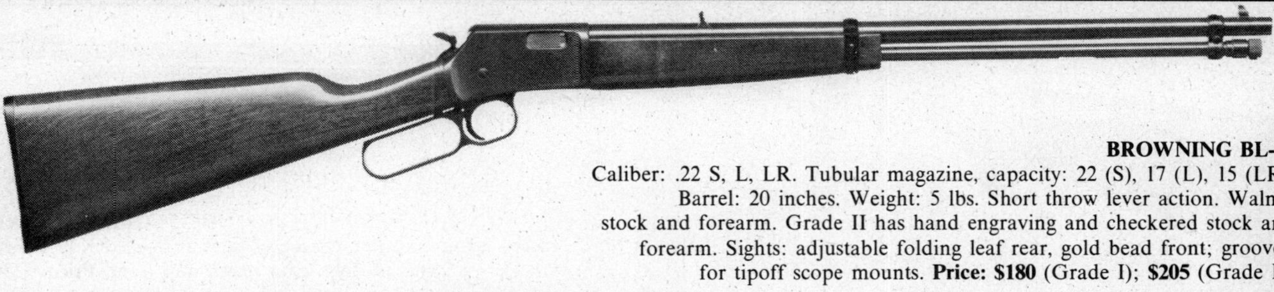

BROWNING BL-22
Caliber: .22 S, L, LR. Tubular magazine, capacity: 22 (S), 17 (L), 15 (LR). Barrel: 20 inches. Weight: 5 lbs. Short throw lever action. Walnut stock and forearm. Grade II has hand engraving and checkered stock and forearm. Sights: adjustable folding leaf rear, gold bead front; grooved for tipoff scope mounts. **Price: $180** (Grade I); **$205** (Grade II)

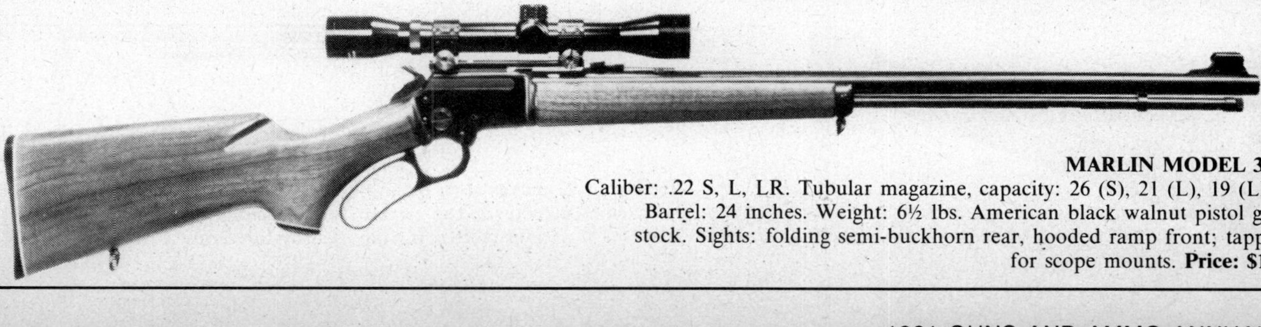

MARLIN MODEL 39A
Caliber: .22 S, L, LR. Tubular magazine, capacity: 26 (S), 21 (L), 19 (LR). Barrel: 24 inches. Weight: 6½ lbs. American black walnut pistol grip stock. Sights: folding semi-buckhorn rear, hooded ramp front; tapped for scope mounts. **Price: $182**

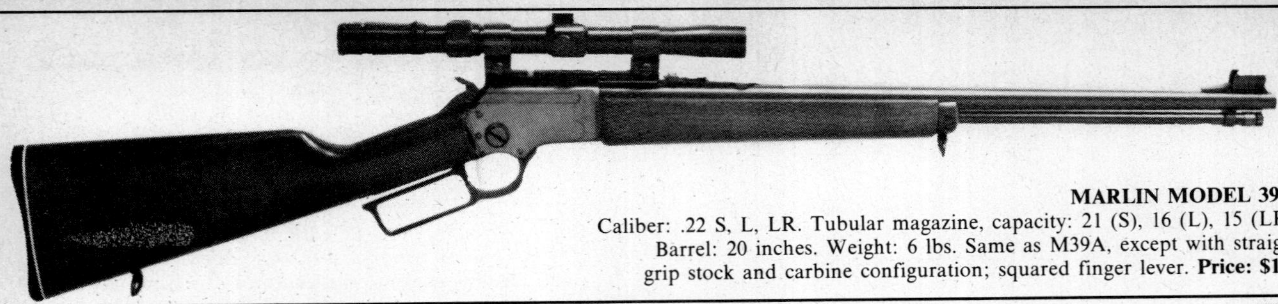

MARLIN MODEL 39M
Caliber: .22 S, L, LR. Tubular magazine, capacity: 21 (S), 16 (L), 15 (LR). Barrel: 20 inches. Weight: 6 lbs. Same as M39A, except with straight grip stock and carbine configuration; squared finger lever. **Price: $182**

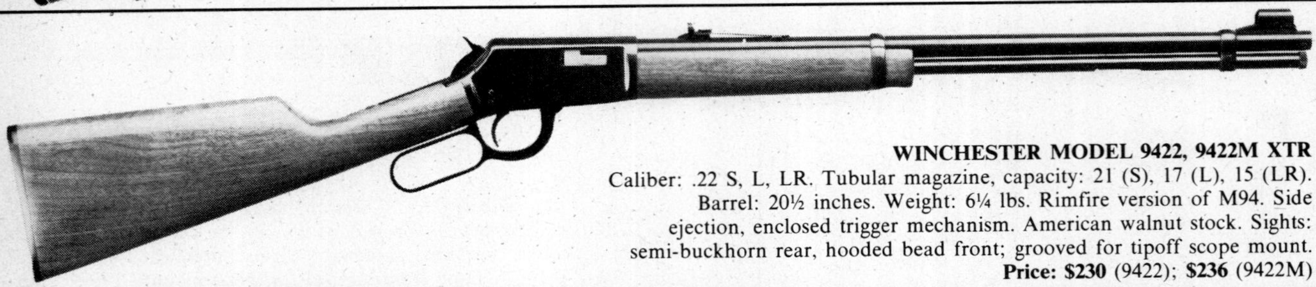

WINCHESTER MODEL 9422, 9422M XTR
Caliber: .22 S, L, LR. Tubular magazine, capacity: 21 (S), 17 (L), 15 (LR). Barrel: 20½ inches. Weight: 6¼ lbs. Rimfire version of M94. Side ejection, enclosed trigger mechanism. American walnut stock. Sights: semi-buckhorn rear, hooded bead front; grooved for tipoff scope mount. **Price: $230** (9422); **$236** (9422M)

Slide Action Rimfire

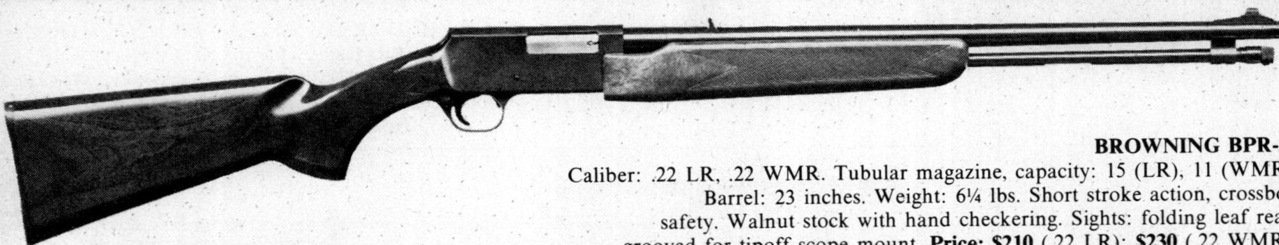

BROWNING BPR-22
Caliber: .22 LR, .22 WMR. Tubular magazine, capacity: 15 (LR), 11 (WMR). Barrel: 23 inches. Weight: 6¼ lbs. Short stroke action, crossbolt safety. Walnut stock with hand checkering. Sights: folding leaf rear; grooved for tipoff scope mount. **Price: $210** (.22 LR); **$230** (.22 WMR); **$310** (.22 LR, Grade II); **$330** (.22 WMR, Grade II)

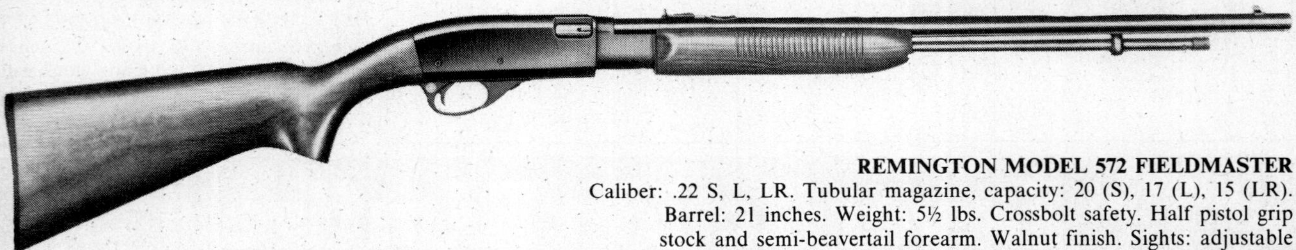

REMINGTON MODEL 572 FIELDMASTER
Caliber: .22 S, L, LR. Tubular magazine, capacity: 20 (S), 17 (L), 15 (LR). Barrel: 21 inches. Weight: 5½ lbs. Crossbolt safety. Half pistol grip stock and semi-beavertail forearm. Walnut finish. Sights: adjustable rear; receiver grooved for tipoff scope mount. **Price: $120**

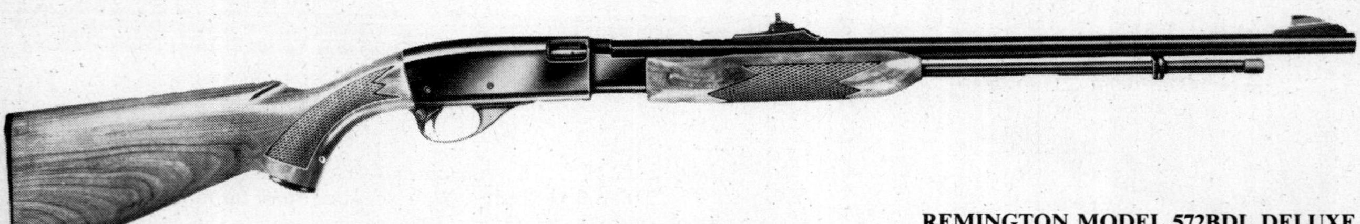

REMINGTON MODEL 572BDL DELUXE
Same as M572, except features custom checkered walnut stock, RKW finish. Sights: sliding ramp rear, ramp front. **Price: $136**

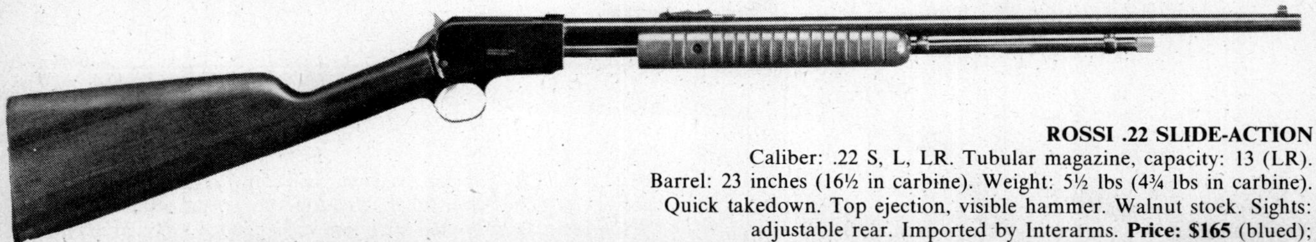

ROSSI .22 SLIDE-ACTION
Caliber: .22 S, L, LR. Tubular magazine, capacity: 13 (LR). Barrel: 23 inches (16½ in carbine). Weight: 5½ lbs (4¾ lbs in carbine). Quick takedown. Top ejection, visible hammer. Walnut stock. Sights: adjustable rear. Imported by Interarms. **Price: $165** (blued); **$175** (nickeled)

Target Rimfire

ANSCHUTZ MODEL 1411 MATCH 54
Caliber: .22 LR. Single-shot. Barrel: 27½. Weight: 11 lbs. Fully adjustable trigger, double locking lugs on bolt. Quick lock time. High Monte Carlo comb with cheekpiece, checkered pistol grip, thumb rest. Sights: tapped for Anschutz sights or scope blocks. **Price: $670; $737** (left hand version)

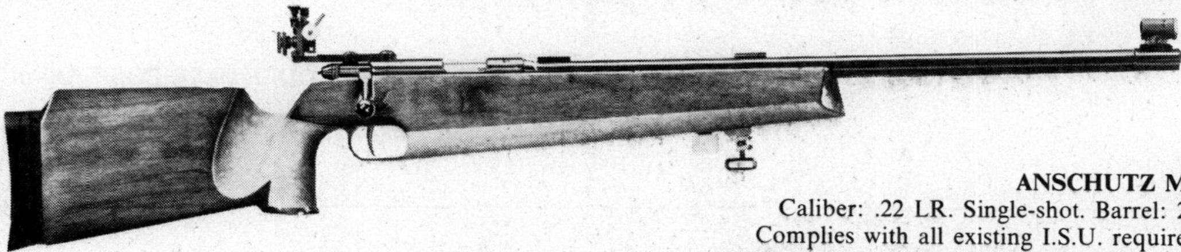

ANSCHUTZ MODEL 1407 MATCH 54
Caliber: .22 LR. Single-shot. Barrel: 26 inches. Weight: 10 lbs. Complies with all existing I.S.U. requirements and is suitable for N.R.A. matches. Features Anschutz 5071 Match Two-Stage trigger. **Price: $617; $679** (left hand version)

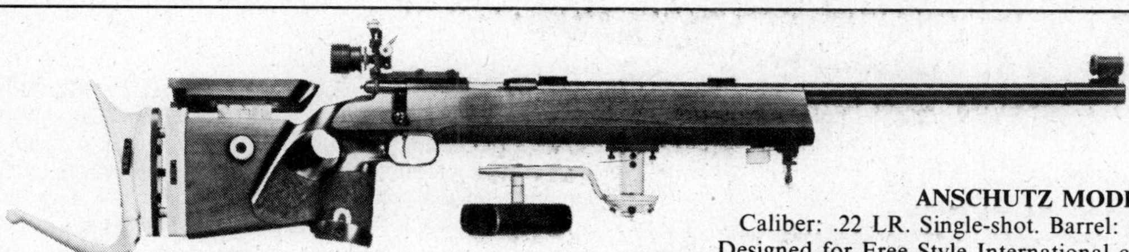

ANSCHUTZ MODEL 1413 SUPER MATCH 54
Caliber: .22 LR. Single-shot. Barrel: 27½ inches. Weight: 15½ lbs. Designed for Free Style International competition. Features reduced ignition time, two-stage trigger, hook buttplate, and adjustable palm rest, cheekpiece. Sights extra. **Price: $1007; $1108** (left hand version)

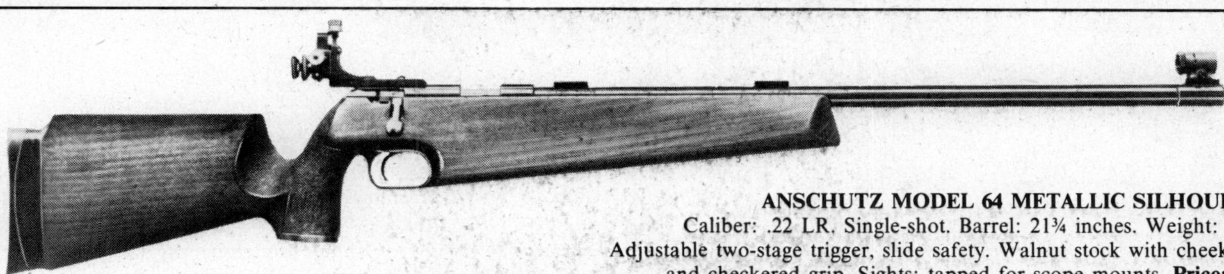

ANSCHUTZ MODEL 64 METALLIC SILHOUETTE
Caliber: .22 LR. Single-shot. Barrel: 21¾ inches. Weight: 8 lbs. Adjustable two-stage trigger, slide safety. Walnut stock with cheekpiece and checkered grip. Sights: tapped for scope mounts. **Price: $311**

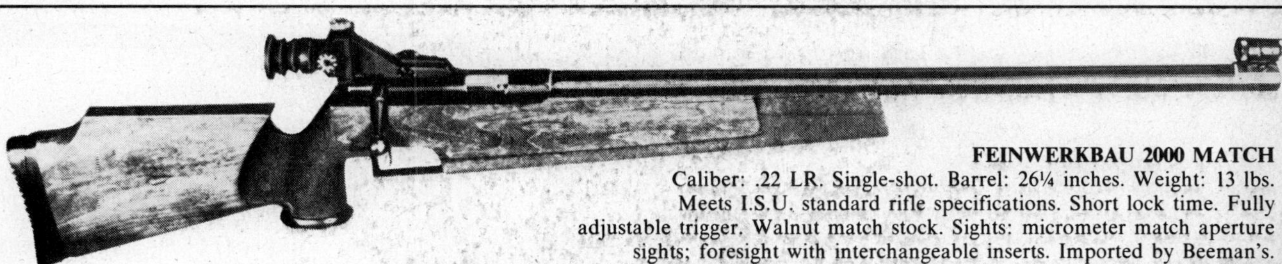

FEINWERKBAU 2000 MATCH
Caliber: .22 LR. Single-shot. Barrel: 26¼ inches. Weight: 13 lbs. Meets I.S.U. standard rifle specifications. Short lock time. Fully adjustable trigger. Walnut match stock. Sights: micrometer match aperture sights; foresight with interchangeable inserts. Imported by Beeman's. **Price: $800**

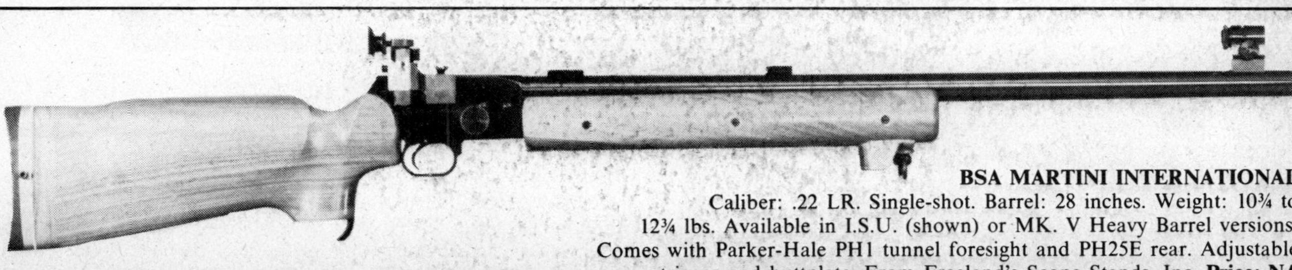

BSA MARTINI INTERNATIONAL
Caliber: .22 LR. Single-shot. Barrel: 28 inches. Weight: 10¾ to 12¾ lbs. Available in I.S.U. (shown) or MK. V Heavy Barrel versions. Comes with Parker-Hale PH1 tunnel foresight and PH25E rear. Adjustable trigger and buttplate. From Freeland's Scope Stands, Inc. **Price: NA**

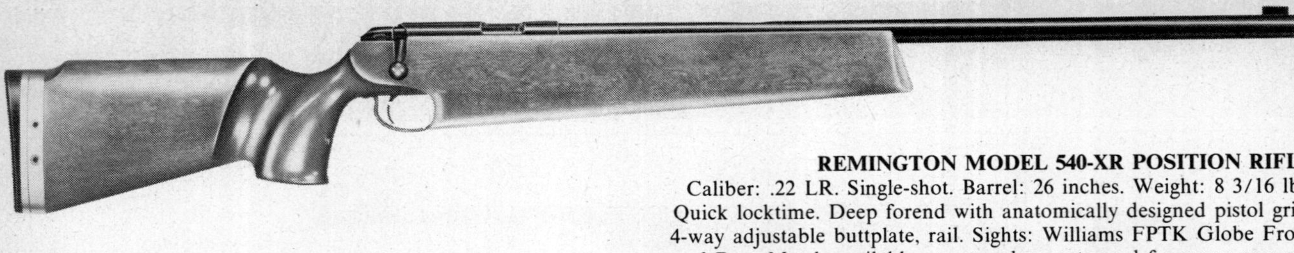

REMINGTON MODEL 540-XR POSITION RIFLE
Caliber: .22 LR. Single-shot. Barrel: 26 inches. Weight: 8 3/16 lbs. Quick locktime. Deep forend with anatomically designed pistol grip. 4-way adjustable buttplate, rail. Sights: Williams FPTK Globe Front and Rear Match available at extra charge; tapped for scope mounts. **Price: $240**

REMINGTON MODEL 40-XR POSITION RIFLE
Caliber: .22 LR. Single-shot. Barrel: 24 inches. Weight: 10 lbs. Position style stock with front swivel block on forend guide rail. Sights: barrel and receiver tapped for iron sights; scope blocks included. **Price: $439**

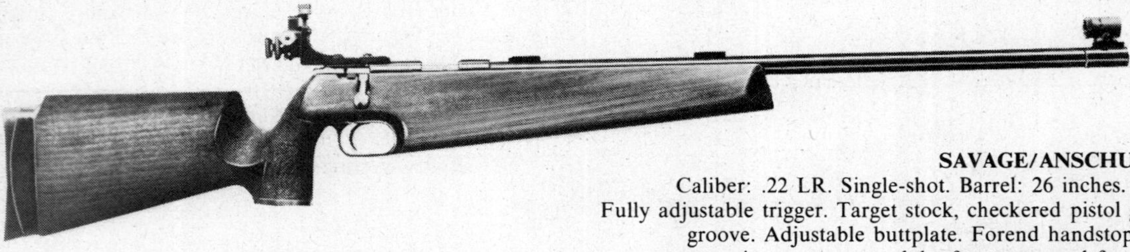

SAVAGE/ANSCHUTZ MODEL 64
Caliber: .22 LR. Single-shot. Barrel: 26 inches. Weight: 7¾ lbs. Fully adjustable trigger. Target stock, checkered pistol grip with thumb groove. Adjustable buttplate. Forend handstop swivels. Sights: micrometer rear, globe front; grooved for Anschutz sights. **Price: $419; $447** (left hand version)

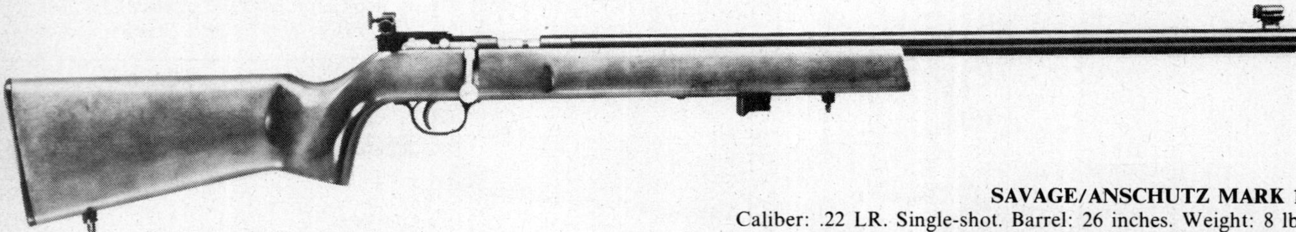

SAVAGE/ANSCHUTZ MARK 12
Caliber: .22 LR. Single-shot. Barrel: 26 inches. Weight: 8 lbs. Slide safety. Wundhammer swell pistol grip. Adjustable hand stop. Sights: micrometer rear, globe front. **Price: $141**

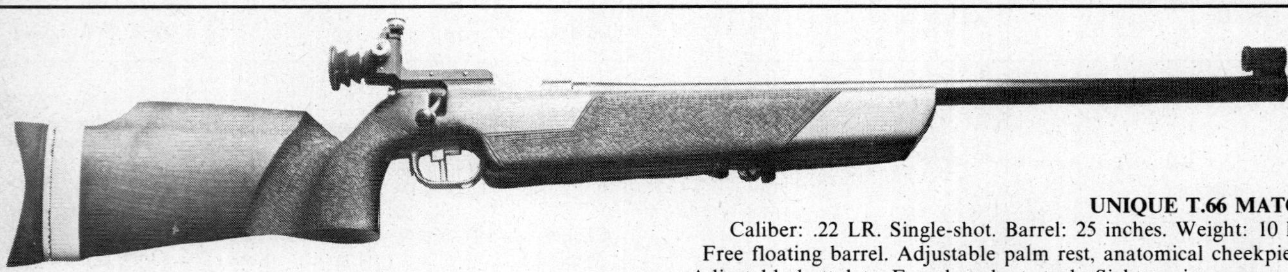

UNIQUE T.66 MATCH
Caliber: .22 LR. Single-shot. Barrel: 25 inches. Weight: 10 lbs. Free floating barrel. Adjustable palm rest, anatomical cheekpiece. Adjustable buttplate. French walnut stock. Sights: micro peep with 5 apertures. Imported by Solersport. **Price: $545**

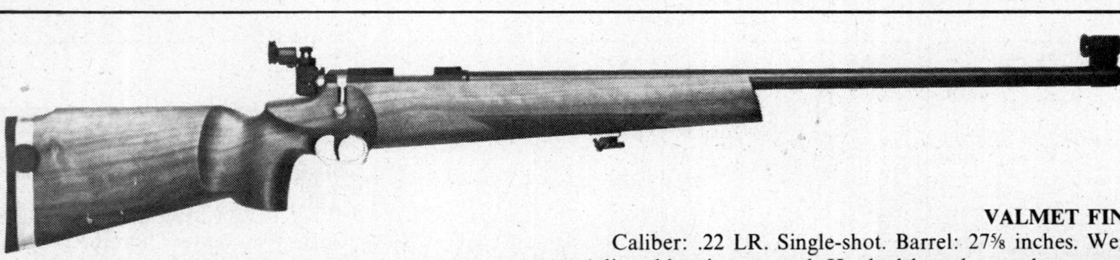

VALMET FINNISH LION
Caliber: .22 LR. Single-shot. Barrel: 27⅝ inches. Weight: 10½ lbs. Adjustable trigger travel. Hooked buttplate, palm rest and hand stop. Sights: detachable globe front, micrometer rear with 3 peep discs. Imported by Mandall Shooting Supplies. **Price: $500; $535** (heavy barrel)

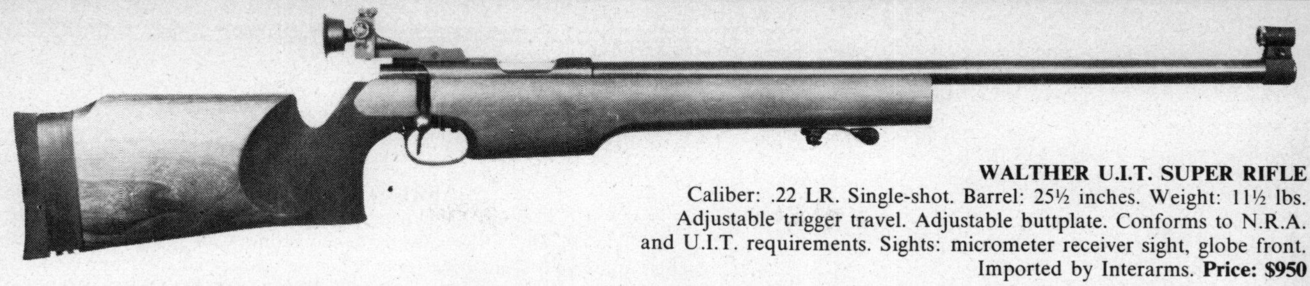

WALTHER U.I.T. SUPER RIFLE
Caliber: .22 LR. Single-shot. Barrel: 25½ inches. Weight: 11½ lbs. Adjustable trigger travel. Adjustable buttplate. Conforms to N.R.A. and U.I.T. requirements. Sights: micrometer receiver sight, globe front. Imported by Interarms. **Price: $950**

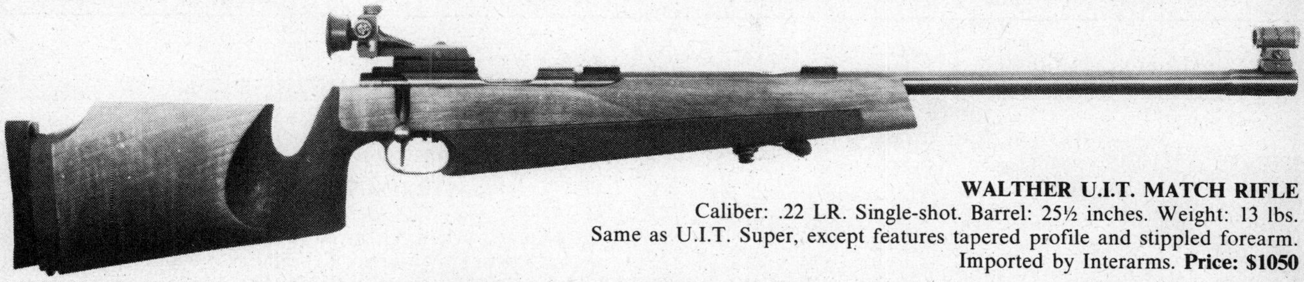

WALTHER U.I.T. MATCH RIFLE
Caliber: .22 LR. Single-shot. Barrel: 25½ inches. Weight: 13 lbs. Same as U.I.T. Super, except features tapered profile and stippled forearm. Imported by Interarms. **Price: $1050**

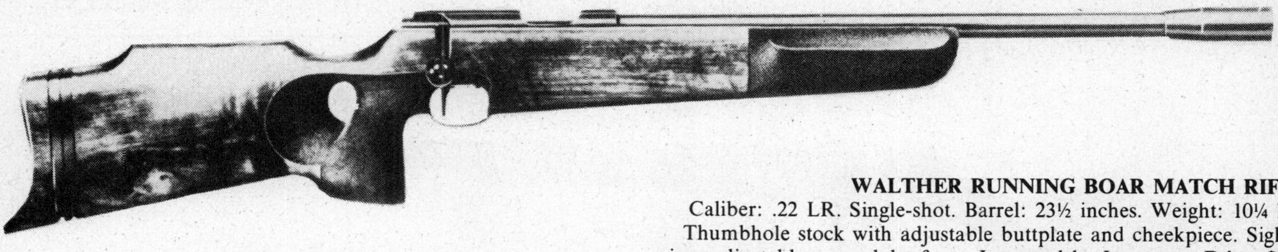

WALTHER RUNNING BOAR MATCH RIFLE
Caliber: .22 LR. Single-shot. Barrel: 23½ inches. Weight: 10¼ lbs. Thumbhole stock with adjustable buttplate and cheekpiece. Sights: micro adjustable rear, globe front. Imported by Interarms. **Price: $900**

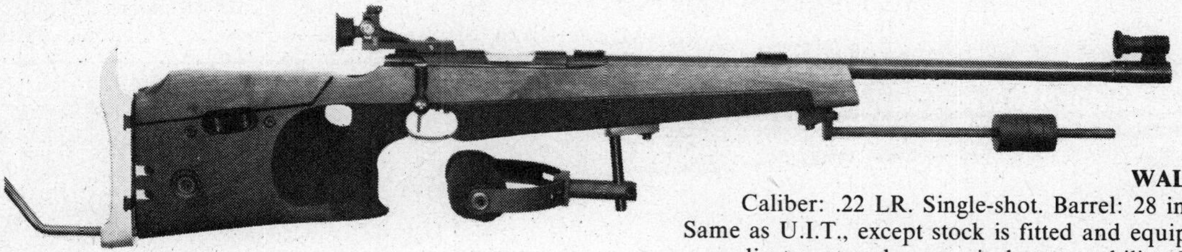

WALTHER MODEL GX-1
Caliber: .22 LR. Single-shot. Barrel: 28 inches. Weight: 16½ lbs. Same as U.I.T., except stock is fitted and equipped for full individual adjustment and anatomical compatability. Imported by Interarms. **Price: $1400**

Single-Shot Centerfire

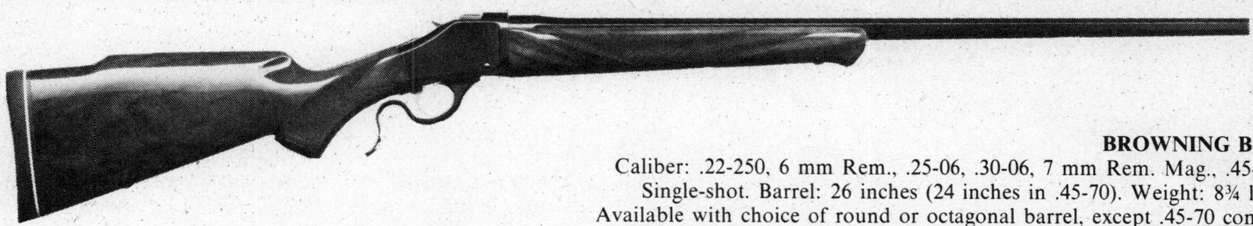

BROWNING B-78
Caliber: .22-250, 6 mm Rem., .25-06, .30-06, 7 mm Rem. Mag., .45-70 Single-shot. Barrel: 26 inches (24 inches in .45-70). Weight: 8¾ lbs. Available with choice of round or octagonal barrel, except .45-70 comes with octagonal bull barrel. Traditional Hi-Wall action. French walnut stock and checkered semi-schnabel forearm. **Price: $400**

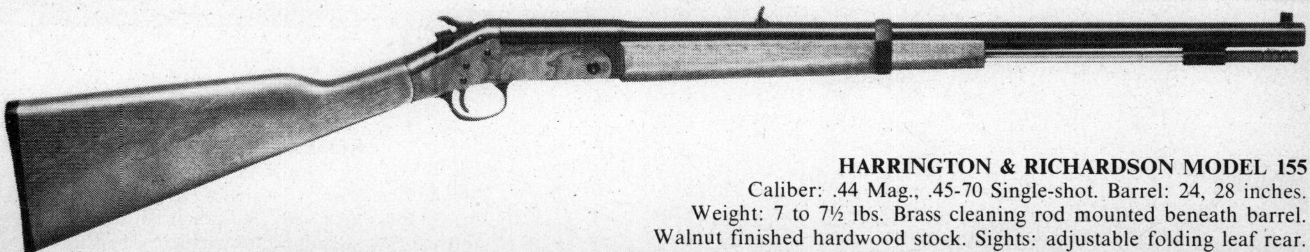

HARRINGTON & RICHARDSON MODEL 155
Caliber: .44 Mag., .45-70 Single-shot. Barrel: 24, 28 inches. Weight: 7 to 7½ lbs. Brass cleaning rod mounted beneath barrel. Walnut finished hardwood stock. Sights: adjustable folding leaf rear. **Price: $100**

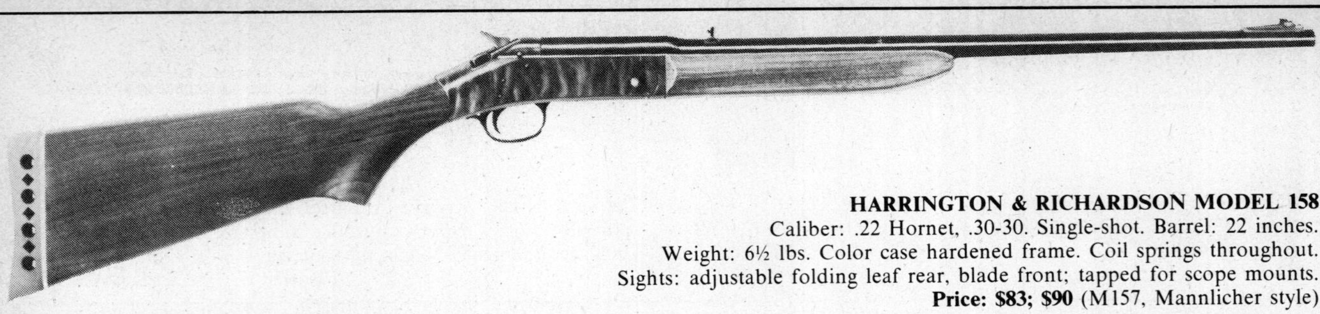

HARRINGTON & RICHARDSON MODEL 158
Caliber: .22 Hornet, .30-30. Single-shot. Barrel: 22 inches. Weight: 6½ lbs. Color case hardened frame. Coil springs throughout. Sights: adjustable folding leaf rear, blade front; tapped for scope mounts. **Price: $83; $90** (M157, Mannlicher style)

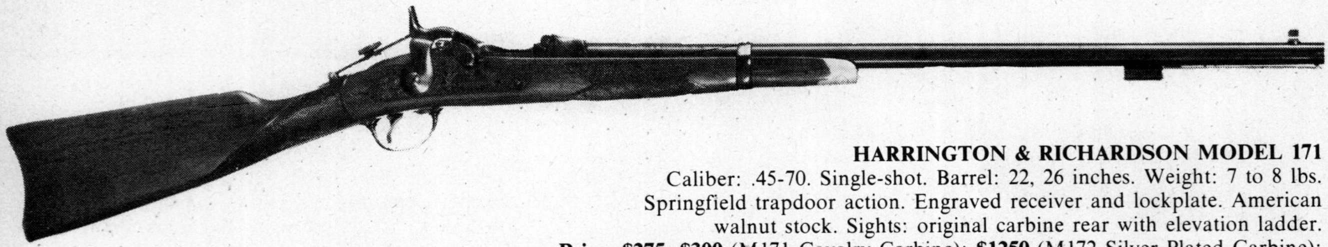

HARRINGTON & RICHARDSON MODEL 171
Caliber: .45-70. Single-shot. Barrel: 22, 26 inches. Weight: 7 to 8 lbs. Springfield trapdoor action. Engraved receiver and lockplate. American walnut stock. Sights: original carbine rear with elevation ladder. **Price: $275; $300** (M171 Cavalry Carbine); **$1250** (M172 Silver Plated Carbine); **$315** (M173 Officer's Model); **$300** (M174 Little Big Horn Commemorative)

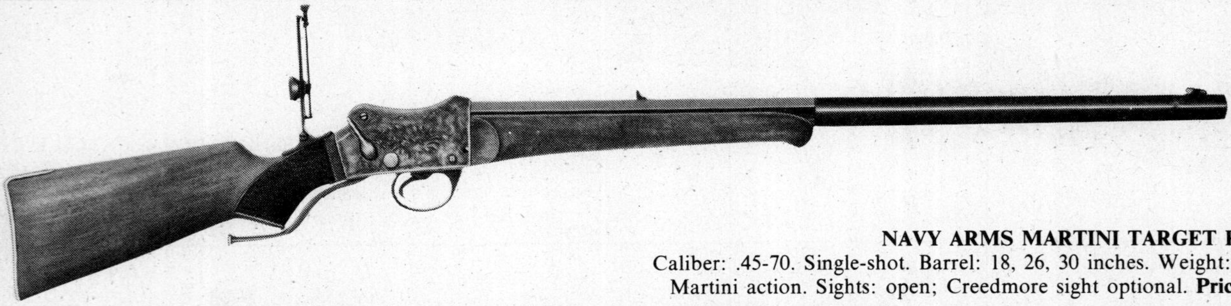

NAVY ARMS MARTINI TARGET RIFLE
Caliber: .45-70. Single-shot. Barrel: 18, 26, 30 inches. Weight: 8 lbs. Martini action. Sights: open; Creedmore sight optional. **Price: NA**

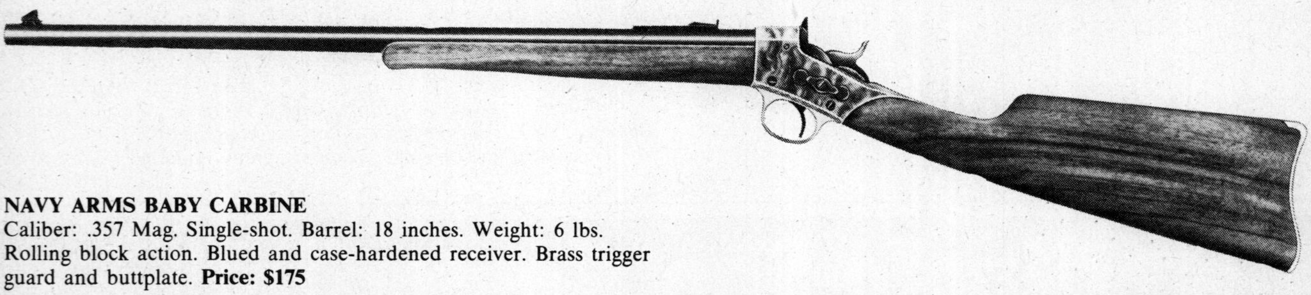

NAVY ARMS BABY CARBINE
Caliber: .357 Mag. Single-shot. Barrel: 18 inches. Weight: 6 lbs. Rolling block action. Blued and case-hardened receiver. Brass trigger guard and buttplate. **Price: $175**

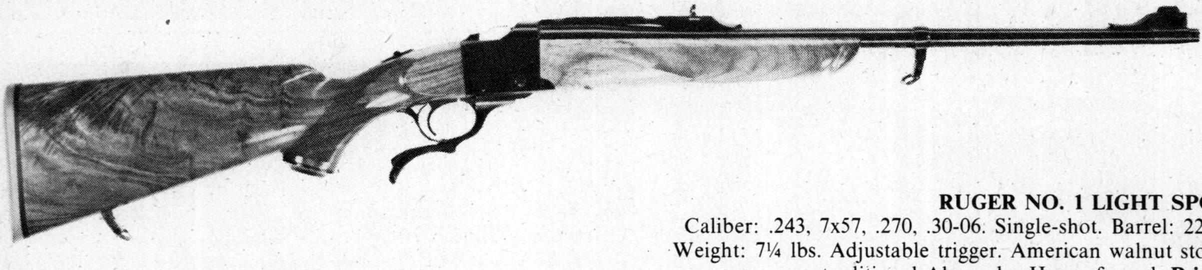

RUGER NO. 1 LIGHT SPORTER
Caliber: .243, 7x57, .270, .30-06. Single-shot. Barrel: 22 inches. Weight: 7¼ lbs. Adjustable trigger. American walnut stock with traditional Alexander Henry forend. **Price: $348**

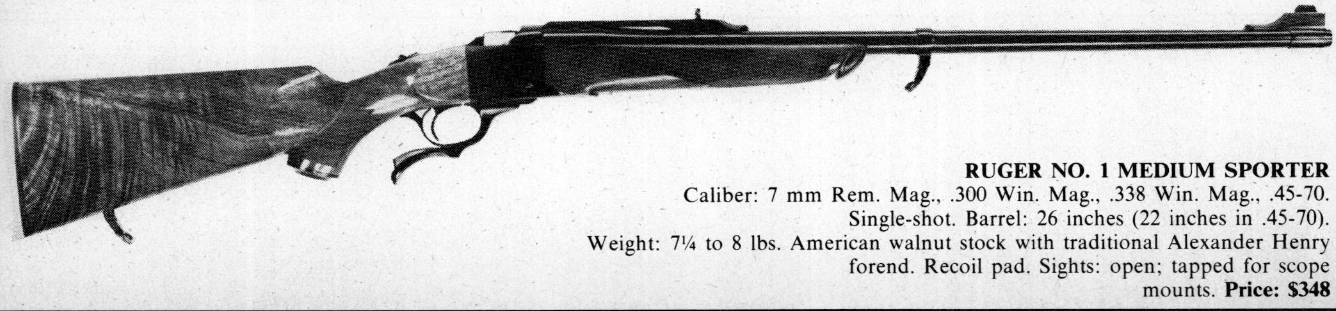

RUGER NO. 1 MEDIUM SPORTER
Caliber: 7 mm Rem. Mag., .300 Win. Mag., .338 Win. Mag., .45-70. Single-shot. Barrel: 26 inches (22 inches in .45-70). Weight: 7¼ to 8 lbs. American walnut stock with traditional Alexander Henry forend. Recoil pad. Sights: open; tapped for scope mounts. **Price: $348**

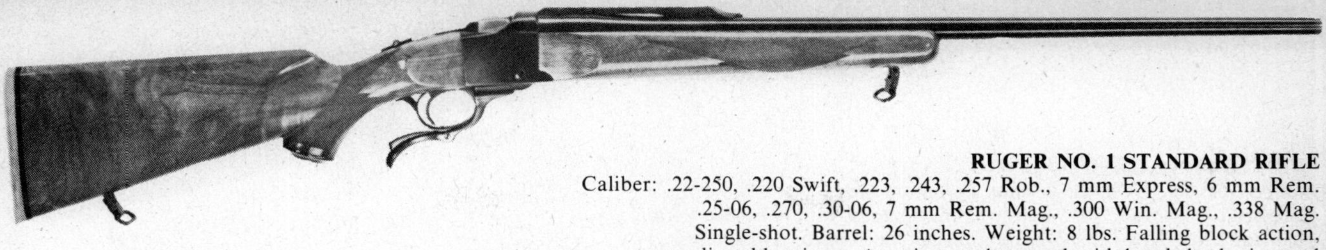

RUGER NO. 1 STANDARD RIFLE
Caliber: .22-250, .220 Swift, .223, .243, .257 Rob., 7 mm Express, 6 mm Rem. .25-06, .270, .30-06, 7 mm Rem. Mag., .300 Win. Mag., .338 Mag. Single-shot. Barrel: 26 inches. Weight: 8 lbs. Falling block action, adjustable trigger. American walnut stock with hand checkering and semi-beavertail forend. Sights: Ruger steel tipoff scope rings furnished. **Price: $348**

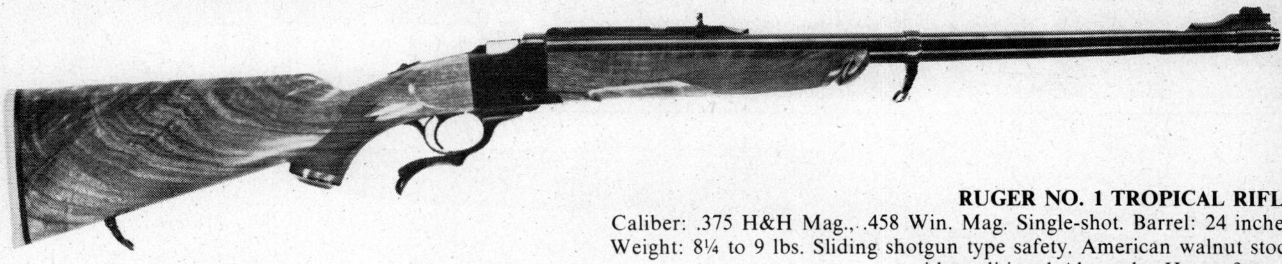

RUGER NO. 1 TROPICAL RIFLE
Caliber: .375 H&H Mag., .458 Win. Mag. Single-shot. Barrel: 24 inches. Weight: 8¼ to 9 lbs. Sliding shotgun type safety. American walnut stock with traditional Alexander Henry forend. Sights: open. **Price: $348**

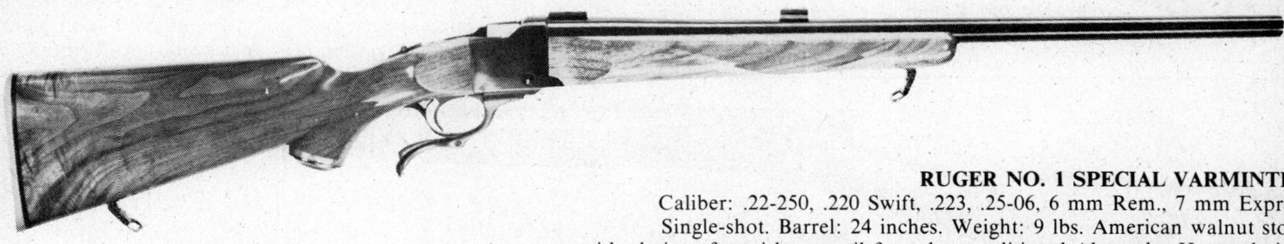

RUGER NO. 1 SPECIAL VARMINTER
Caliber: .22-250, .220 Swift, .223, .25-06, 6 mm Rem., 7 mm Express Single-shot. Barrel: 24 inches. Weight: 9 lbs. American walnut stock with choice of semi-beavertail forend or traditional Alexander Henry design. Sights: Ruger steel block mounts and tipoff scope rings furnished. **Price: $348**

RUGER NO. 3 CARBINE
Caliber: .22 Hornet, .223, .45-70, .375 Win. Single-shot. Barrel: 22 inches. Weight: 6 lbs. Same falling block design as No. 1 models, but with an American style lever. Adjustable trigger. Straight grip American walnut stock. Sights: adjustable folding leaf rear, gold bead front; tapped for scope mounts. **Price: $225**

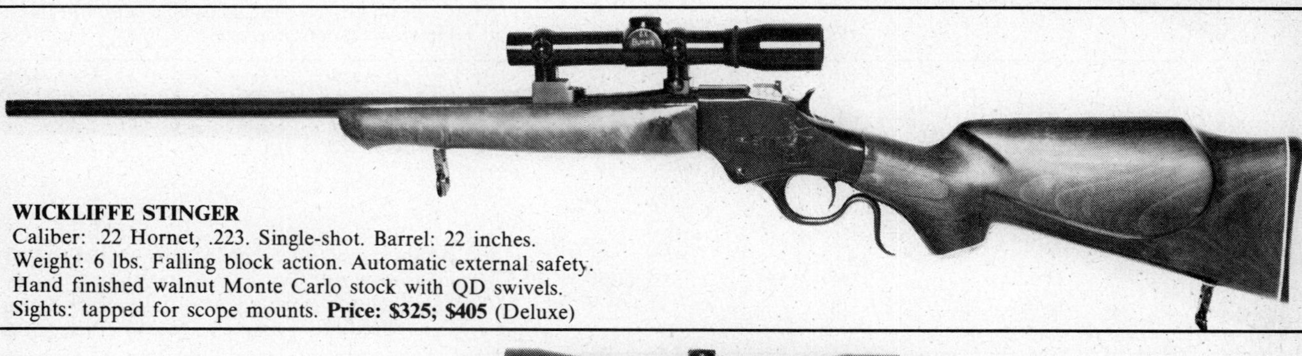

WICKLIFFE STINGER
Caliber: .22 Hornet, .223. Single-shot. Barrel: 22 inches. Weight: 6 lbs. Falling block action. Automatic external safety. Hand finished walnut Monte Carlo stock with QD swivels. Sights: tapped for scope mounts. **Price: $325; $405** (Deluxe)

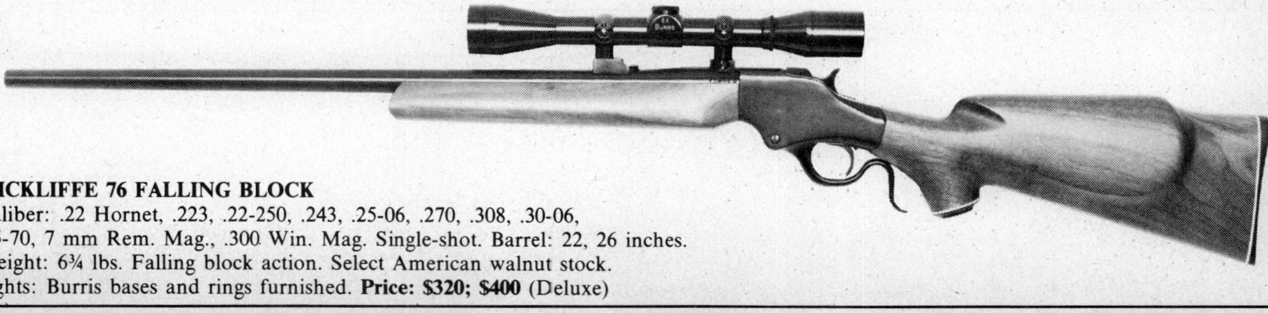

WICKLIFFE 76 FALLING BLOCK
Caliber: .22 Hornet, .223, .22-250, .243, .25-06, .270, .308, .30-06, .45-70, 7 mm Rem. Mag., .300 Win. Mag. Single-shot. Barrel: 22, 26 inches. Weight: 6¾ lbs. Falling block action. Select American walnut stock. Sights: Burris bases and rings furnished. **Price: $320; $400** (Deluxe)

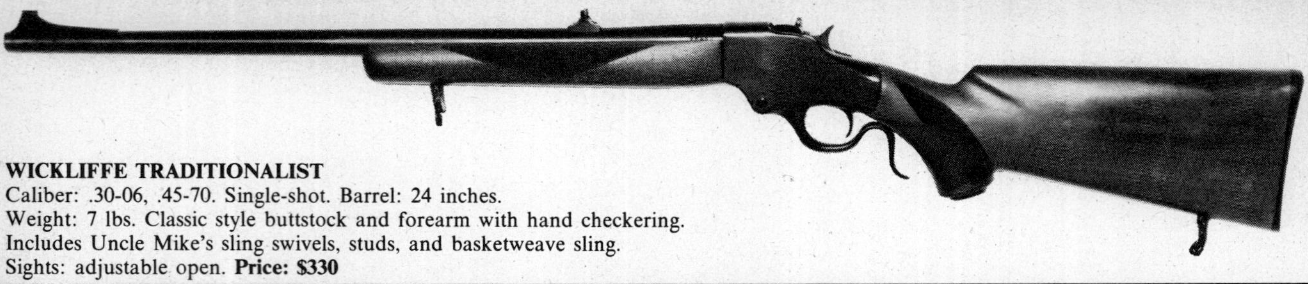

WICKLIFFE TRADITIONALIST
Caliber: .30-06, .45-70. Single-shot. Barrel: 24 inches. Weight: 7 lbs. Classic style buttstock and forearm with hand checkering. Includes Uncle Mike's sling swivels, studs, and basketweave sling. Sights: adjustable open. **Price: $330**

Lever Action Centerfire

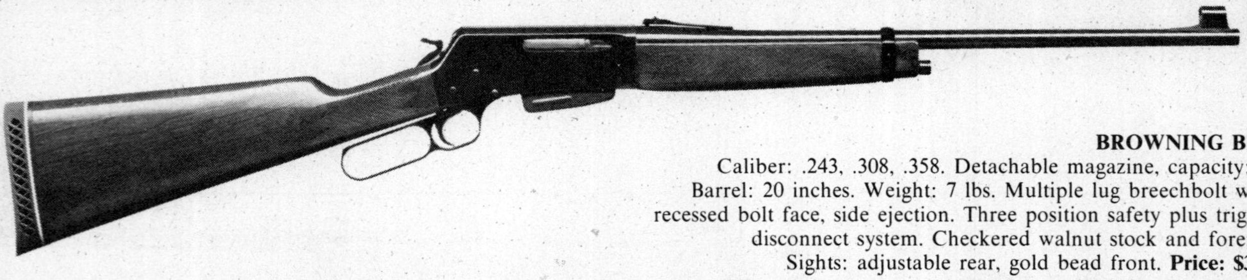

BROWNING BLR
Caliber: .243, .308, .358. Detachable magazine, capacity: 4. Barrel: 20 inches. Weight: 7 lbs. Multiple lug breechbolt with recessed bolt face, side ejection. Three position safety plus trigger disconnect system. Checkered walnut stock and forend. Sights: adjustable rear, gold bead front. **Price: $325**

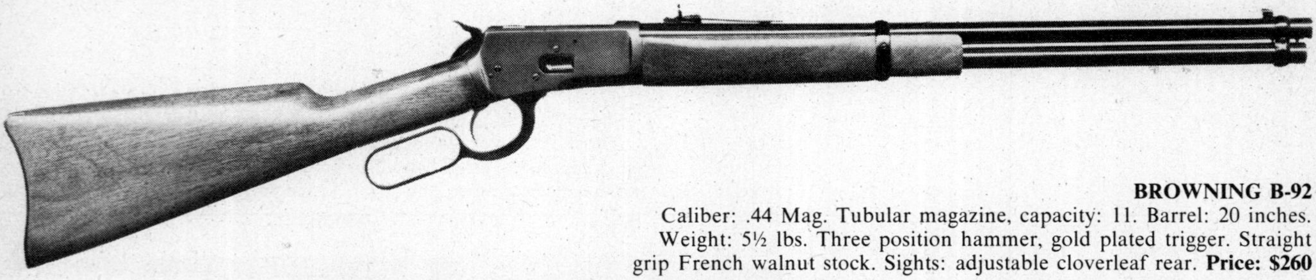

BROWNING B-92
Caliber: .44 Mag. Tubular magazine, capacity: 11. Barrel: 20 inches. Weight: 5½ lbs. Three position hammer, gold plated trigger. Straight grip French walnut stock. Sights: adjustable cloverleaf rear. **Price: $260**

MARLIN MODEL 336C
Caliber: .30-30, .35 Rem. Tubular magazine, capacity: 6. Barrel: 20 inches. Weight: 7 lbs. Side ejection, sand blasted receiver top. American black walnut stock with pistol grip. White spacers at buttplate and pistol grip. Sights: adjustable semi-buckhorn rear, brass bead ramp front; tapped for scope mounts. **Price: $188**

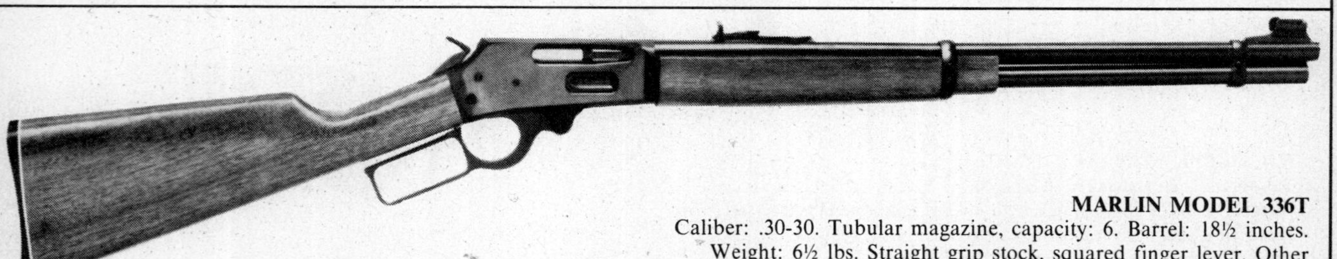

MARLIN MODEL 336T
Caliber: .30-30. Tubular magazine, capacity: 6. Barrel: 18½ inches. Weight: 6½ lbs. Straight grip stock, squared finger lever. Other specifications same as M336C. **Price: $188**

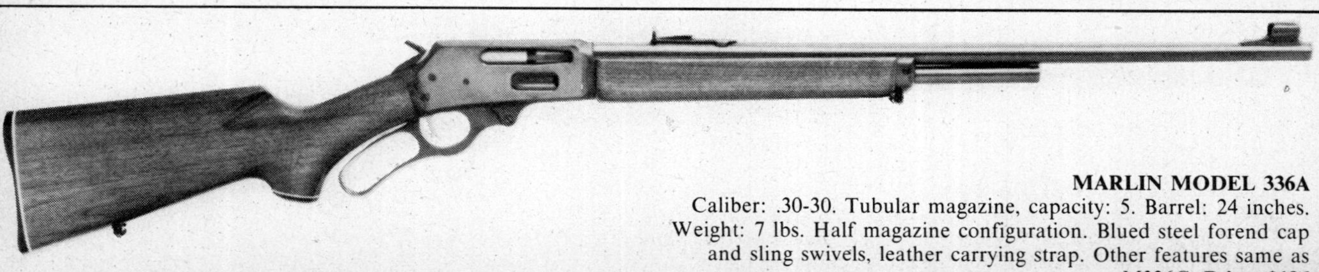

MARLIN MODEL 336A
Caliber: .30-30. Tubular magazine, capacity: 5. Barrel: 24 inches. Weight: 7 lbs. Half magazine configuration. Blued steel forend cap and sling swivels, leather carrying strap. Other features same as M336C. **Price: $196**

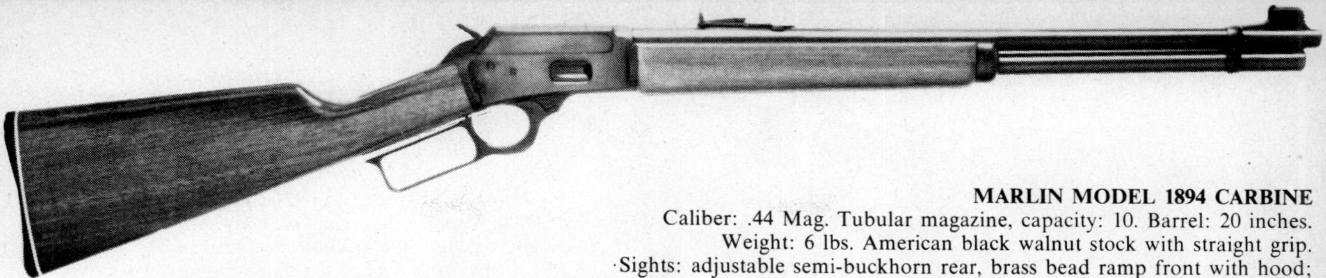

MARLIN MODEL 1894 CARBINE
Caliber: .44 Mag. Tubular magazine, capacity: 10. Barrel: 20 inches. Weight: 6 lbs. American black walnut stock with straight grip. Sights: adjustable semi-buckhorn rear, brass bead ramp front with hood; tapped for scope mounts. **Price: $194**

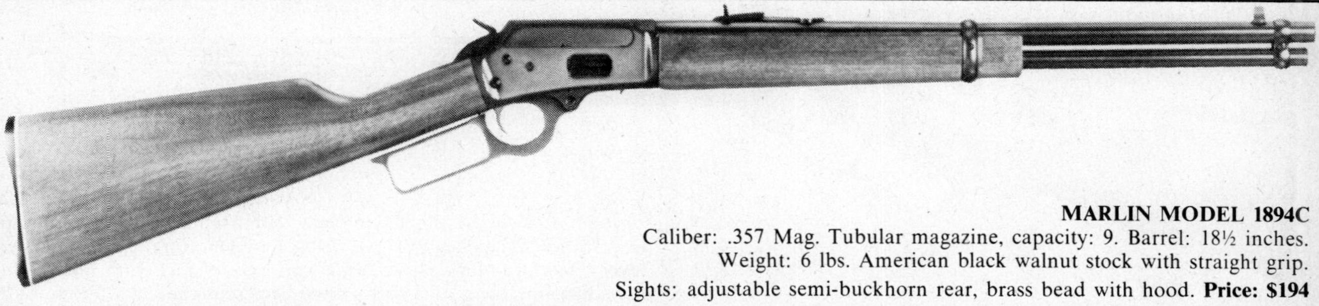

MARLIN MODEL 1894C
Caliber: .357 Mag. Tubular magazine, capacity: 9. Barrel: 18½ inches. Weight: 6 lbs. American black walnut stock with straight grip. Sights: adjustable semi-buckhorn rear, brass bead with hood. **Price: $194**

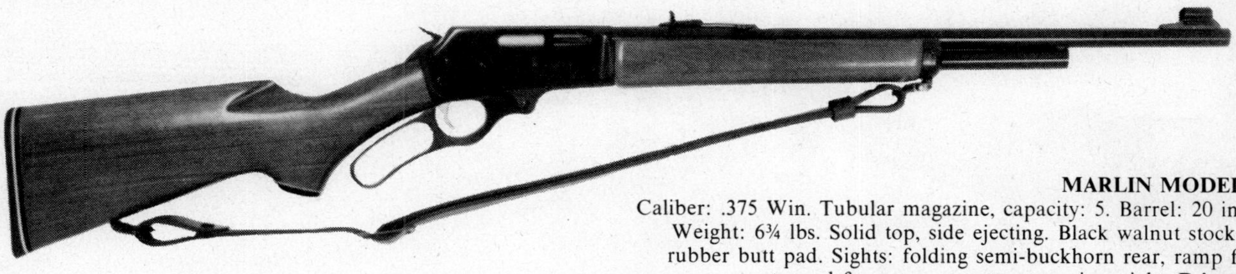

MARLIN MODEL 375
Caliber: .375 Win. Tubular magazine, capacity: 5. Barrel: 20 inches. Weight: 6¾ lbs. Solid top, side ejecting. Black walnut stock with rubber butt pad. Sights: folding semi-buckhorn rear, ramp front; tapped for scope mounts or receiver sight. **Price: $211**

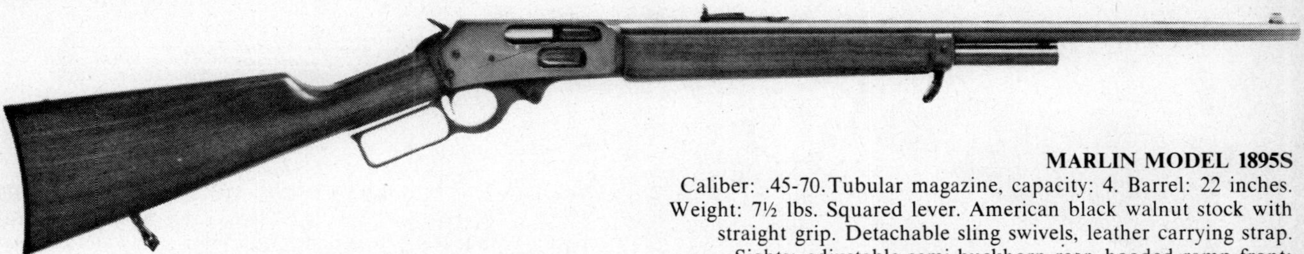

MARLIN MODEL 1895S
Caliber: .45-70.Tubular magazine, capacity: 4. Barrel: 22 inches. Weight: 7½ lbs. Squared lever. American black walnut stock with straight grip. Detachable sling swivels, leather carrying strap. Sights: adjustable semi-buckhorn rear, hooded ramp front; tapped for scope mounts. **Price: $267**

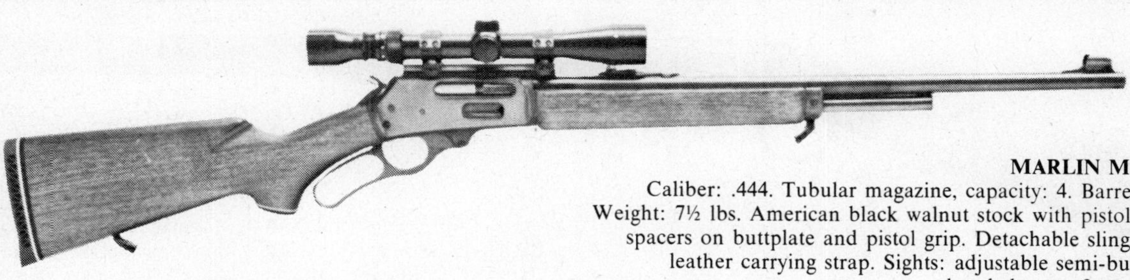

MARLIN MODEL 444S
Caliber: .444. Tubular magazine, capacity: 4. Barrel: 22 inches. Weight: 7½ lbs. American black walnut stock with pistol grip. White spacers on buttplate and pistol grip. Detachable sling swivels and leather carrying strap. Sights: adjustable semi-buckhorn rear, hooded ramp front. **Price: $211**

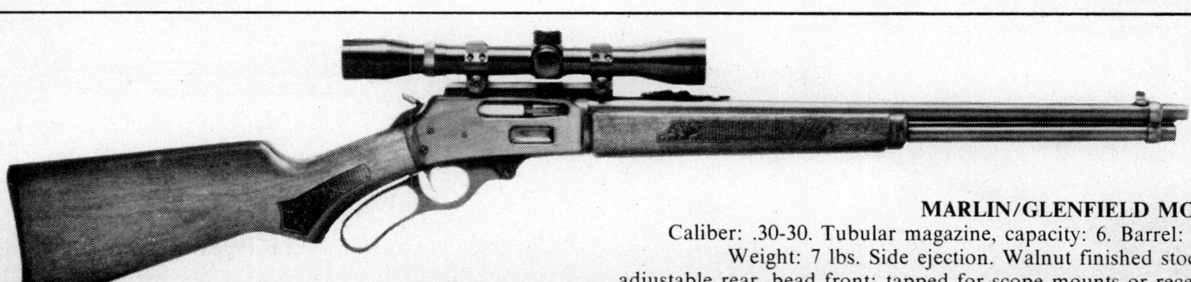

MARLIN/GLENFIELD MODEL 30A
Caliber: .30-30. Tubular magazine, capacity: 6. Barrel: 20 inches. Weight: 7 lbs. Side ejection. Walnut finished stock. Sights: adjustable rear, bead front; tapped for scope mounts or receiver sight. **Price: $175**

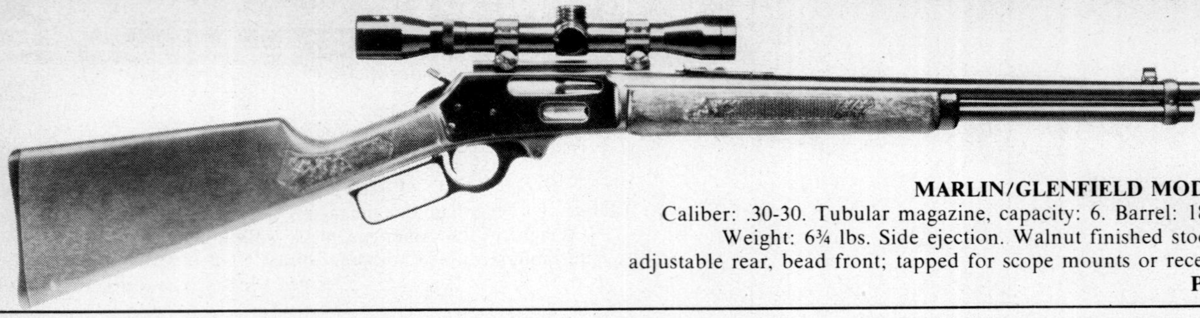

MARLIN/GLENFIELD MODEL 30GT
Caliber: .30-30. Tubular magazine, capacity: 6. Barrel: 18½ inches. Weight: 6¾ lbs. Side ejection. Walnut finished stock. Sights: adjustable rear, bead front; tapped for scope mounts or receiver sight. **Price: $175**

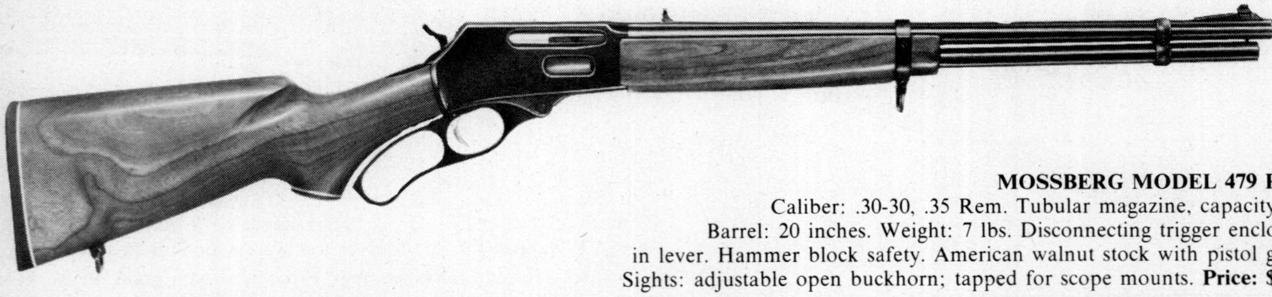

MOSSBERG MODEL 479 PCA
Caliber: .30-30, .35 Rem. Tubular magazine, capacity: 6. Barrel: 20 inches. Weight: 7 lbs. Disconnecting trigger enclosed in lever. Hammer block safety. American walnut stock with pistol grip. Sights: adjustable open buckhorn; tapped for scope mounts. **Price: $175**

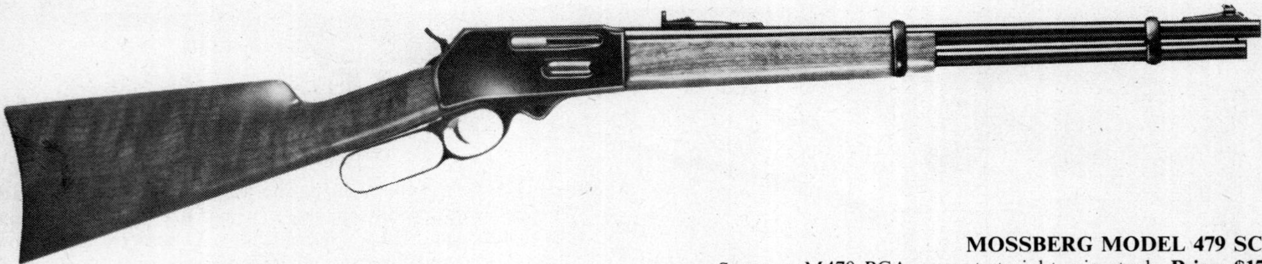

MOSSBERG MODEL 479 SCA
Same as M479 PCA, except straight grip stock. **Price: $175**

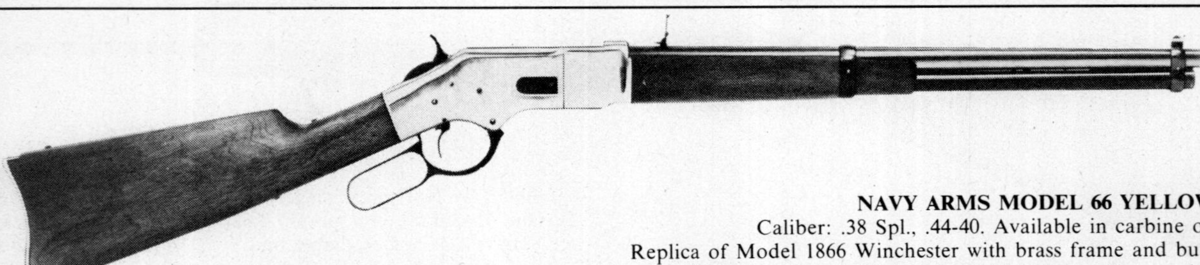

NAVY ARMS MODEL 66 YELLOWBOY
Caliber: .38 Spl., .44-40. Available in carbine or rifle. Replica of Model 1866 Winchester with brass frame and buttplate. Rifle has octagonal barrel; carbine comes with round barrel. **Price: $295; $250** (Carbine)

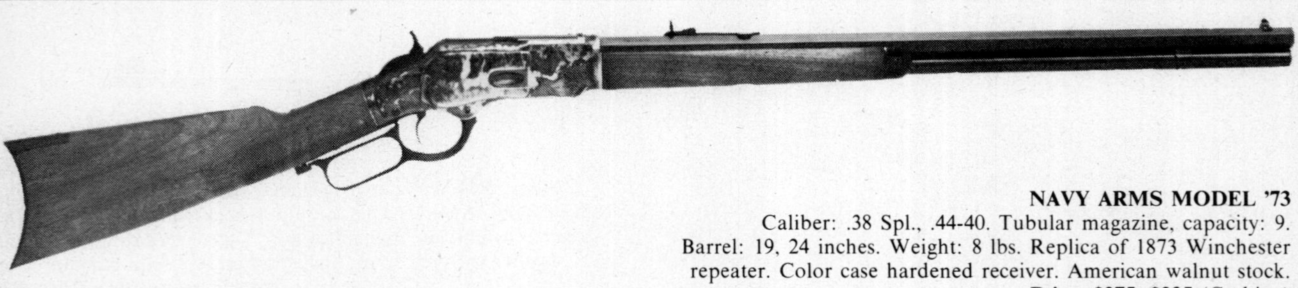

NAVY ARMS MODEL '73
Caliber: .38 Spl., .44-40. Tubular magazine, capacity: 9. Barrel: 19, 24 inches. Weight: 8 lbs. Replica of 1873 Winchester repeater. Color case hardened receiver. American walnut stock. **Price: $275; $235** (Carbine)

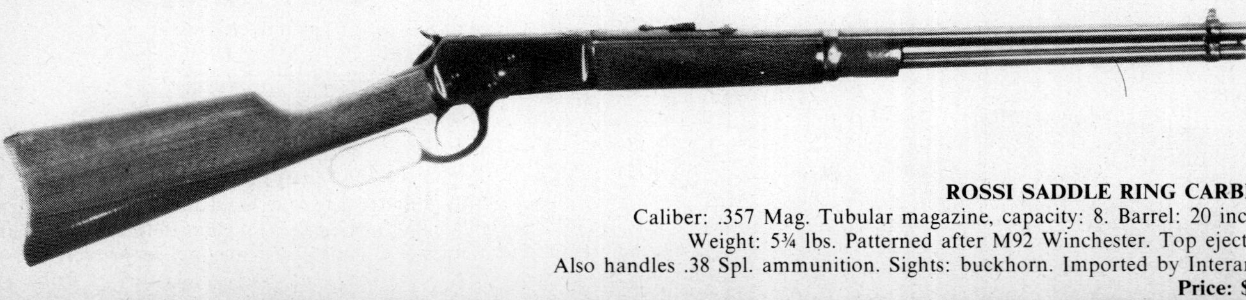

ROSSI SADDLE RING CARBINE
Caliber: .357 Mag. Tubular magazine, capacity: 8. Barrel: 20 inches. Weight: 5¾ lbs. Patterned after M92 Winchester. Top ejection. Also handles .38 Spl. ammunition. Sights: buckhorn. Imported by Interarms. **Price: $310**

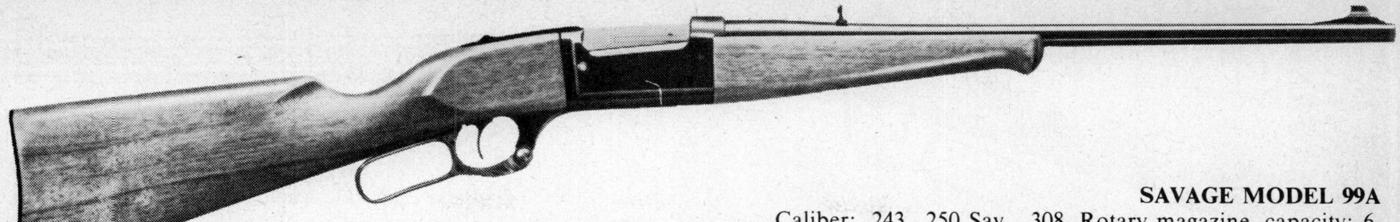

SAVAGE MODEL 99A
Caliber: .243, .250 Sav., .308. Rotary magazine, capacity: 6. Barrel: 22 inches. Weight: 7 lbs. Solid steel receiver. Straight stock configuration, schnabel forend. Sights: folding leaf rear; tapped for scope mounts. **Price: $284**

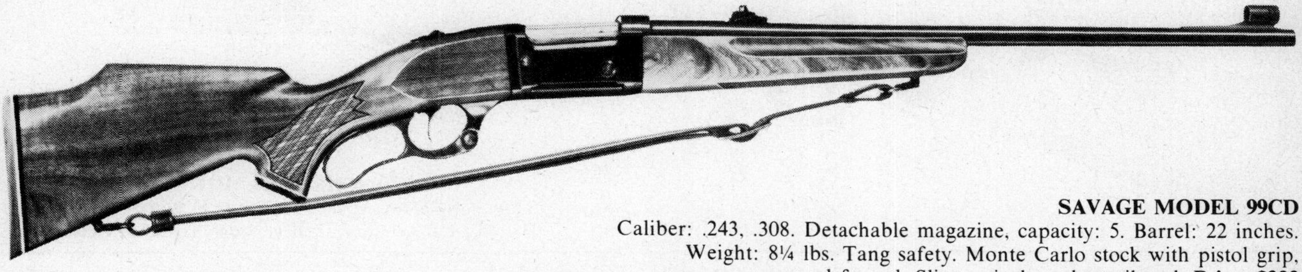

SAVAGE MODEL 99CD
Caliber: .243, .308. Detachable magazine, capacity: 5. Barrel: 22 inches. Weight: 8¼ lbs. Tang safety. Monte Carlo stock with pistol grip, grooved forend. Sling swivels and recoil pad. **Price: $323**

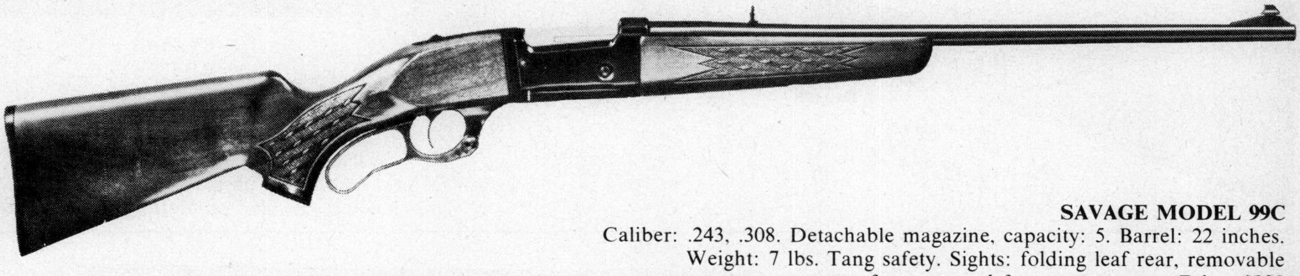

SAVAGE MODEL 99C
Caliber: .243, .308. Detachable magazine, capacity: 5. Barrel: 22 inches. Weight: 7 lbs. Tang safety. Sights: folding leaf rear, removable ramp front; tapped for scope mounts. **Price: $292**

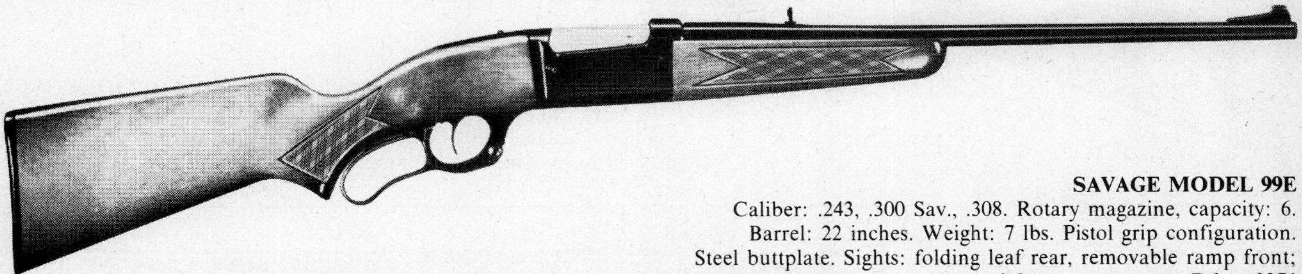

SAVAGE MODEL 99E
Caliber: .243, .300 Sav., .308. Rotary magazine, capacity: 6. Barrel: 22 inches. Weight: 7 lbs. Pistol grip configuration. Steel buttplate. Sights: folding leaf rear, removable ramp front; tapped for scope mounts. **Price: $251**

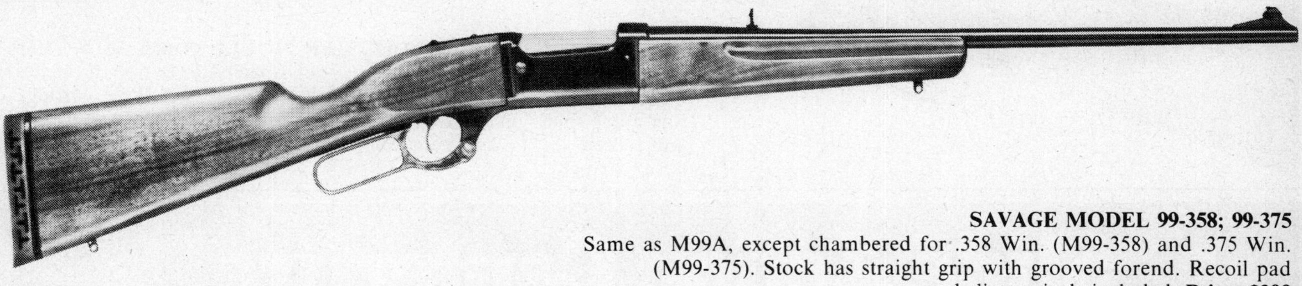

SAVAGE MODEL 99-358; 99-375
Same as M99A, except chambered for .358 Win. (M99-358) and .375 Win. (M99-375). Stock has straight grip with grooved forend. Recoil pad and sling swivels included. **Price: $299**

WINCHESTER MODEL 94 STANDARD
Caliber: .30-30. Tubular magazine, capacity: 6. Barrel: 20 inches. Weight: 6½ lbs. Exposed hammer with half cock safety. Walnut stock with straight grip. Sights: semi-buckhorn rear, hooded blade front; adapted for scope mounts. **Price: $176**

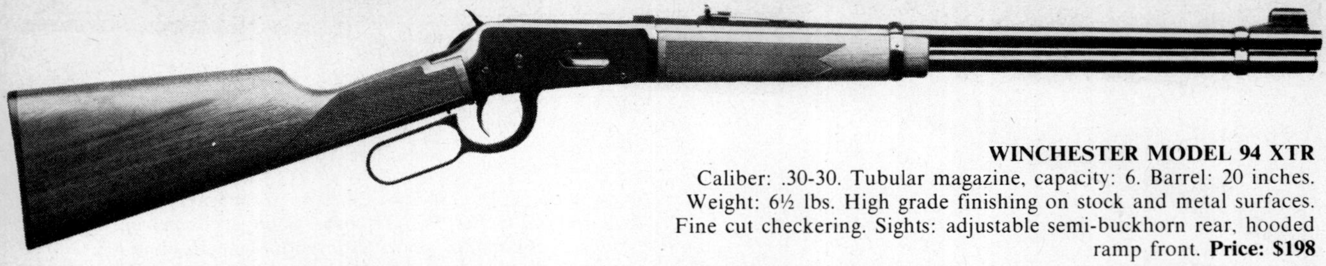

WINCHESTER MODEL 94 XTR
Caliber: .30-30. Tubular magazine, capacity: 6. Barrel: 20 inches. Weight: 6½ lbs. High grade finishing on stock and metal surfaces. Fine cut checkering. Sights: adjustable semi-buckhorn rear, hooded ramp front. **Price: $198**

WINCHESTER MODEL 94 TRAPPER
Caliber: .30-30. Tubular magazine, capacity: 5. Barrel: 16 inches. Weight: 6⅛ lbs. Shortened version of M94. **Price: $176**

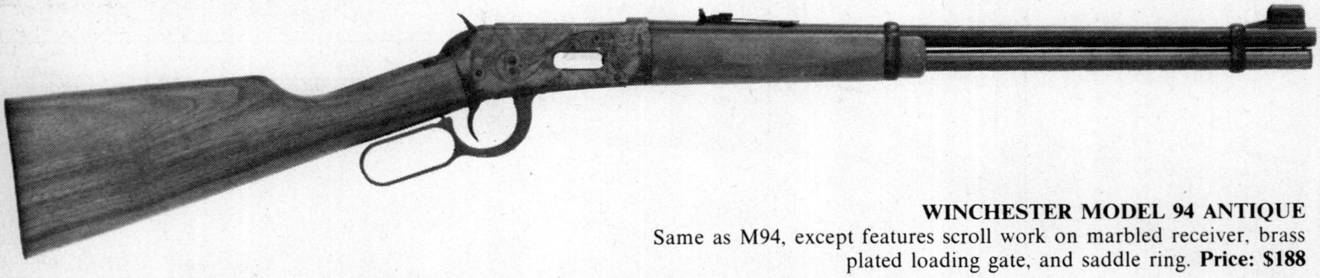

WINCHESTER MODEL 94 ANTIQUE
Same as M94, except features scroll work on marbled receiver, brass plated loading gate, and saddle ring. **Price: $188**

WINCHESTER MODEL 94 OLIVER F. WINCHESTER RIFLE–COMMEMORATIVE
Caliber: .38-55 Win. Barrel: 24 inches, octagonal. Receiver satin gold plated and engraved, lower tang and forend tip satin gold plated. Semi-fancy American walnut stock with high grade cut checkering. Limited edition. **Price: $485**

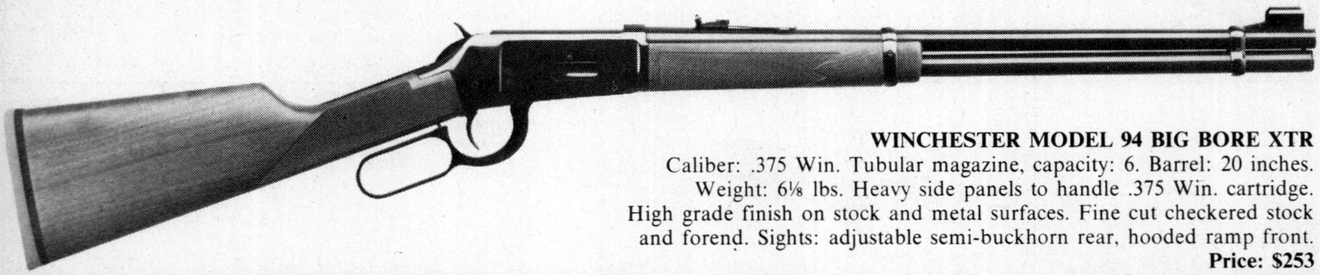

WINCHESTER MODEL 94 BIG BORE XTR
Caliber: .375 Win. Tubular magazine, capacity: 6. Barrel: 20 inches. Weight: 6⅛ lbs. Heavy side panels to handle .375 Win. cartridge. High grade finish on stock and metal surfaces. Fine cut checkered stock and forend. Sights: adjustable semi-buckhorn rear, hooded ramp front. **Price: $253**

Slide Action Centerfire

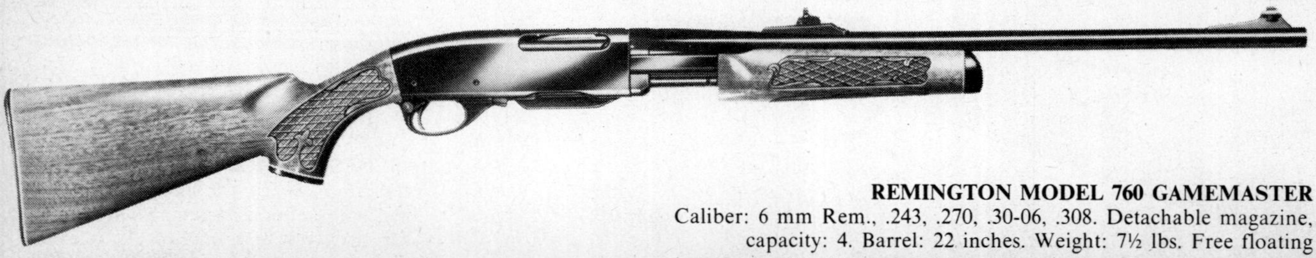

REMINGTON MODEL 760 GAMEMASTER
Caliber: 6 mm Rem., .243, .270, .30-06, .308. Detachable magazine, capacity: 4. Barrel: 22 inches. Weight: 7½ lbs. Free floating barrel. Skip line checkering. Sights: adjustable rear. **Price: $265**

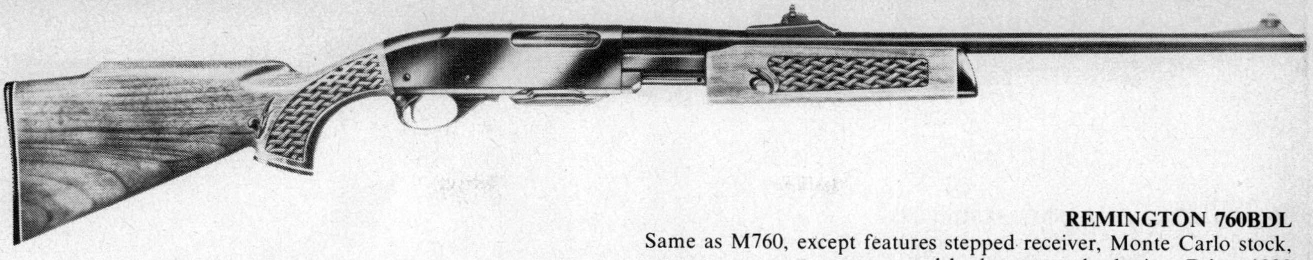

REMINGTON 760BDL
Same as M760, except features stepped receiver, Monte Carlo stock, and basketweave checkering. **Price: $290**

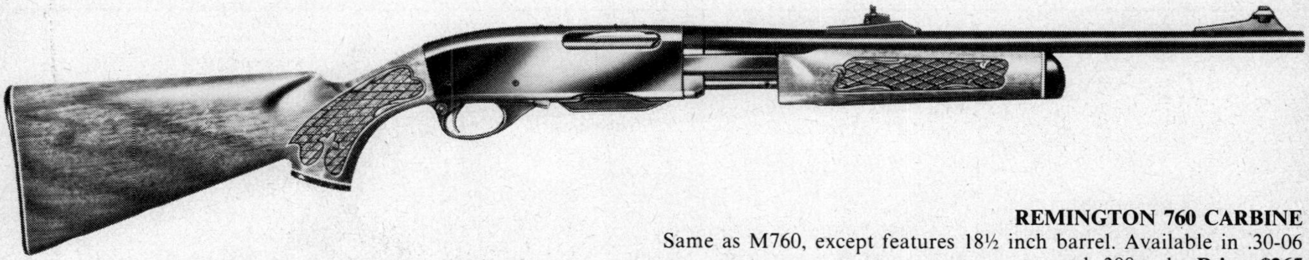

REMINGTON 760 CARBINE
Same as M760, except features 18½ inch barrel. Available in .30-06 and .308 only. **Price: $265**

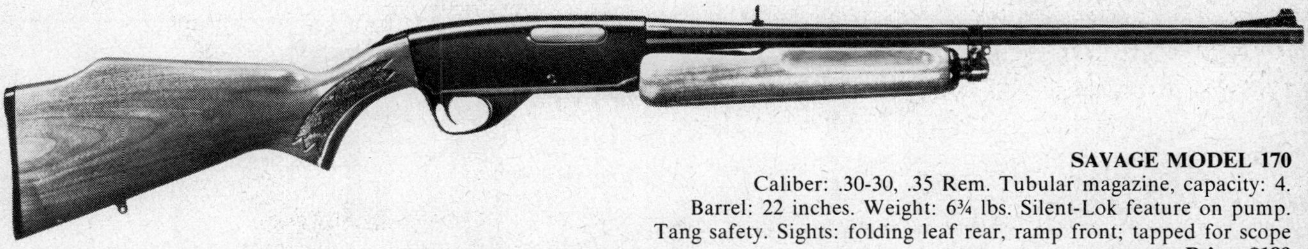

SAVAGE MODEL 170
Caliber: .30-30, .35 Rem. Tubular magazine, capacity: 4. Barrel: 22 inches. Weight: 6¾ lbs. Silent-Lok feature on pump. Tang safety. Sights: folding leaf rear, ramp front; tapped for scope mounts. **Price: $189**

Semiautomatic Centerfire

ARMALITE AR-180
Caliber: .223. Detachable magazine, capacity: 5, 20, 40. Barrel: 18¼ inches. Weight: 7 lbs. All steel receiver and parts Phenolic-fiberglass stock, forend, and grip. Sights: adjustable aperture rear, blade front. **Price: $325; $444** (with Armalite 3X scope)

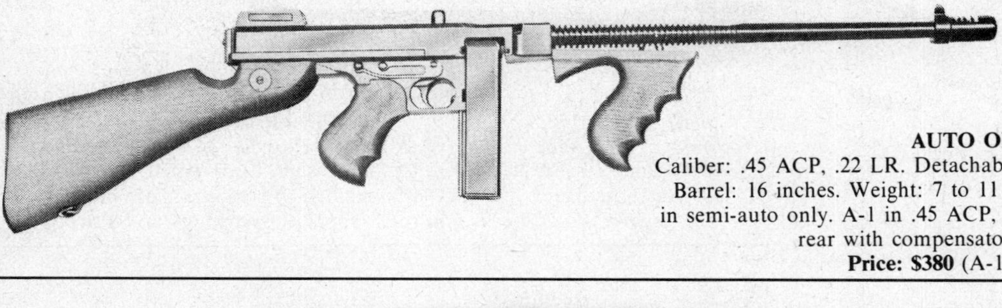

AUTO ORDNANCE MODEL 27 A-1, A-3
Caliber: .45 ACP, .22 LR. Detachable magazine, capacity: 10, 30, 50. Barrel: 16 inches. Weight: 7 to 11 lbs. Replica of Thompson SMG, in semi-auto only. A-1 in .45 ACP, A-3 in .22 LR. Sights: adjustable rear with compensator. Distributed by Numrich Arms. **Price: $380** (A-1); **$420** (A-1 Deluxe); **$420** (A-3)

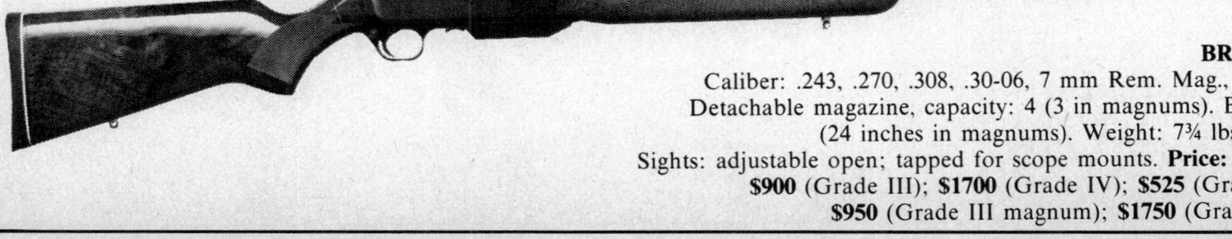

BROWNING BAR
Caliber: .243, .270, .308, .30-06, 7 mm Rem. Mag., .300 Win. Mag. Detachable magazine, capacity: 4 (3 in magnums). Barrel: 22 inches (24 inches in magnums). Weight: 7¾ lbs. Gas operated. Sights: adjustable open; tapped for scope mounts. **Price: $475** (Grade I); **$900** (Grade III); **$1700** (Grade IV); **$525** (Grade I magnum); **$950** (Grade III magnum); **$1750** (Grade IV magnum)

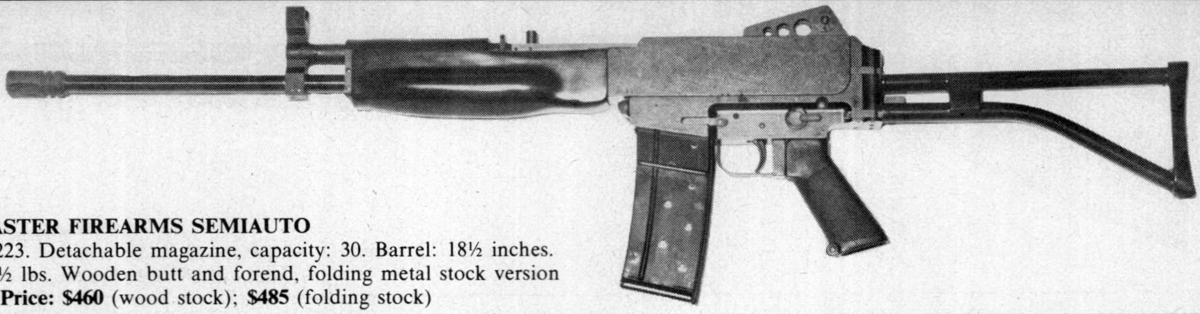

BUSHMASTER FIREARMS SEMIAUTO
Caliber: .223. Detachable magazine, capacity: 30. Barrel: 18½ inches. Weight: 8½ lbs. Wooden butt and forend, folding metal stock version available. **Price: $460** (wood stock); **$485** (folding stock)

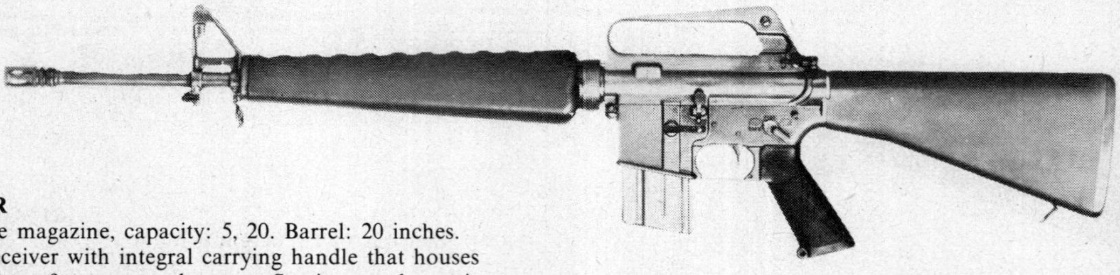

COLT AR-15 SPORTER
Caliber: .223. Detachable magazine, capacity: 5, 20. Barrel: 20 inches. Weight: 7¼ lbs. Alloy receiver with integral carrying handle that houses rear sight and serves as base for scope and mount. Semiauto-only version of the M-16. **Price: $377**; **$466** (with collapsable stock)

DEMRO TAC-1 CARBINE
Caliber: 9 mm, .45 ACP. Detachable magazine, capacity: 30, 32, 50. Barrel: 17 inches. Weight: 8 lbs. Delayed blowback semi-auto. Combination lock safety, thumb safety, and grip safety. Removable stock with built-in magazine loader. Sights: adjustable rear, blade front. **Price: $413**

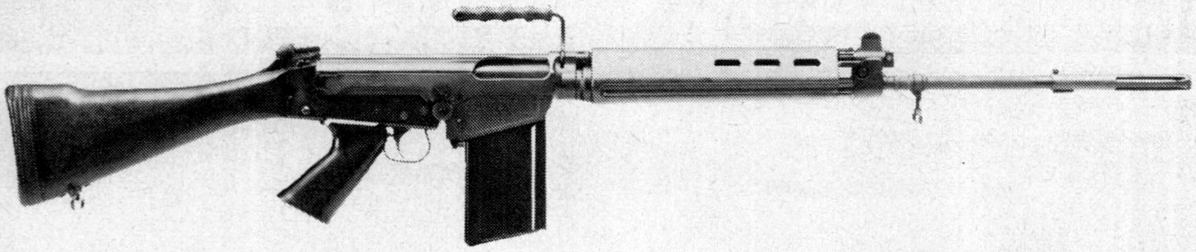

F.N. F.A.L. MATCH SEMI-AUTO
Caliber: .308. Detachable magazine, capacity: 10, 20. Barrel: 21 inches. Weight: 9½ lbs. Gas operated assault rifle. Flash hider optional. Sights: adjustable rear. Imported by Steyr-Daimler-Puch. **Price: $2000**

HECKLER & KOCH 770 SPORTER
Caliber: .243, .308. Magazine capacity: 4. Barrel: 19¾ inches. Weight: 7½ lbs. Delayed roller lock system. Choice of hammer forged, polygonal, or standard rifling. Walnut stock with Monte Carlo cheekpiece. Sights: adjustable folding rear; tapped for scope mounts. **Price: $405**

HECKLER & KOCH MODEL 93 RIFLE
Caliber: .223. Detachable magazine, capacity: 5, 20, 40. Barrel: 16 inches. Weight: 8 to 8½ lbs. Delayed roller lock system. Model A-2 with phenolic stock, Model A-3 with retractable metal stock. Sling included. **Price: $425** (A-2); **$495** (A-3)

HECKLER & KOCH MODEL 91 RIFLE
Caliber: .308. Detachable magazine, capacity: 5, 10, 20.
Barrel: 19 inches. Weight: 10 to 11 lbs. Delayed roller lock system.
Model A-2 with phenolic stock, Model A-3 with retractable metal stock.
Semi-auto version of G-3 rifle. **Price: $428** (A-2); **$515** (A-3)

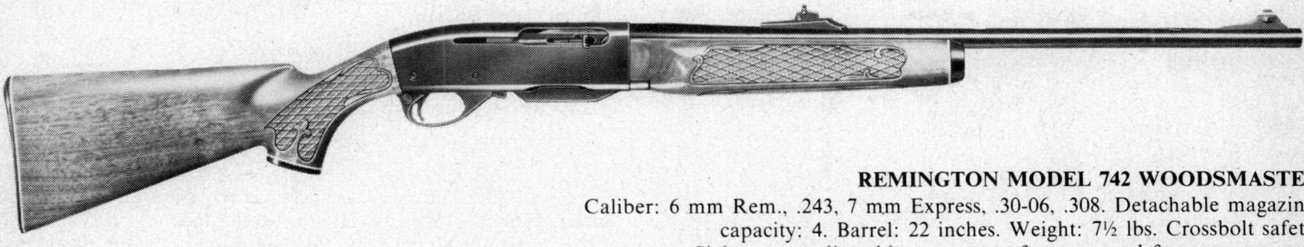

REMINGTON MODEL 742 WOODSMASTER
Caliber: 6 mm Rem., .243, 7 mm Express, .30-06, .308. Detachable magazine, capacity: 4. Barrel: 22 inches. Weight: 7½ lbs. Crossbolt safety. Sights: step adjustable rear, ramp front; tapped for scope mounts. **Price: $305**

REMINGTON MODEL 742BDL
Caliber: .30-06, .308. Same as M742, except basket weave checkering, Monte Carlo stock with cheekpiece, stepped receiver and white line spacers. Left hand version available at no extra charge. **Price: $330**

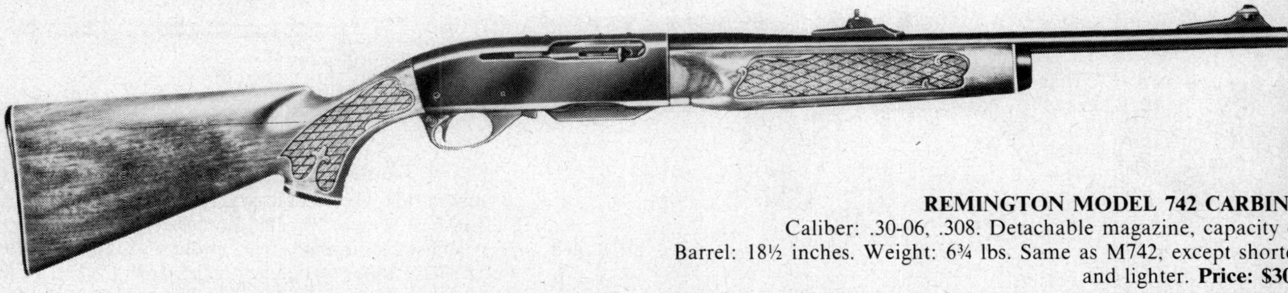

REMINGTON MODEL 742 CARBINE
Caliber: .30-06, .308. Detachable magazine, capacity 4. Barrel: 18½ inches. Weight: 6¾ lbs. Same as M742, except shorter and lighter. **Price: $305**

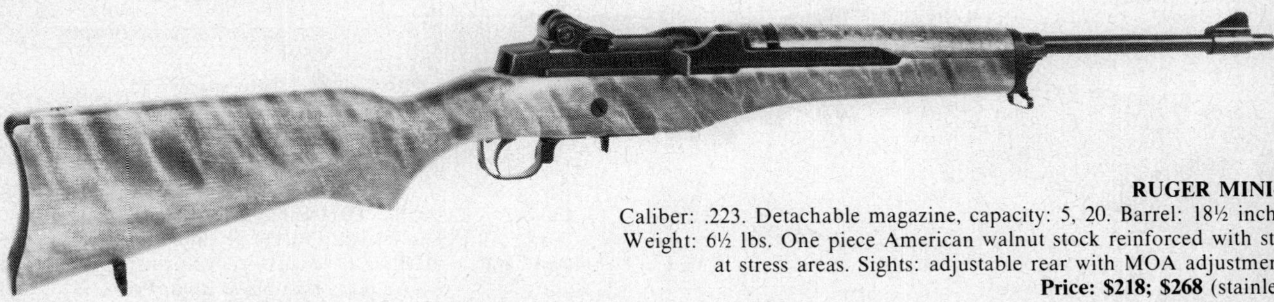

RUGER MINI-14
Caliber: .223. Detachable magazine, capacity: 5, 20. Barrel: 18½ inches. Weight: 6½ lbs. One piece American walnut stock reinforced with steel at stress areas. Sights: adjustable rear with MOA adjustments. **Price: $218; $268** (stainless)

RUGER MODEL 44 CARBINE
Caliber: .44 Mag. Tubular magazine, capacity: 4. Barrel: 18½ inches. Weight: 5¾ lbs. Two-stage trigger pull. Sliding cross-button safety. American walnut stock. Sights: folding leaf rear, gold bead front. **Price: $198**

SIG-AMT

Caliber: .308. Detachable magazine, capacity: 5, 10, 20. Barrel: 18¾ inches. Weight: 9½ lbs. Sights: aperture adjustable rear sight graduated for every 109 yards, adjustable front sight. Bipod, cloth sling, 4X scope and mount are optional accessories. Imported by Mandall Shooting Supplies. **Price: $1500**

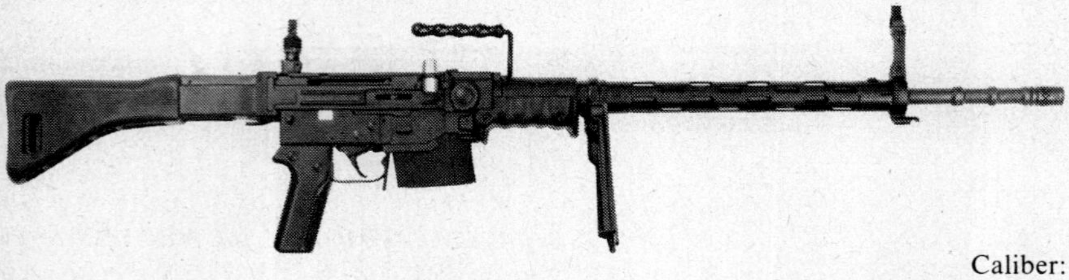

SIG STG-57

Caliber: 7.5 Swiss. Detachable magazine, capacity: 6, 24. Barrel: 23 inches. Weight: 12¼ lbs. Roller lock action. Ventilated barrel sleeve. Straight line stock with bipod. Sights: graduated peep rear. Imported by Mandall Shooting Supplies. **Price: $1700**

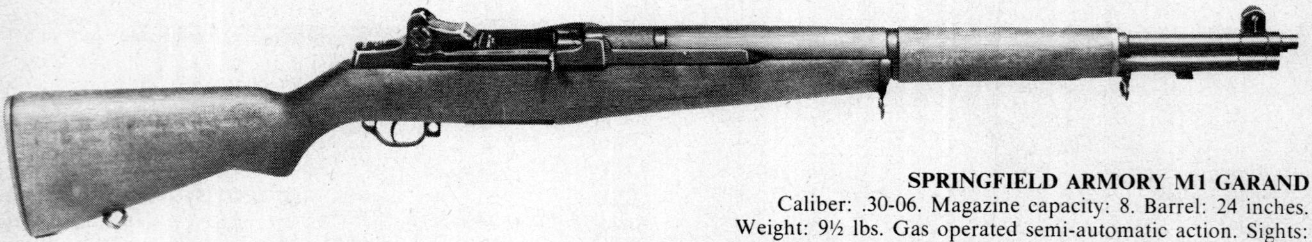

SPRINGFIELD ARMORY M1 GARAND

Caliber: .30-06. Magazine capacity: 8. Barrel: 24 inches. Weight: 9½ lbs. Gas operated semi-automatic action. Sights: adjustable full click aperture rear, military square blade front. **Price: $420; $564** (National Match Grade)

SPRINGFIELD ARMORY M1A RIFLE

Caliber: .308. Detachable magazine, capacity: 5, 10, 20. Barrel: 22 inches. Weight: 8¾ lbs. Rotating bolt, gas operated, air-cooled. Sights: adjustable aperture rear, military square post front. **Price: $420; $580** (National Match); **$690** (Super Match)

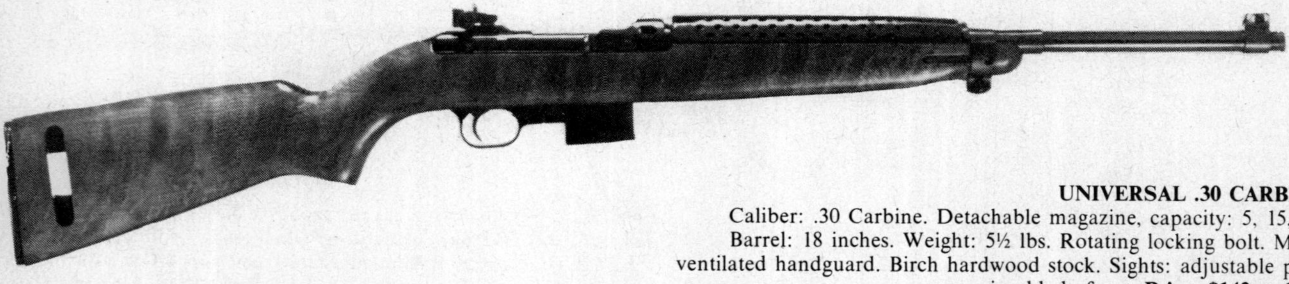

UNIVERSAL .30 CARBINE

Caliber: .30 Carbine. Detachable magazine, capacity: 5, 15, 30. Barrel: 18 inches. Weight: 5½ lbs. Rotating locking bolt. Metal ventilated handguard. Birch hardwood stock. Sights: adjustable peep rear, wing blade front. **Price: $143** to **$214**

VALMET M71S ASSAULT RIFLE

Caliber: .223. Detachable magazine, capacity: 15, 30. Barrel: 16⅝ inches. Weight: 8¾ lbs. Wooden stock. Sights: adjustable open rear. Imported by Interarms. **Price: $695**

Bolt Action Centerfire

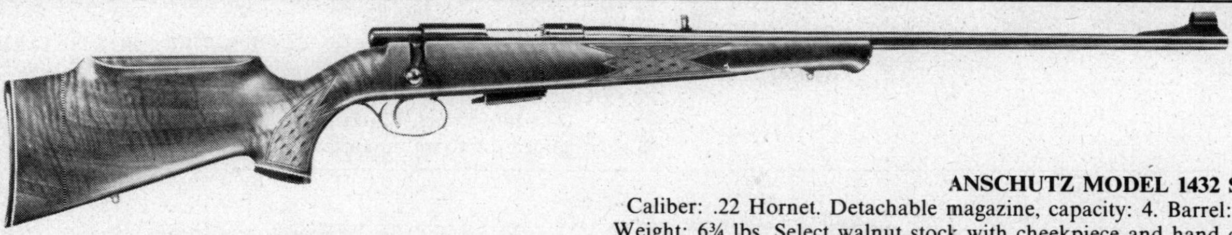

ANSCHUTZ MODEL 1432 SPORTER
Caliber: .22 Hornet. Detachable magazine, capacity: 4. Barrel: 24 inches. Weight: 6¾ lbs. Select walnut stock with cheekpiece and hand checkering. Sights: folding leaf rear, hooded front; tapped for scope mounts. **Price: $633**

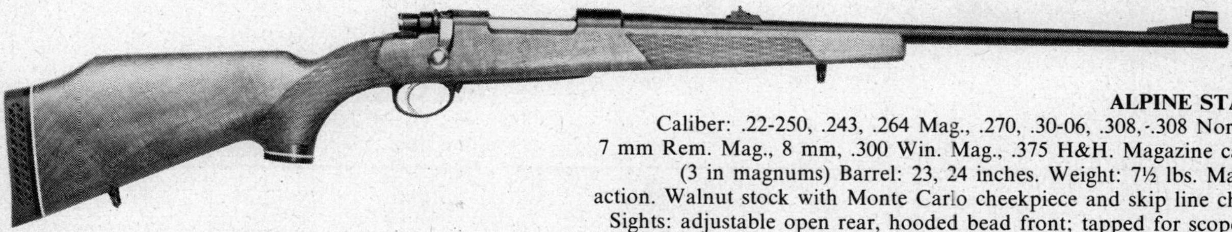

ALPINE STANDARD
Caliber: .22-250, .243, .264 Mag., .270, .30-06, .308, .308 Norma Mag., 7 mm Rem. Mag., 8 mm, .300 Win. Mag., .375 H&H. Magazine capacity: 5 (3 in magnums) Barrel: 23, 24 inches. Weight: 7½ lbs. Mauser type action. Walnut stock with Monte Carlo cheekpiece and skip line checkering. Sights: adjustable open rear, hooded bead front; tapped for scope mounts. Imported by Mandall Shooting Supplies. **Price: $300**

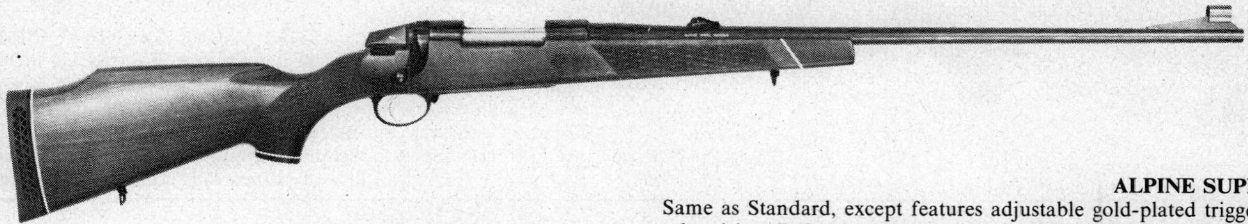

ALPINE SUPREME
Same as Standard, except features adjustable gold-plated trigger, and hand jewelled bolt. Rosewood semi-beavertail forend and grip cap. Imported by Mandall Shooting Supplies. **Price: $400**

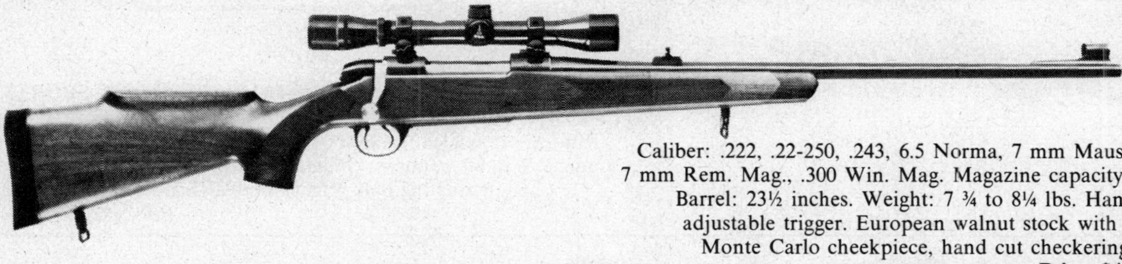

BSA CF-2
Caliber: .222, .22-250, .243, 6.5 Norma, 7 mm Mauser, .270, .308, .30-06 7 mm Rem. Mag., .300 Win. Mag. Magazine capacity: 4 (3 in magnums). Barrel: 23½ inches. Weight: 7 ¾ to 8¼ lbs. Hand finished chamber, adjustable trigger. European walnut stock with forend cap, rollover Monte Carlo cheekpiece, hand cut checkering. Sights: tapped for scope mounts. **Price: $445; $460** (magnums)

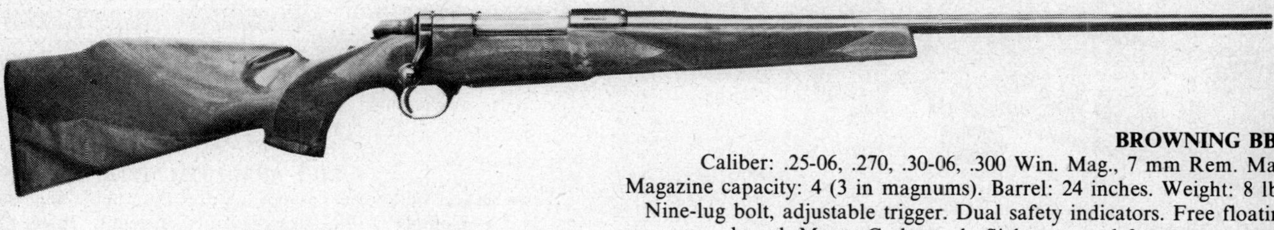

BROWNING BBR
Caliber: .25-06, .270, .30-06, .300 Win. Mag., 7 mm Rem. Mag. Magazine capacity: 4 (3 in magnums). Barrel: 24 inches. Weight: 8 lbs. Nine-lug bolt, adjustable trigger. Dual safety indicators. Free floating barrel. Monte Carlo stock. Sights: tapped for scope mounts. **Price: $400**

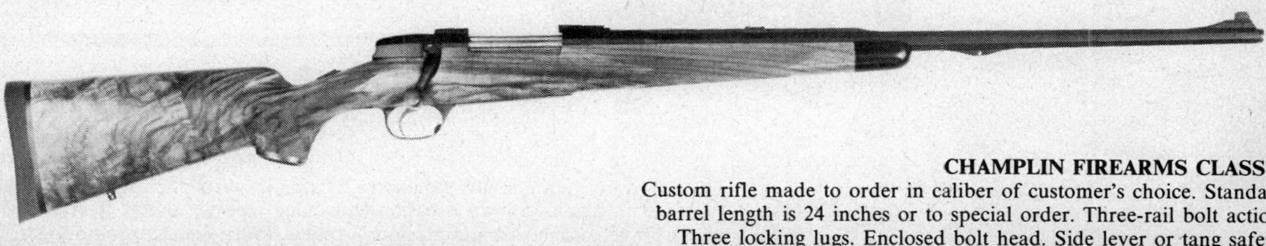

CHAMPLIN FIREARMS CLASSIC
Custom rifle made to order in caliber of customer's choice. Standard barrel length is 24 inches or to special order. Three-rail bolt action. Three locking lugs. Enclosed bolt head. Side lever or tang safety. Adjustable trigger. **Price: $3200** and up

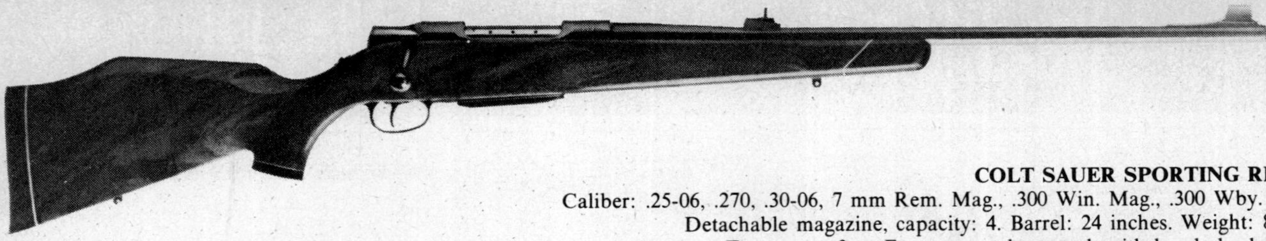

COLT SAUER SPORTING RIFLE
Caliber: .25-06, .270, .30-06, 7 mm Rem. Mag., .300 Win. Mag., .300 Wby. Mag Detachable magazine, capacity: 4. Barrel: 24 inches. Weight: 8 lbs. Top tang safety. European walnut stock with hand checkering. Sights: tapped for scope mounts: **Price: $890; $920** to **$990** (magnums)

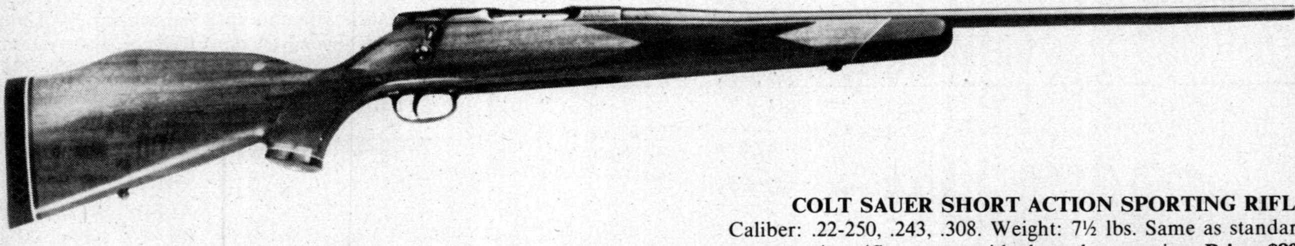

COLT SAUER SHORT ACTION SPORTING RIFLE
Caliber: .22-250, .243, .308. Weight: 7½ lbs. Same as standard sporting rifle, except with short throw action. **Price: $890**

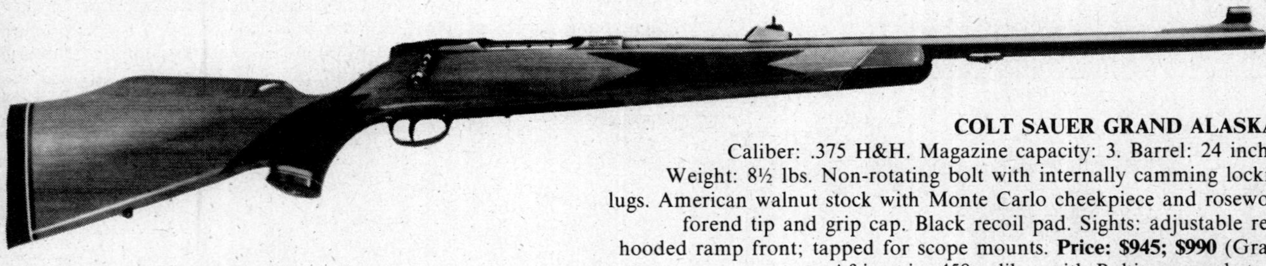

COLT SAUER GRAND ALASKAN
Caliber: .375 H&H. Magazine capacity: 3. Barrel: 24 inches. Weight: 8½ lbs. Non-rotating bolt with internally camming locking lugs. American walnut stock with Monte Carlo cheekpiece and rosewood forend tip and grip cap. Black recoil pad. Sights: adjustable rear, hooded ramp front; tapped for scope mounts. **Price: $945; $990** (Grand African in .458 caliber with Bubinga wood stock)

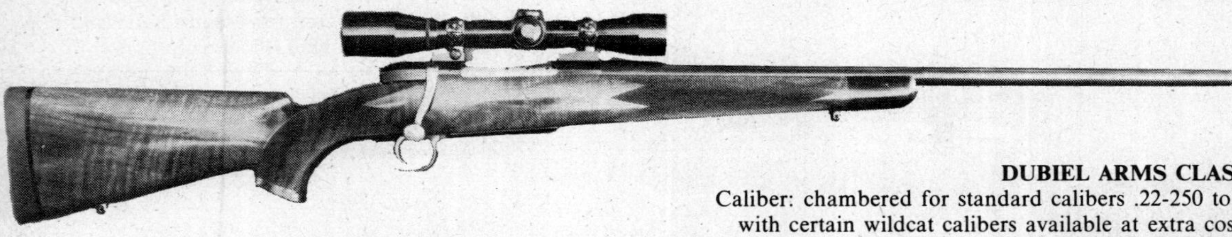

DUBIEL ARMS CLASSIC SPORTER
Caliber: chambered for standard calibers .22-250 to .458 Win. Mag. with certain wildcat calibers available at extra cost. Barrel: length optional. Weight: optional. Douglas Premium barrel, Canjar trigger, short rotating bolt. Right or left handed actions available. **Price: $2250** and up

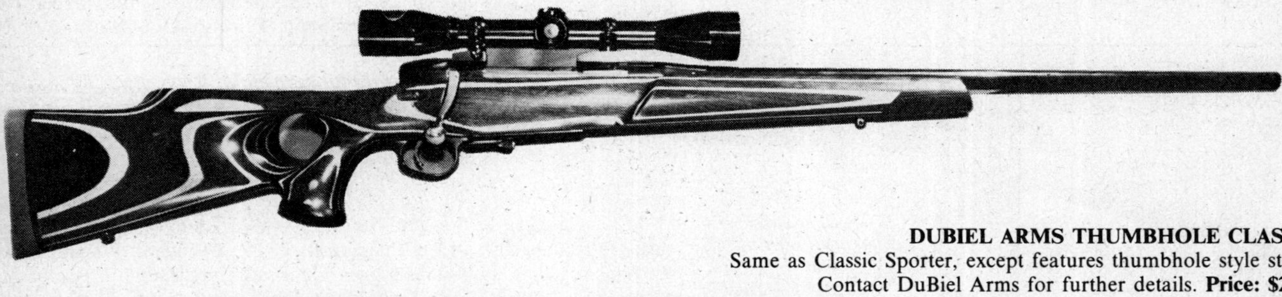

DUBIEL ARMS THUMBHOLE CLASSIC
Same as Classic Sporter, except features thumbhole style stock. Contact DuBiel Arms for further details. **Price: $2250**

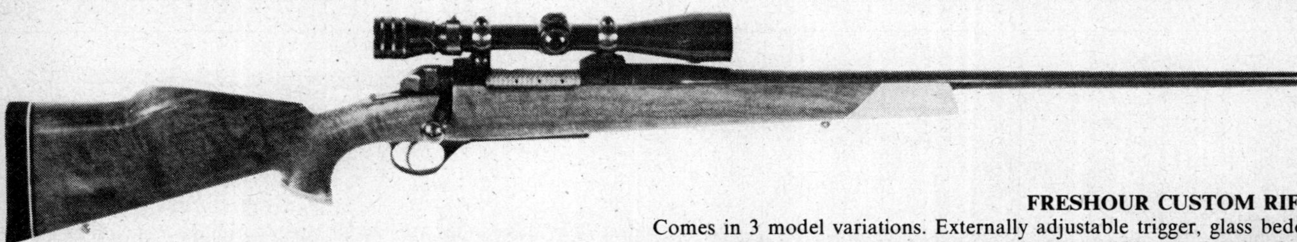

FRESHOUR CUSTOM RIFLE
Comes in 3 model variations. Externally adjustable trigger, glass bedded actions. Shown is basic righthanded repeater. Rifles made to order in most standard and magnum calibers. **Price:** contact Freshour Mfg. Co., 1914 5th Ave. North, Texas City, TX 77590

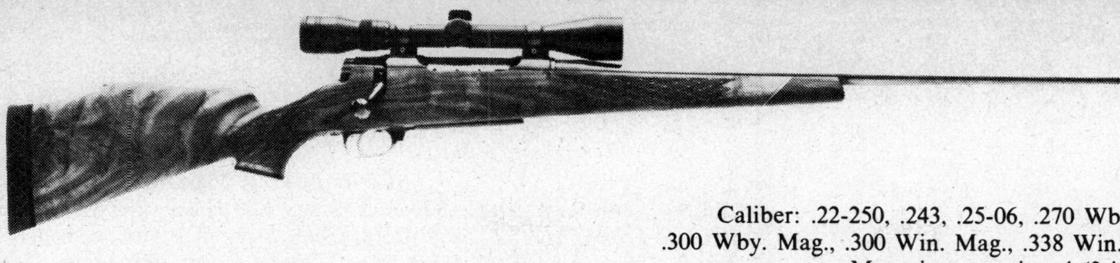

GOLDEN EAGLE MODEL 7000
Caliber: .22-250, .243, .25-06, .270 Wby. Mag., 7 mm Rem. Mag., .30-06, .300 Wby. Mag., .300 Win. Mag., .338 Win. Mag., .375 H&H, .458 Win. Mag. Magazine capacity: 4 (3 in magnums). Barrel: 24, 26 inches. Weight: 7¾ lbs. Walnut stock with checkered pistol grip and forearm. Rosewood forend and grip cap. **Price: $599** to **$749**

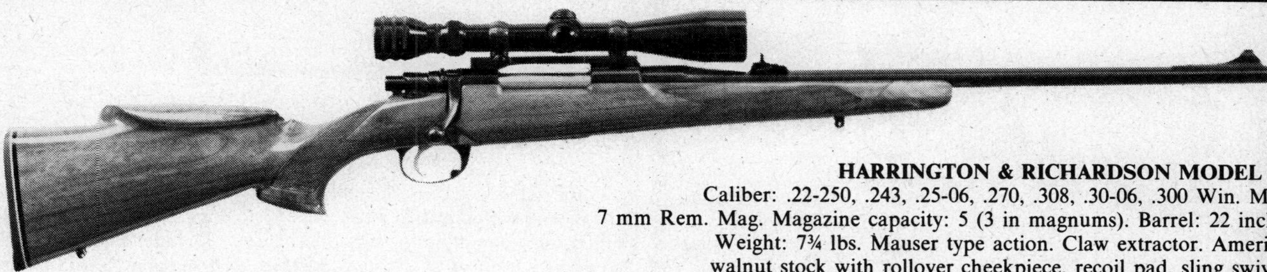

HARRINGTON & RICHARDSON MODEL 300
Caliber: .22-250, .243, .25-06, .270, .308, .30-06, .300 Win. Mag., 7 mm Rem. Mag. Magazine capacity: 5 (3 in magnums). Barrel: 22 inches. Weight: 7¾ lbs. Mauser type action. Claw extractor. American walnut stock with rollover cheekpiece, recoil pad, sling swivels. Sights: adjustable rear; tapped for scope mounts. **Price: $375**

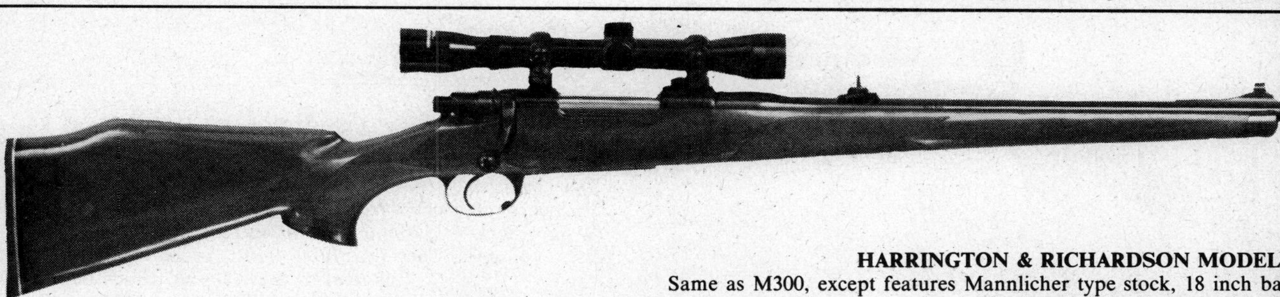

HARRINGTON & RICHARDSON MODEL 301
Same as M300, except features Mannlicher type stock, 18 inch barrel. Not available in .22-250. **Price: $450**

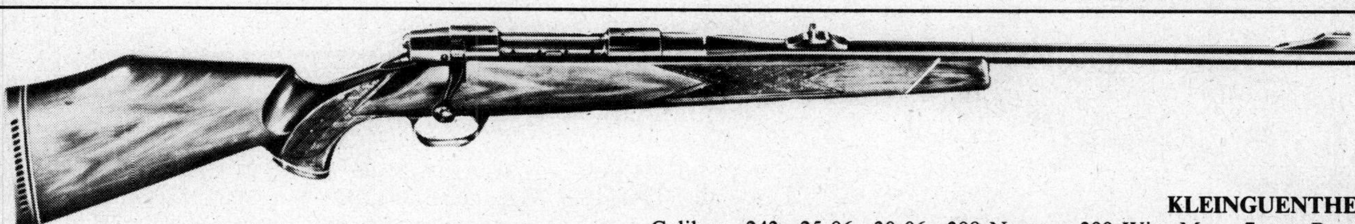

KLEINGUENTHER K-15
Caliber: .243, .25-06, .30-06, .308 Norma, .300 Win. Mag., 7 mm Rem. Mag., .375 H&H, 7x57. Magazine capacity: 4 (3 in magnums). Barrel: 24, 26 inches. Weight: 7⅛ lbs. Short lock time, adjustable trigger. Safety on center of bolt sleeve. Sights: tapped for scope mounts. **Price: NA**

MANNLICHER MODEL M
Caliber: 6.5x57, 6.5x55, 7x64, 7x57, 8x57JS, 9.3x62, 7.5x55, .270 Win., .30-06. Detachable rotary magazine, capacity: 5. Barrel: 20 inches (fullstock); 23½, 25½ (halfstock). Weight: 6¾ lbs. Six rear locking lugs. Choice of single or double set triggers. Monte Carlo cheekpiece, rubber recoil pad. Sights: tapped for scope mounts. Imported by Steyr-Daimler Puch of America. **Price: $703** (fullstock): **$760** (fullstock, left hand): **$650** (halfstock): **$707** (halfstock, left hand)

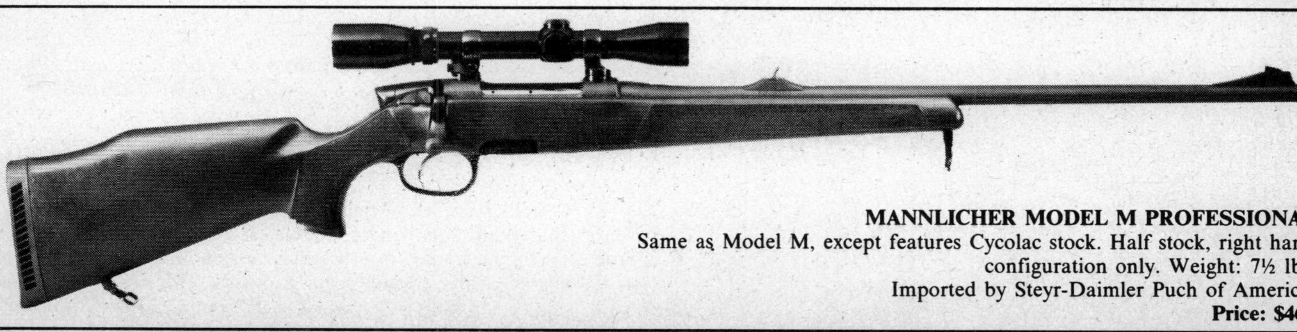

MANNLICHER MODEL M PROFESSIONAL
Same as Model M, except features Cycolac stock. Half stock, right hand configuration only. Weight: 7½ lbs. Imported by Steyr-Daimler Puch of America. **Price: $467**

MANNLICHER MODEL S MAGNUM

Caliber: 7 mm Rem. Mag., .300 Win. Mag., .300 H&H, .338 Win. Mag., .375 H&H, 6.5x68, 8x68S, 9.3x64. Detachable rotary magazine, capacity: 4. Barrel: 25½ inches. Weight: 8½ lbs. Halfstock design with Monte Carlo cheekpiece and hand checkering. Recoil pad. Sights: tapped for scope mounts. Imported by Steyr-Daimler Puch of America. **Price: $756**

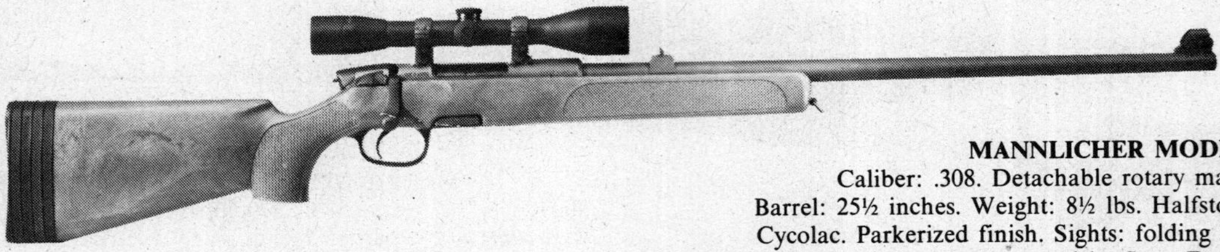

MANNLICHER MODEL SSG MARKSMAN

Caliber: .308. Detachable rotary magazine, capacity: 5, 10. Barrel: 25½ inches. Weight: 8½ lbs. Halfstock design in walnut or Cycolac. Parkerized finish. Sights: folding leaf rear, hooded blade front; tapped for scope mounts. Imported by Steyr-Daimler Puch of America. **Price: $558** (Cycolac); **$613** (walnut); **$1225** (with mounted Kahles scope as shown)

MANNLICHER MODEL ML 79

Caliber: 7x57, 7x64, .270, .30-06, 6.5x57, 6.5x55, 7.5x55. Detachable magazine, capacity: 3, 6. Barrel: 20 inches (fullstock), 23½ inches (halfstock). Weight: 6¾ lbs. Single combination trigger. Walnut stock with Monte Carlo cheekpiece and hand checkering. Sights: adjustable open rear, ramp front; tapped for scope mounts. Imported by Steyr-Daimler Puch of America. **Price: $827** (fullstock); **$774** (halfstock)

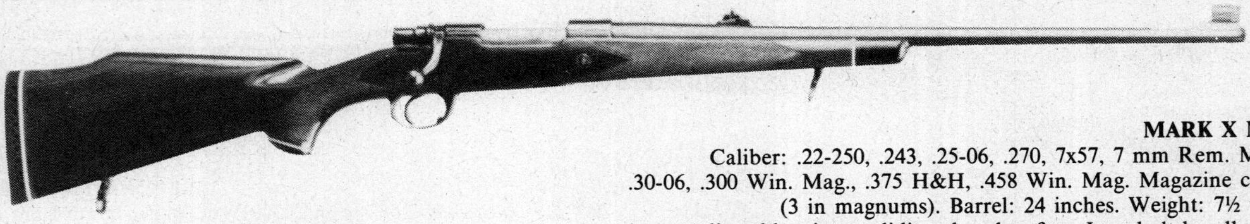

MARK X MAUSER

Caliber: .22-250, .243, .25-06, .270, 7x57, 7 mm Rem. Mag., .308, .30-06, .300 Win. Mag., .375 H&H, .458 Win. Mag. Magazine capacity: 5 (3 in magnums). Barrel: 24 inches. Weight: 7½ to 8¼ lbs. Adjustable trigger, sliding thumb safety. Low bolt handle for scope clearance. European walnut stock. Sights: open leaf rear; tapped for scope mounts. Imported by Interarms. **Price: $319; $329** (magnums); **$389** (.375 and .458)

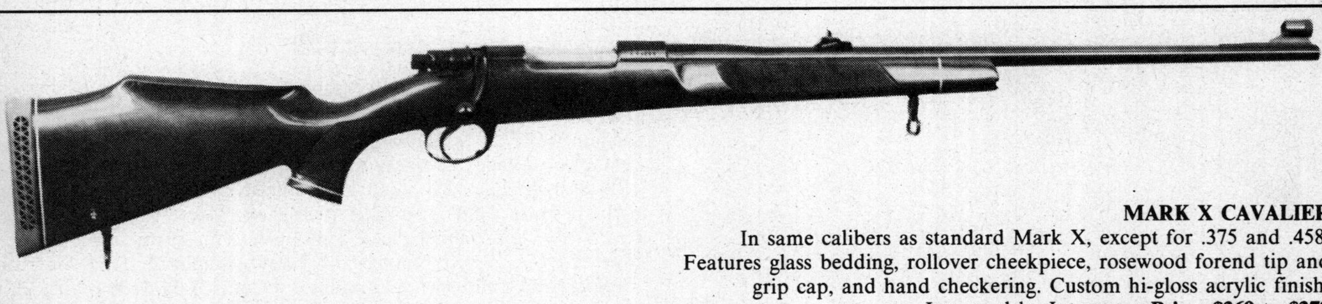

MARK X CAVALIER

In same calibers as standard Mark X, except for .375 and .458. Features glass bedding, rollover cheekpiece, rosewood forend tip and grip cap, and hand checkering. Custom hi-gloss acrylic finish. Imported by Interarms. **Price: $369** to **$379**

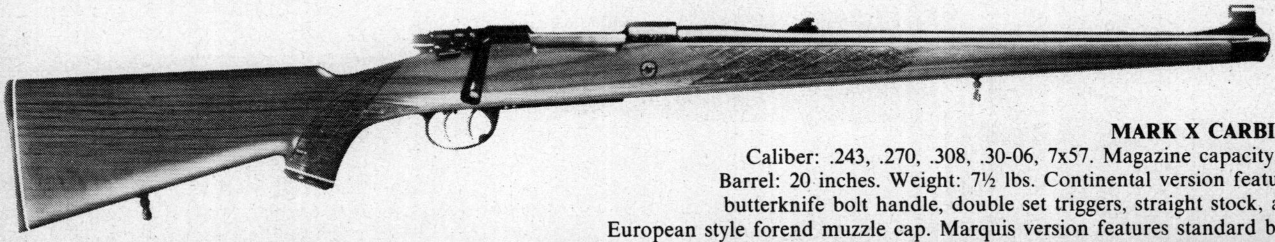

MARK X CARBINE

Caliber: .243, .270, .308, .30-06, 7x57. Magazine capacity: 5. Barrel: 20 inches. Weight: 7½ lbs. Continental version features butterknife bolt handle, double set triggers, straight stock, and European style forend muzzle cap. Marquis version features standard bolt, single adjustable trigger, pistol grip stock. **Price: $369** (Marquis); **$389** (Continental)

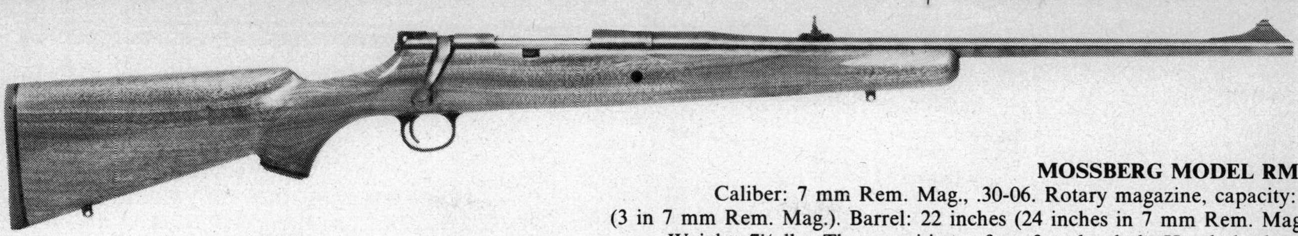

MOSSBERG MODEL RM-7
Caliber: 7 mm Rem. Mag., .30-06. Rotary magazine, capacity: 4 (3 in 7 mm Rem. Mag.). Barrel: 22 inches (24 inches in 7 mm Rem. Mag.) Weight: 7½ lbs. Three position safety, four lug bolt. Hand checkered pistol grip and forend. Sights: folding leaf rear, gold bead front; tapped for scope mounts. **Price: $268**

NAVY ARMS MAUSER
Caliber: .45-70. Magazine capacity: 3. Barrel: 18, 24, 26 inches. Weight: 8 to 8½ lbs. Siamese Mauser action and magazine for use with large, rimmed cartridges. Checkered walnut stock. Carbine (18 inch barrel) has straight grip stock. **Price: $200**

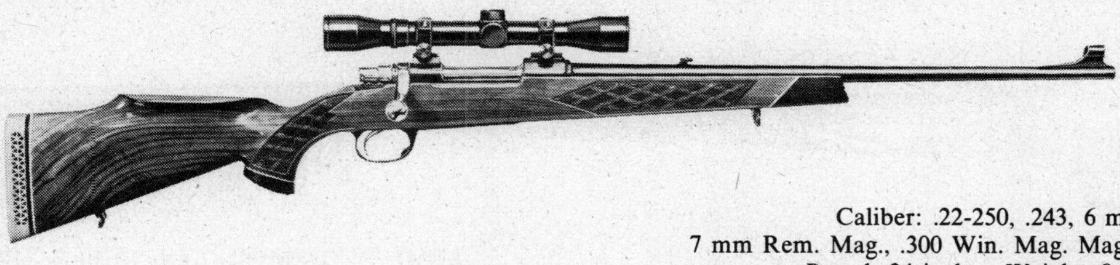

PARKER HALE 1200 SERIES
Caliber: .22-250, .243, 6 mm Rem., .25-06, .270, .30-06, .308, 7 mm Rem. Mag., .300 Win. Mag. Magazine capacity: 4 (3 in magnums). Barrel: 24 inches. Weight: 8 lbs. Adjustable trigger, side safety. Semi-fancy walnut stock. Sights: tapped for scope mounts. Imported by Jana International. **Price: $300; $313** (magnums)

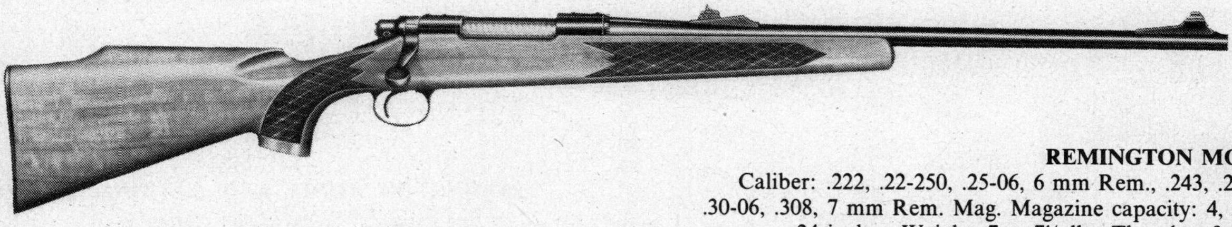

REMINGTON MODEL 700 ADL
Caliber: .222, .22-250, .25-06, 6 mm Rem., .243, .270, 7 mm Exp., .30-06, .308, 7 mm Rem. Mag. Magazine capacity: 4, 5, 6. Barrel: 22, 24 inches. Weight: 7 to 7½ lbs. Thumb safety, jeweled bolt. Recoil pad on magnums. Sights: adjustable rear, blade ramp front. **Price: $298; $313** (magnum)

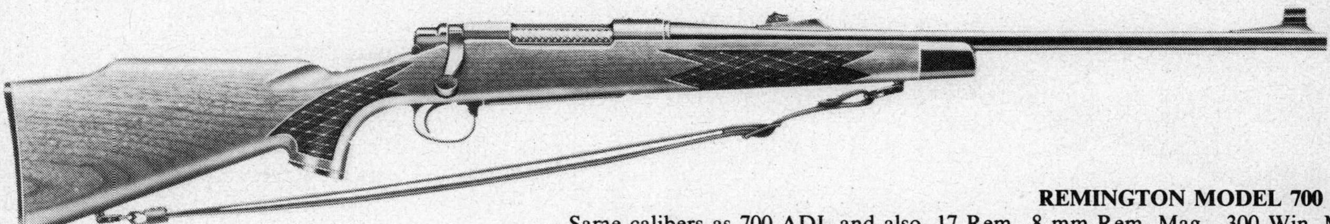

REMINGTON MODEL 700 BDL
Same calibers as 700 ADL and also .17 Rem., 8 mm Rem. Mag., .300 Win. Mag., .375 H&H, .458 Win. Mag. Deluxe version of M700, including sling strap and swivels. Fine line checkering and white line spacers. Left hand version available. **Price: $358** to **$580**

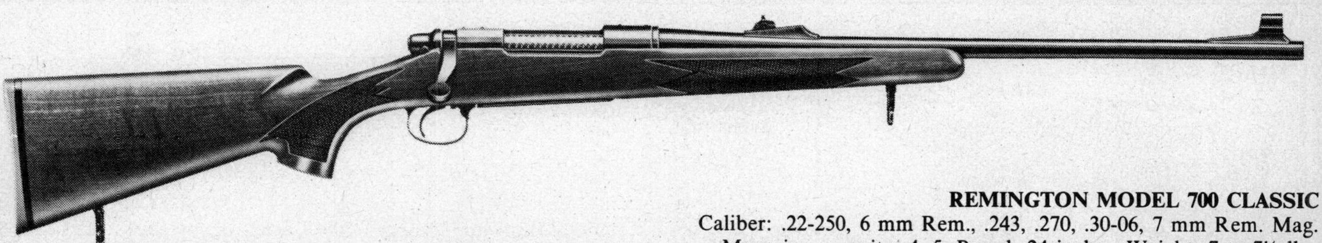

REMINGTON MODEL 700 CLASSIC
Caliber: .22-250, 6 mm Rem., .243, .270, .30-06, 7 mm Rem. Mag. Magazine capacity: 4, 5. Barrel: 24 inches. Weight: 7 to 7½ lbs. Classic styled oil finished stock, round forend tip. Full cut checkering. Sights: tapped for scope mounts. **Price: $328; $343** (magnum)

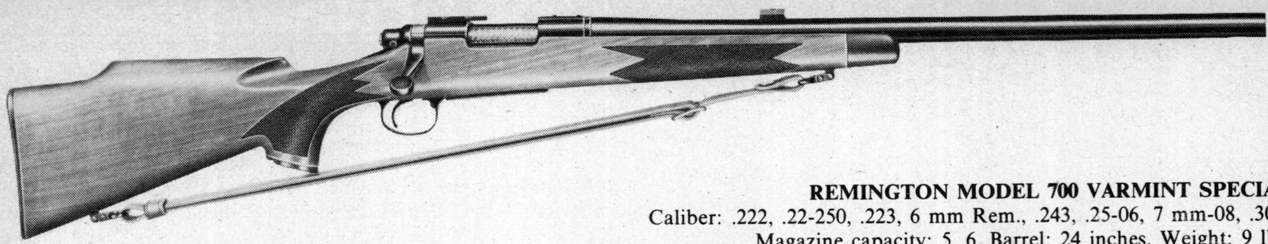

REMINGTON MODEL 700 VARMINT SPECIAL
Caliber: .222, .22-250, .223, 6 mm Rem., .243, .25-06, 7 mm-08, .308. Magazine capacity: 5, 6. Barrel: 24 inches. Weight: 9 lbs. Heavy barrel. Monte Carlo stock with cut checkering, sling and swivels. Sights: front and rear target rifle scope blocks. **Price: $378**

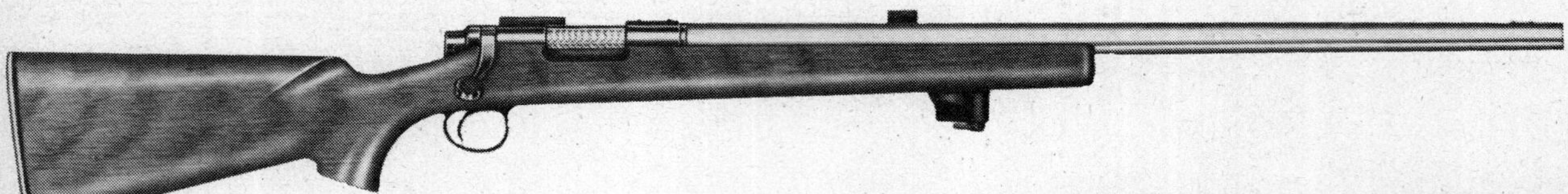

REMINGTON MODEL 40-XB RANGEMASTER
Caliber: .222, .22-250, .243, 6 mm Rem., .25-06, 7 mm Rem. Mag., .308, .30-06, .30-338, .300 Win. Mag. Single-shot. Weight: 9¼ to 11¼ lbs. Match type trigger, free floating stainless steel barrel. Sights: tapped for scope blocks and receiver sights. **Price: $635; $677** (repeater)

REMINGTON MODEL 4C-XB BENCHREST
Caliber: .22 BR Rem., .222, .223, 6x47, 6 mm BR Rem., .308. Single shot. Barrel: 20, 24 inches. Weight: 9¼ to 12 lbs. Match trigger, adjustable from 1½ to 3½ lbs. Stainless steel barrel. **Price: $670**

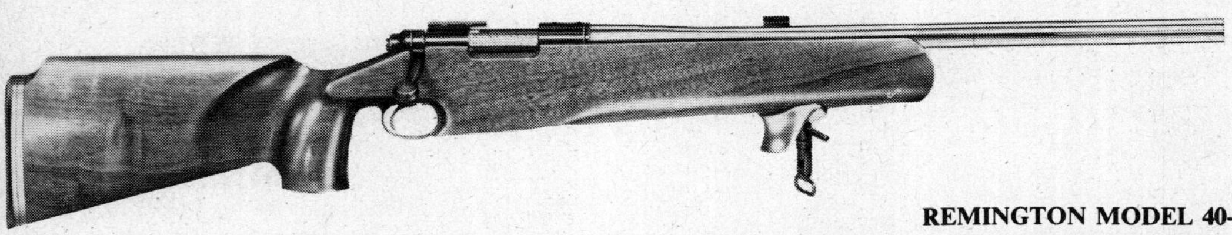

REMINGTON MODEL 40-XC NATIONAL MATCH
Caliber: .308. Magazine capacity: 5. Barrel: 24 inches. Weight: 11 lbs. Trigger adjustable from 2 to 4 lbs. Improved position style stock. Stainless steel barrel. Sights: tapped for metallic sights; scope mount bases installed. **Price: $690**

REMINGTON MODEL 788 VARMINT
Caliber: .223 Rem., .22-250. Detachable magazine, capacity: 3. Barrel: 24 inches. Weight: 7½ lbs. Thumb type safety. Walnut finished stock. Left hand version available. Sights: lock screw adjustable rear, blade ramp front; tapped for scope mounts and receiver sights. **Price: $210**

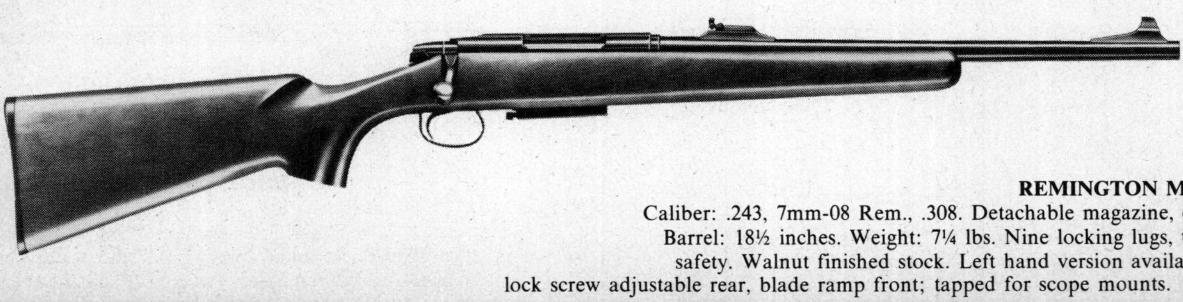

REMINGTON MODEL 788
Caliber: .243, 7mm-08 Rem., .308. Detachable magazine, capacity: 3. Barrel: 18½ inches. Weight: 7¼ lbs. Nine locking lugs, thumb type safety. Walnut finished stock. Left hand version available. Sights: lock screw adjustable rear, blade ramp front; tapped for scope mounts. **Price: $210**

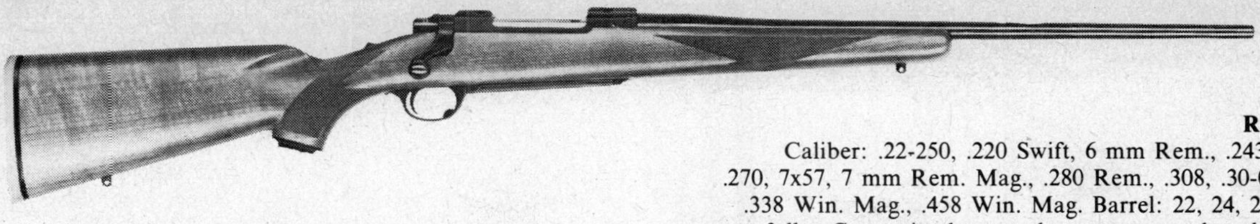

RUGER MODEL 77

Caliber: .22-250, .220 Swift, 6 mm Rem., .243, .257 Rob., .25-06, .270, 7x57, 7 mm Rem. Mag., .280 Rem., .308, .30-06, .300 Win. Mag., .338 Win. Mag., .458 Win. Mag. Barrel: 22, 24, 26 inches. Weight: 7 to 9 lbs. Comes in short stroke or magnum length action. Round top variation comes with iron sights and is tapped for scope mounts. Integral base receiver model comes without iron sights but with bases to accommodate Ruger steel rings. **Price: $297** to **$402**

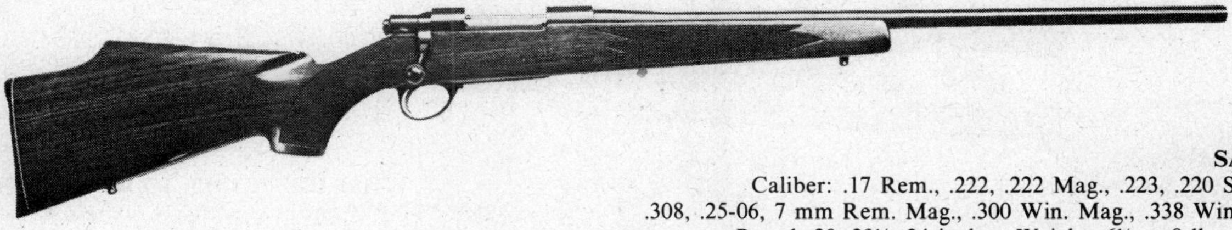

SAKO STANDARD

Caliber: .17 Rem., .222, .222 Mag., .223, .220 Swift, .22-250, .243, .308, .25-06, 7 mm Rem. Mag., .300 Win. Mag., .338 Win. Mag. .375 H&H. Barrel: 20, 23½, 24 inches. Weight: 6½ to 8 lbs. Adjustable trigger. European walnut stock with high gloss finish, fine cut checkering. Imported by Stoeger. **Price: $500** to **$550**

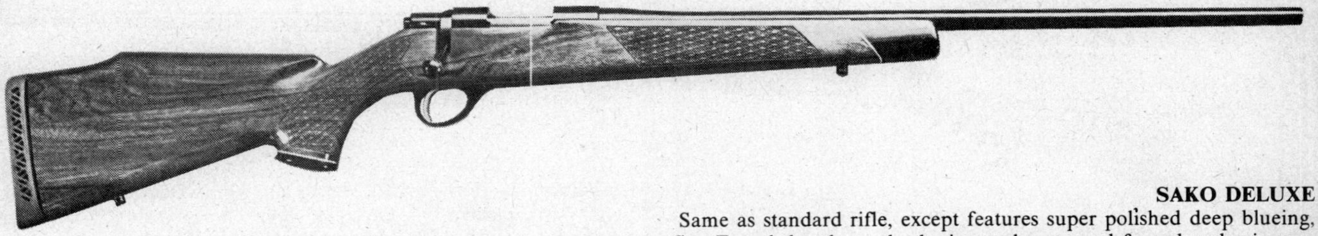

SAKO DELUXE

Same as standard rifle, except features super polished deep blueing, fine French hand cut checkering and rosewood forend and grip cap. Imported by Stoeger. **Price: $717** to **$770**

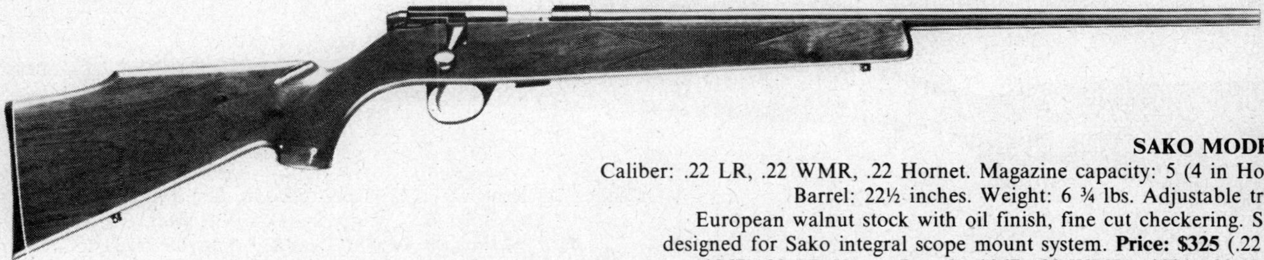

SAKO MODEL 78

Caliber: .22 LR, .22 WMR, .22 Hornet. Magazine capacity: 5 (4 in Hornet). Barrel: 22½ inches. Weight: 6 ¾ lbs. Adjustable trigger. European walnut stock with oil finish, fine cut checkering. Sights: designed for Sako integral scope mount system. **Price: $325** (.22 LR); **$367** (.22 LR Heavy Barrel); **$367** (.22 WMR); **$391** (.22 Hornet)

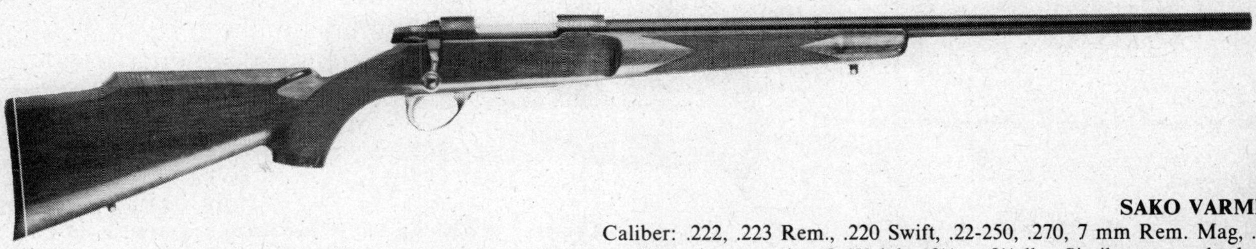

SAKO VARMINT

Caliber: .222, .223 Rem., .220 Swift, .22-250, .270, 7 mm Rem. Mag, .308. Magazine capacity: 5. Weight: 8¼ to 8½ lbs. Similar to standard rifle. **Price: $585**

SAVAGE MODEL 110S SILHOUETTE RIFLE

Caliber: .308. Magazine capacity: 4. Barrel: 22 inches. Weight: 8 ¾ lbs. Heavy tapered, free floating barrel. Wundhammer swell pistol grip. Stippled checkering on grip and forend. **Price: $275**

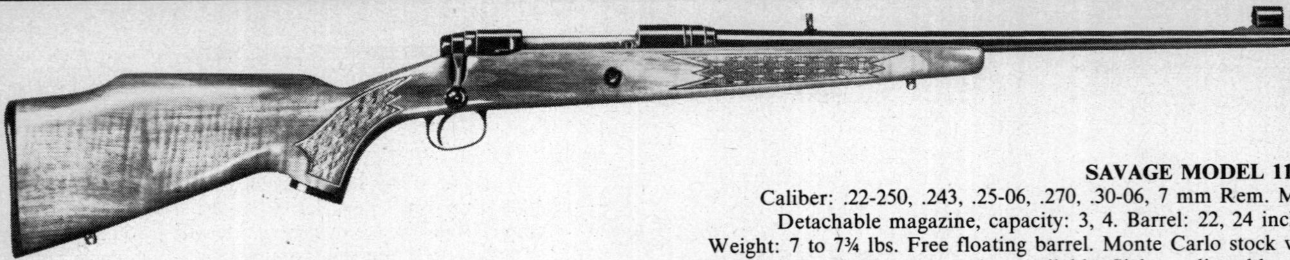

SAVAGE MODEL 110-C
Caliber: .22-250, .243, .25-06, .270, .30-06, 7 mm Rem. Mag. Detachable magazine, capacity: 3, 4. Barrel: 22, 24 inches. Weight: 7 to 7¾ lbs. Free floating barrel. Monte Carlo stock with cheekpiece. Left hand version available. Sights: adjustable rear, hooded ramp front. **Price: $273; $290** (magnum)

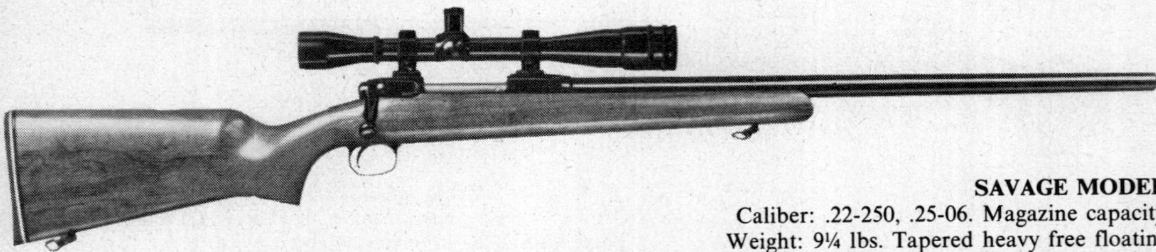

SAVAGE MODEL 112-R VARMINTER
Caliber: .22-250, .25-06. Magazine capacity: 4. Barrel: 26 inches. Weight: 9¼ lbs. Tapered heavy free floating barrel. Select walnut stock with high fluted comb and Wundhammer swell pistol grip. Sights: blocks included for target scopes. **Price: $299**

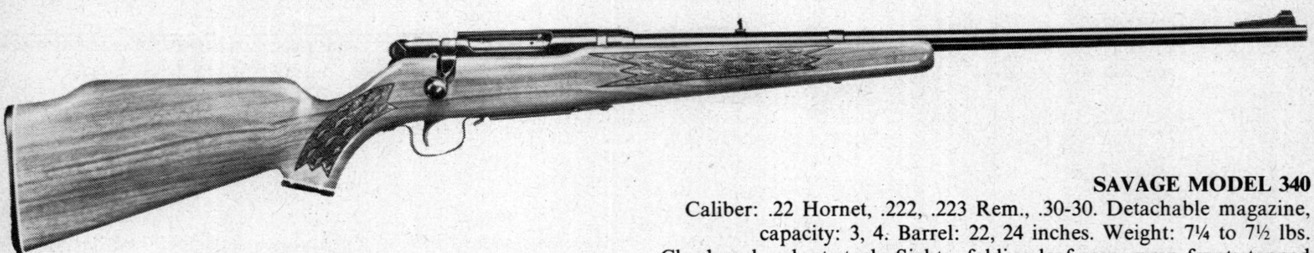

SAVAGE MODEL 340
Caliber: .22 Hornet, .222, .223 Rem., .30-30. Detachable magazine, capacity: 3, 4. Barrel: 22, 24 inches. Weight: 7¼ to 7½ lbs. Checkered walnut stock. Sights: folding leaf rear, ramp front; tapped for scope mounts. **Price: $177**

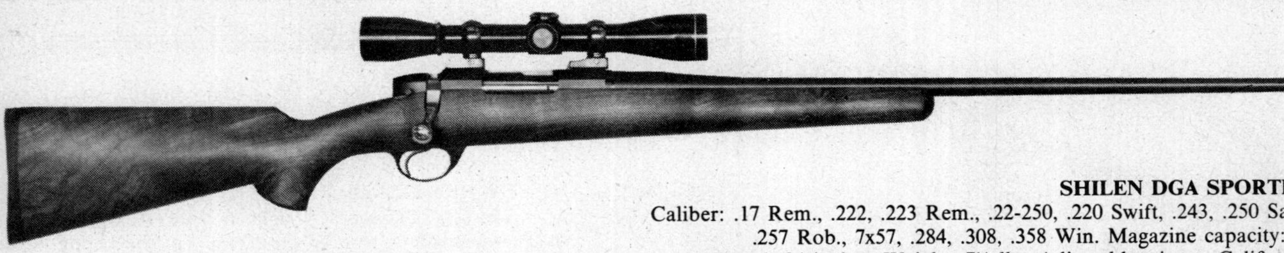

SHILEN DGA SPORTER
Caliber: .17 Rem., .222, .223 Rem., .22-250, .220 Swift, .243, .250 Sav., .257 Rob., 7x57, .284, .308, .358 Win. Magazine capacity: 3. Barrel: 24 inches. Weight: 7½ lbs. Adjustable trigger. California Claro walnut stock. Sights: tapped for scope mounts. **Price: $695**

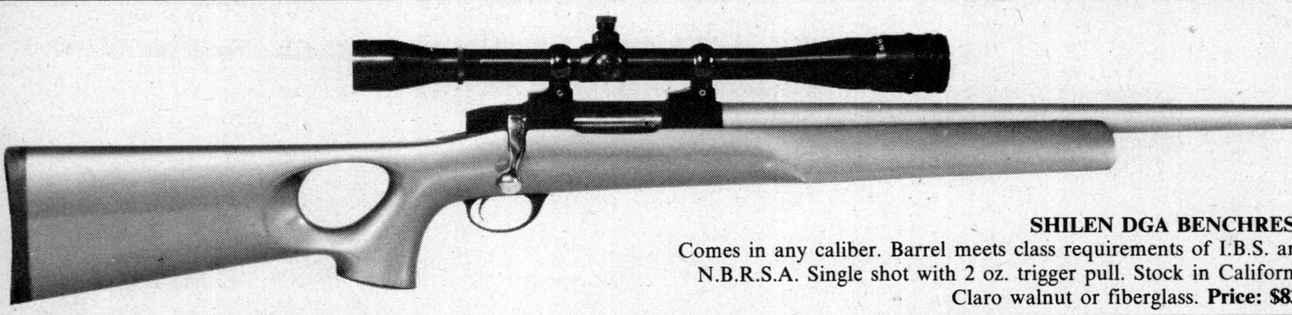

SHILEN DGA BENCHREST
Comes in any caliber. Barrel meets class requirements of I.B.S. and N.B.R.S.A. Single shot with 2 oz. trigger pull. Stock in California Claro walnut or fiberglass. **Price: $835**

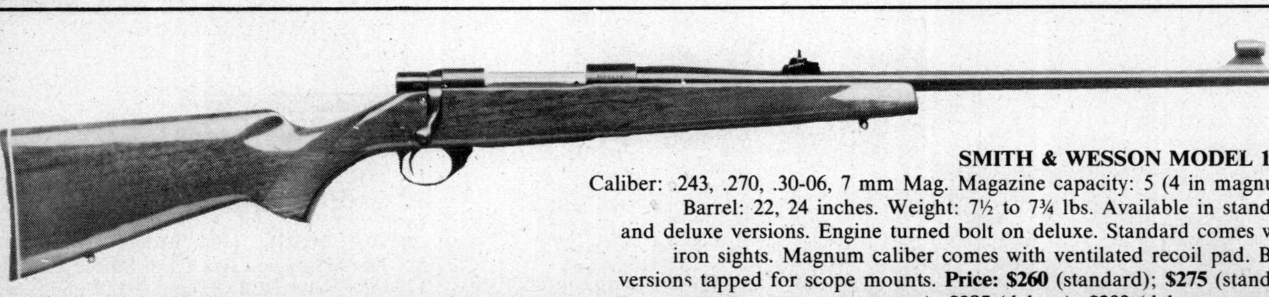

SMITH & WESSON MODEL 1500
Caliber: .243, .270, .30-06, 7 mm Mag. Magazine capacity: 5 (4 in magnum) Barrel: 22, 24 inches. Weight: 7½ to 7¾ lbs. Available in standard and deluxe versions. Engine turned bolt on deluxe. Standard comes with iron sights. Magnum caliber comes with ventilated recoil pad. Both versions tapped for scope mounts. **Price: $260** (standard); **$275** (standard magnum); **$285** (deluxe); **$300** (deluxe magnum)

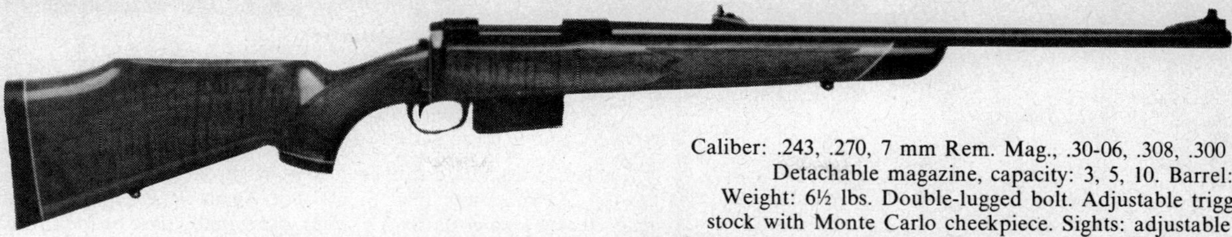

TIKKA
Caliber: .243, .270, 7 mm Rem. Mag., .30-06, .308, .300 Win. Mag. Detachable magazine, capacity: 3, 5, 10. Barrel: 23 inches. Weight: 6½ lbs. Double-lugged bolt. Adjustable trigger. Walnut stock with Monte Carlo cheekpiece. Sights: adjustable open rear, ramp front; tapped for scope mounts. Imported by Armsport. **Price: $495**

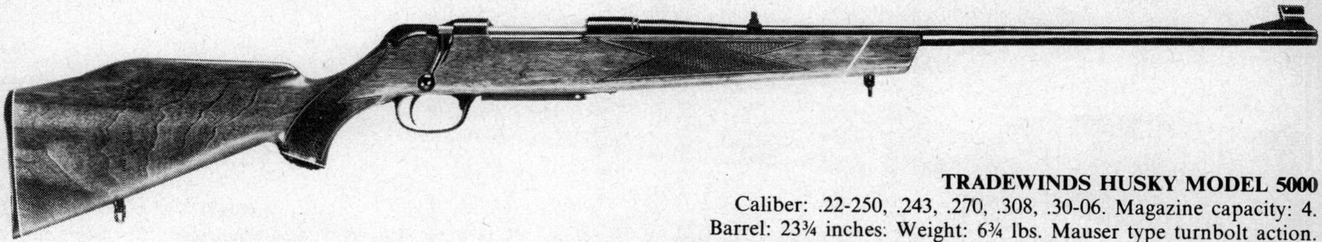

TRADEWINDS HUSKY MODEL 5000
Caliber: .22-250, .243, .270, .308, .30-06. Magazine capacity: 4. Barrel: 23¾ inches: Weight: 6¾ lbs. Mauser type turnbolt action. Fully recessed bolt head. Adjustable trigger. European walnut stock with hand checkering. Sights: tapped for scope mounts. **Price: $295**

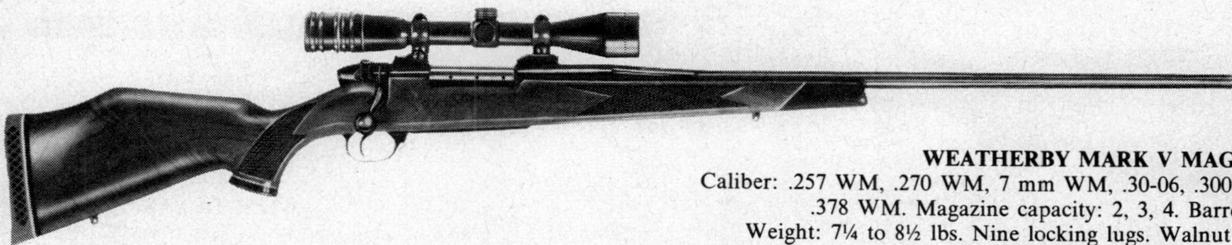

WEATHERBY MARK V MAGNUM DELUXE
Caliber: .257 WM, .270 WM, 7 mm WM, .30-06, .300 WM, .340 WM, .378 WM. Magazine capacity: 2, 3, 4. Barrel: 24, 26 inches. Weight: 7¼ to 8½ lbs. Nine locking lugs. Walnut stock with hand checkering. Sling swivel studs. Left hand version available. **Price: $660** to **$800**

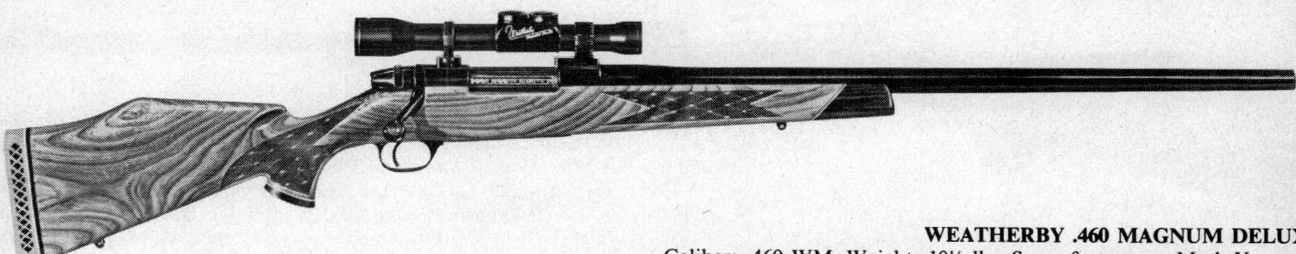

WEATHERBY .460 MAGNUM DELUXE
Caliber: .460 WM. Weight: 10½ lbs. Same features as Mark V, except Pendelton muzzle brake. Hand honed and damascened bolt. **Price: $920**

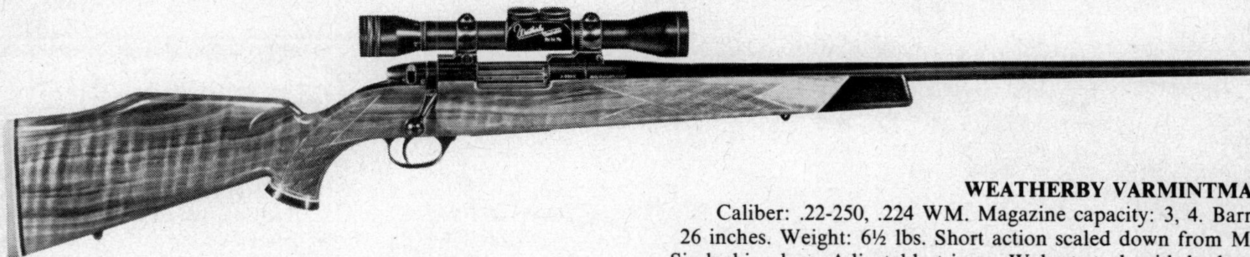

WEATHERBY VARMINTMASTER
Caliber: .22-250, .224 WM. Magazine capacity: 3, 4. Barrel: 24, 26 inches. Weight: 6½ lbs. Short action scaled down from Mark V. Six locking lugs. Adjustable trigger. Walnut stock with basket weave checkering. **Price: $640** to **$650**

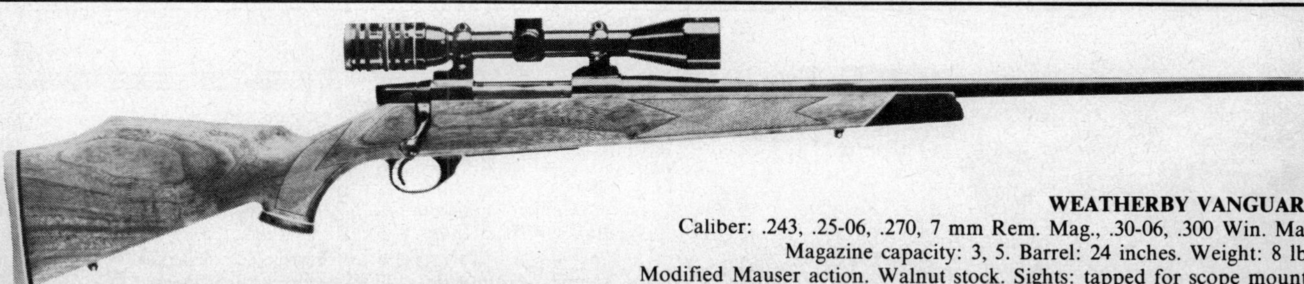

WEATHERBY VANGUARD
Caliber: .243, .25-06, .270, 7 mm Rem. Mag., .30-06, .300 Win. Mag. Magazine capacity: 3, 5. Barrel: 24 inches. Weight: 8 lbs. Modified Mauser action. Walnut stock. Sights: tapped for scope mounts. **Price: $400**

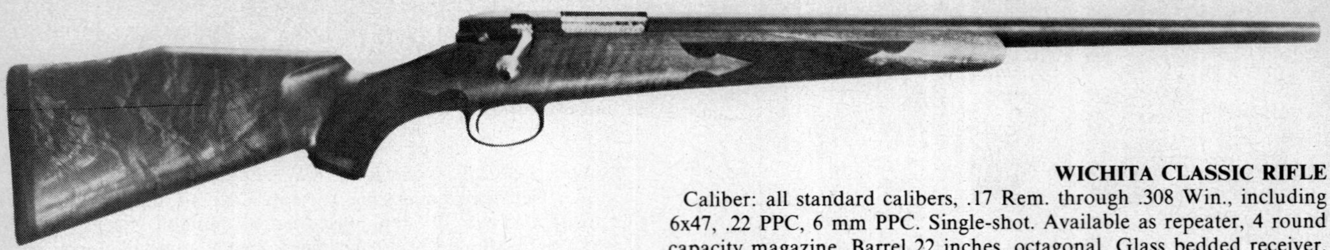

WICHITA CLASSIC RIFLE
Caliber: all standard calibers, .17 Rem. through .308 Win., including 6x47, .22 PPC, 6 mm PPC. Single-shot. Available as repeater, 4 round capacity magazine. Barrel 22 inches, octagonal. Glass bedded receiver. Hand rubbed Claro or American walnut stock. **Price: $1295**

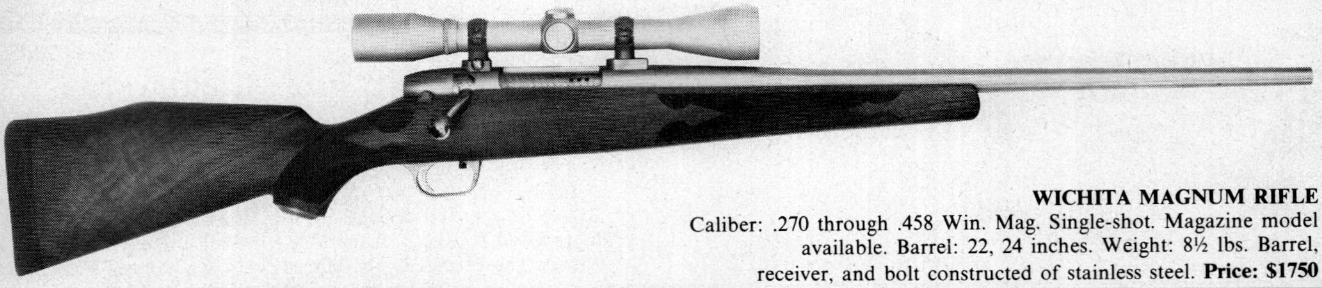

WICHITA MAGNUM RIFLE
Caliber: .270 through .458 Win. Mag. Single-shot. Magazine model available. Barrel: 22, 24 inches. Weight: 8½ lbs. Barrel, receiver, and bolt constructed of stainless steel. **Price: $1750**

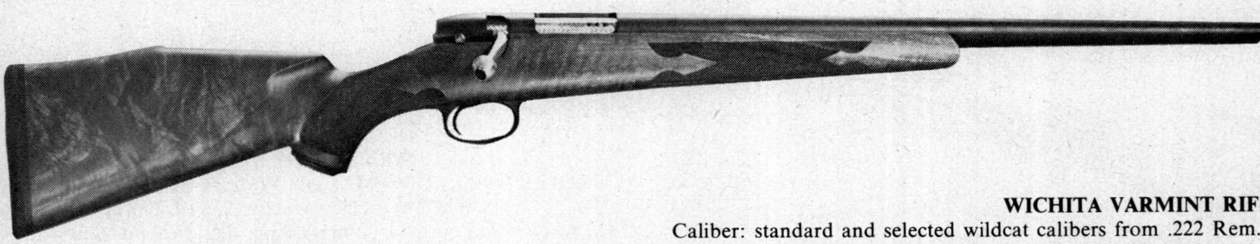

WICHITA VARMINT RIFLE
Caliber: standard and selected wildcat calibers from .222 Rem. to .308. Single-shot. Barrel: 20 inches. Glass bedded receiver. Fancy grade Claro or American walnut stock. **Price: $895**

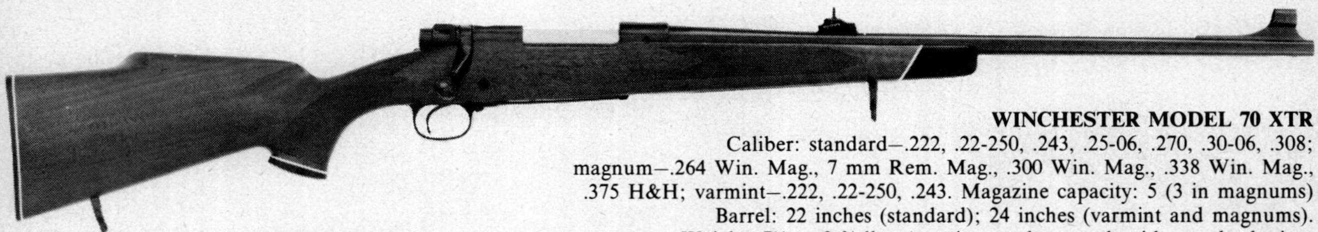

WINCHESTER MODEL 70 XTR
Caliber: standard—.222, .22-250, .243, .25-06, .270, .30-06, .308; magnum—.264 Win. Mag., 7 mm Rem. Mag., .300 Win. Mag., .338 Win. Mag., .375 H&H; varmint—.222, .22-250, .243. Magazine capacity: 5 (3 in magnums) Barrel: 22 inches (standard); 24 inches (varmint and magnums). Weight: 7½ to 9 ¾ lbs. American walnut stock with cut checkering. Sights: leaf rear; tapped for scope mounts. **Price: $354** (standard); **$373** (varmint and magnum); **$547** (.375 H&H)

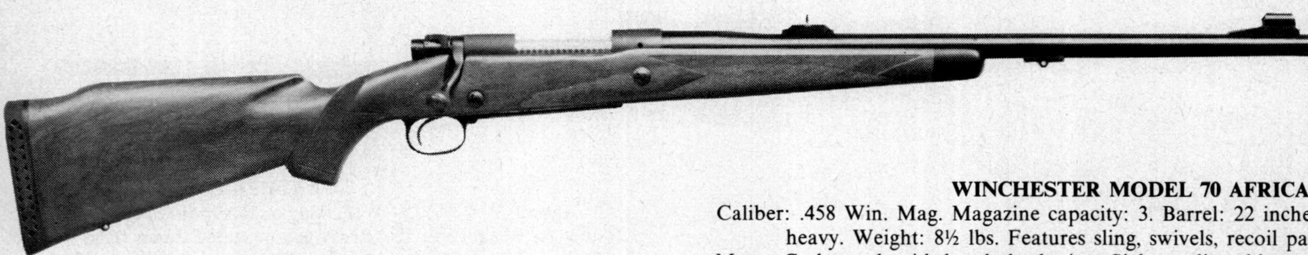

WINCHESTER MODEL 70 AFRICAN
Caliber: .458 Win. Mag. Magazine capacity: 3. Barrel: 22 inches, heavy. Weight: 8½ lbs. Features sling, swivels, recoil pad. Monte Carlo stock with hand checkering. Sights: adjustable rear; tapped for scope mounts. **Price: $607**

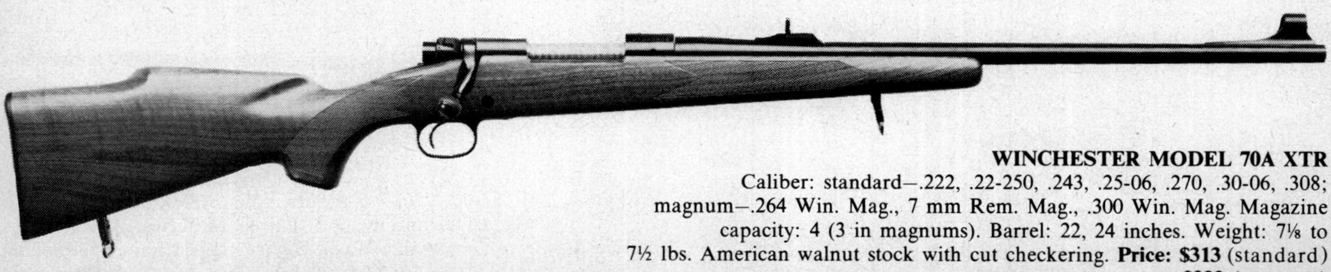

WINCHESTER MODEL 70A XTR
Caliber: standard—.222, .22-250, .243, .25-06, .270, .30-06, .308; magnum—.264 Win. Mag., 7 mm Rem. Mag., .300 Win. Mag. Magazine capacity: 4 (3 in magnums). Barrel: 22, 24 inches. Weight: 7⅛ to 7½ lbs. American walnut stock with cut checkering. **Price: $313** (standard) **$333** (magnum)

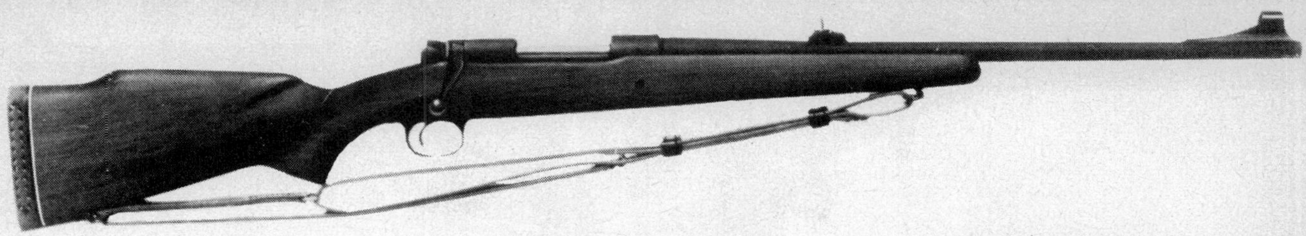

WINCHESTER MODEL 70A POLICE
Caliber: .308, .30-06. Magazine capacity: 4. Barrel: 22 inches. Non-glare finish. Comes with sling, swivels, and recoil pad. **Price: $301**

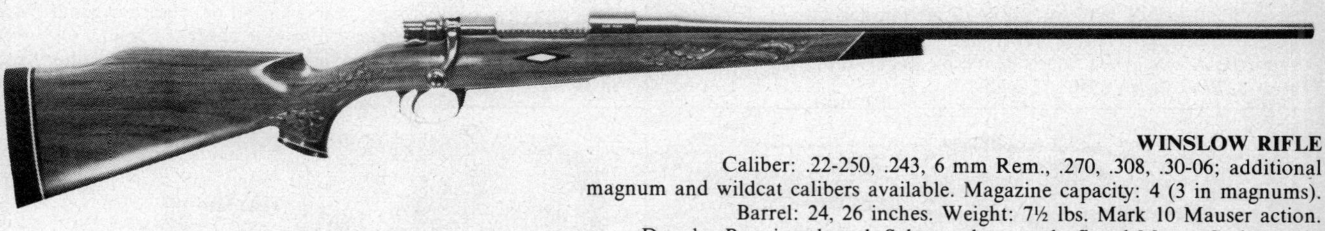

WINSLOW RIFLE
Caliber: .22-250, .243, 6 mm Rem., .270, .308, .30-06; additional magnum and wildcat calibers available. Magazine capacity: 4 (3 in magnums). Barrel: 24, 26 inches. Weight: 7½ lbs. Mark 10 Mauser action. Douglas Premium barrel. Select walnut stock, fluted Monte Carlo comb. Hand skip line checkering. Rosewood grip cap and forend tip. **Price: $1100**

HANDGUN CATALOG

Special Purpose

BUFFALO MODEL-1
Caliber: .357 Mag. Capacity: 2.
Barrel: 3 inches. Weight: 14 oz. Twin barrel derringer, through frame hammer block safety. Laminated birchwood grips. **Price: NA**

HAMMERLI MODEL 120 FREE PISTOL
Caliber: .22 LR. Single-shot.
Barrel: 10 inches. Weight: 44 oz.
Internally adjustable trigger. Maintenance and cleaning tools included. Sights: fully adjustable (one click = 6/10 inch at 50 meters). Imported by Mandall Shooting Supplies. **Price: $700** (M120-1); **$725** (M125-2, contoured grips, palm rest)

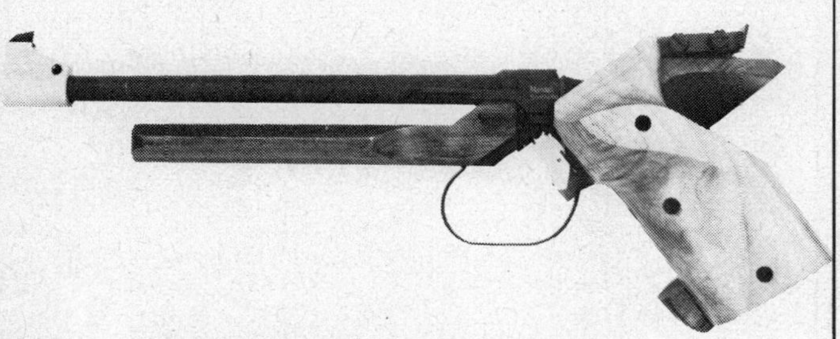

HAMMERLI MODEL 150 FREE PISTOL
Caliber: .22 LR. Single-shot.
Barrel: 11¼ inches. Weight: 42½ oz.
Martini type action. Set trigger. Comes with maintenance tools, cleaning rod and carrying case. Sights: micro rear. Imported by Mandall Shooting Supplies.
Price: $1295

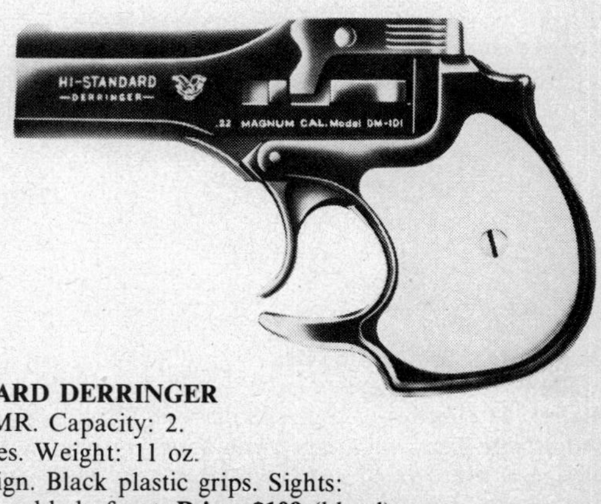

HIGH STANDARD DERRINGER
Caliber: .22 WMR. Capacity: 2.
Barrel: 2½ inches. Weight: 11 oz.
Break open design. Black plastic grips. Sights: square notch rear, blade front. **Price: $109** (blued); **$126** (nickeled)

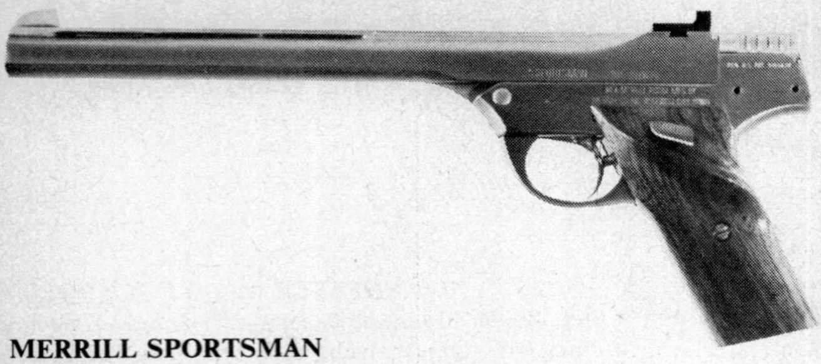

MERRILL SPORTSMAN
Caliber: .22 LR to .30-30, plus Herrett wildcats. Single-shot. Barrel: 9, 12 inches. Weight: 54 oz. (9 inch barrel). Break barrel design, thumb safety. Walnut stocks with thumbrest. Sights: open micro. Made by Rock Pistol Mfg.
Price: $270 (9 inch); **$290** (12 inch)

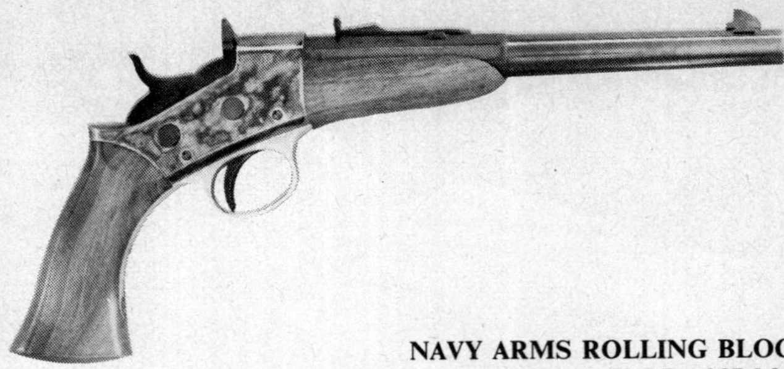

NAVY ARMS ROLLING BLOCK
Caliber: .22 LR, .357 Mag. Single-shot. Barrel: 8 inches. Weight: 43 oz. Half octagonal barrel. Visible hammer. Color case hardened frame. One piece walnut grips. Sights: adjustable semi-buckhorn rear.
Price: $150

REMINGTON XP-100
Caliber: .221 Fireball, 7 mm BR Rem. Single-shot. Barrel: 10½, 15 inches. Weight: 52 to 66 oz. Match type grooved trigger. Stocks suitable for right or left hand shooting. Sights: open rear, blade front; tapped for scope mounts. Case included. **Price: $251**

SEMMERLING LM-4
Caliber: .45 ACP. Detachable magazine, capacity: 5. Barrel: 3½ inches. Weight: 24 oz. Manual operation. Double action trigger mechanism. Plastic stocks. Semi-matte blue finish. Comes with carrying case, spare screws and wrench. Sights: fixed. **Price: $645**

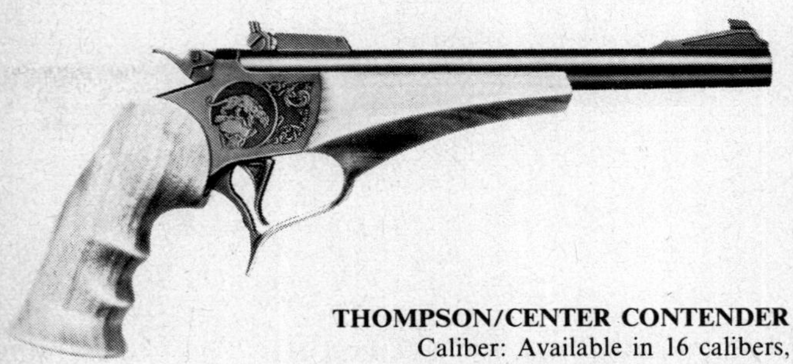

THOMPSON/CENTER CONTENDER
Caliber: Available in 16 calibers, .22 to .45 Win. Mag. Single-shot. Barrel: 10 inches. Weight: 46 oz. Break open action. Dual firing pins in basic frame rotate for centerfire and rimfire barrels. Blued finish, walnut stock and forend. Sights: adjustable open; tapped for scope mounts. **Price: 225** to **$235** (extra barrels approx **$100** each)

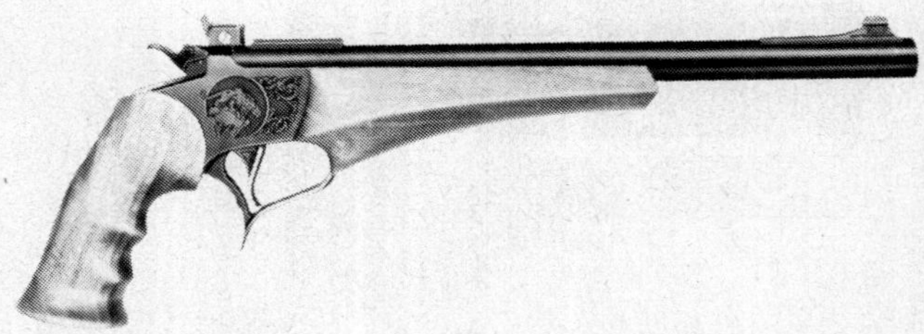

THOMPSON/CENTER CONTENDER SUPER 14
Caliber: Available in 9 calibers, .222 to .45 Win. Mag. including Herrett wildcats. Single-shot. Barrel: 14 inches. Weight: 56 oz. Bull barrel. Sights: adjustable rear, Patridge type front.
Price: $245 (extra barrels approx **$100** each)

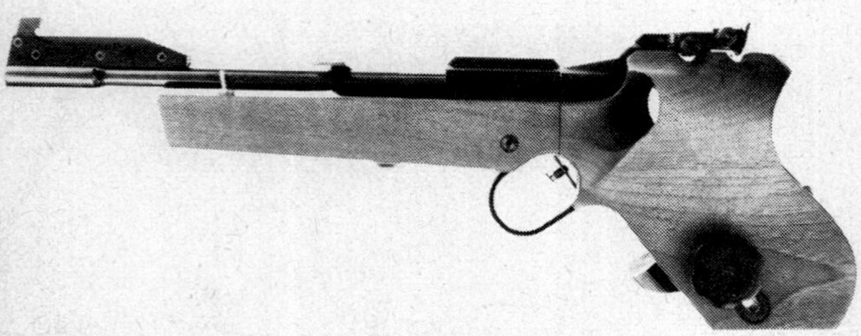

WALTHER FREE PISTOL
Caliber: .22 LR. Single-shot. Barrel: 11¾ inches. Weight: 48 oz. Adjustable electronic trigger with battery. Anatomically designed wraparound grip, adjustable palm rest. Sights: one quarter minute adjustable micro rear. Imported by Interarms. **Price: $1440**

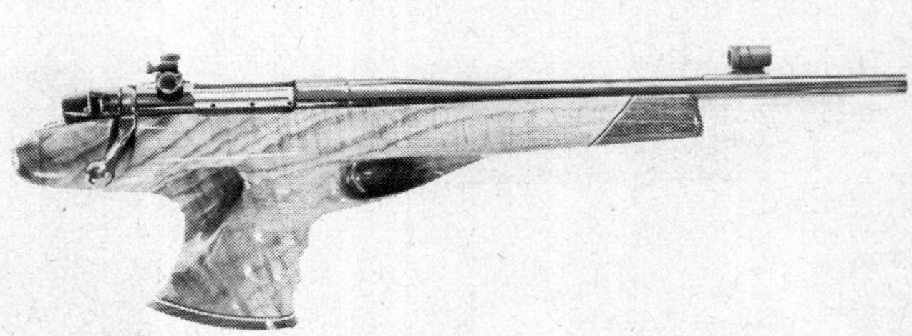

WEATHERBY MK V SILHOUETTE PISTOL
Caliber: .308, .22-250. Single-shot. Barrel: 15 inches. Weight: 5 lbs. Modified Varmintmaster action. Adjustable Canjar trigger. One piece walnut thumbhole stock. Sights: Williams rear peep sight, globe front. **Price: $800**

WICHITA SILHOUETTE PISTOL
Caliber: 7 mm PPC, .308x1½, .308 Win.
Special calibers available on request. Single-shot. Non-glare satin blue finish. Walnut or fiberglass stock.
Price: $595 and up

WICHITA CLASSIC PISTOL
Caliber: up to and including .308 Win.
Single-shot. Barrel: 11¼ inches. Three lug bolts, adjustable trigger. Octagon barrel. Non-glare finish.
Sights: micro open. **Price: $1,295**

Double Action

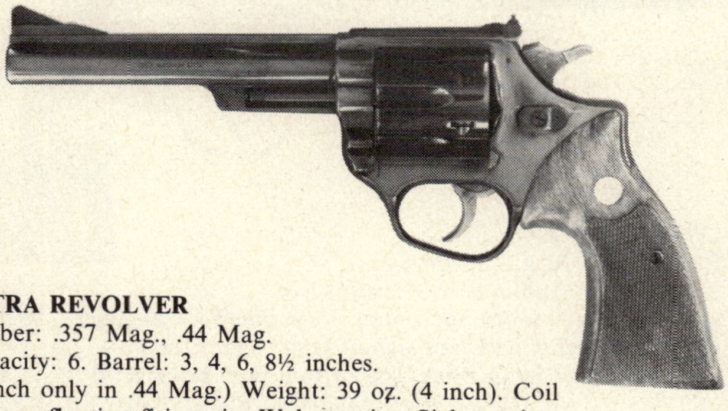

ASTRA REVOLVER
Caliber: .357 Mag., .44 Mag.
Capacity: 6. Barrel: 3, 4, 6, 8½ inches.
(6 inch only in .44 Mag.) Weight: 39 oz. (4 inch). Coil springs, floating firing pin. Walnut grips. Sights: micro click adjustable rear. **Price: $275; $290** (8½ inch); **$390** (.44 Mag)

CHARTER ARMS .357 BULLDOG
Caliber: .357 Mag. Capacity: 5.
Barrel: 6 inches. Weight: 25 oz. Checkered square butt walnut grips. Blued finish. Sights: fixed.
Price: $180

CHARTER ARMS BULLDOG
Caliber: .44 Spl. Capacity: 5.
Barrel: 3 inches. Weight: 19 oz. Blued finish. Checkered walnut round butt. Sights: fixed, Patridge type. **Price: $168**

CHARTER ARMS TARGET BULLDOG
Caliber: .357 Mag., .44 Spl.
Capacity: 5. Barrel: 4 inches. Weight: 20½ oz. Enclosed ejector rod, blued finish. Square butt checkered walnut grips. Sights: adjustable rear.
Price: $189 (.357); **$193** (.44)

CHARTER ARMS POLICE BULLDOG
Caliber: .38 Spl. Capacity: 5.
Barrel: 4 inches. Weight: 20½ oz. Enclosed ejector rod. Blued finish. Square butt checkered walnut grips. Sights: adjustable, snag free. **Price: $175**

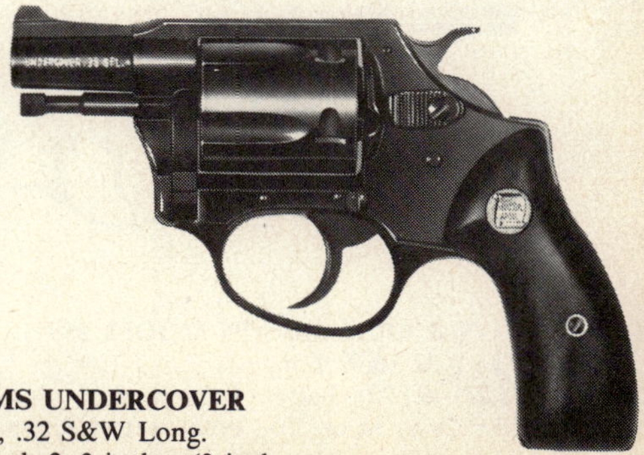

CHARTER ARMS UNDERCOVER
Caliber: .38 Spl., .32 S&W Long.
Capacity: 5. Barrel: 2, 3 inches (3 inch only in .38). Weight: 16½ oz. Square butt in .38 only. Choice of blue or nickel finish. Sights: fixed. **Price: $149**

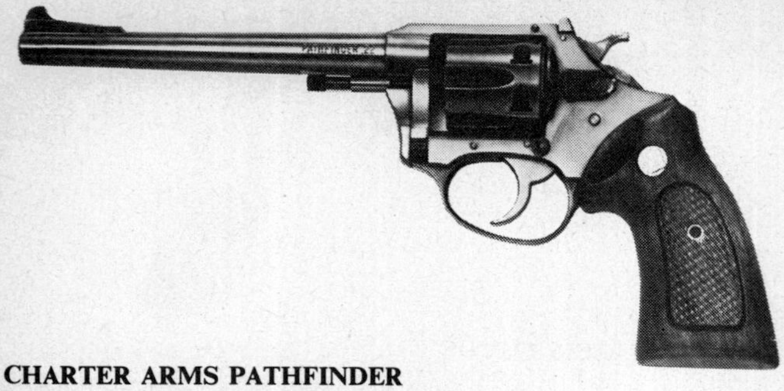

CHARTER ARMS PATHFINDER
Caliber: .22 LR or .22 WMR.
Capacity: 6. Barrel: 3, 6 inches.
Weight: 18½ to 22½ oz. Smooth walnut grips. Blued finish.
Sights: adjustable. **Price: $168** (.22 LR, 3 inch); **$177** (.22 LR, 6 inch); **$182** (.22 WMR)

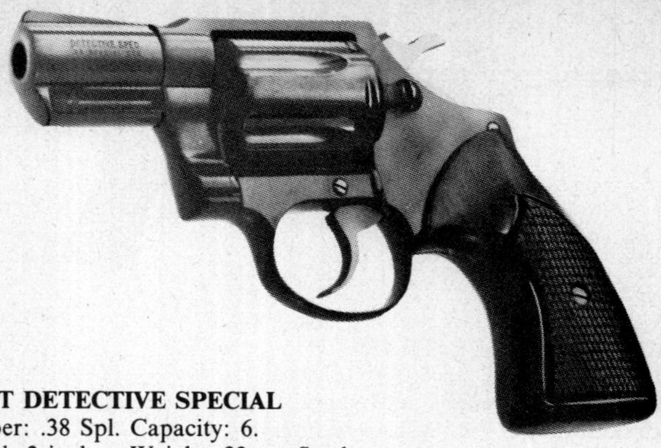

COLT DETECTIVE SPECIAL
Caliber: .38 Spl. Capacity: 6.
Barrel: 2 inches. Weight: 22 oz. Steel frame. Checkered walnut grips. Sights: fixed. **Price: $260** (blued); **$280** (nickel)

COLT LAWMAN MK III
Caliber: .357 Mag. Capacity: 6.
Barrel: 2, 4 inches. Weight: 35 oz. Steel frame, heavy barrel. Walnut grips. Sights: fixed.
Price: $246 (blued); **$261** (nickel)

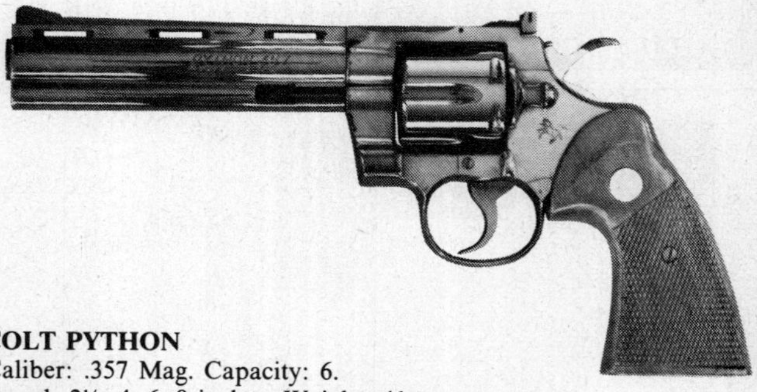

COLT PYTHON
Caliber: .357 Mag. Capacity: 6.
Barrel: 2½, 4, 6, 8 inches. Weight: 41 oz. (4 inch). Heavy frame and barrel. Vent rib. Wide hammer, grooved trigger. Checkered walnut grips. Sights: micro adjustable rear, ramp front. Top of Colt line. **Price: $445** to **$486**

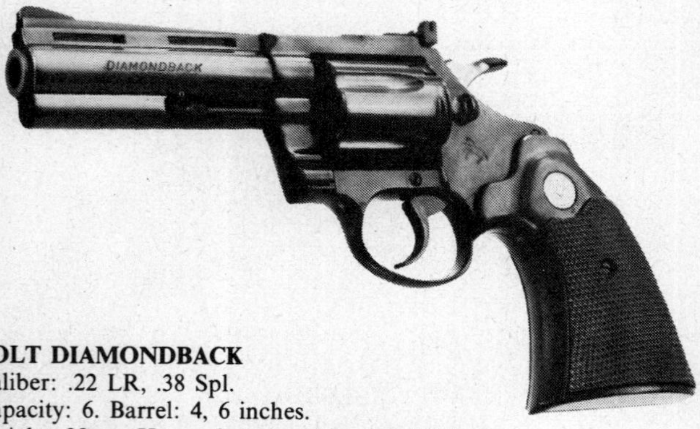

COLT DIAMONDBACK
Caliber: .22 LR, .38 Spl.
Capacity: 6. Barrel: 4, 6 inches.
Weight: 28 oz. Heavy frame and barrel. Wide spur hammer, grooved trigger. Blued. Sights: micro adjustable rear. **Price: $308** to **$315**

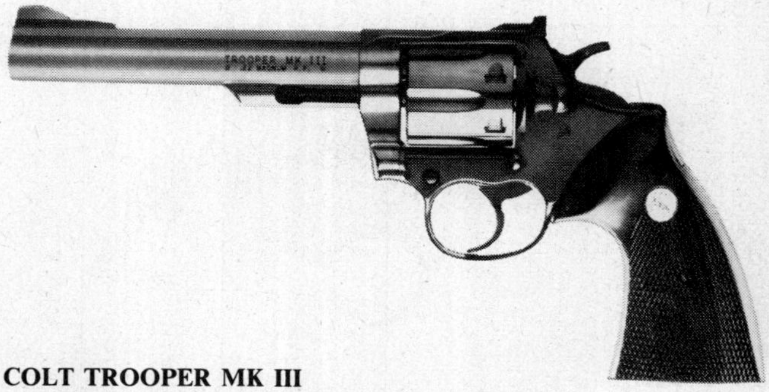

COLT TROOPER MK III
Caliber: .22 LR, .22 WMR, .38 Spl., .357 Mag. Capacity: 6. Barrel: 4, 6, 8 inches.
Weight: 39 to 42 oz. Heavy barrel, target trigger and hammer.
Sights: micro adjustable. **Price: $289** to **$318**

HARRINGTON & RICHARDSON MODEL 999
Caliber: .22 LR, .32 S&W Long.
Capacity: 9 (.22), 6 (.32). Barrel: 4, 6 inches. Weight: 28 to 34 oz. Top break auto ejection. Fluted barrel (4 inch models); integral vent rib barrel (6 inch models). Blued finish. Checkered walnut grips.
Sights: adjustable. **Price: $120**

HARRINGTON & RICHARDSON MODEL 940
Caliber: .22 LR. Capacity: 9.
Barrel: 6 inches. Weight: 36 oz. Swing out cylinder. Vent rib barrel. Available with flat target weight barrel. Sights: adjustable rear, target blade front.
Price: $100

HARRINGTON & RICHARDSON
MODEL 676 CONVERTIBLE
Caliber: .22 LR, .22 WMR.
Capacity: 6. Barrel: 4½, 5½, 7½, 12½ inches.
Weight: 31 to 41 oz. Side loading, comes with two interchangeable cylinders. Color cased frame and ejector rod housing. Coil spring construction. Sights: adjustable.
Price: $100; $125 (12 inch)

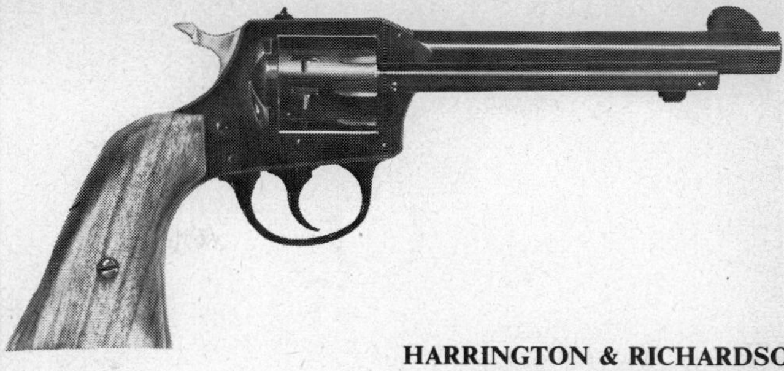

HARRINGTON & RICHARDSON
MODEL 649 CONVERTIBLE
Caliber: .22 LR, .22 WMR.
Capacity: 6. Barrel: 5½ inches. Weight: 32 oz. Interchangeable cylinders, side loading and ejecting. One piece walnut grips. Sights: adjustable rear. **Price: $90; $95** (M60 with nickel finish)

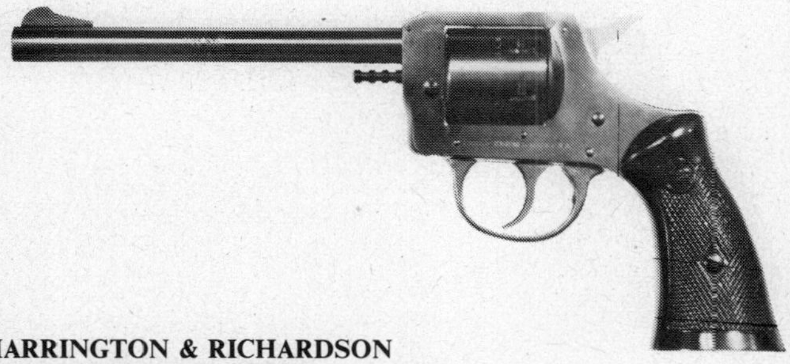

HARRINGTON & RICHARDSON
MODEL 666 CONVERTIBLE
Caliber: .22 LR, .22 WMR.
Capacity: 6. Barrel: 6 inches. Weight: 28 oz. Interchangeable cylinders with pull pin extraction. Black cycolac grips. Sights: fixed. **Price: $75**

HARRINGTON & RICHARDSON MODEL 976
Caliber: .22 LR Capacity: 9.
Barrel: 7½ inches. Weight: 36 oz. Side loading and ejecting. Blued barrel, cylinder, and trigger guard. Color case hardened frame, ejector rod housing, and trigger. Sights: adjustable rear. **Price: $90**

HARRINGTON & RICHARDSON MODEL 929
Caliber: .22 LR. Capacity: 9.
Barrel: 2½, 4, 6 inches. Weight: 22 to 28 oz.
Black cycolac grips. Sights: adjustable rear (4 and 6 inch versions). **Price: $80; $85** (nickel)

HARRINGTON & RICHARDSON MODEL 922
Caliber: .22 LR. Capacity: 9.
Barrel: 2½, 4, 6 inches. Weight: 20 to 26 oz.
Solid frame. Bulldog type grips on 2½ inch version.
Sights: fixed. **Price: $68; $70** (nickel)

HARRINGTON & RICHARDSON MODEL 622
Caliber: .22 LR. Capacity: 6.
Barrel: 2½, 4 inches. Weight: 20 to 26 oz.
Bulldog style black cycolac grips. Sights: fixed.
Price: $60; $65 (nickel)

HARRINGTON & RICHARDSON MODEL 632
Caliber: .32 S&W Long. Capacity: 6.
Barrel: 2½, 4 inches. Weight: 20 to 26 oz.
Sights: fixed. **Price: $68; $73** (nickel)

HARRINGTON & RICHARDSON MODEL 732
Caliber: .32 S&W Long. Capacity: 6.
Barrel: 2½, 4 inches. Weight: 23½ to 26 oz.
Swing out cylinder. Bulldog grips on 2½ inch version. Sights: adjustable rear. **Price: $80; $85** (nickel)

HIGH STANDARD DOUBLE NINE CONVERTIBLE
Caliber: .22 LR, .22 WMR.
Capacity: 9. Barrel: 5½ inches.
Weight: 32 oz. DA swing out dual cylinder.
Steel frame. Smooth walnut grips. Sights: adjustable rear.
Price: $200 (blue)

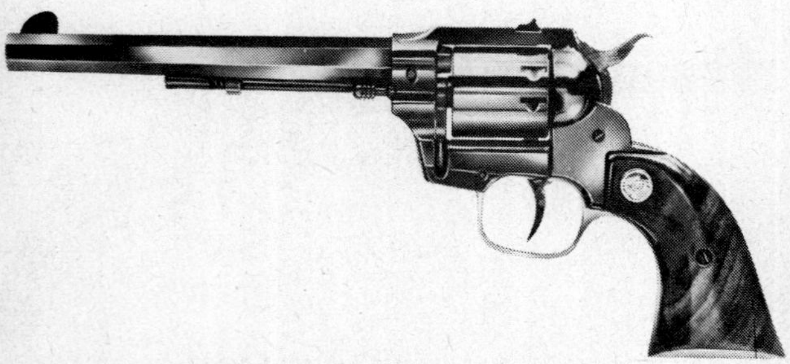

HIGH STANDARD HIGH SIERRA
Caliber: .22 LR, .22 WMR.
Capacity: 9. Barrel: 7 inches. Weight: 35 oz. Octagonal barrel. Swing out dual cylinder, gold plated backstrap and trigger guard. **Price: $204**

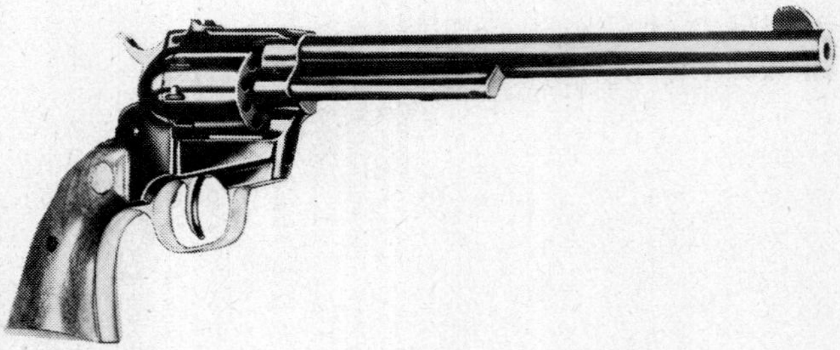

HIGH STANDARD LONGHORN
Caliber: .22 LR, .22 WMR.
Capacity: 9. Barrel: 9½ inches.
Weight: 38 oz. DA swing out dual cylinder.
Steel frame. Smooth walnut grips. **Price: $204** (blue)

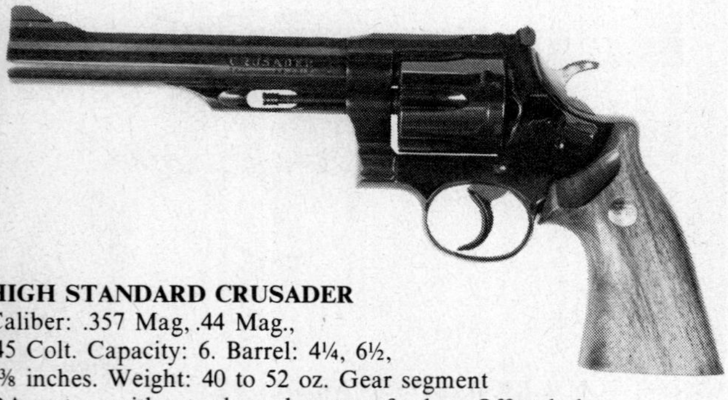

HIGH STANDARD CRUSADER
Caliber: .357 Mag, .44 Mag., .45 Colt. Capacity: 6. Barrel: 4¼, 6½, 8⅜ inches. Weight: 40 to 52 oz. Gear segment DA system without rebound or transfer bar. Offset bolt notches. Cylinder lock. One piece walnut grips checkered in .357 and .45, smooth in .44. Sights: adjustable micro.
Price: $391 to **$407**

IVER JOHNSON SIDEWINDER
Caliber: .22 S, L, LR, .22 WMR. Capacity: 8. Barrel: 4, 6 inches.
Weight: 31 oz. Stag horn style grips. Sight: fixed or adjustable. **Price: $67** to **$90**

LLAMA COMANCHE
Caliber: .22, .38 Spl., .357 Mag.
Capacity: 6. Barrel: 4, 6 inches. Weight: 39 oz. Target type hammer. Floating firing pin. Serrated front and backstrap. Oversize walnut checkered grips. Sights: adjustable rear. Imported by Stoeger. **Price: $267** (blue); **$342** (chrome); **$1500** (gold damascened)

LLAMA SUPER COMANCHE
Caliber: .44 Mag. Capacity: 6.
Barrel: 6 inches. Weight: 50 oz. Two cylinder latches. Coil springs throughout. Heavy ribbed barrel. Broad serrated hammer spur. Sights: adjustable target type. Imported by Stoeger. **Price: $415**

RUGER SPEED-SIX
Caliber: .38 Spl, .357 Mag., 9 mm.
Capacity: 6. Barrel: 2¾, 4 inches.
Weight: 31 oz. Round butt configuration. Blued finish.
Sights: fixed. **Price: $148** to **$165**

RUGER SPEED-SIX STAINLESS
Same specifications as
regular Speed-Six, except in stainless steel.
Price: $162 to **$179**

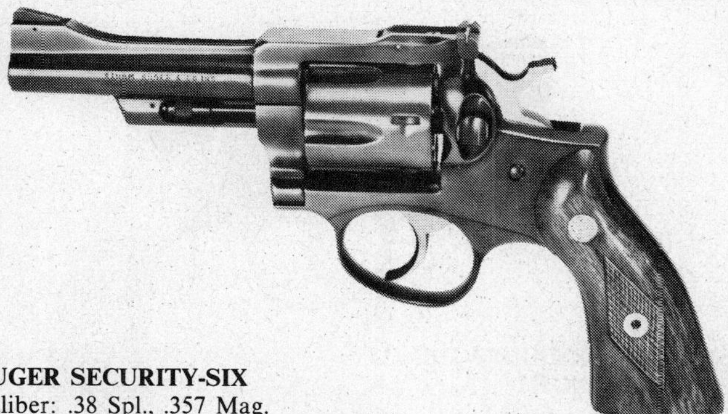

RUGER SECURITY-SIX
Caliber: .38 Spl., .357 Mag.
Capacity: 6. Barrel: 2¾, 4, 6 inches.
Weight: 33½ oz. Spring loaded frame-mounted firing pin.
Blued finish. Checkered walnut grips. Sights: micro
adjustable. **Price: $193** to **$208**

RUGER SECURITY-SIX
Same specifications as
regular Security-Six, except in stainless steel.
Price: $212 to **$228**

RUGER POLICE SERVICE-SIX
Caliber: .38 Spl., .357 Mag.,
9 mm. Capacity: 6. Barrel: 2¾, 4 inches.
Weight: 33 oz. Blued finish. Checkered walnut grips.
Sights: fixed. **Price: $148** to **$165**

RUGER POLICE SERVICE-SIX
Same specifications as
regular Police Service-Six, except in stainless
steel. Not available in 9 mm. **Price: $162**

RUGER REDHAWK .44 MAGNUM
Caliber: .44 Mag. Capacity: 6.
Barrel: 7½ inches. Weight: 52 oz. Stainless
steel construction. Coil springs throughout. Locking crane
feature. Transfer-bar ignition system. Checkered walnut
grips. Sights: adjustable rear, blade front with inter-
changeable inserts. **Price: $325**

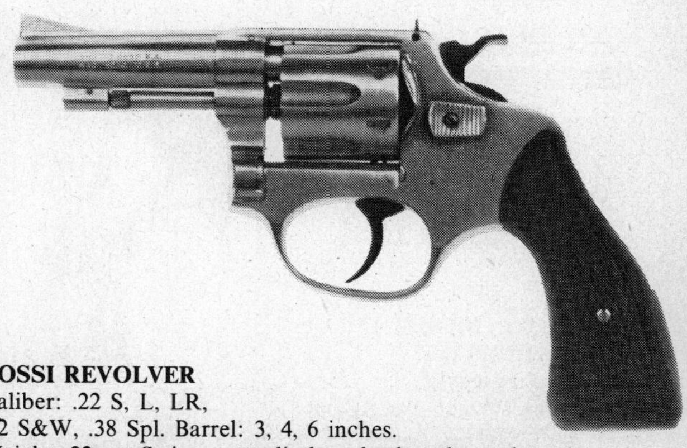

ROSSI REVOLVER
Caliber: .22 S, L, LR,
.32 S&W, .38 Spl. Barrel: 3, 4, 6 inches.
Weight: 22 oz. Swing out cylinder, checkered wood stocks.
Sights: adjustable rear, ramp front. **Price: $130** to **$150**

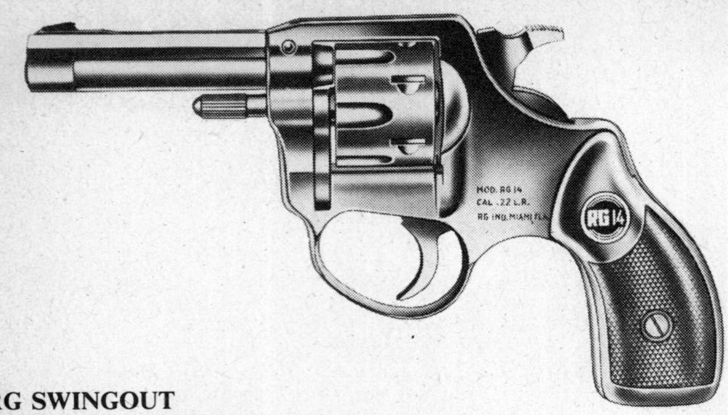

RG SWINGOUT
Caliber: .22 LR, .32 S&W, .38 Spl. Capacity: 6. Barrel: 1¾ to 4 inches. Weight: 15 to 29 oz. Swing out cylinder. Sights: fixed. **Price: $40** to **$100**

SMITH & WESSON MODEL 10
Caliber: .38 Spl. Capacity: 6. Barrel: 2, 4, 5, 6 inches. Weight: 29 to 32 oz. K frame. Checked walnut stocks with choice of round or square butt. Sights: square notch rear, ramp front. Single action kit available. **Price: $143** (blue); **$155** (nickel)

SMITH & WESSON MODEL 10
HEAVY BARREL
Caliber: .38 Spl. Capacity: 6. Barrel: 4 inches. Weight: 34 oz. K frame. Checked walnut stocks, square butt. Sights: square notch rear, ramp front. **Price: $143** (blue); **$155** (nickel)

SMITH & WESSON MODEL 12
M&P AIRWEIGHT
Caliber: .38 Spl. Capacity: 6. Barrel: 2, 4 inches. Weight: 18 oz. (2 inch); 19½ oz. (4 inch). Airweight K frame. Checked walnut stocks with choice of round or square butt. Sights: square notch rear, ramp front. **Price: $185** (blue); **$210** (nickel)

SMITH & WESSON MODEL 13
MILITARY & POLICE .357
Caliber: .357 Mag. Capacity: 6. Barrel: 4 inches. Weight: 34 oz. K frame. Checked walnut stocks, square butt. Sights: square notch rear, ramp front. **Price: $157** (blue); **$171** (nickel)

SMITH & WESSON MODEL 14
K-38 MASTERPIECE
Caliber: .38 Spl. Capacity: 6. Barrel: 6, 8⅜ inches. Weight: 38½ oz. (6 inch); 42½ oz. (8⅜ inch). K frame. Checked walnut stocks, square butt. Sights: micro rear, Patridge front. Blue only. Single action kit available. **Price: $222** to **$264**

SMITH & WESSON MODEL 15
COMBAT MASTERPIECE
Caliber: .38 Spl. Capacity: 6. Barrel: 2, 4 inches. Weight: 28 oz. (2 inch); 34 oz. (4 inch). K frame. Checked walnut stocks, square butt. Sights: micro click rear, Baughman Quick Draw front. **Price: $170** (blue); **$183** (nickel)

SMITH & WESSON MODEL 17
K-22 MASTERPIECE
Caliber: .22 LR. Capacity: 6. Barrel: 6, 8⅜ inches. Weight: 38½ oz. (6 inch); 42½ oz. (8⅜ inch). K frame. Checked walnut stocks, square butt. Sights: micro click rear, Patridge front. Blue only. **Price: $223** to **$264**

SMITH & WESSON MODEL 48
K-22 MASTERPIECE M.R.F.
Caliber: .22 Mag. Rimfire. Capacity: 6. Barrel: 4, 6, 8⅜ inches. Weight: 36 to 43 oz. K frame. Checked walnut stocks. Auxiliary cylinder available in .22 LR. Sights: micro click rear, Baughman Quick Draw front (4 inch), Patridge front (6, 8⅜ inch). Blue only. **Price: $237** to **$248**

SMITH & WESSON MODEL 18
.22 COMBAT MASTERPIECE
Caliber: .22 LR. Capacity: 6.
Barrel: 4 inches. Weight: 36½ oz. K frame. Checked walnut stocks, square butt. Sights: micro click rear, Baughman Quick Draw front. **Price: $205** to **$222**

SMITH & WESSON MODEL 19
COMBAT MAGNUM
Caliber: .357 Mag. Capacity: 6.
Barrel: 2½, 4, 6 inches. Weight: 31 to 40 oz.
K frame. Checked Goncalo Alves Target Stocks. Round butt available in 2½ inch barrel version only. Sights: micro click rear, Baughman Quick Draw front, Patridge front on 6 inch. **Price: $217** to **$246**

SMITH & WESSON MODEL 25
.45 TARGET
Caliber: .45 ACP. Capacity: 6.
Barrel: 6½ inches. Weight: 45 oz. N frame. Checked Goncalo Alves Target Stocks, square butt. Sights: micro click rear, Patridge front. Blue only. **Price: $356**

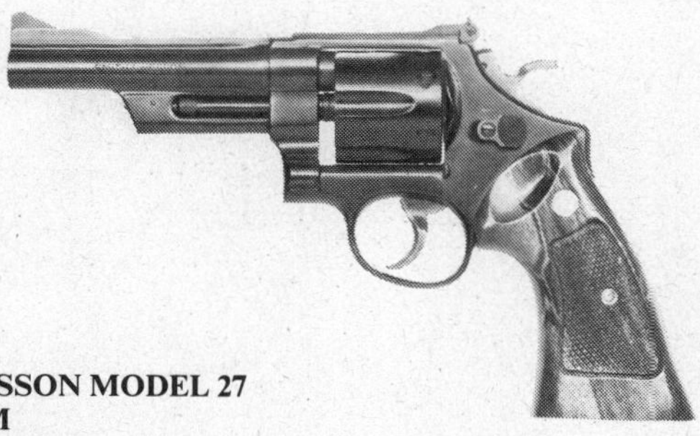

SMITH & WESSON MODEL 27
.357 MAGNUM
Caliber: .357 Mag. Capacity: 6.
Barrel: 3½, 5, 6, 8⅜ inches. Weight: 41 to 47 oz. N frame. Fine checkered top strap and barrel rib. Checked Goncalo Alves Target Stocks, square butt. Sights: micro click rear, choice of any S&W front. Blue or nickel. **Price: $376** to **$388**

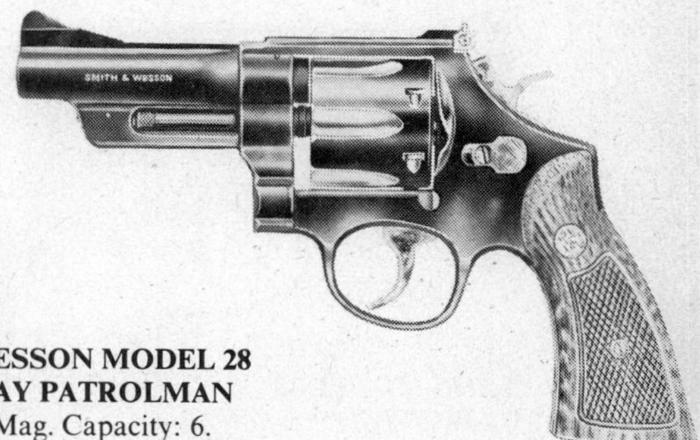

SMITH & WESSON MODEL 28
.357 HIGHWAY PATROLMAN
Caliber: .357 Mag. Capacity: 6.
Barrel: 4, 6 inches. Weight: 41¾ oz. (4 inch); 44 oz. (6 inch). N frame. Sand blast stippling on barrel rib and frame edging. Checked walnut stocks, square butt. Target stocks optional. Sights: micro click rear, Baughman Quick Draw front. Blue only. **Price: $222** to **$239**

SMITH & WESSON MODEL 29
Caliber: .44 Mag. Capacity: 6.
Barrel: 4, 6½, 8⅜ inches. Weight: 43 to 51½ oz. N frame. Oversize Goncalo Alves Target Stocks. Sights: micro click rear, red ramp front. Blue or nickel. **Price: $376** to **$388**

SMITH & WESSON MODEL 57
Caliber: .41 Mag. Capacity: 6
Barrel: 4, 6, 8⅜ inches. Weight: 46 to 54 oz. N frame. Oversize Goncalo Alves Target Stocks. Checked target type hammer. Presentation case included. Sights: micro click rear, red ramp front. Blue or nickel. **Price: $376** to **$388**

SMITH & WESSON MODEL 629
Same as M29, except in stainless steel.
6 inch barrel only. **Price: $421**

SMITH & WESSON MODEL 31
Caliber: .32 S&W Long. Capacity: 6.
Barrel: 2, 3 inches. Weight: 20 oz. (2 inch); 22 oz. (3 inch). J frame. Checked walnut stocks, square butt. Sights: square notch rear, fixed serrated ramp front.
Blue only. **Price: $179**

SMITH & WESSON MODEL 34
.22/.32 KIT GUN
Caliber: .22 LR. Capacity: 6.
Barrel: 2, 4 inches. Weight: 22½ oz. (2 inch); 24¼ oz. (4 inch). Checked walnut stocks with choice of round or square butt. Sights: micro click rear, fixed serrated ramp front. **Price: $186** (blue); **$202** (nickel)

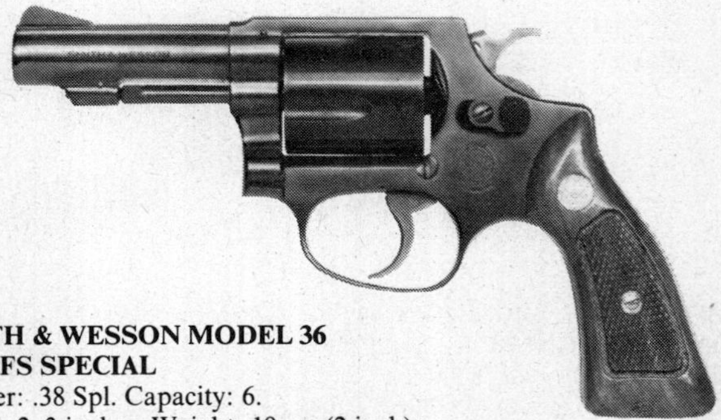

SMITH & WESSON MODEL 36
CHIEFS SPECIAL
Caliber: .38 Spl. Capacity: 6.
Barrel: 2, 3 inches. Weight: 19 oz. (2 inch); 22 oz. (3 inch). J frame. Checked walnut service stocks with choice of round or square butt. Sights: square notch rear, fixed serrated ramp front. **Price: $171** (blue); **$185** (nickel)

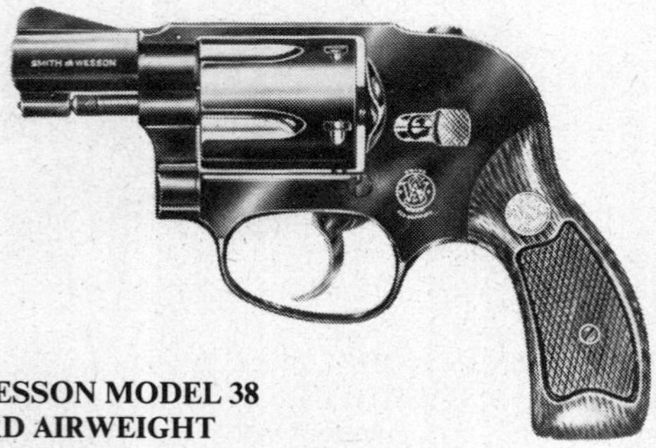

SMITH & WESSON MODEL 38
BODY GUARD AIRWEIGHT
Caliber: .38 Spl. Capacity: 5.
Barrel: 2 inches. Weight: 14½ oz. Airweight J frame. Hammerless design. Checked walnut service stocks. Sights: square notch rear, fixed front. **Price: $198** (blue); **$223** (nickel)

SMITH & WESSON MODEL 60
CHIEFS SPECIAL STAINLESS
Caliber: .38 Spl. Capacity: 5.
Barrel: 2 inches. Weight: 19 oz. J frame.
Checked walnut service grips, round butt. Sights: square notch rear, fixed ramp front. **Price: $208**

SMITH & WESSON MODEL 63
.22/.32 STAINLESS KIT GUN
Caliber: .22 LR. Capacity: 6.
Barrel: 4 inches. Weight: 24½ oz. J frame.
Checked walnut service grips, square butt. Sights: micro click rear, red ramp front. **Price: $216**

SMITH & WESSON MODEL 64
M&P STAINLESS
Caliber: .38 Spl. Capacity: 6.
Barrel: 2, 4 inches. Weight: 29 oz. (2 inch); 34 oz. (4 inch). K frame. Heavy barrel on 4 inch version.
Checked walnut service stocks, round butt on 2 inch, square butt on 4 inch. Sights: square notch rear, fixed ramp front.
Price: $163

SMITH & WESSON MODEL 65 STAINLESS
Caliber: .357 Mag. Capacity: 6.
Barrel: 4 inches. Weight: 34 oz. K frame.
Checked walnut service grips, square butt. Sights: square notch rear, serrated ramp front. **Price: $173**

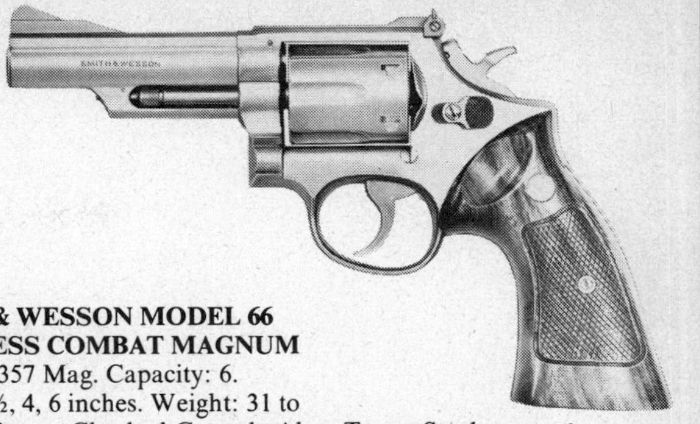

SMITH & WESSON MODEL 66 STAINLESS COMBAT MAGNUM
Caliber: .357 Mag. Capacity: 6.
Barrel: 2½, 4, 6 inches. Weight: 31 to 40 oz. K frame. Checked Goncalo Alves Target Stocks, round butt on 2½ inch, square butt on 4, 6 inch. Sights: micro click rear, red ramp front. **Price: $232**

SMITH & WESSON MODEL 67 STAINLESS COMBAT MASTERPIECE
Caliber: .38 Spl. Capacity: 6.
Barrel: 4 inches. Weight: 34 oz. K frame
Checked walnut service stocks, square butt. Sights: micro click rear, red ramp front. **Price: $213**

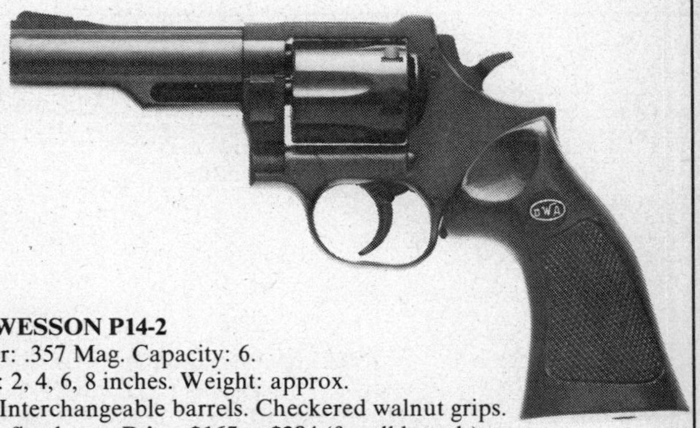

DAN WESSON P14-2
Caliber: .357 Mag. Capacity: 6.
Barrel: 2, 4, 6, 8 inches. Weight: approx. 34 oz. Interchangeable barrels. Checkered walnut grips. Sights: fixed rear. **Price: $165** to **$384** (for all barrels)

Single Action Revolvers

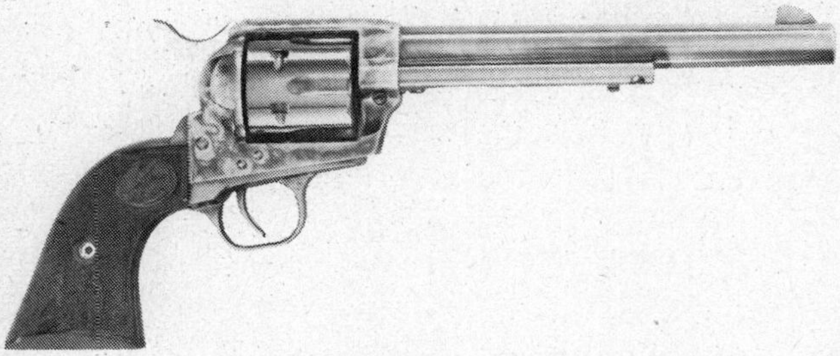

COLT SINGLE ACTION ARMY
Caliber: .357 Mag., .44 Spl., .45 Colt. Capacity: 6. Barrel: 4¾, 5½, 7½, 12 inches. Weight: 40 oz. (5½ inch). Heat colored frame, blued barrel. Checked hard rubber grips. **Price: $400** to **$478**

COLT NEW FRONTIER SINGLE ACTION ARMY
Caliber: .45 Colt, .44 Spl., .44-40. Capacity: 6. Barrel: 7½ inches.
Weight: 44 oz. Heat colored flat top frame. Plain walnut grips. Sights: adjustable rear, ramp front. **Price: $400** to **$476**

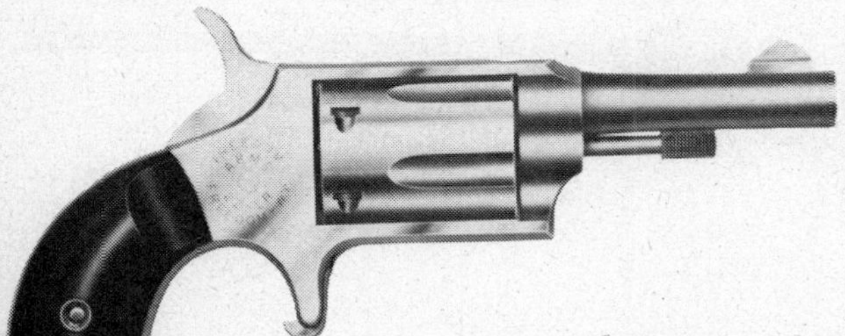

FREEDOM ARMS MINI-REVOLVER
Caliber: .22 S, L, LR, .22 WMR.
Capacity: 5. Barrel: 1, 1¾ inches.
Weight: 4 oz. Hammer notch safety. Stainless steel construction. Sights: fixed. **Price: $125;** (.22 LR); **$145** (.22 WMR)

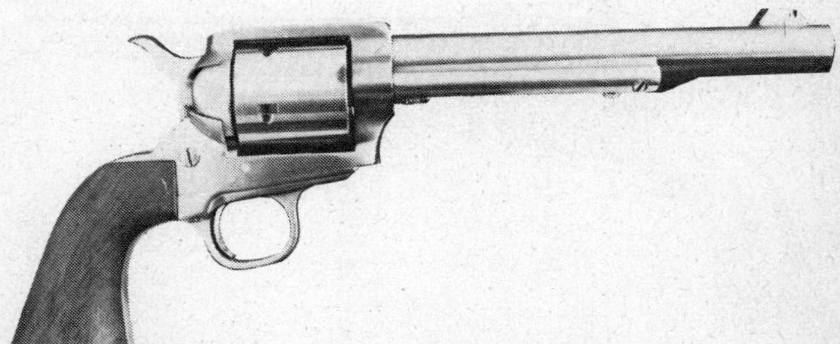

FREEDOM ARMS .454 CASULL
Caliber: .454 Casull Mag.
Capacity: 5. Barrel: 7½ inches. Weight: 50 oz. Stainless steel construction. Sights: fixed rear, blade front. **Price: $500** (available spring, 1981)

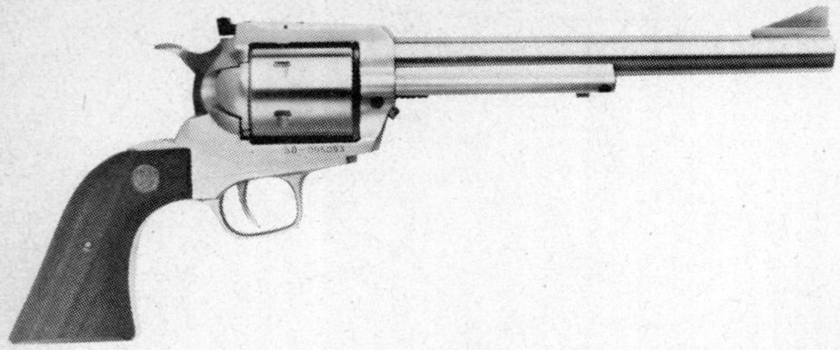

U.S. ARMS ABILENE .44 MAG
Caliber: .44 Mag. Capacity: 6. Barrel: 7½, 8½, 9 inches. Weight: approx. 48 oz. Wide serrated trigger. Smooth walnut grips. Blued or brushed chrome finish. Sights: adjustable rear, ramp front. Distributed by Mossberg. **Price: $275** to **$313**

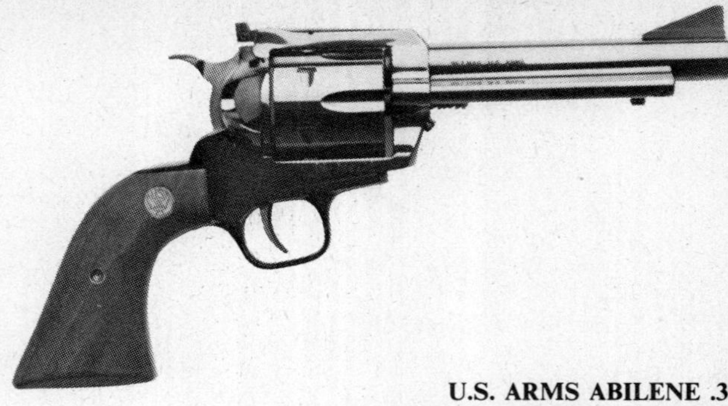

U.S. ARMS ABILENE .357
Caliber: .357 Mag. Capacity: 6. Barrel: 4⅝, 6, 7½ inches. Smooth walnut grips. Blued finish. Sights: adjustable rear, ramp front. Distributed by Mossberg. **Price: $233**

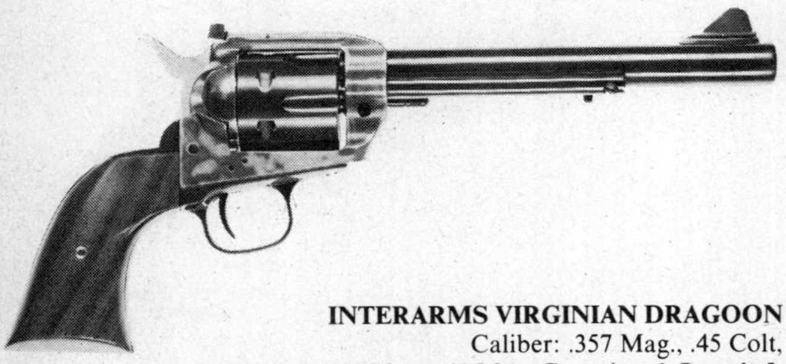

INTERARMS VIRGINIAN DRAGOON
Caliber: .357 Mag., .45 Colt, .44 Mag., .41 Mag. Capacity: 6. Barrel: 5, 6, 7½, 8⅜, 12 inches. Weight: approx. 48 oz. Color case hardened steel frame. Swiss safe feature. Walnut grips. Sights: micro adjustable rear. Made in U.S.A. **Price: $250** ; **$280** (stainless); **$340** (12 inch); **$370** (12 inch, stainless)

MITCHELL SINGLE ACTION ARMY MODEL
Caliber: .357 Mag., .44 Mag., .45 Colt. Capacity: 6. Barrel: 4¾, 6, 7½ inches. Forged steel frame, color case hardened. Brass two piece backstrap. Sights: adjustable rear, ramp front. **Price: $170** to **$190**

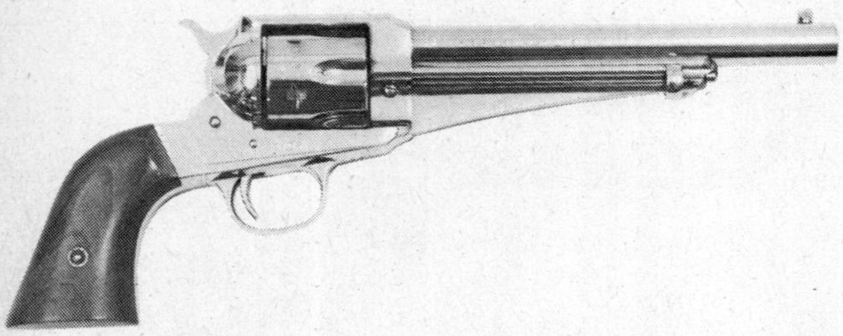

NAVY ARMS MODEL 1875
Caliber: .357 Mag., .45 Colt, .44-40. Capacity: 6. Barrel: 7½ inches. Weight: 44 oz. Replica of Remington revolver with one piece walnut stocks, color case hardened frame, blued barrel and cylinder. **Price: $180** (blue); **$215** (nickel)

NAVY ARMS STANDARD FRONTIER REVOLVER
Caliber: .357 Mag., .45 Colt, .30 Carbine, .38-40, .44-40, 9 mm Luger. Capacity: 6. Barrel: 4½, 5½, 7½ inches. Weight: 46 oz. Color case hardened frame, blued barrel and cylinder. Sights: fixed.

NAVY ARMS FRONTIER TARGET
Same as Standard Revolver, except comes with target adjustable sights. **Price: $150**

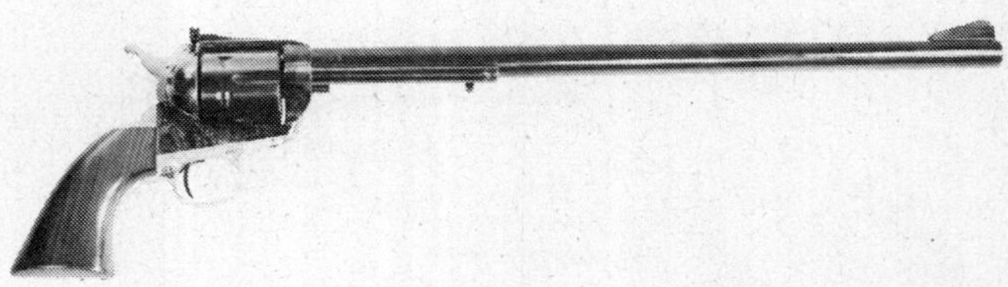

NAVY ARMS BUNTLINE TARGET
Same as Target Revolver, except comes with 16½ inch barrel. **Price: $210**

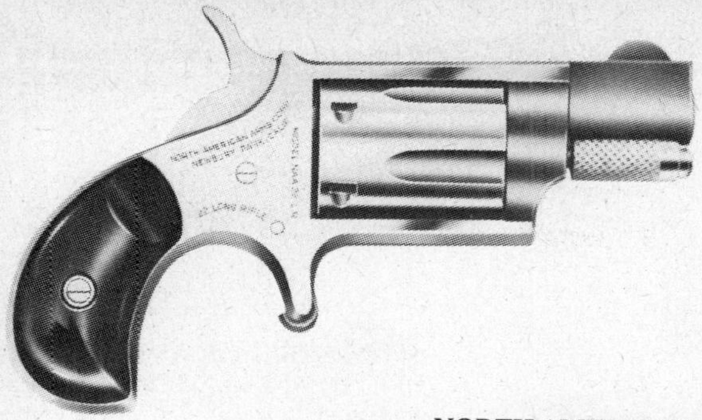

NORTH AMERICAN ARMS MINI-DERRINGER
Caliber: .22 LR. Capacity: 5. Barrel: 1 inch. Weight: 4.5 oz. Stainless steel construction. **Price: $128**

RUGER NEW MODEL SUPER BLACKHAWK
Caliber: .44 Mag. Capacity: 6. Barrel: 7½, 10 inches. Weight: 48 oz. (7½ inch); 50 oz. (10 inch). Broad serrated hammer spur. Non fluted cylinder. Squared trigger guard. Walnut stocks. Blued finish. Sights: micro adjustable rear. **Price: $207**

RUGER NEW MODEL BLACKHAWK
Caliber: .30 Carbine, .357 Mag., .41 Mag., .45 Colt. Capacity: 6. Barrel: 4⅝, 6½, 7½ inches. Weight: 38 to 44 oz. Transfer bar system permits safe carry of fully loaded cylinder. Walnut stocks. Blued finish. Sights: micro adjustable rear. **Price: $180** ; **$225** (stainless)

RUGER NEW MODEL BLACKHAWK CONVERTIBLE
Same as standard Blackhawk, except in .357 Mag. and .45 Colt only. .357 Mag. version comes with interchangeable cylinder for 9 mm Parabellum; .45 Colt version comes with interchangeable cylinder for .45 ACP. **Price: $197**

RUGER NEW MODEL SUPER SINGLE-SIX
Caliber: .22 LR and .22 WMR. Capacity: 6. Barrel: 4⅝, 5½, 6½, 9½ inches. Weight: 31½ to 39 oz. Transfer bar system permits safe carry with fully loaded cylinder. Comes with cylinder for .22 WMR. Music wire springs. Blued finish. Sights: adjustable. **Price: $142**

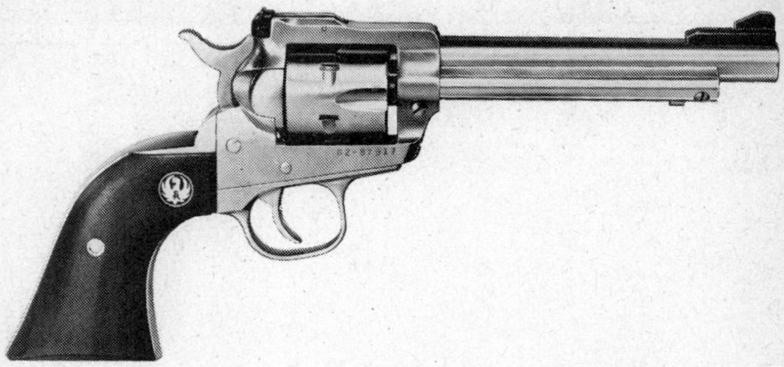

RUGER NEW MODEL SINGLE-SIX STAINLESS
Same as standard Single-Six, except comes in stainless steel. **Price: $201**

UNITED SPORTING ARMS SEVILLE
Caliber: .357 Mag., 9 mm Win. Mag., .41 Mag., .45 Colt, .45 Win. Mag. Capacity: 6. Barrel: 4⅝, 5½, 6½, 7½, 9½ inches. Weight: 48 oz. Walnut grips. Available with interchangeable barrels and cylinders to convert to additional calibers. Sights: adjustable rear, ramp front. Blue finish. **Price: $252**

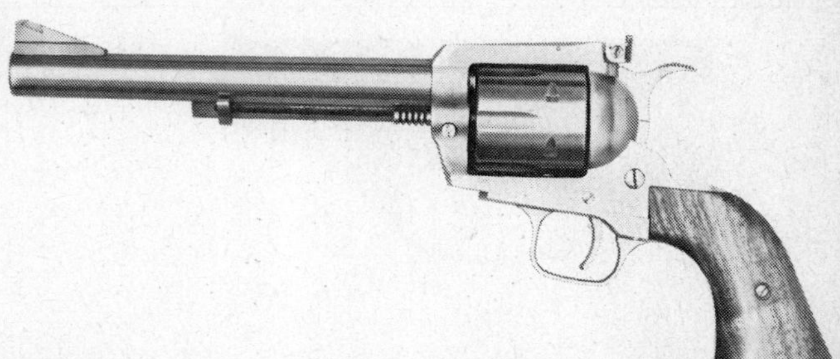

UNITED SPORTING ARMS EL DORADO
Same as Seville, except stainless steel construction. **Price: $350**

WESTERN CATTLEMAN MAGNUM
Caliber: .45 Colt, .357 Mag., .44 Mag., .44-40. Capacity: 6. Barrel: 4¾, 5½, 7¼ inches. Weight: 41 oz. Color case hardened frame. Available in all steel or with brass backstrap and trigger guard. Walnut stocks. Sights: fixed. **Price: $250**

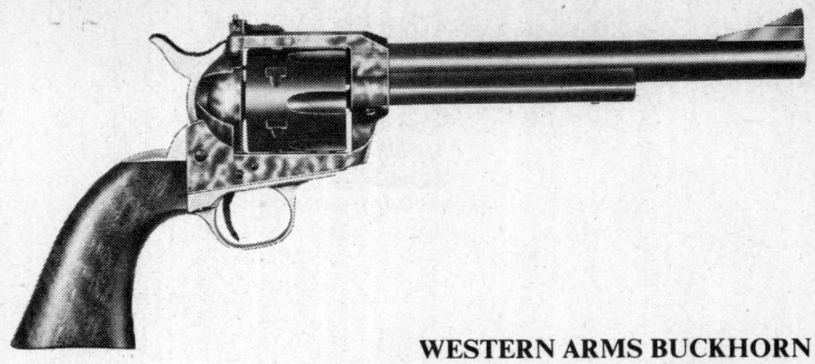

WESTERN ARMS BUCKHORN
Caliber: .45 Colt, .357 Mag., .44 Mag. Capacity: 6. Barrel: 5¾, 6, 7½, 12 inches. Weight: 44 oz. Color case hardened frame. Available in all steel or with brass backstrap and trigger guard. Walnut stocks. Sights: fixed. **Price: $275**

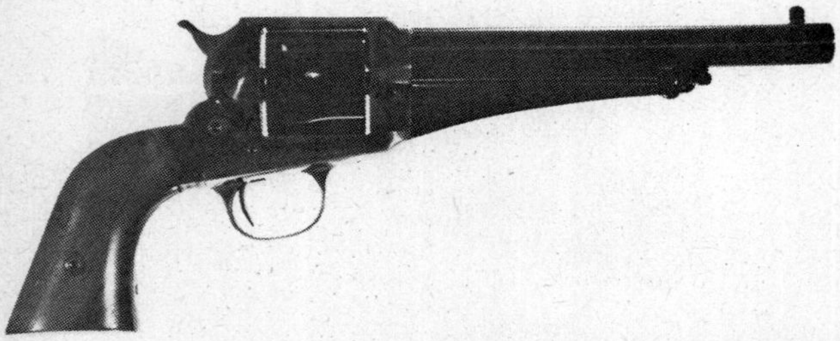

WESTERN ARMS MODEL 1875
Caliber: .357 Mag., .44-40, .45 Colt. Capacity: 6. Barrel: 7½ inches. Weight: 44 oz. Replica of M1875 Remington. Color case hardened frame. Walnut grips. **Price: $250**

WESTERN ARMS MODEL 22
Caliber: .22 LR and .22 WMR. Capacity: 6. Barrel: 4¾, 5½, 6½ inches. All steel construction, color case hardened frame. Available with steel or brass backstrap. Comes with interchangeable cylinders for .22 LR and .22 WMR. Walnut grips. Sights: fixed; target model with adjustable rear. **Price: $160; $180** (target model)

Semiautomatic

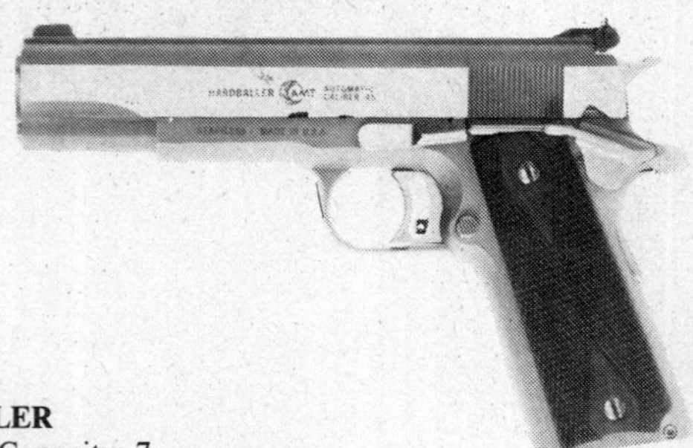

AMT HARDBALLER
Caliber: .45 ACP. Capacity: 7. Barrel: 4⅞ inches. Weight: 39 oz. Stainless steel construction. Combat safety and trigger. Sights: low adjustable. **Price: $495** ; **$695** (7 inch Longside version)

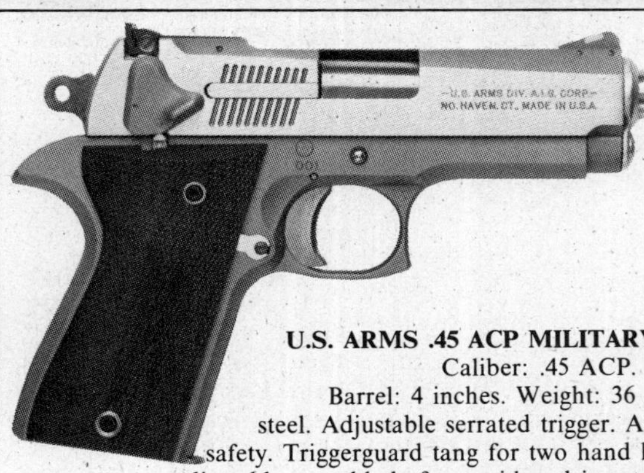

U.S. ARMS .45 ACP MILITARY-COMBAT
Caliber: .45 ACP. Capacity: 7. Barrel: 4 inches. Weight: 36 oz. Stainless steel. Adjustable serrated trigger. Ambidextrous safety. Triggerguard tang for two hand hold. Sights: adjustable rear, blade front with red insert. Distributed by Mossberg. **Price: NA**

ASTRA CONSTABLE
Caliber: .22 LR, .32, .380. Capacity: 10 (8 in .32 and .380. Barrel: 3½ inches. Weight: 26 oz. Double action. Contoured tang. Blued, chromed, or engraved. Sights: adjustable. Imported by Interarms. **Price: $245** ; **$270** (chrome)

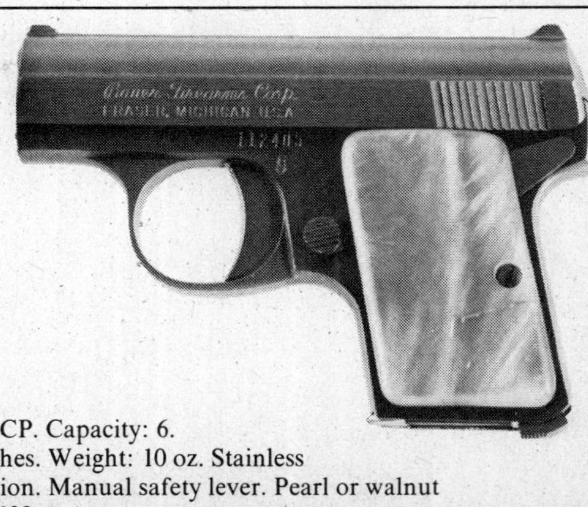

BAUER .25
Caliber: .25 ACP. Capacity: 6. Barrel: 2¼ inches. Weight: 10 oz. Stainless steel construction. Manual safety lever. Pearl or walnut grips. **Price: $100**

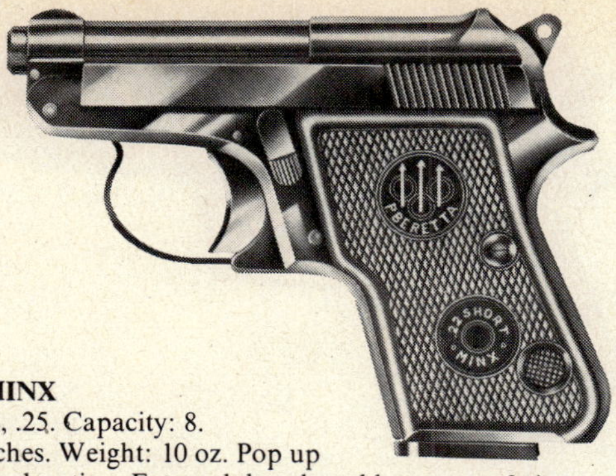

BERETTA MINX
Caliber: .22 S, .25. Capacity: 8.
Barrel: 2½ inches. Weight: 10 oz. Pop up barrel, blowback action. External thumb and hammer safeties. Pushbutton magazine release. Distributed by J.L. Galef & Son.
Price: $125

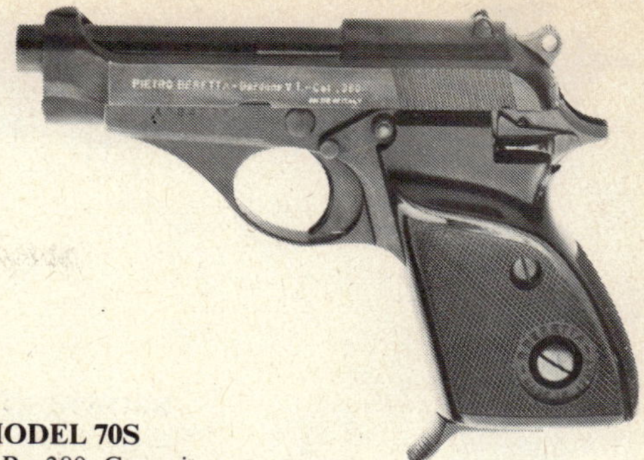

BERETTA MODEL 70S
Caliber: .22 LR, .380. Capacity: 10 (.22), 7 (.380). Barrel: 3⅝ inches. Weight: 23¼ oz. External hammer. Checked plastic stocks. Sights: fixed. **Price: $235**

BERETTA MODEL 76
Caliber: .22 LR. Capacity: 10. Barrel: 6 inches. Weight: 35 oz. External hammer, thumb safety. Sights: adjustable. **Price: $285** (plastic grips); **$320** (wood grips)

BERETTA MODEL 81/84
Caliber: .32 (M81), .380 (M84). Capacity: 13 (.380), 12 (.32). Barrel: 4 inches. Weight: 23 oz. Double action. Thumb magazine release. Blued finish. Sights: fixed. **Price: $335** (plastic grips); **$355** (wood grips)

BERETTA MODEL 92
Caliber: 9 mm. Capacity: 15. Barrel: 5 inches. Weight: 32 oz. Double action. Thumb magazine release. Blued finish. Sights: fixed. **Price: $470** (plastic grips); **$495** (wood grips)

BERNARDELLI MODEL 80
Caliber: .22 LR, .380. Capacity: 7 (.380), 10 (.22). Barrel: 3½ inches. Weight: 26½ oz. Hammer blocking side safety. Loaded round indicator. Dual recoil buffer springs. Sights: adjustable rear. Imported by Interarms. **Price: $200**

BERNARDELLI MODEL 100
Caliber: .22 LR. Capacity: 10. Barrel: 5¾ inches. Weight: 38 oz. Broad barrel rib. Walnut thumbrest grips. Sights: adjustable rear, interchangeable front. Imported by Interarms.
Price: $350 (includes case)

BERSA MODEL 644
Caliber: .22 LR. Capacity: 10. Single action trigger. Visible hammer. Hammer, firing pin, and magazine safety. Contoured nylon grips. Imported by Interarms.
Price: $175

BROWNING HI-POWER
Caliber: 9 mm. Capacity: 14.
Barrel: 4¾ inches. Weight: 32 oz. External hammer. Manual thumb safety. Checkered walnut stocks.
Price: $400 (fixed sights); **$440** (adjustable sights)
Engraved versions available at extra cost.

BROWNING BDA
Caliber: .45 ACP. Capacity: 7.
Barrel: 4¼ inches. Weight: 29 oz. Double action. Black plastic grips. Sights: fixed square notch rear, blade front. **Price: $350**

BROWNING BDA 380
Caliber: .380. Capacity: 12.
Barrel: 3⅞ inches. Weight: 23 oz.
Double action. Safety operates from left or right side. Walnut grips. Sights: fixed. **Price: $300**

BROWNING CHALLENGER
Caliber: .22 LR. Capacity: 10.
Barrel: 6¾ inches. Weight: 38 oz.
All steel hand fitted frame. Side thumb safety. Hardwood grips. Sights: screw adjustable rear. **Price: $190**

WILDEY GAS-ACTION AUTO
Caliber: 9 mm Win. Mag., .45 Win. Mag.
Capacity: 14 (9 mm), 8 (.45). Barrel: 5, 6, 7, 8, 10 inches. Weight: approx. 51 oz. Gas operated.
DA trigger mechanism. Rotary bolt lock. Stainless steel.
Price: $490; $500 (8, 10-inch)

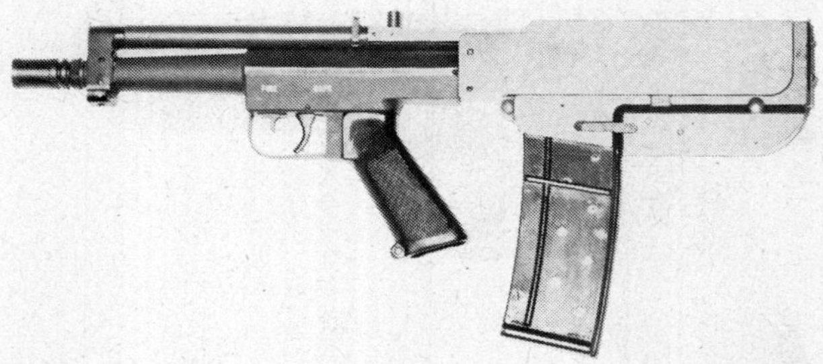

BUSHMASTER PISTOL
Caliber: .223. Detachable magazine, capacity: 30. Barrel: 11½ inches.
Weight: 5¼ lbs. AK-47 type gas system, lightweight alloy receiver. **Price: $340**

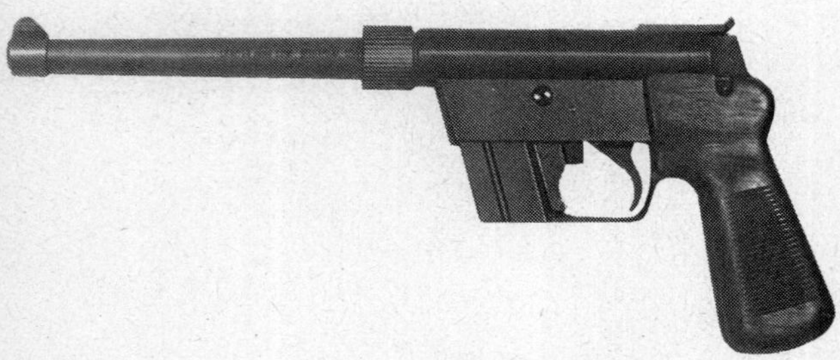

CHARTER ARMS EXPLORER II
Caliber: .22 LR. Capacity: 8.
Barrel: 6, 8, 10 inches. Weight: 28 oz.
Semi-gloss textured enamel finish. Simulated walnut grips. Sights: adjustable rear. Price: **NA**

AMERICAN ARMS TP-70
Caliber: .25. Capacity: 6.
Overall length: 4 ¾ inches. Weight: 12½ oz.
Double action. Stainless steel. **Price: $180**

COLT GOVERNMENT MODEL MK IV
Caliber: .45 ACP, .38 Super,
9 mm. Capacity: 7. Barrel: 5 inches.
Weight: 39 oz. External hammer. Checkered walnut grips.
Sights: dovetail slide windage adjustment rear, low blade front. **Price: $300** to **$320**

COLT COMBAT COMMANDER
Caliber: 9 mm, .38 Super, .45 ACP.
Capacity: 8 (7 in .45). Barrel: 4¼ inches.
Weight: 33 oz. Sandblasted ramp. Walnut grips.
Price: $296 to **$314**

COLT GOLD CUP MK IV
Caliber: .45 ACP. Capacity: 7.
Barrel: 5 inches. Weight: 38½ oz. Adjustable trigger. Hand fitted. Patridge front sight, micro adjustable rear. **Price: $400**

CROWN CITY EAGLE
Caliber: .45 ACP. Capacity: 7.
Barrel: 5 inches. Weight: 39 oz. Grip and thumb safeties. Adjustable trigger. Non-glare black oxide finish. Sights: fixed. **Price: $275** ; **$325** (adjustable sights)

CROWN CITY CONDOR/FALCON COMBO
Commander size interchangeable slide and barrel unit included. Stainless steel construction. Otherwise, features same as Eagle. **Price: $562** ; **$637** (adjustable sights)

DETONICS .45
Caliber: .45 ACP. Capacity: 6.
Barrel: 3¼ inches. Weight: 29 oz. Thumb safety. Walnut grips. Sights: dovetailed windage adjustment rear. **Price: $435** to **$575** (depending on options)

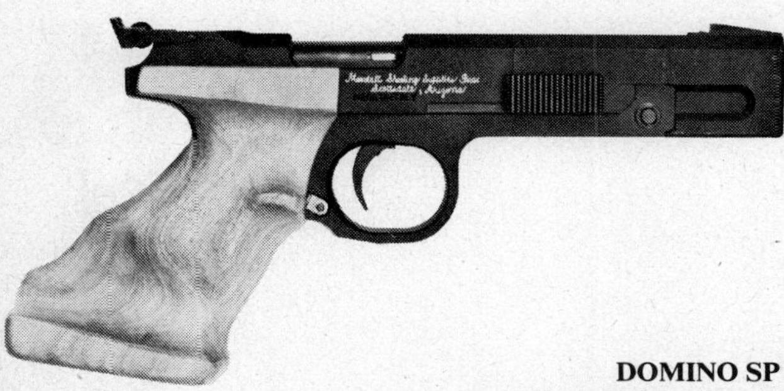

DOMINO SP 602
Caliber: .22 LR. Capacity: 5.
Barrel: 5¾ inches. Weight: 44 oz.
Adjustable and removable trigger mechanism. Magazine inserts from the top. Extremely low line of sight. Adjustable orthopaedic grip. From Mandall Shooting Supplies. **Price: $1050** ; **$1050** (.22 S version)

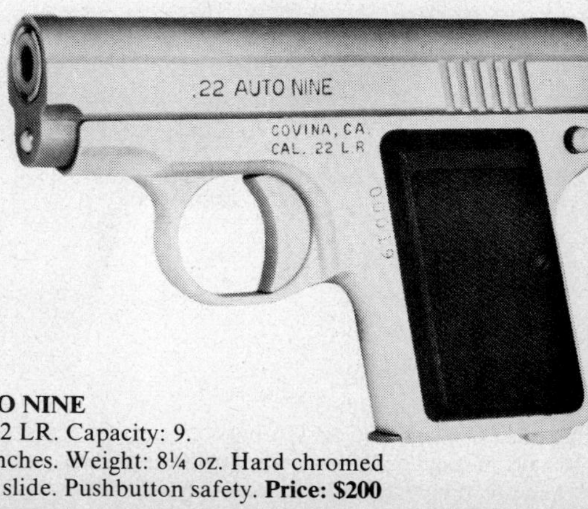

FTL AUTO NINE
Caliber: .22 LR. Capacity: 9.
Barrel: 2 inches. Weight: 8¼ oz. Hard chromed barrel and slide. Pushbutton safety. **Price: $200**

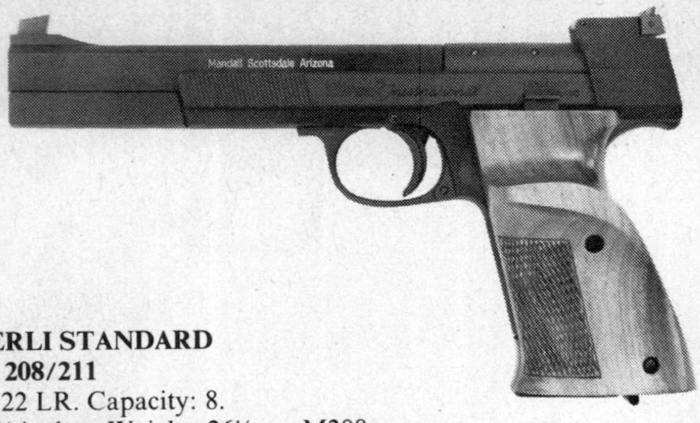

HAMMERLI STANDARD MODEL 208/211
Caliber: .22 LR. Capacity: 8.
Barrel: 4¾ inches. Weight: 26½ oz. M208 features adjustable palm rest grip. M211 (shown) features adjustable thumbrest grip. Extra heavy barrel weight. Sights: micro rear. From Mandall Shooting Supplies.
Price: $1050

HAMMERLI MODEL 230
Caliber: .22 S. Capacity: 5.
Barrel: 6¼ inches. Weight: 44 oz.
Adjustable trigger. Sights: micro rear. From Mandall Shooting Supplies. **Price: $1095**

SIG-SAUER P230
Caliber: .32, .380, 9 mm.
Capacity: 7 (8 in .32). Barrel: 3 ¾ inches.
Weight: 16 oz. Double action. Visible hammer.
Fixed sights. Imported by Mandall Shooting Supplies. **Price: $300**

HECKLER & KOCH P9S
Caliber: 9 mm. Capacity: 9.
Barrel: 4 inches. Weight: 33 oz. Double action.
Thumb safety. Heel magazine release. Sights: adjustable.
Price: $384

HECKLER & KOCH P9S .45
Same as 9 mm P9S, except chambered for .45 ACP.
Price: $384

HECKLER & KOCH HK-4
Caliber: .22 LR, .25, .32, .380.
Capacity: 8, 7 (.380). Weight: 17 oz. DA pocket auto with conversion kit for all calibers listed. Comes with magazines, barrels. Sights: fixed. **Price: $340**

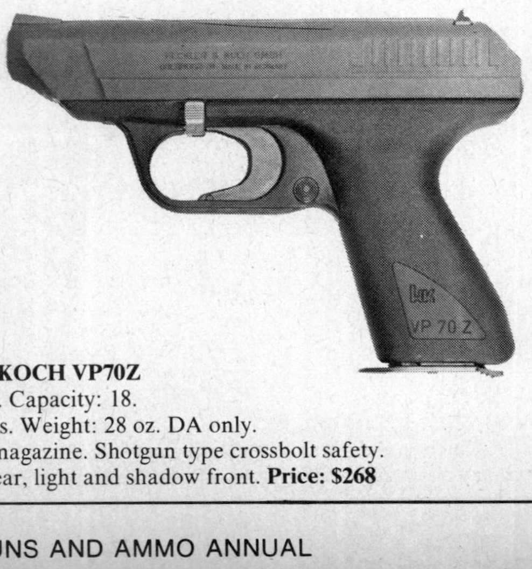

HECKLER & KOCH VP70Z
Caliber: 9 mm. Capacity: 18.
Barrel: 4 inches. Weight: 28 oz. DA only.
Double stack magazine. Shotgun type crossbolt safety.
Sights: fixed rear, light and shadow front. **Price: $268**

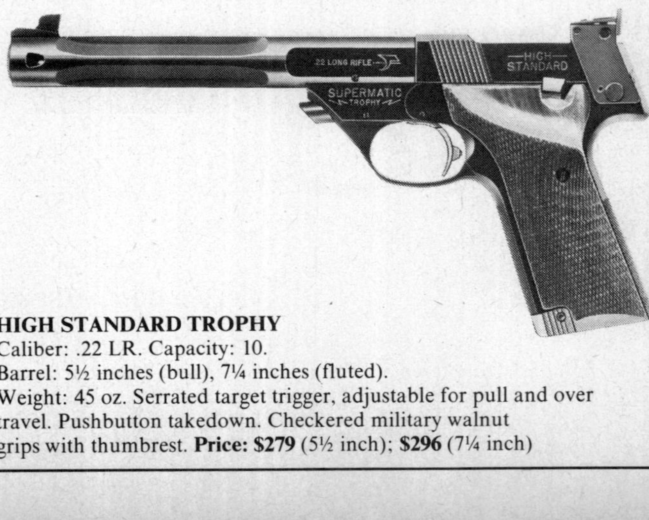

HIGH STANDARD TROPHY
Caliber: .22 LR. Capacity: 10.
Barrel: 5½ inches (bull), 7¼ inches (fluted).
Weight: 45 oz. Serrated target trigger, adjustable for pull and over travel. Pushbutton takedown. Checkered military walnut grips with thumbrest. **Price: $279** (5½ inch); **$296** (7¼ inch)

HIGH STANDARD CITATION
Caliber: .22 LR. Capacity: 10.
Barrel: 5½ inches (bull), 7¼ inches (fluted).
Weight: 45 oz. Pushbutton takedown. Trigger adjustable for over travel. Walnut stocks with thumbrest. Sights: micro click adjustable rear. **Price: $262** (5½ inch); **$279** (7¼ inch)

HIGH STANDARD SHARPSHOOTER
Caliber: .22 LR. Capacity: 10.
Barrel: 5½ inches (bull). Weight: 42 oz. Pushbutton takedown. Wide, serrated target trigger. Checkered walnut grips with thumbrest. Sights: micro click adjustable rear (independent of slide). **Price: $232**

HIGH STANDARD VICTOR
Caliber: .22 LR. Capacity: 10.
Barrel: 5½ inches. Weight: 47 oz. Vent rib barrel. Wide serrated trigger with overtravel adjustment. Stippled backstrap and frontstrap. Checkered walnut grips. Sights: micro click adjustable rear. **Price: $319**

IVER JOHNSON X-300 PONY
Caliber: .380. Capacity: 6.
Barrel: 3 inches. Weight: 20 oz. Thumb safety. All steel construction. Checkered walnut grips. Sights: adjustable rear, blade front. **Price: $230** (blue); **$244** (chrome)

IVER JOHNSON SUPER ENFORCER
Caliber: .30. Capacity: 15, 30.
Barrel: 9½ inches. Walnut stock. **Price: $242**

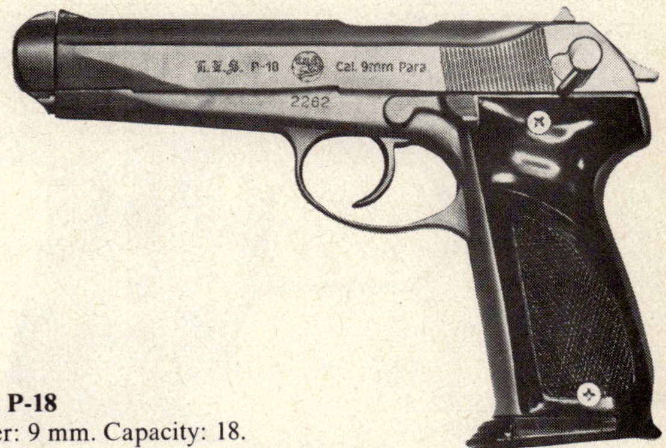

L.E.S. P-18
Caliber: 9 mm. Capacity: 18.
Barrel: 5½ inches. Weight: 36 oz. Inertia firing pin. Stainless steel construction. SA and DA versions available. Sights: fixed. **Price: $290** to **$345**

LLAMA LARGE FRAME AUTO
Caliber: 9 mm, .38 Super, .45 ACP.
Capacity: 9, 7 (.45). Barrel: 5 inches.
Weight: 40 oz. Locked breech action. Wide spur hammer. Side lever safety. Serrated trigger. Sights: adjustable. From Stoeger. **Price: $245** to **$1500** (depending on finish)

LLAMA SMALL FRAME AUTO
Caliber: .22 LR, .32, .380.
Capacity: 7, 8 (.22). Barrel: 3¾ inches.
Weight: 23 oz. Blowback operation. Side lever thumb safety. Black plastic stocks with thumbrest. Sights: screw adjustable rear. From Stoeger. **Price: $245** to **$1500** (depending on finish)

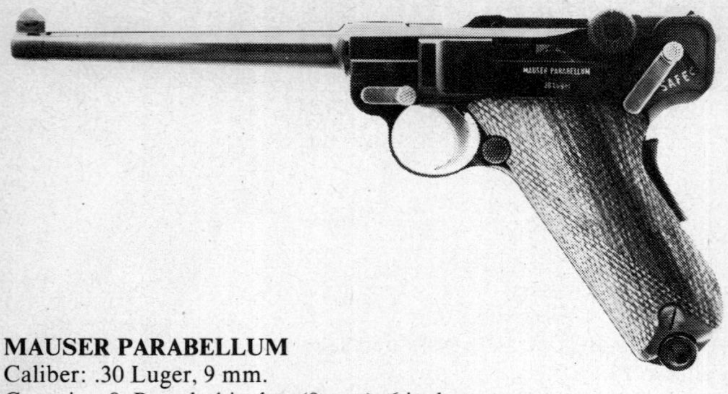

MAUSER PARABELLUM
Caliber: .30 Luger, 9 mm.
Capacity: 8. Barrel: 4 inches (9 mm), 6 inches (.30 Luger). Weight: 32 oz. Swiss pattern American Eagle Luger in limited production run. Comes with velvet lined presentation case. From Interarms. **Price: $625**

MAUSER HSC
Caliber: .32, .380.
Capacity: 7. Barrel: 3¾ inches. Weight: 23 oz. Bright blue. Walnut finish. Sights: fixed. From Interarms. **Price: $250** ; **$275** (one of 5000 series)

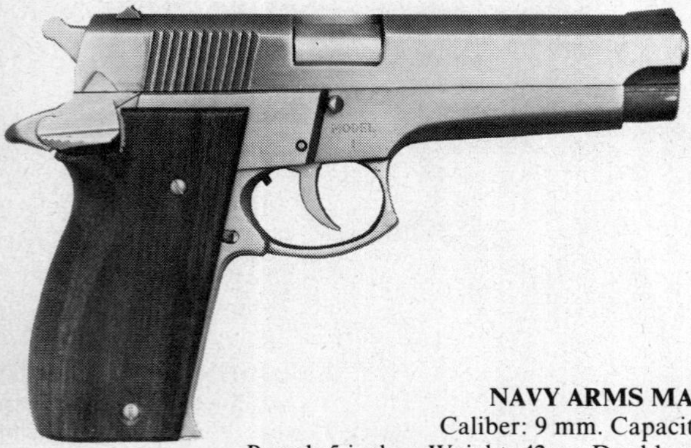

NAVY ARMS MAMBA
Caliber: 9 mm. Capacity: 14.
Barrel: 5 inches. Weight: 42 oz. Double action. Stainless steel construction. Combat style trigger guard. Sights: fixed. **Price: $325**

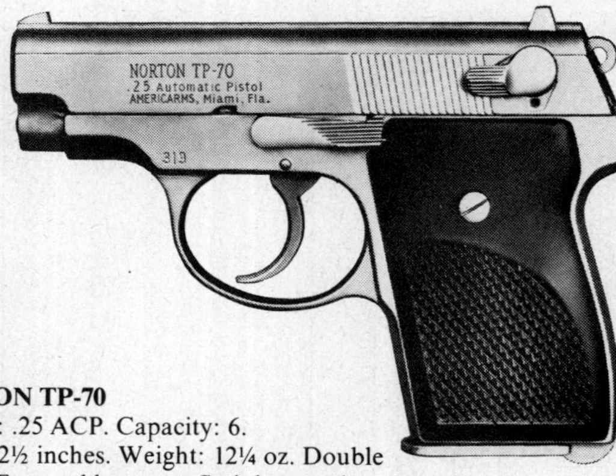

NORTON TP-70
Caliber: .25 ACP. Capacity: 6.
Barrel: 2½ inches. Weight: 12¼ oz. Double action. External hammer. Stainless steel construction. Available from R.E. Shattuck. **Price: $140**

RAVEN ARMS MODEL P-25
Caliber: .25. Capacity: 6.
Barrel: 2½ inches. Weight: 15 oz. Walnut grips. Available in blue, chrome, and nickel finish. **Price: $60**

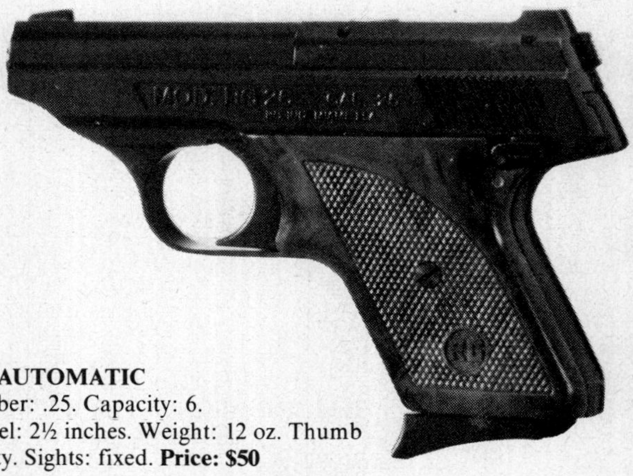

RG AUTOMATIC
Caliber: .25. Capacity: 6.
Barrel: 2½ inches. Weight: 12 oz. Thumb safety. Sights: fixed. **Price: $50**

RUGER STANDARD MODEL .22 AUTO
Caliber: .22 LR. Capacity: 9.
Barrel: 4¾, 6 inches. Weight: 36, 38 oz. Two stage trigger pull. Positive safety. Hard rubber grips. Blued finish. Sights: adjustable rear. **Price: $109**

RUGER MARK I TARGET MODEL
Caliber: .22 LR. Capacity: 9.
Barrel: 6⅞ inches (tapered); 5½ inches (bull). Weight: 42 oz. Anti-backlash trigger. Hard rubber grips. Blued finish. Sights: micro adjustable rear. **Price: $140**

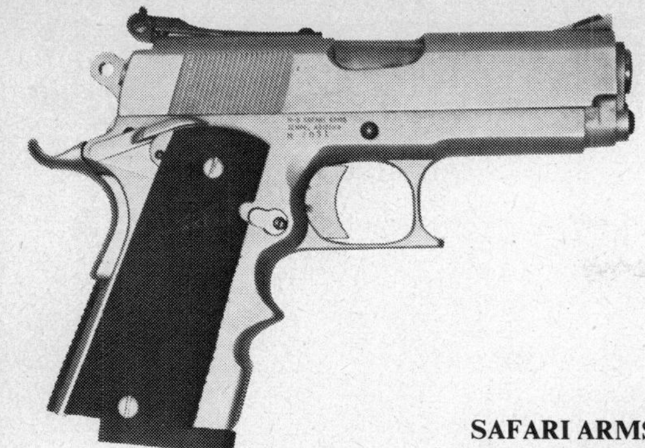

SAFARI ARMS ENFORCER
Caliber: .45 ACP. Capacity: 6. Barrel: 3¾ inches. Weight: 35 oz. Short barrel version of Matchmaster. **Price: $486** (Teflon); **$526** (Armaloy)

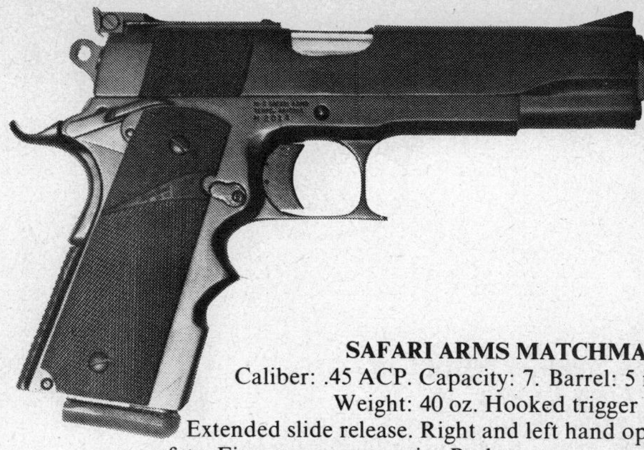

SAFARI ARMS MATCHMASTER
Caliber: .45 ACP. Capacity: 7. Barrel: 5 inches. Weight: 40 oz. Hooked trigger guard. Extended slide release. Right and left hand operated safety. Finger groove on grip. Pachmayr neoprene grips. Sights: Behlert adjustable rear. **Price: $486** (Teflon); **$526** (Chrome)

VEGA COMBAT MATCH .45
Caliber: .45 ACP. Capacity: 7. Barrel: 5 inches. Weight: 39 oz. Wide spur hammer. Oversized combat safety. All stainless steel construction. Sights: fixed or adjustable. From Pacific International. **Price: $329; $359** (adjustable rear sight)

SIG P 210-6
Caliber: 7.65, 9 mm. Capacity: 8. Barrel: 4¾ inches. Weight: 37½ oz. Adjustable trigger stop. Sandblasted finish. Plastic grips. Sights: micro adjustable rear. .22 Cal. Conversion kit available. From Mandall Shooting Supplies. **Price: $1450** ; **$700** (Conversion kit)

SIG-HAMMERLI P-240
Caliber: .32 S&W Long. Capacity: 5. Barrel: 5½ inches. Weight: 40 oz. Adjustable trigger. Meets I.S.U. Centerfire requirements. Sights: micro adjustable. From Mandall Shooting Supplies. **Price: $1450**

SILE/BENELLI B76/B77
Caliber: 9 mm (B76), .32 ACP (B77) Capacity: 8. Barrel: 4¼ inches. Weight: 34 oz. Double action. External hammer, thumb safety. External parts blued, internal parts hard chrome plated. Checkered walnut grips. Sights: low visibility fixed. **Price: $320**

SMITH & WESSON MODEL 39 AUTO
Caliber: 9 mm. Capacity: 8. Barrel: 4 inches. Weight: 26½ oz. Double action. Checkered grips. Sights: micro click adjustable rear, serrated ramp front. **Price: $246** (blue); **$271** (nickel)

SMITH & WESSON MODEL 52 .38 MASTER
Caliber: .38 Spl. Mid Range Wadcutter. Capacity: 5. Barrel: 5 inches. Weight: 41 oz. Target trigger with adjustable trigger stop. Bright blue finish with sandblast stippling around sighting area. Sights: micro click adjustable rear, Patridge front on ramp base. **Price: $471**

SMITH & WESSON MODEL 59 AUTO
Caliber: 9 mm. Capacity: 14.
Barrel: 4 inches. Weight: 27 oz. Checked nylon grips. Sights: micro click adjustable rear (windage only), serrated ramp front. **Price: $294** (blue); **$321** (nickel)

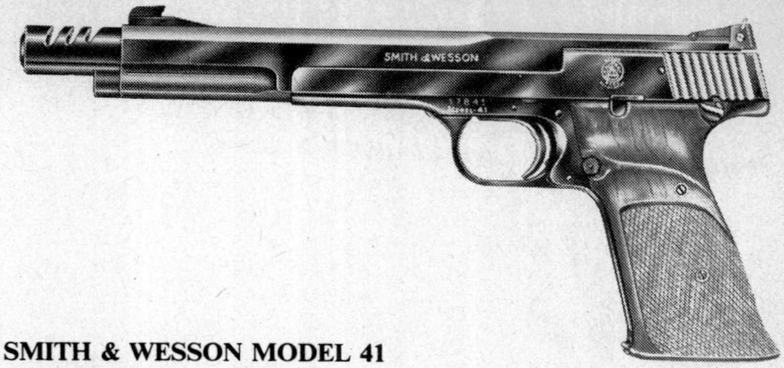

SMITH & WESSON MODEL 41
Caliber: .22 LR. Capacity: 10.
Barrel: 7 inches. Weight: 43½ oz. Target trigger with adjustable stop. Checkered walnut stocks with thumb rest adaptable to left or right hand shooters. Bright blue finish. Sights: micro click adjustable rear, undercut Patridge front. **Price: $294**

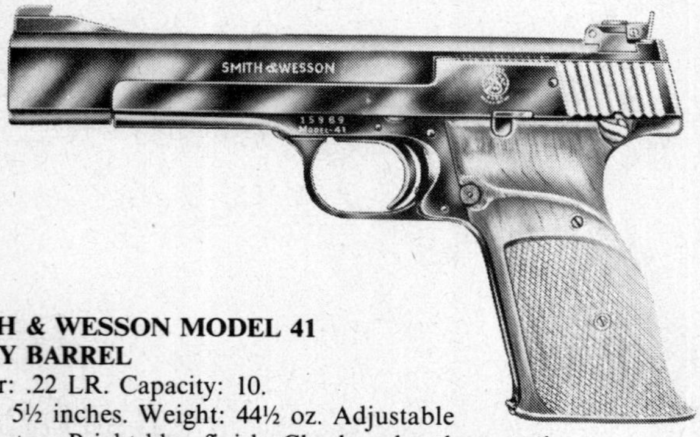

SMITH & WESSON MODEL 41 HEAVY BARREL
Caliber: .22 LR. Capacity: 10.
Barrel: 5½ inches. Weight: 44½ oz. Adjustable trigger stop. Bright blue finish. Checkered walnut stocks with thumb rest adaptable for left or right hand shooters. Sights: micro click rear, undercut Patridge front.
Price: $294

STAR PD
Caliber: .45 ACP. Capacity: 6.
Barrel: 3½ inches. Weight: 24 oz. Duraluminum frame. Thumb safety. Checkered walnut grips. Sights: adjustable rear. Imported by Interarms. **Price: $305**

STAR MODEL BM
Caliber: 9 mm. Capacity: 9.
Barrel: 4¼ inches. Weight: 26 oz. All steel construction. Thumb safety. Auto magazine disconnector. Plastic grips. Imported by Interarms. **Price: $255** (blue); **$275** (chrome)

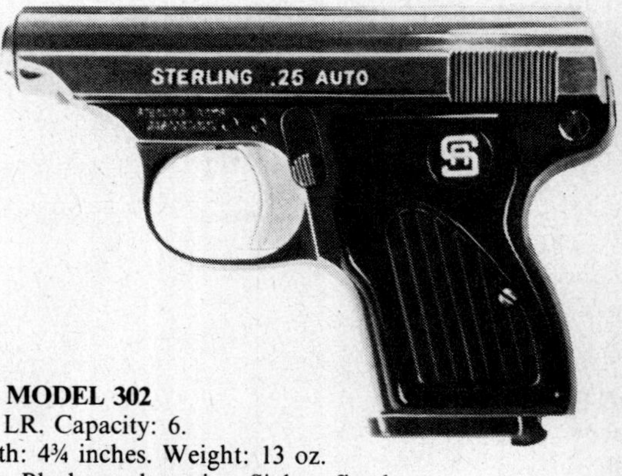

STERLING MODEL 302
Caliber: .22 LR. Capacity: 6.
Overall length: 4¾ inches. Weight: 13 oz.
Single action. Black cycolac grips. Sights: fixed.
Price: $105 (blue); **$125** (stainless)

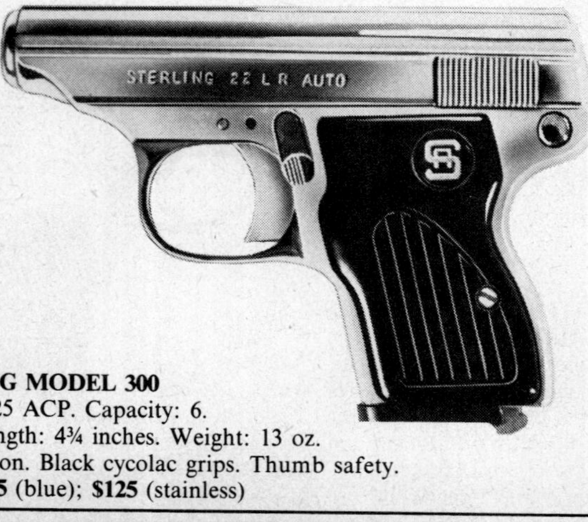

STERLING MODEL 300
Caliber: .25 ACP. Capacity: 6.
Overall length: 4¾ inches. Weight: 13 oz.
Single action. Black cycolac grips. Thumb safety.
Price: $105 (blue); **$125** (stainless)

STERLING MODEL 400 MK II
Caliber: .380. Capacity: 8.
Overall length: 6½ inches. Weight: 26 oz.
Double action. Rolling block safety. Low profile serrated hammer. Pushbutton magazine release. Sights: adjustable rear. **Price: $210**

STERLING MODEL 450 DA .45
Caliber: .45 ACP. Capacity: 7.
Overall length: 7½ inches. Weight: 36 oz.
Double action. Contoured trigger guard. Rolling block safety. Sights: adjustable. Blued finish. **Price: NA**

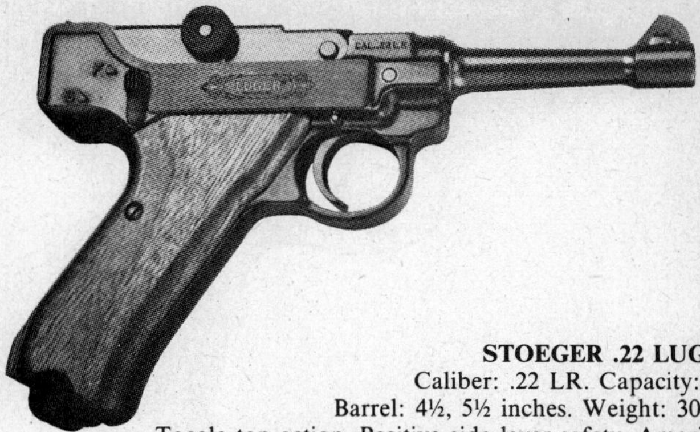

STOEGER .22 LUGER
Caliber: .22 LR. Capacity: 10.
Barrel: 4½, 5½ inches. Weight: 30 oz.
Toggle top action. Positive side lever safety. American black walnut stocks. Sights: fixed. From Stoeger. **Price: $149**

UNIQUE MODEL D.E.S./69
Caliber: .22 LR. Capacity: 5.
Barrel: 6 inches. Weight: 35 oz. Wide trigger. Sights: micro rear, ramp front. From Solersport. **Price: $600**

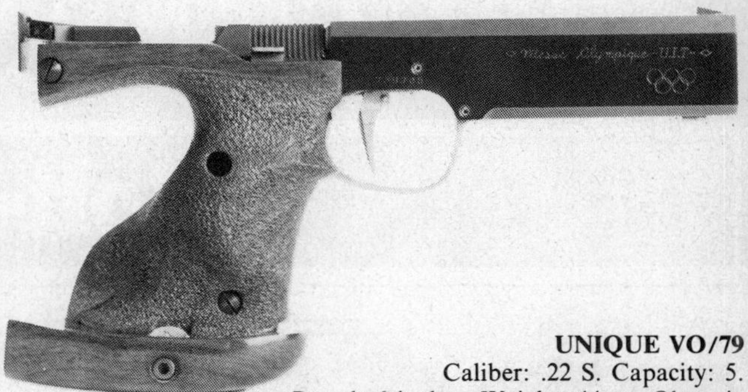

UNIQUE VO/79
Caliber: .22 S. Capacity: 5.
Barrel: 6 inches. Weight: 44 oz. Olympic rapid fire pistol. Equipped with dry firing mechanism. Walnut grip with thumb rest and adjustable palm shelf. From Solersport. **Price: $709**

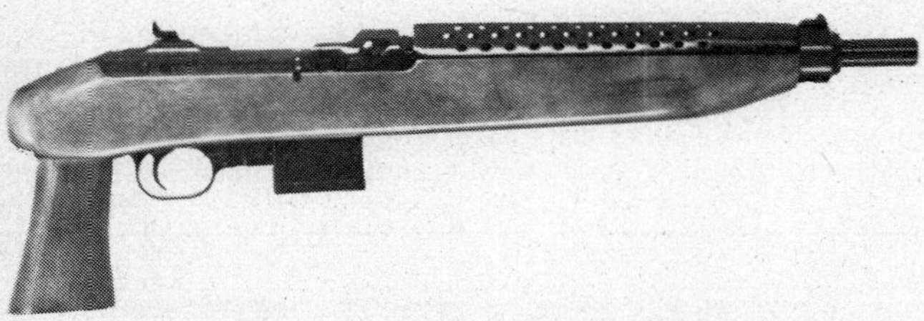

UNIVERSAL ENFORCER
Caliber: .30 Carbine. Capacity: 5, 15, 30. Barrel: 11¼ inches. Weight: 4 lbs. Flip type safety. Gas operated. Metal ventilated handguard. Birch stock and grip. Sights: square notch adjustable rear. **Price: $245**

WALTHER MODEL PP
Caliber: .22 LR, .32, .380.
Capacity: 10 (.22), 8 (.32), 7 (.380).
Barrel: 3⅞ inches. Weight: 23½ oz. DA auto. Plastic grips. Comes with two magazines. Sights: fixed. Imported by Interarms. **Price: $525** (.22); **$500** (.32, .380)

WALTHER MODEL PPK/S AMERICAN
Caliber: .380. Capacity: 7.
Barrel: 3¼ inches. Weight: 23 oz. Shortened version of Walther PP. Blued finish. Sights: fixed. Made in America. **Price: $265**

WALTHER P-38
Caliber: .22 LR, .30 Luger, 9 mm.
Capacity: 8. Barrel: 5 inches. Weight: 28 oz.
DA action. Plastic grips with side lanyard ring. Blue finish in CF calibers, matte finish in .22. Imported by Interarms. **Price: $750** (.22); **$660** (9 mm, .30 Luger)

WALTHER P38-K
Caliber: 9 mm. Same as P-38, except $2\frac{3}{4}$ inch barrel. Weight: 26 oz. **Price: $690**

WALTHER OSP/GSP
Caliber: .22 S, LR, .32 S&W.
Capacity: 5. Barrel: $5\frac{3}{4}$ inches.
Weight: $44\frac{3}{4}$ oz. Conforms to I.S.U. and NRA Match regulations. Imported by Interarms. **Price: $700** to **$1250**

SHOTGUN CATALOG

Bolt-Action

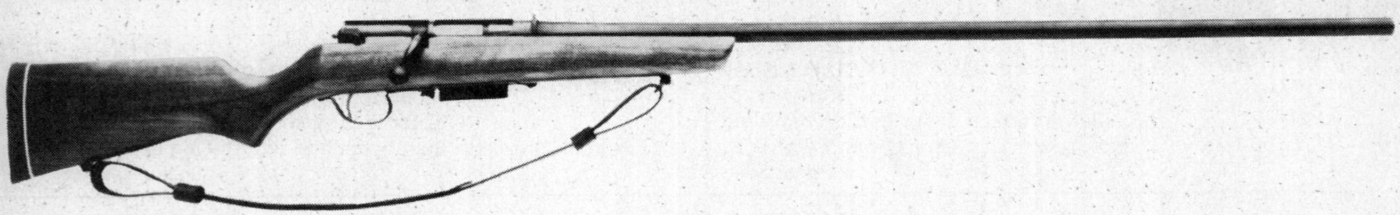

MARLIN MODEL 55 GOOSE GUN
Gauge: 12. Chamber: 3 inch. Detachable magazine, capacity: 2. Barrel: 36 inches. Weight: 8 lbs. Choke: F. American black walnut stock. Gold plated trigger. Thumb safety. Comes with leather carrying strap. **Price: $116**

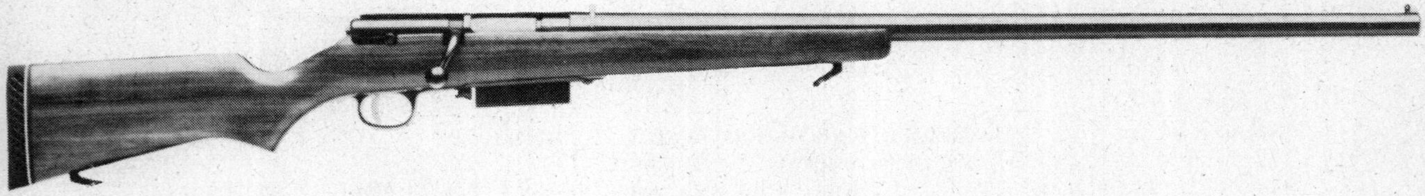

MARLIN SUPER GOOSE 10
Gauge: 10. Chamber: $3\frac{1}{2}$ inch. Detachable magazine, capacity: 2. Barrel: 34 inches. Weight: $10\frac{1}{2}$ lbs. Choke: F. Thumb safety. American black walnut stock with detachable sling swivels and carrying strap. Bead front sight. **Price: $207**

MOSSBERG 395K, 385K, 183K
Gauge: 12 (395K), 20 (385K), .410 (183K). Chamber: 3 inch. Detachable magazine, capacity: 2. Barrel: 26 inches. Weight: $6\frac{3}{4}$ lbs. Choke: selective. Thumb safety. C-Lect-Choke attachment included. Walnut finished hardwood stock with recoil pad. **Price: $104** (395K); **$98** (385K); **$90** (183K)

SAVAGE/STEVENS MODEL 58
Gauge: .410. Chamber: 3 inch. Detachable magazine, capacity: 3. Barrel: 24 inches. Weight: 5½ lbs. Choke: F. Thumb safety. Self cocking bolt action with band extractor. Walnut finished hardwood stock. **Price: NA**

Slide Action

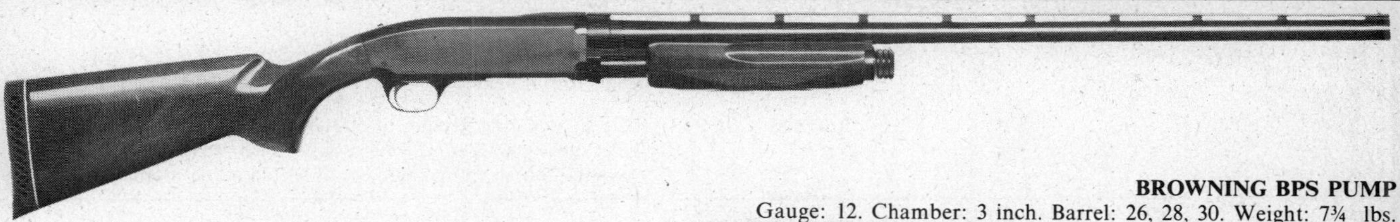

BROWNING BPS PUMP
Gauge: 12. Chamber: 3 inch. Barrel: 26, 28, 30. Weight: 7¾ lbs. Choke: IC, M, F. Bottom feeding and ejecting. High post vent rib. Double action bars. Select walnut stock. **Price: $300** (Grade I hunting); **$310** (Grade I Trap)

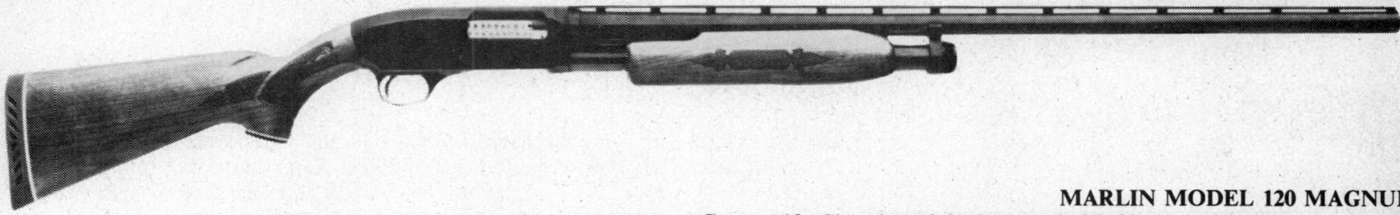

MARLIN MODEL 120 MAGNUM
Gauge: 12. Chamber: 3 inch. Barrel: 26, 28, 30 inches. Weight: 8¼ lbs. Choke: F, M, IC. Crossbolt safety. Checkered walnut stock and forend Semi-beavertail forend. Deluxe recoil pad. **Price: $253**

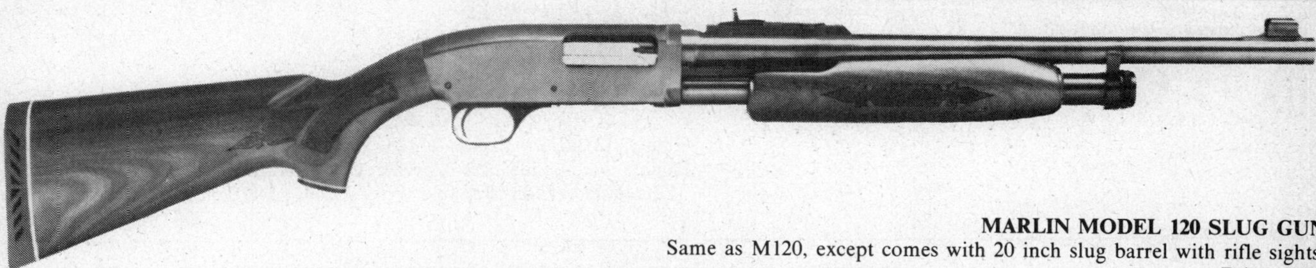

MARLIN MODEL 120 SLUG GUN
Same as M120, except comes with 20 inch slug barrel with rifle sights. **Price: $253**

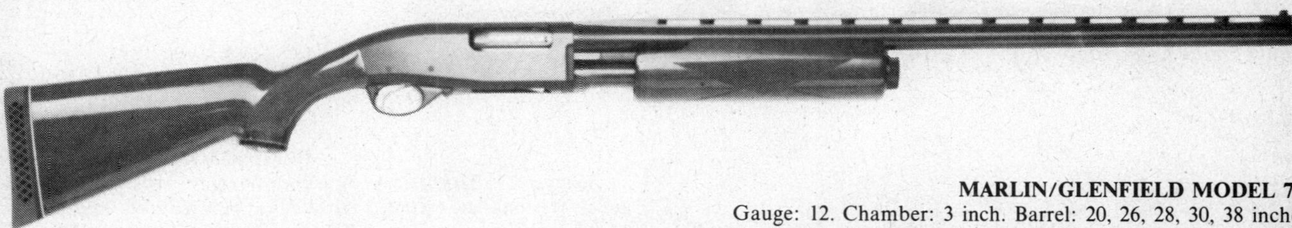

MARLIN/GLENFIELD MODEL 778
Gauge: 12. Chamber: 3 inch. Barrel: 20, 26, 28, 30, 38 inches. Weight: 7¾ lbs. Choke: M, F, IC. Double action bars. Engine turned bolt. Crossbolt safety. Vent rib barrel. **Price: $184** to **$209**

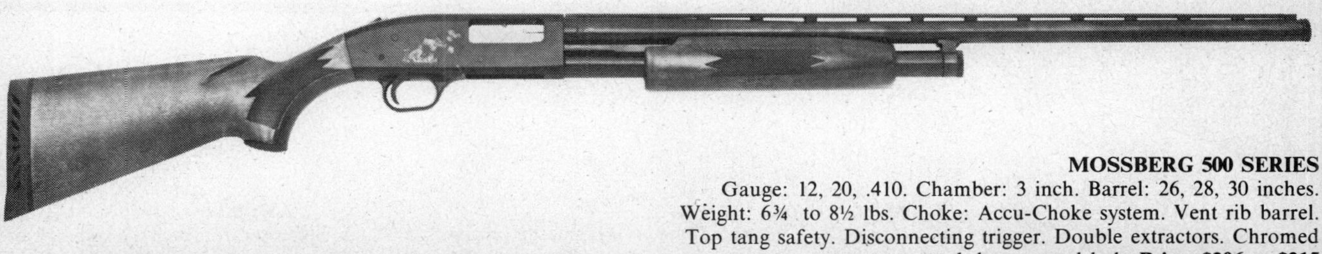

MOSSBERG 500 SERIES
Gauge: 12, 20, .410. Chamber: 3 inch. Barrel: 26, 28, 30 inches. Weight: 6¾ to 8½ lbs. Choke: Accu-Choke system. Vent rib barrel. Top tang safety. Disconnecting trigger. Double extractors. Chromed and damascened bolt. **Price: $206** to **$215**

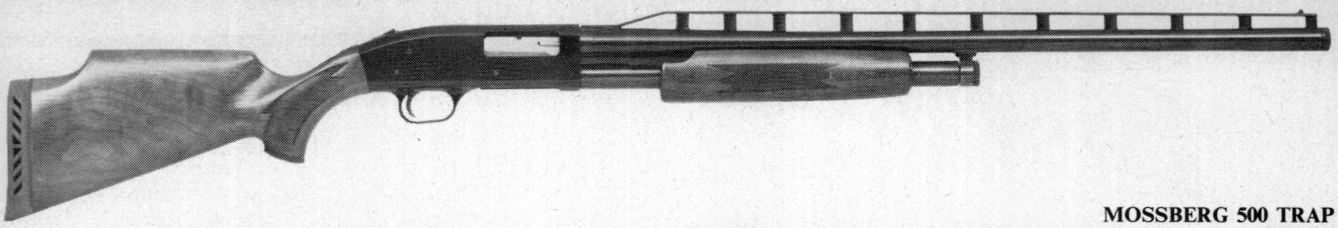

MOSSBERG 500 TRAP
Same as 500 series, except comes with high rib barrel in 28 or 30 inch length. Includes pistol grip cap and heavy duty recoil pad. **Price: $306** to **$315**

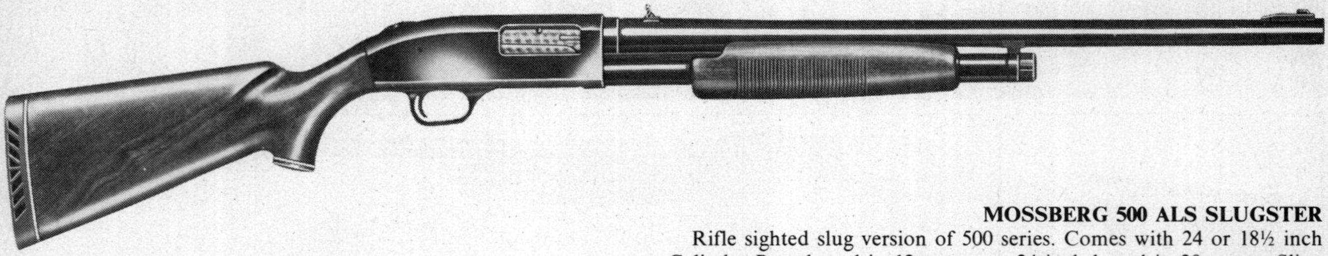

MOSSBERG 500 ALS SLUGSTER
Rifle sighted slug version of 500 series. Comes with 24 or 18½ inch Cylinder Bore barrel in 12 gauge, or 24 inch barrel in 20 gauge. Sling swivels and strap included. Tapped for scope mounts. **Price: $200**

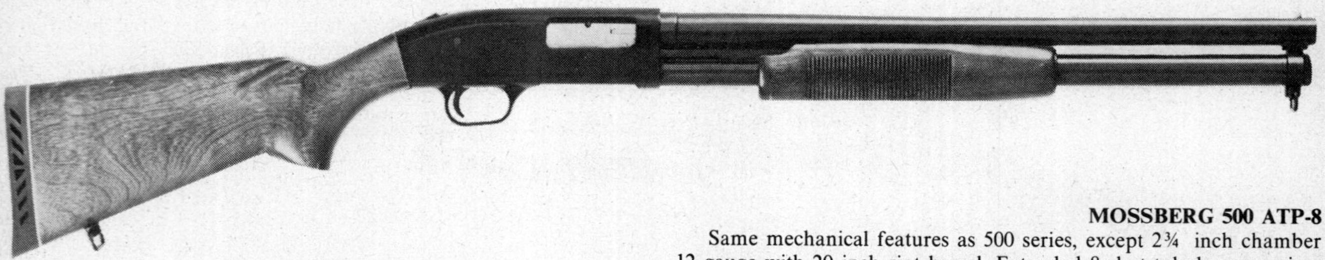

MOSSBERG 500 ATP-8
Same mechanical features as 500 series, except 2¾ inch chamber 12 gauge with 20 inch riot barrel. Extended 8 shot tubular magazine. Recoil pad and sling swivels. **Price: $185**

REMINGTON MODEL 870 WINGMASTER
Gauge: 12, 16, 20. Barrel: 26, 28, 30 inches. Weight: 7 lbs. Choke: F,M,IC. Double action slide bars. Checkered walnut stock and forearm. Recoil pad. Left hand version available. **Price: $240; $278** (vent rib); **$267** (magnum); **$306** (magnum, vent rib)

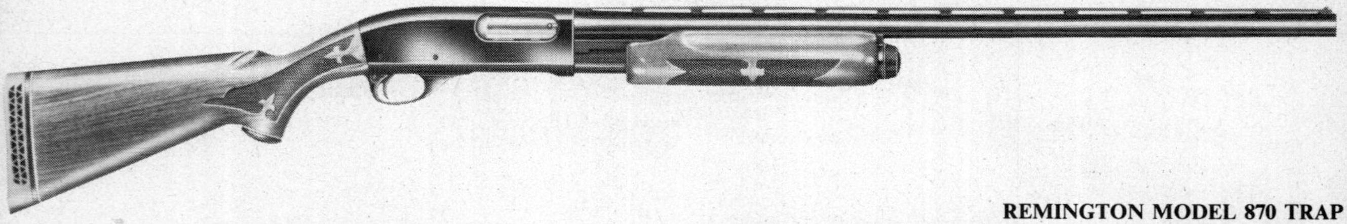

REMINGTON MODEL 870 TRAP
Gauge: 12. Barrel: 26 to 30 inches. Weight: 7 lbs. Choke: F, MT. Same as Wingmaster, except features stock with reduced drop. Vent rib. Checkered walnut stock and forearm. Recoil pad. **Price: $289** to **$491**

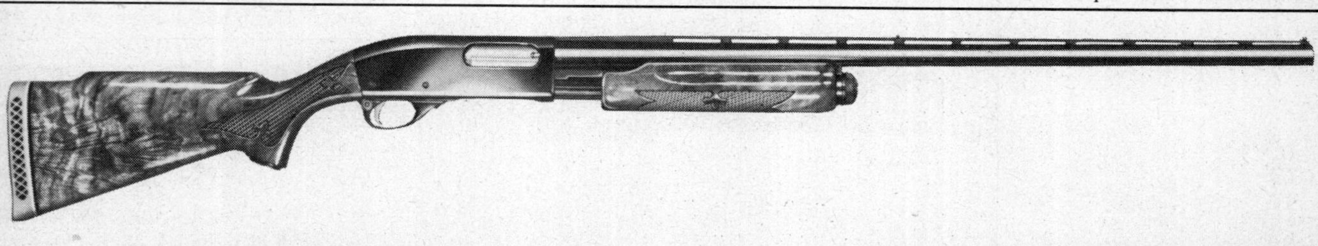

REMINGTON 870 MAGNUM
Same as Wingmaster, except has 3 inch chamber. Comes in 12 and 20 gauge. Barrel: 28, 30 inches. Choke: F, M. **Price: $297; $335** (vent rib)

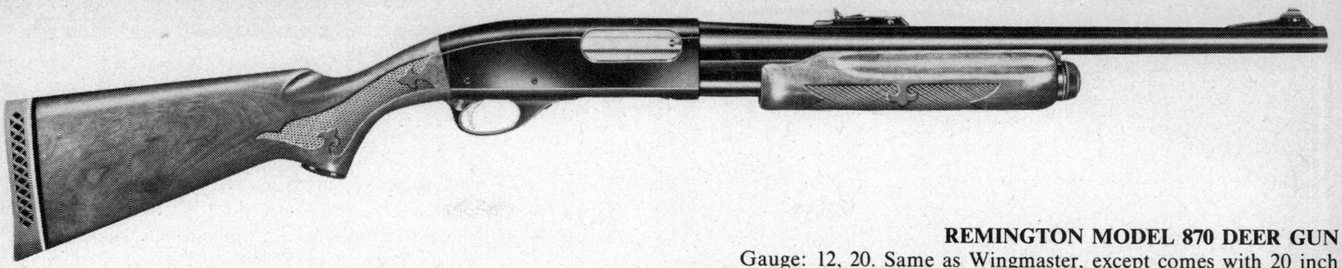

REMINGTON MODEL 870 DEER GUN
Gauge: 12, 20. Same as Wingmaster, except comes with 20 inch Improved Cylinder barrel. Adjustable sights. **Price: $246; $261** (deluxe)

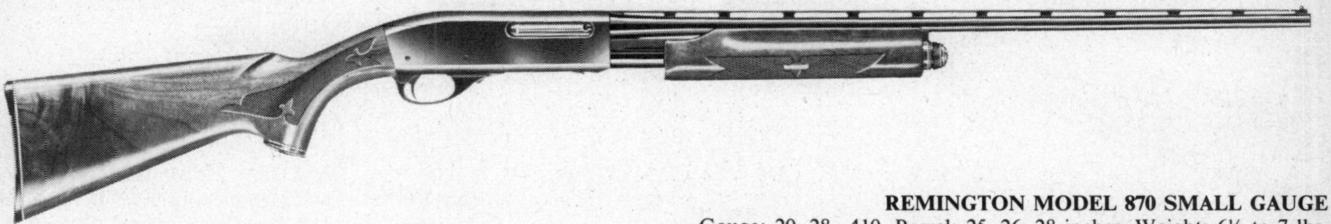

REMINGTON MODEL 870 SMALL GAUGE
Gauge: 20, 28, .410. Barrel: 25, 26, 28 inches. Weight: 6½ to 7 lbs. Choke: F, M, IC. Scaled down version of Wingmaster. **Price: $245; $283** (vent rib); **$273** (magnum); **$311** (magnum, vent rib)

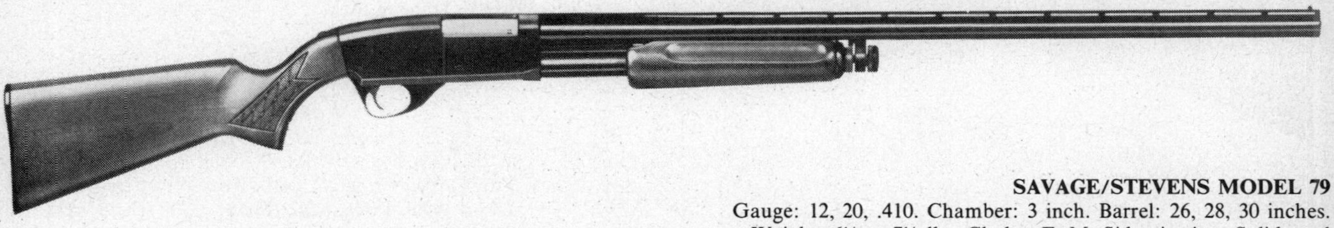

SAVAGE/STEVENS MODEL 79
Gauge: 12, 20, .410. Chamber: 3 inch. Barrel: 26, 28, 30 inches. Weight: 6¼ to 7½ lbs. Choke: F, M. Side ejection. Solid steel receiver. Quick takedown. Top tang safety. Walnut finished hardwood stock. **Price: NA**

SKB MODEL 7300
Gauge: 12, 20. Chamber: 3 inch. Barrel: 26, 28, 30 inches. Choke: F, M, IC. Vent rib barrel. Bolt release on left and right side. Twin action slide bars. French walnut stock and forearm. Slug barrel available. **Price: $310**

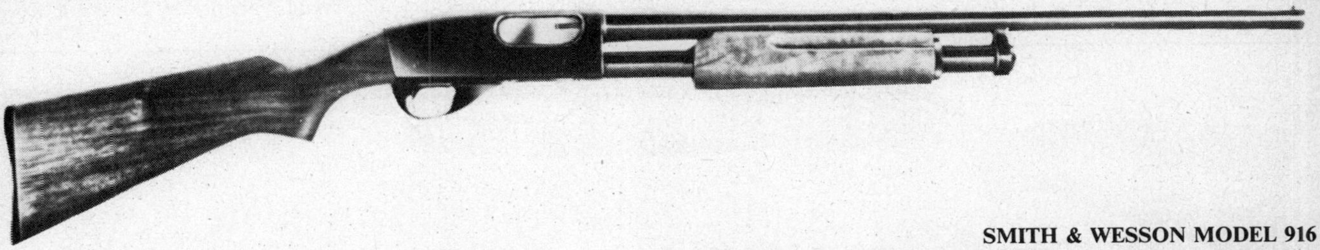

SMITH & WESSON MODEL 916
Gauge: 12, 16, 20. Barrel: 20, 26, 28, 30 inches. Weight: 7¼ lbs. Choke: F, M, IC. Top tang safety. American walnut stock. **Price: NA**

WEATHERBY PATRICIAN PUMP
Gauge: 12. Barrel: 26, 28, 30 inches. Weight: 7¼ lbs. Choke: F, M, IC. Crossbolt safety. Gold plated forearm release. Checkered walnut stock. **Price: $380** (Field); **$450** (Deluxe); **$410** (Trap); **$380** (Buckmaster slug gun)

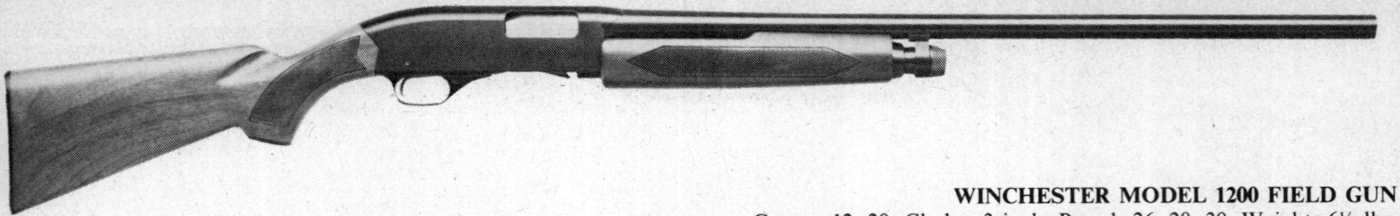

WINCHESTER MODEL 1200 FIELD GUN
Gauge: 12, 20. Choke: 3 inch. Barrel: 26, 28, 30. Weight: 6½ lbs. Choke: F, M, IC. Winchoke and vent rib optional. Crossbolt safety. **Price: $214; $236** (vent rib and Winchoke)

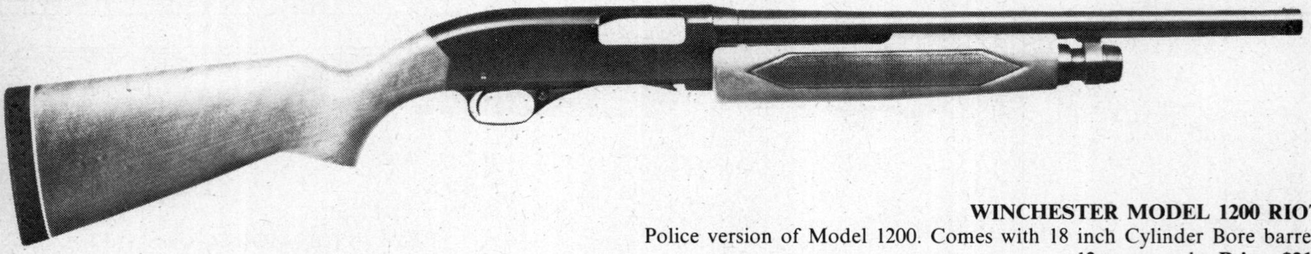

WINCHESTER MODEL 1200 RIOT
Police version of Model 1200. Comes with 18 inch Cylinder Bore barrel. 12 gauge only. **Price: $212**

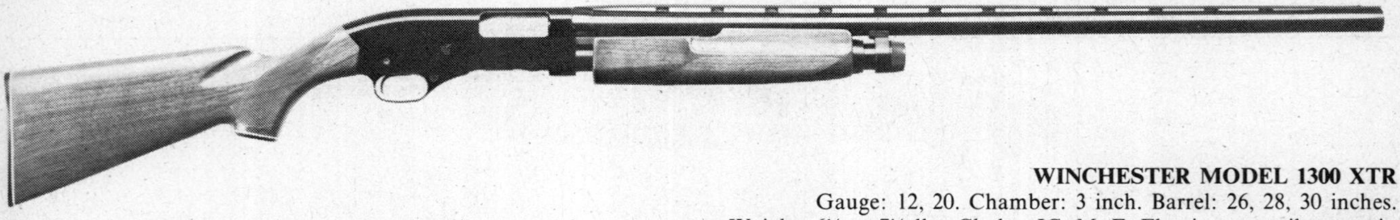

WINCHESTER MODEL 1300 XTR
Gauge: 12, 20. Chamber: 3 inch. Barrel: 26, 28, 30 inches. Weight: 6¼ to 7¼ lbs. Choke: IC, M, F. Floating vent rib, one pin disassembly. Front locking rotating bolt. Crossbolt safety. American walnut stock. **Price: $242** (Field); **$270** (Field, vent rib)

Semiautomatic

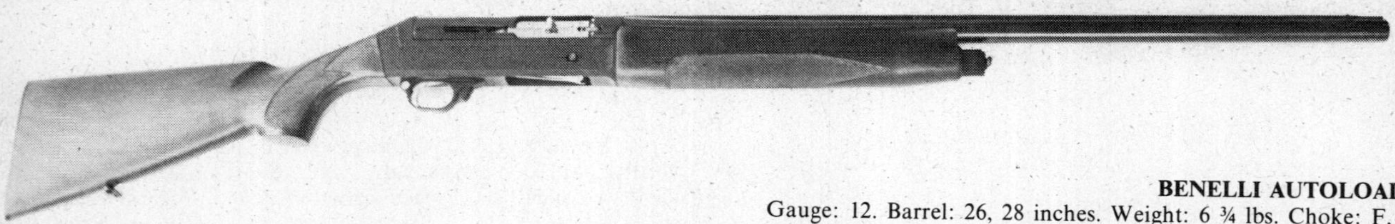

BENELLI AUTOLOADER
Gauge: 12. Barrel: 26, 28 inches. Weight: 6 ¾ lbs. Choke: F, IM, M, IC, S. Crossbolt safety. Quick interchangeable barrels. Hand checkered European walnut stock and forend. Imported by Heckler & Koch. **Price: NA**

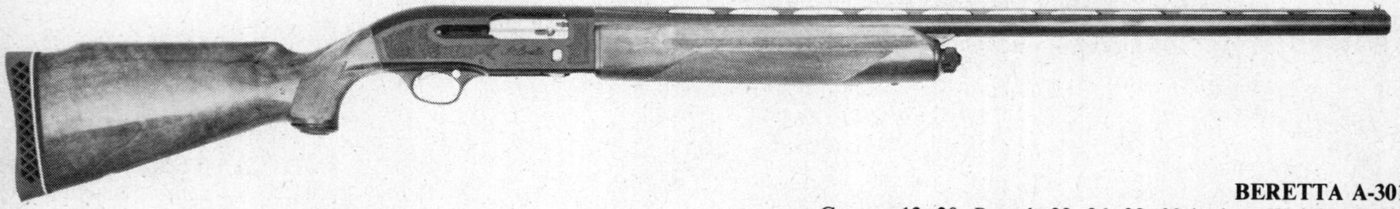

BERETTA A-301
Gauge: 12, 20. Barrel: 22, 26, 28, 30 inches. Weight: 7 lbs. Choke F, M, IC. Gas operated. Crossbolt safety. Vent rib barrel. **Price: $470; $515** (magnum); **$470** (Skeet); **$485** (Trap)

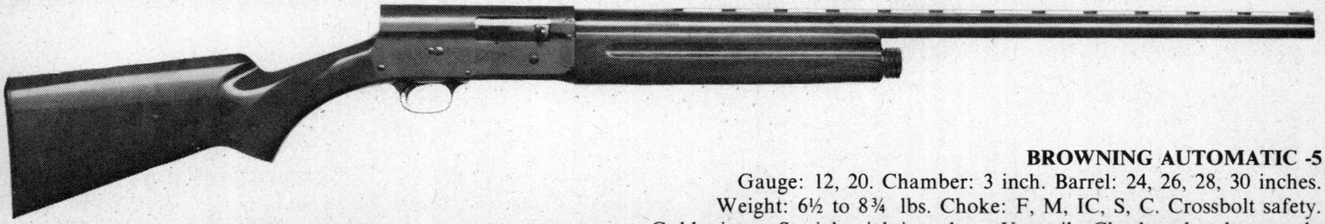

BROWNING AUTOMATIC -5
Gauge: 12, 20. Chamber: 3 inch. Barrel: 24, 26, 28, 30 inches. Weight: 6½ to 8¾ lbs. Choke: F, M, IC, S, C. Crossbolt safety. Gold trigger. Straight sighting plane. Vent rib. Checkered walnut stock. **Price: $475; $485** (magnum)

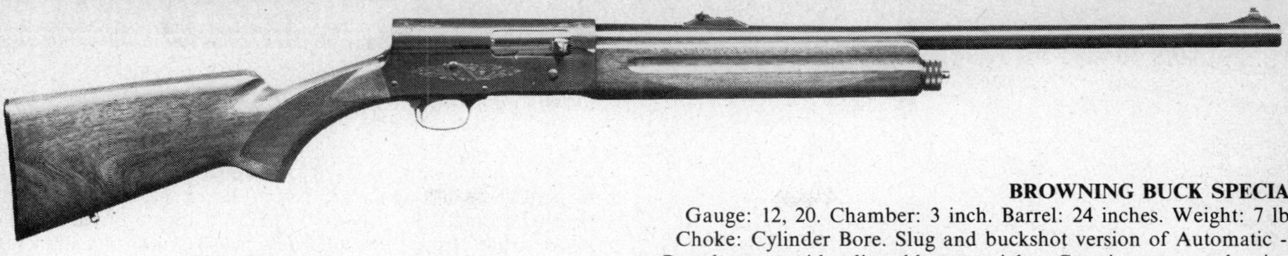

BROWNING BUCK SPECIAL
Gauge: 12, 20. Chamber: 3 inch. Barrel: 24 inches. Weight: 7 lbs. Choke: Cylinder Bore. Slug and buckshot version of Automatic -5. Barrel comes with adjustable open sights. Carrying strap and swivels optional. **Price: $502; $512** (magnum)

BROWNING B2000
Gauge: 12, 20. Chamber: 3 inch. Barrel: 26, 28, 30, 32 inches. Weight: 6½ to 7¾ lbs. Choke: F, M, IC, C. Vent rib barrel. Cold forged receiver. German silver medium sight bead. Checkered French walnut stock and forearm. **Price: $430** (Hunting model); **$475** (Trap and Skeet)

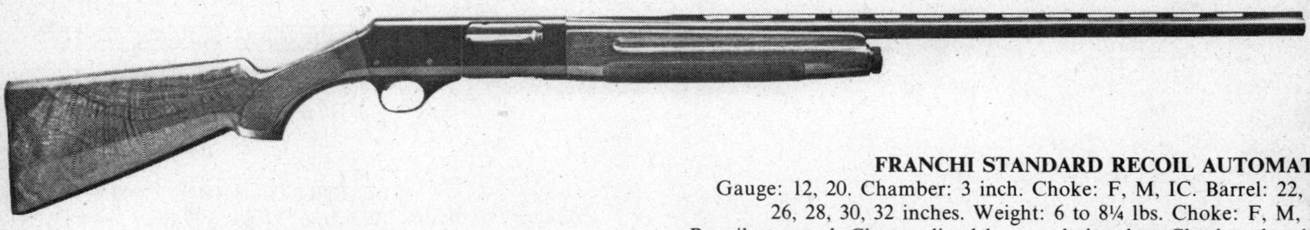

FRANCHI STANDARD RECOIL AUTOMATIC
Gauge: 12, 20. Chamber: 3 inch. Choke: F, M, IC. Barrel: 22, 24, 26, 28, 30, 32 inches. Weight: 6 to 8¼ lbs. Choke: F, M, IC. Recoil operated. Chrome lined bore and chamber. Checkered walnut stock and forearm. Imported by Stoeger. **Price: $335; $350** (magnum); **$350** (Slug gun)

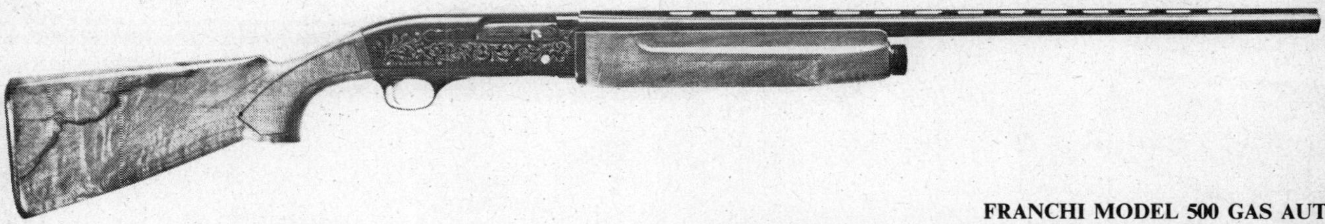

FRANCHI MODEL 500 GAS AUTO
Gauge: 12. Barrel: 26, 28, 30 inches. Weight: 7 lbs. Choke: F, M, S, IC. Gas operated. Pushbutton safety. Vent rib barrel. Checkered walnut stock and pistol grip. Imported by Stoeger. **Price: $349**

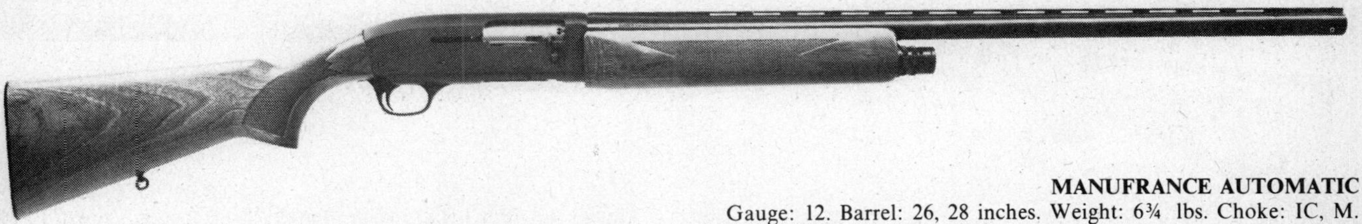

MANUFRANCE AUTOMATIC
Gauge: 12. Barrel: 26, 28 inches. Weight: 6¾ lbs. Choke: IC, M. Lightweight machined receiver. Vent rib barrel with matte sighting plane. Select French walnut stock and forend. Imported by Interarms. **Price: $279**

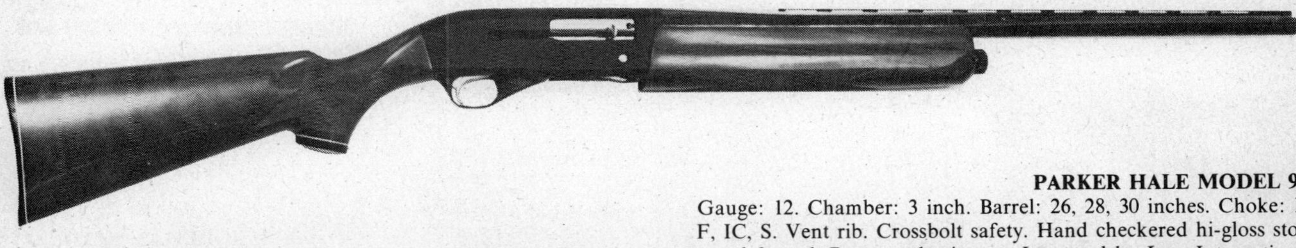

PARKER HALE MODEL 900
Gauge: 12. Chamber: 3 inch. Barrel: 26, 28, 30 inches. Choke: M, F, IC, S. Vent rib. Crossbolt safety. Hand checkered hi-gloss stock and forend. Rosewood grip cap. Imported by Jana International. **Price: $320; $341** (magnum)

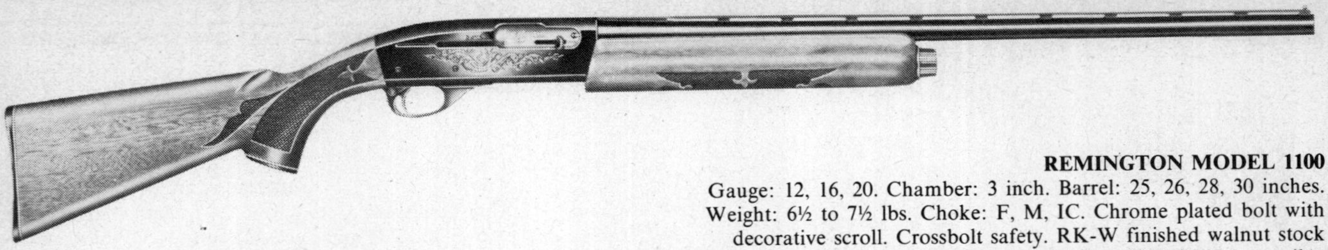

REMINGTON MODEL 1100
Gauge: 12, 16, 20. Chamber: 3 inch. Barrel: 25, 26, 28, 30 inches. Weight: 6½ to 7½ lbs. Choke: F, M, IC. Chrome plated bolt with decorative scroll. Crossbolt safety. RK-W finished walnut stock and forearm. Available in left hand version. **Price: $305; $345** (vent rib); **$338** (magnum); **$376** (magnum, vent rib)

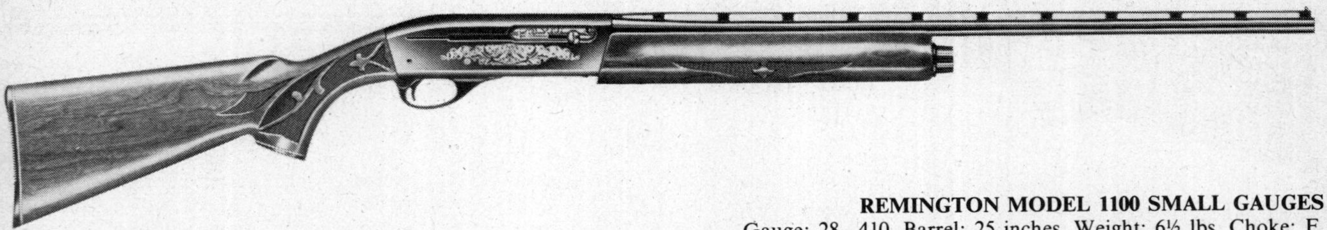

REMINGTON MODEL 1100 SMALL GAUGES
Gauge: 28, .410. Barrel: 25 inches. Weight: 6½ lbs. Choke: F, F, IC. **Price: $310; $348** (vent rib); **$359** (Skeet, vent rib); **$423** (Tournament Skeet, vent rib)

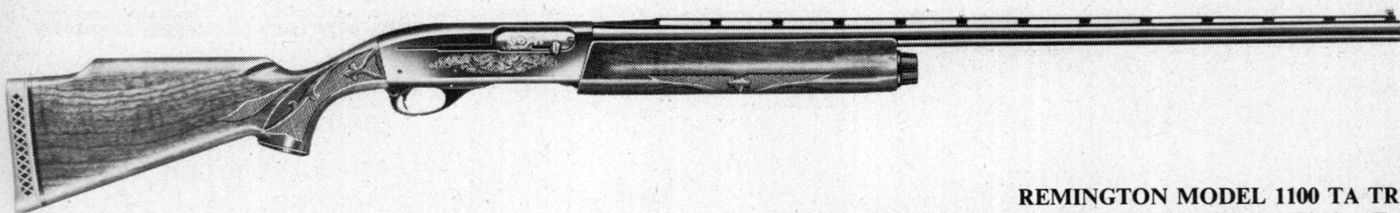

REMINGTON MODEL 1100 TA TRAP
Gauge: 12. Barrel: 30 inches. Weight: 8¼ lbs. Choke: F, M. Available with Trap or Monte Carlo stock; otherwise same as M1100. **Price: $361** (Trap, vent rib); **$371** (Monte Carlo, vent rib); **$428** (Tournament, vent rib); **$439** (Tournament, vent rib, Monte Carlo)

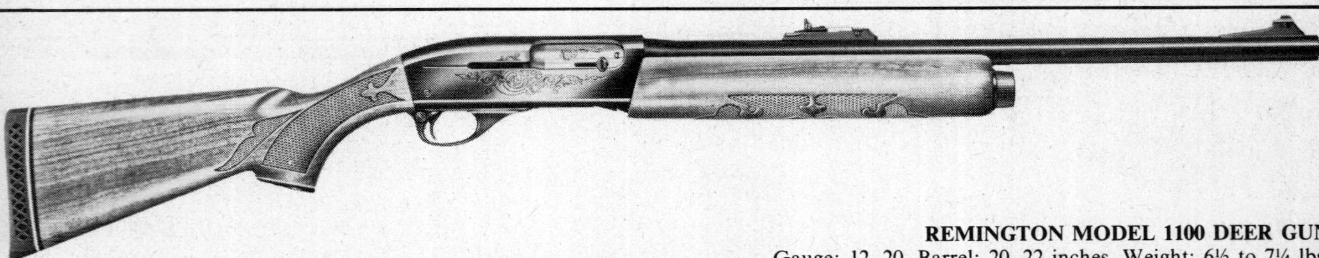

REMINGTON MODEL 1100 DEER GUN
Gauge: 12, 20. Barrel: 20, 22 inches. Weight: 6½ to 7¼ lbs. Choke: IC. Rifled slug version of M1100. Recoil pad. Sights: adjustable rear, ramp front. **Price: $342**

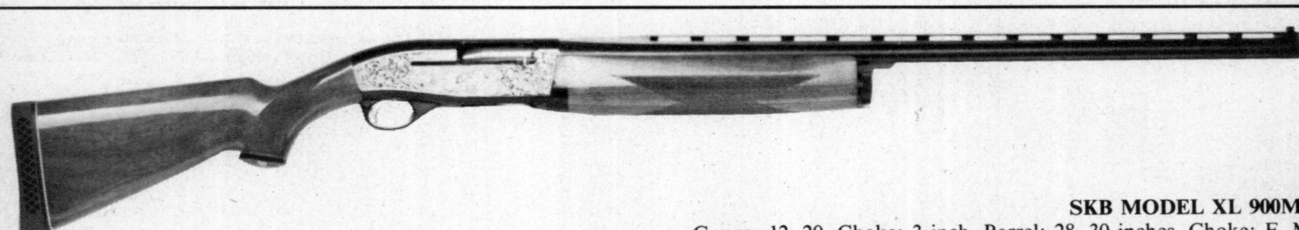

SKB MODEL XL 900MR
Gauge: 12, 20. Choke: 3 inch. Barrel: 28, 30 inches. Choke: F, M, IC. Target grade vent rib barrel. White spacers on buttplate. Flourescent bead front sight. French walnut stock. **Price: $371; $393** (Skeet); **$413** (Trap)

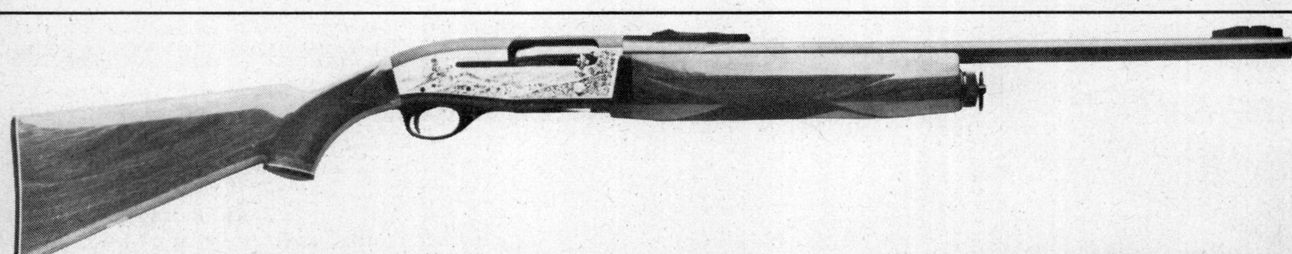

SKB MODEL XL 900MR SLUG
Gauge: 12, 20. Barrel: 24 inches. Choke: cylinder bore. Sights: rifle type. **Price: $371**

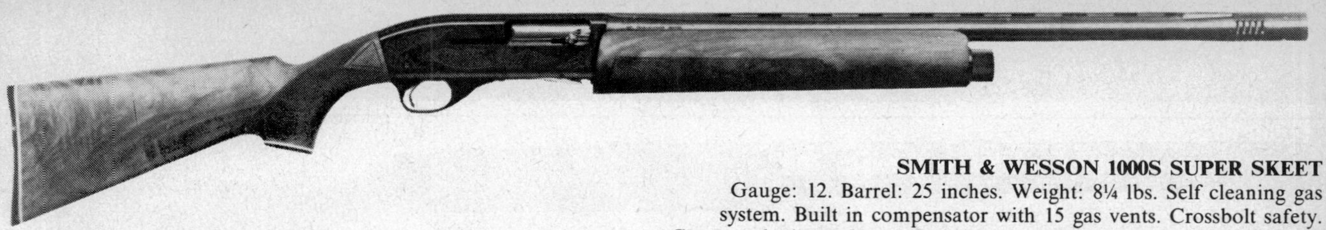

SMITH & WESSON 1000S SUPER SKEET
Gauge: 12. Barrel: 25 inches. Weight: 8¼ lbs. Self cleaning gas system. Built in compensator with 15 gas vents. Crossbolt safety. Contoured trigger. Vent rib. Select grade American walnut stock with right hand plam swell. Double sighting beads with flourescent red front. **Price: $329**

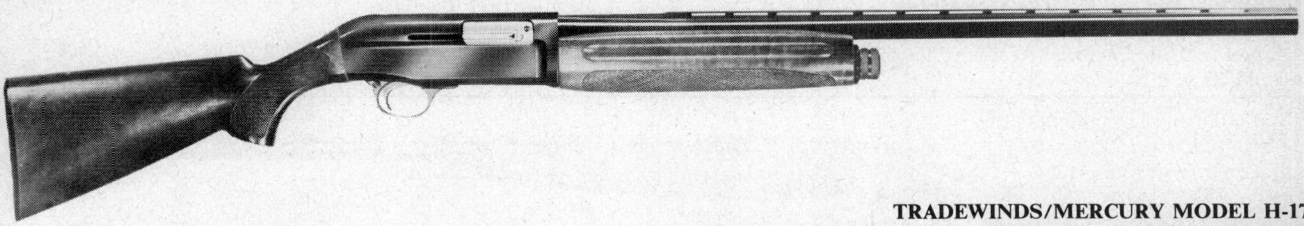

TRADEWINDS/MERCURY MODEL H-170
Gauge: 12. Barrel: 26, 28 inches. Weight: 7 lbs. Choke: optional. Hard chrome-lined bore, semi-floating bolt. Vent rib. Select European walnut stock. **Price: $345**

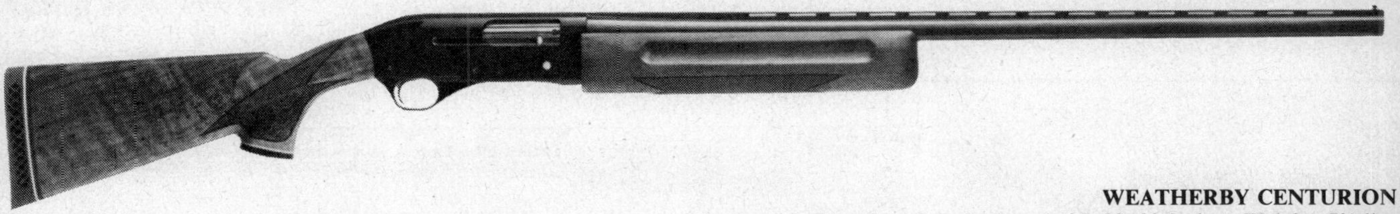

WEATHERBY CENTURION
Gauge: 12. Chamber: 3 inch. Barrel: 26, 28, 30 inches. Weight: 7¾ lbs. Choke: F, M, IC, S. Crossbolt safety. Gas operated, floating piston. Vent rib. Walnut stock with hand checkering. **Price: $400** (Field); **$470** (Deluxe); **$430** (Trap)

WINCHESTER MODEL 1500 XTR
Gauge: 12, 20. Barrel: 26, 28, 30 inches. Weight: 6½ to 7 lbs. Choke: IC, M, F. Self-adjusting gas system, crossbolt safety. One pin disassembly. American walnut stock with cut checkering. **Price: $304** (Field); **$331** (Field, vent rib)

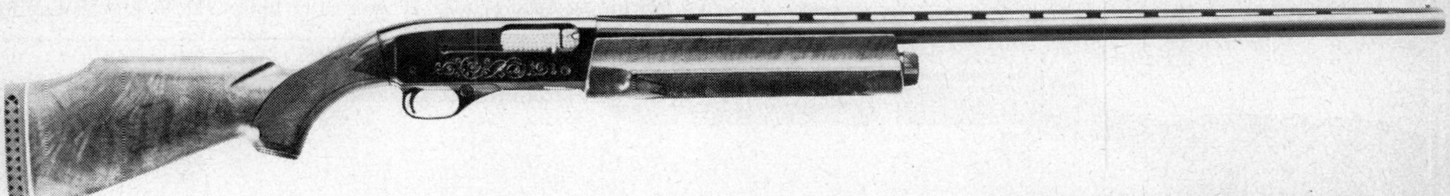

WINCHESTER SUPER-X MODEL 1 XTR
Gauge: 12. Barrel: 26, 28, 30 inches. Weight: 8½ lbs. Choke: F, M, IC, IM, S. Floating vent rib, three-piece engine turned bolt. Crossbolt safety. Monte Carlo or regular stock. **Price: $500** (Field, vent rib); **$576** (Trap, vent rib); **$576** (Skeet, vent rib)

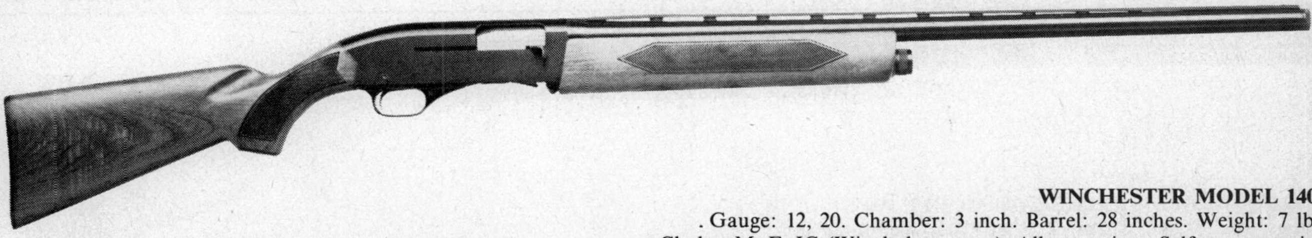

WINCHESTER MODEL 1400
. Gauge: 12, 20. Chamber: 3 inch. Barrel: 28 inches. Weight: 7 lbs. Choke: M, F, IC (Winchoke system). Alloy receiver. Self-compensating valve for standard or magnum loads. Checkered walnut stock and forend. **Price: $275** (Field); **$298** (Field, vent rib)

Single-Shot

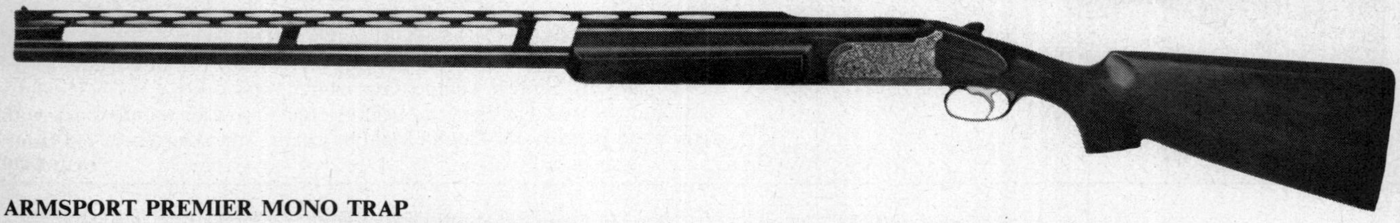

ARMSPORT PREMIER MONO TRAP
Gauge: 12. Barrel: 32, 34 inches. Weight: 8 lbs. Choke: M, F. High post vent rib. Engraved receiver. Fine grained walnut stock. **Price: $1795**

BERETTA MARK II TRAP
Gauge: 12. Barrel: 32, 34 inches. Weight: 8½ lbs. Choke: M, F. Long forend with tapered full beavertail shape. Vent rib. **Price: $530**

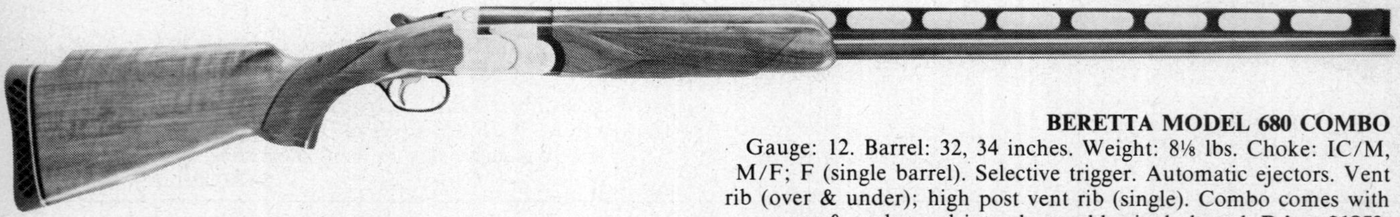

BERETTA MODEL 680 COMBO
Gauge: 12. Barrel: 32, 34 inches. Weight: 8⅛ lbs. Choke: IC/M, M/F; F (single barrel). Selective trigger. Automatic ejectors. Vent rib (over & under); high post vent rib (single). Combo comes with over & under and interchangeable single barrel. **Price: $1850; $1340** (single barrel trap only)

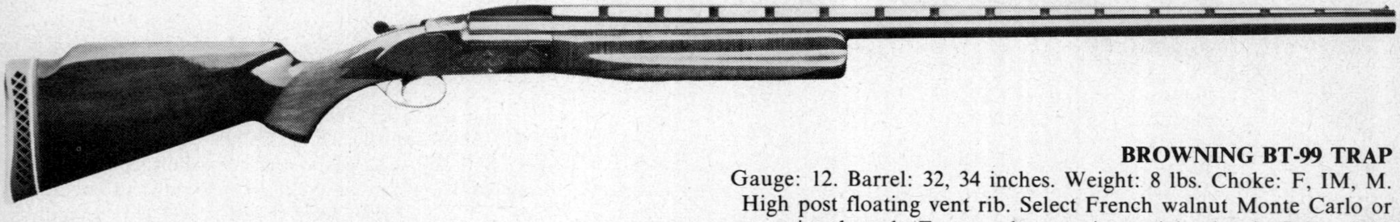

BROWNING BT-99 TRAP
Gauge: 12. Barrel: 32, 34 inches. Weight: 8 lbs. Choke: F, IM, M. High post floating vent rib. Select French walnut Monte Carlo or conventional stock. Front and center ivory sights beads. **Price: $600** (Grade I); **$850** (Grade II with extra bbl.); **$1350** (Pigeon Grade)

HARRINGTON & RICHARDSON MODEL 176
Gauge: 10. Chamber: 3½ inches. Barrel: 32, 36 inches. Weight: 10 lbs. Choke: F. Break open action. Blued barrel with case hardened frame. American walnut stock. **Price: $85**

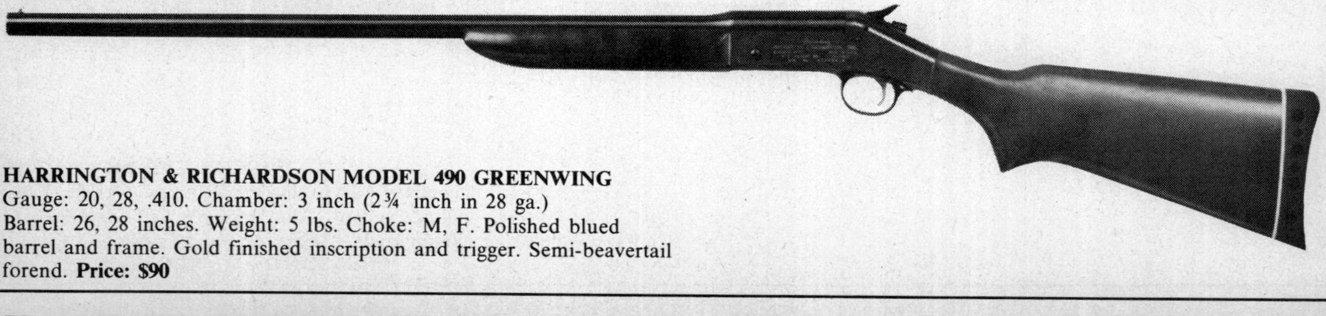

HARRINGTON & RICHARDSON MODEL 490 GREENWING
Gauge: 20, 28, .410. Chamber: 3 inch (2¾ inch in 28 ga.)
Barrel: 26, 28 inches. Weight: 5 lbs. Choke: M, F. Polished blued barrel and frame. Gold finished inscription and trigger. Semi-beavertail forend. **Price: $90**

PERAZZI COMPETITION I
Gauge: 12. Barrel: 32, 34 inches. Weight: 8½ lbs. Choke: F, M. Handfitted parts. Chrome lined chamber. Hand honed choke. Hand checkered walnut stock. Distributed by Winchester. **Price: $2650**

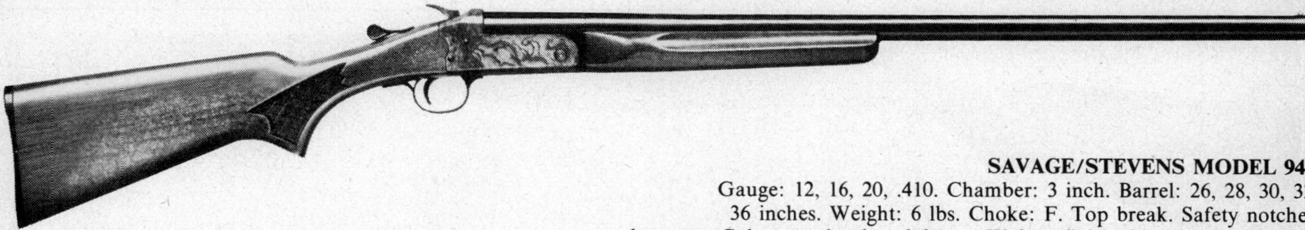

SAVAGE/STEVENS MODEL 94C
Gauge: 12, 16, 20, .410. Chamber: 3 inch. Barrel: 26, 28, 30, 32, 36 inches. Weight: 6 lbs. Choke: F. Top break. Safety notched hammer. Color case hardened frame. Walnut finished stock and forearm. **Price: $73; $83** (Youth's Model)

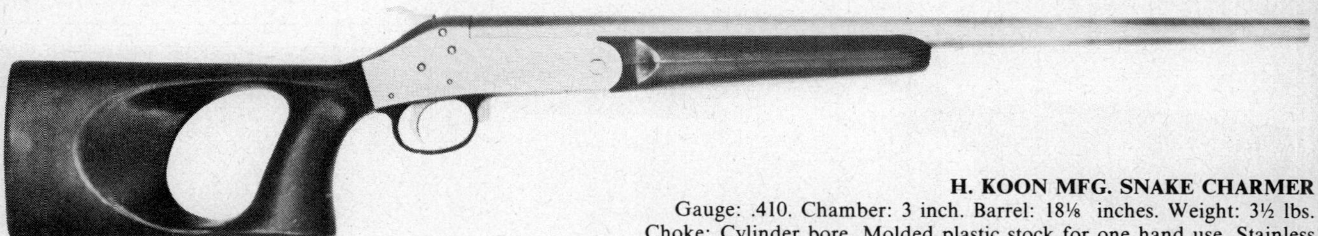

H. KOON MFG. SNAKE CHARMER
Gauge: .410. Chamber: 3 inch. Barrel: 18⅛ inches. Weight: 3½ lbs. Choke: Cylinder bore. Molded plastic stock for one hand use. Stainless steel construction. External hammer. **Price: $90** (vinyl case included)

Over & Under

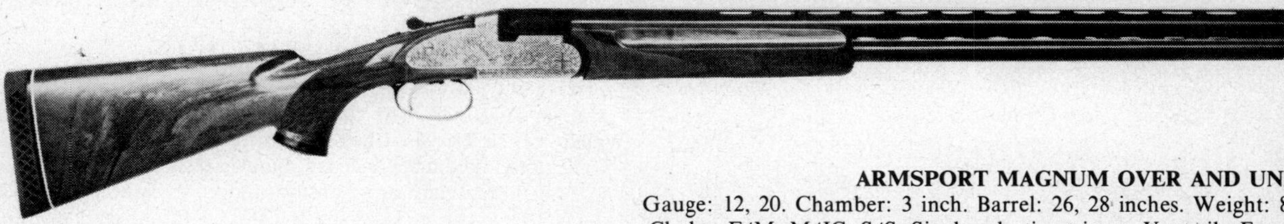

ARMSPORT MAGNUM OVER AND UNDER
Gauge: 12, 20. Chamber: 3 inch. Barrel: 26, 28 inches. Weight: 8 lbs. Choke: F/M, M/IC, S/S. Single selective trigger. Vent rib. Engraved receiver. Automatic ejectors. Recoil pad. **Price: $695**

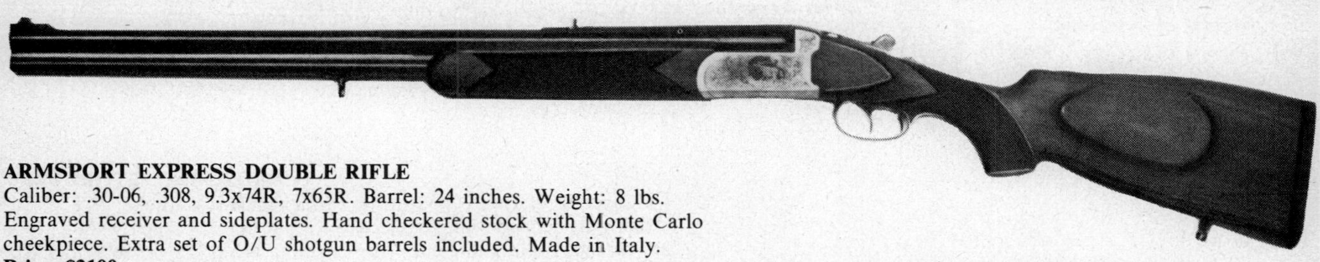

ARMSPORT EXPRESS DOUBLE RIFLE
Caliber: .30-06, .308, 9.3x74R, 7x65R. Barrel: 24 inches. Weight: 8 lbs. Engraved receiver and sideplates. Hand checkered stock with Monte Carlo cheekpiece. Extra set of O/U shotgun barrels included. Made in Italy. **Price: $3100**

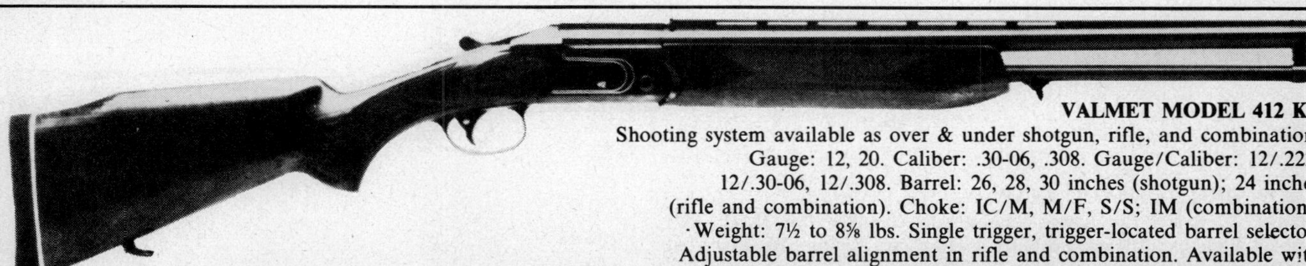

VALMET MODEL 412 KE
Shooting system available as over & under shotgun, rifle, and combination. Gauge: 12, 20. Caliber: .30-06, .308. Gauge/Caliber: 12/.222, 12/.30-06, 12/.308. Barrel: 26, 28, 30 inches (shotgun); 24 inches (rifle and combination). Choke: IC/M, M/F, S/S; IM (combination). Weight: 7½ to 8⅜ lbs. Single trigger, trigger-located barrel selector. Adjustable barrel alignment in rifle and combination. Available with extractors or ejectors. American walnut stock with hand checkering. **Price: $1300** (shotgun with set of rifle barrels and combination barrels)

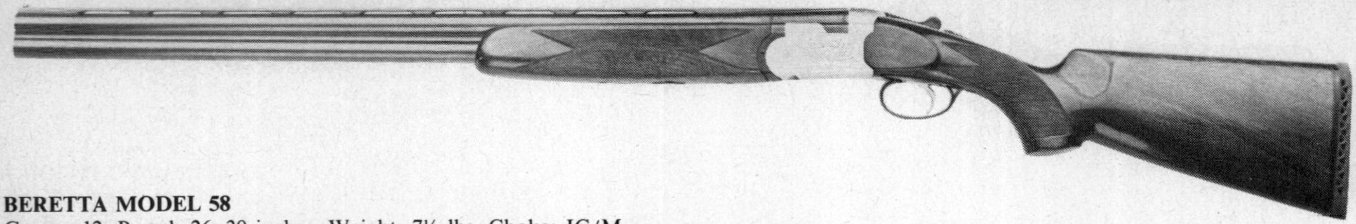

BERETTA MODEL 58
Gauge: 12. Barrel: 26, 30 inches. Weight: 7½ lbs. Choke: IC/M, M/F. Same as 55B, except fitted for trap or skeet competition. Comes with recoil pad. **Price: $829**

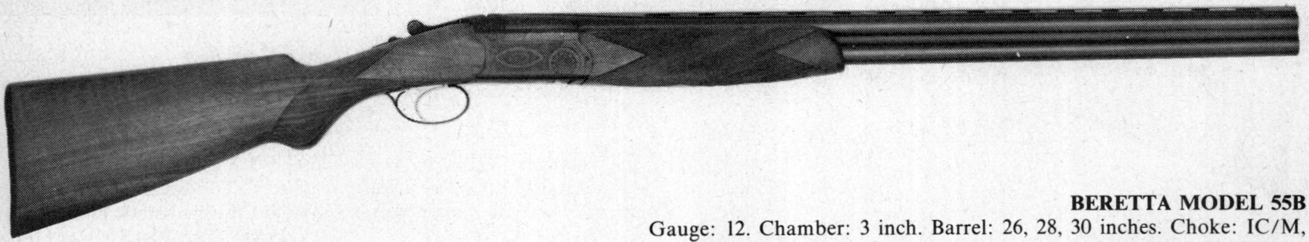

BERETTA MODEL 55B
Gauge: 12. Chamber: 3 inch. Barrel: 26, 28, 30 inches. Choke: IC/M, M/F. Chrome lined bores. Vent rib. Selective automatic ejectors. Tang safety. **Price: $626**

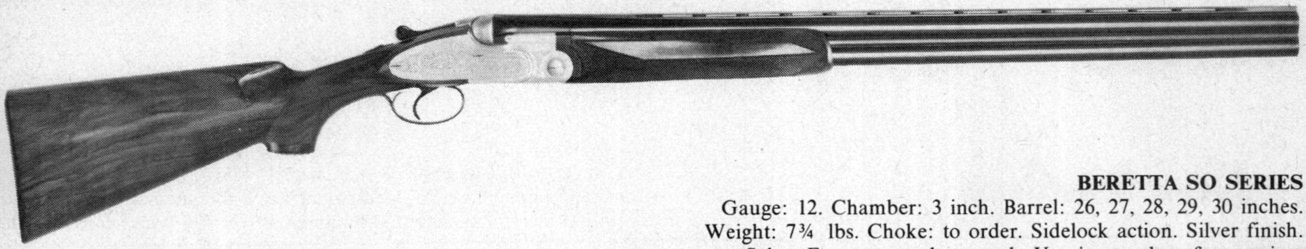

BERETTA SO SERIES
Gauge: 12. Chamber: 3 inch. Barrel: 26, 27, 28, 29, 30 inches. Weight: 7¾ lbs. Choke: to order. Sidelock action. Silver finish. Select European walnut stock. Varying grades of engraving. **Price: $4680** to **$8050**

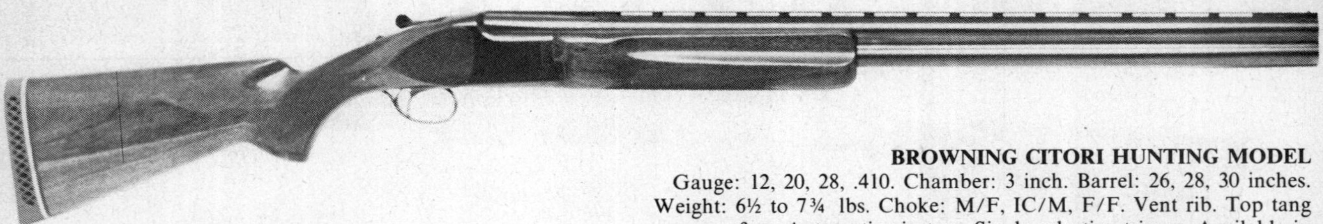

BROWNING CITORI HUNTING MODEL
Gauge: 12, 20, 28, .410. Chamber: 3 inch. Barrel: 26, 28, 30 inches. Weight: 6½ to 7¾ lbs. Choke: M/F, IC/M, F/F. Vent rib. Top tang safety. Automatic ejectors. Single selective trigger. Available in Standard and Sporter Models, in choice of three grades. **Price: $600** to **$1600**

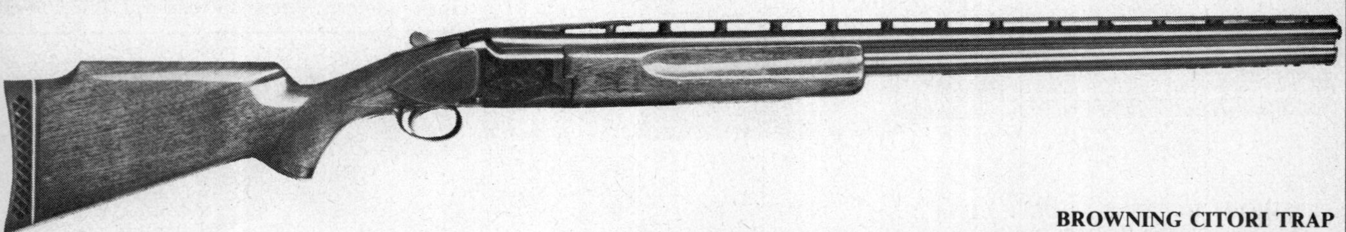

BROWNING CITORI TRAP
Gauge: 12. Barrel: 30, 32 inches. Weight: 8¼ lbs. Choke: F/F, F/M. High post vent rib. Single selective trigger. Select French walnut stock and forend. Front and center ivory sighting beads. **Price: $675** to **$1675**

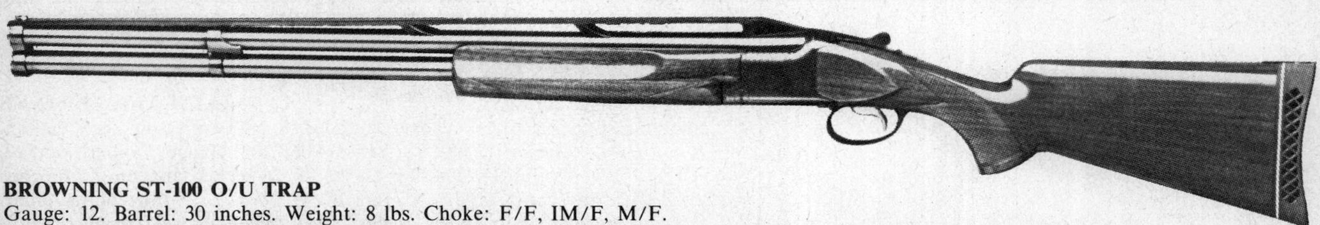

BROWNING ST-100 O/U TRAP
Gauge: 12. Barrel: 30 inches. Weight: 8 lbs. Choke: F/F, IM/F, M/F. Impact adjustment device on barrels. High post floating Broadway rib. Contoured selective trigger. Semi-beavertail forearm. Select hand checkered walnut stock. **Price: $3250**

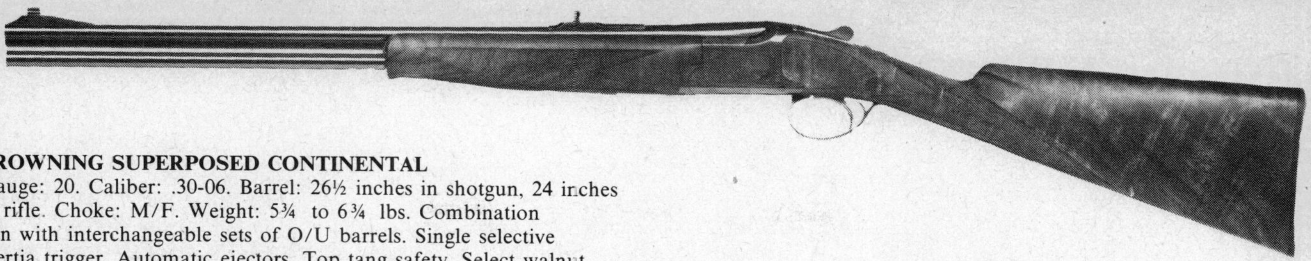

BROWNING SUPERPOSED CONTINENTAL
Gauge: 20. Caliber: .30-06. Barrel: 26½ inches in shotgun, 24 inches in rifle. Choke: M/F. Weight: 5¾ to 6¾ lbs. Combination gun with interchangeable sets of O/U barrels. Single selective inertia trigger. Automatic ejectors. Top tang safety. Select walnut stock with fine line checkering. **Price: $4650**

GEBRUDER MERKEL 203E
Gauge: 12, 16, 20. Chamber: 3 inch. Barrel: by order. Weight: 7 lbs. Choke: by order. Double, single, or selective trigger. English Arabesque or hunting scene engraving. **Price: $5316**

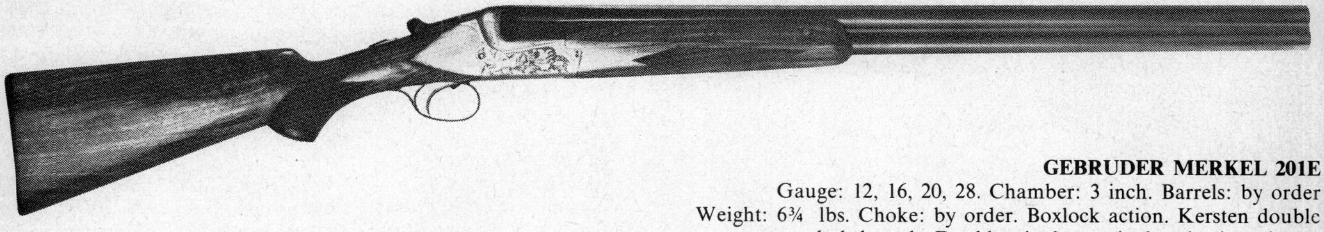

GEBRUDER MERKEL 201E
Gauge: 12, 16, 20, 28. Chamber: 3 inch. Barrels: by order Weight: 6¾ lbs. Choke: by order. Boxlock action. Kersten double crossbolt breech. Double, single, or single selective trigger: **Price: $3124**

F.I.E. OVER AND UNDER
Gauge: 12, 20. Barrel: 26, 28, 30 inches. Weight: 9½ lbs. Choke: M/F, S/S, F/F. Vent rib. Coil spring mechanism. Twin bolt locking system. Double triggers. Brazilian hardwood stock and forend. **Price: $200** to **$218**

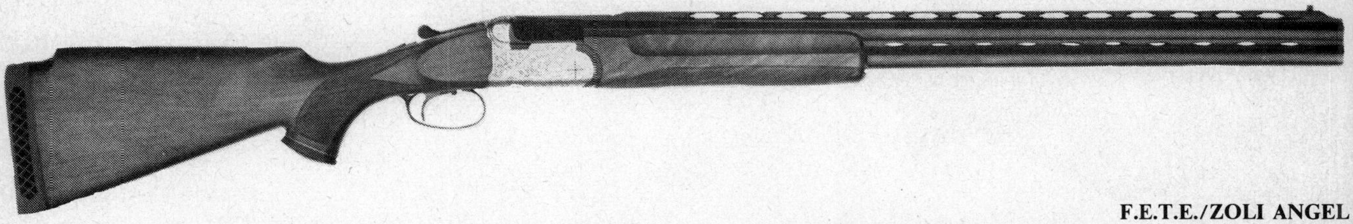

F.E.T.E./ZOLI ANGEL
Gauge: 12, 20. Barrel: 26, 28, 30 inches. Weight: 7½ lbs. Choke: F/M, M/IC, S/S. Engraved receiver. European walnut stock. Gold trigger. **Price: $630** (Field); **$700** (Trap); **$650** (Skeet)

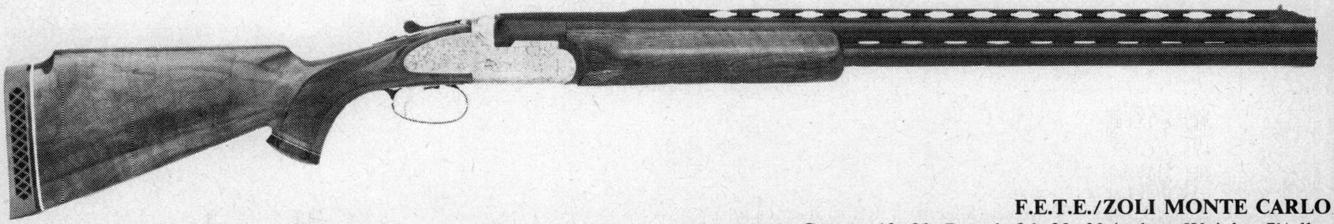

F.E.T.E./ZOLI MONTE CARLO
Gauge: 12, 20. Barrel: 26, 28, 30 inches. Weight: 7½ lbs. Choke: F/M M/IC, S/S. Comes in 3 grades. Single selective trigger. Top tang safety. **Price: $730** (Field); **$750** (Skeet); **$800** (Trap)

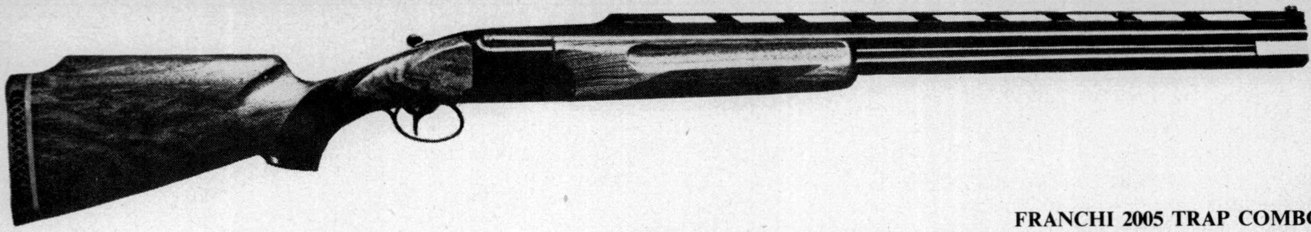

FRANCHI 2005 TRAP COMBO
Gauge: 12. Barrel: 32 inches (O/U); 34 inches (single). Weight: 8½ lbs. Choke: IM/F (O/U), F (single). Mechanical triggers. High post vent rib. Velour lined hard case included. **Price: $1595**

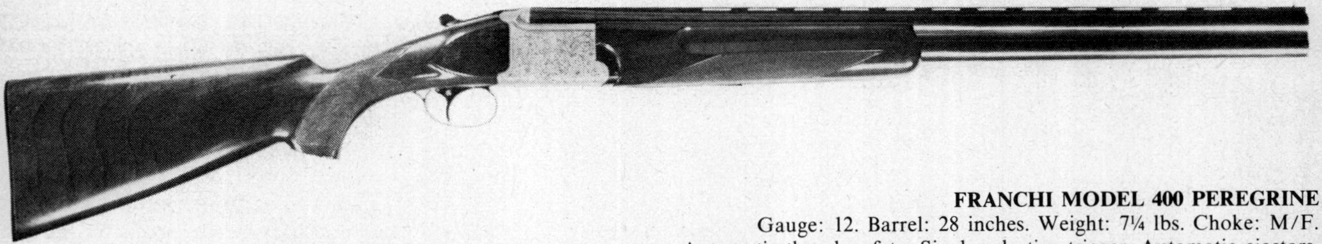

FRANCHI MODEL 400 PEREGRINE
Gauge: 12. Barrel: 28 inches. Weight: 7¼ lbs. Choke: M/F. Automatic thumb safety. Single selective trigger. Automatic ejectors. European walnut stock and forend. **Price: $567**

GOLDEN EAGLE O/U SHOTGUN
Gauge: 12, 20, 28, .410. Chamber: 3 inch. Barrel: 28, 30, 32 inches. Weight: 7 lbs. Single selective trigger. Vent rib. Contoured beavertail walnut forend. White line spacer recoil pad. **Price: $899** to **$3499**

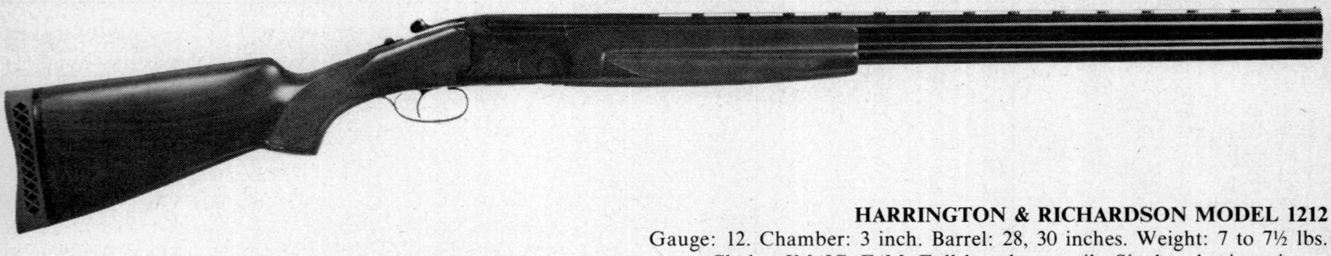

HARRINGTON & RICHARDSON MODEL 1212
Gauge: 12. Chamber: 3 inch. Barrel: 28, 30 inches. Weight: 7 to 7½ lbs. Choke: IM/IC, F/M. Full length vent rib. Single selective trigger. Engraved breech, receiver, and triggerguard. **Price: $475** to **$489**

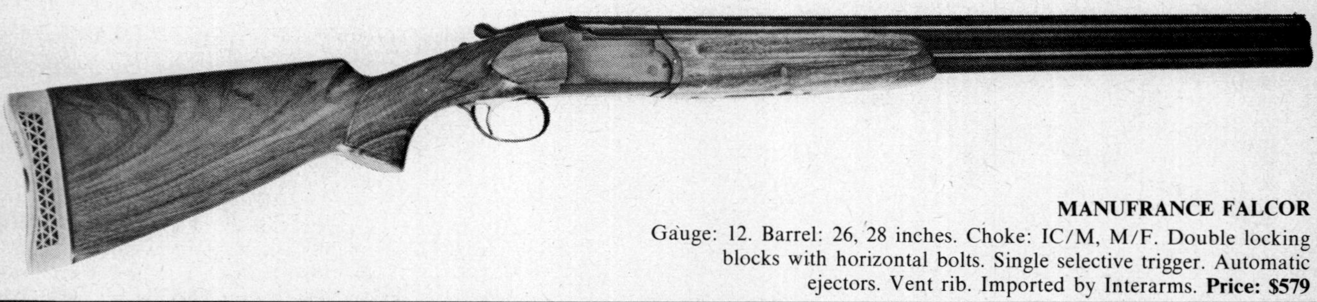

MANUFRANCE FALCOR
Gauge: 12. Barrel: 26, 28 inches. Choke: IC/M, M/F. Double locking blocks with horizontal bolts. Single selective trigger. Automatic ejectors. Vent rib. Imported by Interarms. **Price: $579**

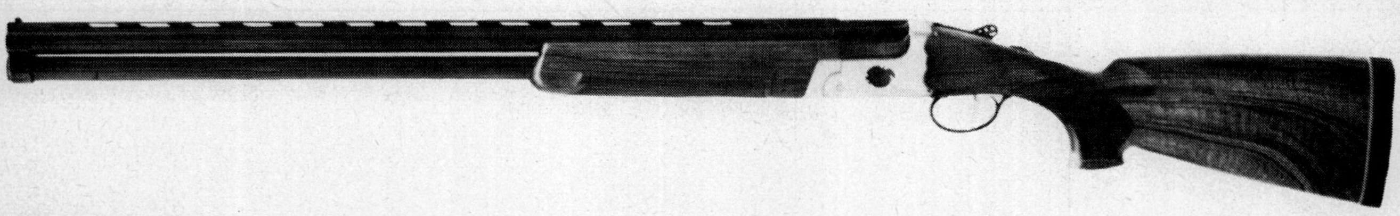

IAB C-300 COMBO
Gauge: 12. Barrel: 31 to 34 inches. Weight: 8 ¾ lbs. Comes with 31 inch double barrels and extra single barrel. Super Combo comes with 2 single barrels. Imported by Puccinelli Co. **Price: $3550**

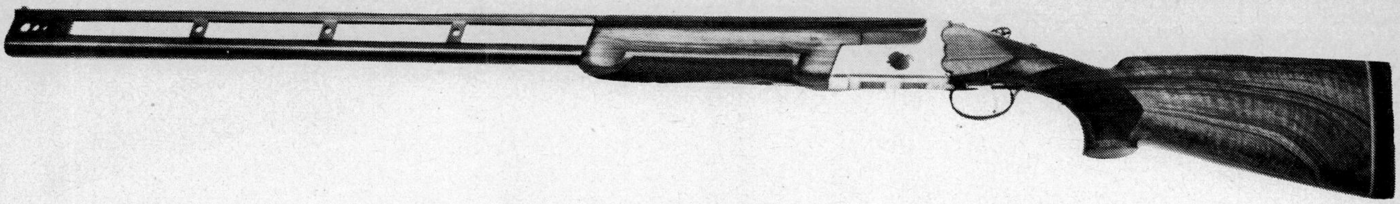

IAB S-300
Gauge: 12. Barrel: 26 to 32 inches. Weight: 8 lbs. Vent rib. Single selective trigger. Imported by Puccinelli Co. **Price: $2750**

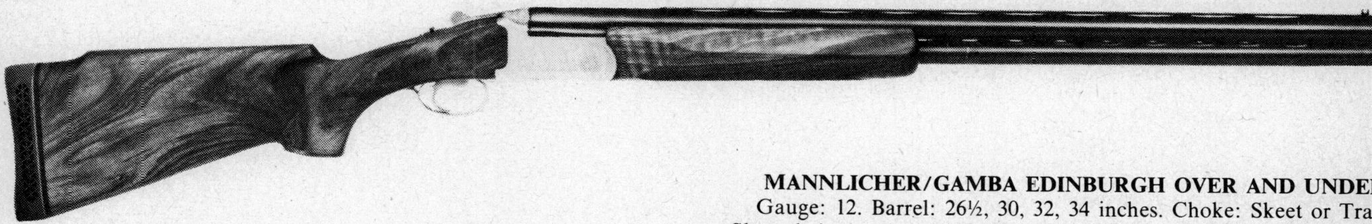

MANNLICHER/GAMBA EDINBURGH OVER AND UNDER
Gauge: 12. Barrel: 26½, 30, 32, 34 inches. Choke: Skeet or Trap Shaped single trigger (selective available). Anson action. Automatic ejectors. Double vent rib. Hand checkered Monte Carlo stock with shaped pistol grip. Imported by Steyr, Daimler. Puch. **Price: $1815**

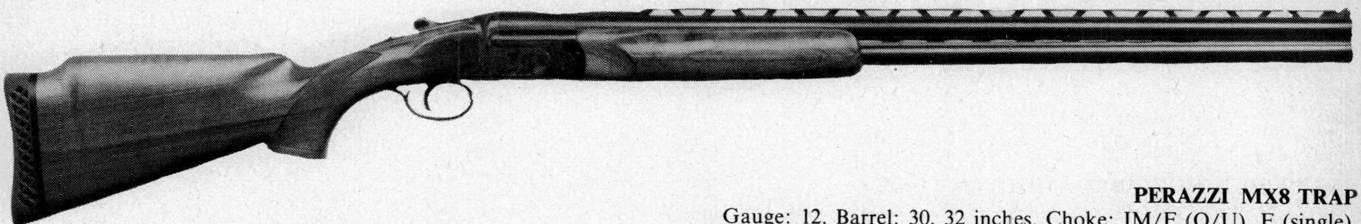

PERAZZI MX8 TRAP
Gauge: 12. Barrel: 30, 32 inches. Choke: IM/F (O/U), F (single). High straight rib, vent side ribs. Detachable trigger mechanism. Case hardened receiver. Single barrel available. Distributed by Winchester. **Price: $3950; $5650** (Combo)

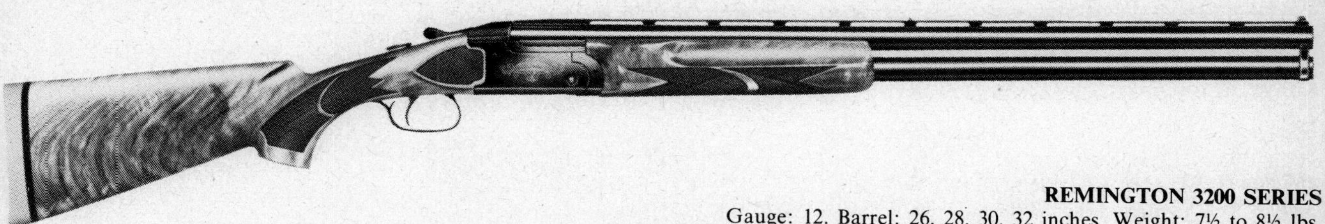

REMINGTON 3200 SERIES
Gauge: 12. Barrel: 26, 28, 30, 32 inches. Weight: 7½ to 8½ lbs. Choke: S/S, IM/F, F/F. Comes in Skeet, Trap, and Special Trap versions. All feature vent rib, selective single trigger, high grade walnut stocks. **Price: $1225** to **$3900**

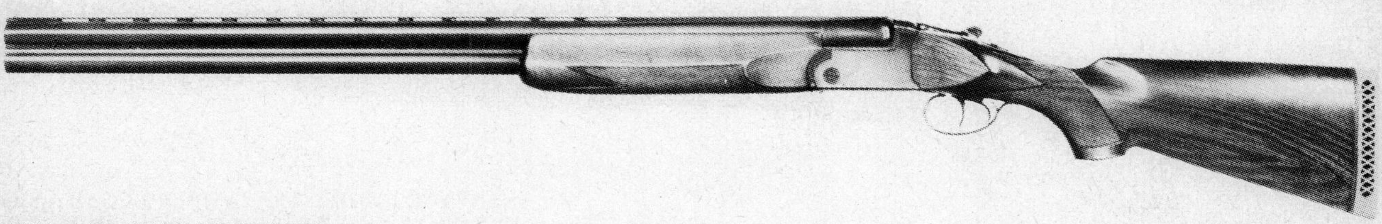

RICHLAND ARMS MODEL 810
Gauge: 10. Chamber: 3½ inch. Barrel: 32 inches. Weight: 10½ lbs. Choke: F/M, F/F. Double triggers. Kersten lock action. European walnut stock. **Price: $500**

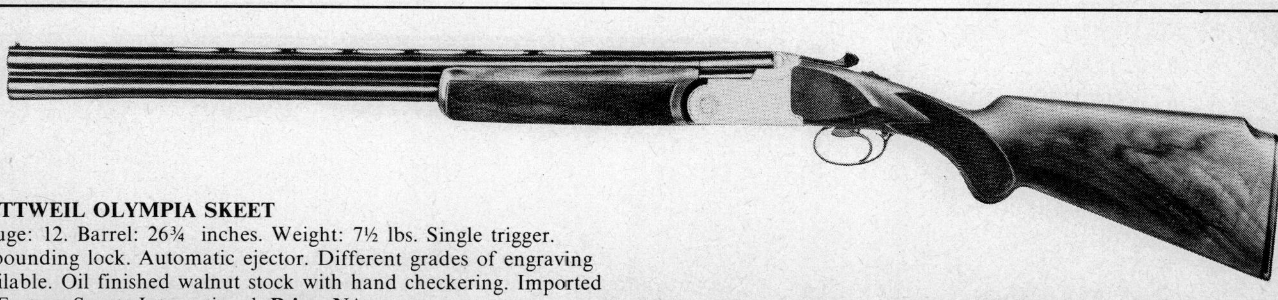

ROTTWEIL OLYMPIA SKEET
Gauge: 12. Barrel: 26¾ inches. Weight: 7½ lbs. Single trigger. Rebounding lock. Automatic ejector. Different grades of engraving available. Oil finished walnut stock with hand checkering. Imported by Eastern Sports International. **Price: NA**

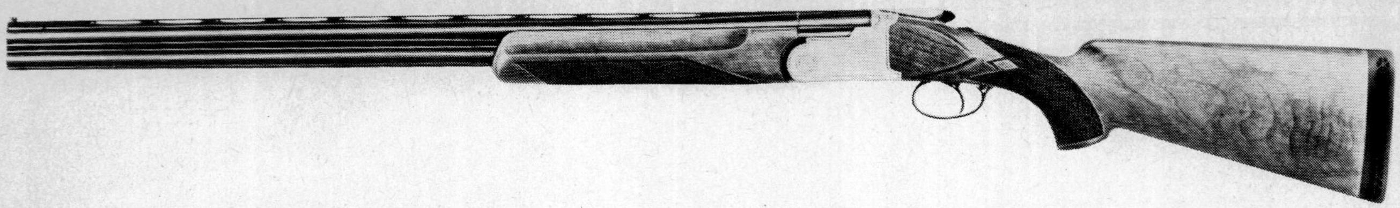

ROTTWEIL TRAP MONTREAL
Gauge: 12. Barrel: 30 inches. Weight: 8¼ lbs. Competition single trigger. Rebounding lock. Auxiliary mid rib sight. Fluted beavertail forend. Imported by Eastern Sports International. **Price: $2,750**

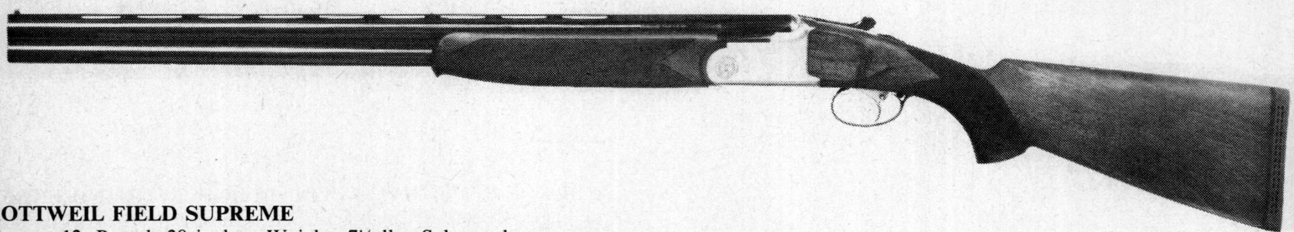

ROTTWEIL FIELD SUPREME
Gauge: 12. Barrel: 28 inches. Weight: 7¼ lbs. Selector button on single trigger. Vent rib. Plastic front sight. Imported by Eastern Sports International. **Price: $2,750**

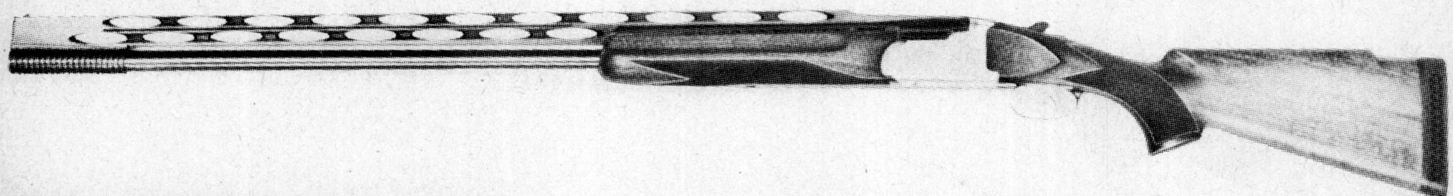

ROTTWEIL ADJUSTABLE AMERICAN TRAP
Gauge: 12. Barrel: 32, 34 inches. Choke: IM/F (O/U), F & IM (single). New variable point of impact system. Chrome lined barrels. Comes with one set O/U barrels and two single. Comes with fitted case. Imported by Eastern Sports International. **Price: $3,495**

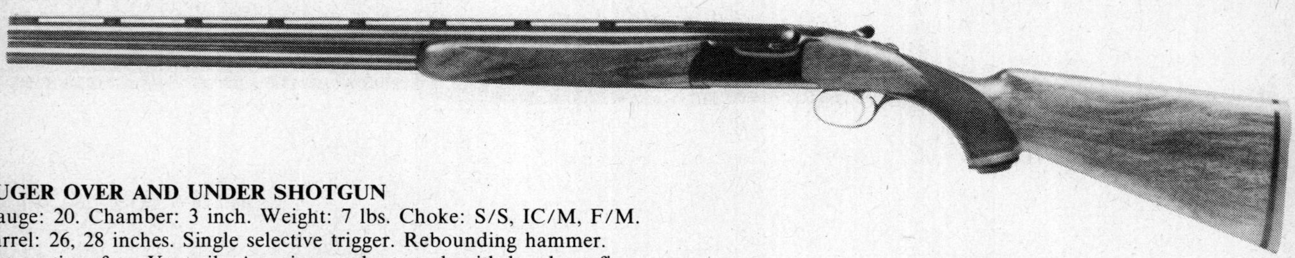

RUGER OVER AND UNDER SHOTGUN
Gauge: 20. Chamber: 3 inch. Weight: 7 lbs. Choke: S/S, IC/M, F/M. Barrel: 26, 28 inches. Single selective trigger. Rebounding hammer. Automatic safety. Vent rib. American walnut stock with hand cut fine line checkering. Semi-beavertail forearm. **Price: $675**

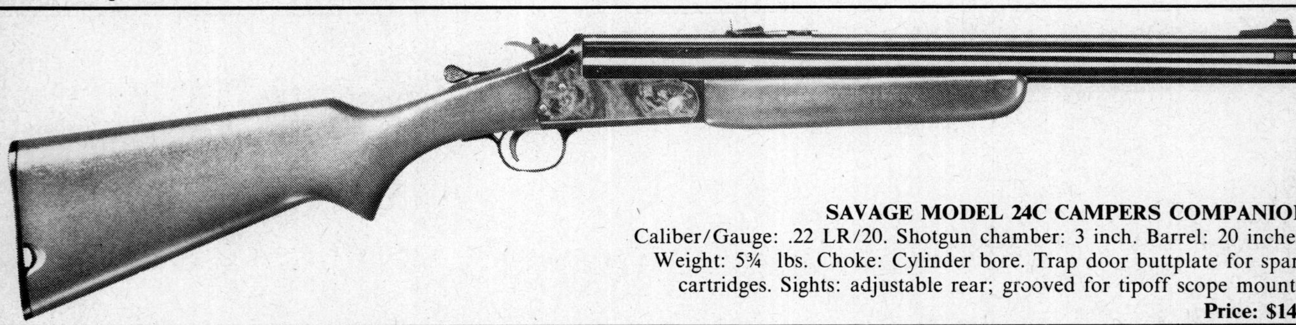

SAVAGE MODEL 24C CAMPERS COMPANION
Caliber/Gauge: .22 LR/20. Shotgun chamber: 3 inch. Barrel: 20 inches. Weight: 5¾ lbs. Choke: Cylinder bore. Trap door buttplate for spare cartridges. Sights: adjustable rear; grooved for tipoff scope mounts. **Price: $140**

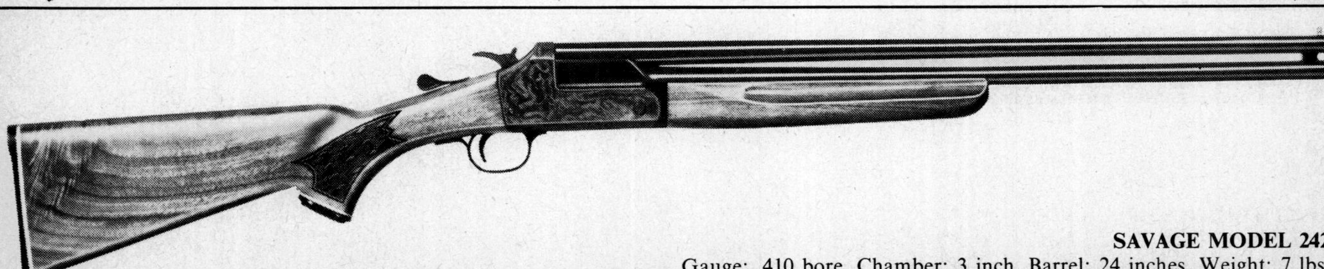

SAVAGE MODEL 242
Gauge: .410 bore. Chamber: 3 inch. Barrel: 24 inches. Weight: 7 lbs. Choke: F. Two way opening lever for left or right hand use. Walnut stock and forend. **Price: $156**

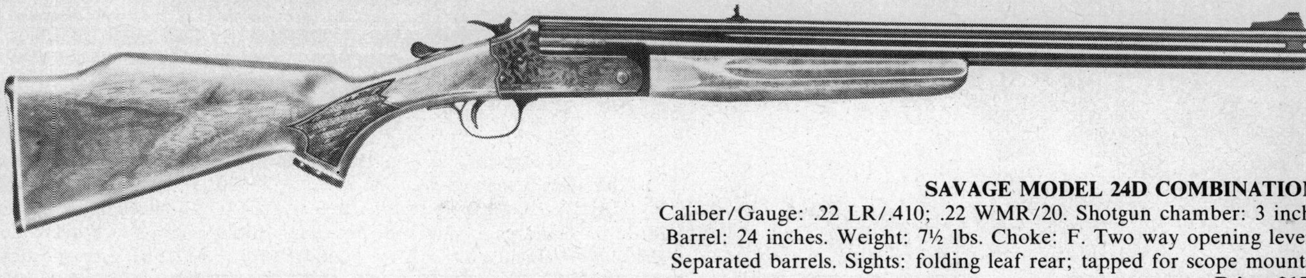

SAVAGE MODEL 24D COMBINATION
Caliber/Gauge: .22 LR/.410; .22 WMR/20. Shotgun chamber: 3 inch. Barrel: 24 inches. Weight: 7½ lbs. Choke: F. Two way opening lever. Separated barrels. Sights: folding leaf rear; tapped for scope mounts. **Price: $156**

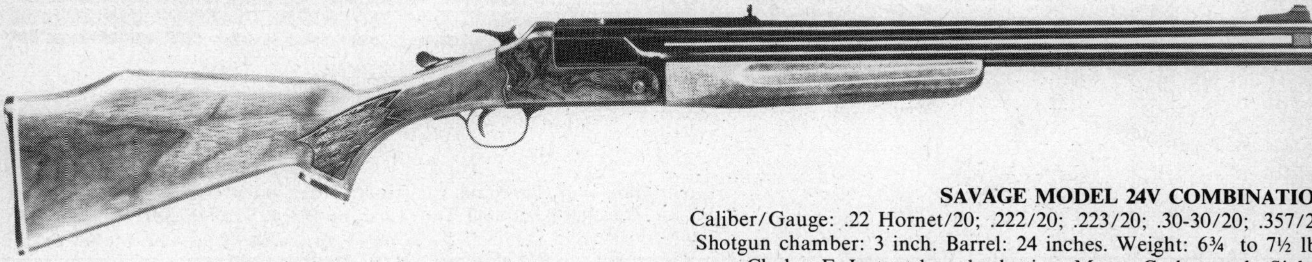

SAVAGE MODEL 24V COMBINATION
Caliber/Gauge: .22 Hornet/20; .222/20; .223/20; .30-30/20; .357/20. Shotgun chamber: 3 inch. Barrel: 24 inches. Weight: 6¾ to 7½ lbs. Choke: F. Instant barrel selection. Monte Carlos stock. Sights: folding leaf rear, ramp front; tapped for scope mounts. **Price: $186**

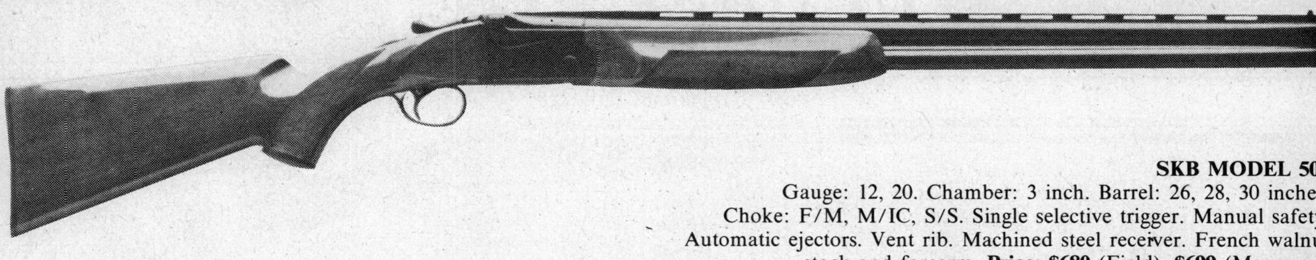

SKB MODEL 500
Gauge: 12, 20. Chamber: 3 inch. Barrel: 26, 28, 30 inches. Choke: F/M, M/IC, S/S. Single selective trigger. Manual safety. Automatic ejectors. Vent rib. Machined steel receiver. French walnut stock and forearm. **Price: $680** (Field), **$699** (Magnum); **$690** (Skeet); **$703** (Trap)

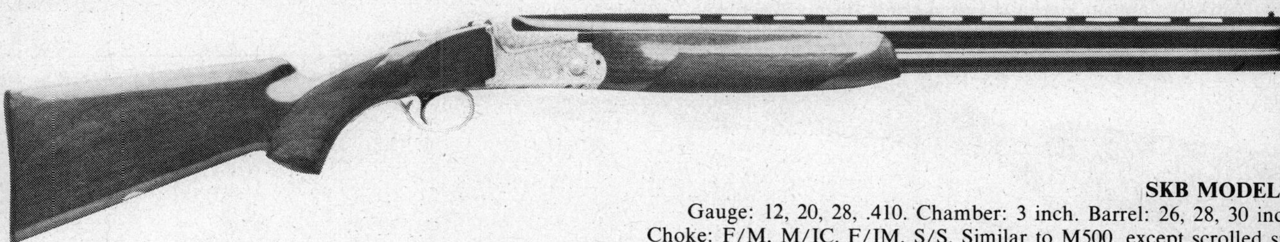

SKB MODEL 600
Gauge: 12, 20, 28, .410. Chamber: 3 inch. Barrel: 26, 28, 30 inches. Choke: F/M, M/IC, F/IM, S/S. Similar to M500, except scrolled silver plated frame and select grade French walnut. **Price: $815** (Field); **$890** (M680 English style); **$839** (Skeet); **$850** (Trap); **$2569** (3 barrel combo)

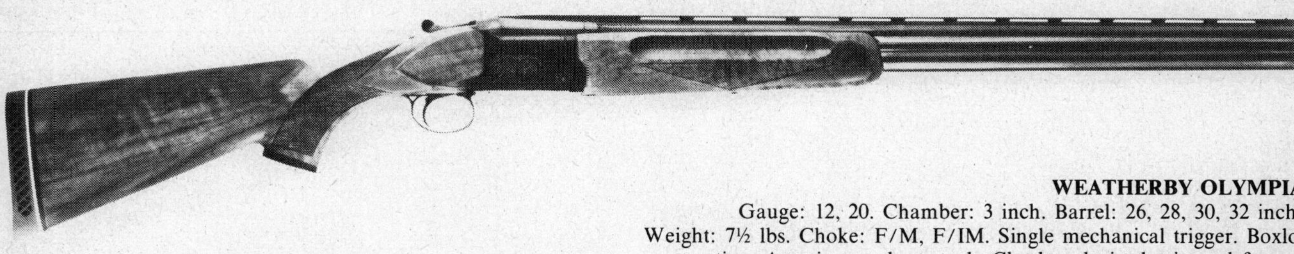

WEATHERBY OLYMPIAN
Gauge: 12, 20. Chamber: 3 inch. Barrel: 26, 28, 30, 32 inches. Weight: 7½ lbs. Choke: F/M, F/IM. Single mechanical trigger. Boxlock action. American walnut stock. Checkered pistol grip and forearm. **Price: $750; $830** (Trap)

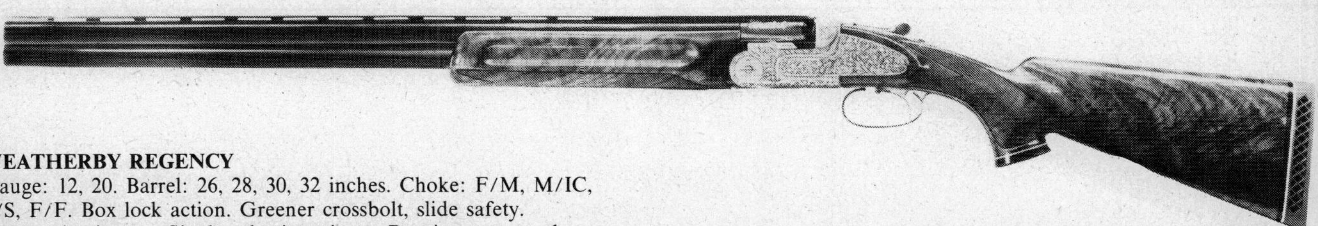

WEATHERBY REGENCY
Gauge: 12, 20. Barrel: 26, 28, 30, 32 inches. Choke: F/M, M/IC, S/S, F/F. Box lock action. Greener crossbolt, slide safety. Automatic ejectors. Single selective trigger. Receiver engraved in scroll work and floral patterns. Walnut stock with hand checkering. **Price: $1100; $1200** (Trap)

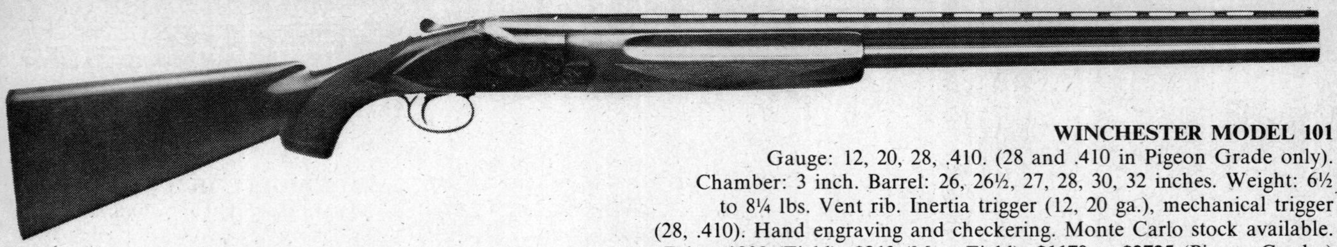

WINCHESTER MODEL 101
Gauge: 12, 20, 28, .410. (28 and .410 in Pigeon Grade only). Chamber: 3 inch. Barrel: 26, 26½, 27, 28, 30, 32 inches. Weight: 6½ to 8¼ lbs. Vent rib. Inertia trigger (12, 20 ga.), mechanical trigger (28, .410). Hand engraving and checkering. Monte Carlo stock available. **Price: $902** (Field); **$913** (Mag. Field); **$1170** to **$2725** (Pigeon Grades)

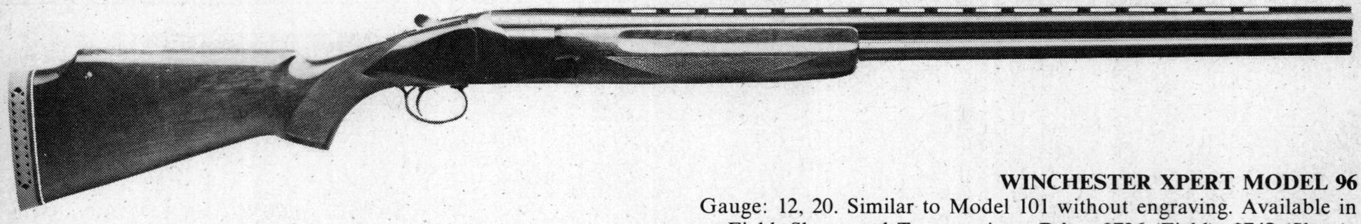

WINCHESTER XPERT MODEL 96
Gauge: 12, 20. Similar to Model 101 without engraving. Available in Field, Skeet, and Trap versions. **Price: $726** (Field); **$748** (Skeet); **$748** (Trap); **$764** (Trap with Monte Carlo)

Double Barrel

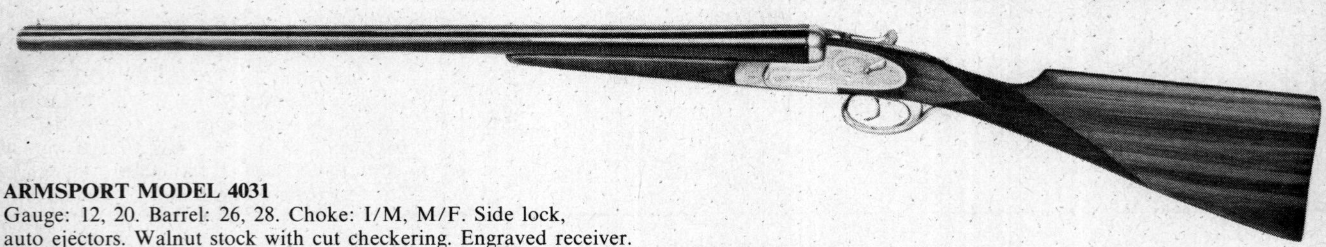

ARMSPORT MODEL 4031
Gauge: 12, 20. Barrel: 26, 28. Choke: I/M, M/F. Side lock, auto ejectors. Walnut stock with cut checkering. Engraved receiver. **Price: $3600**

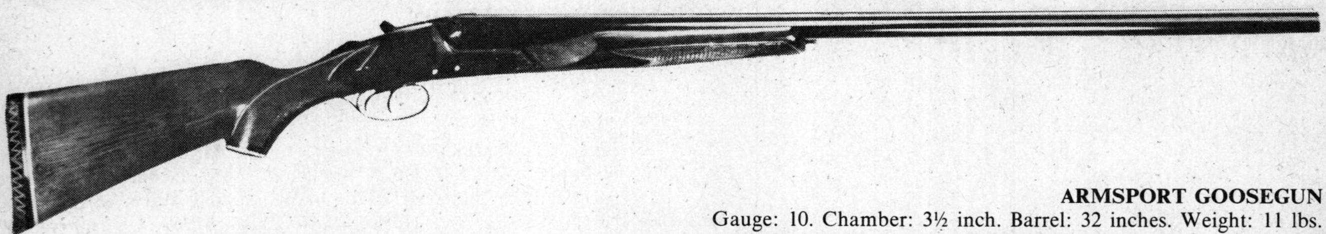

ARMSPORT GOOSEGUN
Gauge: 10. Chamber: 3½ inch. Barrel: 32 inches. Weight: 11 lbs. Choke: F/F. Double triggers. Checkered European walnut stock and forend. Rubber recoil pad. **Price: $595**

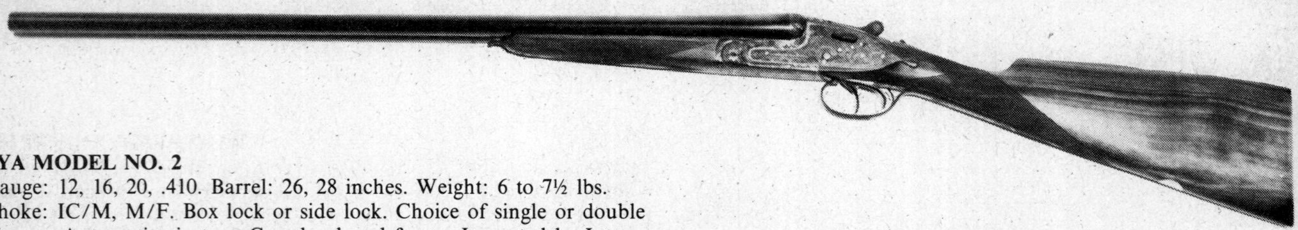

AYA MODEL NO. 2
Gauge: 12, 16, 20, .410. Barrel: 26, 28 inches. Weight: 6 to 7½ lbs. Choke: IC/M, M/F. Box lock or side lock. Choice of single or double triggers. Automatic ejectors. Case hardened frame. Imported by Interarms. **Price: $750** to **$2200**

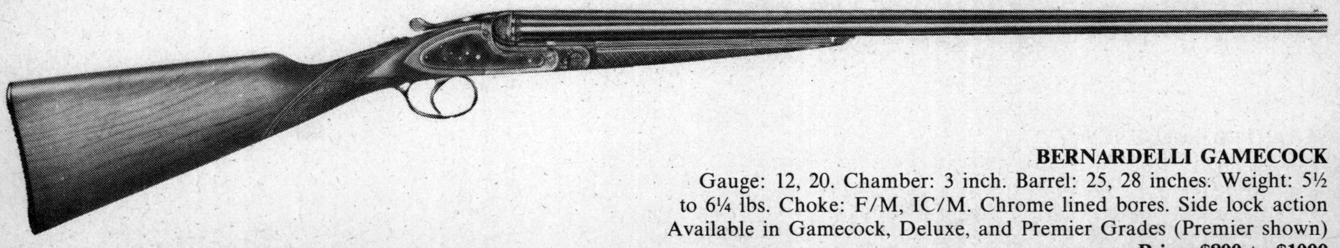

BERNARDELLI GAMECOCK
Gauge: 12, 20. Chamber: 3 inch. Barrel: 25, 28 inches. Weight: 5½ to 6¼ lbs. Choke: F/M, IC/M. Chrome lined bores. Side lock action Available in Gamecock, Deluxe, and Premier Grades (Premier shown) **Price: $800** to **$1000**

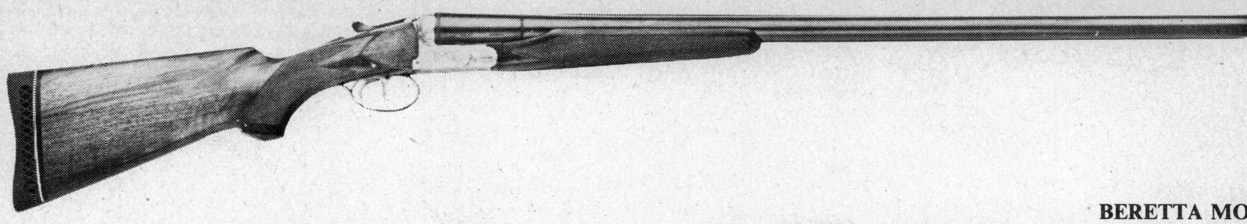

BERETTA MODEL 410
Gauge: 10. Chamber: 3½ inch. Barrel: 32 inches. Weight: 10 lbs. Choke: F/F. Silver gray receiver. Double triggers. European walnut stock and forend. Hand checkering. Matted rib. **Price: $1135**

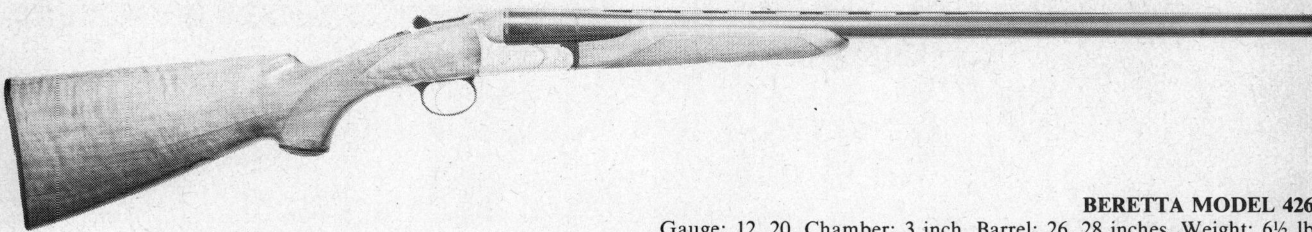

BERETTA MODEL 426E
Gauge: 12, 20. Chamber: 3 inch. Barrel: 26, 28 inches. Weight: 6½ lbs. Choke: IC/M, F/M. Box lock action. Double triggers. European walnut stock and forearm with hand checkering. **Price: $986**

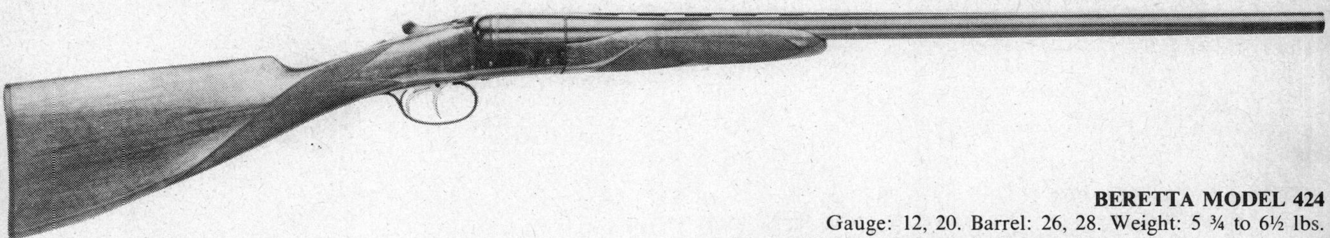

BERETTA MODEL 424
Gauge: 12, 20. Barrel: 26, 28. Weight: 5 ¾ to 6½ lbs. Choke: I/M, M/F. Boxlock design with coil springs. Extractors. European walnut stock with hand checkering. **Price: $757**

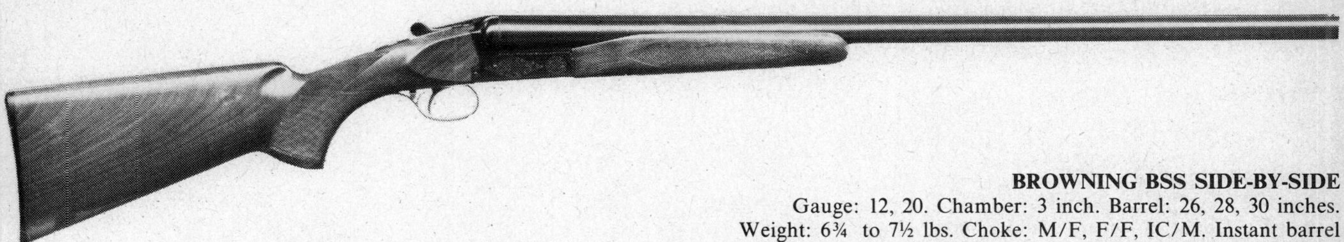

BROWNING BSS SIDE-BY-SIDE
Gauge: 12, 20. Chamber: 3 inch. Barrel: 26, 28, 30 inches. Weight: 6¾ to 7½ lbs. Choke: M/F, F/F, IC/M. Instant barrel selection. Single selective trigger. Top tang safety. Automatic ejection. Checkered French walnut stock and forearm. **Price: $460** to **$825**

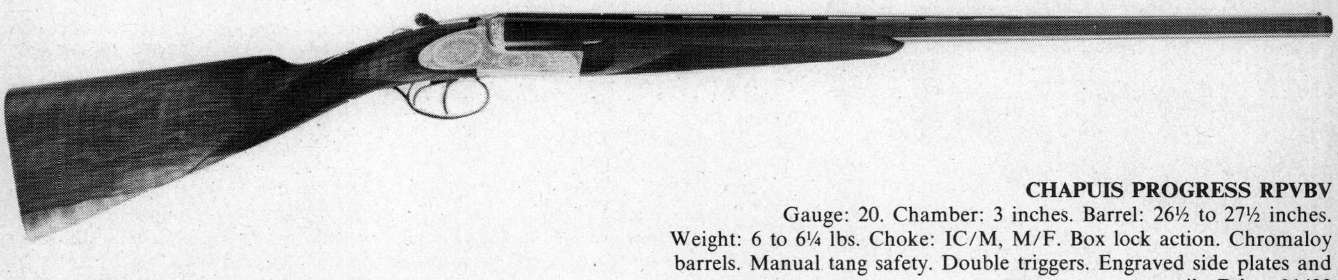

CHAPUIS PROGRESS RPVBV
Gauge: 20. Chamber: 3 inches. Barrel: 26½ to 27½ inches. Weight: 6 to 6¼ lbs. Choke: IC/M, M/F. Box lock action. Chromaloy barrels. Manual tang safety. Double triggers. Engraved side plates and vent rib. **Price: $1400**

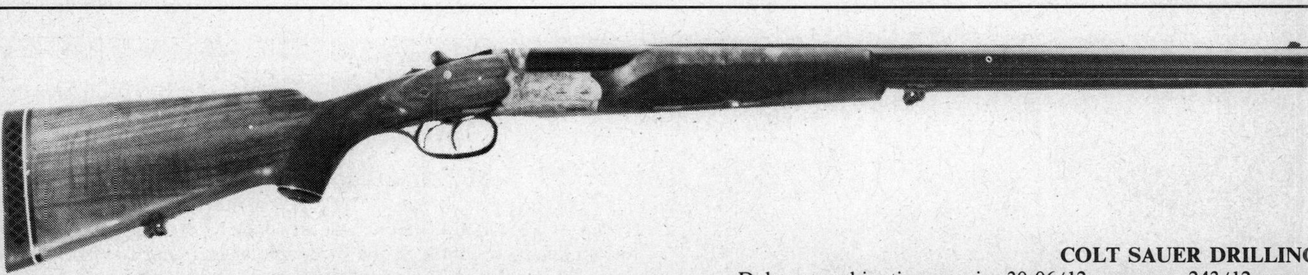

COLT SAUER DRILLING
Deluxe combination gun in .30-06/12 gauge or .243/12 gauge. Barrel: 25 inches. Weight: 8 lbs. **Price: $2995**

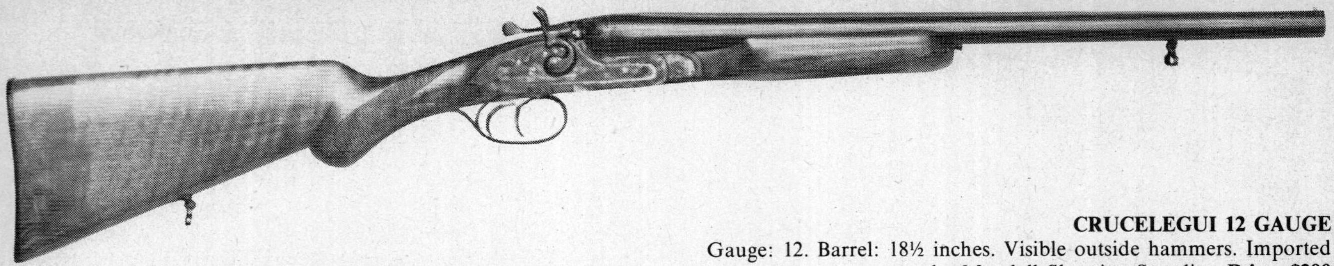

CRUCELEGUI 12 GAUGE
Gauge: 12. Barrel: 18½ inches. Visible outside hammers. Imported by Mandall Shooting Supplies. **Price: $300**

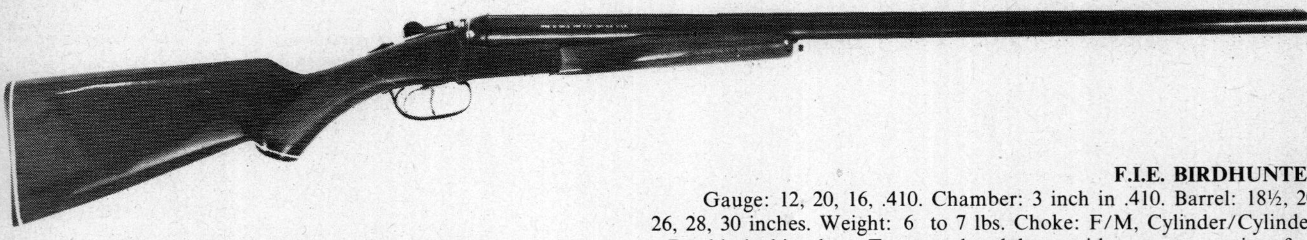

F.I.E. BIRDHUNTER
Gauge: 12, 20, 16, .410. Chamber: 3 inch in .410. Barrel: 18½, 20, 26, 28, 30 inches. Weight: 6 to 7 lbs. Choke: F/M, Cylinder/Cylinder. Double locking lugs. Top tang breakdown with tang automatic safety. Brazilian hardwood stock. **Price: $137** to **$153**

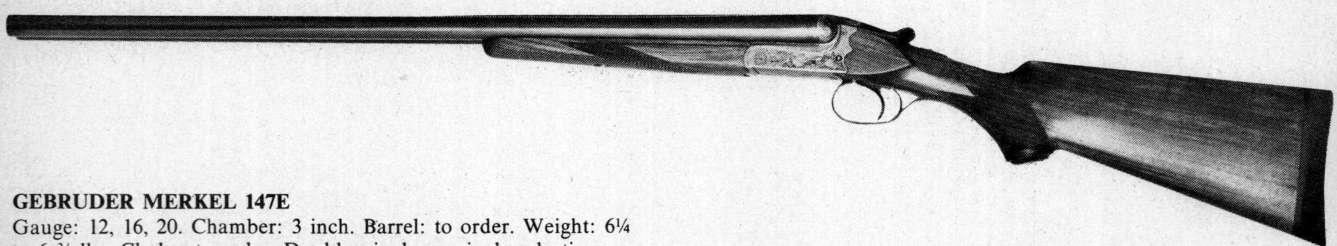

GEBRUDER MERKEL 147E
Gauge: 12, 16, 20. Chamber: 3 inch. Barrel: to order. Weight: 6¼ to 6 ¾ lbs. Choke: to order. Double, single, or single selective trigger. **Price: $1105**

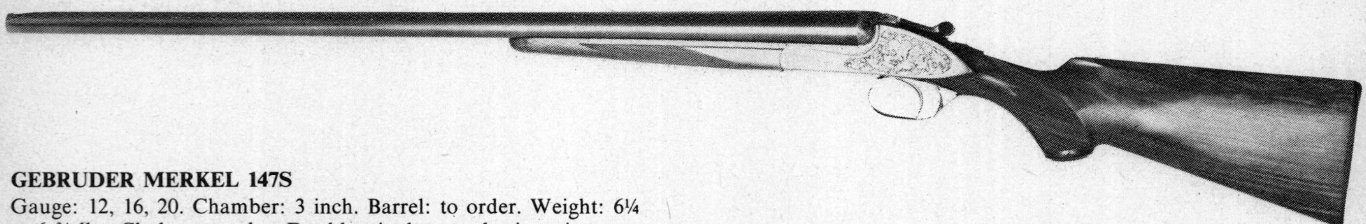

GEBRUDER MERKEL 147S
Gauge: 12, 16, 20. Chamber: 3 inch. Barrel: to order. Weight: 6¼ to 6 ¾ lbs. Choke: to order. Double, single, or selective trigger. Walnut stock in pistol grip or English style. Hunting scene engraving. **Price: $2486**

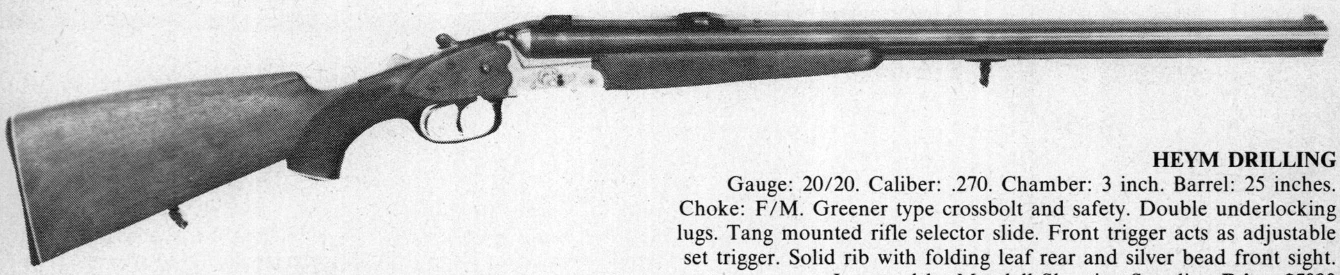

HEYM DRILLING
Gauge: 20/20. Caliber: .270. Chamber: 3 inch. Barrel: 25 inches. Choke: F/M. Greener type crossbolt and safety. Double underlocking lugs. Tang mounted rifle selector slide. Front trigger acts as adjustable set trigger. Solid rib with folding leaf rear and silver bead front sight. Imported by Mandall Shooting Supplies. **Price: $5000**

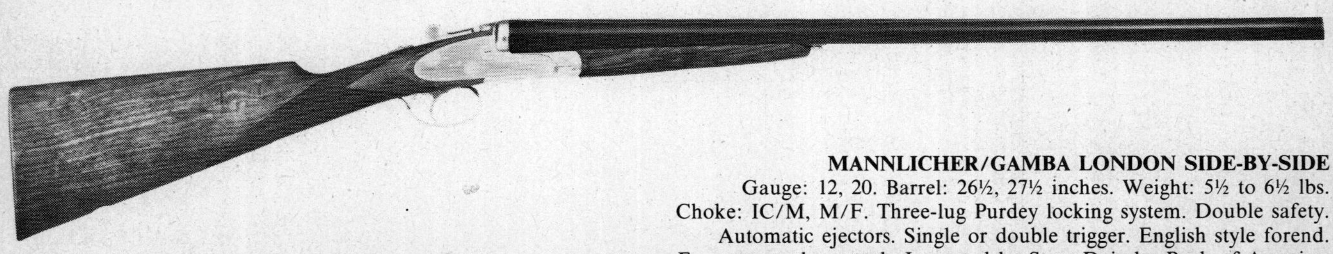

MANNLICHER/GAMBA LONDON SIDE-BY-SIDE
Gauge: 12, 20. Barrel: 26½, 27½ inches. Weight: 5½ to 6½ lbs. Choke: IC/M, M/F. Three-lug Purdey locking system. Double safety. Automatic ejectors. Single or double trigger. English style forend. European walnut stock. Imported by Steyr-Daimler Puch of America. **Price: $3325** (double trigger); **$3475** (single trigger)

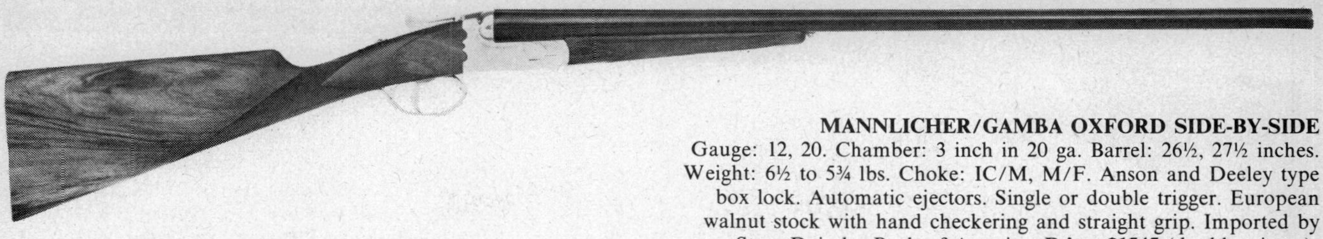

MANNLICHER/GAMBA OXFORD SIDE-BY-SIDE
Gauge: 12, 20. Chamber: 3 inch in 20 ga. Barrel: 26½, 27½ inches. Weight: 6½ to 5¾ lbs. Choke: IC/M, M/F. Anson and Deeley type box lock. Automatic ejectors. Single or double trigger. European walnut stock with hand checkering and straight grip. Imported by Steyr-Daimler Puch of America. **Price: $1545** (double trigger); **$1670** (single trigger)

RICHLAND ARMS MODEL 711 LONG RANGE
Gauge: 10, 12. Chamber: 3½, 3 inches. Barrel: 32, 30 inches. Weight: 7¾ to 11 lbs. Anson and Deely type action with Purdey lock. Double triggers and extractors. Select Spanish walnut stock. **Price: $508** (10 ga.); **$455** (12 ga.)

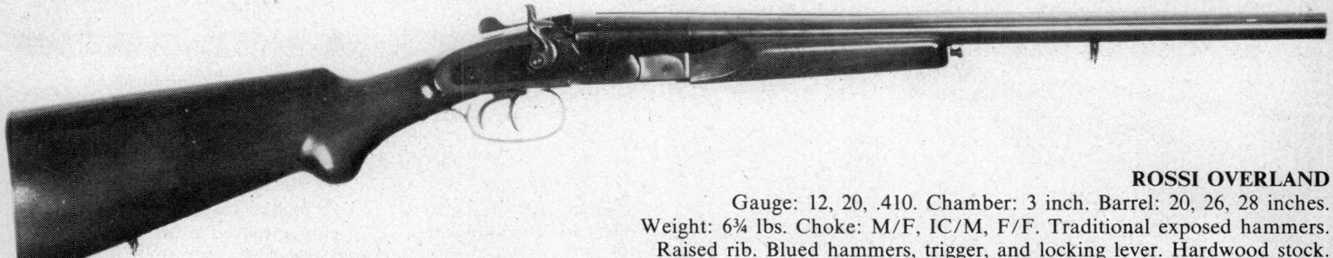

ROSSI OVERLAND
Gauge: 12, 20, .410. Chamber: 3 inch. Barrel: 20, 26, 28 inches. Weight: 6¾ lbs. Choke: M/F, IC/M, F/F. Traditional exposed hammers. Raised rib. Blued hammers, trigger, and locking lever. Hardwood stock. Imported by Interarms. **Price: $275; $285** (.410)

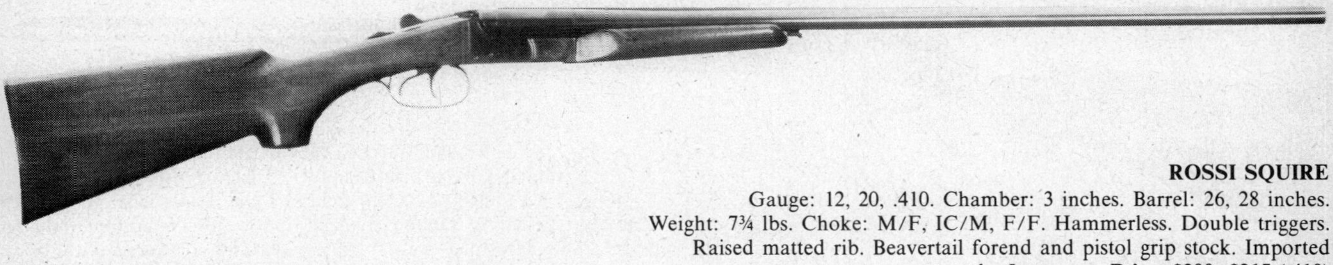

ROSSI SQUIRE
Gauge: 12, 20, .410. Chamber: 3 inches. Barrel: 26, 28 inches. Weight: 7¾ lbs. Choke: M/F, IC/M, F/F. Hammerless. Double triggers. Raised matted rib. Beavertail forend and pistol grip stock. Imported by Interarms. **Price: $300; $315** (.410)

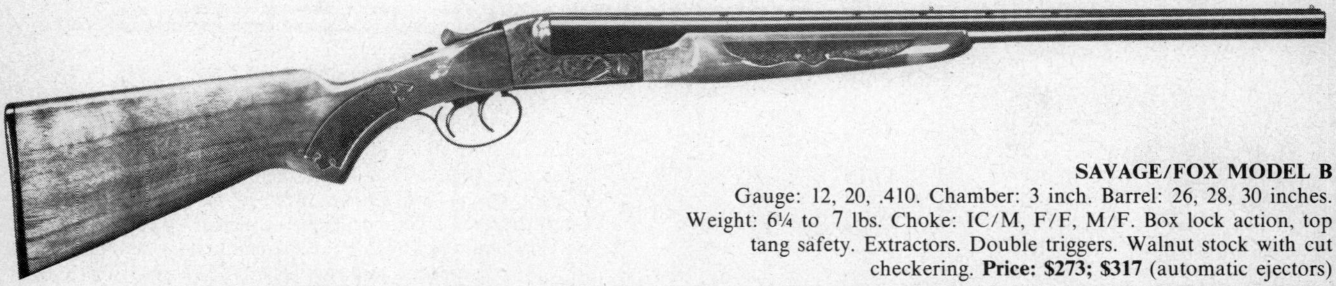

SAVAGE/FOX MODEL B
Gauge: 12, 20, .410. Chamber: 3 inch. Barrel: 26, 28, 30 inches. Weight: 6¼ to 7 lbs. Choke: IC/M, F/F, M/F. Box lock action, top tang safety. Extractors. Double triggers. Walnut stock with cut checkering. **Price: $273; $317** (automatic ejectors)

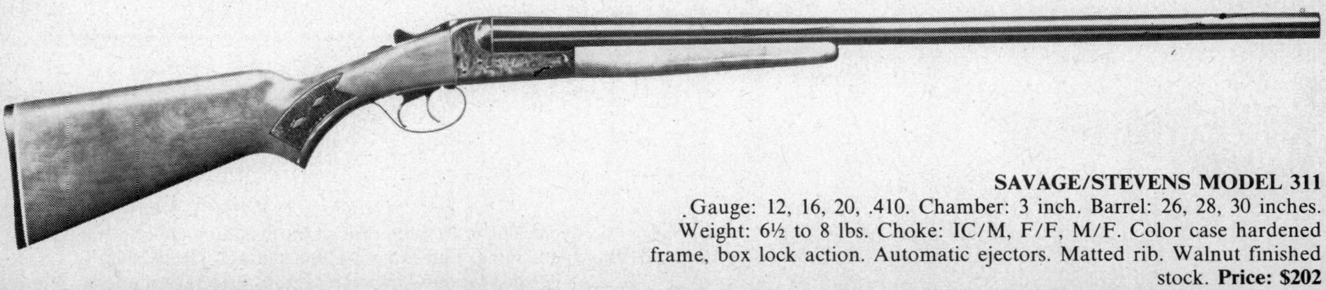

SAVAGE/STEVENS MODEL 311
Gauge: 12, 16, 20, .410. Chamber: 3 inch. Barrel: 26, 28, 30 inches. Weight: 6½ to 8 lbs. Choke: IC/M, F/F, M/F. Color case hardened frame, box lock action. Automatic ejectors. Matted rib. Walnut finished stock. **Price: $202**

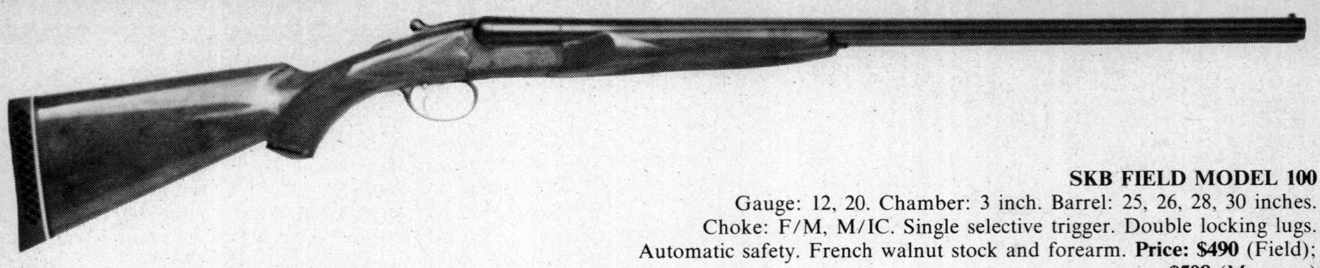

SKB FIELD MODEL 100
Gauge: 12, 20. Chamber: 3 inch. Barrel: 25, 26, 28, 30 inches. Choke: F/M, M/IC. Single selective trigger. Double locking lugs. Automatic safety. French walnut stock and forearm. **Price: $490** (Field); **$508** (Magnum)

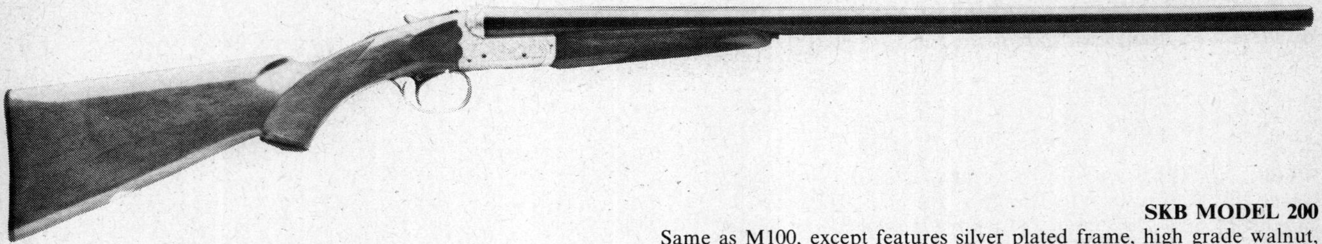

SKB MODEL 200
Same as M100, except features silver plated frame, high grade walnut, beavertail forearm, and gold plated oval nameplate. **Price: $678; $699** (Magnum); **$705** (Skeet); **$718** (M280 English style)

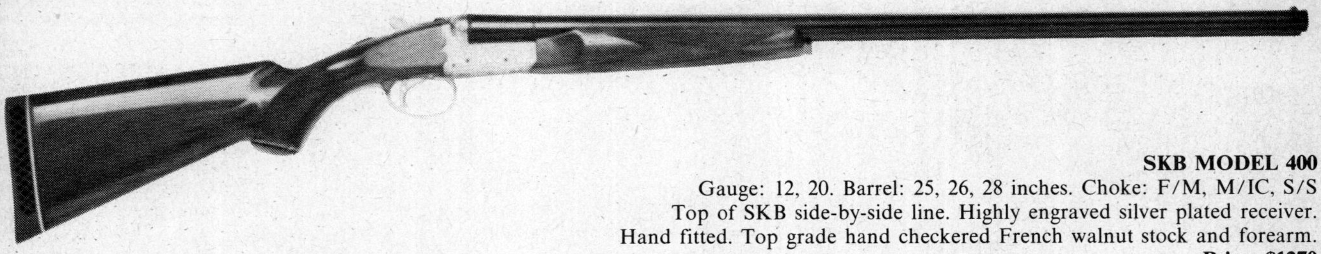

SKB MODEL 400
Gauge: 12, 20. Barrel: 25, 26, 28 inches. Choke: F/M, M/IC, S/S Top of SKB side-by-side line. Highly engraved silver plated receiver. Hand fitted. Top grade hand checkered French walnut stock and forearm. **Price: $1270**

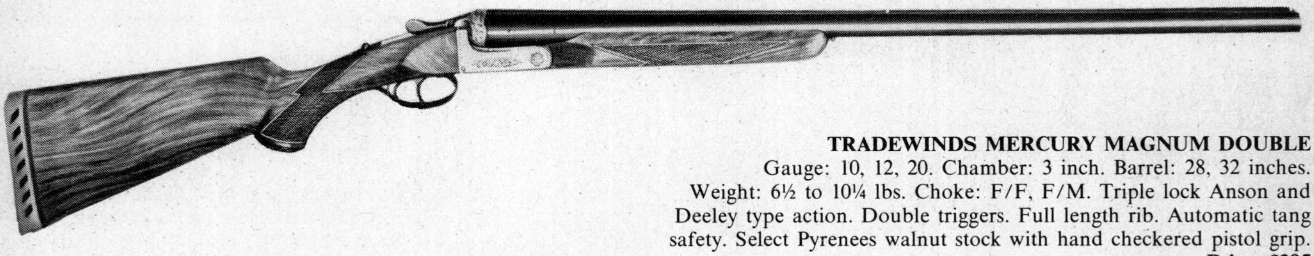

TRADEWINDS MERCURY MAGNUM DOUBLE
Gauge: 10, 12, 20. Chamber: 3 inch. Barrel: 28, 32 inches. Weight: 6½ to 10¼ lbs. Choke: F/F, F/M. Triple lock Anson and Deeley type action. Double triggers. Full length rib. Automatic tang safety. Select Pyrenees walnut stock with hand checkered pistol grip. **Price: $395**

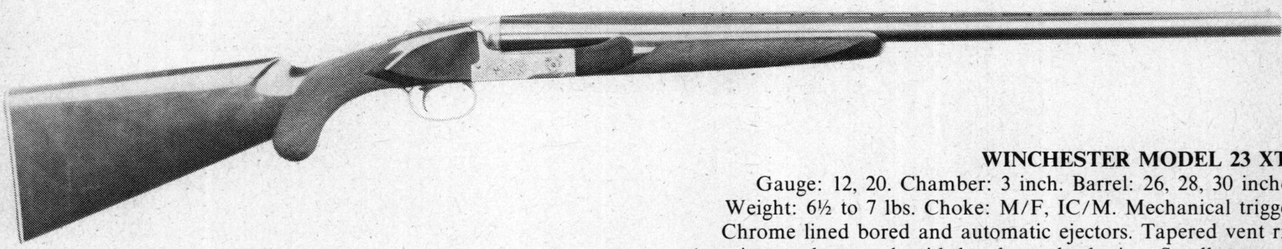

WINCHESTER MODEL 23 XTR
Gauge: 12, 20. Chamber: 3 inch. Barrel: 26, 28, 30 inches. Weight: 6½ to 7 lbs. Choke: M/F, IC/M. Mechanical trigger. Chrome lined bored and automatic ejectors. Tapered vent rib. American walnut stock with hand cut checkering. Scroll engraved trigger guard, top lever, and receiver. **Price: $980**

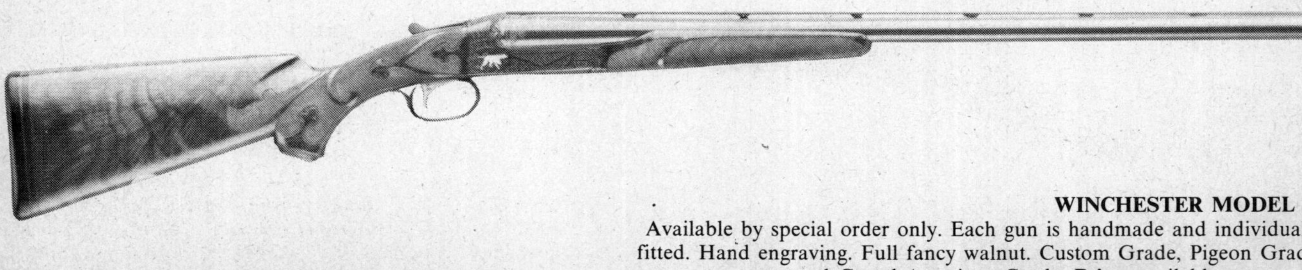

WINCHESTER MODEL 21
Available by special order only. Each gun is handmade and individually fitted. Hand engraving. Full fancy walnut. Custom Grade, Pigeon Grade, and Grand American Grade. **Price:** available on request

BLACKPOWDER CATALOG

Handguns

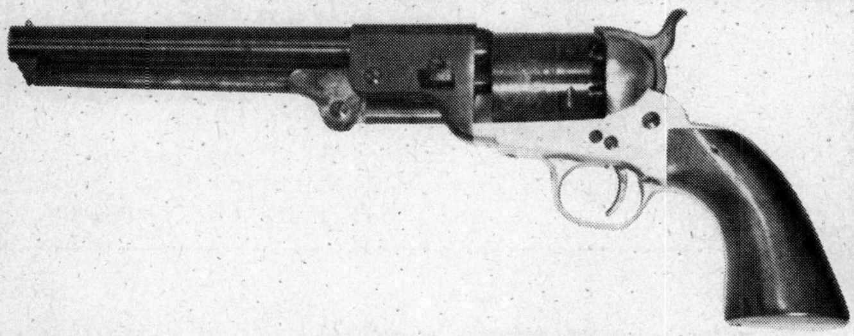

ARMOURY INC. CONFEDERATE REVOLVER
Caliber: .36. Ignition: percussion.
Barrel: 7½ inches. Weight: 41 oz. Brass frame, engraved cylinder. Blued barrel. Also available in .44 caliber as Model CB44. **Price: $115**

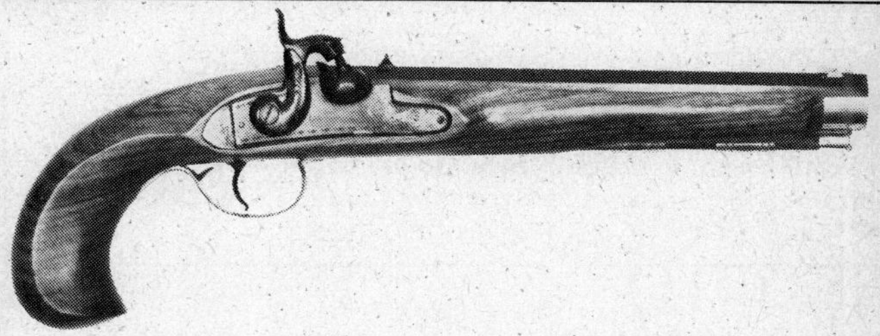

ARMOURY INC. KENTUCKY PISTOL
Caliber: .45. Ignition: percussion, flint. Overall length: 15½ inches. Weight: 35 oz. Octagon rifled barrel. Brass tipped ramrod. Dovetailed rear sight and brass blade front. **Price: $130**

ARMOURY INC. SHERIFF REVOLVER
Caliber: .36, .44. Ignition: percussion.
Barrel: 5 inches. Weight: 39 oz. Brass frame, backstrap, and triggerguard. Octagon rifled barrel with blued finish. Engraved cylinder with battle scene.
Price: $115

ARMSPORT SPILLER AND BURR
Caliber: .36. Ignition: percussion.
Barrel: 7 inches. Weight: 40 oz. One piece brass frame and blued cylinder. Octagon barrel. Walnut stocks. Sights: fixed. **Price: $90**

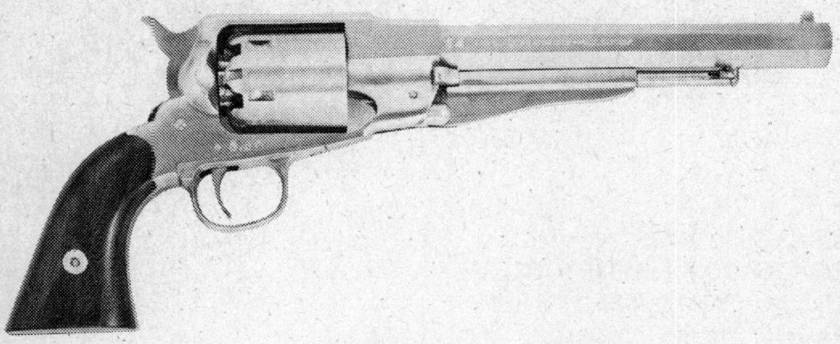

ARMSPORT REMINGTON ARMY STAINLESS
Caliber: .44. Ignition: percussion.
Barrel: 8 inches. Weight: 44 oz. Stainless replica of Remington classic. Walnut stocks. **Price: $175**

ARMSPORT ELI WHITNEY
Caliber: .36. Ignition: percussion.
Barrel: 7½ inches. Weight: 44 oz. One piece frame. Bright blue finish. Octagon barrel. Walnut grips.

ARMSPORT CORSAIR PISTOL
Caliber: .44. Ignition: percussion.
Barrel: 9½ inches. Weight: 40 oz. Twin percussion locks. Brass triggerguard. Select grain walnut stock.
Price: $110; $90 (kit)

ARMSPORT PATRIOT PISTOL
Caliber: .45. Ignition: percussion.
Barrel: 9¼ inches. Weight: 36 oz. Hooked breech. Double set triggers. Walnut stocks. Adjustable open sights.
Price: $125; $95 (kit)

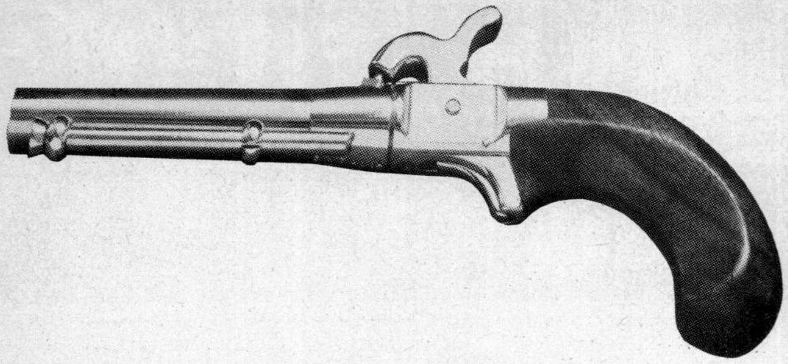

CLASSIC ARMS NEW ORLEANS ACE
Caliber: .44. Ignition: percussion.
Overall length: 9 inches. Weight: 16 oz.
Brass framed replica of early gambler's pistol. **Price: $50; $33** (kit)

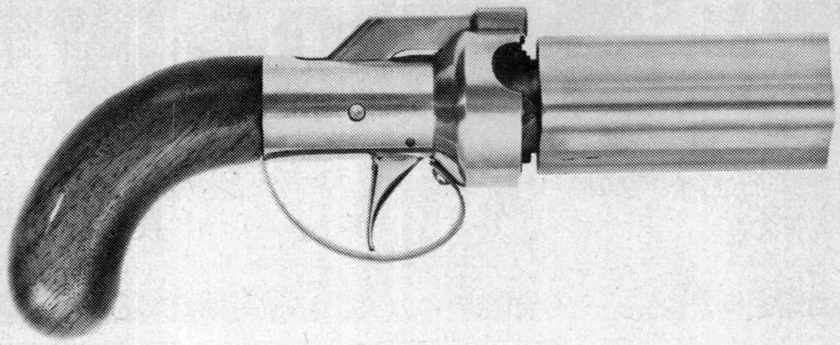

CLASSIC ARMS ETHAN ALLEN PEPPERBOX
Caliber: .36. Ignition: percussion.
Overall length: 9 inches. Weight: 36 oz.
Replica of California Gold Rush Pistol. 4 shot capacity. Solid brass frame. **Price: $70; $48** (kit)

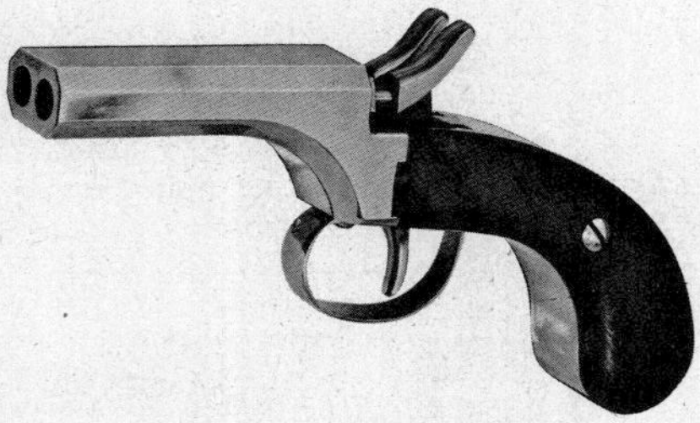

CLASSIC ARMS SNAKE EYES
Caliber: .36. Ignition: percussion.
Overall length: 6¾ inches. Weight: 24 oz.
Replica of double barreled gambler's pistol. Solid brass receiver and pearlite grip. **Price: $60; $44** (kit)

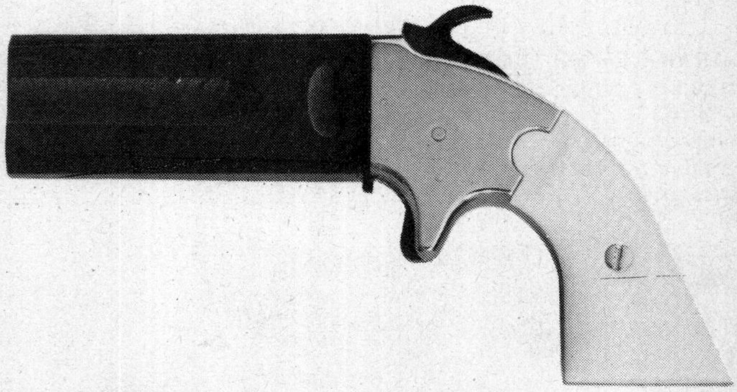

CLASSIC ARMS TWISTER
Caliber: .36. Barrel: 3⅜ inches.
Smoothbore. **Price: $73; $45** (kit)

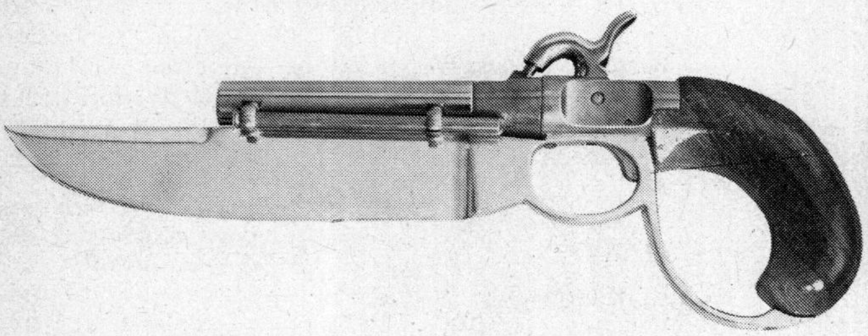

CLASSIC ARMS ELGIN CUTLASS PISTOL
Caliber: .44. Ignition: percussion.
Barrel: 9 inches. Replica of U.S. Navy boarding pistol.
Price: $100; $75 (kit)

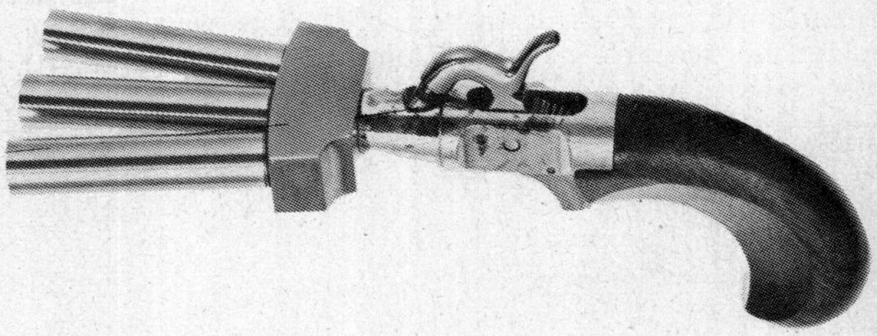

CLASSIC ARMS DUCKFOOT MUTINY GUN
Caliber: .36. Ignition: percussion.
Overall length: 10½ inches. Weight: 32 oz.
Replica of 3 barreled English Admiralty riot gun for shipboard disturbances. **Price: $60; $40** (kit)

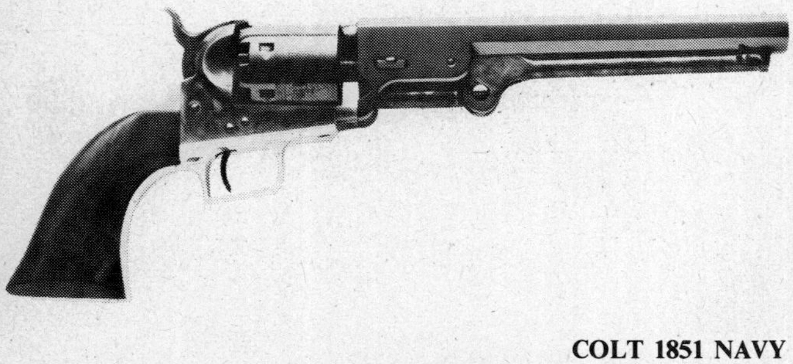

COLT 1851 NAVY
Caliber: .36. Ignition: percussion.
Barrel: 7½ inches. Weight: 42 oz. Capacity: 6.
Engraved cylinder. Color case hardened frame. Octagon barrel. Silver plated brass backstrap and triggerguard. Walnut stocks.
Price: $334

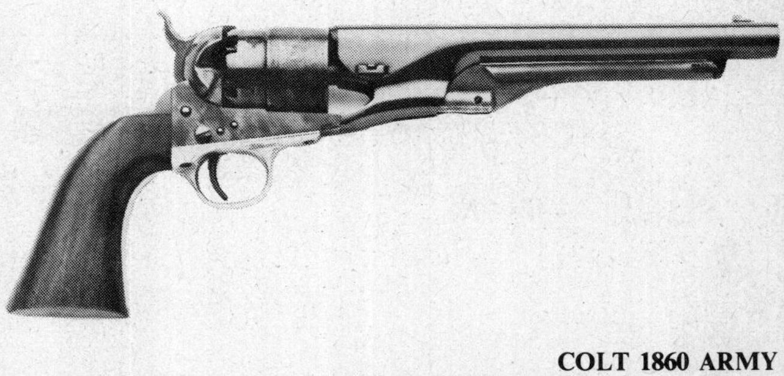

COLT 1860 ARMY
Caliber: .44. Ignition: percussion.
Barrel: 8 inches. Weight: 42 oz. Capacity: 6.
Historic "creeping" loading lever. Roll engraved cylinder and walnut grips. Backstrap cut for shoulder stock. Brass triggerguard. German silver front sight. **Price: $343**

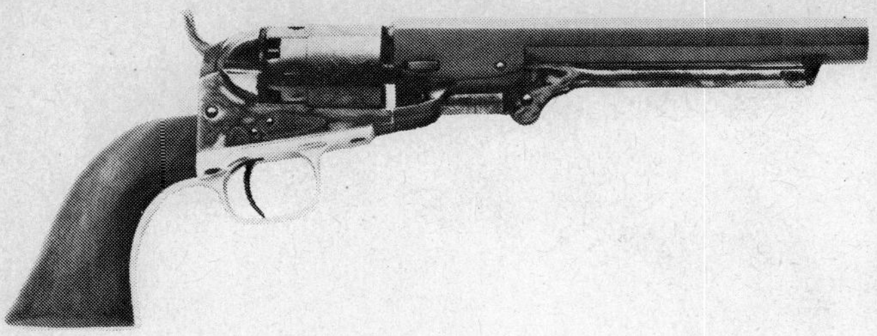

COLT 1862 POCKET NAVY
Caliber: .36. Ignition: percussion.
Barrel: 5½ inches. Weight: 27 oz. Capacity: 5.
Rebated cylinder, hinged loading lever. Silver plated backstrap and triggerguard. Color case hardened frame. Brass pin front sight. **Price: $314**

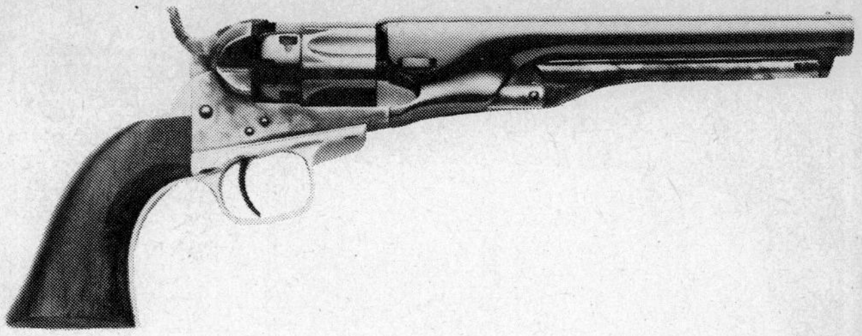

COLT 1862 POCKET POLICE
Caliber: .36. Ignition: percussion.
Barrel: 5½ inches. Weight: 25 oz. Capacity: 5.
Color case hardened frame. Silver plated triggerguard and backstrap. One piece walnut grips. **Price: $314**

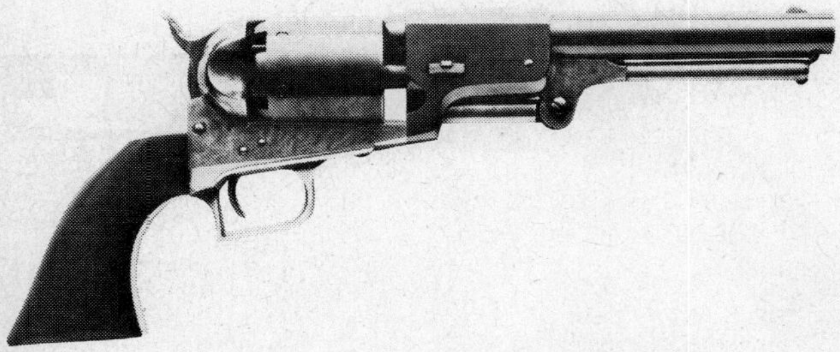

COLT 1ST MODEL DRAGOON
Caliber: .44. Ignition: percussion.
Barrel: 7½ inches. Weight: 66 oz. Capacity: 6.
Engraved cylinder with oval bolt cuts. Flared square back triggerguard. "V" type mainspring and short trigger. **Price: $357**

COLT 2ND MODEL DRAGOON
Caliber: .44. Ignition: percussion.
Barrel: 7½ inches. Weight: 66 oz. Capacity: 6.
Rectangular bolt cuts on engraved cylinder. Straight square back triggerguard. Color case hardened steel frame. **Price: $357**

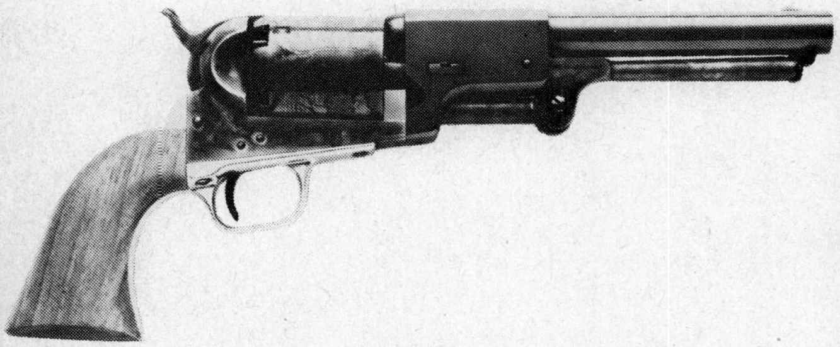

COLT 3RD MODEL DRAGOON
Caliber: .44. Ignition: percussion.
Barrel: 7½ inches. Weight: 66 oz. Capacity: 6.
Similar to other Dragoons, except features oval triggerguard, flat mainspring, and rectangular bolt cuts. **Price: $357**

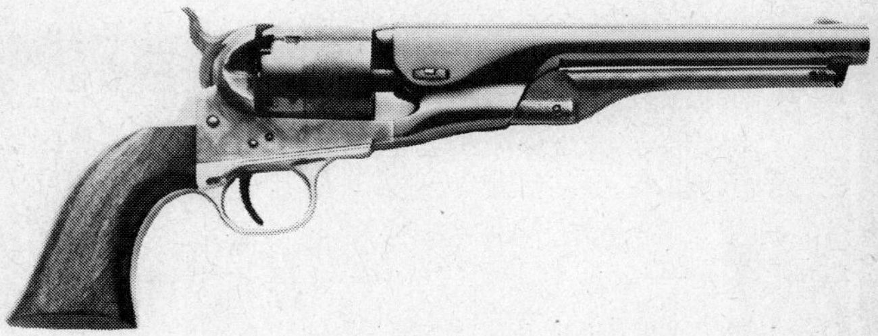

COLT 1861 NAVY
Caliber: .36. Ignition: percussion.
Barrel: 7½ inches. Weight: 42 oz. Capacity: 6.
Engraved cylinder, silver plated backstrap and triggerguard. Color case hardened frame. One piece walnut grips. **Price: $334**

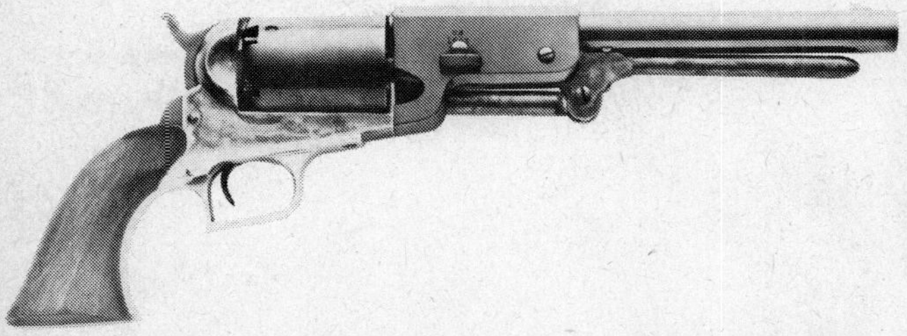

COLT WALKER
Caliber: .44. Ignition: percussion.
Barrel: 9 inches. Weight: 73 oz. Capacity: 6.
Engraved cylinder, brass triggerguard in squared configuration. Checkered spur hammer. German silver front sight. **Price: $447**

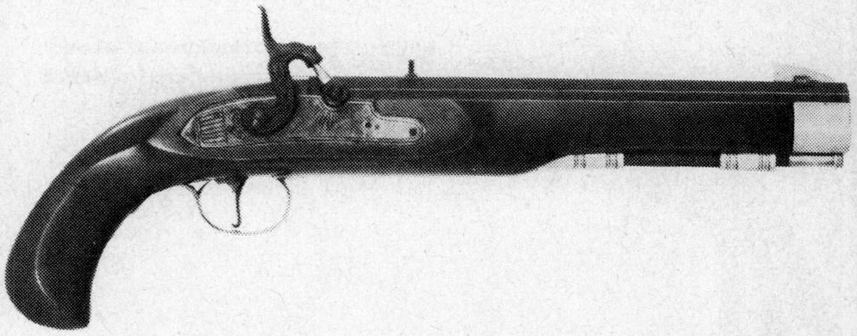

CVA KENTUCKY PISTOL
Caliber: .45. Ignition: percussion or flint. Barrel: 9 inches. Weight: 40 oz.
Blued barrel. "V" type mainspring. Polished brass furniture. Fixed sights. **Price: $84** (percussion); **$52** (kit); **$60** (flint, kit)

CVA MOUNTAIN PISTOL
Caliber: .45. Ignition: percussion. Barrel: 9 inches. Weight: 43 oz. All steel tang bolt and barrel tenon. Inletted American maple stock. Fixed sights. **Price: $103; $80** (kit)

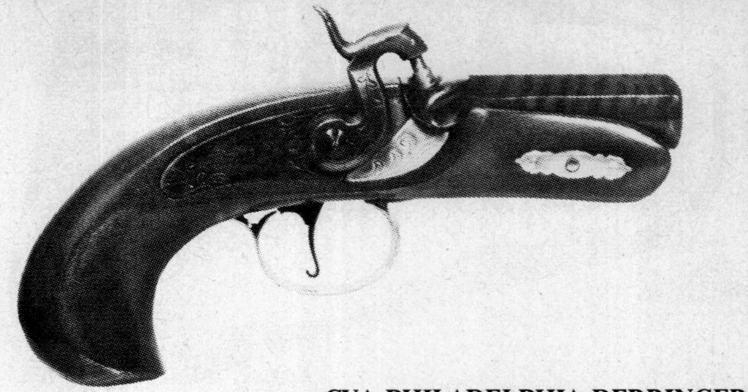

CVA PHILADELPHIA DERRINGER
Caliber: .45. Ignition: percussion. Barrel: 3⅛ inch. Weight: 14 oz. Coil spring percussion lock. Case hardened finish. Brass hardware. Comes with steel nipple wrench. **Price: $49; $32** (kit)

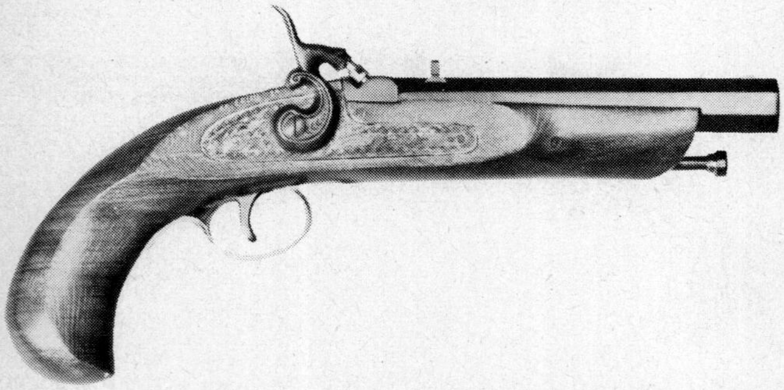

CVA COLONIAL PISTOL
Caliber: .45. Ignition: percussion or flint. Barrel: 7 inches. Weight: 30 oz. Interlocking breech system. Case hardened lock. Blued barrel. **Price: $57** (percussion); **$34** (kit); **$41** (flint, kit)

CVA TOWER PISTOL
Caliber: .45. Ignition: percussion or flint. Barrel: 8 inches. Weight: 39 oz. Brass grip cap, triggerguard, thimbles and nose cap. Color case hardened lock. Blued steel ramrod. **Price: $80** (percussion); **$52** (kit); **$60** (flint, kit)

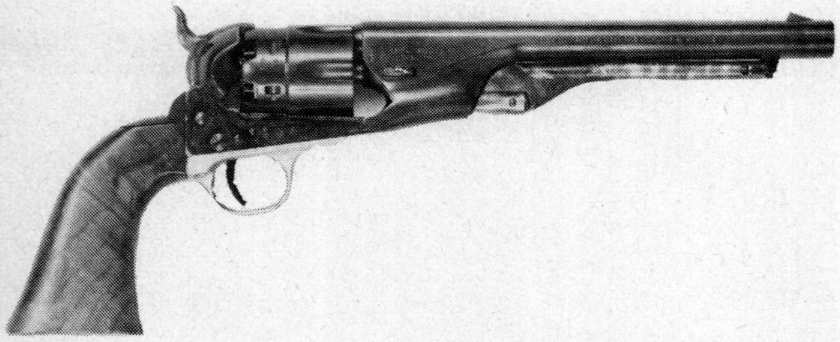

DIXIE GUN WORKS
1860 ARMY
Caliber: .44. Ignition: percussion. Barrel: 8 inches. Weight: 43 oz. Case hardened frame and loading lever. Fluted cylinder. Frame cut for shoulder stock. **Price: $100**

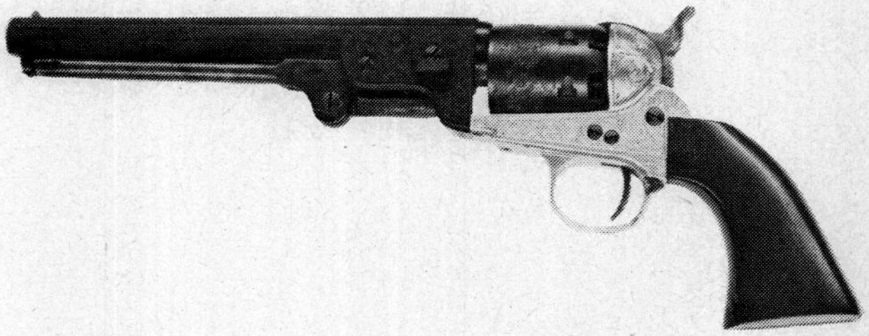

DIXIE GUN WORKS
NAVY REVOLVER
Caliber: .36. Ignition: percussion. Barrel: 7½ inches. Weight: 40 oz. Replica of Civil War era naval sidearm. Brass frame. **Price: $50; $70** (engraved)

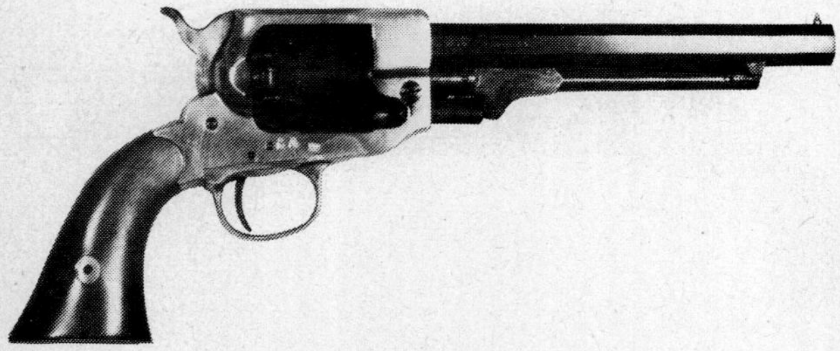

DIXIE GUN WORKS
SPILLER & BURR REVOLVER
Caliber: .36. Ignition: percussion. Barrel: 7 inches. Weight: 40 oz. Brass frame replica of Confederate revolver. Octagon barrel. Case hardened loading lever. Two piece walnut grips. **Price: $80**

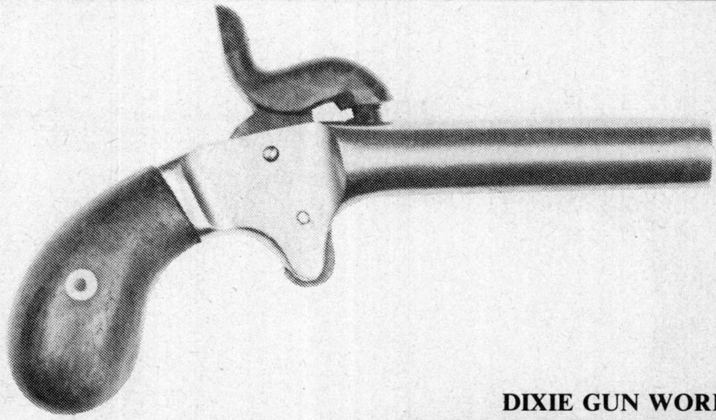

DIXIE GUN WORKS
BRASS FRAME DERRINGER
Caliber: .40. Ignition: percussion. Overall length: 6 inches. Weight: 14 oz. Small center hammer derringer of the type used on the Mississippi River in the last century. **Price: $40; 50** (engraved); **$8** (case)

DIXIE GUN WORKS FRENCH CHARLEVILLE
Caliber: .69. Ignition: flint.
Barrel: 7½ inches. Smoothbore reproduction of 1777 Cavalry pistol from Revolutionary War. Brass frame, buttcap, triggerguard. Steel backstrap. **Price: $95**

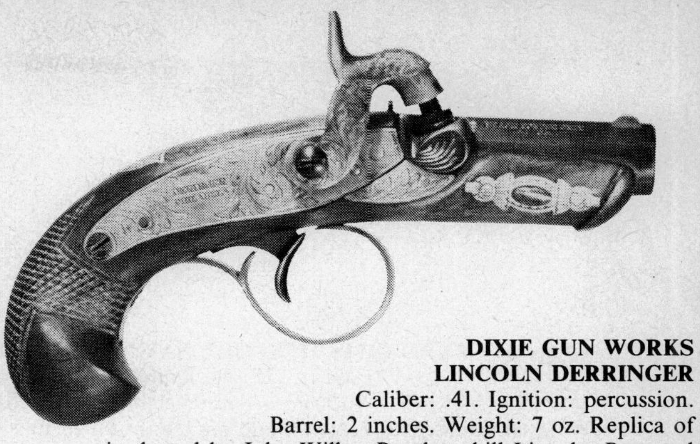

DIXIE GUN WORKS LINCOLN DERRINGER
Caliber: .41. Ignition: percussion.
Barrel: 2 inches. Weight: 7 oz. Replica of pistol used by John Wilkes Booth to kill Lincoln. Browned barrel. Checkered stock. Silver furniture. **Price: $145**

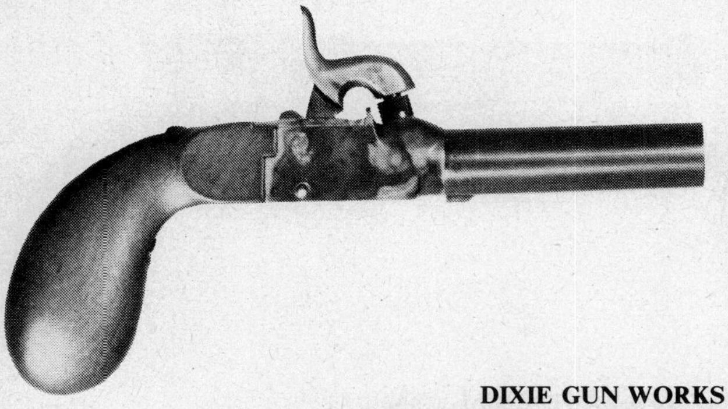

DIXIE GUN WORKS SCREW BARREL DERRINGER
Caliber: .445. Ignition: percussion.
Folding trigger. Screw barrel design makes ram rod unnecessary. Bag type walnut grips. **Price: $50**

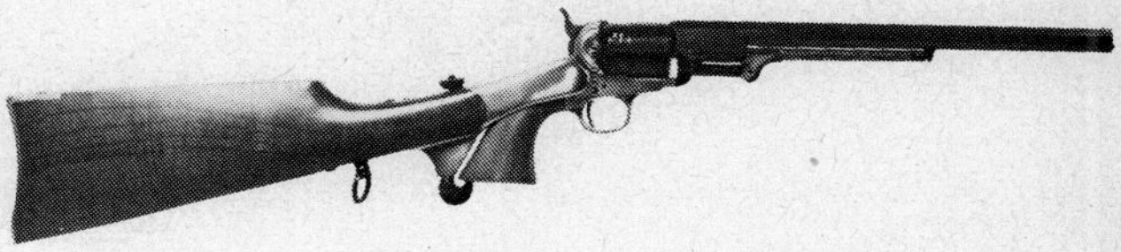

DIXIE GUN WORKS WYATT EARP REVOLVER
Caliber: .44. Ignition: percussion.
Barrel: 12 inches. Rebated cylinder. Brass frame, backstrap, triggerguard. Case hardened hammer, trigger, loading lever. Walnut grips. **Price: $80; $45** (shoulder stock)

EUROARMS 1851 COLT NAVY "SHERIFF"
Same as Colt Navy, except with 5 inch barrel. Non-fluted cylinder. **Price: $107**

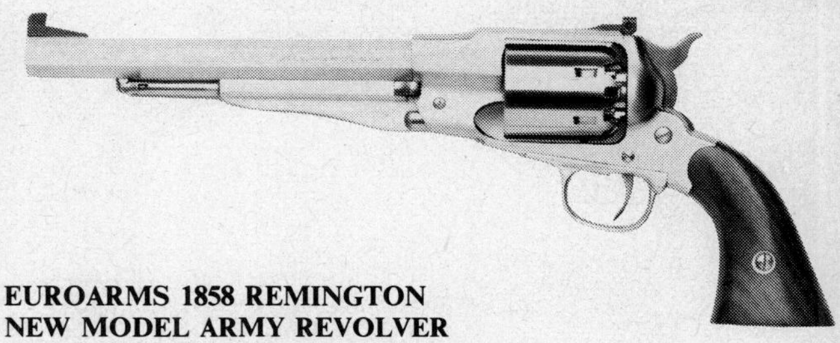

EUROARMS 1858 REMINGTON NEW MODEL ARMY REVOLVER
Caliber: .36, .44. Ignition: percussion.
Barrel: 6½, 8 inches. Weight: 40 oz. Replica of Remington Civil War era classic. Solid frame Blued finish with color case hardened frame. Brass triggerguard. Sight notch in top strap. Stainless steel version and target sights available. **Price: $122** to **$193**

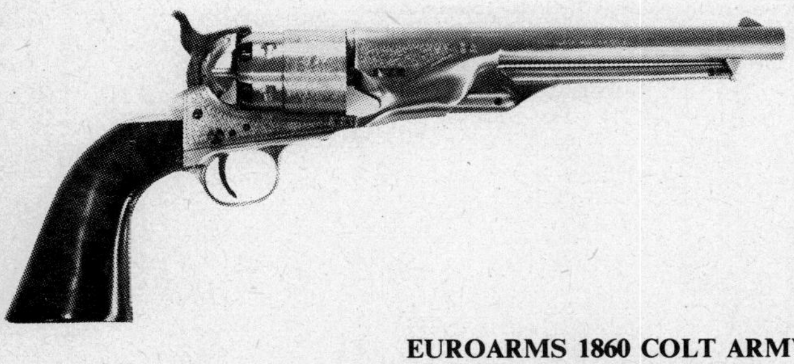

EUROARMS 1860 COLT ARMY
Caliber: .44. Ignition: percussion.
Barrel: 5, 8 inches. Weight: 38, 41 oz. Four screw model with slotted frame for shoulder stock. Blued barrel, cylinder. Color case hardened frame. Walnut stocks. Brass backstrap and triggerguard. **Price: $119** (engraving optional)

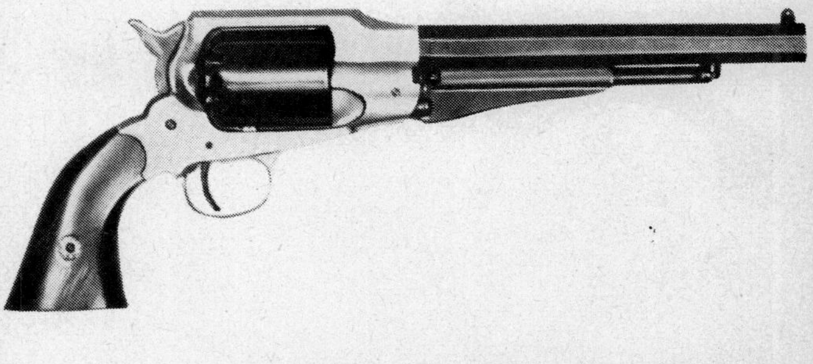

F.I.E. REMINGTON NEW ARMY
Caliber: .44. Ignition: percussion.
Barrel: 7¾ inches. Weight: 42 oz. Polished brass frame. Blued finish. Walnut grips. **Price: $61; $59** (kit)

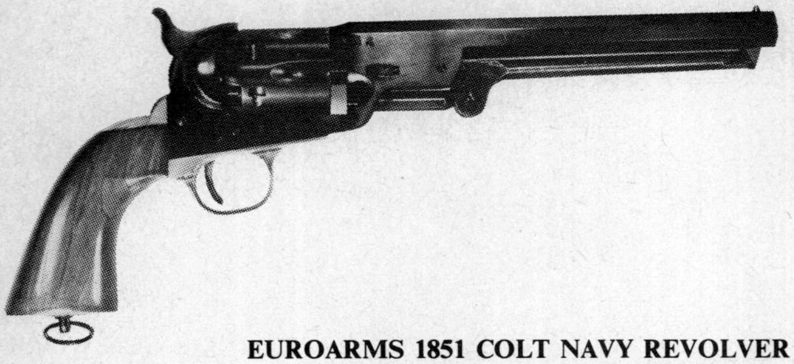

EUROARMS 1851 COLT NAVY REVOLVER
Caliber: .36, .44. Ignition: percussion.
Barrel: 5, 7½ inches. Weight: 38 to 41 oz.
Fluted cylinder. Brass guard and backstrap. Blued finish.
Walnut grips. "V" sight notch in hammer spur. Price:
Price: $88 to **$193** (various finishes and options)

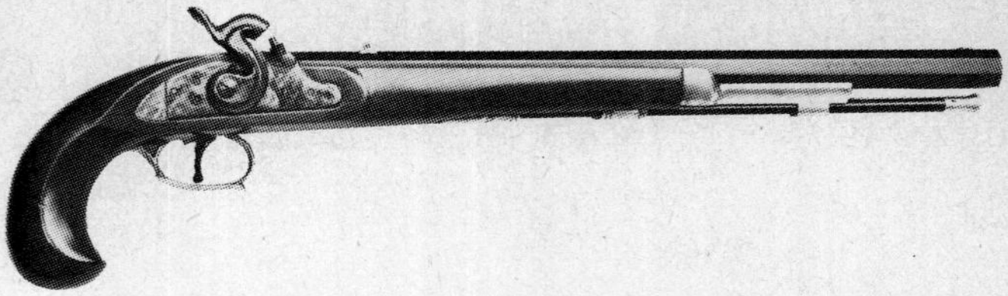

HOPKINS & ALLEN
Caliber: .44. Ignition: percussion.
Barrel: 17 inches. Weight: 56 oz. Long pistol
with hardwood stock and bright brass fittings.
Price: $83; $62 (kit)

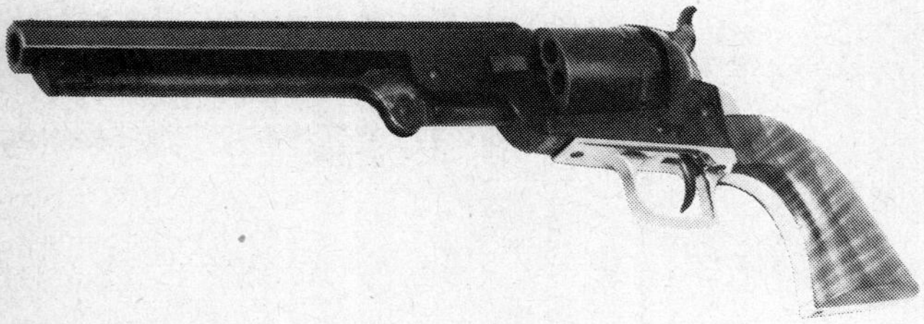

LYMAN 1851 SQUAREBACK NAVY
Caliber: .36. Ignition: percussion.
Barrel: 7½ inches. Weight: 50 oz. Case hardened
steel frame, engraved cylinder. One piece walnut grips.
Price: $150; $110 (kit)

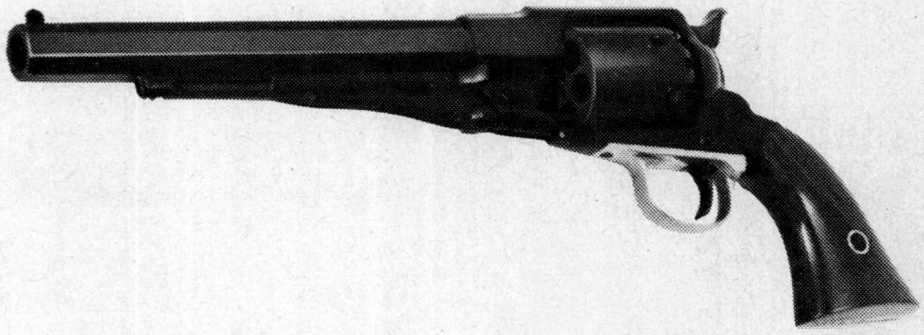

LYMAN NEW MODEL ARMY .44
Caliber: .44. Ignition: percussion.
Barrel: 8 inches. Weight: 50 oz. Machined
steel frame, polished brass triggerguard. European walnut
grips. **Price: $156; $120** (kit)

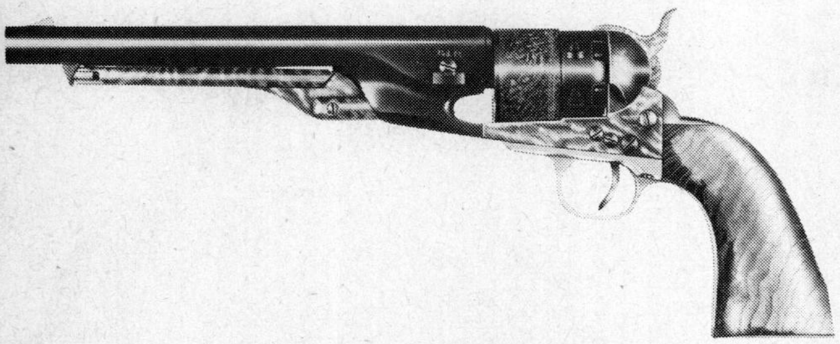

LYMAN 1860 ARMY .44
Caliber: .44. Ignition: percussion.
Barrel: 8 inches. Weight: 50 oz. Case hardened
frame, engraved cylinder. Nickel plated brass triggerguard
and backstrap. **Price: $156; $120** (kit)

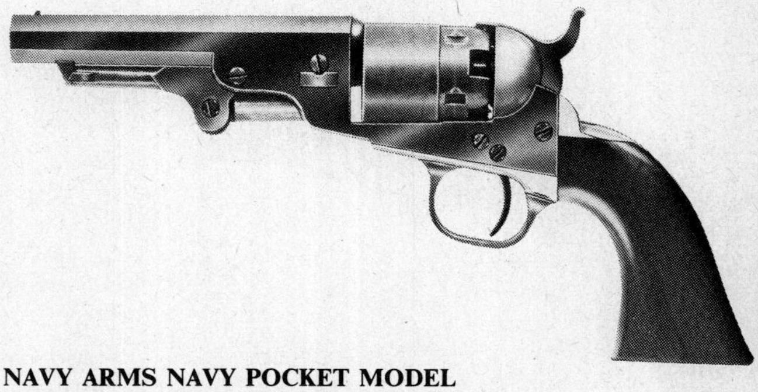

NAVY ARMS NAVY POCKET MODEL
Caliber: .36. Ignition: percussion.
Barrel: 4½, 5½, 6½ inches. Weight: 25 oz.
Replica of one of Colt's rarest revolvers. Case hardened
frame. Walnut stocks. **Price: $125**

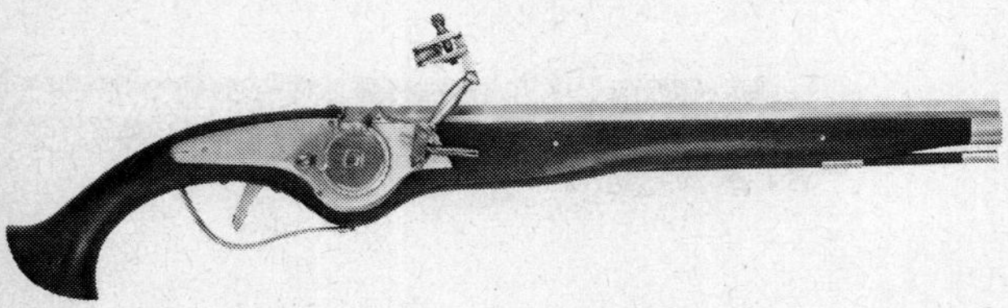

NAVY ARMS WHEELLOCK PISTOL
Caliber: .45. Ignition: flint.
Barrel: 17 inches. Weight: 4½ lbs. Replica
of 1517 wheellock. **Price: $195; $150** (kit)

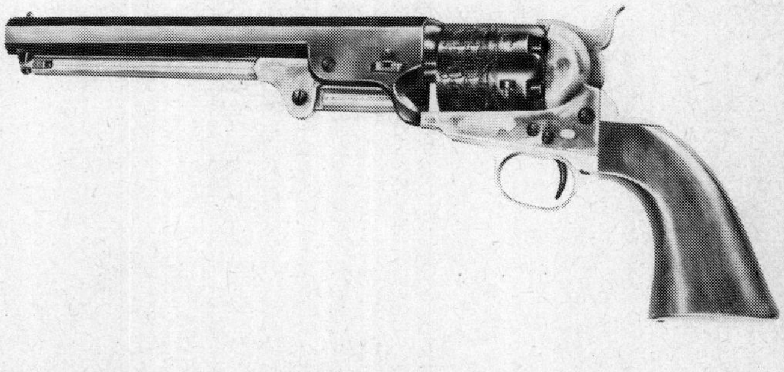

NAVY ARMS YANK NAVY MODEL
Caliber: .36. Ignition: percussion.
Barrel: 7½ inches. Weight: 45 oz. Case hardened
frame. Silver plated triggerguard. Roll engraved cylinder
with naval scene. **Price: $125**

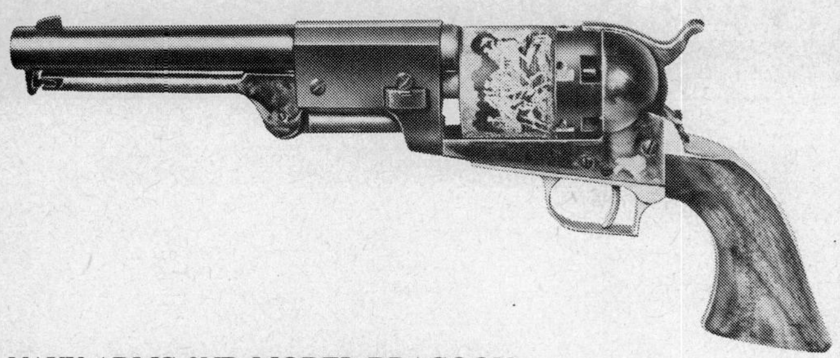

NAVY ARMS 2ND MODEL DRAGOON
Caliber: .44. Ignition: percussion.
Barrel: 7½ inches. Weight: 64 oz. Blued with case hardened frame. Square back triggerguard. Roll engraved cylinder. **Price: $138**
(1st and 3rd Model Dragoons also available)

NAVY ARMS 1860 ARMY
Caliber: .44. Ignition: percussion.
Barrel: 8 inches. Weight: 41 oz. Polished brass triggerguard. Case hardened frame. Roll engraved cylinder.
Price: $110

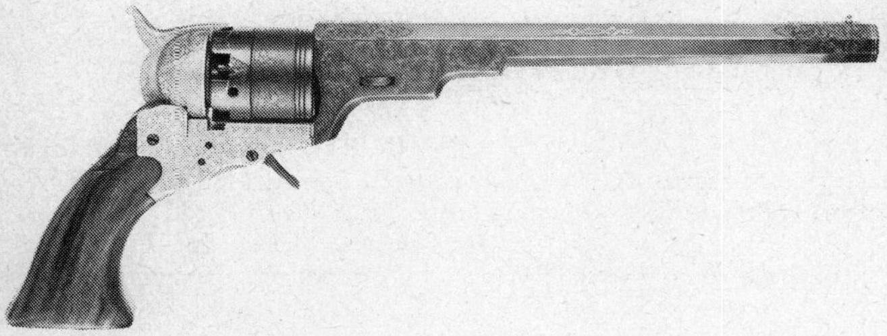

NAVY ARMS PATERSON COLT
Caliber: .36. Ignition: percussion.
Barrel: 9 inches. Weight: 40 oz. Limited edition replica of Colt's first successful revolver. **Price: $195; $500** (deluxe, engraved)

TEN-RING PRECISION TARGET PISTOL
Caliber: .44. Ignition: percussion or flint. Barrel: 9 inches. Weight: 48 oz.
Adjustable trigger, matte finish. Price: $215 (percussion); **$250** (flint)

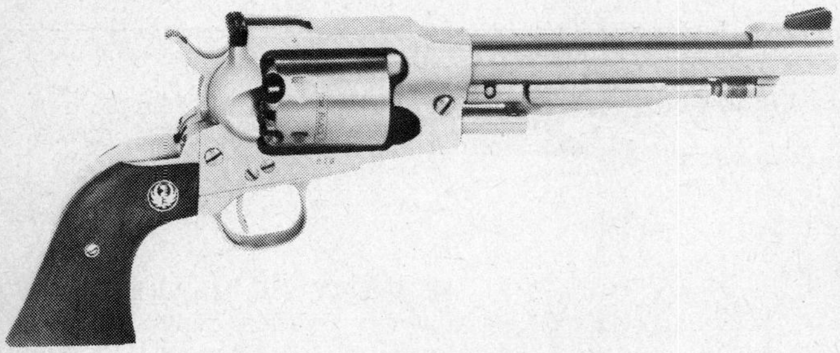

RUGER OLD ARMY
Caliber: .44. Ignition: percussion.
Barrel: 7½ inches. Weight: 46 oz. Capacity: 6.
American walnut grips. Blued finish. Adjustable micro sights. **Price: $173; $225** (stainless)

THOMPSON/CENTER PATRIOT
Caliber: .45. Ignition: percussion.
Barrel: 9 inches. Weight: 36 oz. Hooked breech system.
Twist rifled target type barrel. Patridge type sights.
Price: $155

WESTERN ARMS TEXAS DRAGOON MODEL 1862
Caliber: .44. Ignition: percussion.
Barrel: 7½ inches. Weight: 66 oz. Capacity: 6.
Brass triggerguard and backstrap. Roll engraved cylinder.
Walnut grips. Case hardened and blued finish. **Price: $195**

WESTERN ARMS NEW MODEL POCKET PISTOL 1862
Caliber: .36. Ignition: percussion.
Barrel: 4½, 5½, 6½ inches. Weight: 26 oz.
Hand fitted frame. Octagon barrel. Silver triggerguard and backstrap. Engraved cylinder. Walnut grips. **Price: $175**

Longarms

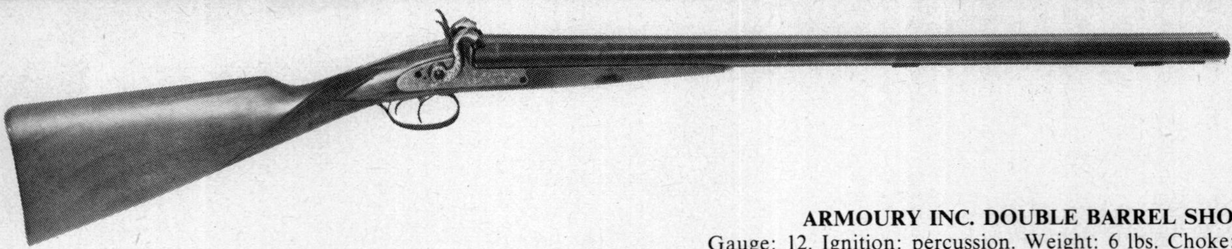

ARMOURY INC. DOUBLE BARREL SHOTGUN
Gauge: 12. Ignition: percussion. Weight: 6 lbs. Choke: M, F. Engraved side locks, browned barrels. Straight grip. **Price: $220**

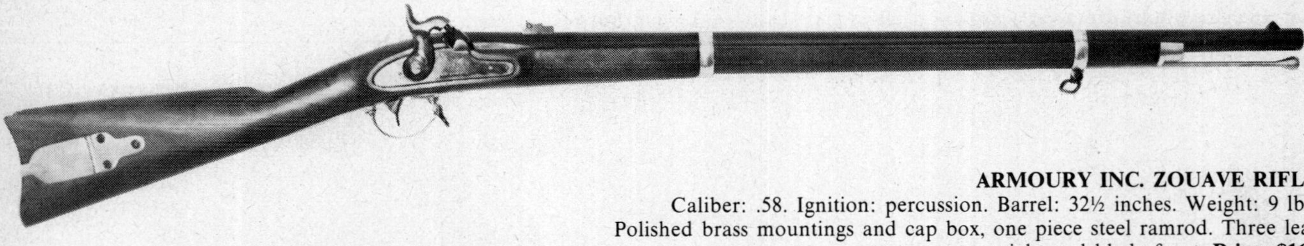

ARMOURY INC. ZOUAVE RIFLE
Caliber: .58. Ignition: percussion. Barrel: 32½ inches. Weight: 9 lbs. Polished brass mountings and cap box, one piece steel ramrod. Three leaf rear sight and blade front. **Price: $165**

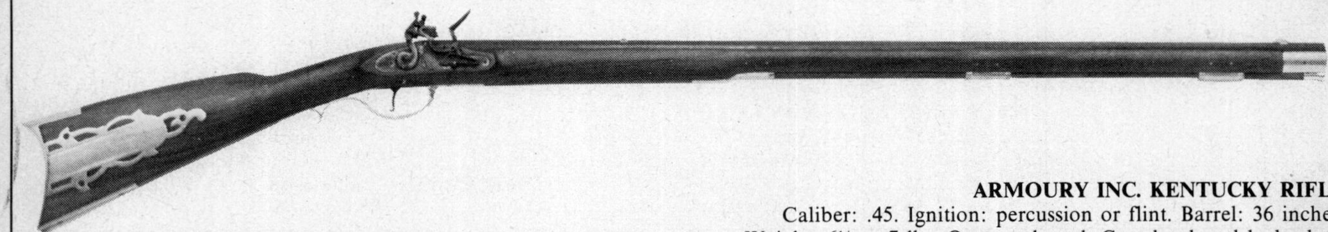

ARMOURY INC. KENTUCKY RIFLE
Caliber: .45. Ignition: percussion or flint. Barrel: 36 inches. Weight: 6½ to 7 lbs. Octagon barrel. Case hardened lock plate. Brass-tipped ramrod. Walnut stock. **Price: $270**

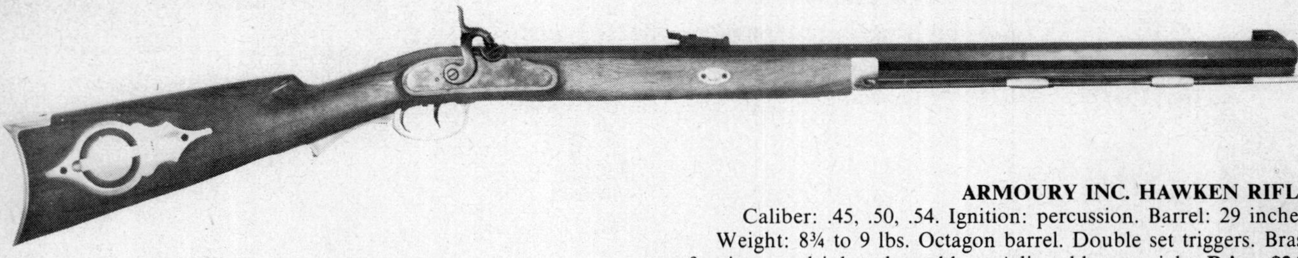

ARMOURY INC. HAWKEN RIFLE
Caliber: .45, .50, .54. Ignition: percussion. Barrel: 29 inches. Weight: 8¾ to 9 lbs. Octagon barrel. Double set triggers. Brass furniture and inletted patchbox. Adjustable rear sight. **Price: $245**

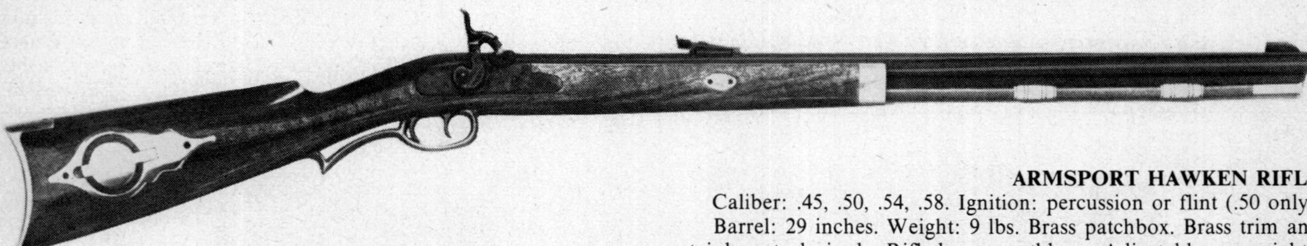

ARMSPORT HAWKEN RIFLE
Caliber: .45, .50, .54, .58. Ignition: percussion or flint (.50 only). Barrel: 29 inches. Weight: 9 lbs. Brass patchbox. Brass trim and stainless steel nipple. Rifled or smoothbore. Adjustable open sights. **Price: $215; $240** (deluxe)

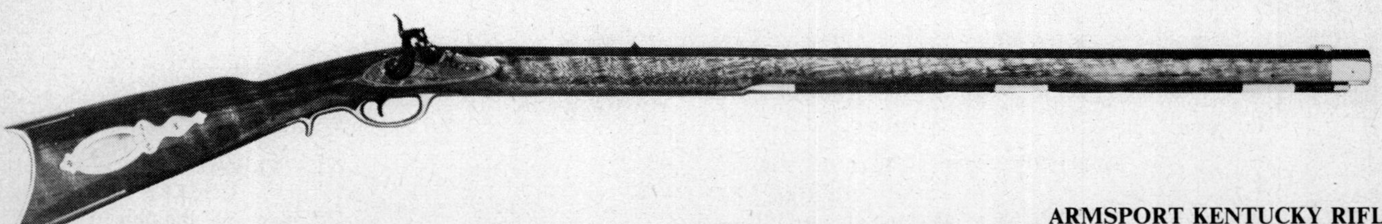

ARMSPORT KENTUCKY RIFLE
Caliber: .45, .50. Ignition: percussion or flint. Barrel: 36 inches. Weight: 7½ lbs. Brass patchbox, buttplate, triggerguard. Select walnut stock. Fixed sights. **Price: $199; $275** (deluxe); **$178** (kit)

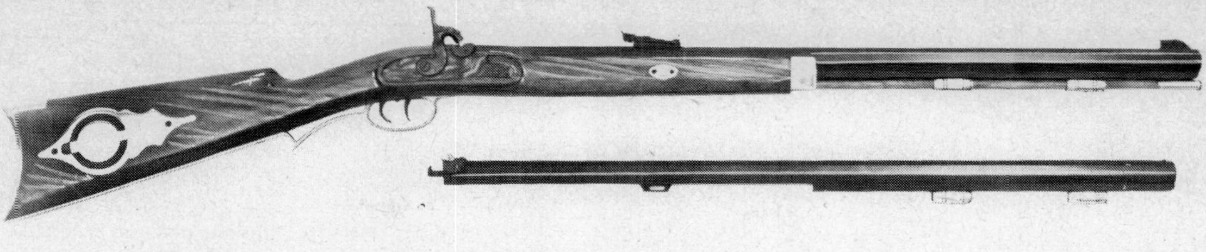

ARMSPORT HAWKEN RIFLE/SHOTGUN COMBO
Same specifications as regular Hawken, except comes with interchangeable .50 cal. rifle barrel and 20 gauge shotgun barrel. Percussion only. **Price: $299**

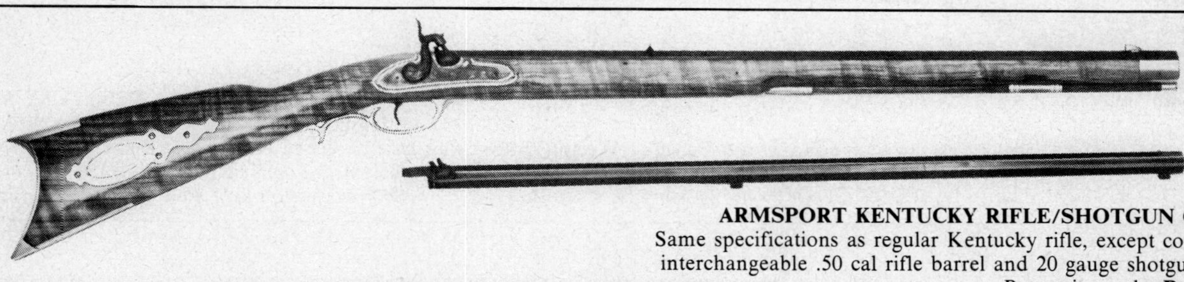

ARMSPORT KENTUCKY RIFLE/SHOTGUN COMBO
Same specifications as regular Kentucky rifle, except comes with interchangeable .50 cal rifle barrel and 20 gauge shotgun barrel. Percussion only. **Price: $260**

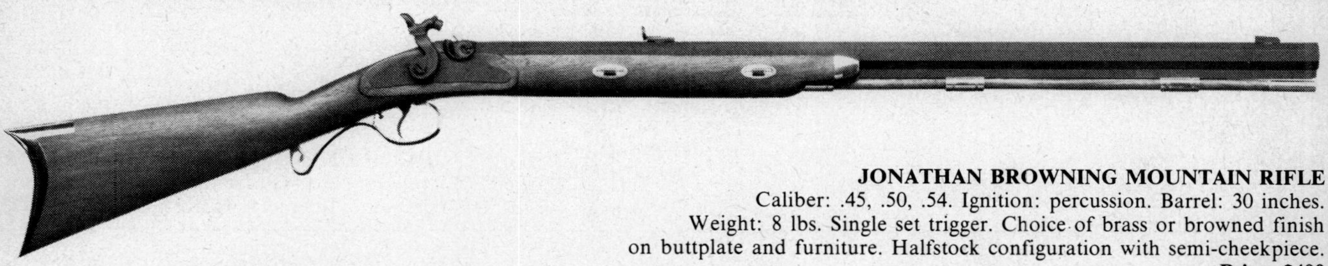

JONATHAN BROWNING MOUNTAIN RIFLE
Caliber: .45, .50, .54. Ignition: percussion. Barrel: 30 inches. Weight: 8 lbs. Single set trigger. Choice of brass or browned finish on buttplate and furniture. Halfstock configuration with semi-cheekpiece. **Price: $400**

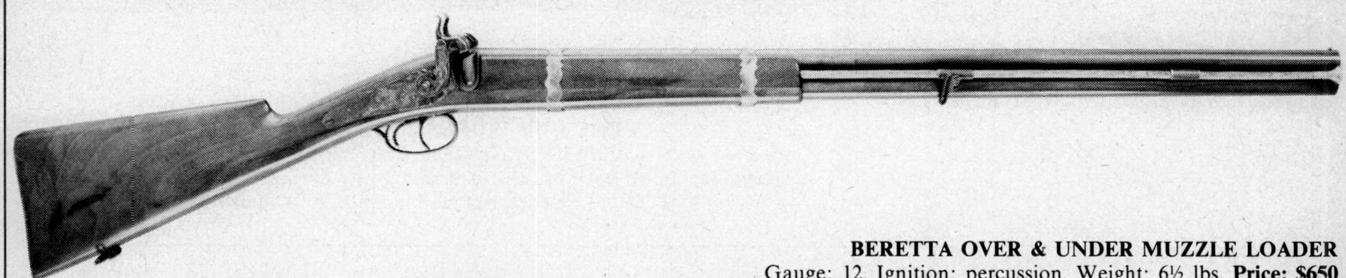

BERETTA OVER & UNDER MUZZLE LOADER
Gauge: 12. Ignition: percussion. Weight: 6½ lbs. **Price: $650**

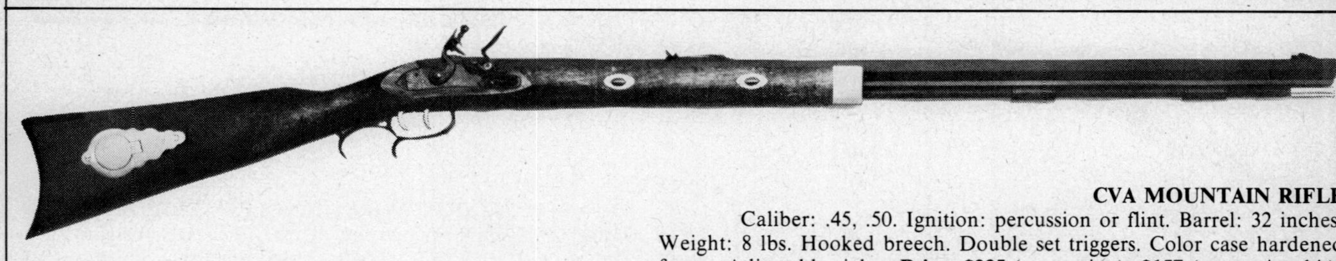

CVA MOUNTAIN RIFLE
Caliber: .45, .50. Ignition: percussion or flint. Barrel: 32 inches. Weight: 8 lbs. Hooked breech. Double set triggers. Color case hardened frame. Adjustable sights. **Price: $235** (percussion); **$157** (percussion kit); **$250** (flint); **$170** (flint kit)

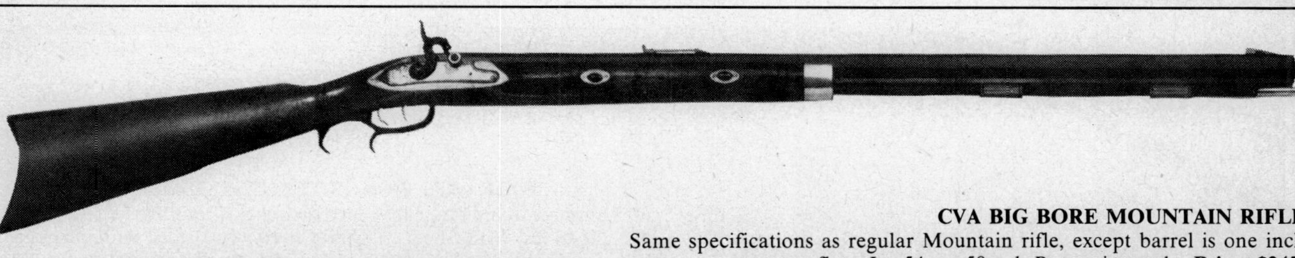

CVA BIG BORE MOUNTAIN RIFLE
Same specifications as regular Mountain rifle, except barrel is one inch across flats. In .54 or .58 cal. Percussion only. **Price: $245; $165** (kit)

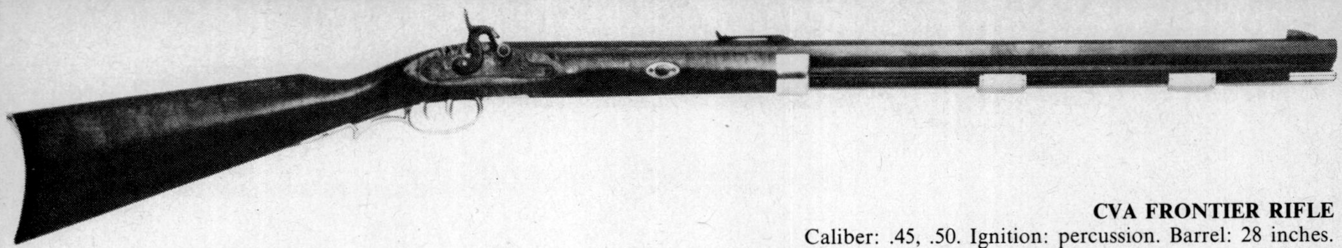

CVA FRONTIER RIFLE
Caliber: .45, .50. Ignition: percussion. Barrel: 28 inches. Weight: 7½ lbs. Hooked breech. Double set triggers. Polished brass furniture. Walnut stock. **Price: $180; $140** (kit)

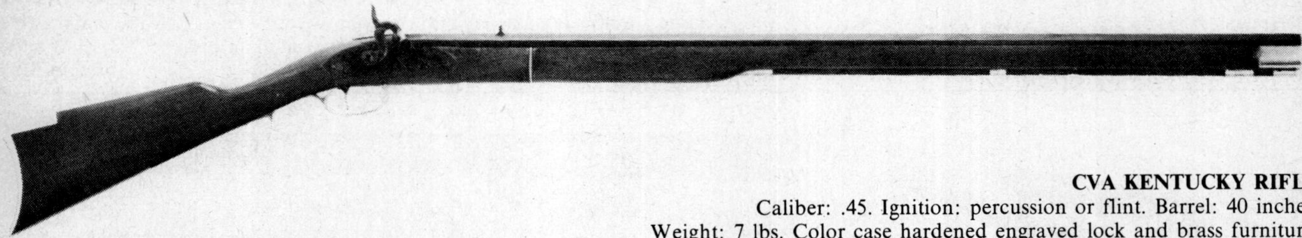

CVA KENTUCKY RIFLE
Caliber: .45. Ignition: percussion or flint. Barrel: 40 inches. Weight: 7 lbs. Color case hardened engraved lock and brass furniture. Walnut finished stock. Fixed sights. **Price: $150** (percussion); **$90** (percussion kit); **$158** (flint); **$100** (flint kit)

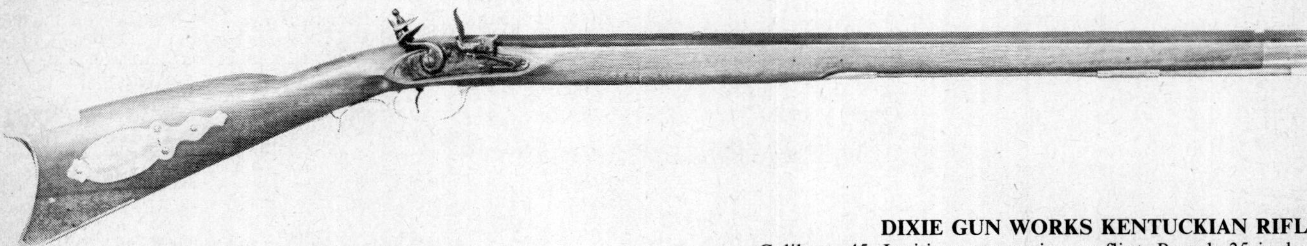

DIXIE GUN WORKS KENTUCKIAN RIFLE
Caliber: .45. Ignition: percussion or flint. Barrel: 35 inches. Weight: 6¼ lbs. Brass furniture. European walnut stock. **Price: $165**

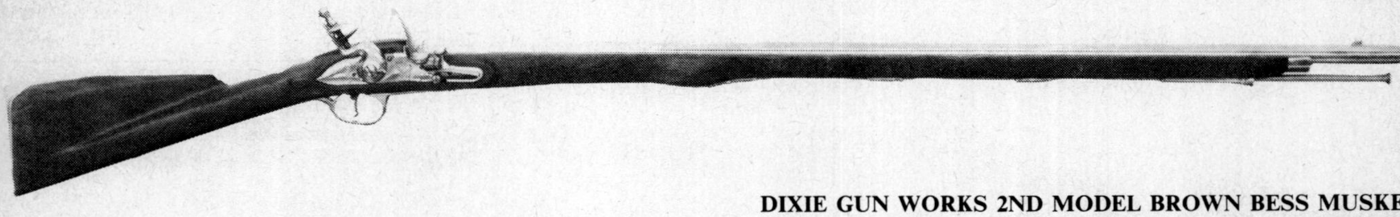

DIXIE GUN WORKS 2ND MODEL BROWN BESS MUSKET
Caliber: .75. Ignition: flint. Barrel: 41¾ inches. Weight: 9½ lbs. Reproduction of issue musket of British Army during the American Revolution. Bright finished barrel. Smoothbore. Walnut stock. Brass furniture. **Price: $265**

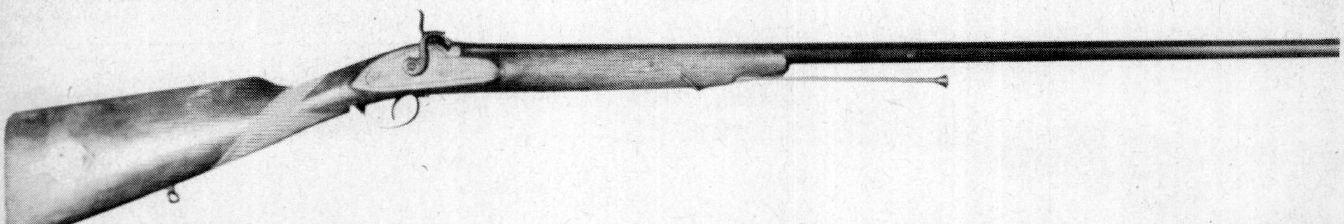

DIXIE GUN WORKS SINGLE BARREL SHOTGUN
Gauge: 28. Ignition: percussion. Barrel: 32 inches. Weight: 7 lbs. Walnut stock. Steel ramrod. **Price: $60**

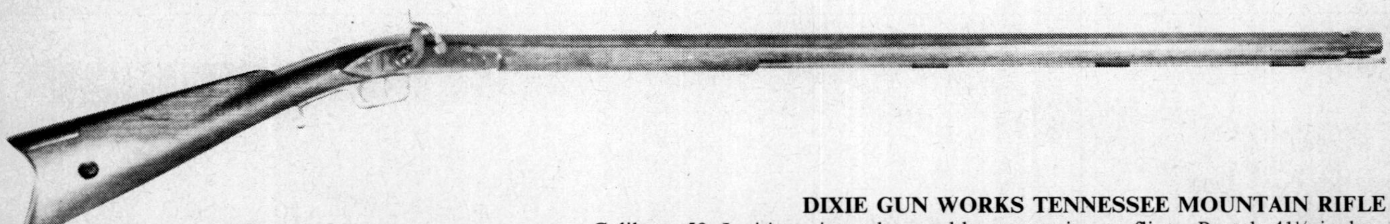

DIXIE GUN WORKS TENNESSEE MOUNTAIN RIFLE
Caliber: .50. Ignition: interchangeable percussion or flint. Barrel: 41½ inches. Weight: 8 lbs. Double set triggers. Cherrywood stock with cheekpiece. Grease hole. Adjustable buckhorn rear sight. **Price: $225; $245** (left hand version); **$175** (kit, right hand only)

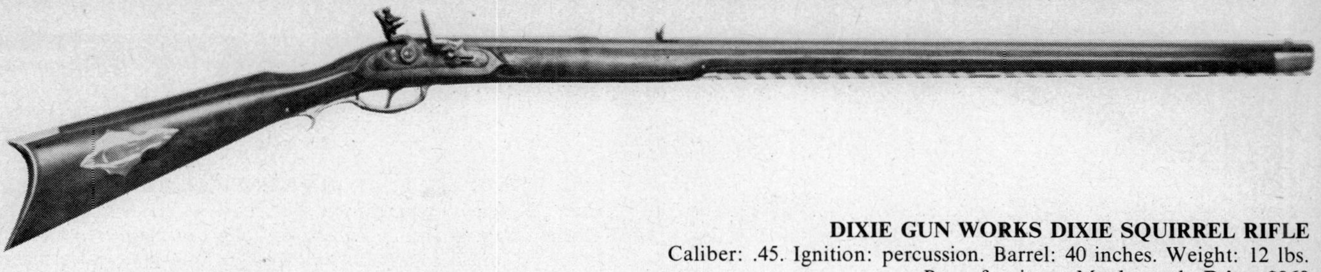

DIXIE GUN WORKS DIXIE SQUIRREL RIFLE
Caliber: .45. Ignition: percussion. Barrel: 40 inches. Weight: 12 lbs. Brass furniture. Maple stock. **Price: $260**

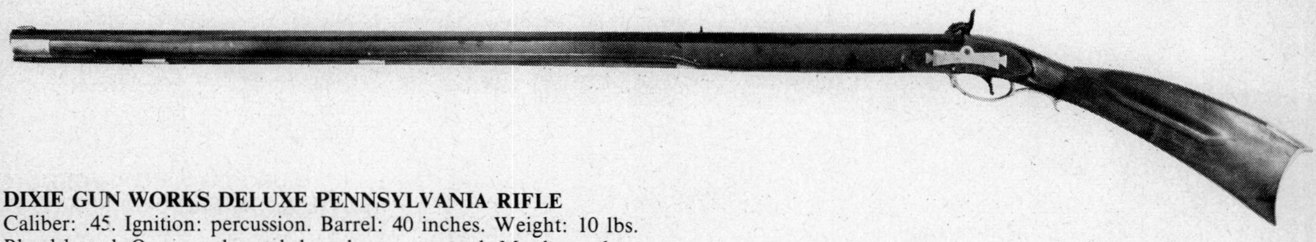

DIXIE GUN WORKS DELUXE PENNSYLVANIA RIFLE
Caliber: .45. Ignition: percussion. Barrel: 40 inches. Weight: 10 lbs. Blued barrel. Open work patch box, brass mounted. Maple stock. **Price: $298**

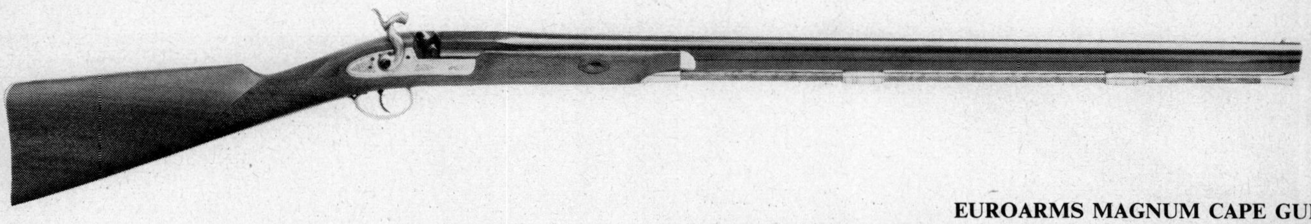

EUROARMS MAGNUM CAPE GUN
Gauge: 12. Ignition: percussion. Barrel: 32 inches. Weight: 7½ lbs. Single barrel shotgun. Blued triggerguard, buttplate, barrel, and underrib. European walnut stock. **Price: $240**

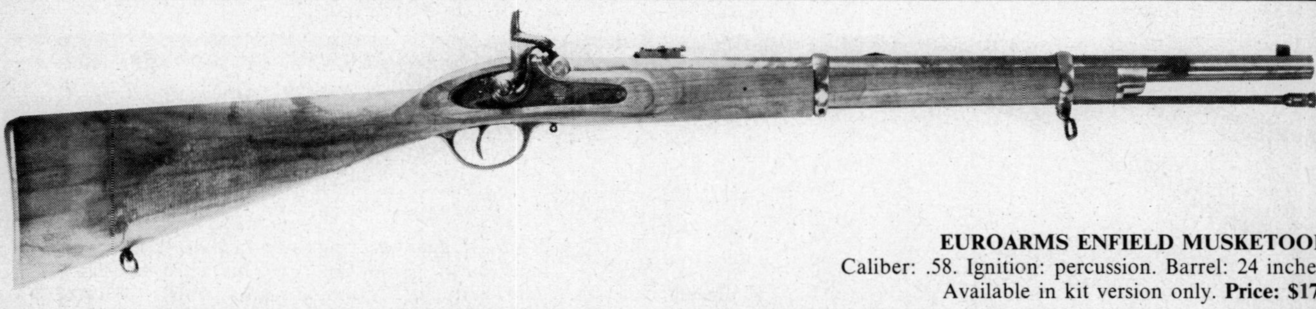

EUROARMS ENFIELD MUSKETOON
Caliber: .58. Ignition: percussion. Barrel: 24 inches. Available in kit version only. **Price: $179**

EUROARMS PENNSYLVANIA RIFLE
Caliber: .45, .50. Ignition: percussion or flint. Barrel: 36 inches. Weight: 7 lbs. Color case hardened locks with border engraving. Rifled octagon barrel. Brass furniture. Wooden ramrod. **Price: $192** (percussion); **$222** (flint)

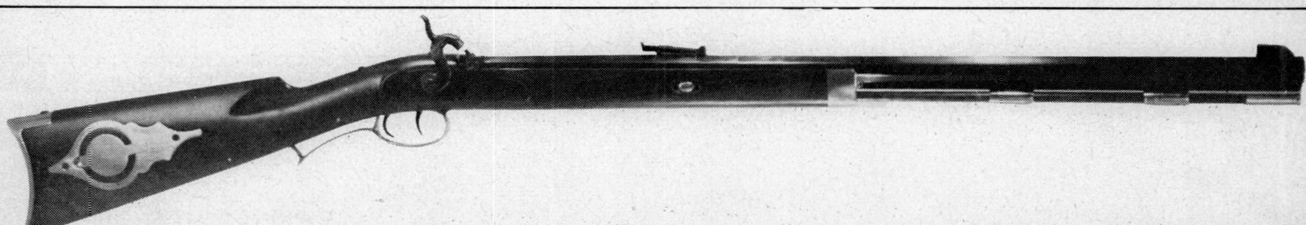

EUROARMS HAWKEN RIFLE
Caliber: .45, .50. .54, .58. Ignition: percussion or flint. Barrel: 28 inches. Weight: 9½ lbs. Brass furniture. Double set triggers. Select walnut stock. **Price: NA**

EUROARMS ENFIELD RIFLED MUSKET
Caliber: .58. Ignition: percussion. Barrel: 39 inches. Weight: 9½ lbs. Brass buttplate, triggerguard, and nosecap. Color case hardened lock and hammer. Replica of 1853 Enfield Musket. Adjustable barrel bands. **Price: $223** (musketoon); **$253** (musket)

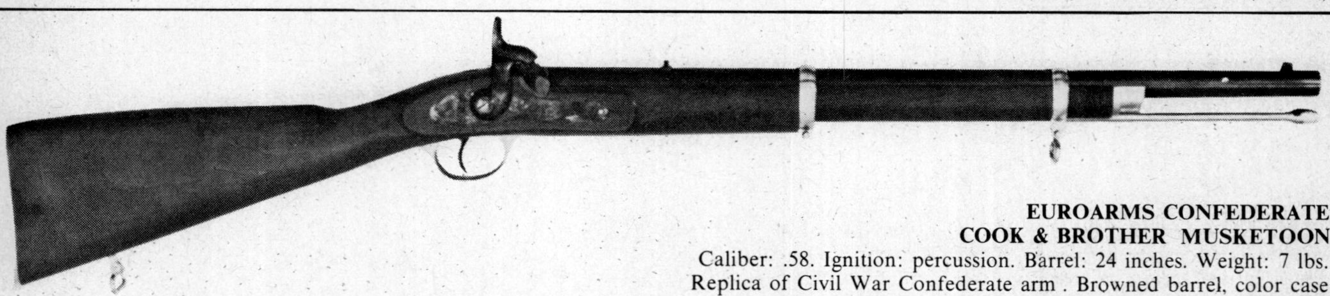

EUROARMS CONFEDERATE COOK & BROTHER MUSKETOON
Caliber: .58. Ignition: percussion. Barrel: 24 inches. Weight: 7 lbs. Replica of Civil War Confederate arm . Browned barrel, color case hardened frame. Brass fittings, walnut stock. **Price: $223**

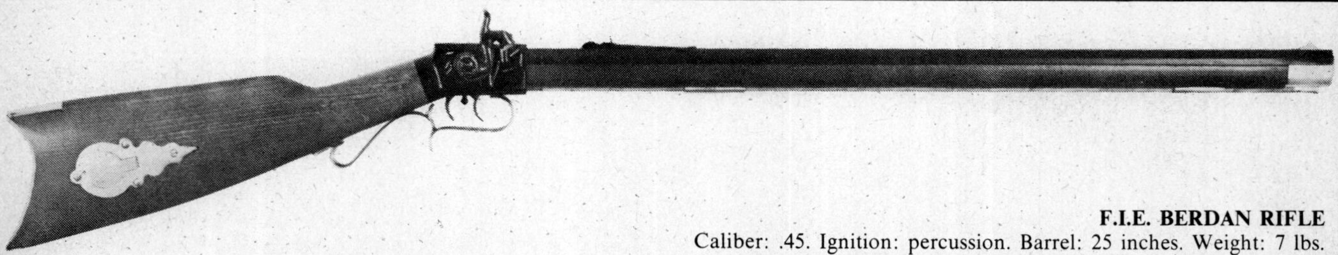

F.I.E. BERDAN RIFLE
Caliber: .45. Ignition: percussion. Barrel: 25 inches. Weight: 7 lbs. Double set adjustable triggers. Brass triggerguard. Walnut finished hardwood stock. Adjustable rear sight. **Price: $98**

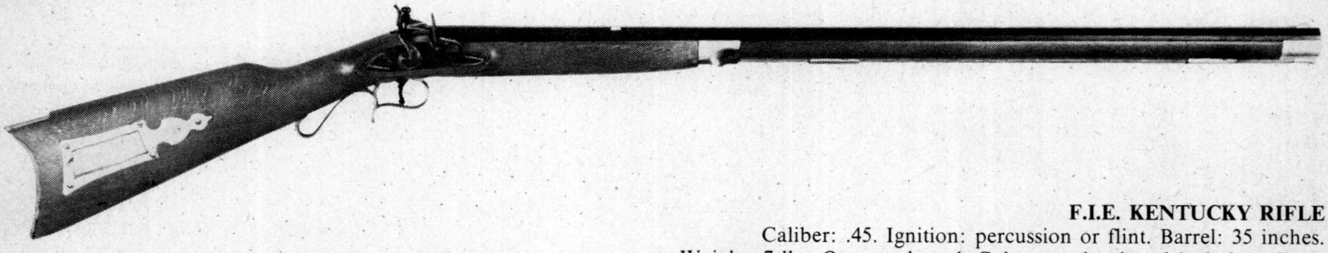

F.I.E. KENTUCKY RIFLE
Caliber: .45. Ignition: percussion or flint. Barrel: 35 inches. Weight: 7 lbs. Octagon barrel. Color case hardened lockplate. Brass patchbox, triggerguard, buttplate. Hardwood ramrod. **Price: $75; $107** (kit)

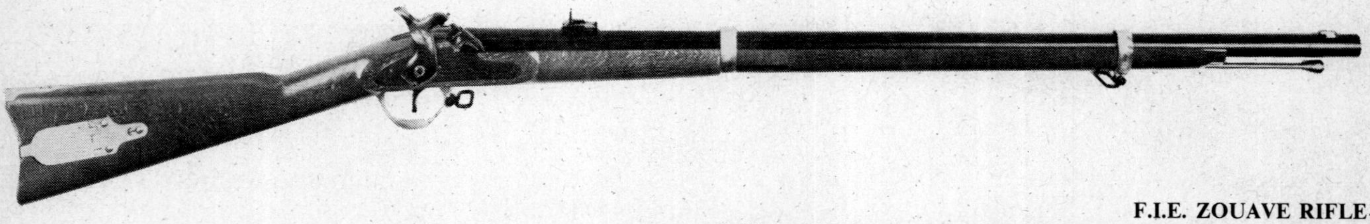

F.I.E. ZOUAVE RIFLE:
Caliber: .58. Ignition: percussion. Barrel: 33 inches. Weight: 9 lbs. Spring loaded removable barrel bands. Comes with two nipples—one for percussion cap and one for winged musket cap. **Price: $117**

HOPKINS & ALLEN MINUTEMAN
Caliber: .36, .45, .50. Ignition: percussion or flint. Barrel: 39 inches. Weight: 10 lbs. Octagon barrel. Brass furniture. American maple stock. Fixed sights. **Price: $275** (percussion); **$290** (flint); **$160** (percussion kit); **$170** (flint kit)

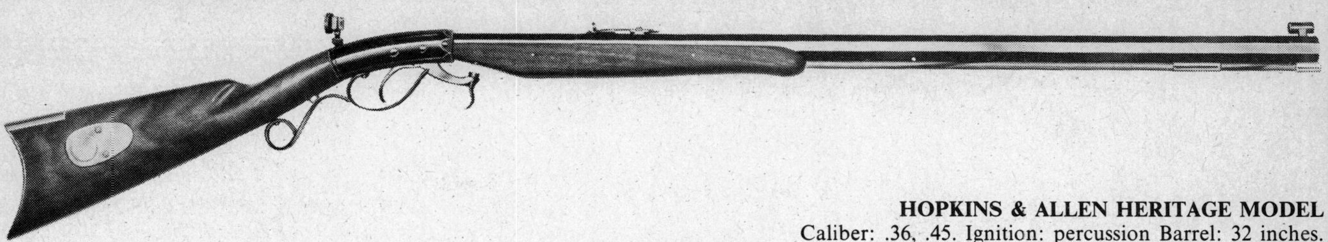

HOPKINS & ALLEN HERITAGE MODEL
Caliber: .36, .45. Ignition: percussion Barrel: 32 inches. Weight: 8½ lbs. Underhammer design features uninterrupted sighting plane. American walnut stock. **Price: $200**

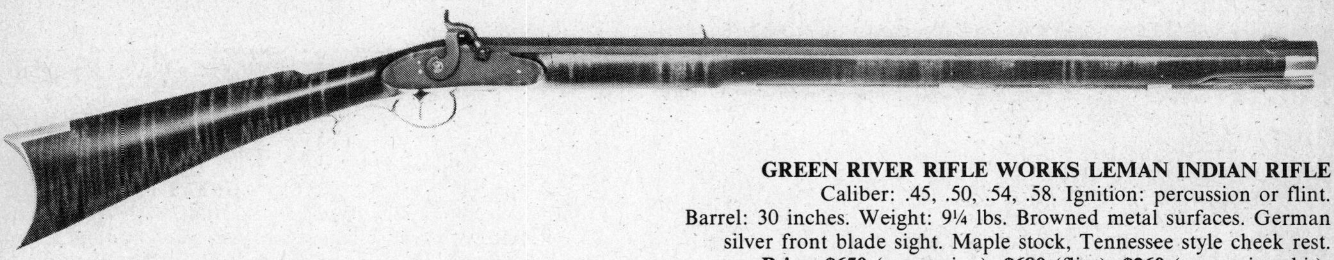

GREEN RIVER RIFLE WORKS LEMAN INDIAN RIFLE
Caliber: .45, .50, .54, .58. Ignition: percussion or flint. Barrel: 30 inches. Weight: 9¼ lbs. Browned metal surfaces. German silver front blade sight. Maple stock, Tennessee style cheek rest. **Price: $650** (percussion); **$680** (flint); **$260** (percussion, kit); **$290** (flint, kit)

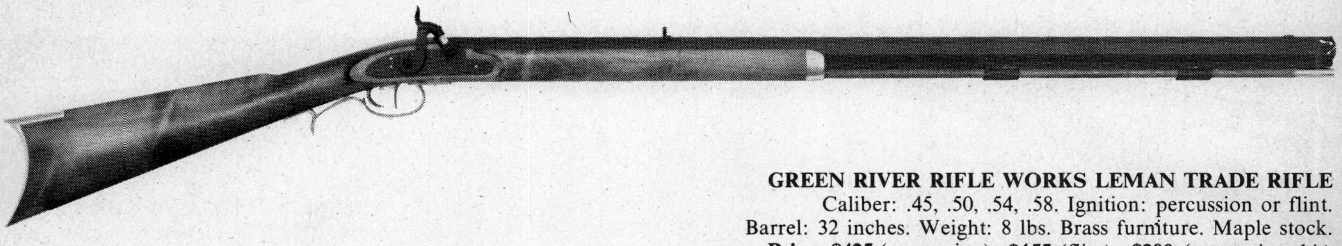

GREEN RIVER RIFLE WORKS LEMAN TRADE RIFLE
Caliber: .45, .50, .54, .58. Ignition: percussion or flint. Barrel: 32 inches. Weight: 8 lbs. Brass furniture. Maple stock. **Price: $425** (percussion); **$455** (flint); **$200** (percussion, kit) **$230** (flint, kit)

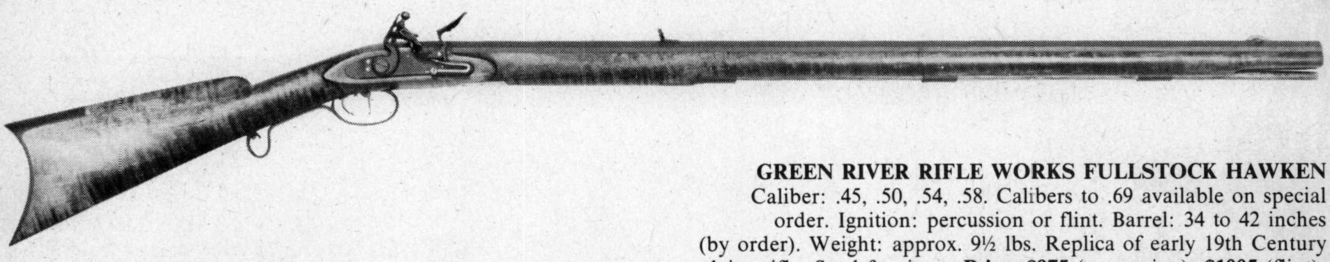

GREEN RIVER RIFLE WORKS FULLSTOCK HAWKEN
Caliber: .45, .50, .54, .58. Calibers to .69 available on special order. Ignition: percussion or flint. Barrel: 34 to 42 inches (by order). Weight: approx. 9½ lbs. Replica of early 19th Century plains rifle. Steel furniture. **Price: $975** (percussion); **$1005** (flint); **$300** (percussion, kit); **$330** (flint, kit)

GREEN RIVER RIFLE WORKS HALFSTOCK HAWKEN
Same as Hawken, except comes in halfstock configuration. **Price: $850** (percussion); **$880** (flint); **$275** (percussion, kit); **$305** (flint, kit)

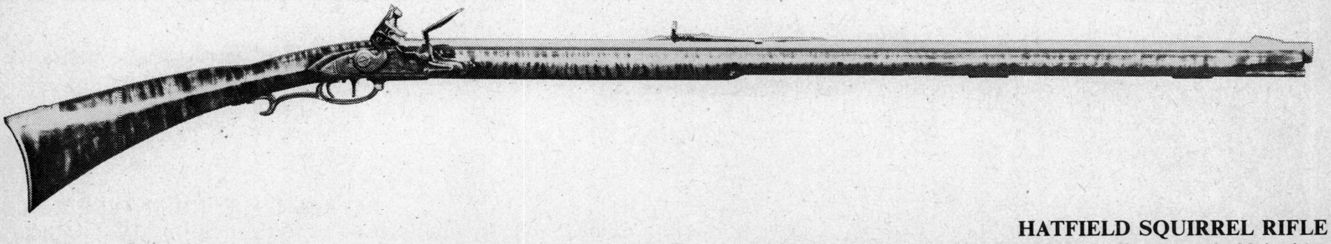

HATFIELD SQUIRREL RIFLE
Caliber: .32, .36, .45, .50. Ignition: flint. Barrel: 39 inches. Weight: 7½ to 8 lbs. Double set triggers. Select American fancy maple stock. Buckhorn rear sight. **Price: $436**

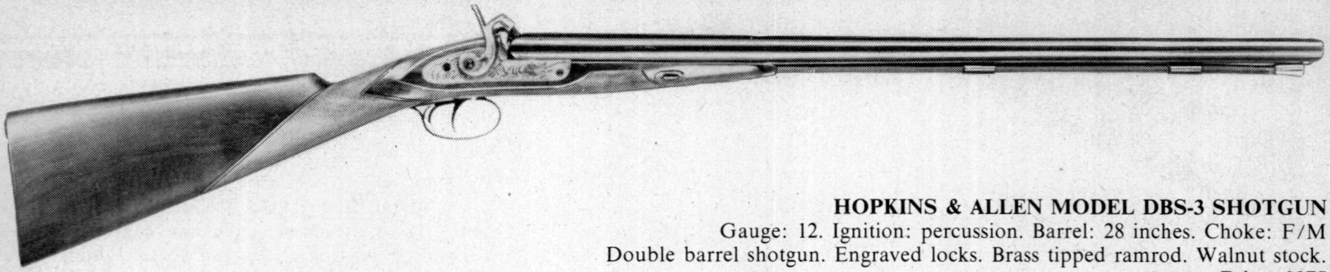

HOPKINS & ALLEN MODEL DBS-3 SHOTGUN
Gauge: 12. Ignition: percussion. Barrel: 28 inches. Choke: F/M
Double barrel shotgun. Engraved locks. Brass tipped ramrod. Walnut stock.
Price: $275

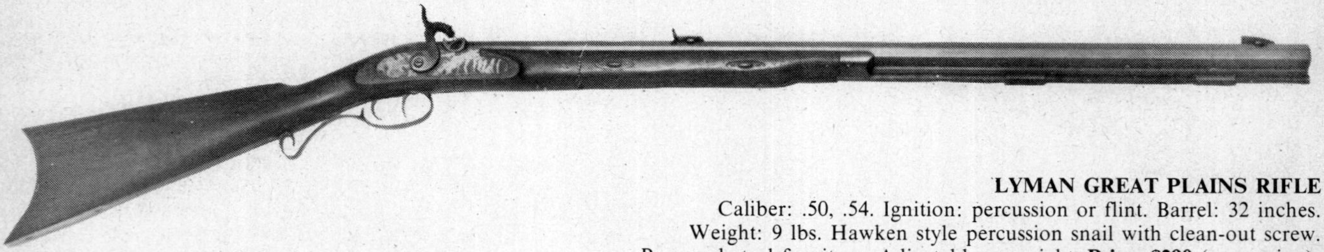

LYMAN GREAT PLAINS RIFLE
Caliber: .50, .54. Ignition: percussion or flint. Barrel: 32 inches.
Weight: 9 lbs. Hawken style percussion snail with clean-out screw.
Browned steel furniture. Adjustable rear sight. **Price: $290** (percussion);
$325 (flint); **$225** (percussion, kit)

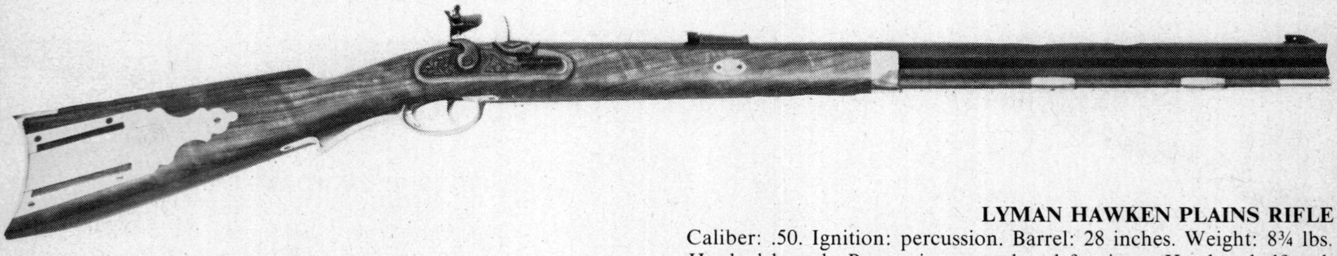

LYMAN HAWKEN PLAINS RIFLE
Caliber: .50. Ignition: percussion. Barrel: 28 inches. Weight: 8¾ lbs.
Hooked breech. Brass triggerguard and furniture. Hawken halfstock
configuration in European walnut. **Price: $240; $190** (kit)

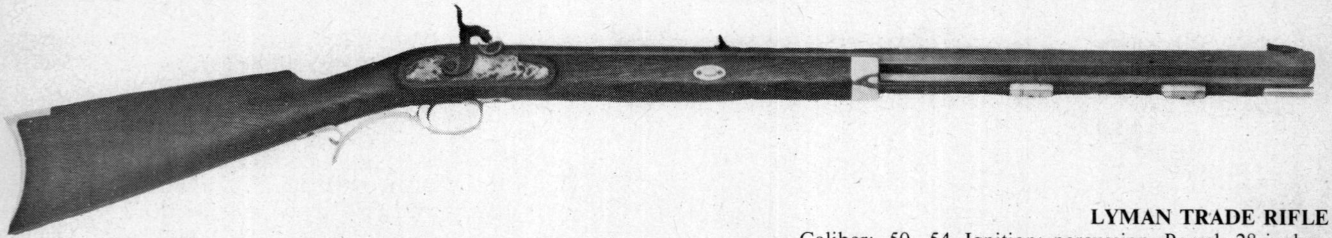

LYMAN TRADE RIFLE
Caliber: .50, .54. Ignition: percussion. Barrel: 28 inches.
Weight: 8¾ lbs. Hooked breech. Brass triggerguard and furniture.
European walnut stock. Fixed sights. **Price: $225; $180** (kit)

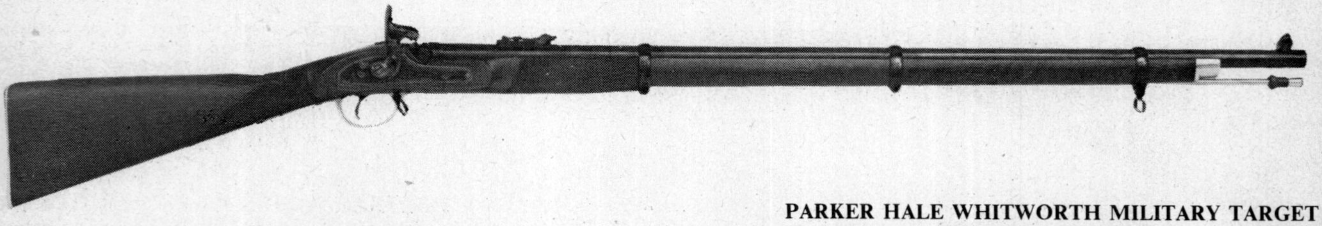

PARKER HALE WHITWORTH MILITARY TARGET
Caliber: .451. Ignition: percussion. Barrel: 36 inches.
Weight: 9½ lbs. Hexagonal bore. Detented lock. Adjustable sights.
Available from Navy Arms. **Price: $575**

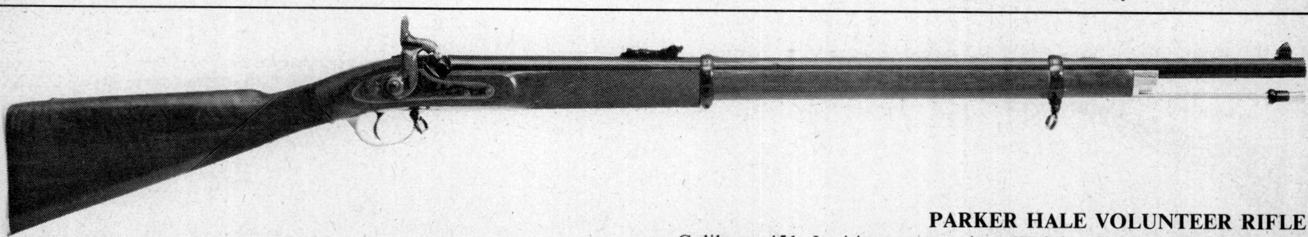

PARKER HALE VOLUNTEER RIFLE
Caliber: .451. Ignition: percussion. Weight: 9½ lbs. Detented lock.
Walnut stock with hand cut checkering. Comes complete with accessory kit
including bullet mould, sizing die and punch, wad punch, nipple key.
Available from Navy Arms. **Price: $575**

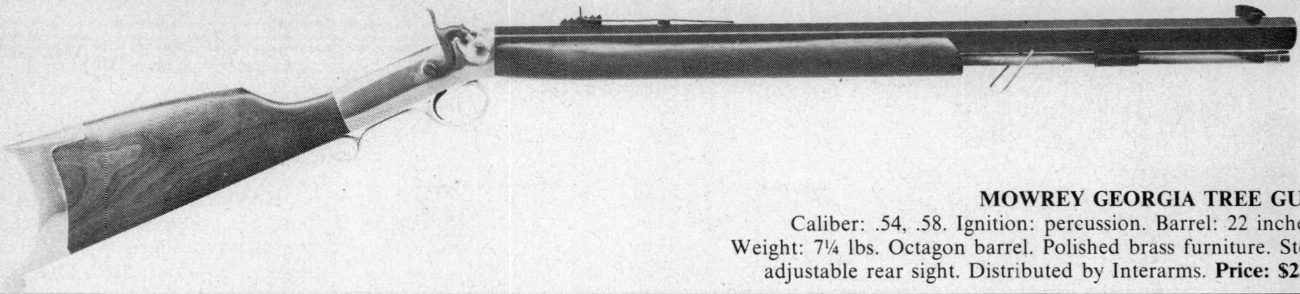

MOWREY GEORGIA TREE GUN
Caliber: .54, .58. Ignition: percussion. Barrel: 22 inches. Weight: 7¼ lbs. Octagon barrel. Polished brass furniture. Step adjustable rear sight. Distributed by Interarms. **Price: $239**

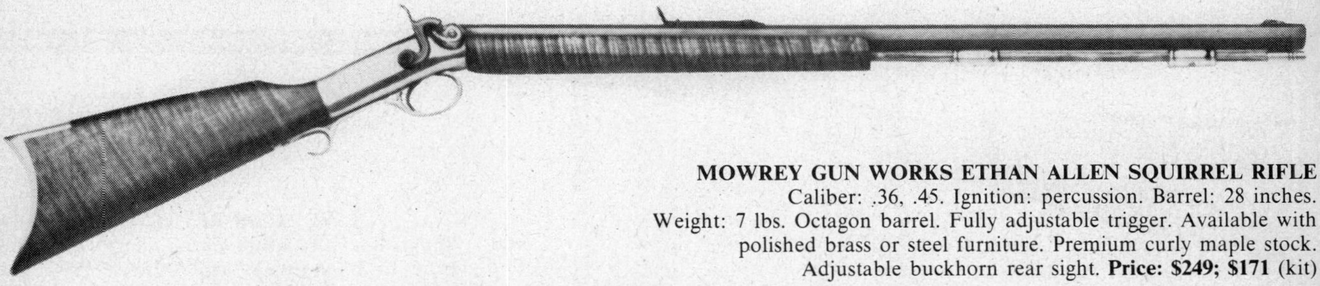

MOWREY GUN WORKS ETHAN ALLEN SQUIRREL RIFLE
Caliber: .36, .45. Ignition: percussion. Barrel: 28 inches. Weight: 7 lbs. Octagon barrel. Fully adjustable trigger. Available with polished brass or steel furniture. Premium curly maple stock. Adjustable buckhorn rear sight. **Price: $249; $171** (kit)

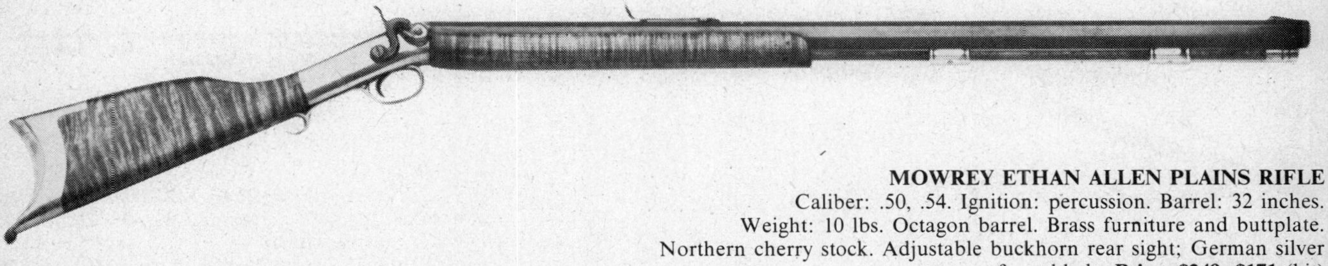

MOWREY ETHAN ALLEN PLAINS RIFLE
Caliber: .50, .54. Ignition: percussion. Barrel: 32 inches. Weight: 10 lbs. Octagon barrel. Brass furniture and buttplate. Northern cherry stock. Adjustable buckhorn rear sight; German silver front blade. **Price: $249; $171** (kit)

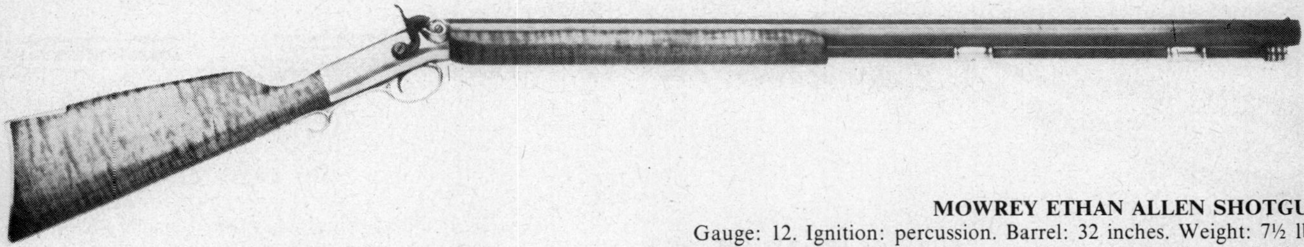

MOWREY ETHAN ALLEN SHOTGUN
Gauge: 12. Ignition: percussion. Barrel: 32 inches. Weight: 7½ lbs. Octagon barrel. Modified choke. Polished brass furniture. Northern cherry stock. **Price: $249; $171** (kit)

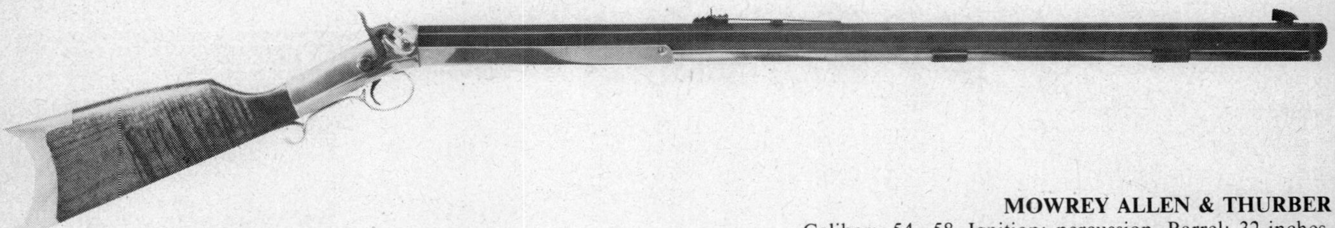

MOWREY ALLEN & THURBER
Caliber: .54, .58. Ignition: percussion. Barrel: 32 inches. Weight: 10¼ lbs. Octagon barrel. Polished brass furniture, forend, and ramrod. Distributed by Interarms. **Price: $229**

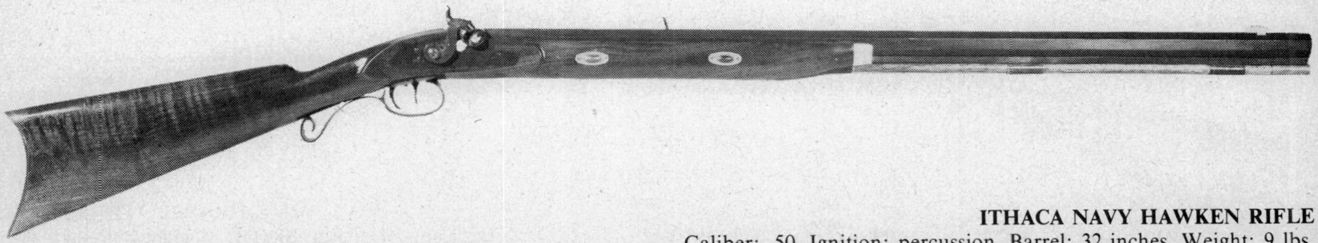

ITHACA NAVY HAWKEN RIFLE
Caliber: .50. Ignition: percussion. Barrel: 32 inches. Weight: 9 lbs. Octagon barrel. Double set triggers, hooked breech. Silver barrel key inlays. Black walnut half stock. Available from Navy Arms. **Price: $395; $275** (kit)

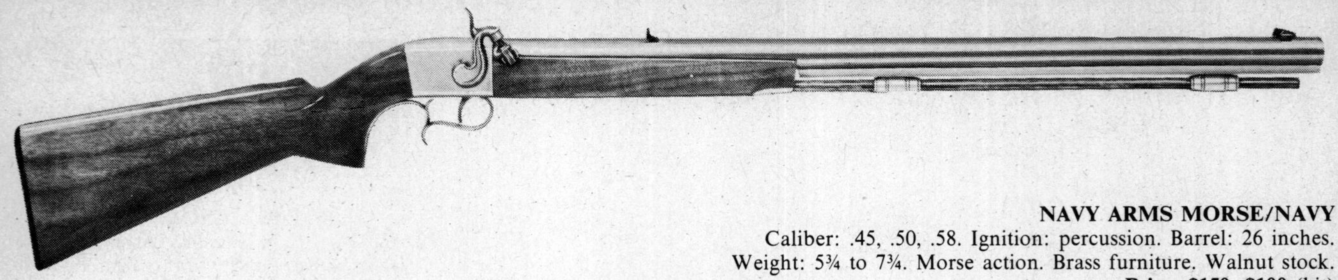

NAVY ARMS MORSE/NAVY
Caliber: .45, .50, .58. Ignition: percussion. Barrel: 26 inches. Weight: 5¾ to 7¾. Morse action. Brass furniture. Walnut stock. **Price: $150; $100** (kit)

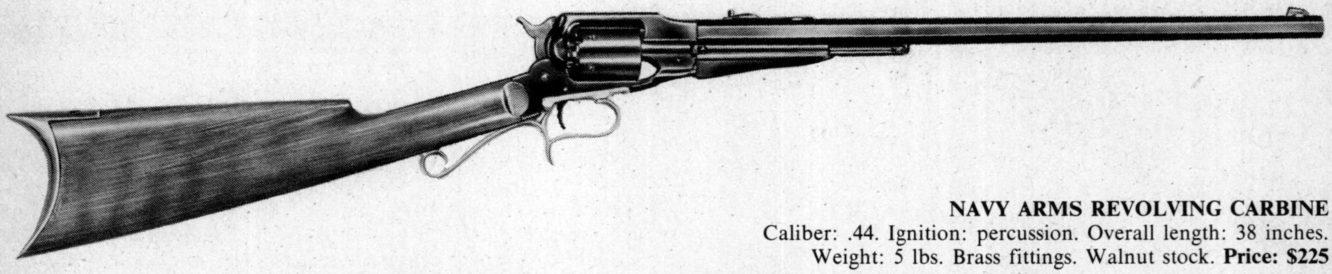

NAVY ARMS REVOLVING CARBINE
Caliber: .44. Ignition: percussion. Overall length: 38 inches. Weight: 5 lbs. Brass fittings. Walnut stock. **Price: $225**

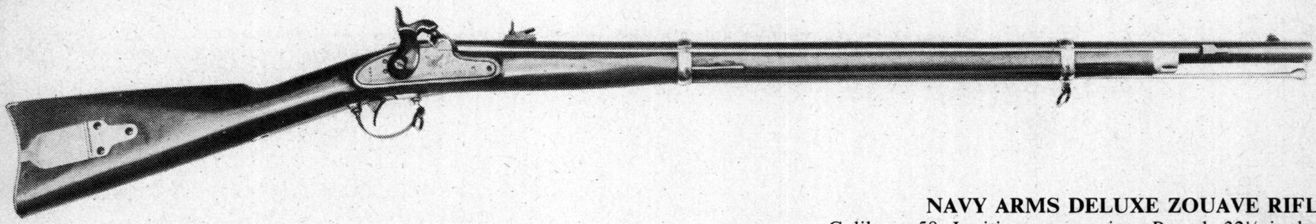

NAVY ARMS DELUXE ZOUAVE RIFLE
Caliber: .58. Ignition: percussion. Barrel: 32½ inches. Replica of Civil War era rifle. Blued barrel, case hardened lock. Brass fittings. Deluxe hardwood stock. **Price: $325**

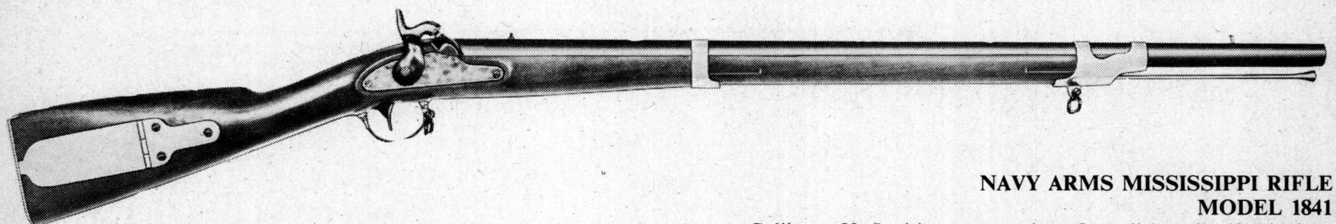

NAVY ARMS MISSISSIPPI RIFLE MODEL 1841
Caliber: .58. Ignition: percussion. Overall length: 48½ inches. Replica of U.S. Issue longarm of Mexican War period. Brass fittings. Walnut stock. **Price: $250**

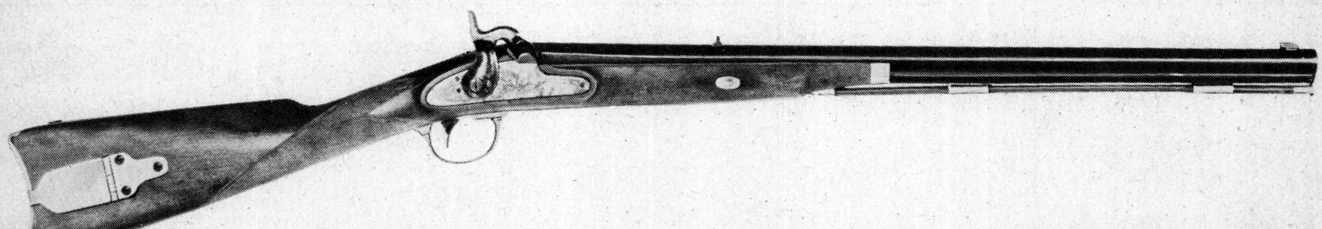

NAVY ARMS BUFFALO HUNTER
Caliber: .58. Ignition: percussion. Overall length: 41½ inches. Weight: 8 lbs. **Price: $250**

NAVY ARMS HARPERS FERRY RIFLE
Caliber: .58. Ignition: flint. Barrel: 35 inches. Weight: 9 lbs. Replica of Lewis and Clark 1806 expedition rifle. Browned finish. **Price: $265**

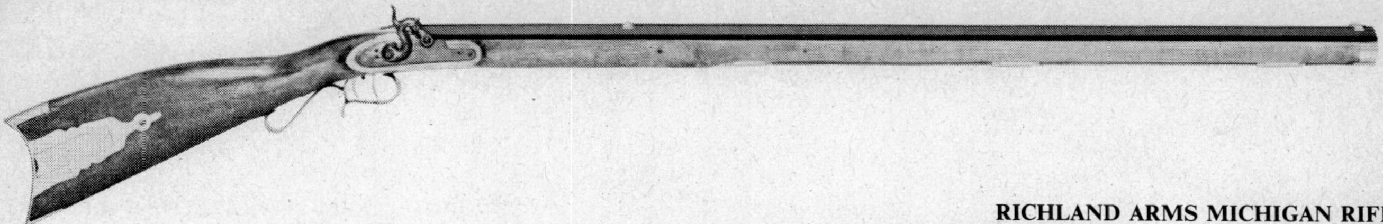

RICHLAND ARMS MICHIGAN RIFLE
Caliber: .45, .50. Ignition: percussion or flint. Barrel: 36 inches. Weight: 9 lbs. Brass triggerguard, patchbox, and buttplate. Case hardened lockplate. Select maple stock. **Price: $158**

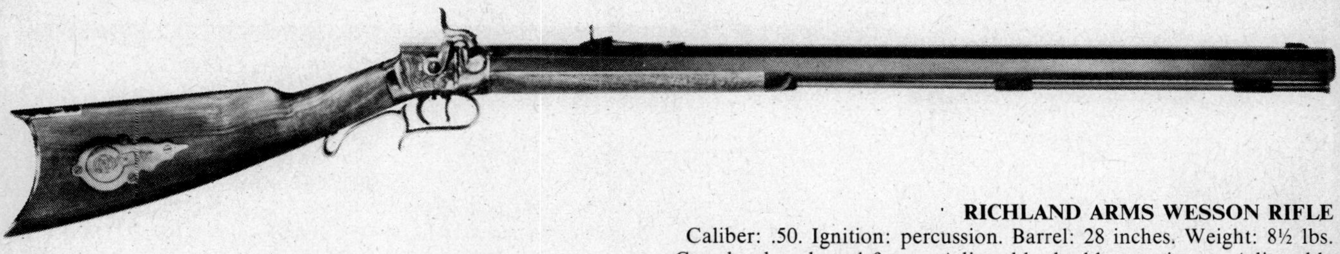

RICHLAND ARMS WESSON RIFLE
Caliber: .50. Ignition: percussion. Barrel: 28 inches. Weight: 8½ lbs. Case hardened steel frame. Adjustable double set triggers. Adjustable sights. **Price: $225**

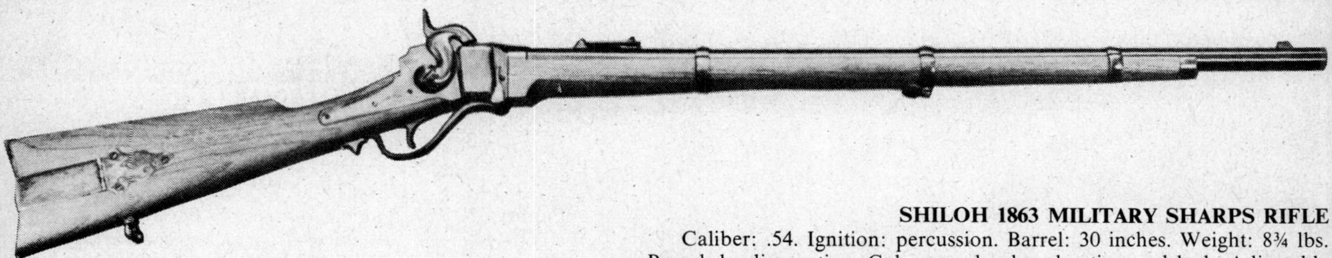

SHILOH 1863 MILITARY SHARPS RIFLE
Caliber: .54. Ignition: percussion. Barrel: 30 inches. Weight: 8¾ lbs. Breech loading action. Color case hardened action and lock. Adjustable sights. **Price: $499**

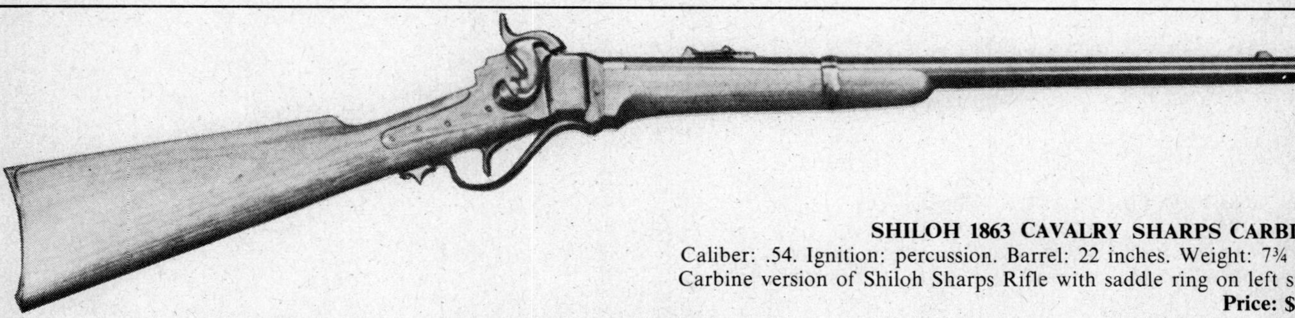

SHILOH 1863 CAVALRY SHARPS CARBINE
Caliber: .54. Ignition: percussion. Barrel: 22 inches. Weight: 7¾ lbs. Carbine version of Shiloh Sharps Rifle with saddle ring on left side. **Price: $399**

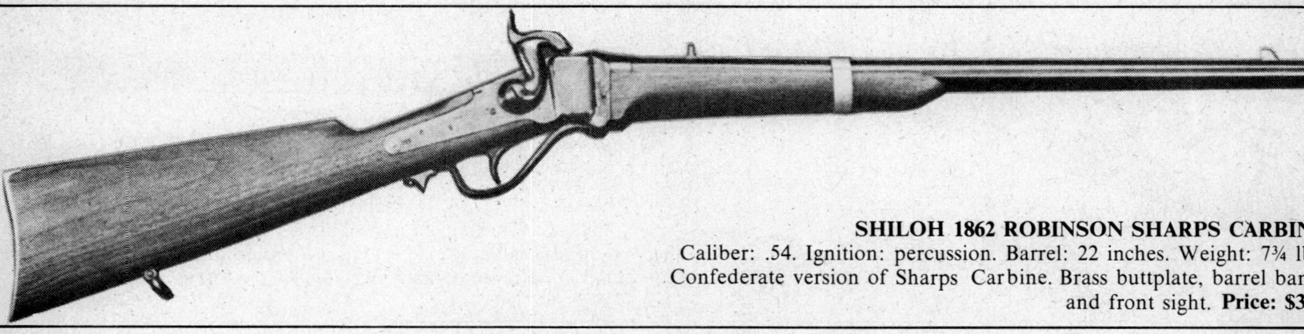

SHILOH 1862 ROBINSON SHARPS CARBINE
Caliber: .54. Ignition: percussion. Barrel: 22 inches. Weight: 7¾ lbs. Confederate version of Sharps Carbine. Brass buttplate, barrel band, and front sight. **Price: $399**

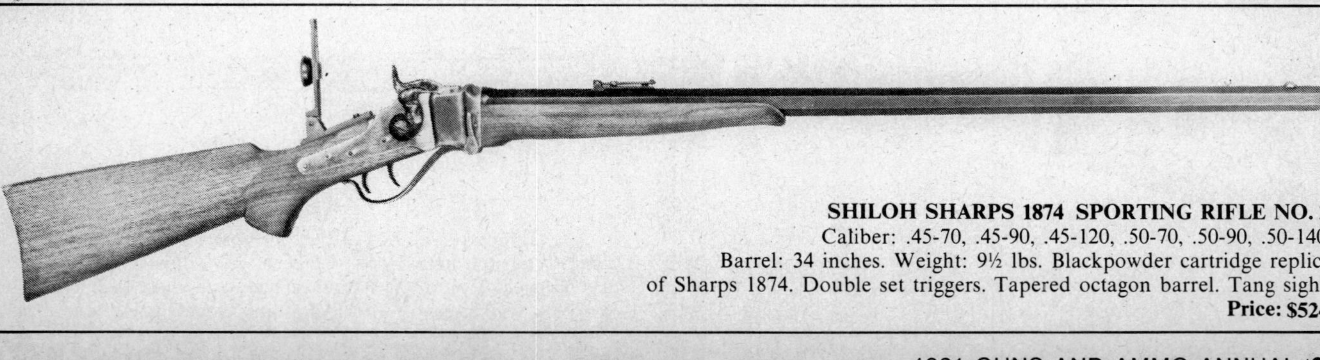

SHILOH SHARPS 1874 SPORTING RIFLE NO. 2
Caliber: .45-70, .45-90, .45-120, .50-70, .50-90, .50-140. Barrel: 34 inches. Weight: 9½ lbs. Blackpowder cartridge replica of Sharps 1874. Double set triggers. Tapered octagon barrel. Tang sight. **Price: $524**

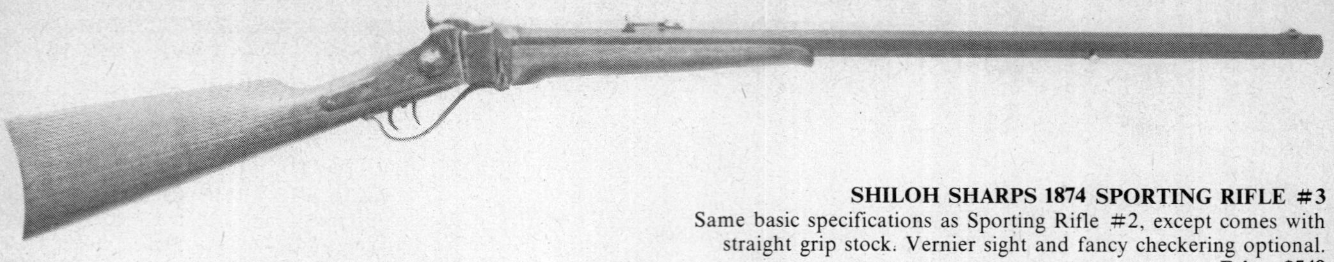

SHILOH SHARPS 1874 SPORTING RIFLE #3
Same basic specifications as Sporting Rifle #2, except comes with straight grip stock. Vernier sight and fancy checkering optional. **Price: $549**

SHILOH SHARPS BUSINESS RIFLE
Caliber: .45-70, .45-90, .45-120, .50-70, .50-90, .50-140. Barrel: 28 inches. Weight: 9½ lbs. Heavy tapered barrel. Double set triggers. Replica of 1874 Sharps rifle. American walnut stock. **Price: $490**

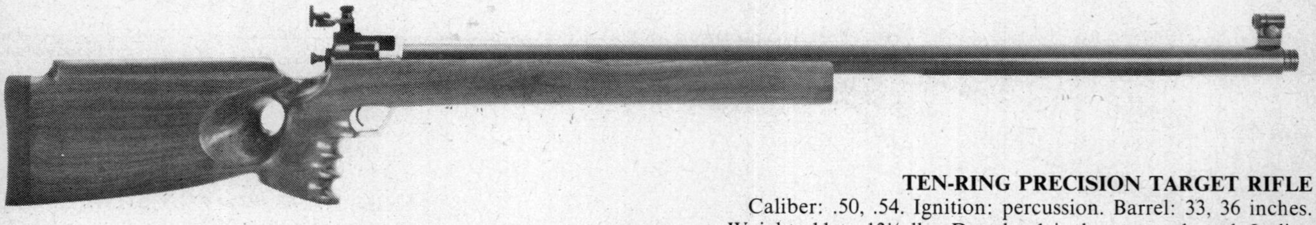

TEN-RING PRECISION TARGET RIFLE
Caliber: .50, .54. Ignition: percussion. Barrel: 33, 36 inches. Weight: 11 to 13½ lbs. Douglas 1 inch octagon barrel. In-line percussion action. Adjustable trigger. Redfield Olympic aperture rear sight. Walnut freestyle stock. **Price: $485**

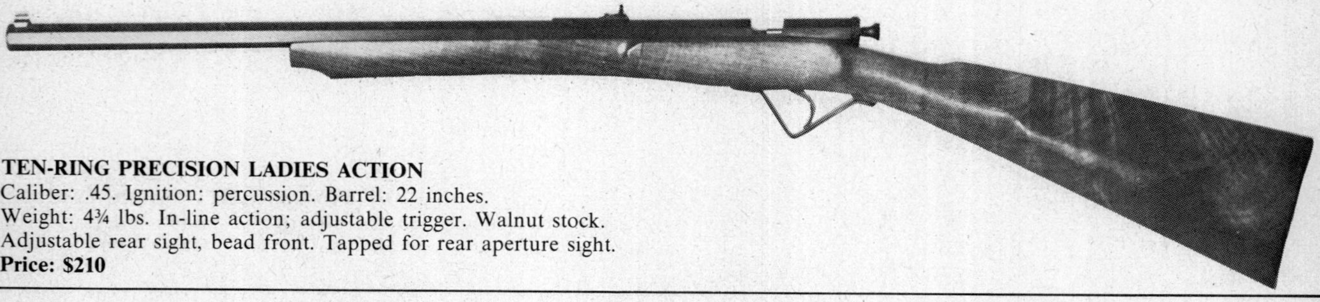

TEN-RING PRECISION LADIES ACTION
Caliber: .45. Ignition: percussion. Barrel: 22 inches. Weight: 4¾ lbs. In-line action; adjustable trigger. Walnut stock. Adjustable rear sight, bead front. Tapped for rear aperture sight. **Price: $210**

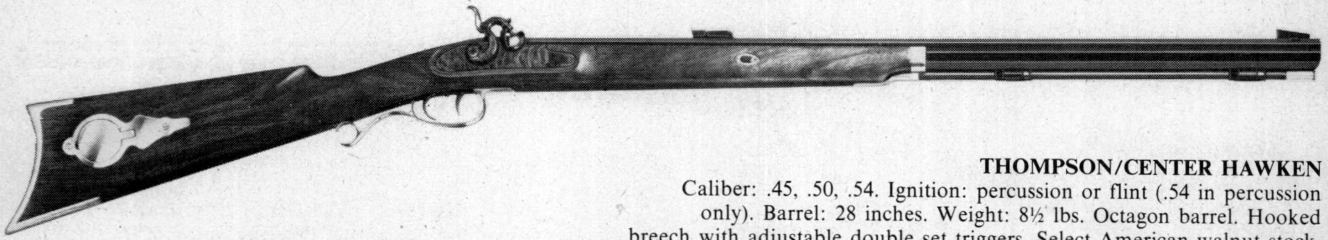

THOMPSON/CENTER HAWKEN
Caliber: .45, .50, .54. Ignition: percussion or flint (.54 in percussion only). Barrel: 28 inches. Weight: 8½ lbs. Octagon barrel. Hooked breech with adjustable double set triggers. Select American walnut stock. with contoured cheekpiece. Open adjustable sights. **Price: $245** (percussion). **$255** (flint)

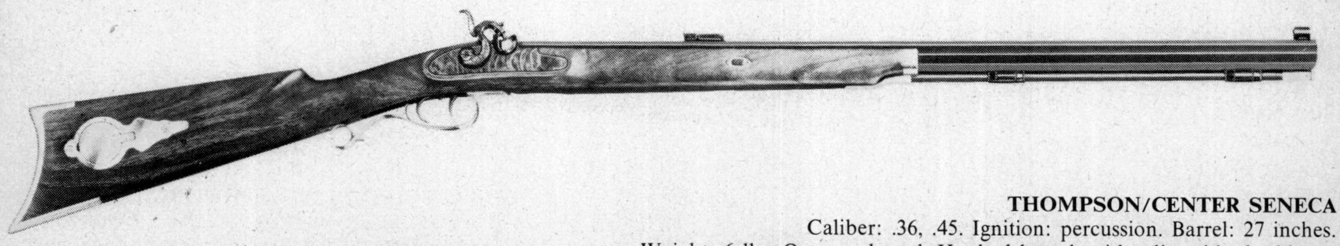

THOMPSON/CENTER SENECA
Caliber: .36, .45. Ignition: percussion. Barrel: 27 inches. Weight: 6 lbs. Octagon barrel. Hooked breech with adjustable double set triggers. Color case hardened lock. American walnut stock. Adjustable open sights. **Price: $245**

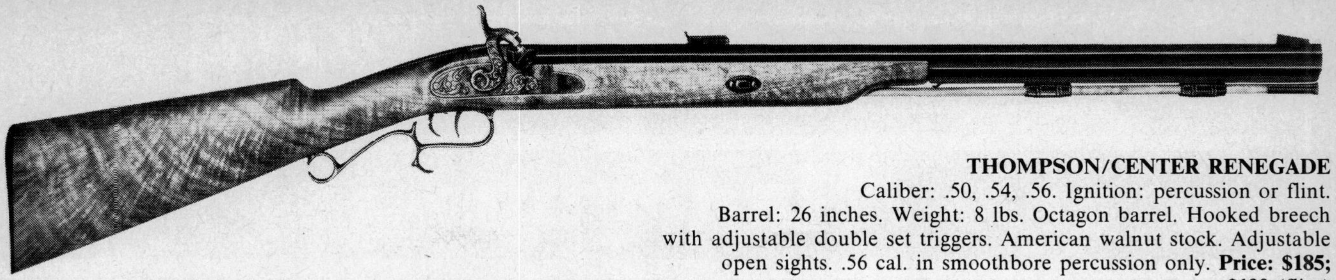

THOMPSON/CENTER RENEGADE
Caliber: .50, .54, .56. Ignition: percussion or flint. Barrel: 26 inches. Weight: 8 lbs. Octagon barrel. Hooked breech with adjustable double set triggers. American walnut stock. Adjustable open sights. .56 cal. in smoothbore percussion only. **Price: $185; $190** (flint)

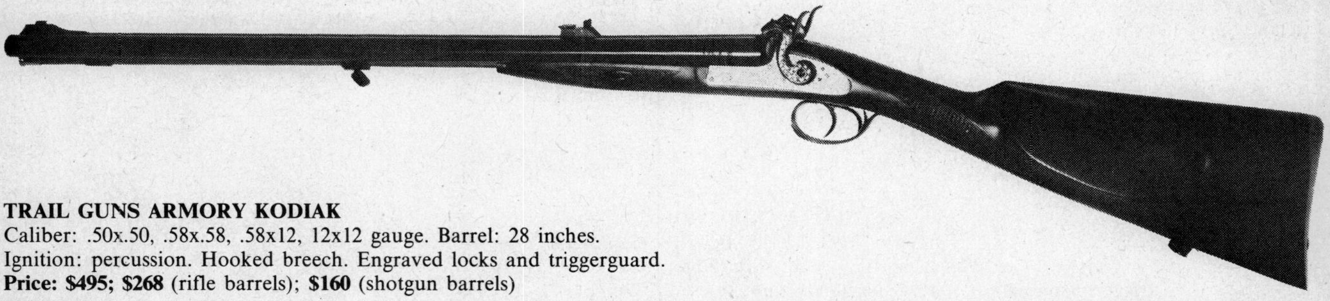

TRAIL GUNS ARMORY KODIAK
Caliber: .50x.50, .58x.58, .58x12, 12x12 gauge. Barrel: 28 inches. Ignition: percussion. Hooked breech. Engraved locks and triggerguard. **Price: $495; $268** (rifle barrels); **$160** (shotgun barrels)

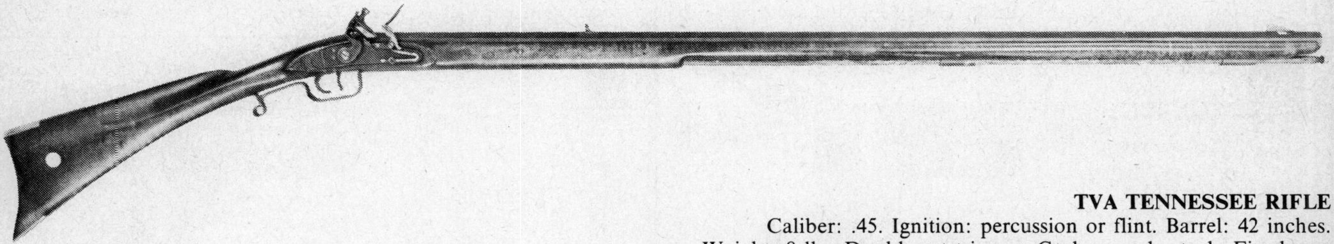

TVA TENNESSEE RIFLE
Caliber: .45. Ignition: percussion or flint. Barrel: 42 inches. Weight: 8 lbs. Double set triggers. Curley maple stock. Fixed rear sight, brass based silver blade front. **Price: $295**

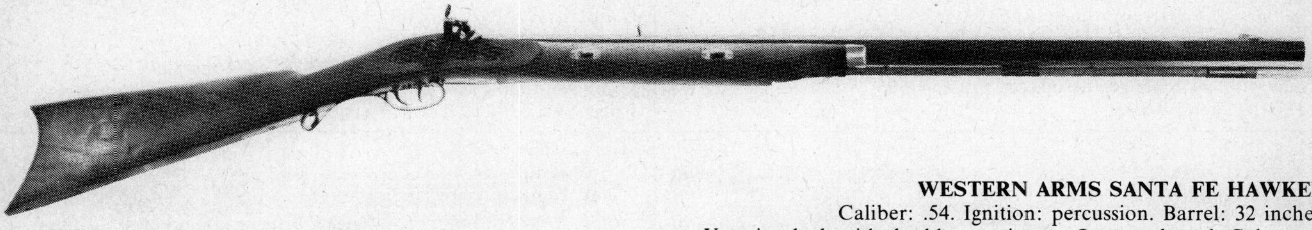

WESTERN ARMS SANTA FE HAWKEN
Caliber: .54. Ignition: percussion. Barrel: 32 inches. V-spring lock with double set triggers. Octagon barrel. Color case hardened lock. Walnut stock in halfstock configuration with cheekpiece. **Price: $325**

WESTERN ARMS MODEL 1873
Caliber: .22 LR, .22 WMR, .38 Spl., .357 Mag., .44-40. Barrel: 24 inches. Octagon barrel. Walnut stock. Stepped rear sight. Replica of Winchester Model 1873. **Price: $350** (carbine version available)

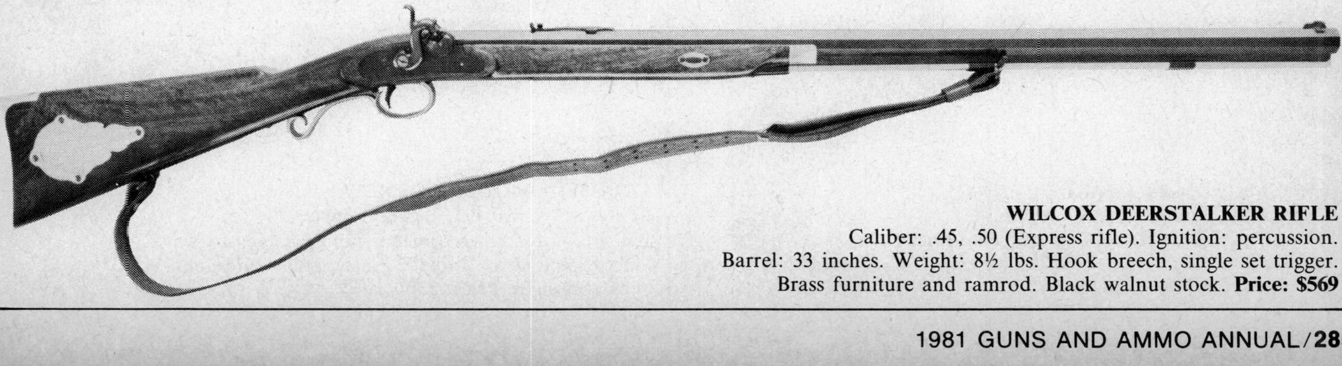

WILCOX DEERSTALKER RIFLE
Caliber: .45, .50 (Express rifle). Ignition: percussion. Barrel: 33 inches. Weight: 8½ lbs. Hook breech, single set trigger. Brass furniture and ramrod. Black walnut stock. **Price: $569**

AIRGUN CATALOG

Handguns

BEEMAN MODEL 900
Caliber: .177. Barrel: 7 inches.
Weight: 51 oz. Power: spring piston. Adjustable trigger. Automatic safety. Adjustable walnut stocks. Micro click rear sight with notch inserts. **Price: $345; $355** (left hand version).

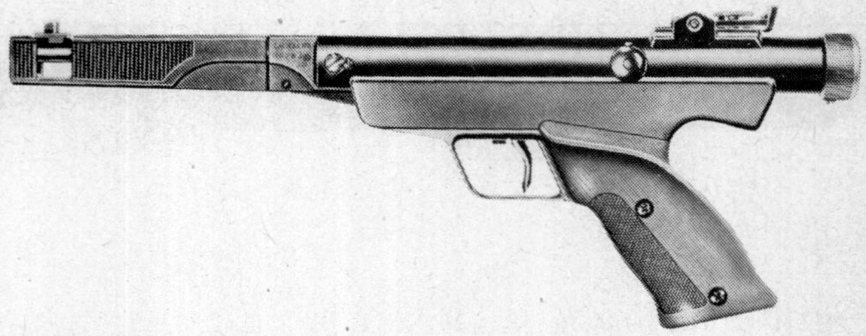

BEEMAN MODEL 850
Caliber: .177. Barrel: 7 inches.
Weight: 50 oz. Power: spring piston. Recoilless action. Rotating barrel housing. Adjustable sights. Optional muzzle weight. **Price: $180; $185** (left hand version)

BEEMAN MODEL 800
Caliber: .177. Barrel: 7 inches.
Weight: 50 oz. Power: spring piston. Recoilless action. Synthetic grips. Micro click rear sight, hooded front with interchangeable inserts. **Price: $150; $155** (left hand version)

BEEMAN MODEL 700
Caliber: .177. Barrel: 7 inches.
Weight: 49 oz. Power: spring piston. Adjustable double pull trigger. Automatic safety. Checked synthetic grips. Adjustable rear sight, hooded front.
Price: $93; $103 (left hand version)

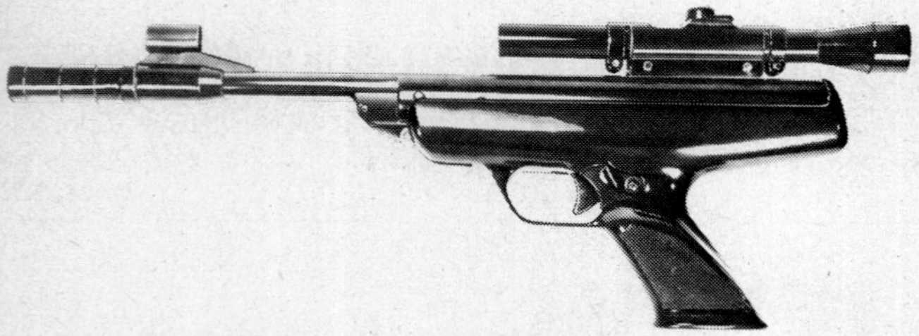

BSA SCORPION AIR PISTOL
Caliber: .177, .22. Barrel: 7⅞ inches.
Weight: 56 oz. Power: barrel cocking. Velocity: 510 fps (.177), 380 fps (.22). Adjustable trigger. Adjustable rear sight. **Price: $100** (without scope)

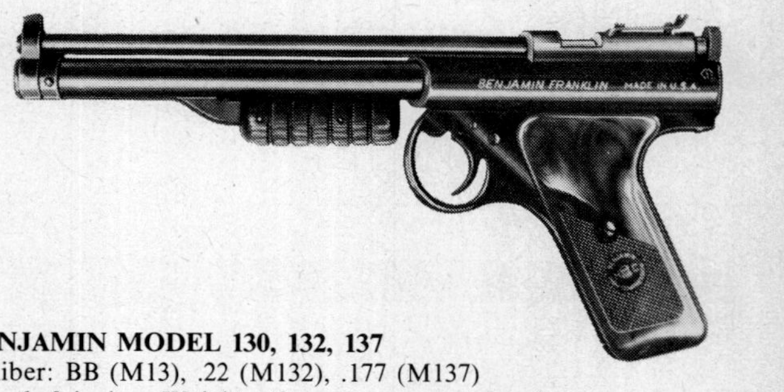

BENJAMIN MODEL 130, 132, 137
Caliber: BB (M13), .22 (M132), .177 (M137)
Barrel: 8 inches. Weight: 48 oz. Power: pneumatic. Pump action. Button rifled barrel in pellet version. Plastic grips. Adjustable open rear sight. **Price: $62**

CROSMAN MARK I TARGET PISTOL
Caliber: .177, .22. Barrel: 7¼ inches.
Weight: 42 oz. Power: gas. Adjustable velocity from 300 to 400 fps. Checked thumbrest grips. Adjustable rear sight, Patridge front. **Price: $42**

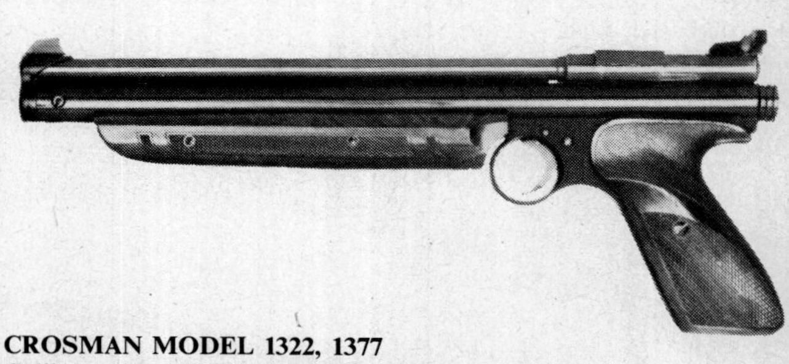

CROSMAN MODEL 1322, 1377
Caliber: .177, BB, .22. Barrel: 8 inches.
Weight: 37 oz. Power: pneumatic. Crossbolt safety. Slide breech cover. Molded plastic grips. Adjustable open rear sight. **Price: $36**

CROSMAN MODEL 38C, 38T
Caliber: .177. Barrel: 3½, 6 inches.
Weight: 38, 43 oz. Power: gas. 6 shot, single or double action. Die-cast receiver. Blue-black finish.
Adjustable rear sight. **Price: $37**

CROSMAN 1600 AIR PISTOL
Caliber: BB. Barrel: 7¾ inches.
Weight: 29 oz. Power: gas. All steel barrel
Slide action safety. **Price: $22**

CROSMAN MODEL 454
Caliber: BB. Barrel: 7 inches.
Weight: 30 oz. Power: gas. 16-shot repeater.
Slide action safety. Adjustable rear sight. **Price: $27**

CROSMAN MODEL 44 PEACEMAKER
Caliber: .177. Barrel: 4¾ inches.
Weight: 34 oz. Power: gas. 6-shot repeater.
Revolving cylinder. 265 fps velocity. **Price: $21**

DAISY MODEL 1200 TARGET PISTOL
Caliber: BB. Barrel: 7½ inches.
Weight: 30 oz. Power: gas. Gravity feed.
Vent rib. Molded woodgrain grips. Adjustable sights.
Price: $35

DAISY MODEL 717, 722
Caliber: .177, .22. Barrel: 13½ inches.
Weight: 48 oz. Power: pneumatic. 360 fps velocity (.177), 290 fps (.22) Molded grips. Adjustable micro sights. **Price: $51; $2** (left hand grips)

FEINWERKBAU MODEL 80 MATCH
Caliber: .177. Barrel: 7½ inches.
Weight: 2¾ to 3¼ lbs. Power: spring piston.
Recoilless operation. Two-stage adjustable trigger.
Walnut stocks with adjustable palm shelf. From Beeman's and Air Rifle Headquarters. **Price: $495**

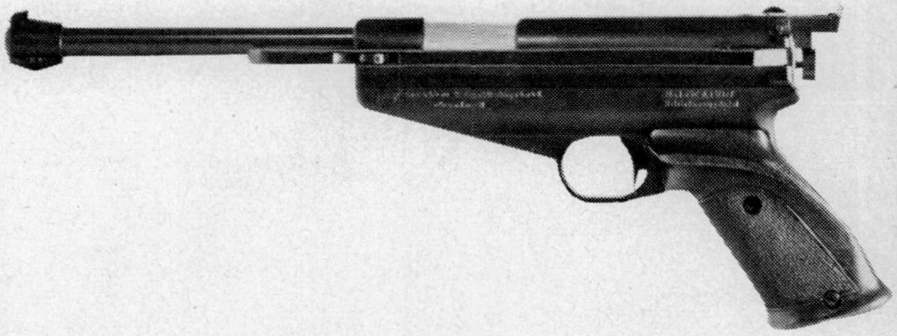

FEINWERKBAU MODEL 65 MK I
Caliber: .177. Barrel: 7½ inches.
Weight: 42 to 50 oz. Power: spring piston.
Recoilless action. Sidelever cocking. 500 fps velocity.
Available with adjustable grip. Left hand version available.
From Beeman's and Air Rifle Headquarters. **Price: $400** to **$469**

HAMMERLI MASTER
Caliber: .177. Barrel: 6 inches.
Weight: 43 oz. Power: gas. Match trigger with 3-way adjustment. Wood target stocks. Micro adjustable sights. From Mandall Shooting Supplies.
Price: $300

HEALTHWAYS MODEL 9100 TOPSCORE
Caliber: .177. Barrel: 6½ inches.
Weight: 28 oz. Power: spring. Bull barrel.
225 fps muzzle velocity. Adjustable sights. **Price: $23**

HEALTHWAYS MODEL 9401
Caliber: .177. Barrel: 5⅞ inches.
Weight: 28 oz. Power: gas. Double action repeater. 400 fps velocity. Molded grips. **Price: $32**

HY-SCORE MODEL 819 SM
Caliber: .177. Barrel: 7 1/16 inches.
Weight: 53 oz. Power: spring. Recoilless action.
Six-way adjustable trigger. Shaped European walnut stock.
Interchangeable inserts for rear sight. **Price: $315**

MARKSMAN MODEL 1010
Caliber: .177, BB. Barrel: 2 inches.
Weight: 24 oz. Power: pneumatic. Repeater, 20-round capacity. Fires darts in single shot mode.
Two safeties. **Price: $15**

SHERIDAN CO2 PISTOL
Caliber: .20. Barrel: 6½ inches.
Weight: 27 oz. Power: gas. Simulated walnut grips. Blued finish. Adjustable rear sight, blade front.
Price: $48

SMITH & WESSON MODEL 78G, 79G
Caliber: .177, .22. Barrel: 8½ inches.
Weight: 42 oz. Power: gas. Pull bolt action.
Checkered walnut stocks. Micro adjustable sights.
Price: $65

WALTHER MODEL LP-3
Caliber: .177. Barrel: 9⅜ inches.
Weight: 45½ oz. Power: pneumatic. 4-way adjustable trigger. Plastic thumbrest grips. Micro click adjustable rear sight. From Interarms and Beeman's. **Price: $430**

WALTHER LP-3 MATCH
Same basic specifications as standard LP-3, except has ISU regulation adjustable grips. From Interarms and Beeman's. **Price: $540**

WALTHER MODEL LP-53
Caliber: .177. Barrel: 9⅜ inches. Weight: 40½ oz. Power: pneumatic. Target grips, barrel weights available. Micro rear sight with interchangeable blade front. From Interarms and Beeman's. **Price: $250**

WEBLEY HURRICANE
Caliber: .177, .22. Barrel: 8 inches. Weight: 36 oz. Power: spring piston. Manual safety. Forward tilting barrel. Checked plastic stocks. Micro click rear sight. Imported by Beeman's. **Price: $90**

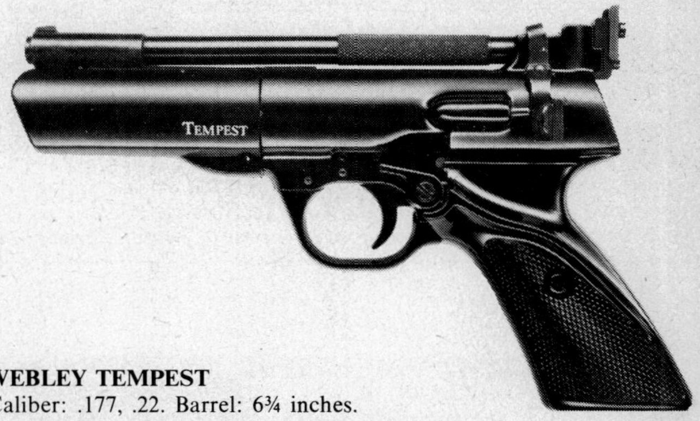

WEBLEY TEMPEST
Caliber: .177, .22. Barrel: 6¾ inches. Weight: 32 oz. Power: spring piston. Manual safety. Adjustable trigger pull. 530 fps velocity (.177) Adjustable rear sight. Imported by Beeman's. **Price: $80**

WEIRAUCH HW70
Caliber: .177. Barrel: 6¼ inches. Weight: 38 oz. Power: pneumatic. 2-stage trigger. Molded grip with thumbrest. Adjustable rear sight with hooded front. From Beeman's and Air Rifle Headquarters. **Price: $80**

WISHCO CUSTOM MATCH
Caliber: .177. Barrel: 7 inches. Weight: 34 oz. Power: spring. Walnut stocks with thumbrest. Adjustable sights. From Beeman's and Air Rifle Headquarters **Price: $113**

WISCHO S-20 STANDARD
Caliber: .177. Barrel: 7 inches. Weight: 34 lbs. Power: spring piston. 450 fps velocity. Walnut stock suitable for right or left hand. From Beeman's. **Price: $99**

FEINWERKBAU F-65 CM-X
Caliber: .177. Barrel: 7½ inches. Weight: 42 oz. Power: spring piston. Two-stage trigger. Anatomically designed grips. From Beeman's and Air Rifle Headquarters. **Price: $499**

Longarms

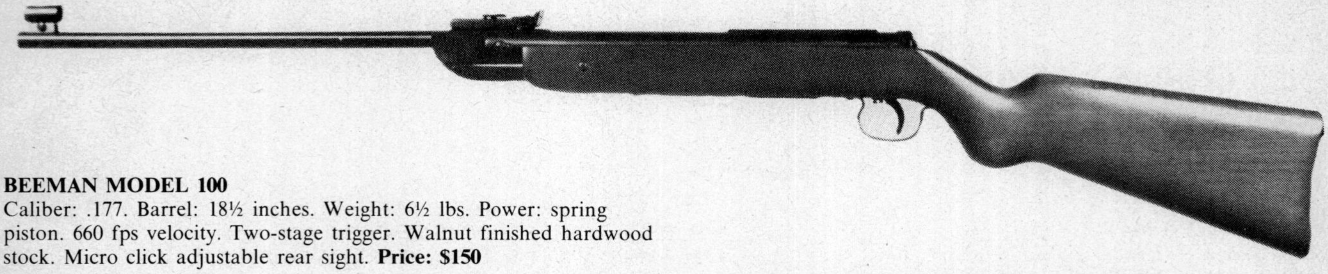

BEEMAN MODEL 100
Caliber: .177. Barrel: 18½ inches. Weight: 6½ lbs. Power: spring piston. 660 fps velocity. Two-stage trigger. Walnut finished hardwood stock. Micro click adjustable rear sight. **Price: $150**

BEEMAN MODEL 250
Caliber: .177. Barrel: 20½ inches. Weight: 7¾ lbs. Power: spring piston. 800 fps velocity. Automatic safety. Adjustable trigger. Micro rear sight, front inserts: grooved for scope mounts. **Price: $210**

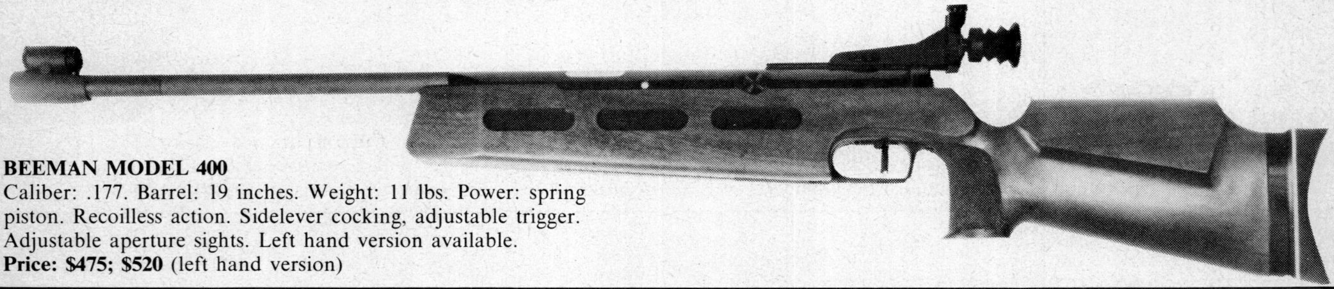

BEEMAN MODEL 400
Caliber: .177. Barrel: 19 inches. Weight: 11 lbs. Power: spring piston. Recoilless action. Sidelever cocking, adjustable trigger. Adjustable aperture sights. Left hand version available.
Price: $475; $520 (left hand version)

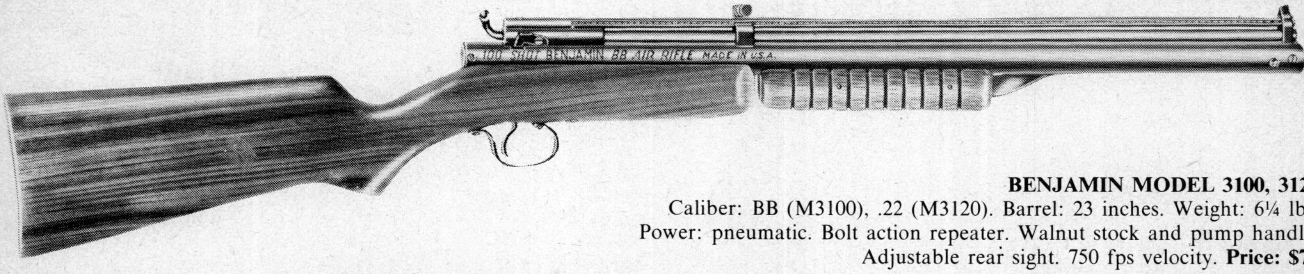

BENJAMIN MODEL 3100, 3120
Caliber: BB (M3100), .22 (M3120). Barrel: 23 inches. Weight: 6¼ lbs. Power: pneumatic. Bolt action repeater. Walnut stock and pump handle. Adjustable rear sight. 750 fps velocity. **Price: $76**

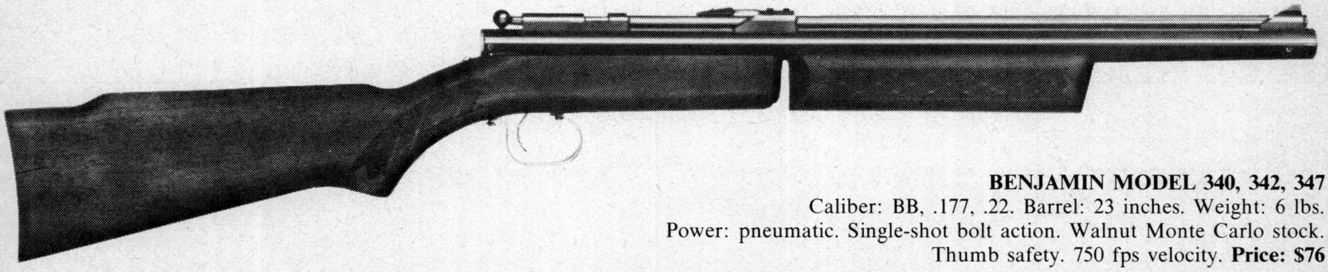

BENJAMIN MODEL 340, 342, 347
Caliber: BB, .177, .22. Barrel: 23 inches. Weight: 6 lbs. Power: pneumatic. Single-shot bolt action. Walnut Monte Carlo stock. Thumb safety. 750 fps velocity. **Price: $76**

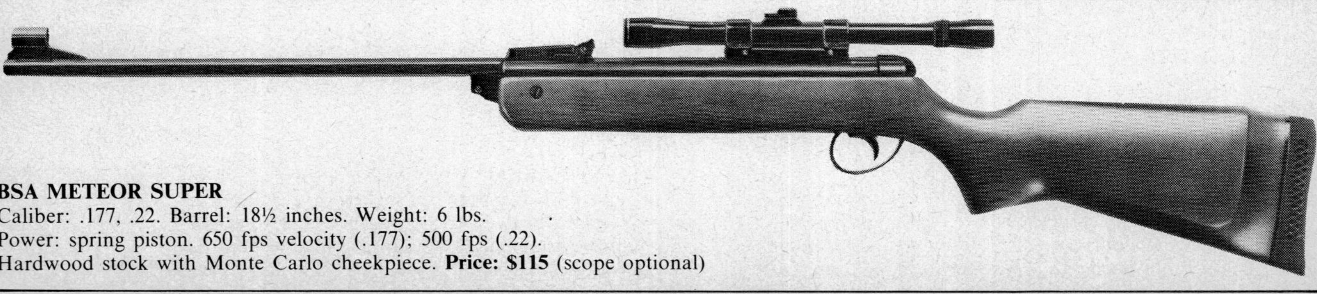

BSA METEOR SUPER
Caliber: .177, .22. Barrel: 18½ inches. Weight: 6 lbs. Power: spring piston. 650 fps velocity (.177); 500 fps (.22). Hardwood stock with Monte Carlo cheekpiece. **Price: $115** (scope optional)

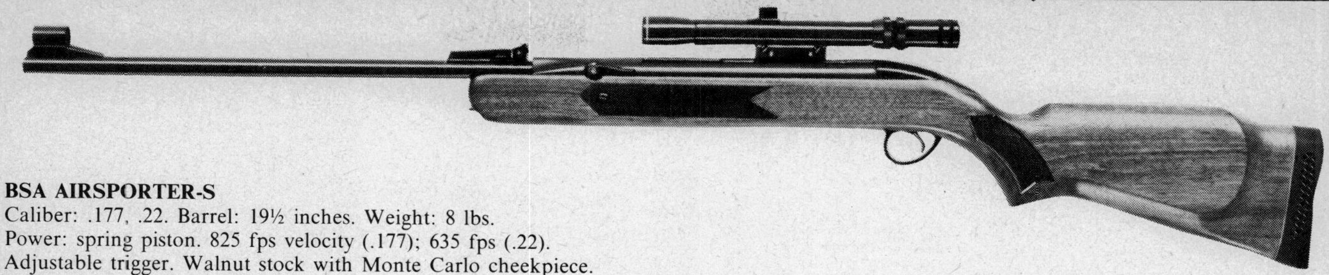

BSA AIRSPORTER-S
Caliber: .177, .22. Barrel: 19½ inches. Weight: 8 lbs.
Power: spring piston. 825 fps velocity (.177); 635 fps (.22).
Adjustable trigger. Walnut stock with Monte Carlo cheekpiece.
Price: $250 (scope optional)

BSA BUCCANEER
Caliber: .177, .22. Barrel: 18½ inches. Weight: 6 lbs. Power: spring piston. 510 fps velocity (.177); 400 fps (.22). Thumb safety. Adjustable sights. Thumb hole stock. **Price: $125** (scope optional)

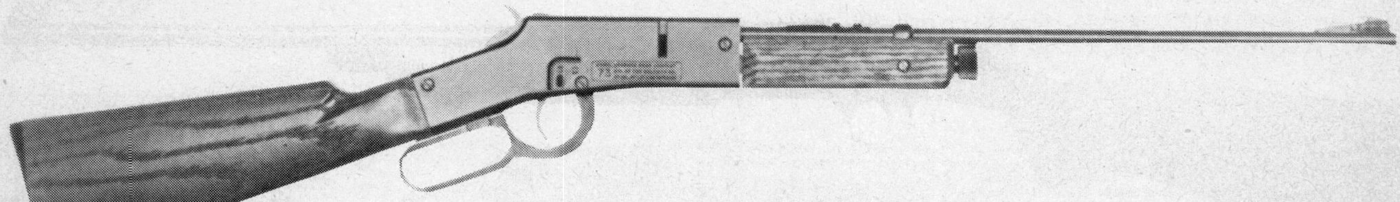

CROSMAN MODEL 73 SADDLE PAL
Caliber: .177. BB. Barrel: 18 inches. Weight: 3¼ lbs. Power: gas. Lever safety. 16 shot repeater. Adjustable rear sight. **Price: $24**

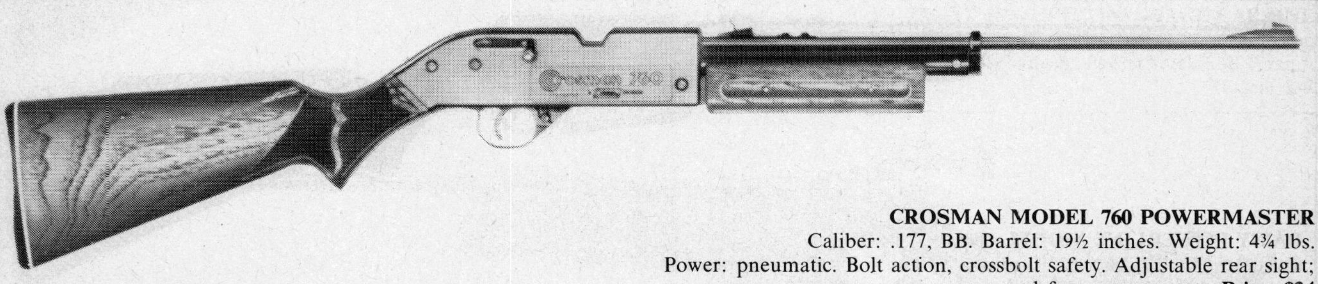

CROSMAN MODEL 760 POWERMASTER
Caliber: .177, BB. Barrel: 19½ inches. Weight: 4¾ lbs. Power: pneumatic. Bolt action, crossbolt safety. Adjustable rear sight; grooved for scope mounts. **Price: $34**

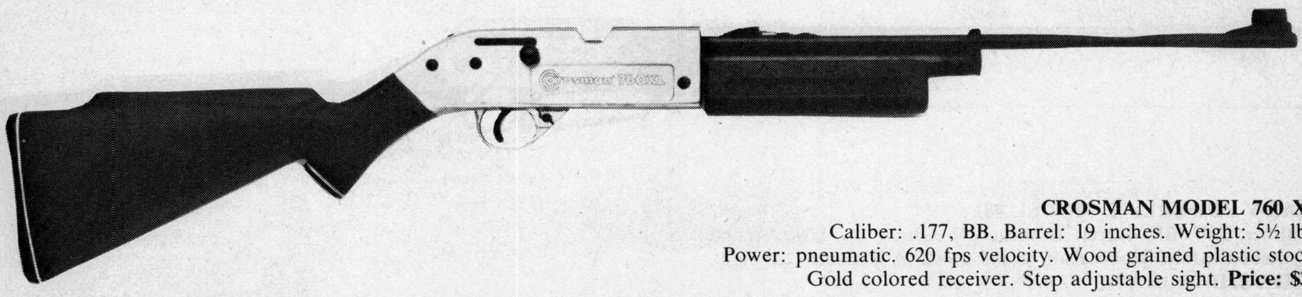

CROSMAN MODEL 760 XL
Caliber: .177, BB. Barrel: 19 inches. Weight: 5½ lbs. Power: pneumatic. 620 fps velocity. Wood grained plastic stock. Gold colored receiver. Step adjustable sight. **Price: $39**

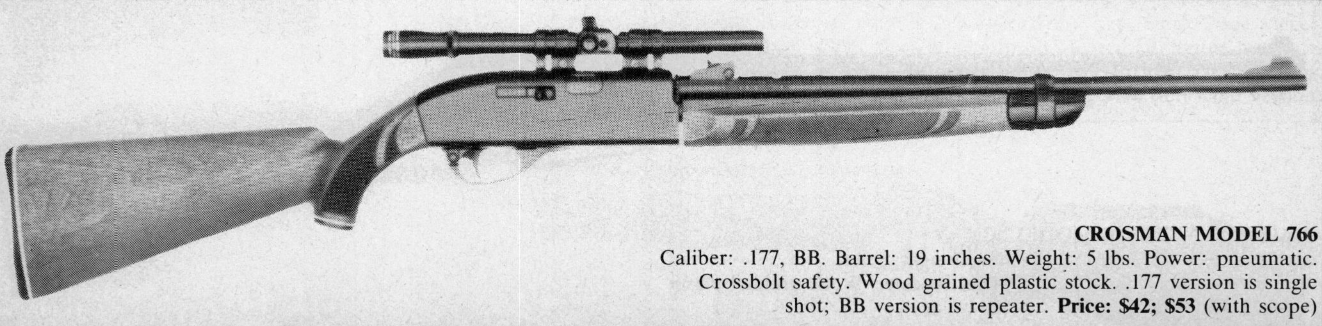

CROSMAN MODEL 766
Caliber: .177, BB. Barrel: 19 inches. Weight: 5 lbs. Power: pneumatic. Crossbolt safety. Wood grained plastic stock. .177 version is single shot; BB version is repeater. **Price: $42; $53** (with scope)

Air Rifle Headquarters. **Price: $210; $270** (deluxe)

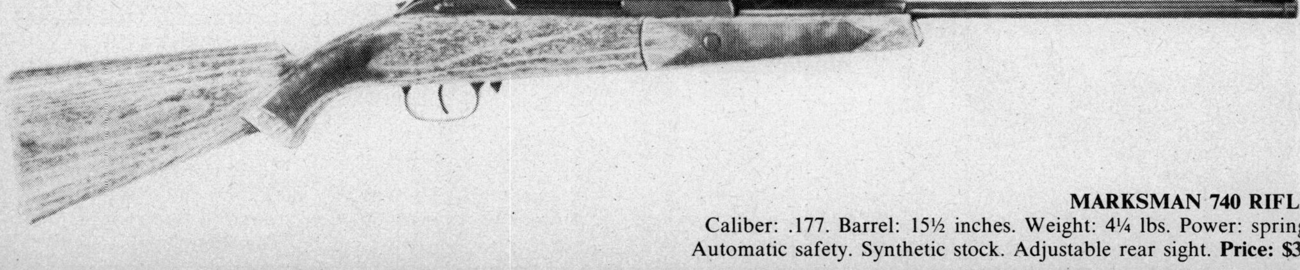

MARKSMAN 740 RIFLE
Caliber: .177. Barrel: 15½ inches. Weight: 4¼ lbs. Power: spring. Automatic safety. Synthetic stock. Adjustable rear sight. **Price: $30**

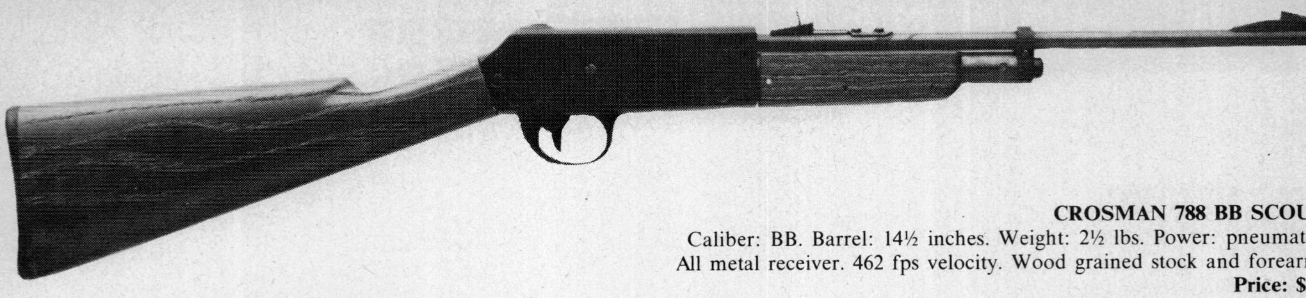

CROSMAN 788 BB SCOUT
Caption: BB. Barrel: 14½ inches. Weight: 2½ lbs. Power: pneumatic. All metal receiver. 462 fps velocity. Wood grained stock and forearm. **Price: $21**

NORICA 80 G
Caliber: .177. Barrel: 18 inches. Weight: 6¾ lbs. Power: spring piston. 647 fps velocity. Two-stage adjustable trigger. Automatic safety. Beechwood stock with cheekpiece. Adjustable rear sight, tunnel front with inserts. From Air Rifle Headquarters. **Price: $169**

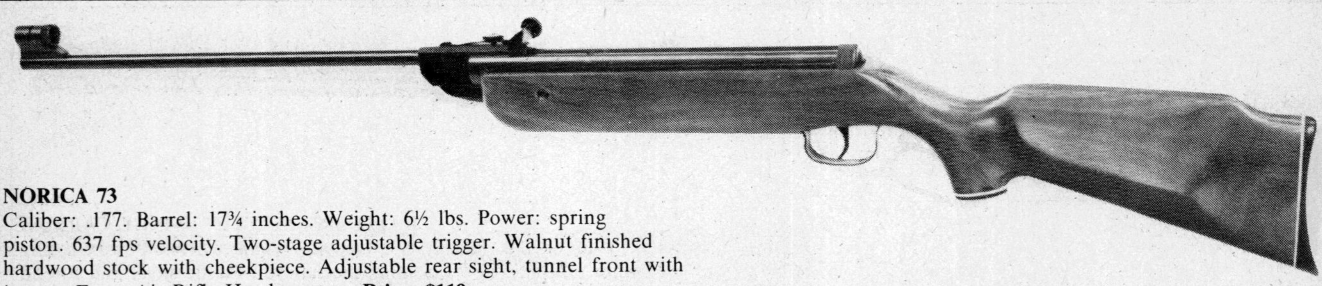

NORICA 73
Caliber: .177. Barrel: 17¾ inches. Weight: 6½ lbs. Power: spring piston. 637 fps velocity. Two-stage adjustable trigger. Walnut finished hardwood stock with cheekpiece. Adjustable rear sight, tunnel front with inserts. From Air Rifle Headquarters. **Price: $119**

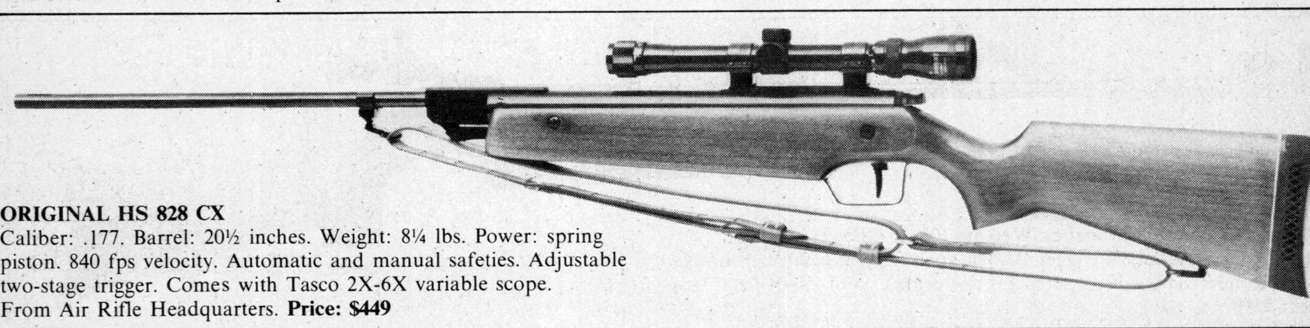

ORIGINAL HS 828 CX
Caliber: .177. Barrel: 20½ inches. Weight: 8¼ lbs. Power: spring piston. 840 fps velocity. Automatic and manual safeties. Adjustable two-stage trigger. Comes with Tasco 2X-6X variable scope. From Air Rifle Headquarters. **Price: $449**

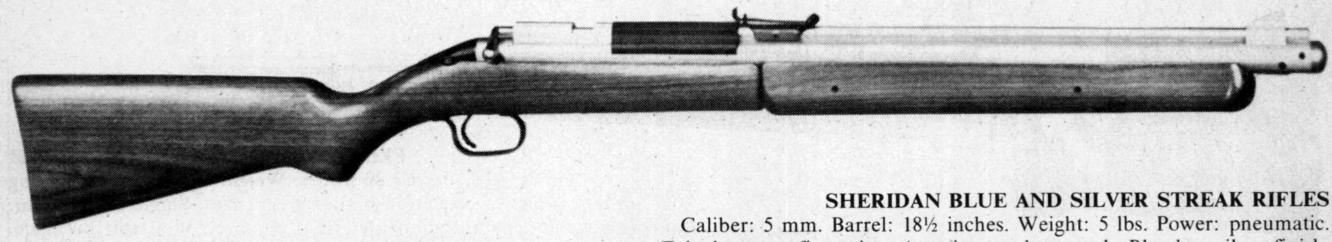

SHERIDAN BLUE AND SILVER STREAK RIFLES
Caliber: 5 mm. Barrel: 18½ inches. Weight: 5 lbs. Power: pneumatic. Takedown configuration. American walnut stock. Blued or silver finish. Adjustable sights. Left hand version available. **Price: $80** to **$83**

SHERIDAN CO2 RIFLE
Caliber: 5 mm. Barrel: 18½ inches. Weight: 6 lbs. Power: gas. Bolt action single shot. Thumb safety. Walnut stock. Blued or silver finish. Left hand version available. **Price: $80** to **$83**

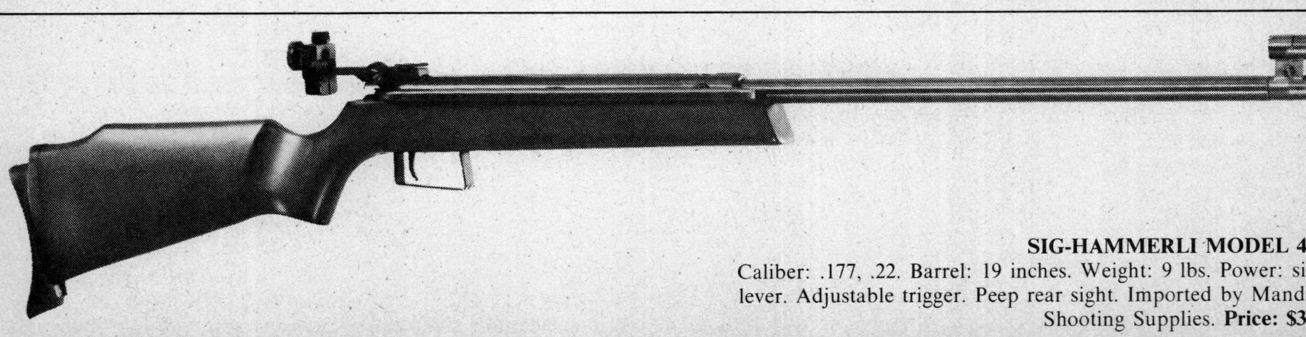

SIG-HAMMERLI MODEL 403
Caliber: .177, .22. Barrel: 19 inches. Weight: 9 lbs. Power: side lever. Adjustable trigger. Peep rear sight. Imported by Mandall Shooting Supplies. **Price: $300**

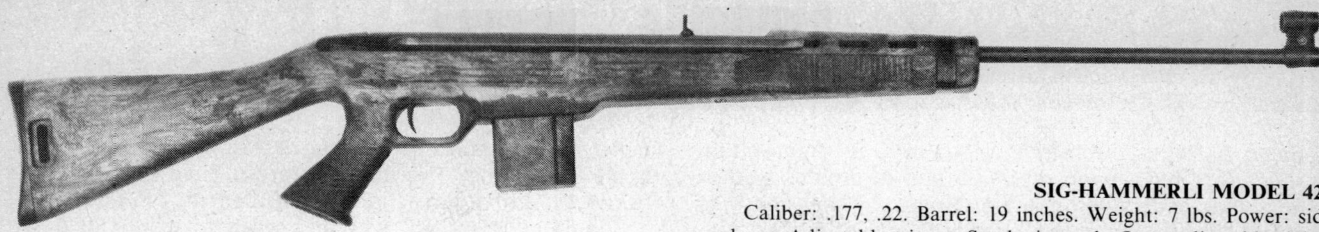

SIG-HAMMERLI MODEL 420
Caliber: .177, .22. Barrel: 19 inches. Weight: 7 lbs. Power: side lever. Adjustable trigger. Synthetic stock. Open adjustable sights. Imported by Mandall Shooting Supplies. **Price: $200**

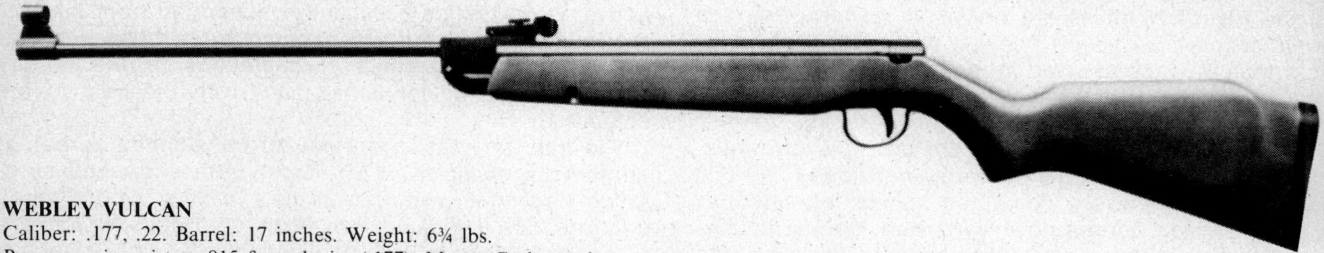

WEBLEY VULCAN
Caliber: .177, .22. Barrel: 17 inches. Weight: 6¾ lbs. Power: spring piston. 815 fps velocity (.177). Monte Carlo stock. Grooved for scope mounts. From Beeman's. **Price: $140**

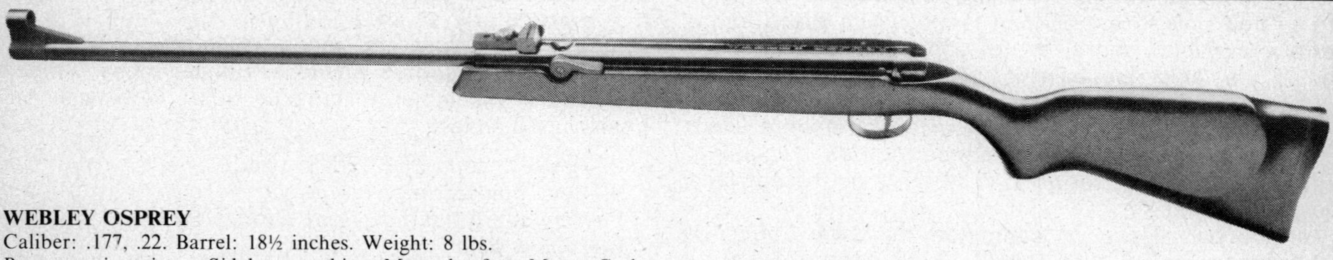

WEBLEY OSPREY
Caliber: .177, .22. Barrel: 18½ inches. Weight: 8 lbs. Power: spring piston. Sidelever cocking. Manual safety. Monte Carlo stock with non-slip rubber buttplate. From Beeman's. **Price: $180**

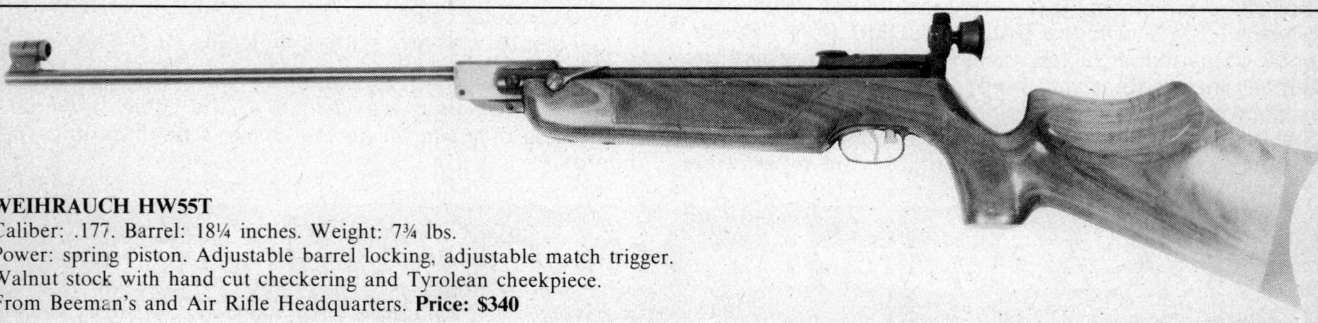

WEIHRAUCH HW55T
Caliber: .177. Barrel: 18¼ inches. Weight: 7¾ lbs. Power: spring piston. Adjustable barrel locking, adjustable match trigger. Walnut stock with hand cut checkering and Tyrolean cheekpiece. From Beeman's and Air Rifle Headquarters. **Price: $340**

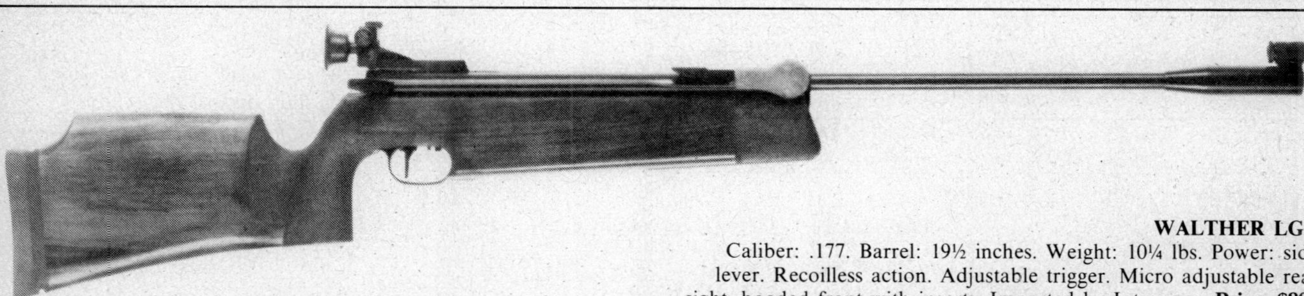

WALTHER LGR
Caliber: .177. Barrel: 19½ inches. Weight: 10¼ lbs. Power: side lever. Recoilless action. Adjustable trigger. Micro adjustable rear sight, hooded front with inserts. Imported by Interarms. **Price: $800**

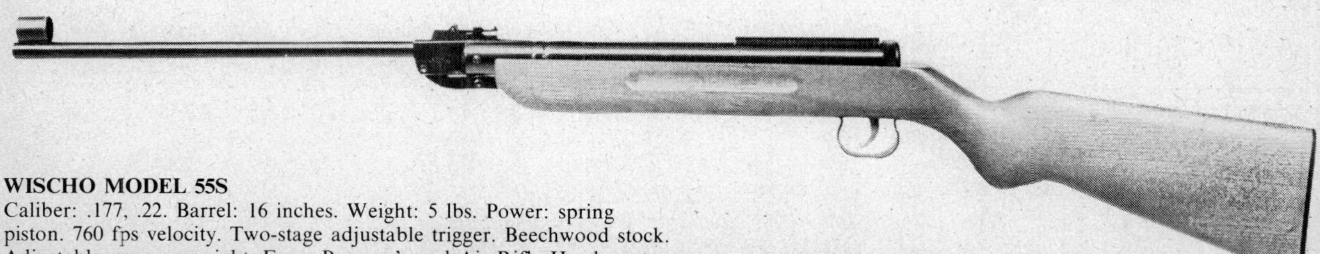

WISCHO MODEL 55S
Caliber: .177, .22. Barrel: 16 inches. Weight: 5 lbs. Power: spring piston. 760 fps velocity. Two-stage adjustable trigger. Beechwood stock. Adjustable open rear sight. From Beeman's and Air Rifle Headquarters. **Price: $190** (.177); **$200** (.22)

BALLISTICS TABLES

Catalog ballistics published by major U.S. ammunition manufacturers have been modified and supplemented to produce this consolidated table. Some of the changes are merely error corrections or extensions of limited catalog data. Others are adaptations intended to present more realistic figures of greater practical use. For example, the trajectory tables are based, in general, upon a normal scope line-of-sight 1½-inch above the bore axis, rather than 0.9-inch as used in much of the factory literature these days. The 0.9-inch height is used here only for iron sights.

For a more drastic example of the changes, the muzzle velocities shown for Weatherby cartridges in .300 and smaller calibers are computed values, obtained by adjusting catalog figures downward with the LeDuc formulas to get speeds for 24-inch, rather than 26-inch barrels. No change has been made in muzzle velocities for the .340 and larger Weatherby calibers, since they are offered only in the 26-inch length. With these long-barrel exceptions, and a few for short barrels, all identified in the table, the ballistics shown are based upon 24-inch barrels. In one case, the .350 Rem. Mag., estimated muzzle velocities for a 24-inch barrel were used to recalculate ballistic performance figures, replacing published 20-inch barrel data, to fit the most recent factory chambering.

All rifle ballistics were computed for G&A Specialty Publications by BALLISTICOMP, 2203 Stratford Lane, Colorado Springs, CO 80909.

Ballistic coefficients (B.C.) were calculated from catalog velocity data with the BALLISTICOMP Bullet Performance Computer. In a few instances, catalog tables are internally inconsistent with respect to the drag deceleration shown for different cartridges loaded with the same bullet. In such cases, the most representative of the conflicting coefficients was used to refigure the downrange bullet performance for those rounds which were obviously out of line. The manufacturers have been notified; hopefully, some corrections will appear in the catalogs.

The BALLISTICOMP was also used to work out downrange velocities and trajectories for Weatherby and Federal loads. Except for these, and the corrections noted above, velocities are essentially unchanged from the factory tables, and the corresponding trajectories are simple (line-of-sight geometry) adaptations of the factory numbers, checked for consistency with the help of the computer.

The trajectory tables are set up for sighting zeroes at appropriate multiples of 50 yards. Both short and long-range sighting-in distances (if the two differ by 100 yards or more) are indicated for those cartridges which are reasonably useful in both applications.

The various factory bullet designations are anything but standardized and ballistically, are pretty nondescriptive, particularly when trademarks are used. For instance, "soft point" can mean anything from a fairly sharp tip to a round or flat nose. Hence, these symbols are used in the tables to indicate bullet type, with emphasis upon shape:

P—Pointed, as in PSP, Spitzer, etc.
S—Semi-pointed; e.g., some Win. ST's
V—Very sharp tip; e.g., Rem. Bronze Pt.
H—Hollow point
R—Round nose
F—Flat nose
M—Full metal case
*—Asterisk with any symbol indicates Boat Tail

Some cartridges dropped by the factories in the not-too-distant past are carried in the table because a few boxes may still be on the dealer's shelf or in private hoards.

U.S. Centerfire Rifle Cartridges

CARTRIDGE	BULLET			VELOCITY—fps (ENERGY—ft-lb)						TRAJECTORY inches above (+), (-) below line of sight 0 indicates sight-in (zeroed) range							
	Wt (gr)	Type	B.C.	Muzzle	100 yd	200 yd	300 yd	400 yd	500 yd	50 yd	100 yd	150 yd	200 yd	250 yd	300 yd	400 yd	500 yd
.17 REMINGTON	25	H	.150	4040	3284	2644	2086	1606	1235	-0.3	+0.3	0	-1.3	-3.8	-7.9		
				(906)	(599)	(388)	(242)	(143)	(85)		+1.7	+2.3	+1.8	0	-3.3	-17	
.218 BEE	46	H	.130	2760	2102	1550	1156	961	850	0	0	-2.0	-6.6	-15	-28		
				(778)	(451)	(245)	(136)	(94)	(74)								
.22 HORNET	45	S,H	.130	2690	2042	1502	1128	948	840	0	0	-2.1	-7.1	-16	-30		
				(723)	(417)	(225)	(127)	(90)	(70)								
	46	H	.130	2690	2042	1502	1128	948	841	0	0	-2.1	-7.1	-16	-30		
				(739)	(426)	(230)	(130)	(92)	(72)								
.222 REMINGTON	50	P,M	.177	3140	2602	2123	1700	1350	1107	+0.1	+0.7	0	-2.3	-6.5	-13	-36	
				(1094)	(752)	(500)	(321)	(202)	(136)								
	50	H	.188	3140	2635	2182	1777	1432	1172	+0.1	+0.7	0	-2.2	-6.2	-13	-34	
				(1094)	(771)	(529)	(351)	(228)	(152)								
	55	M*	.273	3020	2675	2355	2057	1783	1537	+0.1	+0.7	0	-2.0	-5.6	-11	-28	
				(1114)	(874)	(677)	(517)	(388)	(288)								

P=pointed, S=semi-pointed, V=very sharp, H=hollow point, R=round nose, F=flat nose, M=metal case, *=boattail.

CARTRIDGE	BULLET			VELOCITY—fps ENERGY—ft-lb						TRAJECTORY (+) inches above, (-) below line of sight 0 indicates sight-in (zeroed) range							
	Wt (gr)	Type	B.C.	Muzzle	100 yd	200 yd	300 yd	400 yd	500 yd	50 yd	100 yd	150 yd	200 yd	250 yd	300 yd	400 yd	500 yd
.223 REMINGTON	55	P	.197	3240	2747	2304	1905	1554	1270	0	+0.6	0	-2.0	-5.6	-11	-30	
(5.56mm)				(1282)	(921)	(648)	(443)	(295)	(197)								
	55	M	.201	3240	2759	2326	1933	1587	1301	0	+0.6	0	-1.9	-5.5	-11	-28	
				(1282)	(929)	(661)	(456)	(308)	(207)								
	55	H	.208	3240	2773	2352	1969	1627	1341	0	+0.6	0	-1.9	-5.4	-11	-29	
				(1282)	(939)	(675)	(473)	(323)	(220)								
	55	M*	.273	3240	2877	2543	2231	1943	1679	0	+0.6	0	-1.7	-4.7	-9.3	-24	
				(1282)	(1011)	(790)	(608)	(461)	(344)								
.222 REM. MAG.	55	P,H	SAME EXTERIOR BALLISTICS AS .223 REMINGTON														
.30-30/.224 REM.	55	S	.140	3400	2693	2085	1570	1187	986	+0.5	+1.7	+1.6	0	-3.6	-10	-34	
ACCELERATOR				(1411)	(885)	(531)	(301)	(172)	(119)								
.224 WEATHERBY	50	P	.183	3645	3056	2546	2091	1692	1369	-0.2	+0.3	0	-1.7	-4.5	-9.2		
MAGNUM				(1475)	(1037)	(720)	(485)	(318)	(208)		+2.1	+2.6	+1.9	0	-3.8	-18	
	55	P	.211	3555	3054	2614	2212	1845	1527	-0.2	+0.3	0	-1.5	-4.4	-8.7		
				(1543)	(1139)	(834)	(597)	(416)	(285)		+2.1	+2.7	+2.0	0	-3.6	-17	
.225 WINCHESTER	55	P	.197	3570	3039	2580	2156	1772	1445	-0.2	+0.4	0	-1.7	-4.4	-8.7		
				(1556)	(1128)	(813)	(568)	(383)	(255)		+2.1	+2.7	+2.0	0	-3.5	-16	-40
.22-250 REM.	55	P	.197	3730	3180	2695	2257	1863	1519	-0.2	+0.3	0	-1.3	-3.9	-7.8		
				(1699)	(1235)	(887)	(622)	(424)	(282)		+1.8	+2.4	+1.8	0	-3.2	-15	-37
	55	H	.208	3730	3206	2740	2324	1940	1603	-0.2	+0.3	0	-1.3	-3.8	-7.6		
				(1699)	(1255)	(917)	(659)	(460)	(314)		+1.8	+2.3	+1.7	0	-3.1	-15	-36
	55	H*	.278	3730	3326	2965	2626	2315	2025	-0.3	+0.3	0	-1.1	-3.3	-6.5		
				(1699)	(1351)	(1073)	(842)	(654)	(501)		+1.5	+2.0	+1.5	0	-2.6	-12	-27
.308 WIN./.224	55	P	.197	3770	3215	2726	2286	1888	1541	-0.2	+0.3	0	-1.3	-3.8	-7.6		
ACCELERATOR				(1735)	(1262)	(907)	(638)	(435)	(290)		+1.8	+2.3	+1.7	0	-3.1	-15	-36
.30-06/.224 REM.	55	P	.197	4080	3485	2965	2502	2083	1709	-0.3	+0.2	0	-1.0	-2.8	-6.2		
ACCELERATOR				(2033)	(1983)	(1074)	(764)	(530)	(356)		+1.4	+1.9	+1.4	0	-2.6	-12	-30
.220 SWIFT	50	P	.220	4110	3611	3133	2681	2298	1941	-0.4	+0.1	0	-0.9	-2.7	-5.6	-15	-31
				(1877)	(1448)	(1090)	(799)	(586)	(418)								
.243 WINCHESTER	80	P,H	.255	3350	2955	2593	2259	1951	1670	-0.1	+0.5	0	-1.6	-4.5	-8.8		
				(1993)	(1551)	(1194)	(906)	(676)	(495)		+2.2	+2.7	+2.0	0	-3.5	-16	-37
	85	H*	.400	3320	3066	2827	2599	2382	2176	-0.2	+0.4	0	-1.4	-3.9	-7.5		
				(2080)	(1774)	(1508)	(1275)	(1071)	(893)		+1.9	+2.4	+1.7	0	-2.9	-13	-28
	100	P	.350	2960	2697	2449	2215	1993	1786	+0.6	+1.6	+1.4	0	-3.0	-7.5	-22	
				(1945)	(1615)	(1332)	(1089)	(882)	(708)								
	100	P*	.469	2960	2761	2567	2385	2208	2040	+0.5	+1.5	+1.4	0	-2.7	-6.8	-20	-40
				(1945)	(1692)	(1463)	(1263)	(1082)	(924)								
6 mm REMINGTON	80	P,H	.255	3470	3064	2694	2352	2036	1747	-0.1	+0.4	0	-1.4	-4.1	-8.1		
(.244 REM.)				(2139)	(1667)	(1289)	(982)	(736)	(542)		+2.0	+2.5	+1.8	0	-3.2	-15	-34
	90	P	.297	3190	2863	2558	2273	2007	1760	0	+0.5	0	-1.7	-4.8	-9.3		
				(2033)	(1638)	(1307)	(1032)	(805)	(619)		+2.5	+2.8	+2.1	0	-3.5	-16	-37
	100	P	.350	3130	2857	2600	2357	2127	1911	0	+0.5	0	-1.7	-4.7	-9.1		
				(2175)	(1812)	(1501)	(1233)	(1004)	(811)		+2.4	+2.8	+2.0	0	-3.4	-15	-35
.240 WEATHERBY	70	P	.247	3720	3265	2870	2505	2172	1858	-0.2	+0.3	0	-1.2	-3.5	-7.0		
MAGNUM				(2150)	(1657)	(1280)	(975)	(733)	(536)		+1.7	+2.1	+1.6	0	-2.9	-13	-30
	87	P	.311	3390	3059	2756	2466	2196	1949	-0.1	+0.3	0	-1.5	-4.1	-7.3		
				(2220)	(1807)	(1467)	(1175)	(931)	(734)		+2.0	+2.5	+1.9	0	-3.1	-14	-32
	100	P	.366	3300	3022	2762	2515	2282	2062	0	+0.4	0	-1.6	-4.3	-8.4		
				(2418)	(2027)	(1694)	(1404)	(1156)	(944)		+2.2	+2.7	+1.9	0	-3.2	-14	-31
.250 SAVAGE	87	P	.263	3030	2673	2342	2036	1755	1504	+0.7	+1.7	+1.6	0	-3.2	-8.1	-25	
				(1773)	(1380)	(1059)	(801)	(595)	(437)								
	100	S	.252	2820	2467	2140	1839	1569	1339	+1.0	+2.1	+1.9	0	-3.8	-9.8	-30	
				(1765)	(1351)	(1017)	(751)	(547)	(398)								
	100	P	.284	2820	2504	2210	1936	1684	1461	+1.0	+2.0	+1.7	0	-3.6	-9.2	-28	
				(1765)	(1392)	(1084)	(832)	(630)	(474)								

P=pointed, S=semi-pointed, V=very sharp, H=hollow point, R=round nose, F=flat nose, M=metal case, *=boattail.

CARTRIDGE	BULLET			VELOCITY—fps (ENERGY—ft-lb)						TRAJECTORY (+) inches above, (-) below line of sight 0 indicates sight-in (zeroed) range							
	Wt (gr)	Type	B.C.	Muzzle	100 yd	200 yd	300 yd	400 yd	500 yd	50 yd	100 yd	150 yd	200 yd	250 yd	300 yd	400 yd	500 yd
.256 WIN. MAG.	60	H	.122	2760	2097	1542	1149	957	846	0	0	-2.0	-6.7	-15	-28		
				(1015)	(586)	(317)	(176)	(122)	(95)								
.257 ROBERTS	87	P	.263	3170	2802	2462	2147	1857	1594	+0.5	+1.5	+1.4	0	-2.9	-7.2	-22	
				(1941)	(1516)	(1171)	(890)	(666)	(491)								
	100	S	.252	2900	2541	2210	1904	1627	1387	+0.8	+2.0	+1.8	0	-3.6	-9.1	-28	
				(1867)	(1433)	(1084)	(805)	(588)	(427)								
	117	R	.240	2650	2291	1961	1663	1404	1199	+1.3	+2.6	+2.3	0	-4.5	-12	-36	
				(1824)	(1363)	(999)	(718)	(512)	(373)								
.257 WEATHERBY MAGNUM	87	P	.215	3695	3186	2741	2333	1965	1630	-0.2	+0.3	0	-1.3	-3.9	-8.3		
				(2637)	(1960)	(1451)	(1051)	(746)	(513)		+1.8	+2.3	+1.7	0	-3.1	-15	-36
	100	P	.280	3440	3065	2729	2412	2117	1845	-0.1	+0.4	0	-1.5	-4.1	-8.2		
				(2627)	(2086)	(1653)	(1292)	(995)	(756)		+2.0	+2.4	+1.8	0	-3.1	-14	-32
	117	S	.256	3200	2819	2469	2144	1847	1578	+0.5	+1.5	+1.4	0	-2.8	-7.0	-22	
				(2660)	(2064)	(1583)	(1194)	(886)	(547)								
.25-06 REMINGTON	87	H	.232	3440	2995	2591	2222	1884	1583	-0.1	+0.4	0	-1.5	-4.4	-8.7		
				(2286)	(1733)	(1297)	(954)	(686)	(484)		+2.1	+2.7	+2.0	0	-3.5	-16	-39
	90	H	.260	3440	3043	2680	2344	2034	1749	-0.1	+0.4	0	-1.5	-4.1	-8.2		
				(2364)	(1850)	(1435)	(1098)	(827)	(611)		+2.0	+2.5	+1.9	0	-3.3	-15	-35
	100	P	.294	3230	2893	2580	2287	2014	1762	0	+0.5	0	-1.7	-4.6	-9.1		
				(2316)	(1858)	(1478)	(1161)	(901)	(689)		+2.7	+3.0	+2.2	0	-3.7	-16	-37
	117	P	.346	3060	2788	2530	2281	2051	1832	+0.1	+0.6	0	-1.8	-5.0	-9.7		
				(2432)	(2019)	(1663)	(1351)	(1093)	(872)		+2.6	+3.0	+2.2	0	-3.7	-16	-37
	117	P*	.438	3060	2837	2626	2425	2237	2055	+0.1	+0.5	0	-1.8	-4.9	-9.5		
				(2432)	(2091)	(1791)	(1527)	(1300)	(1097)		+2.6	+2.9	+2.1	0	-3.6	-15	-34
	120	P,H	.358	3010	2749	2502	2269	2048	1840	+0.6	+1.6	+1.4	0	-2.7	-7.1	-21	
				(2414)	(2013)	(1668)	(1372)	(1117)	(902)								
.25-20 WINCHESTER	86	F	.193	1460	1194	1030	931	858	798	+2.0	0	-8.2	-23				
				(407)	(272)	(203)	(165)	(141)	(122)						(IRON SIGHTS)		
.25-35 WINCHESTER	117	F	.216	2230	1866	1545	1282	1097	984	+1.2	+1.9	0	-4.9	-13	-26		
				(1292)	(904)	(620)	(427)	(313)	(251)								
	117	R	.240	2230	1905	1613	1363	1169	1041	+1.2	+1.8	0	-4.6	-13	-25		
				(1292)	(943)	(676)	(483)	(355)	(281)								
6.5 mm REM. MAG.	100	P	.256	3390	2998	2639	2308	2000	1720	-0.1	+0.4	0	-1.5	-4.3	-8.5		
				(2551)	(1995)	(1546)	(1183)	(888)	(657)		+2.1	+2.6	+1.9	0	-3.4	-15	-36
	120	P	.323	3210	2905	2621	2353	2102	1867	0	+0.5	0	-1.6	-4.5	-8.8		
				(2745)	(2248)	(1830)	(1475)	(1177)	(929)		+2.3	+2.8	+2.0	0	-3.4	-15	-35
.264 WIN. MAG.	100	P	.253	3320	2926	2565	2231	1923	1644	-0.1	+0.5	0	-1.6	-4.6	-9.1		
				(2447)	(1901)	(1461)	(1105)	(821)	(600)		+2.3	+2.8	+2.1	0	-3.5	-16	-38
	140	P	.385	3030	2782	2548	2326	2114	1914	+0.1	+0.6	0	-1.8	-5.0	-10		
				(2853)	(2405)	(2018)	(1682)	(1389)	(1139)		+2.6	+3.0	+2.2	0	-3.6	-16	-36
.270 WINCHESTER	100	P	.250	3480	3067	2690	2343	2023	1730	-0.1	+0.4	0	-1.4	-4.1	-8.1		
				(2689)	(2088)	(1606)	(1219)	(909)	(664)		+2.0	+2.5	+1.8	0	-3.2	-15	-35
	130	P	.335	3110	2823	2554	2300	2061	1837	0	+0.6	0	-1.8	-4.9	-9.4		
				(2791)	(2300)	(1883)	(1527)	(1226)	(974)		+2.5	+2.9	+2.2	0	-3.6	-16	-37
	30	V	.370	3110	2849	2604	2371	2150	1941	0	+0.5	0	-1.7	-4.7	-9.1		
				(2791)	(2343)	(1957)	(1622)	(1334)	(1087)		+2.4	+2.8	+2.2	0	-3.4	-15	-35
	130	P*	.426	3110	2882	2666	2456	2262	2073	0	+0.5	0	-1.6	-4.5	-8.7		
				(2791)	(2397)	(2051)	(1741)	(1477)	(1240)		+2.4	+2.8	+2.1	0	-3.0	-14	-32
	150	R	.261	2900	2550	2225	1926	1653	1415	+0.8	+1.9	+1.8	0	-3.5	-9.0	-28	
				(2801)	(2165)	(1649)	(1235)	(910)	(667)								
	150	P	.345	2900	2632	2380	2142	1918	1709	+0.7	+1.8	+1.6	0	-3.2	-7.9	-23	
				(2801)	(2307)	(1886)	(1528)	(1225)	(973)								
	150	P*	.483	2900	2708	2523	2346	2176	2013	+0.6	+1.6	+1.4	0	-2.8	-7.0	-20	
				(2801)	(2442)	(2120)	(1833)	(1577)	(1349)								

P=pointed, S=semi-pointed, V=very sharp, H=hollow point, R=round nose, F=flat nose, M=metal case, *=boattail.

CARTRIDGE	BULLET			VELOCITY—fps (ENERGY—ft-lb)						TRAJECTORY (+) inches above, (-) below line of sight 0 indicates sight-in (zeroed) range							
	Wt (gr)	Type	B.C.	Muzzle	100 yd	200 yd	300 yd	400 yd	500 yd	50 yd	100 yd	150 yd	200 yd	250 yd	300 yd	400 yd	500 yd
.270 WEATHERBY MAGNUM	100	P	.226	3640	3157	2728	2340	1995	1669	-0.2	+0.3	0	-1.4	-4.0	-8.2		
				(2941)	(2213)	(1652)	(1216)	(884)	(618)		+1.9	+2.4	+1.8	0	-3.1	-14	-34
	130	P	.318	3275	2959	2668	2390	2132	1894	-0.1	+0.5	0	-1.6	-4.5	-8.8		
				(3095)	(2527)	(2054)	(1649)	(1312)	(1035)		+2.2	+2.7	+2.0	0	-3.3	-15	-35
	150	P	.343	3160	2881	2612	2361	2121	1901	0	+0.5	0	-1.7	-4.6	-9.1		
				(3325)	(2764)	(2272)	(1856)	(1498)	(1203)		+2.4	+2.8	+2.0	0	-3.4	-15	-36
.280 REMINGTON	150	P	.344	2890	2624	2373	2135	1912	1705	+0.7	+1.8	+1.5	0	-3.1	-8.0	-24	
				(2781)	(2293)	(1875)	(1518)	(1217)	(968)								
	165	R	.289	2820	2510	2220	1950	1701	1479	+0.9	+2.0	+1.7	0	-3.5	-9.1	-29	
				(2913)	(2308)	(1805)	(1393)	(1060)	(801)								
.284 WINCHESTER	125	P	.300	3140	2810	2512	2238	1972	1728	+0.6	+1.5	+1.4	0	-2.8	-7.4	-21	
				(2736)	(2191)	(1751)	(1390)	(1079)	(829)								
	150	P	.344	2860	2595	2344	2108	1886	1680	+0.7	+1.8	+1.6	0	-3.2	-8.2	-24	
				(2724)	(2243)	(1830)	(1480)	(1185)	(940)								
7 mm MAUSER	139	P	.331	2660	2398	2151	1918	1699	1505	-0.1	0	-1.4	-4.6	-9.6	-17		
				(2183)	(1774)	(1428)	(1135)	(891)	(699)		+2.3	+2.1	0	-3.9	-10	-29	
	175	R	.275	2440	2137	1857	1603	1382	1204	+0.8	+1.3	0	-3.5	-9.6	-19		
				(2313)	(1774)	(1340)	(998)	(742)	(563)								
7 mm-08 REM.	140	P	.388	2860	2625	2402	2189	1988	1798	+0.7	+1.8	+1.5	0	-3.1	-7.8	-23	
				(2542)	(2142)	(1793)	(1489)	(1228)	(1005)								
7 mm REM. MAG.	125	P	.300	3310	2976	2666	2376	2105	1852	-0.1	+0.4	0	-1.5	-4.3	-8.5		
				(3040)	(2458)	(1972)	(1567)	(1230)	(952)		+2.1	+2.6	+1.9	0	-3.3	-15	-34
	150	P	.344	3110	2830	2568	2320	2085	1866	0	+0.6	0	-1.7	-4.8	-9.3		
				(3221)	(2667)	(2196)	(1792)	(1448)	(1160)		+2.5	+2.9	+2.1	0	-3.5	-16	-36
	150	P*	.518	3110	2925	2748	2575	2407	2246	0	+0.5	0	-1.6	-4.3	-8.2		
				(3221)	(2849)	(2515)	(2208)	(1929)	(1680)		+2.3	+2.6	+1.9	0	-3.1	-13	-29
	160	P	.423	2950	2730	2518	2321	2127	1945	+0.6	+1.6	+1.4	0	-2.7	-7.0	-21	
				(3091)	(2647)	(2252)	(1913)	(1607)	(1344)								
	165	P*	.606	2860	2709	2562	2422	2288	2164	+0.6	+1.6	+1.4	0	-2.8	-6.9	-20	-39
				(2996)	(2688)	(2404)	(2149)	(1918)	(1715)								
	175	P	.430	2860	2645	2440	2244	2057	1879	+0.7	+1.7	+1.5	0	-3.0	-7.6	-22	
				(3178)	(2718)	(2313)	(1956)	(1644)	(1372)								
	175	P*	.654	2860	2718	2580	2443	2316	2186	+0.6	+1.6	+1.4	0	-2.7	-6.8	-19	-38
				(3178)	(2870)	(2586)	(2319)	(2084)	(1856)								
7 mm EXPRESS REM.	150	P	.344	2970	2699	2444	2203	1975	1763	+0.6	+1.6	+1.4	0	-2.9	-7.5	-22	
				(2937)	(2426)	(1989)	(1616)	(1299)	(1035)								
7 mm WEATHERBY MAGNUM	139	P	.334	3205	2909	2633	2374	2130	1903	0	+0.5	0	-1.6	-4.5	-8.8		
				(3170)	(2611)	(2139)	(1739)	(1400)	(1117)		+2.3	+2.8	+2.0	0	-3.4	-15	-35
	154	P	.364	3075	2811	2561	2329	2106	1896	0	+0.6	0	-1.7	-4.7	-9.1		
				(3233)	(2701)	(2242)	(1854)	(1516)	(1229)		+2.5	+2.9	+2.1	0	-3.4	-14	-34
.30 CARBINE	110	R,H	.170	1990	1567	1236	1035	923	842	+0.9	0	-4.5	-14	-28			
															(IRON SIGHTS)		
(20″ BARREL)				(967)	(600)	(373)	(262)	(208)	(173)								
	110	M	.183	1990	1596	1278	1070	952	870	+0.9	0	-4.3	-13	-27			
				(967)	(622)	(399)	(280)	(221)	(185)						(IRON SIGHTS)		
.30 REMINGTON	170	R,F	.254	2120	1822	1555	1328	1153	1036	+1.4	+2.0	0	-5.1	-14	-27		
				(1696)	(1253)	(913)	(666)	(502)	(405)								
.303 SAVAGE	180	R	.248	2120	1813	1539	1308	1134	1021	+1.4	+2.0	0	-5.1	-14	-27		
				(1796)	(1314)	(946)	(684)	(514)	(417)								
	190	F	.250	1940	1657	1410	1211	1073	982	+1.9	+2.5	0	-6.2	-17	-33		
				(1588)	(1158)	(839)	(619)	(486)	(407)								
.303 BRITISH	180	R	.246	2460	2124	1817	1542	1311	1137	+0.8	+1.3	0	-3.6	-10	-19		
				(2418)	(1803)	(1319)	(950)	(687)	(517)								
	180	P	.369	2460	2233	2018	1816	1629	1459	+0.6	+1.2	0	-3.1	-8.4	-16	-39	
				(2418)	(1993)	(1627)	(1318)	(1060)	(851)								
	215	R	.287	2170	1899	1652	1432	1250	1113	+1.2	+1.8	0	-4.6	-12	-24		
				(2248)	(1721)	(1303)	(979)	(746)	(591)								

P=pointed, S=semi-pointed, V=very sharp, H=hollow point, R=round nose, F=flat nose, M=metal case, *=boattail.

CARTRIDGE	BULLET			VELOCITY—fps (ENERGY—ft-lb)						TRAJECTORY (+) inches above, (-) below line of sight 0 indicates sight-in (zeroed) range							
	Wt (gr)	Type	B.C.	Muzzle	100 yd	200 yd	300 yd	400 yd	500 yd	50 yd	100 yd	150 yd	200 yd	250 yd	300 yd	400 yd	500 yd
.30-30 WINCHESTER	125	H	.175	2570	2089	1660	1323	1092	942	+0.9	+1.6	0	-4.0	-11	-22		
				(1833)	(1211)	(765)	(486)	(331)	(246)								
	150	R	.193	2390	1973	1605	1303	1095	973	+1.0	+1.6	0	-4.4	-12	-24		
				(1902)	(1296)	(858)	(565)	(399)	(315)								
	150	H,F	.219	2390	2018	1684	1398	1177	1036	+0.8	+1.5	0	-4.1	-11	-22		
				(1902)	(1356)	(944)	(651)	(461)	(357)								
	170	H,F,R	.254	2200	1895	1619	1381	1191	1061	+1.2	+1.8	0	-4.6	-13	-26		
				(1827)	(1355)	(989)	(720)	(535)	(425)								
.300 SAVAGE	150	R	.224	2630	2247	1897	1585	1324	1131	0	0	-1.7	-5.5	-12	-21		
				(2303)	(1681)	(1198)	(837)	(584)	(426)		+2.7	+2.4	0	-4.8	-13	-39	
	150	S	.271	2630	2311	2015	1743	1500	1295	0	0	-1.6	-5.1	-11	-19		
				(2303)	(1779)	(1352)	(1012)	(749)	(558)		+2.5	+2.2	0	-4.3	-11	-34	
	150	P	.314	2630	2354	2095	1853	1631	1433	0	0	-1.5	-4.8	-10	-18		
				(2303)	(1845)	(1462)	(1143)	(886)	(684)		+2.4	+2.1	0	-4.1	-10	-31	
	180	R	.248	2350	2025	1728	1467	1252	1098	+1.1	+1.5	0	-4.0	-11	-21		
				(2207)	(1639)	(1193)	(860)	(626)	(482)								
	180	P	.384	2350	2137	1935	1745	1570	1413	+0.9	+1.3	0	-3.4	-9.2	-18		
				(2207)	(1825)	(1496)	(1217)	(985)	(798)								
.30-40 KRAG	180	R	.248	2430	2099	1795	1525	1298	1128	+0.9	+1.4	0	-3.7	-10	-20		
				(2360)	(1761)	(1288)	(929)	(673)	(508)								
	180	P	.384	2430	2213	2007	1813	1632	1468	+0.1	0	-1.8	-5.6	-12	-20		
				(2360)	(1957)	(1610)	(1314)	(1064)	(861)		+2.8	+2.4	0	-4.5	-11	-34	
	220	S	.380	2160	1956	1765	1587	1427	1287	+1.3	+1.7	0	-4.2	-11	-21		
				(2279)	(1869)	(1522)	(1230)	(995)	(809)								
.308 WINCHESTER	110	P	.186	3180	2666	2206	1795	1444	1178	+0.6	+1.7	+1.6	0	-3.3	-9.0	-29	
				(2469)	(1736)	(1188)	(787)	(509)	(339)								
	125	P	.268	3050	2697	2370	2067	1788	1537	+0.6	+1.7	+1.5	0	-3.1	-7.9	-24	
				(2581)	(2018)	(1559)	(1186)	(887)	(656)								
	150	S	.271	2820	2488	2179	1893	1633	1405	+1.0	+2.1	+1.8	0	-3.6	-9.5	-29	
				(2648)	(2061)	(1581)	(1193)	(888)	(657)								
	150	P	.314	2820	2533	2263	2009	1774	1560	+0.9	+2.0	+1.7	0	-3.4	-8.8	-26	
				(2648)	(2137)	(1705)	(1344)	(1048)	(810)	0.9							
	165	P*	.470	2700	2515	2329	2156	1989	1831	+0.9	+2.0	+1.7	0	-3.3	-8.4	-24	
				(2670)	(2317)	(1987)	(1703)	(1449)	(1228)								
	180	R	.248	2620	2274	1955	1666	1414	1212	0	0	-1.7	-5.3	-11	-20		
				(2743)	(2066)	(1527)	(1109)	(799)	(587)		+2.6	+2.3	0	-4.6	-12	-36	
	180	P	.384	2620	2393	2178	1974	1782	1604	-0.1	0	-1.5	-4.6	-9.5	-17		
				(2743)	(2288)	(1896)	(1557)	(1269)	(1028)		+2.3	+2.0	0	-3.9	-9.6	-28	
	200	S	.343	2450	2208	1980	1767	1572	1397	+0.1	0	-1.7	-5.7	-12	-20		
				(2665)	(2165)	(1741)	(1386)	(1097)	(867)		+2.8	+2.4	0	-4.6	-12	-35	
.300 H & H MAG.	150	P	.314	3130	2822	2534	2264	2011	1776	0	+0.6	0	-1.8	-4.9	-9.5		
				(3262)	(2652)	(2138)	(1707)	(1347)	(1050)		+2.5	+2.9	+2.1	0	-3.7	-16	-38
	180	P	.384	2880	2640	2412	2196	1990	1798	+0.7	+1.8	+1.5	0	-3.0	-7.7	-23	
				(3315)	(2785)	(2325)	(1927)	(1583)	(1292)								
	220	S	.380	2580	2350	2138	1931	1743	1567	0	0	-1.6	-4.8	-10	-18		
				(3251)	(2697)	(2232)	(1821)	(1484)	(1199)		+2.4	+2.0	0	-4.0	-10	-29	
.300 WIN. MAG.	150	S	.271	3290	2921	2585	2269	1975	1710	-0.1	+0.5	0	-1.7	-4.6	-8.9		
				(3604)	(2841)	(2225)	(1714)	(1299)	(974)		+2.2	+2.7	+2.0	0	-3.4	-15	-35
	150	P	.314	3290	2965	2670	2392	2131	1890	-0.1	+0.5	0	-1.6	-4.4	-8.7		
				(3604)	(2927)	(2374)	(1905)	(1512)	(1190)		+2.2	+2.6	+2.0	0	-3.3	-15	-34
	180	R	.248	2960	2593	2254	1932	1651	1405	+0.8	+2.1	+1.9	0	-3.4	-8.9	-27	
				(3501)	(2867)	(2030)	(1492)	(1089)	(789)								
	180	P	.384	2960	2720	2489	2270	2059	1864	+0.6	+1.6	+1.5	0	-2.9	-7.2	-21	
				(3501)	(2956)	(2476)	(2059)	(1694)	(1388)								
	200	P*	.596	2830	2671	2524	2377	2236	2103	+0.7	+1.7	+1.5	0	-2.9	-7.2	-20	-40
				(3556)	(3168)	(2829)	(2509)	(2220)	(1964)								
	220	S	.380	2680	2448	2228	2020	1823	1640	-0.1	0	-1.4	-4.3	-9.0	-16		
				(3508)	(2927)	(2424)	(1993)	(1623)	(1314)		+2.2	+1.8	0	-3.6	-9.2	-27	

P=pointed, S=semi-pointed, V=very sharp, H=hollow point, R=round nose, F=flat nose, M=metal case, *=boattail.

CARTRIDGE	BULLET			VELOCITY—fps (ENERGY—ft-lb)						TRAJECTORY (+) inches above, (-) below line of sight 0 indicates sight-in (zeroed) range							
	Wt (gr)	Type	B.C.	Muzzle	100 yd	200 yd	300 yd	400 yd	500 yd	50 yd	100 yd	150 yd	200 yd	250 yd	300 yd	400 yd	500 yd
.30-06 SPRINGFLD	110	P	.186	3380	2843	2365	1936	1561	1261	+0.4	+1.4	+1.3	0	-2.9	-7.7	-25	
				(2790)	(1974)	(1366)	(915)	(595)	(388)								
	125	P	.268	3140	2780	2447	2138	1853	1595	+0.5	+1.5	+1.3	0	-2.8	-7.4	-22	
				(2736)	(2145)	(1662)	(1269)	(953)	(706)								
	150	S	.271	2920	2580	2265	1972	1704	1466	+0.7	+1.9	+1.6	0	-3.3	-8.7	-26	
				(2839)	(2217)	(1708)	(1295)	(967)	(716)								
	150	P	.314	2910	2617	2342	2083	1843	1622	+0.7	+1.8	+1.6	0	-3.1	-8.2	-24	
				(2820)	(2281)	(1827)	(1445)	(1131)	(876)								
	150	V	.362	2910	2656	2416	2189	1974	1773	+0.7	+1.7	+1.5	0	-3.0	-7.7	-23	
				(2820)	(2349)	(1944)	(1596)	(1298)	(1047)								
	165	P	.339	2800	2534	2283	2047	1825	1621	+0.9	+2.0	+1.7	0	-3.4	-8.7	-26	
				(2872)	(2352)	(1909)	(1535)	(1220)	(963)								
	165	P*	.470	2800	2607	2421	2244	2073	1911	+0.7	+1.8	+1.6	0	-3.1	-7.7	-22	
				(2872)	(2489)	(2148)	(1844)	(1575)	(1338)								
	180	R	.248	2700	2348	2023	1727	1466	1251	-0.1	0	-1.5	-4.6	-10	-18		
				(2913)	(2203)	(1635)	(1192)	(859)	(625)		+2.4	+2.1	0	-4.2	-11	-34	
	180	P	.384	2700	2469	2250	2042	1846	1663	-0.1	0	-1.3	-4.2	-8.8	-15		
				(2913)	(2436)	(2023)	(1666)	(1362)	(1105)		+2.1	+1.8	0	-3.5	-9.0	-26	
	180	V	.407	2700	2485	2280	2084	1899	1725	-0.1	0	-1.3	-4.1	-8.7	-15		
				(2913)	(2468)	(2077)	(1736)	(1441)	(1189)		+2.1	+1.8	0	-3.4	-8.8	-26	
	200	P*	.596	2550	2401	2260	2126	1993	1867	0	0	-1.5	4.5	-9.3	-16		
				(2887)	(2560)	(2268)	(2007)	(1764)	(1548)		+2.2	+1.9	0	-3.6	-9.0	-26	
	220	R	.296	2410	2130	1870	1632	1422	1246	+0.8	+1.3	0	-3.5	-9.5	-18		
				(2837)	(2216)	(1708)	(1301)	(988)	(758)								
	220	S	.380	2410	2192	1985	1791	1611	1448	+0.7	+1.3	0	-3.3	-8.7	-17		
				(2837)	(2347)	(1924)	(1567)	(1268)	(1024)								
.300 WEATHERBY MAGNUM	110	P	.260	3785	3342	2958	2601	2275	1967	-0.3	+0.2	0	-1.2	-3.2	-6.4		
				(3498)	(2727)	(2137)	(1652)	(1264)	(945)		+1.4	+1.9	+1.4	0	-2.5	-12	-30
	150	P	.313	.3445	3113	2805	2519	2248	1993	-0.1	+0.3	0	-1.3	-3.8	-7.5		
				(3952)	(3227)	(2620)	(2113)	(1683)	(1323)		+1.9	+2.3	+1.7	0	-3.0	-13	-30
	180	P	.358	3165	2894	2635	2392	2163	1947	0	+0.5	0	-1.7	-4.5	-8.9		
				(4003)	(3347)	(2775)	(2286)	(1870)	(1515)		+2.4	+2.8	+2.0	0	-3.4	-15	-34
	220	R	.300	2830	2530	2250	1988	1744	1527	+0.9	+2.0	+1.7	0	-3.5	-9.0	-27	
				(3912)	(3126)	(2473)	(1930)	(1485)	(1139)								
.32 REMINGTON	170	F	.210	2140	1785	1475	1228	1064	963	+1.4	+2.1	0	-5.4	-15	-29		
				(1728)	(1203)	(821)	(569)	(427)	(350)								
	170	R	.242	2140	1822	1538	1301	1125	1012	+1.4	+2.0	0	-5.1	-14	-27		
				(1728)	(1253)	(893)	(639)	(478)	(387)								
.32 WIN. SPEC'L	170	F	.210	2250	1870	1537	1267	1082	972	+1.2	+1.8	0	-4.9	-13	-27		
				(1911)	(1320)	(892)	(606)	(442)	(357)								
	170	H,R	.242	2250	1921	1626	1372	1174	1044	+1.2	+1.7	0	-4.5	-12	-24		
				(1911)	(1393)	(998)	(710)	(520)	(411)								
.32-20 WINCHESTER	100	F	.146	1210	1021	913	834	769	712	0	-6.3	-21					
				(325)	(231)	(185)	(154)	(131)	(113)						(IRON SIGHTS)		
8mm MAUSER	170	R	.205	2360	1969	1622	1333	1123	997	+1.0	+1.6	0	-4.3	-12	-24		
				(2102)	(1463)	(993)	(671)	(476)	(375)								
8mm REM. MAG.	185	P	.300	3080	2761	2464	2186	1926	1687	+0.6	+1.5	+1.4	0	-2.8	-7.3	-22	
				(3896)	(3132)	(2494)	(1963)	(1524)	(1170)								
	220	P	.377	2830	2581	2346	2123	1913	1716	+0.8	+1.9	+1.6	0	-3.2	-8.2	-24	
				(3912)	(3255)	(2688)	(2201)	(1787)	(1439)								
.338 WIN. MAG.	200	S	.307	2960	2658	2375	2110	1862	1635	+0.6	+1.7	+1.5	0	-3.0	-7.9	-24	
				(3890)	(3137)	(2505)	(1977)	(1539)	(1187)								
	250	S	.330	2660	2395	2145	1910	1693	1497	+1.0	+2.3	+1.9	0	-3.8	-10	-29	
				(3927)	(3184)	(2554)	(2025)	(1591)	(1244)								
	300	R	.299	2430	2152	1893	1655	1443	1265	+1.6	+3.0	+2.5	0	-5.1	-13	-38	
				(3933)	(3084)	(2387)	(1824)	(1387)	(1066)								

P=pointed, S=semi-pointed, V=very sharp, H=hollow point, R=round nose, F=flat nose, M=metal case, *=boattail.

CARTRIDGE	BULLET Wt (gr)	Type	B.C.	VELOCITY—fps (ENERGY—ft-lb) Muzzle	100 yd	200 yd	300 yd	400 yd	500 yd	TRAJECTORY (+) inches above, (-) below line of sight 0 indicates sight-in (zeroed) range 50 yd	100 yd	150 yd	200 yd	250 yd	300 yd	400 yd	500 yd
.340 WEATHERBY	200	P	.321	3210	2905	2615	2345	2094	1862	+1.0	+2.3	+2.9	+2.1	0	-3.5	-15	-35
MAGNUM				(4575)	(3747)	(3036)	(2442)	(1947)	(1539)								
(26″ BARREL)	250	S	.339	2850	2580	2325	2090	1865	1655	+0.8	+1.9	+1.6	0	-3.4	-8.6	-25	
				(4508)	(3694)	(3000)	(2424)	(1930)	(1520)								
.348 WINCHESTER	200	R	.227	2520	2149	1812	1514	1270	1097	+0.7	+1.3	0	-3.5	-9.5	-19		
				(2820)	(2051)	(1458)	(1018)	(716)	(534)								
	200	S	.277	2520	2215	1931	1672	1443	1253	+0.6	+1.2	0	-3.2	-8.8	-17		
				(2820)	(2178)	(1656)	(1241)	(925)	(697)								
.35 REMINGTON	150	S	.184	2300	1874	1506	1218	1039	935	+1.2	+1.8	0	-4.9	-14	-27		
				(1762)	(1169)	(755)	(494)	(359)	(291)								
	200	R	.195	2020	1646	1335	1114	985	901	+1.9	+2.5	0	-6.5	-18	-35		
				(1812)	(1203)	(791)	(551)	(431)	(360)								
.350 REM. MAG.	200	P	.292	2770	2468	2163	1911	1666	1451	+1.0	+2.2	+1.9	0	-3.7	-9.6	-29	
				(3407)	(2704)	(2077)	(1621)	(1232)	(935)								
	250	P	.377	2450	2230	2016	1823	1636	1471	+1.6	+2.8	+2.4	0	-4.4	-11	-33	
				(3331)	(2760)	(2256)	(1844)	(1485)	(1201)								
.358 WINCHESTER	200	S	.263	2490	2171	1876	1610	1379	1194	+0.7	+1.3	0	-3.4	-9.3	-18		
				(2753)	(2093)	(1563)	(1151)	(844)	(633)								
	250	S	.328	2230	1988	1763	1557	1375	1224	+1.0	+1.6	0	-4.1	-11	-21		
				(2760)	(2194)	(1725)	(1346)	(1049)	(832)								
.375 H & H MAG.	270	S	.326	2690	2420	2166	1928	1707	1507	+1.0	+2.2	+1.9	0	-3.7	-9.7	-29	
				(4337)	(3510)	(2812)	(2228)	(1747)	(1361)								
	300	M	.236	2530	2171	1843	1552	1307	1126	+0.6	+1.3	0	-3.4	-9.4	-19		
				(4263)	(3139)	(2262)	(1604)	(1138)	(844)								
	300	S	.325	2530	2268	2022	1793	1584	1397	+1.3	+2.6	+2.2	0	-4.3	-11	-33	
				(4263)	(3426)	(2723)	(2141)	(1671)	(1300)								
.375 WINCHESTER	200	F	.214	2200	1841	1526	1268	1089	980	+1.3	+1.9	0	-5.0	-14	-27		
				(2149)	(1505)	(1034)	(714)	(527)	(426)								
	250	F	.279	1900	1647	1424	1239	1103	1011	+1.9	+2.5	0	-6.3	-17	-32		
				(2004)	(1505)	(1125)	(852)	(675)	(567)								
.378 WEATHERBY	270	S	.323	3180	2872	2592	2327	2076	1843	0	+0.4	0	-1.8	-4.9	-10		
MAGNUM				(6061)	(4944)	(4027)	(3246)	(2583)	(2036)		+2.5	+2.9	+2.1	0	-3.6	-16	-36
(26″ BARREL)	300	S	.328	2925	2640	2380	2125	1896	1680	+0.7	+1.8	+1.6	0	-3.2	-8.0	-24	
				(5698)	(4642)	(3772)	(3007)	(2394)	(1880)								
.38-40 WINCHESTER	180	F	.167	1160	999	901	827	764	710	0	-6.6	-22					
				(538)	(399)	(324)	(273)	(233)	(201)						(IRON SIGHTS)		
44 REM. MAG.	240	H	.159	1760	1362	1094	953	861	789	+1.4	0	-6.1	-18	-37			
(20″ BARREL)				(1650)	(988)	(638)	(484)	(395)	(332)						(IRON SIGHTS)		
	240	F	.168	1760	1380	1114	970	878	805	+1.4	0	-5.9	-18	-36			
				(1650)	(1015)	(661)	(501)	(411)	(345)						(IRON SIGHTS)		
.444 MARLIN	240	F	.150	2350	1815	1377	1087	941	845	+1.3	+1.9	0	-5.3	-16	-32		
				(2942)	(1755)	(1010)	(630)	(472)	(380)								
	265	F	.194	2120	1733	1405	1160	1012	920	+1.5	+2.2	0	-5.8	-16	-32		
				(2644)	(1768)	(1162)	(791)	(603)	(498)								
.45-70 GOV'T	405	F,R	.250	1330	1168	1055	977	918	869	+2.4	0	-8.7	-25				
				(1590)	(1227)	(1001)	(858)	(758)	(679)						(IRON SIGHTS)		
	300	H	.160	1810	1406	1123	971	875	802	+1.3	0	-5.8	-17	-36			
															(IRON SIGHTS)		
				(2182)	(1317)	(840)	(628)	(510)	(428)								
.458 WIN. MAG.	500	M	.346	2040	1823	1623	1442	1284	1161	+1.4	+2.0	0	-5.0	-13	-25		
				(4619)	(3689)	(2924)	(2308)	(1839)	(1496)								
	510	R	.278	2040	1770	1527	1319	1157	1046	+1.6	+2.2	0	-5.4	-14	-28		
				(4712)	(3547)	(2640)	(1970)	(1516)	(1239)								
.460 WEATHERBY	500	R	.250	2700	2350	2030	1730	1478	1267	-0.1	0	-1.5	-4.6	-10	-18		
MAG. (26″ BARREL)				(8092)	(6030)	(4574)	(3322)	(2425)	(1782)		+2.4	+2.1	0	-4.2	-11	-34	

P=pointed. S=semi-pointed. V=very sharp. H=hollow point. R=round nose. F=flat nose. M=metal case. *=boattail.

METRIC C.F. RIFLE CARTRIDGES

CARTRIDGE	BULLET WT. GRS.	BULLET TYPE	VELOCITY MUZZLE	VELOCITY 100 YDS.	VELOCITY 200 YDS.	VELOCITY 300 YDS.	ENERGY (FT. LBS.) MUZZLE	ENERGY 100 YDS.	ENERGY 200 YDS.	ENERGY 300 YDS.	MID-RANGE TRAJ. 100 YDS.	MID-RANGE TRAJ. 200 YDS.	MID-RANGE TRAJ. 300 YDS.
5.6 X 35R VIERLING	46	SP	2030	1500	1140	–––	418	224	130	–––	1.2	7.5	–––
5.6 X 50R MAG.	50	PSP	3510	2903	2420	2000	2480	843	650	445	0.4	2.2	6.2
5.6 X 52R (SAV. H.P.)	71	PSP	2850	2460	2320	2200	1280	947	864	766	0.3	2.3	6.5
5.6 X 61 VOM HOFE SE	77	PSP	3700	3360	3060	2790	2350	1920	1605	1345	0.1	1.1	3.4
5.6 X 61R	77	PSP	3480	3140	2840	2560	2070	1690	1370	1120	0.1	1.3	4.0
6.5 X 54 MANN. SCH.	139	PSPBT	2580	2420	2270	2120	2056	1808	1591	1388	0.2	2.4	6.5
	156	SPRN	2461	2240	2033	1840	2098	1738	1432	1173	0.3	3.0	8.2
	159	SP	2170	1925	1705	1485	1660	1300	1025	810	0.5	4.1	11.5
	160	SP	2160	1950	1750	1570	1660	1350	1090	875	0.5	4.2	11.7
6.5 X 55 SWEDISH MAUS.	139	PSPBT	2789	2630	2470	2320	2402	2136	1883	1662	0.1	2.0	5.6
	156	SPRN	2493	2271	2062	1867	2153	1787	1473	1208	0.3	2.9	7.9
	160	SP	2420	2190	1960	1760	2080	1700	1360	1110	0.3	3.0	8.0
6.5 X 57 MAUSER	93	PSP	3350	2930	2570	2260	2300	1760	1350	1040	0.1	1.7	4.8
6.5 X 57R	93	PSP	3350	2930	2570	2260	2300	1760	1350	1040	0.1	1.7	4.8
6.5 CARCANO	156	SPRN	2000	1810	1640	1484	1386	1135	932	764	1.0	4.8	12.5
6.5 JAPANESE	139	PSPBT	2428	2280	2130	1990	1820	1605	1401	1223	0.3	2.8	7.7
	156	SPRN	2067	1871	1692	1529	1481	1213	992	810	0.6	4.4	11.9
7 X 57 MAUSER	103	PSP	3330	2865	2450	2060	2550	1890	1380	977	0.1	1.7	5.2
	162	TIG	2785	2480	2250	2060	2780	2200	1820	1520	0.3	2.4	6.7
7 X 57R	103	PSP	3260	2810	2390	2000	2430	1820	1320	920	0.1	1.8	5.3
7 X 57 MAUSER	110	PSP	3068	2792	2428	2277	2300	1904	1561	1267	0.0	1.6	5.0
	139	SP	2550	2240	1960	1720	2000	1540	1190	910	0.3	2.9	8.6
7 X 57R	150	PSPBT	2756	2539	2331	2133	2530	2148	1810	1516	0.1	2.2	6.2
7 X 57 MAUSER	162	TIG	2710	2420	2210	2020	2640	2120	1750	1460	0.3	2.4	6.9
7 X 57R	175	SPRN	2490	2170	1900	1680	2410	1830	1403	1097	0.4	3.3	9.0
7 X 61 S&H (26 IN.)	160	PSPBT	3100	2927	2757	2595	3415	3045	2701	2393	0.0	1.5	4.3
w/NOSLER	160	SPN	3150	2927	2757	2595	3415	3045	2701	2393	0.0	1.5	4.3
7 X 64 BRENNEKE	103	PSP	3572	3110	2685	2283	2930	2230	1670	1190	0.1	1.4	4.4
	139	SP	3000	2570	2260	1980	2780	2040	1570	1200	0.2	2.2	6.4
	162	TIG	2960	2603	2375	2200	3150	2440	2030	1740	0.2	2.0	6.0
	177	TIG	2880	2665	2490	2325	3270	2820	2440	2130	0.2	2.0	5.6
7 X 65R	103	PSP	3480	3010	2590	2200	2770	2100	1540	1120	0.1	1.5	4.7
7 X 65R BRENNEKE	139	SP	3000	2570	2260	1980	2780	2040	1570	1200	0.2	2.2	6.4
	162	TIG	2887	2540	2320	2140	3000	2320	1930	1650	0.2	2.2	6.3
	177	TIG	2820	2600	2420	2255	3120	2660	2300	2000	0.2	2.1	5.9
7mm VOM HOFE SE	169	ToSto	3300	3045	2825	2620	4090	3480	3010	2600	0.1	1.4	3.9
7 X 75R VOM HOFE SE	169	ToSto	3070	2840	2630	2430	3550	3050	2620	2240	0.1	1.6	4.5
7.5 X 55 SWISS	180	SPBT	2650	2441	2248	2056	3060	2380	2020	1690	1.0	2.6	13.3
7.62 RUSSIAN	180	PSPBT	2624	2415	2222	2030	2749	2326	1970	1644	0.2	2.5	6.6
7.65 ARGENTINE	150	PSP	2920	2630	2355	2105	2841	2304	1848	1476	0.1	2.0	5.8
7.7 JAPANESE	130	PSP	2950	2635	2340	2065	2513	2004	1581	1231	0.1	2.0	5.9
	180	SPBT	2493	2292	2101	1922	2484	2100	1765	1477	0.3	2.8	7.7
8 X 56 MANN. SCH.	200	SP	2181	1863	1640	1509	2119	1547	1200	1005	0.6	4.3	12.4
8 X 57 JS	123	SP	2968	2339	1805	1318	2415	1497	897	477	0.2	2.7	8.8
	198	TIG	2732	2415	2181	1985	3276	2560	2083	1736	0.3	2.5	7.1
8 X 57 JR	196	SP	2391	1991	1742	1565	2488	1736	3116	1056	0.5	3.9	11.2
	196	SPRN	2362	2045	1761	1513	2428	1820	1530	996	0.4	3.7	10.6
8 X 57 JRS	123	SP	2970	2340	1805	1318	2415	1497	897	477	0.2	2.7	8.8
8 X 57 JS	123	PSP	2887	2515	2170	1857	2277	1728	1286	942	0.1	2.3	6.8
	159	SPRN	2723	2362	2030	1734	2618	1970	1455	1062	0.2	2.6	7.9
	196	SP	2480	2140	1870	1640	2680	2000	1510	1165	0.4	3.3	9.4
	196	SPRN	2526	2195	1894	1627	2778	2097	1562	1152	0.3	3.1	9.1
8 X 57 JRS	198	TIG	2600	2320	2105	1930	2970	2350	1950	1620	0.3	2.7	7.6
8 X 60	196	SP	2585	2162	1890	1690	2905	2030	1560	1245	0.4	3.2	9.2
8 X 60 F	198	TIG	2660	2360	2140	1950	3110	2440	2000	1660	0.3	2.6	7.4
8 X 60 S	196	SP	2585	2162	1890	1690	2905	2030	1560	1245	0.4	3.2	9.2
	198	TIG	2780	2450	2205	2010	3390	2625	2130	1770	0.3	2.4	6.9
8 X 64 BRENNEKE	198	TIG	2830	2480	2230	2030	3520	2700	2180	1790	0.3	2.4	6.8
8 X 64 S BRENNEKE	198	TIG	2830	2480	2230	2030	3520	2700	2180	1790	0.3	2.4	6.8
8 X 65 RS BRENNEKE	198	TIG	2830	2480	2230	2030	3520	2700	2180	1790	0.3	2.4	6.8
8.15 X 46 R	151	FP	1920	1610	1370	1160	1240	870	630	500	0.9	6.0	17.4
8 X 68 S MAGNUM	187	HP	3280	2940	2660	2430	4460	3590	2930	2430	0.3	1.5	6.7
9 X 57 MAUSER	245	SP	2310	1940	1710	1550	2930	2050	1605	1320	0.5	4.0	11.5
9.3 X 62 MAUSER	285	SP	2360	2010	1775	1605	3540	2570	2006	1620	0.4	3.7	10.6
	293	TUG	2515	2310	2150	2020	4110	3480	3010	2634	0.3	2.8	7.5
9.3 X 64 BRENNEKE	293	TUG	2640	2450	2290	2145	4550	3900	3410	3000	0.3	2.4	6.0
9.3 X 72 R	193	FP	1925	1600	1400	1245	1590	1090	835	666	0.5	5.7	16.6
9.3 X 74 R	293	TUG	2360	2160	1998	1870	3580	3000	2560	2250	0.3	3.1	8.7
11.2 X 60R (43 MAUSER)	385	L	1360	1150	1030	940	1580	1130	910	750	3.2	15.0	36.5
10.75 X 68 MAUSER	347	SP	2190	1950	1750	1580	3690	2920	2370	1950	0.5	4.1	11.3
10.75 X 73 (404 JEFFERY)	401	SP	2330	2010	1800	1650	4840	3585	2885	2430	0.4	3.7	10.6

RIMFIRE CARTRIDGES

CARTRIDGE	BULLET WT. (GR)	BULLET TYPE	VELOCITY (FPS) MUZZLE	VELOCITY (FPS) 100 YD	ENERGY (FT-LB) MUZZLE	ENERGY (FT-LB) 100 YD	MID-RANGE TRAJECTORY (IMPACT AT 100 YD)	HANDGUN MUZZLE VEL.	HANDGUN MUZZLE ENERGY
				SPECIALTY LOADS					
.22 SHORT BLANK	–	–	–	–	–	–	–	–	–
.22 SHORT GALLERY	15	F	1690	929	95	29	3.3″	(NOT FOR	
.22 SHORT GALLERY	27	F	1120	910	75	50	4.4″	HANDGUN	
.22 SHORT GALLERY	29	S	1045	872	70	49	4.8″	USE)	
.22 SHORT CB	29	S	715	600	33	23	–	–	–
.22 SHORT CB	29	S	727*	610	34	24	–	706	32
.22 LONG CB	29	S	727*	610	34	24	–	706	32
.22 L.R. SHOT	25	#12	1025	–	58	–		–	–
.22 L.R. SHOT	30	#12	1047*	–	73	–	–	950	–
.22 WIN. MAG. SHOT	50	#11	1126*	–	140	–	–	1000	–
				"STANDARD" VELOCITY					
.22 SHORT TARGET	29	S	830*	695	44	31	8.2″	786	40
.22 SHORT	29	S	1045	872	70	49	4.8″	865	48
.22 L.R. MATCH	40	S	1120	950	111	80	4.2″	–	–
.22 L.R. MATCH	40	S	1138*	975	115	84	4.2″	1027	94
.22 LONG RIFLE	40	S	1150	975	117	84	4.0″	950	80
.22 WIN. AUTO.	45	S	1035	917	107	84	4.6″	–	–
				HIGH VELOCITY					
.22 SHORT	27	H	1120	904	75	49	4.4″	1033	64
.22 SHORT	27	H	1164*	920	81	51	4.6″	1077	70
.22 SHORT	29	S	1095	903	77	52	4.5″	1010	66
.22 SHORT	29	H	1120	905	81	53	4.4″	1033	69
.22 SHORT	29	S	1132*	920	82	54	4.6″	1065	73
.22 LONG	29	S	1180*	946	90	58	4.3″	1031	68
.22 LONG	29	S	1240	962	99	60	3.9″	1095	77
.22 LONG RIFLE	36	H	1280	1010	131	82	3.5″	1080	93
.22 LONG RIFLE	37	H	1280	1013	135	84	3.5″	1080	96
.22 LONG RIFLE	37	H	1370*	1040	154	89	3.5″	1190	116
.22 LONG RIFLE	38	H	1280	1020	138	88	3.5″	1080	98
.22 LONG RIFLE	40	S,D	1255	1017	140	92	3.6″	1060	100
.22 LONG RIFLE	40	S	1341*	1045	160	97	3.5″	1150	117
.22 W.R.F.(REM SPL)	45	S	1320	1055	174	111	3.3″	–	–
				HYPER VELOCITY					
.22 LR XPEDITER	29	C	1680	1079	182	75	2.5″	–	–
.22 LR STINGER	32	H	1687*	1158	202	95	1.9″	1430	145
.22 LR YELLOW JACKET	33	H	1500	1075	165	85	2.8″	–	–
				MAGNUM					
5mm REM. MAGNUM	38	H	2105	1609	374	218	1.3″	–	–
.22 WIN. MAG. RF	40	H,M	1910	1326	324	156	1.7″	1480**	195
.22 WIN. MAG. RF	40	H,S	2025*	1407	364	176	1.7″	1339	159

Bullet Type: F=Frangible, S=Solid, H=Hollow point, D=Dynapoint, C=Cavity, M=Full metal case.
*CCI data (18½″ barrel except WMRF 20″ barrel)—Other rifle MVs (Fed, W-W, R-P) 24″ barrel.
**6½″ handgun barrel—other handgun MVs 6″ barrel.
Some figures derived by BALLISTICOMP projections (where practicable.)
Catalog MVs accepted without change.

CENTERFIRE HANDGUN CARTRIDGES

CENTERFIRE CARTRIDGE	BULLET GRS.	BULLET STYLE	MUZZLE VELOCITY	MUZZLE ENERGY	BARREL INCHES
22 JET	40	SP	2100	390	8⅜
221 FIREBALL	50	SP	2650	780	10½
25 (6.35mm) AUTO	50	MC	810	73	2
256 WIN. MAG.	60	HP	2350	735	8½
30 (7.65mm) LUGER	93	MC	1220	307	4½
32 COLT NEW POLICE	100	L	680	100	4
32 (7.65mm) AUTO	71	MC	960	145	4
32 (7.65mm) AUTO	77	MC	900	162	4
32 S&W	88	L	680	90	3
32 S&W LONG	98	L	705	115	4
32-20 WIN.	100	L	1030	271	6
32-20 WIN.	100	SP	1030	271	6
357 MAGNUM POLICE	110	JHP/SP	1690	697	6
357 MAGNUM POLICE	137	JSP	1630	808	6
357 MAG.	158	SP	1550	845	8⅜
357 MAG.	158	MP	1410	695	8⅜
357 MAG.	158	L	1410	696	8⅜
357 MAG.	158	JSP	1450	735	8⅜
9mm LUGER	90	JHP	1485	441	5
9mm LUGER	112	JSP	1325	428	5
9mm LUGER	116	MC	1165	349	4

CARTRIDGE	BULLET GRS.	BULLET STYLE	MUZZLE VELOCITY	MUZZLE ENERGY	BARREL INCHES
38 SPECIAL	200	L	730	236	6
38 SPECIAL	158	MP	855	256	6
38 SPECIAL WC	148	L	770	195	6
38 SPECIAL MATCH	148	HBWC	775	196	6
38 SPECIAL MATCH, IL	148	L	770	195	6
38 SPECIAL MATCH, IL	158	L	855	256	6
38 SPECIAL HI-SPEED	158	L	1090	425	6
38 SPECIAL	158	SWC,L	1050	—	6
38 SPECIAL	158	RN	900	320	6
38 SHORT COLT	125	L	730	150	6
38 LONG COLT	150	L	730	175	6
38 SUPER	107	JHP	1430	486	5
38 SUPER	112	JSP	1435	504	5
38 SUPER AUTO	130	MC	1280	475	5
38 AUTO	130	MC	1040	312	4½
380 ACP	88	JHP	1065	227	4
380 AUTO	95	MC	955	192	3¾
38-40 WIN.	180	SP	975	380	5
41 MAGNUM	170	JHP/JSP	1775	1188	8⅜
41 REM. MAG.	210	L	1050	515	8¾
41 REM. MAG.	210	SP	1500	1050	8¾
44 S&W SPECIAL	246	L	755	311	6½

CARTRIDGE	BULLET GRS.	STYLE	MUZZLE VELOCITY	MUZZLE ENERGY	BARREL INCHES
44 MAG.	180	JHP/SP	2005	1607	6
44 REM. MAG.	240	SP	1470	1150	6½
9mm LUGER	115	JHP	1140	330	4
9mm LUGER	124	MC	1120	345	4
38 S&W	146	L	685	150	4
38 S&W	146	L	730	172	4
38 SPECIAL POLICE	110	JHP/SP	1282	399	6
38 SPECIAL POLICE	125	JHP	1370	520	6
38 SPECIAL, IL	150	L	1060	375	6
38 SPECIAL INTERNATIONAL	158	SWC KK	1110	439	6
38 SPECIAL	158	JHP	1150	465	6
38 SPECIAL	158	L	855	256	6
38 SPECIAL INT.	158	L	1090	414	6
44 REM. MAG.	240	L	1470	1150	6½
44-40 WIN.	200	SP	975	420	7½

CARTRIDGE	BULLET GRS.	STYLE	MUZZLE VELOCITY	MUZZLE ENERGY	BARREL INCHES
45 COLT	250	L	860	410	5½
45 COLT	255	L	860	410	5½
45 AUTO	190	JHP	1060	473	5
45 AUTO	230	MC	850	369	5
45 ACP	230	JHP	850	370	5
45 AUTO WC	185	MC	775	245	5
45 AUTO MC	230	MC	850	369	5
45 AUTO MATCH	185	MC	775	247	5
45 AUTO MATCH	210	L	710	235	5
45 AUTO RIM	230	L	810	335	5½

IL—INSIDE LUB. JSP—JACKETED SOFT POINT WC—WAD CUTTER
RN—ROUND NOSE HP—HOLLOW POINT LUB—LUBRICATED
MC—METAL CASE SP—SOFT POINT MP—METAL POINT
JHP—JACKETED HOLLOW POINT L—LEAD

FEDERAL SHOTSHELL LOADS

HI-POWER LOADS

Gauge	Federal Load Number	Shell Length Inches	Powder Drams Equiv.	Ounces of Shot	Shot Sizes
12	F127	2¾	3¾	1¼	BB, 2, 4, 5, 6, 7½, 9
16	F164	2¾	3¼	1⅛	4, 5, 6, 7½, 9
20	F203	2¾	2¾	1	4, 5, 6, 7½, 9
28	F283	2¾	2¼	⅞	6, 7½, 8
410	F412	2½	Max.	½	6, 7½, 9
410	F413	3	Max.	11/16	4, 5, 6, 7½, 9

MAGNUM LOADS

Gauge	Federal Load Number	Shell Length Inches	Powder Drams Equiv.	Ounces of Shot	Shot Sizes
12	F130	2¾	3¾	1½	2, 4, 5, 6
12	F131	3	4	1⅞	BB, 2, 4
12	F129	3	4	1⅝	2, 4, 6
16	F165	2¾	3¼	1¼	2, 4, 6
20	F205	2¾	2¾	1⅛	4, 6, 7½
20	F207	3	3	1¼	4, 6, 7½

12, 16 and 20 Gauge maximum loads have Triple-Plus wad column. 28 Gauge load has a plastic shot cup and conventional "Altite" filler wads. The F412 and F413 have all-plastic wad column. Magnums have plastic shot cup and conventional "Altite" filler wads.

12, 16, 20, 28 And 410 Gauge Target Loads

PAPER TUBE CHAMPION

Gauge	Federal Load Number	Shell Length Inches	Powder Drams Equiv.	Ounces of Shot	Shot Sizes
12	C117	2¾	2¾	1⅛	7½, 8, 8½, 9
12	C118	2¾	3	1⅛	7½, 8, 8½, 9

PLASTIC TUBE CHAMPION

Gauge	Federal Load Number	Shell Length Inches	Powder Drams Equiv.	Ounces of Shot	Shot Sizes
12	F117	2¾	2¾	1⅛	7½, 8, 8½, 9
12	F118	2¾	3	1⅛	7½, 8, 8½, 9
12*	F125	2¾	3¼	1⅛	7½, 8, 9
16	F167	2¾	2¾	1⅛	7½, 8, 9
20	S206	2¾	2½	⅞	8, 9
20	F206	2¾	2½	⅞	8, 9
28	F280	2¾	Max.	¾	9
410	F412	2½	Max.	½	9

16 and 20 gauge loads have the Federal "Triple-Plus" plastic wad Column. 28 gauge load has a plastic shot cup and conventional "Altite" filler wads. The F412 has an all-plastic wad column. S206 has paper case; all others have plastic case.

*For International competition only.

FIELD LOADS

Gauge	Federal Load Number	Shell Length Inches	Powder Drams Equiv.	Ounces of Shot	Shot Sizes
12	F124	2¾	3¼	1¼	7½, 8, 9
12	FN124	2¾	3¼	1¼	7½, 8
12	F123	2¾	3¼	1⅛	4, 5, 6, 7½, 8, 9
16	F162	2¾	2¾	1⅛	4, 5, 6, 7½, 8, 9
20	F202	2¾	2½	1	4, 5, 6, 7½, 8, 9

LIGHT FIELD LOADS

Gauge	Federal Load Number	Shell Length Inches	Powder Drams Equiv.	Ounces of Shot	Shot Sizes
12	F120	2¾	3¼	1	6, 8
16	F161	2¾	2½	1	6, 8
20	F201	2¾	2½	⅞	6, 8

"F" prefix load number indicates load in plastic case. "FN" prefix load number indicates nickeled shot.

PREMIUM MAGNUM

Gauge	Federal Load Number	Shell Length Inches	Powder Drams Equiv.	Ounces of Shot	Shot Sizes
10	P108	3½	4¼	2	2,4
12	P158	3	4	1⅞	BB, 2,4
12	P156	2¾	4	1½	BB, 2,4
20	P258	3	3	1¼	2,4

PREMIUM HI-POWER

Gauge	Federal Load Number	Shell Length Inches	Powder Drams Equiv.	Ounces of Shot	Shot Sizes
12	P154	2¾	3¾	1¼	2,4,6,7½
20	P254	2¾	2¾	1	4,6

PREMIUM FIELD LOADS

Gauge	Federal Load Number	Shell Length Inches	Powder Drams Equiv.	Ounces of Shot	Shot Sizes
12	P152	2¾	3¼	1⅛	7½, 8
20	P252	2¾	2½	1	7½, 8

Buckshot & Rifled Slugs

Gauge		Federal Load Number	Shell Length Inches	Powder Drams Equiv.	Shot Sizes
Buck	12	F131	3	Sup. Mag.	00 Buck–15 Pellets
Shot	12	F131	3	Sup. Mag.	4 Buck–41 Pellets
5	12	F130	2¾	Magnum	00 Buck–12 Pellets
Per	12	F130	2¾	Magnum	1 Buck–20 Pellets
Box	12	F130	2¾	Magnum	4 Buck–34 Pellets
	12	F127	2¾	Max.	00 Buck– 9 Pellets
	12	F127	2¾	Max.	0 Buck–12 Pellets
	12	F127	2¾	Max.	1 Buck–16 Pellets
	12	F127	2¾	Max.	4 Buck–27 Pellets
	16	F164	2¾	Max.	1 Buck–12 Pellets
	20	F203	2¾	Max.	3 Buck–20 Pellets
Rifled	12	F127	2¾	Max.	⅞ oz. Rifled Slug
Slugs	16	F164	2¾	Max.	4/5 oz. Rifled Slug
5	20	F203	2¾	Max.	⅝ oz. Rifled Slug
Per	410	F412	2½	Max.	⅓ oz. Rifled Slug
Box					

Buckshot and Slugs do not have shot cup. All have plastic cases.

REMINGTON–PETERS SHOT GUN GAME LOADS & RECOMMENDATIONS

TYPE OF LOAD	GAUGE	LENGTH SHELL INCHES	POWDER EQUIV. DRAMS	OUNCES OF SHOT	SHOT SIZES
LONG RANGE LOADS "EXPRESS" "HIGH VELOCITY"	* 10	2⅞	8	—	Yacht Gun Blank **
	* 10	2⅞	4¾	1⅝	4 **
	12	2¾	3¾	1¼	BB, 2, 4, 5, 6, 7½, 9 **#5 shot**
	16	2¾	3¼	1⅛	4, 5, 6, 7½, 9 **Remington brand only.**
	20	2¾	2¾	1	4, 5, 6, 7½, 9
	28	2¾	2¼	¾	6, 7½, 9†
	410	2½	Max.	½	4, 6, 7½, 9†
	410	3	Max.	11/16	4, 5, 6, 7½, 9†
MAGNUM LOADS "EXPRESS" "HIGH VELOCITY"	* 10	3½	Max.	2	2, (Mag.) **
	12	2¾	Max.	1½	2, 4, 5, 6 (Mag.)
	12	3	Max.	1⅞	2, 4, 6 (Mag.) **#5 shot**
	12	3	4	1⅝	BB, 2, 4 (Mag.) **Remington brand only.**
	16	2¾	Max.	1¼	2, 4, 6 (Mag.)
	20	2¾	Max.	1⅛	4, 6, 7½ (Mag.)
	20	3	Max.	1¼	4, 6, 7½ (Mag.)
"POWER-PAKT" ● BUCKSHOT LOADS	12	2¾	3¾	—	00 Buck- 9 Pellets
	12	2¾	3¾	—	0 Buck-12 Pellets
	12	2¾	3¾	—	1 Buck-16 Pellets
	12	2¾	3¾	—	4 Buck-27 Pellets
	16	2¾	3	—	1 Buck-12 Pellets
	20	2¾	2¾	—	3 Buck-20 Pellets
"POWER-PAKT" ● MAGNUM BUCKSHOT LOADS	12	2¾	4	—	00 Buck-12 Pellets
	12	3	4½	—	00 Buck-15 Pellets
	12	2¾	4	—	1 Buck-20 Pellets
	12	3	4½	—	4 Buck-41 Pellets
					Velocity (Muz. / 50 Yds.) — Energy—F.P. (Muz. / 50 Yds.)
RIFLED SLUG LOADS	12	2¾	3¾	⅞	Rifled Slug 1600 1175 2175 1175
	16	2¾	3	4/5	Rifled Slug 1600 1175 1990 1075
	20	2¾	2¾	⅝	Rifled Slug 1600 1175 1555 840
	410	2½	Max.	1/5	Rifled Slug 1830 1335 650 345
FIELD LOADS "SHUR SHOT" "VICTOR"	12	2¾	3¼	1	4, 5, 6, 8
	12	2¾	3¼	1⅛	4, 5, 6, 7½, 8, 9
	12	2¾	3¼	1¼	7½, 8
		May also be used where heavier trap loads are permissible			
	16	2¾	2½	1	6, 8
	16	2¾	2¾	1⅛	4, 5, 6, 7½, 8, 9
	20	2¾	2½	⅞	6, 8
	20	2¾	2½	1	4, 5, 6, 7½, 8, 9
SCATTER OADS	***12	2¾	3	1⅛	8

†28 Ga. and 410 Ga. 9 shot marked "Skeet Loads". *Standard wad column **Remington Brand ***Special wad column
* **"POWER-PAKT" BUCKSHOT LOADS.** Controlled distribution of a granulated polyethylene filler material helps to cushion the shot pellets . . . reducing shot distortion, and improving shot patterns.

GAME	SHELL	SHOT SIZES	SUGGESTED CHOKES	WHAT EXPERIENCED HUNTERS SAY . . .
DUCKS	"Express" "High Velocity"	**4, 5, 6**	**Full**—For Pass Shooting **Modified**—Over Decoys	Use No. 4 shot for long range and pass shooting. For normal range—No. 5 or No. 6 shot while some hunters use No. 7½ shot for closer range shooting over decoys.
GEESE	"Express" "High Velocity"	**BB, 2, 4**	Full	Goose hunters need wallop to fold up their birds so they use the big loads with large shot. Many hunters prefer No. 4 shot for a denser pattern at shorter ranges over decoys.
PHEASANTS	"Express" "High Velocity"	**5, 6**	**Improved Cylinder**—Close Cover **Modified or Full**—For Long Cornfield Shots	For cornfield shooting where long shots are usual—better use No. 5. On a normal rise over dogs and for all around use, No. 6 is the favorite. Bigger shot may be dangerous when hunting in large groups.
GROUSE OR PARTRIDGE	"Express" "High Velocity" or "Shur Shot" "Victor"	**5, 6 7½, 8**	**Improved Cylinder or Modified**—For Brush Work **Full**—For Open Ranges	On the smaller birds such as ruffed grouse or Hungarian Partridge use the smaller shot. The big western grouse (sage, sooty, and blue) call for heavier loads and larger shot.
QUAIL	"Shur Shot" "Victor"	**7½, 8, 9**	**Improved Cylinder or Modified**	For early season shooting on bob-whites when feathers are light some hunters use No. 9 shot. Later they switch to No. 7½ or No. 8. On the running and wild flushing type of quail, such as the Gambel's, larger shot is sometimes used.
DOVES & PIGEONS	"Express" "High Velocity" or "Shur Shot" "Victor"	**6, 7½, 8**	**Modified**	You can do a good job on mourning doves at normal ranges with the lighter loads and No. 7½ or No. 8 shot—but for longer ranges use the heavy loads and No. 6 or No. 7½. Use the same load on band-tailed pigeons and white wings.
WOODCOCK	"Shur Shot" "Victor"	**7½, 8, 9**	**Improved Cylinder or Modified**	Your choice of shot size here will depend on ranges at which your game is shot. For fast shooting in the alder thickets, No. 8 shot is a good choice.
RABBITS	"Express" "High Velocity" or "Shur Shot" "Victor"	**4, 5, 6**	**Improved Cylinder or Modified**—For Brush **Full**—For Long Open Shots	For cottontail rabbits at normal range, the lighter loads ar suitable, but for larger game such as jack rabbits and snow shoe rabbits use heavy loads.
SQUIRRELS	"Express" "High Velocity"	**5, 6**	**Modified**	Most hunters use 5's or 6's and prefer the heavy loads particularly in the tall timber.
RAIL	"Shur Shot" "Victor"	**7½, 8, 9**	**Improved Cylinder**	For the little sora rail No. 8 or No. 9 does the job while many hunters use No. 7½ on the marsh hen or clapper rail.
TURKEY	"Express" "High Velocity"	**BB, 2, 4 5, 6, 7½**	**Full**	Choice of shot size depends on your range. If you're a good caller No. 6 or No. 7½ shot makes a clean kill. BB's, No. 2's, 4's and 5's are best for long shots.
FOX	"Express" "High Velocity"	**BB, 2, 4**	**Full**	It's a toss-up between BB's and No. 2 shot. But remember—the smaller the shot, the denser the pattern.
DEER, BLACK BEAR & WOLF	"Express" "High Velocity"	**Rifled Slug and Buck Shot**	For rifled slugs and buckshot any choke may be used. Best results are obtained with Improved Cylinder	For deer and black bear 12 and 16 gauge slugs are the best. 0 and 00 Buckshot are the most popular sizes for deer hunters who use buckshot.

WINCHESTER-WESTERN SHOTGUN LOADS

SUPER SPEED GAME LOADS

Gauge	Length Of Shell Inches	Powder Dram Equiv.	Oz. Shot	Standard Shot Sizes
10†	2⅞	4¾	1⅝	4
12	2¾	3¾	1¼	BB, 2, 4, 5, 6, 7½, 9
12	2¾ Mag.	4	1½	2, 4, 5, 6
12	3 Mag.	4	1⅜	2, 4, 6
12	3 Mag.	4¼	1⅝	2, 4, 6
12	3 Mag.	Max.	1⅞	BB, 2, 4
16	2¾	3¼	1⅛	4, 5, 6, 7½, 9
16	2¾ Mag.	3½	1¼	2, 4, 6
20	2¾	2¾	1	4, 5, 6, 7½, 9
20	2¾ Mag.	3	1⅛	4, 6, 7½
20	3 Mag.	Max.	1¼	4, 6, 7½
28	2¾	2¼	¾	6, 7½, 9†
410	2½	Max.	½	4, 6, 7½, 9†
410	3	Max.	¾	4, 5, 6, 7½, 9

SUPER-X GAME LOADS

10‡	2⅞	Max.	1⅝	4
10‡	3½ Mag.	Max.	2	2
12	2¾	Max.	1¼	BB, 2, 4, 5, 6, 7½, 9
12	2¾ Mag.	Max.	1½	2, 4, 5, 6
12	3 Mag.	Max.	1⅜	2, 4, 6
12	3 Mag.	Max.	1⅝	2, 4, 6
12	3 Mag.	Max.	1⅞	BB, 2, 4
16	2¾	3¼	1⅛	4, 5, 6, 7½, 9
16	2¾ Mag.	Max.	1¼	2, 4, 6
20	2¾	Max.	1	4, 5, 6, 7½, 9
20	2¾ Mag.	Max.	1⅛	4, 6, 7½
20	3 Mag.	Max.	1¼	4, 6, 7½
28	2¾	Max.	¾	6, 7½, 9†
410	2½	Max.	½	4, 6, 7½, 9†
410	3	Max.	¾	4, 5, 6, 7½, 9

†NO. 9 SHELLS MARKED "SKEET LOADS."

SUPER-X WITH LUBALOY (COPPERIZED SHOT)

12	2¾	Max.	1¼	2, 4, 5, 6, 7½
12	2¾ Mag.	Max.	1½	2, 4
12	3 Mag.	Max.	1⅜	2, 4, 6
12	3 Mag.	Max.	1⅝	2, 4, 6
20	2¾	Max.	1	4, 5, 6, 7½
20	3 Mag.	Max.	1⅛	6
20	3 Mag.	Max.	1³⁄₁₆	4

SUPER-X AND WINCHESTER XX MAGNUM

12	2¾ Mag.	4	1½	2, 4

SUPER-X AND SUPER-SPEED SUPER BUCKSHOT LOADS

Gauge	Length Of Shell Inches	Pellets Total	Buckshot Size
12	2¾	9	00 Buck
12	2¾ Mag.	12	00 Buck
12‡	3 Mag.	15	00 Buck
12	2¾	12	0 Buck
12	2¾	16	1 Buck
12	2¾ Mag.	20	1 Buck
12	2¾	27	4 Buck
12‡	3 Mag.	41	4 Buck
16	2¾	12	1 Buck
20	2¾	20	3 Buck

‡Plastic, not compression-formed.

SUPER-X AND SUPER-SPEED RIFLED SLUG LOADS

Gauge	Length Of Shell Inches	Powder Dram Equiv.	Slug Wt. Oz.	
12	2¾	Max.	1	Rifled Slug
16	2¾	Max.	⅞	Rifled Slug
20	2¾	Max.	⅜	Rifled Slug
410	2½	Max.	⅕	Rifled Slug

UPLAND PLASTIC MARK 5 FIELD LOADS

■ Winchester Brand Only.

Gauge	Length Of Shell Inches	Powder Dram Equiv.	Oz. Shot	Standard Shot Sizes
■ 10‡	2⅞	8	Black Powder	Blank
12	2¾	3	1	4, 5, 6, 8
12	2¾	3¼	1⅛	4, 5, 6, 7½, 8, 9
12	2¾	3¼	1¼	7½, 8
12‡	2¾	6	Black Powder	Blank
16	2¾	2½	1	6, 8
16	2¾	2¾	1⅛	4, 5, 6, 7½, 8, 9
20	2¾	2¼	⅞	6, 8
20	2¾	2½	1	4, 5, 6, 7½, 8, 9

UPLAND BRUSH LOAD

12‡	2¾	3	1⅛	8

WESTERN FIELD TRIAL POPPER LOAD

Gauge	Length Of Shell Inches	Powder Dram Equiv.	Oz. Shot	Standard Shot Sizes
12	2¾	—	—	Blank

WINCHESTER AND WESTERN DOUBLE A TRAP LOADS

Gauge	Length Of Shell Inches	Powder Dram Equiv.	Oz. Shot	Standard Shot Sizes
12	2¾	2¾	1⅛	7½, 8
12	2¾	3	1⅛	7½, 8

WINCHESTER DOUBLE A INTERNATIONAL TRAP LOADS

N – Nickel Plated Shot. B – Black Shot.

12	2¾	3¼	1¼	7½, 8N
12	2¾	3¼	1¼	7½, 8B

WINCHESTER AND WESTERN DOUBLE A SKEET LOADS

Gauge	Length Of Shell Inches	Powder Dram Equiv.	Oz. Shot	Standard Shot Sizes
12	2¾	2¾	1⅛	9
12	2¾	3	1⅛	9
20	2¾	2¼	⅞	9
28	2¾	Max.	¾	9
410	2¼	Max.	½	9

BLACK POWDER LOADING DATA

STANDARD MUZZLE LOADING CHARGES

TYPE AND CALIBER	SIZE BALL USED	POWDER CHARGE IN GRAINS
.69 U.S. Muskets Models 1808-1812-1821-1831-1840-1842	680	80 FFG
.69 U.S. Muskets Models 1821-1840-1842, rerifled from smooth bore	.680	75 FFG
.69 Whitneyville Plymouth Navy Rifle	.680	70 FFG
.58 Zouave	.570 to .575	60 FFG
.58 U.S. Civil War Rifles Models 1855-61-63	.570 to .575	60 FFG
.67 Tower Flint Pistol	.650	35 FFFG
.58 U.S. Springfield Pistols Model 1855	.570 to .575	40 FFG
.577 English Enfield Rifle	.570 to .575	60 FFG
.58 Fayetteville, Richmond, Cook Bros., Columbus and other Confederate Civil War Muskets	.570 to .575	60 FFG
.54 1804-1814-1817 U.S. Rifles	.535	75 FFFG
.54 U.S. Mississippi Rifle	.535	75 FFFG
.52 Sharp's Carbines & Rifles	.535 to .555	60 FFFG
.54 U.S. Pistol Models 1819-21-26-36-42	.535	35 FFG
.54 Halls Carbine	.535	85 FFG
.54 Halls Rifle	.535	100 FFFG
.53 Burnside	.558	60 FFFG
.50 Maynard U.S. Issue	.517	50 FFFG
.54 Gallager	.535	55 FFFG
.54 Greene Carbine	.565 to .569	68 FFFG
.54 Starr	.560	60 FFFG
.50 Smith Carbine	.515	50 FFFG
.56 Colt Root Rifle	.562	70 FFG
.44 Colt Dragoon	.453	40 FFFG
.44 Colt & Remington Army	.453	28 FFFG
.44 Modern made Remington and Colt Revolver	450	28 FFFG
.44 Modern made Colt Dragoon	.451	40 FFFG
.36 Modern made Remington and Colt Revolver	.376	22 FFFG
.36 Colt & Remington Navy, Whitney, Pettingill, Savage, Cooper, Bacon and most .36 Revolvers	.376	22 FFFG
.31 Colt & Remington Pocket Revolver and other .31 American Revolvers	.321	15 FFFG
.40 Dixie Flint & Percussion Rifles	.395	65 FFFG
.45 Dixie Flint & Percussion Rifles	.445	65 FFFG
.40 Dixie Flint & Percussion Pistols	.395	25 FFFG
.45 Dixie Flint & Percussion Pistols	.445	25 FFFG
.41 Dixie brass framed Derringer	.395	10 FFFG
.40 Dixie Half Stock Target Rifle	.395	55 FFFG
.45 Dixie Half Stock	.445	75 FFFG
Pepperboxes—no standard caliber	.30 to .38	15 to 20 gr.

RECOMMENDATIONS

FOR KENTUCKY RIFLES: One grain weight of powder for each caliber, i.e.: .41 caliber rifle will use about 41 grains of black powder.

FOR SINGLE SHOT PISTOLS: For most accurate shooting, use the lightest load possible and still get the ball to target without a big drop in trajectory.

FOR PERCUSSION RIFLES: Same directions as the single shot pistol.

TABLE FOR CONVERTING DRAMS TO GRAINS

1	Dram equals	27½	grains
1¼	Drams equal	34	grains
1½	Drams equal	41	grains
1¾	Drams equal	48	grains
2	Drams equal	55	grains
2¼	Drams equal	61	grains
2½	Drams equal	68	grains
2¾	Drams equal	75	grains
3	Drams equal	82	grains
3¼	Drams equal	89	grains
3½	Drams equal	96	grains
3¾	Drams equal	102	grains
4	Drams equal	109	grains

MUZZLE VELOCITIES OF ROUND BALLS IN KENTUCKY RIFLES BASED ON RATIO OF BALL WEIGHT TO POWDER WEIGHT

Lead		Feet per Sec.
1	1	2570
1½	1	2325
2	1	2135
2½	1	1975
3	1	1850
3½	1	1750
4	1	1670
4½	1	1600
5	1	1535
5½	1	1480
6	1	1425
6½	1	1375
7	1	1325
7½	1	1280
8	1	1235
8½	1	1200
9	1	1155
9½	1	1120
10	1	1035

PERCUSSION CAP SIZES

	Inside Diameter	Length Of Cap
No. 10 Alcan	.167	.178
No. 11 Alcan (Italy)	.168	.153
No. 1075 German	.170	.170
Eley No. F4-12	.170	—
No. 1055 German	.170	.220
No. 11 Remington	.170	.190
No. 11 Italian (Dixie)	.172	.206
Eley No. F4-21	.175	—
No. 11 Winchester	.175	.200
Eley No. F4-25	.177	—
No. 12 Alcan	.178	.195
No. 12 Remington	.178	.190

HANDY CONVERSION TABLES FOR THE SHOOTER

COMPARATIVE WEIGHTS OF POWDER CHARGES, BULLETS AND SPHERICAL BALLS

Powder Weights		Bullet Weights		Weights of Spherical Balls	
Drams	Grains	Ounces	Grains	Bore and Diam.	Grains
0¼	7	0⅛	55	4-bore (1.052")	1750
0½	14	0¼	109	6-bore (.919")	1166
0¾	20½	0⅜	164	8-bore (.835")	875
1	27⅛	0½	219	10-bore (.775")	700
2	55	0⅝	273	11-bore (.751")	636
3	82	0¾	328	12-bore (.729")	583
4	109	0⅞	383	14-bore (.693")	500
5	137	1	437½	16-bore (.662")	437½
6	164	1⅛	492	18-bore (.637")	388
7	191	1¼	547	20-bore (.615")	350
8	219	1⅜	602	24-bore (.579")	292
9	246	1½	656	28-bore (.550")	250
10	273	1¾	766	30-bore (.537")	233
11	301	2	875	32-bore (.526")	219
12	328	4	1750	36-bore (.506")	194
13	355	6	2625	40-bore (.488")	175
14	383	8	3500	44-bore (.473")	159
15	410	12	5250	48-bore (.459")	146
16 or 1 oz.	437½	16 or 1 lb.	7000	50-bore (.453")	140

VALUES FOR CONVERSION OF VELOCITIES IN DIFFERENT LENGTHS OF BARRELS

This table enables one to calculate the approximate velocity of his rifle making allowance for a shorter barrel than the published velocity. It is based on British velocities from 28" bbls. expressed as 1.000. If your barrel is 24" and published velocity taken from a 28" barrel multiply published velocity by .969. If published velocity was taken from a 26" or 24" bbl. use 1.000 as the value and use value corresponding to number of inches under data length.

Length of Barrel in Inches	Value
30	1.014
29	1.007
28	1.000
27	0.993
26	0.985
25	0.977
24	0.969
23	0.964
22	0.954
21	0.944
20	0.937

Comparative Tables of Weight

1 grain = 64.9 milligrams = .0649 gram
1 dram = 1770 milligrams = 1.77 gram
1 ounce = 28350 milligrams = 28.35 gram
1 pound = = 453.59 gram = .453 kg.
1 long ton = = 1016 kg. = 1.016 tonne
1 short ton = = 907 kg. = .907 tonne
1 milligram = .0154 grain
1 gram = 15.432 grain = .0353 oz.
1 kilogram = 15432 grain = 35.3 oz. = 2.2046 lb.
1 tonne = 2204.6 lb. = .9842 long ton
1 tonne = 2204.6 lb. = 1.023 short ton
1 lb. per sq. inch = .0703 kg. per sq. cm.
1 ton per sq. inch = 157.49 kg. per sq. cm.
1 foot pound = .1382 kg. meter
1 foot ton = 309.7 kg. meter = .3097 meter ton
1 atmosphere (14.7 lbs. per sq. in.) = 1.033 technical atmosphere
1 kg. per sq. cm. = 14.233 lbs. per sq. in. = .00635 tons per sq. in. = .967 atm.
1 kg. meter = 7.234 ft. lbs.
1 meter tonne = 3.229 ft. tons

Conversion Factors

To convert				
Grains	to Milligrams	multiply by	64.9	or divide by .0154
Grains	to Grams	multiply by	.0649	or divide by 15.43
Ounces	to Grams	multiply by	28.35	or divide by .0353
Pounds	to Kilograms	multiply by	.453	or divide by 2.205
Long Tons	to Kilograms	multiply by	1016	or divide by .000984
Short Tons	to Kilograms	multiply by	907	or divide by .001023
Milligrams	to Grains	multiply by	.0154	or divide by 64.9
Grams	to Grains	multiply by	15.432	or divide by .0649
Kilograms	to Pounds	multiply by	2.205	or divide by .453
Tonnes	to Long Tons	multiply by	.9842	or divide by 1.016
Tonnes	to Short Tons	multiply by	1.023	or divide by .907
Ft. per sec.	to Meters per sec.	multiply by	.3048	or divide by 3.28
Foot pounds	to Kilogram Meters	multiply by	.1382	or divide by 7.234
Lbs. per sq. in.	to Kgs. per sq. cm.	multiply by	.0703	or divide by 14.223
Meters per sec.	to Feet per sec.	multiply by	3.28	or divide by .3048
Kilogram Meters	to Foot pounds	multiply by	7.234	or divide by .1382
Kgs. per sq. cm.	to Lbs. per sq. in.	multiply by	14.223	or divide by .0703

sportsman's DIRECTORY

Over 3,000,000 GUNS & AMMO men read SPORTSMAN'S DIRECTORY every month as an important source of information on products for the shooting enthusiast. Every ad in the section gets reader attention because of uniform size. At the lowest rate available for display ads of this size, this is your most efficient way to introduce and sell your product(s) to the leading shooters' market available in the country today. For further information, contact "Sam" at GUNS & AMMO, 8490 Sunset Blvd., Los Angeles, CA 90069. Phone 213/657-5100, extension 356.

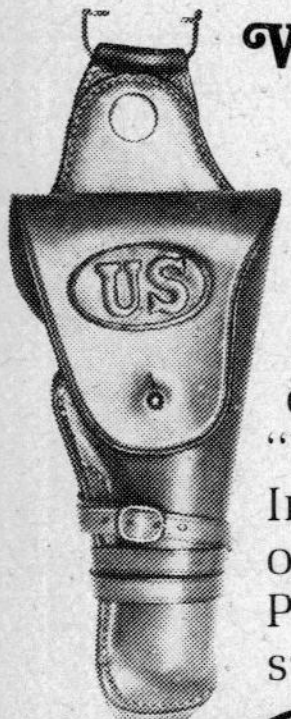

World's Finest Holsters

MODEL 1911 FOR .45 AUTO. $27.95 + $2.00 POST. AVAILABLE RIGHT OR LEFT, BLACK OR RUSSET. "U.S.", "U.S.M.C." OR "U.S.N.". ILLUSTRATED CATALOGUE OF MODERN AND REPRODUCTION HOLSTERS $1.50

El Paso Saddlery Co.

P.O. BOX 27194
EL PASO, TEXAS 79926 915/544-2233

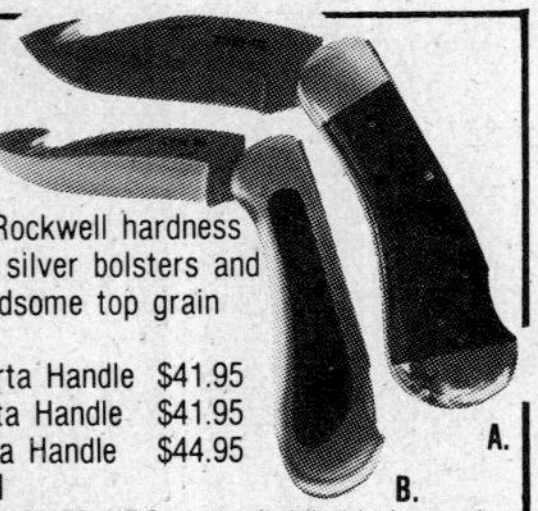

A. GUTHOOK HUNTSMAN
Lockback Folding Knife. Satin finish blade made of Vanadium Molybdinum AUS-8 stainless steel with a Rockwell hardness of C56-58. Solid nickel silver bolsters and liners. Comes with handsome top grain leather sheath.
A-102-BR Brown Micarta Handle $41.95
A-102-BL Black Micarta Handle $41.95
A-102-IV Ivory Micarta Handle $44.95

B. GUTHOOK FISHERMAN
Lockback Folding Knife. 56-58 HRC satin finish blade made of Vanadium Molybdinum AUS-8 stainless steel. Brass liners. Satin finish stainless steel handle with Ebony wood inlay. 2-7/8" locking blade; 3-7/8" closed.
A-101 $26.95
Custom Quality Without The Custom Price Makes Great Christmas Gifts!
Dealer Inquiries Invited—Mastercharge/Visa Welcome
Send $1 for full color catalog and complete information.
CAM III ENTERPRISES.
243 Millbrook Way, Vacaville, CA 95688 707/446-8549

MAKER OF CUSTOM LEATHER PRODUCTS FOR THE PRACTICAL PISTOL SHOOTER

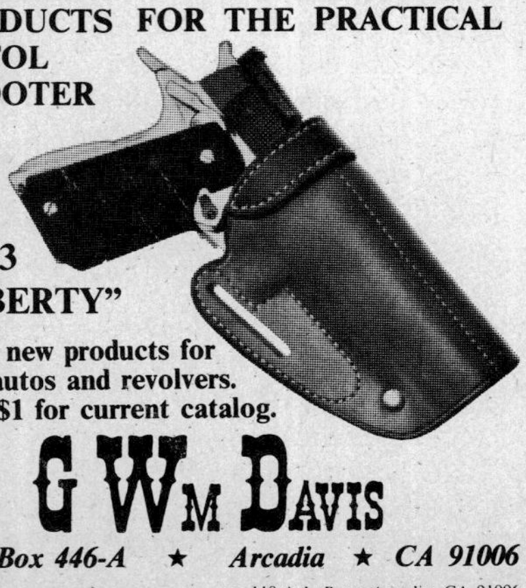

#453 "LIBERTY"

Many new products for both autos and revolvers. Send $1 for current catalog.

G WM DAVIS

P.O. Box 446-A ★ Arcadia ★ CA 91006

118-A LaPorte Arcadia, CA 91006

COMBAT GRIPS

FOR COLT .45 AUTOS AND SIMILAR FRAMES

Fingergrooves eliminate slippage —give total control —gun points naturally at target

$24.95 ppd.

SOLID WALNUT — HAND CRAFTED

BULLSHOOTERS' SUPPLY

Dept GA2 /9625 E. 3rd Street/P.O. Box 13446
Tucson, Arizona 85732/602-298-6924
Bullshooters' Supply is a division of the B.E.W.B. Corp.

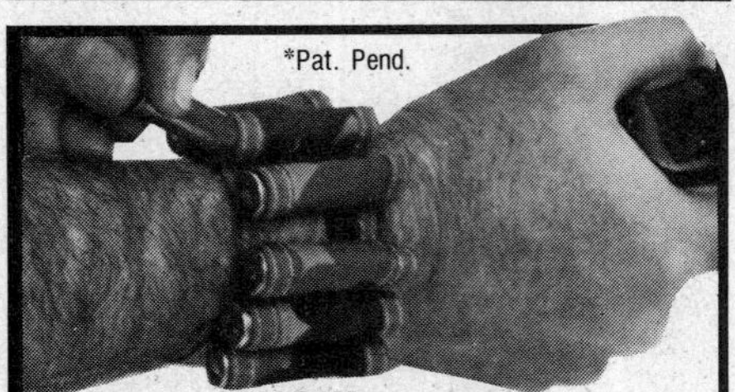

*Pat. Pend.

AMMO-NOW CARTRIDGE STRAP*

Carry 6 rds. on your wrist or rifle stock. Velcro fastening for adjustability. Constructed of durable nylon in a variety of popular calibers and a wide range of colors, including camoflage. $5.50 ea.

CAL DRAKE ENTERPRISES
P.O. BOX 9217, Marina Del Rey, CA 90291

BANKAMERICARD/VISA NO. ____________

EXPIRATION DATE ____/____/____

211 Culver Blvd., Marina Del Rey, CA 90291

KINGSBERY STEEL GUN SAFES

America's Leading Manufacturer of Steel Gun Safes

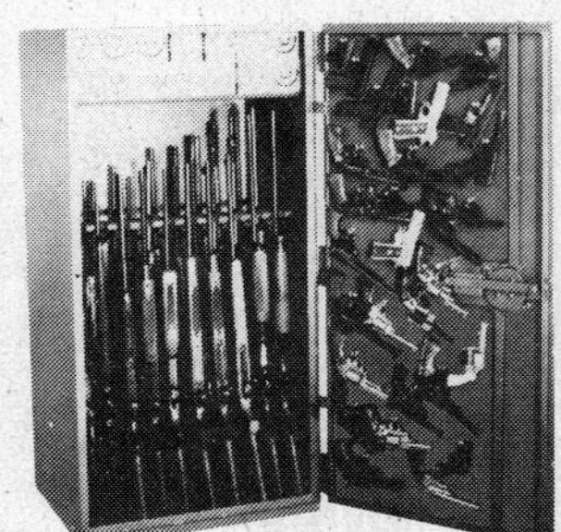

COLLECTORS SAFE
CAN CUSTOM BUILD ANY SIZE

Direct from Factory to Customer

Shown with 30 guns and pistol board for 30 pistols

Weight: 520 lbs.

Price: $790.00

Write or call for brochure: **Toll Free 1-800-531-7153**
Texas Res. Call 512-374-3473
KINGSBERY MFG. CORP. Crystal City, TX 78839

BARGAIN PRICES

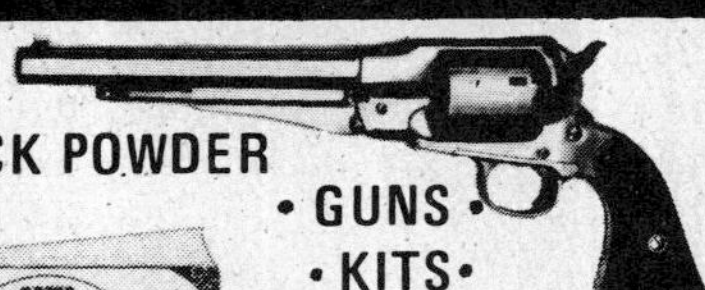

BLACK POWDER
• GUNS • KITS •
TREMENDOUS STOCK
- Revolvers • Pistols
- Rifles • Shotguns
- Hard-to-Find-Items

GUN CATALOG
SEND $1.00

EMF Co. Inc.
(213) 843-7777
Dept. GAA
2911 W. Olive Ave.
Burbank, CA 91505

Electronic
"HEAR MUFFS"®

Offers features of *BOTH* muffs or plugs. If you want something effective, This is it
You can hear your companions conversation while reducing the muzzle blasts to a tolerable level. Or just flat turn off the set. Volume Control adjustment on side.

Tested & proven Semi-Conductor (solid-state) technology and an attractive set of muffs for **74.50** plus **1.50** handling. Battery included.

Mail check or money order to:
Long Branch Enterprises, Inc.
P.O. Box 87 Norco, CA 91760
(Calif. Residents add 6%)
NO C.O.D.
Dealer Inquiries Invited
BankAmericard/Visa Expiration Date ______
No. ____________
If not satisfied, return within 30 days for full refund.

THIS IS THE FIRST CROSSBOW ORIENTED SURVIVAL BOOK IN THE WORLD!
by Howard Steele

DESTINED TO BE A CLASSIC!
DON'T WAIT TOO LONG!

HARDBOUND $15.00
(PERSONALLY AUTOGRAPHED FOR COLLECTORS)
PAPERBACK $10.00
U.L.C., Dept. G-AA
Box 477, Toledo, OR 97391

AMAZING DERRINGER ELECTRIC GUN LIGHTER

An impressive conversation piece. Looks just like the real thing. Never needs batteries, never needs flints. Comes with its own stand. Can be worn on your belt. Available in silver, black, gold, all with pearl handle. Makes a great Christmas gift! Send check or money order for $14.95 plus $2.00 postage and handling to:

SUNSHINE ENTERPRISES,
P.O. Box 9651, Ft. Lauderdale,
FL 33310

Allow 4-6 weeks delivery. One-year money back guarantee. Extra bonus offer: Order by November 30th and receive a free holster. Order After November 30, Each Holster $3.00.

Visa # ____________ Expiration Date ______
Mastercard # ____________ Expiration Date ______

Try
Muzzleloading!

Slip back a generation or so for the time of your life! Find out about the fun, the nostalgia, the challenge. Return this coupon and $1 for our illustrated booklet, **Black Powder—White Smoke... an introduction to muzzleloading.** AIMM—Assn. of Importers/Manufacturers for Muzzleloading.

Mail with $1 to: AIMM Booklet
Suite 300—1221 S.W. Yamhill St.,
Portland, OR 97205

NAME ____________
ADDRESS ____________
CITY/STATE/ZIP ____________

DIRECTORY

ANTIQUE ARMS DEALERS

Bob's Military Antiques
and War Museum
208 Santa Monica Blvd.
Santa Monica, CA 90401

Classic Collections Corp.
1639 E. Del Amo
Carson, CA 90745

Frank Costanza
3740 W. Bilby Rd.
Tucson, AZ 85706

Dixie Gun Works
Gunpowder Lane
Union City, TN 38261

Early & Modern
Firearms Co., Inc.
2911 W. Olive
Burbank, CA 91505

Federal Ordnance
1443 Potrero Ave.
So. El Monte, CA 91733

N. Flayderman & Co., Inc.
Squash Hollow
New Milford, CT 06776

Hawken Shop
3028 N. Lindburgh
St. Louis, MO 63074

Pony Express Sport Shop, Inc.
17460 Ventura Blvd.
Encino, CA 91316

Purple Sage Gun Shop
5417 Central N.W.
Albuquerque, NM 87105

Martin B. Retting, Inc.
11029 Washington Blvd.
Culver City, CA 90230

Navy Arms
689 Bergen Blvd.
Ridgefield, NJ 07657

Santa Ana Gunroom
P.O. Box 1777
Santa Ana, CA 92701

Ralph Shattuck
6661 Castle Dr.
Birmingham, MI 48010

Sotheby Park Bernet
7660 Beverly Blvd.
Los Angeles, CA 90036

ARMS IMPORTERS

Air Rifle Headquarters
Box 327,247 Court St.
Grantsville, WV 26147

American Research & Devel.
P.O. Box 27163
Salt Lake City, UT 84127

Arms International
P.O. Box 95
Malverne, NY 11565

Armsport, Inc.
3590 NW 49th St.
Miami, FL 33142

B&B Sales
12521-3 Oxnard St.
N. Hollywood, CA 91606

Beeman's Precision Airguns
47 Paul Dr.
San Rafael, CA 94903

Canadian Sauer Co.
314 Judson St.
Toronto, Ontario
Canada M8Z 5T6

Century Arms
5 Federal St.
St. Albans, VT 05478

Champlin Firearms
Woodring Airport, Box 3191
Enid, OK 73701

Continental Arms Corp.
697 Fifth Ave.
New York, NY 10022

Dixie Gun Works
Gunpowder Lane
Union City, TN 38261

Dynamit Nobel of America
105 Stonehurst Ct.
Northvale, NJ 07647

Ellwood Epps Orillia, Ltd.
Rt. 3, Highway 11
Orillia, Ontario
Canada L3V 6H3

E.M.F., Inc.
2911 W. Olive Ave.
Burbank, CA 91505

Euroarms of America
1501 Lenoir Dr., Box 3277
Winchester, VA 22601

R.C. Fessler & Co.
1643 Colorado Blvd.
Los Angeles, CA 90041

Firearms Center
308 Leisure Lane
Victoria, TX 77901

F.I.E. Corp.
P.O. Box 4866
Hialeah Lakes
Hialeah, FL 33014

Flaig's
Millvale, PA 15209

J.L. Galef & Sons, Inc.
85 Chambers St.
New York, NY 10007

Gil Hebard Guns
125-129 Public Square
Knoxville, IL 61448

Griffin & Howe, Inc.
589 Broadway
New York, NY 10012

Heckler & Koch
933 N. Kenmore St.
Arlington, VA 22201

Herter's
Waseca, MN 56093

Interarms
10 Prince St.
Alexandria, VA 22313

International Dist., Inc.
P.O. Box 557566
Miami, FL 33155

Jana International
P.O. Box 1107
Denver, CO 80201

Kassner Imports
5480 Linglestoem Rd.
Harrisburg, PA 17112

Kleinguenther, Inc.
P.O. Box 1261
Seguin, TX 78155

S.E. Laszlo
200 Tillary St.
Brooklyn, NY 11201

Lever Arms Service
771 Dunsmuir St.
Vancouver, B.C.
Canada, V6C IM9

London Guns
1528 20th St.
Santa Monica, CA 90404

Mandall Shooting Supply, Inc.
7150 E. 4th St.
Scottsdale, AZ 85252

Mitchell Arms Corp.
116 E. 16th St.
Costa Mesa, CA 92627

William Larkin Moore & Co.
31360 Via Colinas
Westlake Village, CA 91361

Navy Arms
689 Bergen Blvd.
Ridgefield, NJ 07657

Norma Precision & Precision
Sports
798 Cascadilla St.
Ithaca, NY 14850

Richland Arms Co.
321 W. Adrian St.
Blissfield, MI 49228

Sile Distributors
7 Centre Market Pl.
New York, NY 10013

Steyr Daimler Puch of America
85 Metro Way
Secaucus, NJ 07094

Stoeger Industries
55 Ruta Ct.
S. Hackensack, NJ 07606

Tradewinds, Inc.
P.O. Box 1191
Tacoma, WA 98401

Trail Guns Armory
1634 E. Main
League City, TX 77573

Western Arms Corp.
1107 Pen Rd.
Santa Fe, NM 87501

ARMS MFG. & SUPPLIERS

Air Pistol
4535 Huntington Dr. So.
Los Angeles, CA 90032

American Arms Intl'
P.O. Box 27163
Salt Lake City, UT 84127

Armalite, Inc.
118 E. 16th St.
Costa Mesa, CA 92627

Auto Ordnance Corp.
Williams Lane
West Hurley, NY 12491

Bar-Sto Precision Machine
633 So. Victory
Burbank, CA 91502

Bauer Firearms Corp.
34750 Klein Ave.
Fraser, MI 48026

B&B Sales
12521 Oxnard St.
North Hollywood, CA 91606

Beretta U.S.A.
17801 Indian Head Hwy.
Accokeek, MD 20607

Benjamin Air Rifles
1525 S. 8th St.
St. Louis, MO 63104

Bill's House of Guns
P.O. Box 424
Merrill, WI 54452

Browning Arms
Rt.1
Morgan, UT 84050

Caraville Arms
1355 Lawrence Dr.
Newbury Park, CA 91320

Century Mfg., Inc.
910 Keck Ave.
Evansville, IN 47711

Charter Arms
430 Sniffins Lane
Stratford, CT 06497

Colt Firearms
150 Huyshope Ave.
Hartford, CT 06102

Crosman Airguns
980 Turk Hill Rd.
Fairport, NY 14440

Daisy Mfg. Co.
P.O. Box 220
Rogers, AR 72756

Dan Wesson Arms
293 Main St.
Monson, MA 01057

Detonics .45 Associates
2500 Seattle Tower
Seattle, WA 98101

DuBiel Arms
1724 Baker Rd.
Sherman, TX 75090

E.M.F.
2911 W. Olive Ave.
Burbank, CA 91505

Excam, Inc.
4480 E. 11th Ave.
Hialeah, FL 33013

F.I.E. Corp.
P.O. Box 4866
Hialeah Lakes
Hialeah, FL 33014

Freedom Arms
1 Freedom Lane
Freedom, WY 83120

FTL Marketing
12521-3 Oxnard St.
N. Hollywood, CA 91606

Golden Eagle Firearms
5750 Ranchester Dr., Box 42139
Houston, TX 77036

Harrington & Richardson
Industrial Row
Gardner, MA 01440

Heckler & Koch
933 N. Kenmore St.
Arlington, VA 22201

High Standard, Inc.
31 Prestige Park Circle
E. Hartford, CT 06108

J&G Sales, Inc.
720 Elrod Rd.
Prescott, AZ 86301

Iver Johnson Arms
Wilton Ave.
Middlesex, NJ 08846

King's Gun Works
1837 W. Glenoaks Blvd.
Glendale, CA 91201

Klienguenther, Inc.
P.O. Box 1261
Seguin, TX 78155

L.E.S. Co.
2301 Davis St.
N. Chicago, IL 60064

Liberty Org., Inc.
4116 La Crescenta Ave.
La Crescenta, CA 91214

Ljutic Industries
P.O. Box 2117
Yakima, WA 98407

Marlin Firearms
100 Kenna Dr.
N. Haven, CT 06473

McGowen Rifle Barrels
Rt.3
St. Anne, IL 60964

M&N Distributors
3040 W. Lomita Blvd.
Torrance, CA 90505

M-S Safari Arms
PO Box 23370
Phoenix, AZ 85063

O.F. Mossberg & Sons
7 Grasso Ave.
No. Haven, CT 06473

Navy Arms
689 Bergen Blvd.
Ridgefield, NJ 07657

North American Mfg. Corp.
310 West 700 South
Provo, UT 84601

Numrich Arms
Williams Lane
West Hurley, NY 12491

Puccunelli Co.
11 Belle Ave.
San Anselmo, CA 94960

Remington Arms Co.
939 Barnum Ave.
Bridgeport, CT 06602

R.G. Industries Inc.
2485 NW 20th St.
Miami, FL 33135

Rock Pistol Mfg.
704 E. Commonwealth
Fullerton, CA 92631

Rogak, Inc.
2301 Davis St.
No. Chicago, IL 60064

Savage Arms
Springdale Rd.
Westfield, MA 01085

C. Sharps Arms Co.
25 Fallwood Pkwy.
Farmingdale, NY 11735

Sheridan Products
3205 Sheridan Rd.
Racine, WI 53403

Shilen Rifles Inc.
205 Metro Park Blvd.
Ennis, TX 75119

Smith & Wesson
2100 Roosevelt Ave.
Springfield, MA 01101

Sterling Arms Corp.
211 Grand St.
Lockport, NY 14094

Sturm, Ruger & Co.
Lacey Pl.
Southport, CT 06490

Taser Systems Inc.
1100 S. Azusa Ave.
Industry, CA 91748

Tennessee Valley Arms
225 DePot, Box 2022
Union City, TN 38261

Thompson/Center Arms
Farmington Rd.
Rochester, NH 03867

Universal Firearms
3740 E. 10th Ct.
Hialeah, FL 33013

Valor Corp.
5555 NW 36th Ave.
Miami, FL 33142

Weatherby's, Inc.
2781 Firestone Blvd.
South Gate, CA 90280

Wichita Arms
333 Lulu, Box 11371
Wichita, KS 67211

Wildey Firearms Co.
P.O. Box 447, 340 Industrial Rd.
Cheshire, CT 06410

Winchester-Western
275 Winchester Ave.
New Haven, CT 06511

BLACKPOWDER & MUZZLELOADING

A&K Mfg. Co.
1651 N. Nancy Rose Ave.
Tucson, AZ 85712

American Import Co.
1167 Mission St.
San Francisco, CA 94103

Bill's House of Guns
P.O. Box 424
Merrill, WI 54452

Blue & Gray Products
R.D. 6, Box 348
Willsboro, PA 16901

Bruner Shooter Supply
207 Center Park Dr.
Knoxville, TN 37922

Buffalo Robe Indian Trading Post
18555 Sherman Way
Reseda, CA 91335

E. Christopher Firearms
8007 St. Rt. 128
Miamitown, OH 45041

Classic Arms Co.
20 Wilbraham St.
Palmer, MA 01069

Connecticut Valley Arms
Saybrook Rd.
Haddam, CT 06430

Continential Arms Corp.
697 Fifth Ave.
New York, NY 10022

Dixie Gun Works
Gunpowder Lane
Union City, TN 38261

E.M.F. Co.
2911 W. Olive Ave.
Burbank, CA 91505

Euroarms of America
1501 Lenoir Dr.
Winchester, VA 22601

Excam, Inc.
4480 E. 11th Ave.
Hialeah, FL 33013

Federal Ordnance
1442 Potrero Ave.
So. El Monte, CA 91733

Flintlock
1238 So. Beach
Anaheim, CA 92804

Forster Products
82 E. Lanark Ave.
Lanark, IL 61046

J.L. Galef & Sons
85 Chambers St.
New York, NY 10007

Golden Age Arms
14 W. Winter St.
S. Delaware, OH 43015

Green River Forge, Ltd.
Box 715
Roosevelt, UT 84066

Gussert Bullet & Cartridge Co.
P.O. Box 3945
Green Bay, WI 54303

Robert W. Hart & Son
401 Montgomery St.
Nescopeck, PA 18635

The Hawken Shop
3028 N. Lindbergh
St. Louis, MO 63074

Hodgdon Powder Co.
7710 W. 63rd.
Shawnee Mission, KS 66202

Kolpin Mfg., Co.
P.O. Box 231
Berlin, WI 54923

R.K. Lodges
Box 567
Hector, MN 55342

Log Cabin Shop
Box 275
Lodi, OH 44254

Lyman Products Corp.
Rt. 147
Middlefield, CT 06455

Navy Arms
689 Bergen Blvd.
Ridgefield, NJ 07657

Michaels of Oregon
P.O. Box 13010
Portland, OR 97213

Mountain State Muzzleloading Supplies
P.O. Box 154-1
Williamstown, WV 26187

Ox-Yoke Originals
130 Griffin Rd.
West Suffield, CT. 06093

Red River
P.O. Box 3114
Burbank, CA 91504

Rig Products
21320 Deering Ct.
Canoga Park, CA 91303

South Bend Replicas
61650 Oak Rd.
South Bend, IN 46614

Superior Arms Supply
P.O. Box 2008
Toluca Lake, CA 91602

Tennessee Valley Arms
225 DePot, Box 2022
Union City, TN 38261

Track of the Wolf
7335 Zane Way Shopping Center
Brooklyn Park, MN 55429

Trail Guns Armory
1634 E. Main
League City, TX 77573

Valor Corp.
5555 NW 36th Ave.
Miami, FL 33142

Western Arms Corp.
1107 Pen Rd.
Santa Fe, NM 87501

BULLETS

Barnes Bullets
P.O. Box 215
American Fork, UT 84003

B&B Sales
12521 Oxnard St.
N. Hollywood, CA 91606

Bruner Shooter Supply
207 Center Park Dr.
Knoxville, TN 37922

Canadian Sauer Co.
314 Judson St.
Toronto, Ontario
Canada M8Z 5T6

Continental Arms Corp.
697 Fifth Ave.
New York, NY 10022

Federal Ordnance
1443 Potrero Ave.
So. El Monte, CA 91733

Hornady/Pacific/Frontier
P.O. Box 1848
Grand Island, NE 68801

Paul Jones "Fitz"
653 N. Hager St.
San Fernando, CA 91340

Kassner Imports
5480 Linglestown Rd.
Harrisburg, PA 17112

Markell Inc.
4115 Judah St.
San Francisco, CA 94122

Nosler Bullets
P.O. Box 671
Bend, OR 97701

Omark-CCI-Speer-RCBS
Box 856
Lewiston, ID 83501

Reliance Corp.
50 Margin St.
Salem, MA 01970

Remington Arms Co.
939 Barnum Ave.
Bridgeport, CT 06602

Shore Galleries
3318 W. Devon Ave.
Lincolnwood, IL 60659

Sierra Bullet Co.
10532 S. Painter Ave.
Santa Fe Springs, CA 90670

Thompson/Center Arms
Farmington Rd.
Rochester, NH 03867

3-D Investment Inc.
Box J, Main St.
Doniphan, NE 68832

CARTRIDGES

B&B Sales
12521 Oxnard St.
N. Hollywood, CA 91606

Canadian Sauer Co.
314 Judson St.
Toronto, Ontario
Canada M8Z 5T6

Century Arms
5 Fererdal St.
St. Albans, VT 05478

Continental Arms
697 Fifth Ave.
New York, NY 10022

Dynamit Nobel of America
105 Stonehurst Ct.
Northvale, NJ 07647

Federal Cartridge Corp.
2700 Foshay Tower
Minneapolis, MN 55402

Federal Ordnance
1443 Potrero Ave.
So. El Monte, CA 91733

J.L. Galef & Son
85 Chambers St.
New York, NY 10007

H&H Cartridge Corp.
County Rd. 200 W., Box 104
Greensburg, IN 47240

Hornady/Pacific/Frontier
P.O. Box 1848
Grand Island, NE 68801

Hydra Shok Corp.
RD#2, Kuykendall Rd.
Watkins Glen, NY 14891

Kassner Imports
5480 Linglestown Rd.
Harrisburg, PA 17112

KTW Inc.
710 Foster Park Rd.
Lorain, OH 44053

Markell Inc.
4115 Judah St.
San Francisco, CA 94122

Numrich Arms
Williams Lane
West Hurley, NY 12491

Omark-CCI-Speer-RCBS
Box 856
Lewiston, ID 83501

Remington Arms
939 Barnum Ave.
Bridgeport, CT 06602

Shore Galleries
3318 W. Devon Ave.
Lincolnwood, IL 60659

CUSTOM GUNS

Accurate Gun Works
25878 E. Baseline
San Bernadino, CA 92410

Ahlman's
Rt.1, Box 20
Morristown, MN 55052

B&B Sales
12521 Oxnard St.
N. Hollywood, CA 91606

Behlert Custom Guns
725 Lehigh Ave.
Union, NJ 07083

Al Biesen
W. 2039 Sinto Ave.
Spokane, WA 99201

Bo-Mar Tool & Mfg. Co.
P.O. Box 168
Carthage, TX 75633

Champlin Firearms
Woodring Airport, Box 3191
Enid, OK 73701

Continential Arms
697 Fifth Ave.
New York, NY 10022

Davis Co.
2793 Del Monte St.
W. Sacramento, CA 95691

Ellwood Epps, Ltd.
Rt. 3, Highway 11 North
Orillia, Ontario
Canada L3V 6H3

Five Corners Sports
Box 22151
Lyndon, KY 40222

Freeland's Scope Stands
3737 14th Ave.
Rock Island, IL 61201

Griffin & Howe
589 Broadway
New York, NY 10012

Harry Lawson
3328 N. Richey Blvd.
Tucson, AZ 85716

Robert W. Hart & Son
401 Montgomery St.
Nescopeck, PA 18635

Iver Henriksen
1211 So. 2nd St. W.
Missoula, MT 58901

Hoag Gun Works
8523 Canoga Ave.
Canoga Park, CA 91304

Paul Jaeger, Inc.
211 Leedom St.
Jenkintown, PA 19046

Kahles of America
Main St.
Margaretville, NY 12455

Ljutic Industries
P.O. Box 2117
Yakima, WA 98407

Jim Lofland
2275 Larkin Rd.
Boothwyn, PA 19061

London Guns
1528 29th St.
Santa Monica, CA 90404

Mag-na-port Arms
30016 So. River Rd.
Mt. Clemens, MI 48045

Geo. E. Mathews & Son
10224 S. Paramount Blvd.
Downey, CA 90241

McGowen Rifle Barrels
Rt. 3
St. Anne, IL 60964

Navy Arms
689 Bergen Blvd.
Ridgefield, NJ 07657

Pachmayr Gun Works
1220 So. Grand Ave.
Los Angeles, CA 90015

Don Nygord Percision Pistols
6278 Hamilton
La Crescenta, CA 91214

Power Custom Inc.
Box 1604
Independence, MO 64055

Remington Arms
939 Barnum Ave.
Bridgeport, CT 06602

L.W. Seecamp Co.
P.O. Box 255
New Haven, CT 06460

Shaw's
9447 W. Lilac Rd.
Escondido, CA 92025

Shilen Rifles
205 Metro Park Blvd.
Ennis, TX 74119

Silver Dollar Guns
(Geo. Sheldon)
P.O. Box 475
Franklin, NH 03235

Southern Gun Exchange
4311 NE Expressway
Atlanta, GA 30340

Swenson's .45 Shop
P.O. Box 606
Fallbrook, CA 92028

"300" Gunsmith Service
4655 Washington St.
Denver, CO 80216

Trapper Gun
28019 Harper
St. Clair Shores, MI 48081

Weatherby's
2781 Firestone Blvd.
South Gate, CA 90280

Wichita Arms
333 Lulu, Box 11371
Wichita, KS 67211

John B. Williams
P.O. Box 2038
Yorba Linda, CA 92686

CUSTOM STOCKS & GRIPS

Bianchi Gunleather
100 Calle Cortez
Temecula, CA 92390

Al Biesen
W. 2039 Sinto Ave.
Spokane, WA 99201

Black Star
P.O. Box 2038
Yorba Linda, CA 92686

Bullshooter's Supply
P.O. Box 13446
Tucson, AZ 85732

Bullet Industries
1617 E. Seventeen
Santa Ana, CA 92701

Continental Arms
697 Fifth Ave.
New York, NY 10022

Reinhart Fajen
P.O. Box 338
Warsaw, MO 65355

Fitz Co.
653 N. Hager St.
San Fernando, CA 91340

Five Corners Sports
Box 22151
Lyndon, KY 40222

Freeland's Scope Stands
3737 14th Ave.
Rock Island, IL 61201

Fuzzy Farrant & Sons
1235 W. Vine
W. Covina, CA 91790

J.L. Galef & Son
85 Chambers St.
New York, NY 10007

Paul Jaeger
211 Leedom St.
Jenkintown, PA 19046

Harry Lawson
3328 N. Richey Blvd.
Tucson, AZ 85716

The Hawken Shop
3028 N. Lindburgh
St. Louis, MO 63074

Herrett's Stocks
P.O. Box 741
Twin Falls, ID 83301

Hogue Combat Grips
P.O. Box 460
Morro Bay, CA 93402

Art Jewel Enterprises
P.O. Box 819
Berkeley, IL 60163

Kassner Imports
5480 Linglestown
Harrisburg, PA 17112

Magnum Grips
P.O. Box 801
Payson, AZ 85541

Markell Inc.
4115 Judah St.
San Francisco, CA 94122

Mustang Pistol Grips
1334 E. Katella Ave.
Anaheim, CA 92805

Geo. E. Mathews & Son
10224 S. Paramount Blvd.
Downey, CA 90241

Bob Newell
55 Coyote St.
Los Alamos, NM 87544

Martin B. Retting
11029 W. Washington Blvd.
Culver City, CA 90230

Richards Microfit Stocks
8331 N. San Fernando Rd.
Sun Valley, CA 91352

Rogers Holster Co.
10601 Theresa Dr.
Jacksonville, FL 32216

Royal Arms
1210 Bert Acosta St.
El Cajon, CA 92020

Schiermeier Custom Handgun Stocks
306 No. 1st
Kent, WA 98031

Shaw's
9447 W. Lilac Rd.
Escondido, CA 92025

Sile Distributors
7 Centre Market Pl.
New York, NY 10013

Six Enterprises
6564 Hidden Creek Dr.
San Jose, CA 95120

Southern Gun Exchange
4311 NE Expressway
Atlanta, GA 30340

Ed Sowers
8331 DeCelis Pl.
Sepulveda, CA 91343

Sports, Inc.
1871 DeCook Ave.
Park Ridge, IL 60068

LEATHERGOODS

Alessi Holsters
2465 Niagara Falls Blvd.
Tonawanda, NY 14150

American Sales & Mfg.
P.O. Box 677
Laredo, TX 78040

Gunfighter-Andy Anderson
P.O. Box 225
N. Hollywood, CA 91603

Austin-Hall Boot Co.
P.O. Box 12367 A
El Paso, TX 79912

Baird Police Equip.
9889 Alondra Blvd.
Bellflower, CA 90706

B&B Sales
12521 Oxnard St.
N. Hollywood, CA 91606

Behlert Custom Guns
725 Lehigh Ave.
Union, NJ 07083

Belt Slide
1301 Brushy Bend Dr.
Round Rock, TX 78664

Bianchi Gunleather
100 Calle Cortez
Temecula, CA 92390

Ted Blocker Custom Holsters
3559 Lashbrook
Rosemead, CA 91770

Bo-Mar Tool & Mfg.
P.O. Box 168
Carthage, TX 75633

Brauer Bros. Mfg.
2012 Washington Ave.
St. Louis, MO 63103

J.M. Bucheimer
Box 280, Airport Rd.
Frederick, MD 21701

F. Burgess & Co.
Hillcrest Ave.
Meriden, CT 06450

Cobra Gunskin
1865 New Highway
Farmingdale, NY 11735

Davis Leather Co.
P.O. Box 446
Arcadia, CA 91006

D&E Magazines
P.O. Box 4579
Downey, CA 90241

J.C. Decker
Thomas Ave., Box 127
Montgomery, PA 17752

Dixie Gun Works
Gunpowder Lane
Union City, TN 38261

El Dorado
8406 N. Magnolia
Santee, CA 92071

El Paso Saddlery
P.O. Box 27194
El Paso, TX 79926

E.M.F. Co.
2911 W. Olive Ave.
Burbank, CA 91505

The Gun Room
8606 Telegraph
Dearborn Hgts., MI 48127

The Hawken Shop
3028 N. Lindbergh
St. Louis, MO 63074

Don Hume Leather
Box 351
Miami, OK 74354

The Hunter Corp.
P.O. Box 467
Westminster, CO 80030

Jackass Leather Co.
7383 N. Rogers Ave.
Chicago, IL 60626

Kolpin Mfg.
P.O. Box 231
Berlin, WI 54923

Lawman Leather
P.O. Box 447
Katy, TX 77449

George Lawrence Co.
306 SW 1st Ave.
Portland, OR 97204

Liberty Org.
4116 La Crescenta Ave.
La Crescenta, CA 91214

Markell Inc.
4115 Judah St.,
San Francisco, CA 94122

Michaels Of Oregon
P.O. Box 13010
Portland, OR 97213

Mountain State
Muzzleloading
P.O. Box 154-1
Williamstown, WV 26187

Navy Arms
689 Bergen Blvd.
Ridgefield, NJ 07657

Old West
2244-2 Main St.
Chula Vista, CA 92011

Red River
P.O. Box 3114
Burbank, CA 91504

Rogers Holsters
10601 Theresa Dr.
Jacksonville, FL 32216

Roy's Custom Leather Goods
Hwy. 132 & Rawhide Rd.
Magnilia, AR 71753

Safariland
1941 S. Walker Ave.
Monrovia, CA 91016

Salish House
Box 383
Lakeside, MT 59922

Sile Distributors
7 Centre Market Pl.
New York, NY 10013

Smith & Wesson
2100 Roosevelt Ave.
Springfield, MA 01101

Milt Sparks
Box 187
Idaho City, ID 83631

Sports, Inc.
1871 DeCook Ave.
Park Ridge, IL 60068

Tandy Leather Co.
P.O. Box 791
Ft. Worth, TX 76101

Triple K Mfg.
568 6th Ave.
San Diego, CA 92101

MISCELLANEOUS

A-1 Mfg. Co.
108 N. Waggoner
Electra, TX 76360

Accuracy Printing
(firearm log books)
339 Richmond St.
El Segundo, CA 90245

Ahlman's
Rt. 1
Morristown, MN 55052

Alco Carrying Cases, Inc.
601 W. 26th St
New York, NY 10001

Bob Allen Sportswear
(hunting & shooting clothes)
214 SW Jackson
Des Moines, Iowa 50302

Amerex Laboratories
214 W. Rhapsody
San Antonio, TX 78216

Anderson Mfg.
(ignitions, stock finishes)
P.O. Box 3120
Yakima, WA 98903

American Import Co.
(gun racks)
1167 Mission St.
San Francisco, CA 94103

Bachman
(shooting glasses)
1400 E. Erie Ave.
Philadelphia, PA 19124

Eddie Bauer
(clothing)
P.O. Box 3700, 5th & Union
Seattle, WA 98124

Ballisticomp Bullet
Performance
Computers
2203 Stratford Lane
Colorado Springs, CO 80909

Birchwood Casey
(gun care products)
7900 Fuller Rd.
Eden Prairie, MN 55344

Black Star Mfg.
(custom parts)
P.O. Box 2038
Yorba Linda, CA 92686

Break Free Corp.
(cleaners)
17422 Pullman
Irvine, CA 92714

Buck Stop Lure
3015 Grow Rd. NW
Stanton, MI 48888

Buenger Enterprises
(dehumidifiers)
22 Carriage Square, Box 5286
Oxnard, CA 93031

Burnham Bros.
(game calls)
912 Main, Box 669
Marbel Falls, TX 78654

Butler Creek Corp.
(blackpowder)
P.O. Box GG
Jacksonhole, WY 83001

B-Square Co.
(gunsmith tools)
P.O. Box 11281
Ft. Worth, TX 76109

Cherry Corners
(gunsmith supply)
11136 Congress Rd.
Lodi, OH 44254

David Clark Co.
(hearing protection)
364 Franklin St.
Worchester, MA 01602

Colorado School of Trades
(gunsmith school)
1575 Hoyt St.
Lakewood, CO 80215

Continental Shooting Clothes
1014 E. Olympic Blvd.
Los Angeles, CA 90021

Decker Shooting Products
(rifle vise)
1729 Laguna Ave.
Schofield, WI 54476

Del-Ray Products
P.O. Box 91561
Los Angeles, CA 90009

Detroit Bullet Trap Corp.
(range equip.)
2233 N. Palmer Dr.
Schaumburg, IL 60195

D&H Products
(deer hunting accessories)
P.O. Box 22
Glenshaw, PA 15116

Federal Ordnance
1443 Potrero Ave.
So. El Monte, CA 91733

Frielich Police Equip.
396 Broome St.
New York, NY 10013

Firearms & Metal Engraving
Mel Wood
3901 Crestmont Dr.
Santa Maria, CA 93454

Guardian Electronics
31133 W. Via Colinas
Westlake Village, CA 91361

Greentree Enterprises
(body warmers)
P.O. Box 4212
Woodland Park, CO 80863

Harbor House
(gun cases)
12508 Center St.
South Gate, CA 90280

The Hawken Shop
3028 N. Lindbergh
St. Louis, MO 63074

H&D Products
(handgun accessories)
8523 Canoga Ave.
Canoga Park, CA 91304

H&H Associates
P.O. Box 447
Strathmore, CA 93267

Inco Safety Products Co.
(hearing protectors)
P.O. Box 1733
Reading, PA 19603

J. Curtis Earl
(parts)
5512 N. 6th St.
Phoenix, AZ 85012

Kingsbery Mfg., Corp.
(gun safes)
715 W. Zauala St.
Crystal City, TX 78839

Mag-na-Port Arms
(recoil reducers)
30016 So. River Rd.
Mt. Clemens, MI 48045

Robert Maki
(engraver)
814 Revere Rd.
Glenview, IL 60025

Match Grade Ordnance
(cleaning supplies)
Box 2178
Edison, NJ 08817

Merit Gunsight Co.
(accessories)
P.O. Box 995
Sequim, WA 98382

Michaels of Oregon
P.O. Box 13010
Portland, OR 97213

Moore Supply Co.
(game accessories)
3000 S. Main
Salt Lake City, UT 84115

Norton Co., Safety Products
(hearing protector)
16624 Edwards Rd.
Cerritos, CA 90701

Penquin Industries
(cleaning supplies)
Airport Industrial Mall
Coatesville, PA 19320

Perry's Ammo Sling
3143 Hwy. 81 East
McDonough, GA 30253

Poly-Choke Co.
Box 2496
Hartford, CT 06101

Provo Steel & Supply
1400 S. State St.
Provo, UT 84061

Ranch Products
(.45 ACP clips)
P.O. Box 145
Malinta, OH 43535

Remington Arms
(clay targets, traps)
939 Barnum Ave.
Bridgeport, CT 06602

Rig Products
21320 Deering Ct.
Canoga Park, CA 91304

R.S.R. Corp.
(lead shot)
1111 Mockingbird Lane
Dallas, TX 75247

Safariland
(loader clips)
1941 S. Walker Ave.
Monrovia, CA 91016

Saf-T-Case
P.O. Box 5472
Irving, TX 75062

Safety Direct/Silencio
(hearing protectors)
23 Snider Way
Sparks, NV 89431

Se-Kure Controls
5685 N. Lincoln Ave.
Chicago, IL 60654

707 Co. Div. of TGN
(gun care products)
155 Lafayette Ave.
N. White Plains, NY 10603

Sherwood Int'l Export Corp.
(pistol & rifle parts)
18714 Parthenia St.
Northridge, CA 91324

Skunk Screen
315 Church St., P.O. Box Drawer CB
College Station, TX 77841

Sloan's Sporting Goods
(wholesalers)
10 South St.
Ridgefield, CT 06877

Sports Haven
168 W. 1st, Box 30-06
Deer Park, WA 99006

Sports Specialities
(adapters, conversions)
Box 5337 GAN
Hacienda Heights, CA 91745

Johnny Stewart Game Call
P.O. Box 7594,5100 Fort Ave.
Waco, TX 76710

Stowline Inc.
(cases, vaults)
811 S. First
Kent, WA 98031

Swenson's .45 Shop
(safeties)
P.O. Box 606
Fallbrook, CA 92028

Tasco
(binoculars)
7600 NW 26th St.
Miami, FL 33152

Taser Systems Inc.
(non-lethal)
1100 S. Azusa Ave.
Industry, CA 91748

Telepacific Electronics Co.
(chronograph)
772 N. Twin Oaks Valley Rd.
San Marcos, CA 92069

10X Mfg. Co.
(shooting apparel)
316 Lexington Ave.
Cheyenne, WY 82001

Tread Corp.
(gun vaults)
P.O. Box 13207
Roanoke, VA 24032

Trius Products Inc.
(clay target traps)
221 S. Miami Ave.
Cleves, OH 45002

Unique Imports, Inc.
800 Slaters Lane
Alexandria, VA 22314

WD-40 Co.
0161 Cudahy Pl.
San Diego, CA 92110

Weather Shield Sports Equip.
(gun cases, trap & skeet)
Rt. 3, Petoskey Rd., P.O. Box 227
Charlevoix, MI 49720

Ballistic Products
17610 19th Ave. No.
Plymouth, MN 55447

Barnes Bullets
P.O. Box 215
American Fork, UT 84003

B&B Sales
12521 Oxnard St.
N. Hollywood, CA 91606

Canadian Sauer
314 Judson St.
Toronto, Ontario
Canada M8Z 5T6

Carbide Die & Mfg. Co.
15615 E. Arrow Hwy.
Irwindale, CA 91707

C-H Tool & Die Corp.
106 N. Harding St.
Owen, WI 54460

Continental Arms
697 Fifth Ave.
New York, NY 10022

Corbin Mfg. & Supply
P.O. Box 758
Phoenix, OR 97535

Dixie Gun Works
Gunpowder Lane
Union City, TN 38261

DuPont Co.
Explosive Products Div.
10th & Market St.
Wilmington, DE 19898

Elk Mountain Shooters Supply
1719 Marie
Pasco, WA 99301

Framer Bros. Mfg.
1102 N. Washington St.
Eldora, Iowa 50627

Paul Jones "Fitz"
653 N. Hager St.
San Fernando, CA 91340

Forster Products
82 E. Lanark Ave.
Lanark, IL 61046

Hodgdon Powder Co.
7710 W. 63rd.
Shawnee Mission, KS 66202

Hornady/Pacific/Frontier
P.O. Box 1848
Grand Island, ME 68801

Hulme Firearm Service
P.O. Box 83
Millbrae, CA 94030

Lee Precision
4275 Hwy. U
Hartford, WI 53027

Lyman Products
RT. 147
Middlefield, CT 06455

Markell Inc.
4115 Judah St.
San Francisco, CA 94122

Marmel Products
P.O. Box 97
Utica, MI 48087

Marquart Precision
Rear 136 Grove Ave.
P.O. Box 1740
Prescott, AZ 86301

Mr. Paul McLean
2670 Lakeshore Blvd.
Toronto, Ontario
Canada M8V 1G8

MEC Inc.
715 South St.
Mayville, WI 53050

Mequon Reloading Corp.
P.O. Box 253
Mequon, WI 53092

MTM
5680 Webster St.
Dayton, OH 45414

Multi-Scale Charge LTD.
3269 Niagara Falls Blvd.
N. Tonawanda, NY 14120

Oehler Research
P.O. Box 9135
Austin, TX 78766

Omark-CCI-Speer-RCBS
Box 856
Lewiston, ID 83501

Omark-RCBS
605 Oro Dam Blvd.
Oroville, CA 95965

Pak-Tool Co.
4411 SW 100 St
Seattle, WA 98146

Penquin Industries
Airport Industrial Mall
Coatesville, PA 19320

Ponsness-Warren
P.O. Box 8
Rathdrum, ID 83858

Quinetics Corp.
5731 Kenwick
San Antonio, TX 78238

Redding Reloading Equip.
114 Starr Rd.
Cortland, NY 13045

Remington Arms
939 Barnum Ave.
Bridgeport, CT 06602

Saeco Reloading
525 Maple Ave., Box 778
Carpinteria, CA 93013

Shore Galleries
3318 W. Devon Ave.
Lincolnwood, IL 60659

Sile Distributors
7 Centre Market Pl.
New York, NY 10013

Texan Reloaders
444 Cips St.
Watseka, IL 60970

Thompson/Center
Farmington Rd.
Rochester, NH 03867

Whitney Sales
Hollywood Center, X Products
P.O. Box 875
Reseda, CA 91335

L.E. Wilson Inc.
Box 324
Cashmere, WA 98815

SHOOTING ORGANIZATIONS

Alabama Gun Collectors Assn.
P.O. Box 5548
Tuscaloosa, AL 35405

Amateur Trapshooting Assn.
Vandalia, OH 45377

American Pistol Institute
Jeff Cooper
P.O. Box 401
Paulden, AZ 86334

American Pistol & Revolver Assn.
512 E. Wilson Ave.
Glendale, CA 91206

Association of Importers & Mfg. for Muzzle Loading (AIMM)
1221 SW Yamhill, Ste 300
Portland, OR 97205

Ray Chapman Academy
609 E. Broadway
Columbia, MO 65201

International Handgun Metallic Silhouette Assn.
512 Lomak

Armsport
3590 NW 49th St.
Miami, FL 33142

Behlert Custom Guns
725 Lehigh Ave.
Union, NJ 07083

Bill's House of Guns
P.O. Box 424
Merrill, WI 54452

Bo-Mar Tool & Mfg.
P.O. Box 168
Carthage, TX 75633

B-Square
P.O. Box 11281
Ft. Worth, TX 76109

Burnham Bros.
912 Main, Box 669
Marble Falls, TX 78654

Burris Co.
331 E. 8th St., Box 1747
Greeley, CO 80632

Buehler Scope Mounts
17 Orinda Way
Orinda, CA 94563

Bullshooter's Supply
P.O. Box 13446
Tucson, AZ 85732

Bushnell Optics Co.
2828 E. Foothill Blvd.
Pasadena, CA 91107

Clear View Mfg. Co.
20821 Grand River
Detroit, MI 48219

Conetrol Scope Mounts
Hwy. 123 South
Sequin, TX 78155

Danley's
P.O. Box 1-GA
Ft. Johnson, NY 12070

D&H Products
P.O. Box 22
Glenshaw, PA 15116

Federal Ordnance
1443 Potrero Ave.
So. El Monte, CA 91733

F.I.E. Corp.
P.O. Box 4886
Hialeah Lakes
Hialeah, FL 33014

Freeland's Scope Stands
3737 14th Ave.
Rock Island, IL 61201

Fujinon Optical
672 White Plains Rd.
Scarsdale, NY 10583

Bushnell Optics Co.
2828 E. Foothill Blvd.
Pasadena, CA 91107

Clear View Mfg. Co.
20821 Grand River
Detroit, MI 48219

Conetrol Scope Mounts
Hwy. 123 South
Sequin, TX 78155

Danley's
P.O. Box 1-GA
Ft. Johnson, NY 12070

D&H Products
P.O. Box 22
Glenshaw, PA 15116

Federal Ordnance
1443 Potrero Ave.
So. El Monte, CA 91733

F.I.E. Corp.
P.O. Box 4886
Hialeah Lakes
Hialeah, FL 33014

Freeland's Scope Stands
3737 14th Ave.
Rock Island, IL 61201

Fujinon Optical
672 White Plains Rd.
Scarsdale, NY 10583

J.L. Galef & Son
85 Chambers St.
New York, NY 10007

Griffin & Howe
589 Broadway
New York, NY 10012

The Hawken Shop
3028 N. Lindbergh
St. Louis, MO 63074

Heckler & Koch
933 N. Kenmore St
Arlington, VA 22201

H&H Associates
P.O. Box 447
Strathmore, CA 93267

J.B. Holden Co.
295 W. Pearl
Plymouth, MI 48170

Paul Jaeger
211 Leedom St.
Jenkintown, PA 19046

Jason/Empire
9200 Cody
Overland Park, KS 66212

Kahles of America
Main St.
Margaterville, NY 12455

Kwik-Site Co.
5555 Treadwell
Wayne, MI 48184

Lee's Red Ramps
7252 E. Ave. U-3
Littlerock, CA 93543

Leupold & Stevens
P.O. Box 688
Beaverton, OR 97075

Lyman Products
Rt. 147
Middlefield, CT 06455

Mag-na-port Arms
30016 So. River Rd.
Mt. Clemens, MI 48045

Marble Arms
P.O. Box 111
Gladstone, MI 49837

Millett Industries
16131 Gothard St.
Huntington Beach, CA 92647

Miniature Machine Co.
210 E. Poplar St.
Deming, NM 83030

Numrich Arms
Williams Lane
West Hurly, NY 12491

Poly-Choke Co.
Box 2496
Hartford, CT 06101

Ranch Products
P.O. Box 145
Malinta, OH 43535

Ranging Inc.
90 N. Lincoln Rd.
East Rochester, NY 14445

Redfield
5800 E. Jewell Ave.
Denver, CO 80224

Sherwood Int'l Export
18714 Parthenia St.
Northridge, CA 91324

S&K Mfg.
Box 247
Pittsfield, PA 16340

Steyr Daimler Puch of America
85 Metro Way
Secaucus, NJ 07094

Swift Instruments
952 Dorchester Ave.
Boston, MA 02125

Swift Instruments
P.O. Box 562
San Jose, CA 95106

Tasco
7600 NW 26th St.
Miami, FL 33152

Thompson/Center
Farmington Rd.
Rochester, NH 03867

John Unertl Optical
3551-55 East St.
Pittsburgh, PA 15214

Weatherby's
2781 Firestone Blvd.
South Gate, CA 90280

W.R. Weaver Co.
7125 Industrial Ave.
El Paso, TX 79915

Williams Gun Sight Co.
7300 Lapeer Rd.
Davidson, MI 48423

Wichita Arms
333 Lulu, Box 11371
Wichita, KS 67211

Carl Zeiss, Inc.
444 Fifth Ave.
New York, NY 10018

KNIFE IMPORTERS AND MANUFACTURERS

Al Mar (knives)
5861 SW Benfield Ct.
Lake Oswego, OR 97034

The American Import Company
1167 Mission Street
San Francisco, CA 94103

Armament Systems Products, Unltd.
P.O. Box 18595
Atlanta, GA 30326

Bali Song, In.
3039 Roswell St.
Los Angeles, CA 90065

Ballard Cutlery
P.O. Box 97
Golf, IL 60029

B&B Sales
12521 Oxnard St.
North Hollywood, CA 91606

Bill's House of Guns
P.O. Box 424
Merrill, WI 54452

Buck Knives, Inc
P.O. Box 1267
El Cajon, CA 92022

Bucklin Blades
243 Millbrook Wy.
Vacaville, CA 95688

Burnham Bros. Inc
912 Main Box 669
Marble Falls, TX 78654

Cam III Enterprises
243 Millbrook Way
Vacaville, CA. 95688

Camillus Cutlery Company
Camillus, NY 13031

W.R. Case & Sons Cutlery Co.
20 Russell Blvd.
Bradford, PA 16701

Chicago Cutlery
Consumer Products
5420 N. Country Rd. 18
Minneapolis, MN 55428

E. Christopher Firearms, Inc.
8007 St. Rt. 128
Miamitown, OH 45041

Classic Arms Co.
20 Wilbraham St.
Palmer, MS. 01069

Continental Arms Corporation
697 Fifth Avenue
New York, NY 10022

Dixie Gun Works, Inc.
Gunpowder Lane
Union City, TN 38261

D.L. R. Knives Inc.
P.O. Box 25352
S.L.C., UT 84125

F.I.E. Corporation
P.O. Box 4866
Hialeah Lakes
Hialeah, FL 33014

The Hawken Shop
3028 N. Lindbergh
St. Louis, MO 63074

Imperial Knives
1776 Broadway
NY, NY 10019

Jet-aer Corp.
100 Sixth Ame
Paterson, NJ 07524

Ka-Bar Cutlery
5777 Grant Ave.
Cleveland, OH 44105

Lakota Corp.
30961 Agoura Rd.
Suite 311
Westlake Village, CA 91361

Liberty ORG. Inc
4116 La Crescenta Ave.
La Crescenta, CA 91214

Jimmy Lile Handmade Knives
Route 1
Russellville, AR 72801

Markell Inc.
4115 Judah St.
San Francisco, CA 94122

Martini Knives
P.O. Box 48866
Chicago, IL 60648

Nordic Knives
1634 Copenhagen Dr.
Solvang, CA 93463

Pony Express
Sport Shop, Inc.
17460 Ventura Boulevard
Encino, CA 91316

Precise International
3 Chestnut St.
Suffern, NY 10901

Schrade Cutlery Corporation
1776 Broadway
New York, NY 10019

Shore Galleries, Inc.
3318 West Devon Avenue
Lincolnwood, IL 60659

Smith & Wesson
2100 Roosevelt Ave.
Springfield, MA 01101

Westbury Sales Company
373 Maple Ave. P.O. Box 434
Westbury, NY 11590

Western Cutlery Co.
1800 Pike Rd.
Longmont, CO 80501